Trigonometry

Second Edition

eText Reference

Kirk Trigsted

University of Idaho

PEARSON

Boston Columbus Indianapolis New York San Francisco Upper Saddle River
Amsterdam Cape Town Dubai London Madrid Milan Munich Paris Montreal Toronto
Delhi Mexico São Paulo Sydney Hong Kong Seoul Singapore Taipei Tokyo

Editor in Chief: Anne Kelly
Acquisitions Editor: Dawn Murrin
Editorial Assistant: Joseph Colella
Design Manager: Andrea Nix
Art Director: Heather Scott
Project Manager: Ron Hampton
Senior Math Media Producer: Tracy Menoza
Project Manager, MathXL: Kristina Evans
Marketing Manager: Peggy Lucas
Marketing Assistant: Justine Goulart
Manufacturing Buyer: Debbie Rossi

**For my wife, Wendy, and our children—
Benjamin, Emily, Gabrielle, and Isabelle.**

ScoutAutomatedPrintCode

PEARSON

ISBN-13: 978-0-321-86978-4
ISBN-10:   0-321-86978-8

Contents

Chapter Five Polar Equations, Complex Numbers, and Vectors 5-1

Preface

Introduction

This *eText Reference* contains the pages of Kirk Trigsted's *Trigonometry* in a portable, bound format. The structure of this *eText Reference* helps student organize their notes by providing them with space to summarize the videos and animations. Students can also use the *eText Reference* to review the eText material anytime, anywhere.

A Note to Students

This eText was created for you and was specifically designed to be read online. Unlike a traditional textbook, I have created content that gives you, the reader, the ability to be an active participant in your learning. The eText pages have large, readable fonts and were designed to avoid scrolling. Throughout the material, I have carefully integrated hundreds of hyperlinks to definitions, previous chapters, interactive videos, animations, and other important content. Many of the videos and animations require you to actively participate. Take some time to "click around" and get comfortable with the navigation of the eText and explore its many features. To log into your ecourse powered by MyMathLab, you must first register at www.mymathlab.com. This eText Reference is a bound, printed version of the eText that provides a place for you to do practice work and summarize key concepts from the online videos and animations.

Before you attempt each homework assignment, read the appropriate section(s) of the eText. At the beginning of each section (starting in Chapter One), you will encounter a feature called Things to Know. This feature includes all the prerequisite objectives from previous sections that you will need to successfully learn the material presented in the new section. If you do not think that you have a basic understanding of any of these objectives, click on any of the hyperlinks and rework those objectives, taking advantage of the videos or animations.

Try testing yourself by working through the corresponding You Try It exercises. Remember, you learn math by *doing* math! The more time you spend working through the videos, animations, and exercises, the more you will understand. If your instructor assigns homework in MyMathLab or MathXL, rework the exercises until you get them right. Be sure to go back and read the eText at anytime while you are working on your homework. This eText is catered to your educational needs, and I hope you enjoy the experience.

A Note to Instructors

I have taught with MyMathLab for many years and have experienced first-hand how fewer and fewer students are using their traditional textbooks. As the use of technology plays an ever increasing role in how we are teaching our students, it is only natural to have a textbook that mirrors the way our students are learning. I am excited to have written an eText from the ground up to be used as an online, interactive tool for students to read while working in MyMathLab. I wrote this eText entirely from an online perspective, keeping MyMathLab and its existing functionality specifically in mind. Every hyperlink, video, and animation was strategically integrated within the context of each page to maximize the student learning experience. All of the interactive media was designed so students could actively participate while they learn math.

I am a proponent of students learning terms and definitions. Therefore, I have created hyperlinks throughout the text to the definitions of important mathematical terms. I have also inserted a significant amount of just-in-time review throughout the text by creating links to prerequisite topics. Students have the ability to reference these review materials with just a click of the mouse.

Each section has five reading assessment questions for those instructors who like to assign reading. These questions are conceptual in nature and were designed to test students on their reading comprehension. Each question was specifically designed to give specific feedback that directs the back to the appropriate pages of the eText. Note that these questions are static. Students are given two chances to obtain the correct answer. Students will not be able to regenerate a similar exercise. The reading assessment questions can be identified in the Homework/Test Manager by the code "RA."

<table>
<tr><td>Preview and Add to Homework</td><td># Items in your Homework: 0
Preview Item: 73 of 77 | Item #: 3.1.RA-1</td></tr>
</table>

Section 3.1 | Objective: Reading Assessment
Availability: Homework, Tests and Quizzes, Study Plan
Origin: Publisher

Which of the following statements defines a relation?

○ A. A relation is a correspondence between two sets A and B such that each element of set A equals one or more elements in set B.

★ B. A relation is a correspondence between two sets A and B such that each element of set A corresponds to one or more elements in set B.

○ C. A relation is a correspondence between two sets A and B such that each element of set A corresponds to exactly one element in set B.

○ D. A relation is a correspondence between two sets A and B such that each element of set A equals exactly one element in set B.

Connect to a Tutor
Ask My Instructor
Print
Ask the Publisher
Add Instructor Tip

Question is complete.

All parts showing   Close

● Show completed problem   ○ Work problem as student   ☐ Student to show work   Question points: 1   Scoring options

Show Answer | Reload   Copy and Edit   ◄ Previous | Add | Next ►

The first edition of *Trigonometry* included several multipart exercises. I received feedback from many users who liked to use these multipart exercises for homework but often wished that they could assign only one of the parts for testing. This led to the creation of a new type of exercise called brief exercises. The multipart exercises are now called step-by-step exercises.

Step-by-Step Exercises

The step-by-step exercises were designed to use the power of MathXL to systematically walk the student through some of the more complex, conceptual topics. For example, instead of simply asking for the graph of quadratic function, the step-by-step exercise walks the student through the entire graphing process by asking for all of the important aspects of a parabola. The step-by-step exercises can be readily identified in the Homework/Test Manager by the code "SbS" that precedes the exercise number. An example of such an exercise follows.

21 **22** **23** **24** **25** **26** **27** **28** **29** **30** Question 4.1.SbS-23

1 correct | 1 of 106 complete

Use the quadratic function $f(x) = 3x^2 + 6x - 4$ to address the following questions.

a) Use the vertex formula to determine the vertex.

The vertex is $(-1, -7)$.
(Type an ordered pair. Simplify your answer.)

b) Does the graph "open up" or "open down"?

○ Down

● Up

c) What is the equation of the axis of symmetry?

$x = -1$
(Simplify your answer.)

d) Find any x-intercepts. Select the correct choice below and, if necessary, fill in the answer box within your choice.

● A. $x = \dfrac{-3 + \sqrt{21}}{3}, \dfrac{-3 - \sqrt{21}}{3}$

(Type an exact answer, using radicals as needed. Use a comma to separate answers as needed.)

○ B. There is no x-intercept.

e) Find the y-intercept. Select the correct choice below and, if necessary, fill in the answer box within your choice.

● A. The y-intercept is -4. (Type an integer or a fraction.)

○ B. There is no y-intercept.

f) Sketch the graph. Which of the following is the graph of $f(x) = 3x^2 + 6x - 4$?

● A. ○ B. ○ C. ○ D.

g) State the domain and range in interval notation.

The domain is $(-\infty, \infty)$.   The range is $[-7, \infty)$.
(Use integers or fractions for any numbers in the expression.)

Question is complete.

All parts showing   Similar Exercise   Close

Help Me Solve This
View an Example
Textbook
Ask My Instructor
Print

In the exercise sets at the end of each section of this eText Reference, all step-by-step exercises are labelled with an SbS icon.

You Try It In Exercises 21–30, use the quadratic function to address the following.

   a. Use the vertex formula to determine the vertex.

   b. Does the graph "open up" or "open down"?

   c. What is the equation of the axis of symmetry?

   d. Find any x-intercepts.

   e. Find the y-intercept.

   f. Sketch the graph.

   g. State the domain and range in interval notation.

SbS 21. $f(x) = x^2 - 8x$ SbS 22. $f(x) = -x^2 - 4x + 8$

SbS 23. $f(x) = 3x^2 + 6x - 4$ SbS 24. $f(x) = 2x^2 - 5x - 3$

SbS 25. $f(x) = -x^2 + 2x - 6$ SbS 26. $f(x) = \dfrac{1}{2}x^2 + 6x + 1$

SbS 27. $f(x) = -\dfrac{1}{3}x^2 - 9x + 5$ SbS 28. $f(x) = -3x^2 + 7x + 5$

SbS 29. $f(x) = x^2 + \dfrac{8}{3}x - 1$ SbS 30. $f(x) = -\dfrac{1}{4}x^2 + 6x - 1$

Brief Exercises

The brief exercises are copies of the step-by-step exercises and are designed to test one concept of a multistep problem and were designed for instructors who may want to pinpoint a specific skill for a quiz and testing purposes. The brief exercises can be identified in the Homework/Test Manager by the code "BE" that precedes the exercise number. Below is the brief exercise that corresponds to the step-by-step exercise seen on the previous page.

Objective: Graphing Quadratic Functions Using the Vertex Formula

«‹ ‹ 71 72 73 74 75 76 77 78 79 80 › ›› Question 4.1.BE-75

2 correct | 2 of 106 complete

Use the quadratic function $f(x) = 3x^2 + 6x - 4$ to address the following questions.

a) Find any x-intercepts. Select the correct choice below and, if necessary, fill in the answer box within your choice.

A. $x = \dfrac{-3 + \sqrt{21}}{3}, \dfrac{-3 - \sqrt{21}}{3}$

(Type an exact answer, using radicals as needed. Use a comma to separate answers as as needed.)

B. There is no x-intercept.

b) Find the y-intercept. Select the correct choice below and, if necessary, fill in the answer box within your choice.

A. The y-intercept is -4. (Type an integer or a fraction.)

B. There is no y-intercept.

Help Me Solve This
View an Example
Textbook
Ask My Instructor
Print

Question is complete.

All parts showing

Similar Exercise | Close

Skill Check Exercises

The Skill Check Exercises (SCE) were created based on comments submitted by users of the Trigsted precalculus series. Skill Check Exercises help students make algebra connections as they work through exercises that present new concepts. These exercises appear at the beginning of most exercises sets and are assignable in MyMathLab.

1.1 Exercises

Skill Check Exercises

For exercises SCE-1 through SCE-6, determine the Least Common Denominator (LCD) of the given expression.

SCE-1. $\dfrac{2}{9} + \dfrac{1}{3} - \dfrac{1}{6}$

SCE-2. $\dfrac{1}{8}(2p - 1) - \dfrac{7}{3}p - \dfrac{p - 4}{6}$

SCE-3. $\dfrac{a - 3}{6} - \dfrac{3(a - 1)}{10} + \dfrac{2a + 1}{5}$

SCE-4. $\dfrac{3x}{x + 1} - \dfrac{5x + 7}{x - 1}$

SCE-5. $\dfrac{1}{2x} - \dfrac{1}{4} + \dfrac{6}{8x^2}$

SCE-6. $\dfrac{w}{w - 3} - \dfrac{2w}{2w - 1} - \dfrac{w - 3}{2w^2 - 7w + 3}$

SCE-7. $\dfrac{3}{x - 1} + \dfrac{4}{x + 1} - \dfrac{8x}{x^2 - 1}$

SCE-8. $\dfrac{6}{x^2 - x} - \dfrac{2}{x} + \dfrac{3}{x - 1}$

All Skill Check Exercises will be labeled as SCE-1, SCE-2, SCE-3, and so forth in the Assignment Manager in MyMathLab and MathXL.

Resources for Success

MyMathLab® Online Course (access code required)

MyMathLab delivers **proven results** in helping individual students succeed. It provides **engaging experiences** that personalize, stimulate, and measure learning for each student. And, it comes from an **experienced partner** with educational expertise and an eye on the future. MyMathLab helps prepare students and gets them thinking more conceptually and visually through the following features:

Adaptive Study Plan

The Study Plan makes studying more efficient and effective for every student. Performance and activity are assessed continually in real time. The data and analytics are used to provide personalized content-reinforcing concepts that target each student's strengths and weaknesses.

Getting Ready

Students refresh prerequisite topics through assignable skill review quizzes and personalized homework integrated in MyMathLab.

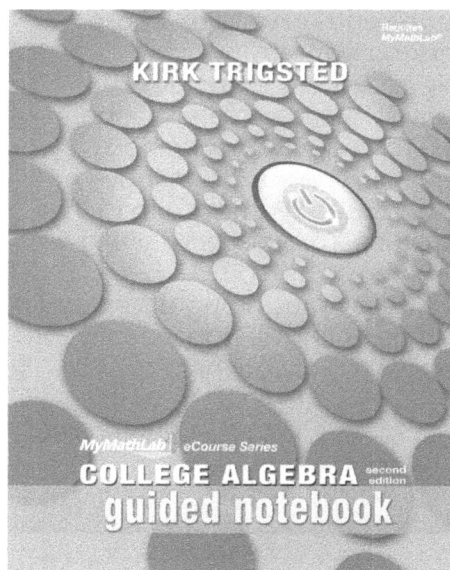

Guided Notebook

The *Guided Notebook* is a printed, interactive workbook created by the author to guide students through the eText. The workbook asks students to write down key definitions and work through important examples before they start their homework.

Skills for Success Modules

MyMathLab's new Skills for Success modules are designed to help students succeed in college and perpare for the future professions. Activities and assignments are available for topics such as "Time Management," "Stress Management," and "Financial Literacy."

Support and Resources

MyMathLab for Trigonometry Student Access Kit, by Kirk Trigsted
0-321-86933-8/978-0-321-86933-3

Guided Notebook

0-321-86915-X/978-0-321-86915-9

The *Guided Notebook* is a printed, interactive workbook created by the author to guide students through the eText. The *Guided Notebook* asks students to write down key definitions and work through important examples before they start their homework. This resource is available in a loose-leaf, three-hole-punched format to provide the foundation for a personalized course note-book. Instructors can also customize the *Guided Notebook* files found within MyMathLab®.

Instructor Resources

PowerPoint® slides present key concepts and definitions from the text. Slides are available for download from within MyMathLab and from Pearson Education's online catalog.

TestGen® software enables instructors to build, edit, print, and administer tests using a computerized bank of questions. TestGen is algorithmically based, allowing instructors to create multiple but equivalent versions of the same question or test. Instructors can also modify test bank questions or add new questions. The software and test bank are available for download from Pearson Education's online catalog.

Acknowledgments

There are so many wonderful, dedicated people that I would like to thank. First of all, I must thank my beautiful wife, Wendy, for her continued loving support. I am also so grateful to my loving children, Benjamin, Emily, Gabrielle, and Isabelle for being so patient and understanding when daddy was locked in his basement office. Thank you for making daddy take breaks from time to time!

There are so many extraordinary and talented people who have contributed to this project. From Pearson, I would like to thank Anne Kelly, Dawn Murrin, and Joe Colella for their editorial superiority; Greg Tobin and Chris Hoag for their continued support; Eileen Moore, Kristina Evans, and Phil Oslin and the rest of the MathXL development team for a truly incredible job; Tracy Menoza and Ruth Berry for taking care of the eText details; the entire art team for their exceptional eye for detail; and Ron Hampton, our project manager, for all his support.

There are so many people from so many colleges and universities whose work on this project was immeasurable. I must personally thank my friend Phoebe Rouse from LSU whose contributions to this book were enormous— "Miss Phoebe" has officially become one of the family! I appreciate everyone else who made this book a reality: Tracy Duff from PreMedia Global helped me in so many ways; Alice Champlin and Anthony T. J. Kilian at Magnitude Entertainment for creating all of the awesome media assets; Pamela Trim for her continued dedication to detail; and Phil Veer and Bruce Yarbrough for sitting through the video shoots.

The list on the following pages includes all reviewers, focus group attendees, and class testers. Please accept my deepest apologies if I have inadvertently omitted anyone. I am humbled and so very grateful for all of your help and I thank you from the bottom of my heart.

—Kirk Trigsted

Teri Barnes, McLennan Community College
Linda Barton, Ball State University
Sam Bazzi, Henry Ford Community College
Molly Beauchman, Yavapai College
Brian Beaudrie, Northern Arizona University
Annette Benbow, Tarrant County College
Patricia Blus, National Louis University
Nina Bohrod, Anoka-Ramsey Community College
Barbara Boschmans, Northern Arizona University
David Bramlet, Jackson State University
Densie Brown, Collin County Community College
Connie Buller, Metropolitan Community College
Joe Castillo, Broward Community College
Mariana Coanda, Broward Community College
Alicia Collins, Mesa Community College
Earl W. Cook, Chattahoochee Valley Community College
Kemba C. Countryman, Chattahoochee Valley Community College
Douglas Culler, Midlands Technical College
Momoyo Dahle, Las Positas Community College
Diane Daniels, Mississippi State University
Emmett Dennis, Southern Connecticut State University
Donna Densmore, Bossier Parish Community College
Debbie Detrick, Kansas City Kansas Community College
Holly Dickin, University of Idaho
Timothy Doyle, University of Illinois at Chicago
Christina Dwyer, Manatee Community College
Stephanie Edgerton, Northern Arizona University
Jeanette Eggert, Concordia University
Brett Elliott, Southeastern Oklahoma State University
Amy Erickson, Georgia Gwinnett College
Nicki Feldman, Pulaski Technical College
Catherine Ferrer, Valencia Community College
Gerry Fitch, Louisiana State University
Cynthia Francisco, Oklahoma State University
Robert Frank, Westmoreland County Community College
Jim Frost, St. Louis Community College–Meramec
Angelito Garcia, Truman College
Lee Gibson, University of Louisville
Charles B. Green, University of North Carolina–Chapel Hill
Jeffrey Hakim, American University
Mike Hall, Arkansas State University
Melissa Hardeman, University of Arkansas, Little Rock
Celeste Hernandez, Richland College
Pamela Howard, Boise State University
Jeffrey Hughes, Hinds Community College
Eric Hutchinson, College of Southern Nevada
Robert Indrihovic, Florence-Darlington Technical College
Philip Kaatz, Mesalands Community College
Cheryl Kane, University of Nebraska-Lincoln

Robert Keller, Loras College
Mike Kirby, Tidewater Community College
Susan Knights, Boise State University
Marie Kohrmann, North Lake College
Debra Kopcso, Louisiana State University
Stephanie Kurtz, Louisiana State University
Jennifer LaRose, Henry Ford Community College
Jeff Laub, Central Virginia Community College
Jennifer Legrand, St. Charles Community College
Kurt Lewandowski, Clackamas Community College
Oscar Macedo, University of Texas at El Paso
Shanna Manny, Northern Arizona University
Pamela Spurlock Mills, University of Arkansas
Peggy L. Moch, Valdosta State University
Susan Moosai, Florida Atlantic University
Rebecca Morgan, Wayne State University
Rebecca Muller, Southeastern Louisiana University
Veronica Murphy, Saint Leo University
Charlie Naffziger, Central Oregon Community College
Prince Raphael A. Okojie, Fayetteville State University
Enyinda Onunwor, Stark State University
Shahla Peterman, University of Missouri–St. Louis
Sandra Poinsett, College of Southern Maryland
Steve Proietti, Northern Essex Community College
Ray Purdom, University of North Carolina at Greensboro
Nancy Ressler, Oakton Community College
Mary Revels, Southeast Community College
Joe Rody, Arizona State University
Cheryl Roddick, San Jose State University
Amy Rouse, Louisiana State University Laboratory School
Phoebe Rouse, Louisiana State University
Patricia Rowe, Columbus State Community College
Amy Rushall, Northern Arizona University
Chie Sakabe, University of Idaho
Jorge Sarmiento, County College of Morris
John Savage, Montana State University–College of Technology
Julie Sawyer, University of Idaho
Victoria Seals, Gwinnett Technical College
Susan P. Sherry, Northern Virginia Community College
Randell Simpson, Temple College
Rita Sowell, Volunteer State Community College
John Squires, Cleveland State Community College
Pam Stogsdill, Bossier Parish Community College
Eleanor Storey, Front Range Community College
Robert Strozak, Old Dominion University
Lalitha Subramanian, Potomac State College of West Virginia University
Mary Ann Teel, University of North Texas
Gwen Terwilliger, University of Toledo
Jamie Thomas, University of Wisconsin–Manitowoc

Keith B. Thompson, Davidson County Community College
Terry Tiballi, State University of New York at Oswego
Suzanne Topp, Salt Lake Community College
Diann Torrence, Delgado Community College
Philip Veer, Johnson County Community College
Marcia Vergo, Metropolitan Community College
Kimberly Walters, Mississippi State University
Aimee Welch, Louisiana State University Laboratory School
Jacci White, Saint Leo University
Ralph L. Wildy Jr., Georgia Military College
Sherri M. Wilson, Fort Lewis College
Xuezheng Wu, Madison College
Janet Wyatt, Metropolitan Community College–Longview
Ghidei Zedingle, Normandale Community College
Brian Zimmerman, University of Mississippi

CHAPTER ONE

An Introduction to Trigonometric Functions

CHAPTER ONE CONTENTS

1.1 An Introduction to Angles: Degree and Radian Measure

THINGS TO KNOW

Before working through this section, be sure that you are familiar with the following concept:

VIDEO   ANIMATION   INTERACTIVE

You Try It

1. Sketching the Graph of a Circle

OBJECTIVES

1 Understanding Degree Measure

2 Finding Coterminal Angles Using Degree Measure

3 Understanding Radian Measure

4 Converting between Degree Measure and Radian Measure

5 Finding Coterminal Angles Using Radian Measure

Introduction to Section 1.1

An angle is made up of two rays that share a common endpoint called the **vertex**. An angle is created by rotating one ray away from a fixed ray. The fixed ray is called the **initial side** of the angle and the rotated ray is called the terminal side of the angle. When we draw an angle, we show the direction in which the **terminal side** is rotated and the amount of rotation. Greek letters such as θ, α, and β (theta, alpha, and beta) or capital letters such as A, B, and C are used to label angles. When we rotate the ray in a *counterclockwise* fashion, we say that the angle has a *positive amount of rotation* (the angle has positive measure). An angle formed from a *clockwise rotation* is said to *have negative measure*. In Figure 1a, angle θ has positive measure. In Figure 1b, angle α has negative measure.

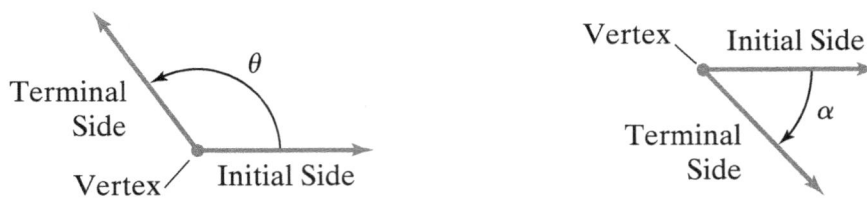

(a) An angle with positive measure   (b) An angle with negative measure

Figure 1

An angle is in **standard position** if the vertex is at the **origin** of a rectangular coordinate system and the initial side of the angle is along the positive x-axis. The terminal side of the angle will always lie in one of the **four quadrants** or on either axis. Angle θ and angle α are sketched again in Figure 2. These angles are now said to be in standard position.

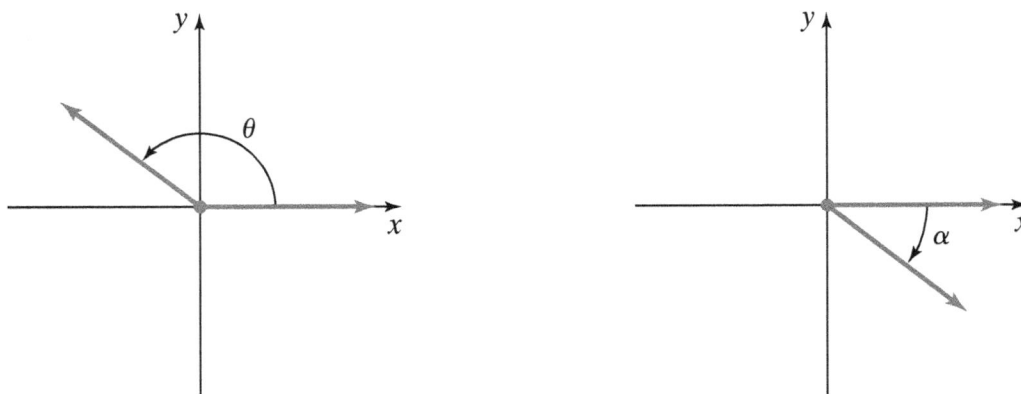

(a) Angle θ has **positive measure** because the direction of rotation is **counterclockwise**. The terminal side of θ lies in Quadrant II.

(b) Angle α has **negative measure** because the direction of rotation is **clockwise**. The terminal side of α lies in Quadrant IV.

Figure 2 Angles in standard position

OBJECTIVE 1   UNDERSTANDING DEGREE MEASURE

DEGREE MEASURE

There are two ways that we will measure angles. We start by introducing **degree measure**. The notation for degrees is the ° symbol. The angle formed by rotating the terminal side one complete counterclockwise rotation has a measure of 360°. See Figure 3. We can draw an angle by knowing the desired amount of rotation necessary. For example, an angle of 90° is equal to $\frac{1}{4}$ of one complete rotation because $\frac{90°}{360°} = \frac{1}{4}$. So, we can draw an angle of 90° by rotating the terminal side of an angle $\frac{1}{4}$ of one complete counterclockwise rotation. See Figure 4. Similarly, an angle of −45° is $\frac{1}{8}$ of one complete clockwise rotation because $\frac{45°}{360°} = \frac{1}{8}$. See Figure 5.

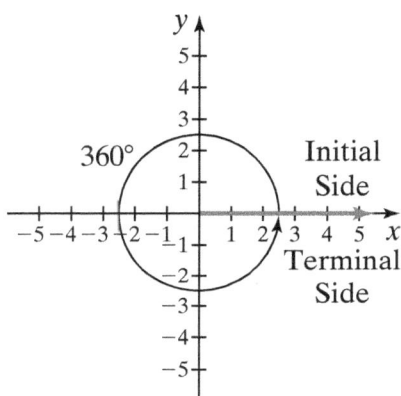

Figure 3 A 360° angle is one complete counterclockwise rotation.

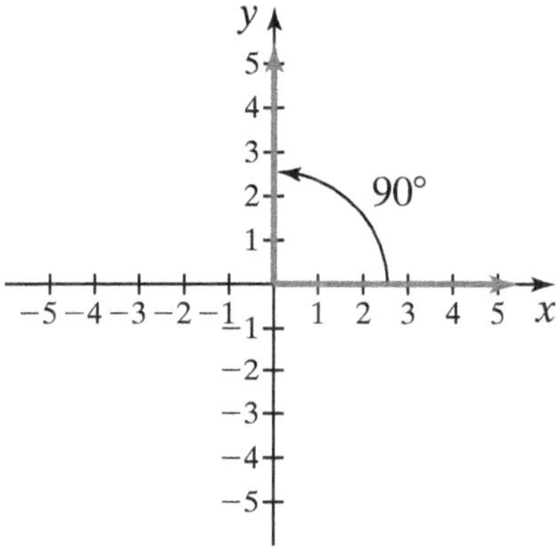

Figure 4 An angle of 90° is $\frac{1}{4}$ of one complete counterclockwise rotation.

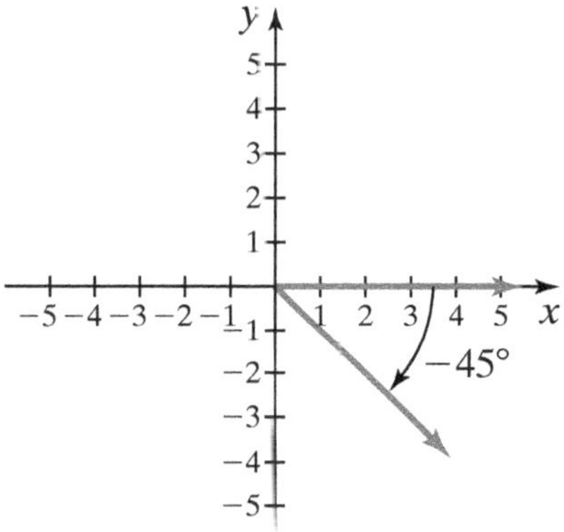

Figure 5 An angle of −45° is $\frac{1}{8}$ of one complete clockwise rotation.

Eventually you should be comfortable sketching the positive and negative angles measured in degrees, as seen in Figure 6. Notice in Figure 6a that all angles that appear in Quadrant I are between 0° and 90°. These are called acute angles. All angles that appear in Quadrant II are between 90° and 180° and are called obtuse angles. Angles whose terminal sides lie along an axis are called quadrantal angles. The quadrantal angle of exactly 90° is called a **right angle**. The quadrantal angle of exactly 180° is called a **straight angle**.

The angles sketched in Figure 6a are formed by rotating the terminal side of the angle in a counterclockwise direction and are thus positive angles. The angles sketched in Figure 6b are negative angles because they are formed by rotating the terminal side of the angle in a clockwise fashion. It is important to point out that even though two angles in standard position may have identical terminal sides, the measure of the two angles may be different. For example, notice in Figure 6a that the angle with a measure of 45° has the exact same terminal side as an angle of −315° sketched in Figure 6b. Angles in standard position having the same terminal side are called **coterminal angles**. Coterminal angles will be discussed in detail in Objective 3 of this section.

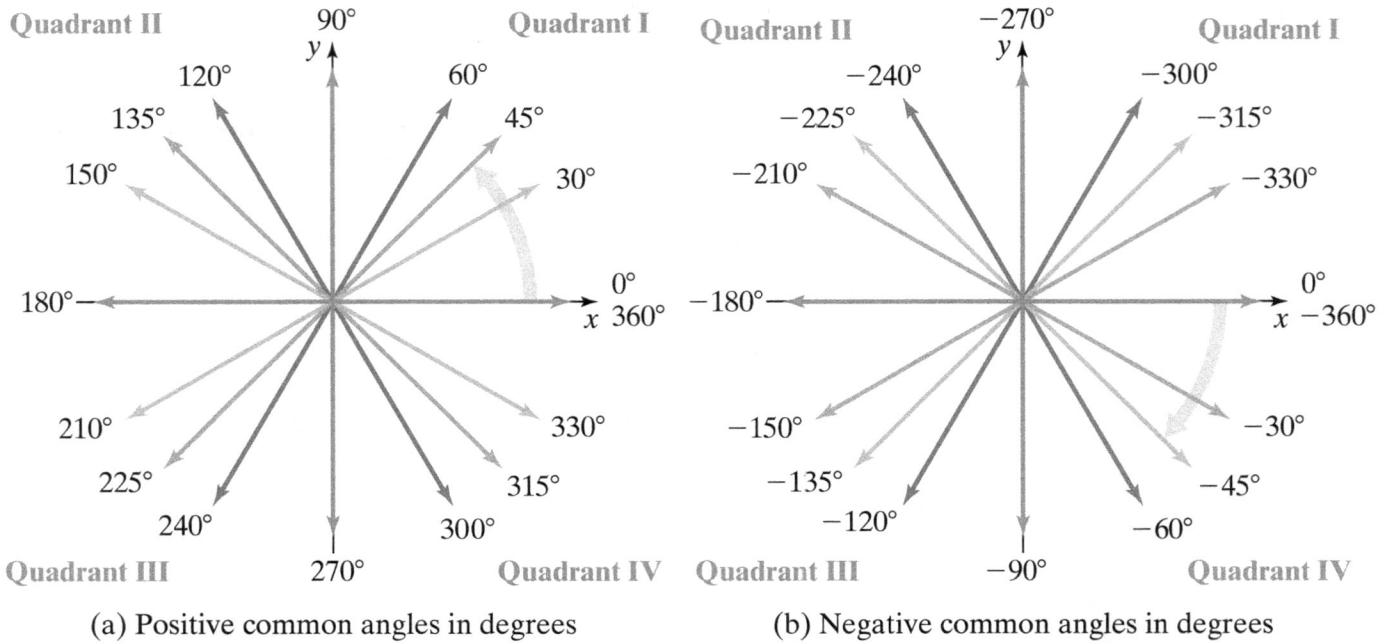

(a) Positive common angles in degrees (b) Negative common angles in degrees

Figure 6

My video summary ▶ **Example 1** Drawing Angles Given in Degree Measure

Draw each angle in standard position and state the quadrant in which the terminal side of the angle lies or the axis on which the terminal side of the angle lies.

a. $\theta = 60°$ **b.** $\alpha = -270°$ **c.** $\beta = 420°$

Solution Watch this video to see how to draw each angle seen below.

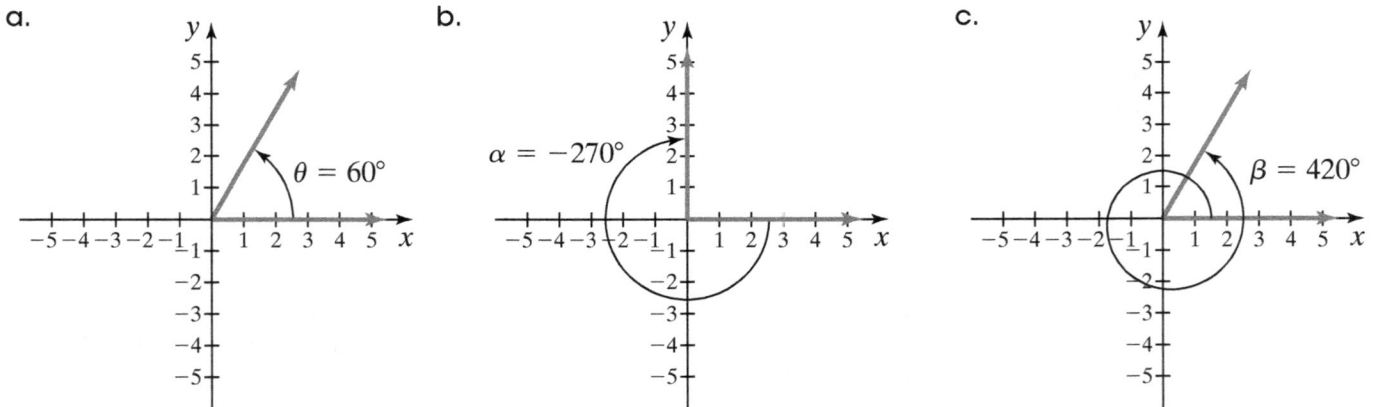

a.

b.

c.

Terminal side lies in Quadrant I. Terminal side lies on the positive y-axis. Terminal side lies in Quadrant I.

Note Not every angle written in degrees has nice integer measurements as in Example 1. An angle can have any decimal measurement, such as $\theta = 41.23°$ or $\alpha = -0.375°$. These angles are said to be written in **degree decimal form**. Angles can also be written in **degrees, minutes, seconds form**. These two forms are discussed in Appendix A.

You Try It Work through this You Try It problem.

Work Exercises 1–5 in this textbook or in the MyMathLab Study Plan.

OBJECTIVE 2   FINDING COTERMINAL ANGLES USING DEGREE MEASURE

Consider the 45°, 405°, and −315° angles in standard position seen in Figure 7. Notice that each angle has the exact same terminal side. Angles in standard position having the same terminal side are called **coterminal angles.**

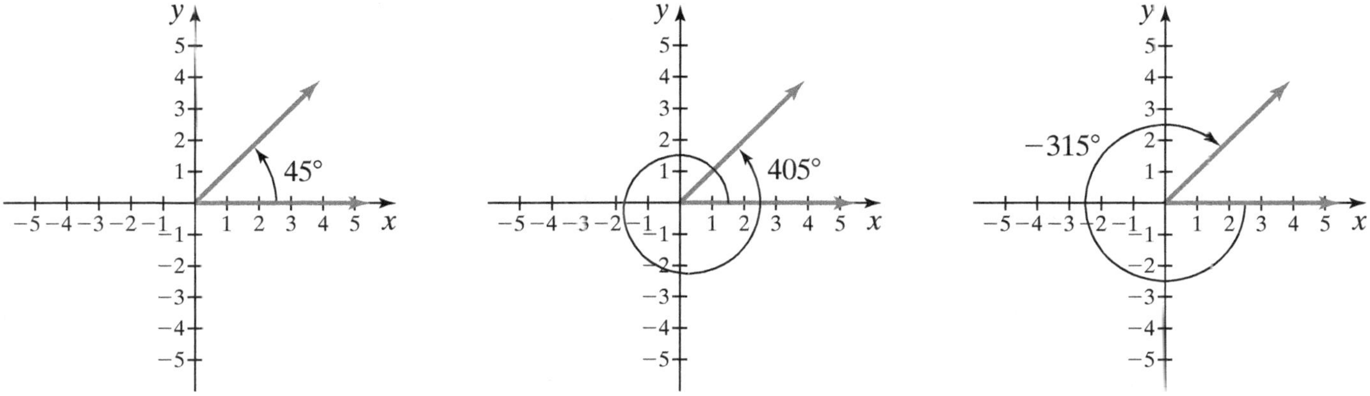

Figure 7
Coterminal angles in standard position have the same terminal side.

Definition   Coterminal Angles

Coterminal angles are angles in standard position having the same terminal side but different measures.

Coterminal angles can be obtained by adding any nonzero integer multiple of 360° to a given angle. Given any angle θ and any nonzero integer k, the angles θ and $\theta + k \cdot 360°$ are coterminal angles.

Below are several examples of angles that are coterminal with the angle 45°.

$$k = 1: \quad 45° + 1 \cdot 360° \quad = 405°$$

$$k = -1: \ 45° + (-1) \cdot 360° = -315°$$

$$k = 2: \quad 45° + 2 \cdot 360° \quad = 765°$$

$$k = -2: \ 45° + (-2) \cdot 360° = -675°$$

The number of angles coterminal to a given angle is infinite. However, we will often be interested in determining the angle of least nonnegative measure that is coterminal with a given angle. Every angle has a coterminal angle of least nonnegative measure. If θ is a given angle, then we will use the notation θ_C to denote the angle of least nonnegative measure coterminal with θ. Note that $0° \leq \theta_C < 360°$ (or $0 \leq \theta_C < 2\pi$.) See Example 2.

My video summary ▷ **Example 2** Finding Coterminal Angles Using Degree Measure

Find the angle of least nonnegative measure, θ_C, that is coterminal with $\theta = -697°$.

Solution The angles coterminal with $\theta = -697°$ have the form $-697° + k(360°)$, where k is an integer. Because θ is a negative angle, we choose positive values of k until we find the angle of least nonnegative measure.

$$k = 1: \ -697° + (1)(360°) = -337°$$

The angle $-337°$ has negative measure, so we now choose $k = 2$.

$$k = 2: \ -697° + (2)(360°) = 23°$$

The angle $23°$ is the smallest *nonnegative* angle coterminal with $\theta = -697°$. Therefore, the angle of least nonnegative measure that is coterminal with $\theta = -697°$ is $\theta_C = 23°$. For the solution to this example, watch this video.

You Try It Work through this You Try It problem.

Work Exercises 6–9 in this textbook or in the My MathLab **Study Plan.**

OBJECTIVE 3 UNDERSTANDING RADIAN MEASURE

My animation summary

You Try It Before we introduce radian measure, it is important that you can sketch a circle centered at the origin whose equation is given in standard form. To practice sketching a circle centered at the origin, work this exercise. Consider the circle $x^2 + y^2 = r^2$ and angle θ seen in Figure 8. An angle whose vertex is at the center of a circle is called a **central angle**. Every central angle intercepts a portion of a circle called an **intercepted arc**. We typically use the variable s to represent the length of an intercepted arc.

When a central angle has an intercepted arc whose length is equal to the radius of the circle, then the central angle is said to have a measurement equal to **1 radian**. It is worth pointing out that 1 radian has a degree measure of about $57.3°$. See Figure 9. Watch this animation for a further explanation of radian measure and to see how to obtain a relationship between radians and degrees.

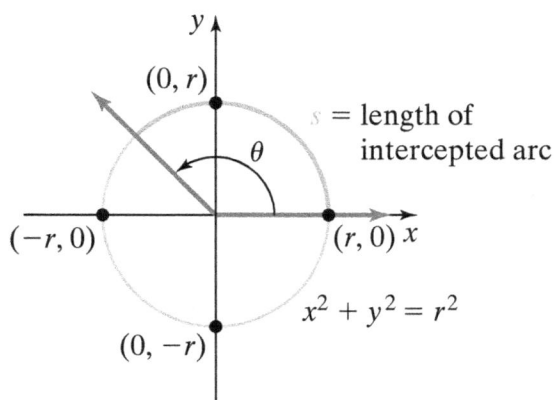

Figure 8 A central angle θ and the corresponding intercepted arc

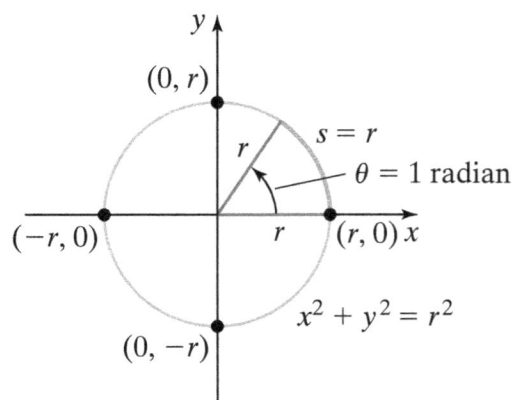

Figure 9 A central angle of $\theta = 1$ radian has a corresponding intercepted arc equal to the length of the radius. An angle of 1 radian is approximately $57.3°$.

Definition Radian

One **radian** is the measure of a central angle that has an intercepted arc equal in length to the radius of the circle.

My animation summary

☐ Make sure that you have carefully watched this animation to verify that there are exactly 2π radians in a circle. This allows us to establish the relationship that $360° = 2\pi$ radians or $180° = \pi$ radians.

Relationship between Degrees and Radians

$$360° = 2\pi \text{ radians}$$

$$180° = \pi \text{ radians}$$

Although we can draw angles that are given in radians by converting the radian measure to degrees, it is important to be able to draw angles given in radian measure by knowing the amount of rotation necessary to sketch them. For example, the angle formed by rotating the terminal side of an angle one complete counterclockwise rotation has a measure of 2π radians. See Figure 10. An angle formed by rotating its terminal side $\frac{1}{4}$ of one complete counterclockwise rotation has a radian measure of $\theta = \frac{\pi}{2}$ because $\frac{1}{4}(2\pi) = \frac{\pi}{2}$. So, we draw an angle of $\theta = \frac{\pi}{2}$ by rotating the terminal side $\frac{1}{4}$ of one complete counterclockwise rotation. See Figure 11. Similarly, an angle formed by rotating its terminal side $\frac{1}{8}$ of one complete clockwise rotation has a radian measure of $\theta = -\frac{\pi}{4}$ because $\frac{1}{8}(-2\pi) = -\frac{\pi}{4}$. See Figure 12.

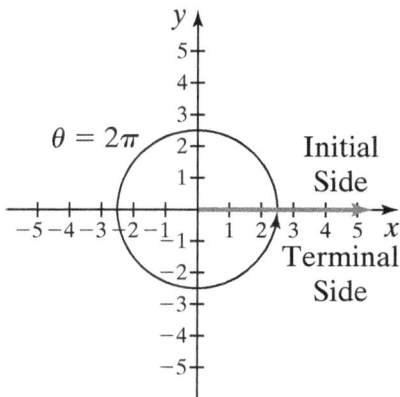

Figure 10 A angle of 2π radians is one complete counterclockwise rotation.

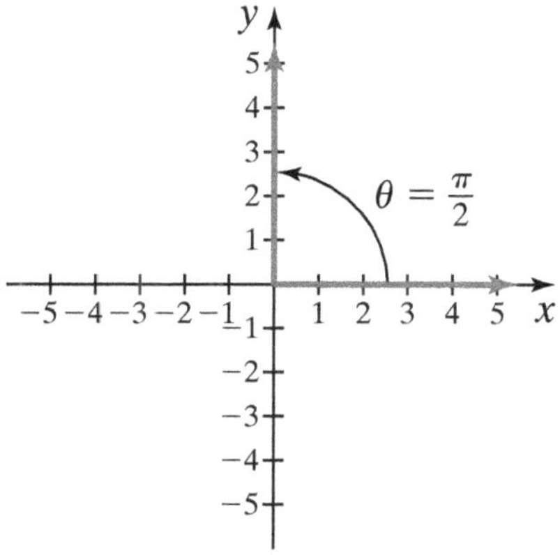

Figure 11 An angle of $\frac{\pi}{2}$ radians is $\frac{1}{4}$ of one complete counterclockwise rotation.

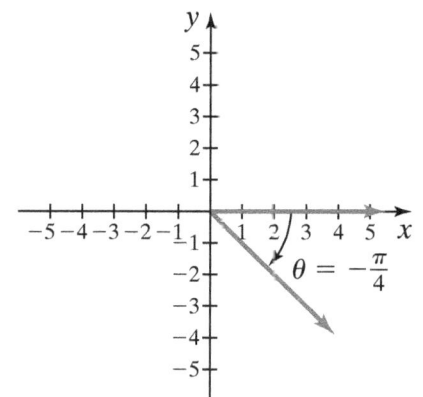

Figure 12 An angle of $-\frac{\pi}{4}$ is $\frac{1}{8}$ of one complete clockwise rotation.

It is important to point out that there is no symbol to indicate radian measure. Although we may use the abbreviation "rad" to represent radians, we will typically omit this abbreviation when angles measured in radians are written in terms of π. For example, we will simply use $\dfrac{\pi}{4}$ to represent $\dfrac{\pi}{4}$ radians. At this point, you should be comfortable sketching angles that are given in degrees or radians. You should be able to sketch each angle seen in Figure 13 as well as determine the quadrant or axis on which the terminal side of the angle lies.

(a) Positive common angles in radians   (b) Negative common angles in radians

Figure 13

My interactive video summary

⊙ **Example 3** Drawing Angles Given in Radian Measure

Draw each angle in standard position and state the quadrant in which the terminal side of the angle lies or the axis on which the terminal side of the angle lies.

a. $\theta = \dfrac{\pi}{3}$ **b.** $\alpha = -\dfrac{3\pi}{2}$ **c.** $\beta = \dfrac{7\pi}{3}$

Solution Watch this interactive video to see how to draw each angle seen below.

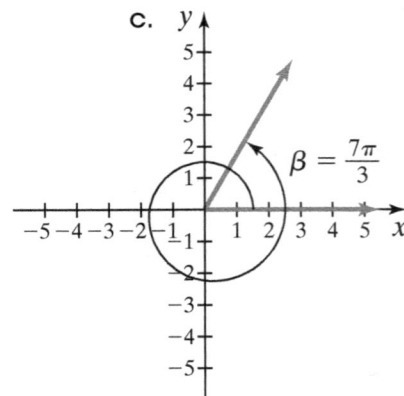

Terminal side lies in Quadrant I.   Terminal side lies on the positive y-axis.   Terminal side lies in Quadrant I.

At first glance it may appear that the angles $\theta = \dfrac{\pi}{3}$ and $\beta = \dfrac{7\pi}{3}$ from Example 3 are the same. Both angles are in standard position and have the same terminal side. However, the angles are quite different because different amounts of rotation are required to sketch the two angles. Recall that angles in standard position that have the same terminal side are called **coterminal angles**. We discuss how to find coterminal angles using radian measure at length in Objective 5. ●

You Try It   Work through this You Try It problem.

Work Exercises 10–15 in this textbook or in the MyMathLab Study Plan.

OBJECTIVE 4   CONVERTING BETWEEN DEGREE MEASURE AND RADIAN MEASURE

My animation summary

We established in this animation that $180° = \pi$ radians. Dividing both sides of this equation 180° or by π radians gives the following rules for converting between degrees and radians.

Converting between Degrees and Radians

To convert degrees to radians, multiply by $\dfrac{\pi \text{ radians}}{180°}$.

To convert radians to degrees, multiply by $\dfrac{180°}{\pi \text{ radians}}$.

My interactive video summary

⊙ **Example 4** Converting from Degree Measure to Radian Measure

Convert each angle given in degree measure into radians.

a. $45°$ b. $-150°$ c. $56°$

Solution To convert from degrees to radians, we multiply by $\dfrac{\pi \text{ radians}}{180°}$.

a. $45° = 45° \cdot \dfrac{\pi \text{ radians}}{180°} \qquad = \dfrac{45\pi}{180} \text{ radians} \quad = \dfrac{\pi}{4} \text{ radians}$

b. $-150° = -150° \cdot \dfrac{\pi \text{ radians}}{180°} = -\dfrac{150\pi}{180} \text{ radians} = -\dfrac{5\pi}{6} \text{ radians}$

c. $56° = 56° \cdot \dfrac{\pi \text{ radians}}{180°} \qquad = \dfrac{56\pi}{180} \text{ radians} \quad = \dfrac{14\pi}{45} \text{ radians}$ ●

You Try It   Work through this You Try It problem.

Work Exercises 16–20 in this textbook or in the MyMathLab Study Plan.

My interactive video summary

⊙ **Example 5** Converting from Radian Measure to Degree Measure

Convert each angle given in radian measure into degrees. Round to two decimal places if needed.

a. $\dfrac{2\pi}{3}$ radians   b. $-\dfrac{11\pi}{6}$ radians   c. 3 radians

Solution To convert from degrees to radians, we multiply by $\dfrac{180°}{\pi \text{ radians}}$.

a. $\quad \dfrac{2\pi}{3} \text{ radians} = \dfrac{2\pi}{3} \text{ radians} \cdot \dfrac{180°}{\pi \text{ radians}} = \dfrac{360°}{3} = 120°$

b. $\quad -\dfrac{11\pi}{6} \text{ radians} = -\dfrac{11\pi}{6} \text{ radians} \cdot \dfrac{180°}{\pi \text{ radians}} = -\dfrac{1980°}{6} = -330°$

c. $\quad 3 \text{ radians} = 3 \text{ radians} \cdot \dfrac{180°}{\pi \text{ radians}} = \dfrac{540°}{\pi} \approx 171.89°$

You Try It Work through this You Try It problem.

Work Exercises 21–26 in this textbook or in the MyMathLab Study Plan.

OBJECTIVE 5   FINDING COTERMINAL ANGLES USING RADIAN MEASURE

Recall that angles in standard position having the same terminal side are called **coterminal angles**. Coterminal angles in degrees can be obtained by adding any nonzero integer multiple of 360° to a given angle. Similarly, coterminal angles given in radians can be obtained by adding a nonzero integer multiple of 2π to a given angle. Therefore, for any angle θ and for any nonzero integer k, we can find a coterminal angle using the expression below:

$$\theta \qquad + \qquad k \cdot 2\pi$$

Original angle   plus   an integer multiple of 2π

Example 6 Finding Coterminal Angles Using Radian Measure

Find three angles that are coterminal with $\theta = \dfrac{\pi}{3}$ using $k = 1$, $k = -1$, and $k = -2$.

Solution Using the expression $\dfrac{\pi}{3} + k \cdot 2\pi$ with $k = 1$, $k = -1$, and $k = -2$, we get the following.

$$k = 1: \quad \dfrac{\pi}{3} + (1)2\pi = \dfrac{\pi}{3} + 2\pi = \dfrac{\pi}{3} + \dfrac{6\pi}{3} = \dfrac{7\pi}{3}$$

$$k = -1: \dfrac{\pi}{3} + (-1)2\pi = \dfrac{\pi}{3} - 2\pi = \dfrac{\pi}{3} - \dfrac{6\pi}{3} = -\dfrac{5\pi}{3}$$

$$k = -2: \dfrac{\pi}{3} + (-2)2\pi = \dfrac{\pi}{3} - 4\pi = \dfrac{\pi}{3} - \dfrac{12\pi}{3} = -\dfrac{11\pi}{3}$$

These angles are sketched in Figure 14.

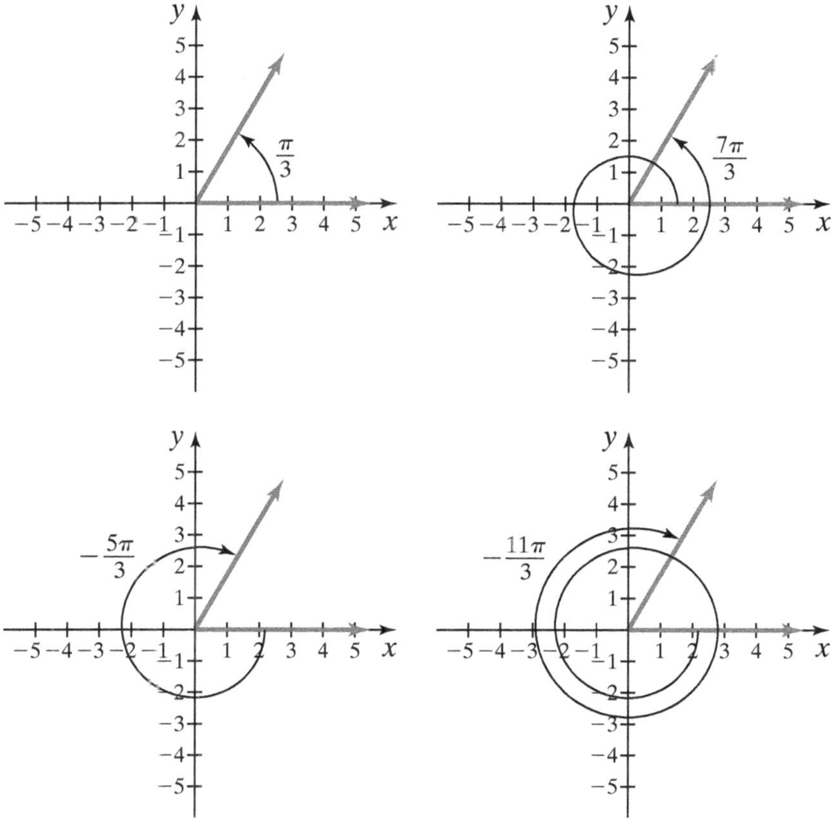

Figure 14 The angles $\dfrac{\pi}{3}, \dfrac{7\pi}{3}, -\dfrac{5\pi}{3}$, and $-\dfrac{11\pi}{3}$ are coterminal angles.

My video summary ▷ **Example 7** Finding Coterminal Angles Using Radian Measure

Find the angle of least nonnegative measure, θ_C, that is coterminal with $\theta = -\dfrac{21\pi}{4}$.

Solution The angles coterminal with $\theta = -\dfrac{21\pi}{4}$ have the form $-\dfrac{21\pi}{4} + k \cdot 2\pi$, where k is an integer. Because θ is a negative angle, we choose positive integer values of k until we find the angle of least nonnegative measure.

$$k = 1: -\frac{21\pi}{4} + (1)2\pi = -\frac{21\pi}{4} + 2\pi = -\frac{21\pi}{4} + \frac{8\pi}{4} = -\frac{13\pi}{4}$$

$$k = 2: -\frac{21\pi}{4} + (2)2\pi = -\frac{21\pi}{4} + 4\pi = -\frac{21\pi}{4} + \frac{16\pi}{4} = -\frac{5\pi}{4}$$

$$k = 3: -\frac{21\pi}{4} + (3)2\pi = -\frac{21\pi}{4} + 6\pi = -\frac{21\pi}{4} + \frac{24\pi}{4} = \frac{3\pi}{4}$$

The angle of least nonnegative measure that is coterminal with $\theta = -\dfrac{21\pi}{4}$ is $\theta_C = \dfrac{3\pi}{4}$.

⚠ **Do not write** $-\dfrac{21\pi}{4} = \dfrac{3\pi}{4}$. **These angles are coterminal but they are not equal.**

You Try It Work through this You Try It problem.

Work Exercises 27–32 in this textbook or in the MyMathLab Study Plan.

1.1 Exercises

Skill Check Exercises

For exercises SCE-1 and SCE-2, simplify the expression.

SCE-1. $(-252)\left(\dfrac{\pi}{180}\right)$

SCE-2. $\left(\dfrac{5\pi}{6}\right)\left(\dfrac{180}{\pi}\right)$

For exercises SCE-3 and SCE-4, rewrite the expression 2π as a fraction using the given denominator.

SCE-3. $2\pi = \dfrac{\square\,\pi}{2}$

SCE-4. $2\pi = \dfrac{\square\,\pi}{5}$

For exercises SCE-5 and SCE-6, determine the numerator on the right-hand side of the equation for the specified value of k.

SCE-5. $\dfrac{8\pi}{4}\cdot k = \dfrac{\square}{4}$ when $k = 2$

SCE-6. $\dfrac{10\pi}{5}\cdot k = \dfrac{\square}{5}$ when $k = -3$

In Exercises 1–5, draw each angle in standard position and state the quadrant in which the terminal side of the angle lies or the axis on which the terminal side of the angle lies.

1. $15°$ **2.** $270°$ **3.** $-150°$ **4.** $480°$ **5.** $-570°$

6. Find the angle of least nonnegative measure, θ_C, that is coterminal with $\theta = 848°$.

7. Find the angle of least nonnegative measure, θ_C, that is coterminal with $\theta = -622°$.

8. Find the angle of least nonnegative measure that is coterminal with $-51°$. Then find the measure of the negative angle that is coterminal with $-51°$ such that the angle lies between $-720°$ and $-360°$.

9. Find two positive angles and two negative angles that are coterminal with the quadrantal angle $\theta = -630°$ such that each angle lies between $-1440°$ and $720°$.

In Exercises 10–15, draw each angle in standard position and state the quadrant in which the terminal side of the angle lies or the axis on which the terminal side of the angle lies.

10. 5π **11.** $\dfrac{4\pi}{3}$ **12.** $\dfrac{5\pi}{2}$ **13.** $-\dfrac{13\pi}{6}$ **14.** $\dfrac{9\pi}{4}$ **15.** $-\dfrac{14\pi}{3}$

In Exercises 16–20, convert each angle given in degree measure into radians.

16. $15°$ **17.** $150°$ **18.** $-540°$ **19.** $96°$ **20.** $-357°$

In Exercises 21–26, convert each angle given in radian measure into degrees. Round to two decimal places if needed.

21. $\dfrac{\pi}{12}$ radians   **22.** $\dfrac{3\pi}{5}$ radians   **23.** $-\dfrac{5\pi}{2}$ radians

24. $\dfrac{8\pi}{7}$ radians   **25.** $-\dfrac{29\pi}{15}$ radians   **26.** 3 radians

27. Find the angle of least nonnegative measure, θ_C, that is coterminal with $\theta = \dfrac{27\pi}{8}$.

28. Find the angle of least nonnegative measure, θ_C, that is coterminal with $\theta = \dfrac{41\pi}{6}$.

29. Find the angle of least nonnegative measure, θ_C, that is coterminal with $\theta = -\dfrac{29\pi}{2}$.

30. Find the angle of least nonnegative measure, θ_C, that is coterminal with $\theta = -\dfrac{\pi}{70}$.

31. Find the angle of least nonnegative measure, θ_C, that is coterminal with $-\dfrac{11\pi}{12}$. Then find the

measure of the negative angle that is coterminal with $-\dfrac{11\pi}{12}$ such that the angle lies between -4π and -2π.

32. Find two positive angles and two negative angles that are coterminal with the quadrantal angle $\theta = -\dfrac{7\pi}{2}$ such that each angle lies between -8π and 4π.

1.2 Applications of Radian Measure

THINGS TO KNOW

Before working through this section, be sure that you are familiar with the following concept:

You Try It

1. Converting between Degree Measure and Radian Measure (Section 1.1)

VIDEO   ANIMATION   INTERACTIVE

OBJECTIVES

1 Determining the Area of a Sector of a Circle

2 Computing the Arc Length of a Sector of a Circle

3 Understanding Linear Velocity and Angular Velocity

Introduction to Section 1.2

My animation summary

▣ In this section, we will look at various applications of radian measure. You may want to watch this animation, which was introduced in the previous section, to review the definition of a radian and to verify that you understand the relationship between radians and degrees.

OBJECTIVE 1   DETERMINING THE AREA OF A SECTOR OF A CIRCLE

A **sector** of a circle is a portion of a circle bounded by two radii and an arc of the circle. See Figure 15.

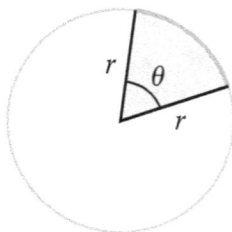

Figure 15 A sector of a circle

Recall that the area of a circle is $A = \pi r^2$. The area of a sector of a circle is simply a portion of the total area. To establish a formula for the area of a sector of a circle, let's look at a concrete example. Suppose that the central angle of a sector of a circle is $\theta = \dfrac{\pi}{4}$. We know that there are 2π radians in an entire circle. Thus, the sector with a central angle of $\theta = \dfrac{\pi}{4}$ represents a portion of the circle equal to $\dfrac{\frac{\pi}{4}}{2\pi} = \dfrac{\pi}{4} \cdot \dfrac{1}{2\pi} = \dfrac{1}{8}$.

Therefore, the area of the sector of the circle must be $\dfrac{1}{8}$ of the area of the entire circle. So, the area of a sector of a circle with a central angle of $\theta = \dfrac{\pi}{4}$ is $A = \dfrac{1}{8}\pi r^2$.

In general, if θ is the central angle of a sector of a circle in radians, then the sector represents a portion of the circle equal to $\dfrac{\theta}{2\pi}$ of the entire circle. Thus, the area, A, of a sector of a circle is given by $A = \left(\dfrac{\theta}{2\pi}\right)(\pi r^2) = \dfrac{1}{2}\theta r^2$.

Area of a Sector of a Circle

The area, A, of a sector of a circle with radius, r, and central angle θ radians is given by $A = \dfrac{\theta}{2\pi}(\pi r^2) = \dfrac{1}{2}\theta r^2$.

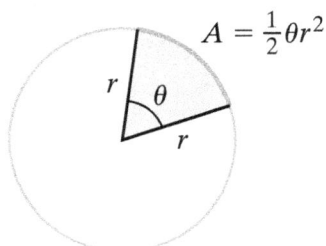

$$A = \tfrac{1}{2}\theta r^2$$

Example 1 Computing the Area of a Sector of a Circle

Find the area of the sector of a circle of radius 15 inches formed by a central angle of $\theta = \dfrac{\pi}{10}$ radians. Round the answer to two decimal places.

Solution We can use the formula $A = \dfrac{1}{2}\theta r^2$ with $\theta = \dfrac{\pi}{10}$ and $r = 15$ inches.

$$A = \frac{1}{2}\theta r^2 \qquad \text{Write the formula for the area of a sector of a circle.}$$

$$= \frac{1}{2}\left(\frac{\pi}{10}\right)(15\text{ in.})^2 \qquad \text{Substitute } \theta = \frac{\pi}{10} \text{ and } r = 15 \text{ in.}$$

$$= \left(\frac{\pi}{20}\right)(225\text{ in.}^2) \qquad \text{Multiply.}$$

$$= \frac{45\pi}{4}\text{ in.}^2 \qquad \text{Simplify.}$$

$$\approx 35.34\text{ in.}^2 \qquad \text{Round to two decimal places.}$$

The area of the sector of the circle is approximately 35.34 square inches.

> ⚠ **The formula for the area of a sector of a circle, $A = \dfrac{1}{2}\theta r^2$, is only valid if the angle θ is in radians. An angle given in degrees must first be converted to radians as in Example 2. To review how to convert degree measure to radian measure, watch this video.**

My video summary

⊙ Example 2 Computing the Area of a Sector of a Circle

Find the area of the sector of a circle of diameter 21 meters formed by a central angle of $135°$. Round the answer to two decimal places.

Solution Because the radius of the circle is half the diameter, we know that $r = \dfrac{21}{2} = 10.5$ meters. We must now convert the central angle of $135°$ to radians.

$$135° = 135° \frac{\pi \text{ radians}}{180°} = \frac{135\pi}{180} \text{ radians} = \frac{3\pi}{4} \text{ radians}$$

We can now find the area of the sector of the circle using $r = 10.5$ and $\theta = \dfrac{3\pi}{4}$.

Therefore, the area of the sector of the circle is

$$A = \frac{1}{2}\theta r^2 = \frac{1}{2}\left(\frac{3\pi}{4}\right)(10.5)^2 \approx 129.89 \text{ square meters.}$$

Watch this video to see every step of this solution.

You Try It Work through this You Try It problem.

Work Exercises 1–6 in this textbook or in the MyMathLab Study Plan.

OBJECTIVE 2 COMPUTING THE ARC LENGTH OF A SECTOR OF A CIRCLE

The arc length of a sector of a circle depends on the corresponding central angle that intercepts the arc and the length of the radius of the circle. See Figure 16.

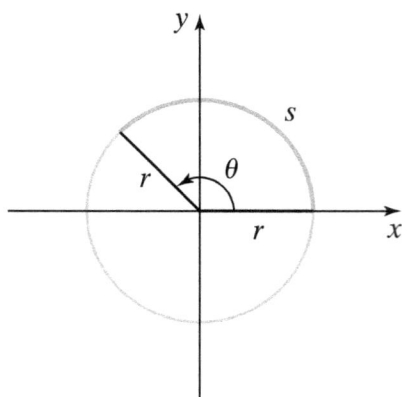

Figure 16 The arc length of a sector of a circle s, depends on the corresponding central angle θ and the length of the radius r.

Recall that the circumference of an entire circle is equal to $2\pi r$. The arc length of a sector of a circle is simply a portion of the circumference. We established that a sector represents a portion of the circle equal to $\dfrac{\theta}{2\pi}$ of the entire circle.

Thus, the arc length, s, of a sector of a circle is given by $s = \left(\dfrac{\theta}{2\pi}\right)(2\pi r) = r\theta$.

Arc Length of a Sector of a Circle

If θ is a central angle (in radians) on a circle of radius r, then the length of the intercepted arc, s, is given by $s = r\theta$.

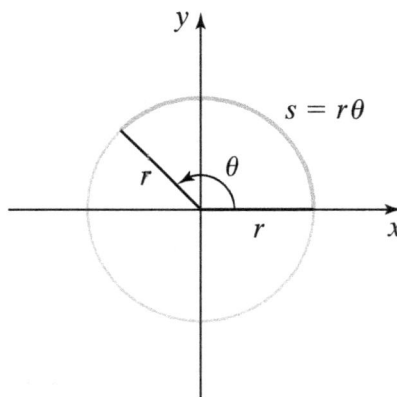

Example 3 Computing the Arc Length of a Sector of a Circle

Find the length of the arc intercepted by a central angle of $\theta = \dfrac{\pi}{6}$ in a circle of radius $r = 22$ centimeters. Round the answer to two decimal places.

Solution We can use the arc length formula $s = r\theta$ with $\theta = \dfrac{\pi}{6}$ and $r = 22$ cm.

Therefore, the length of the arc is $s = r\theta = (22)\left(\dfrac{\pi}{6}\right) = \dfrac{22\pi}{6} = \dfrac{11\pi}{3} \approx 11.52$ cm.

⚠ **The arc length formula $s = r\theta$ is only valid if the angle θ is in radians. An angle given in degrees must first be converted to radians. To review how to convert degree measure to radian measure, watch this video.**

✎ *My video summary* ▷ **Example 4** Calculate the Radius of a Circle

A 120° central angle intercepts an arc of 23.4 inches. Calculate the radius of the circle. Round the answer to two decimal places.

Solution First, we must convert the angle to radian measure.

$$120° = 120° \cdot \frac{\pi \text{ radians}}{180°} = \frac{120\pi}{180} \text{ radians} = \frac{2\pi}{3} \text{ radians}$$

Now, substitute $s = 23.4$ in. and $\theta = \dfrac{2\pi}{3}$ into the arc length formula $s = r\theta$ and solve for r.

$s = r\theta$	Write the formula for the arc length of a sector of a circle.
$23.4 \text{ in.} = r\left(\dfrac{2\pi}{3}\right)$	Substitute $s = 23.4$ in. and $\theta = \dfrac{2\pi}{3}$.
$\dfrac{3}{2\pi} \cdot 23.4 \text{ in.} = r\left(\dfrac{2\pi}{3}\right) \cdot \dfrac{3}{2\pi}$	Multiply both sides by the reciprocal of $\dfrac{2\pi}{3}$.
$11.17 \text{ in.} \approx r$	Simplify and round to two decimal places.

The radius is $r \approx 11.17$ inches. For the solution to this example, watch this video. ●

✎ *My video summary* ▷ **Example 5** Calculate the Angle of Rotation of a Mechanical Gear

Two gears are connected so that the smaller gear turns the larger gear. When the smaller gear with a radius of 3.6 centimeters rotates 300°, how many degrees will the larger gear with a radius of 7.5 cm rotate?

Solution We will first determine how far the smaller gear rotates. The smaller gear has a radius of $r_1 = 3.6$ cm and an angle of rotation of $\theta_1 = 300°$, which is equivalent to $\theta_1 = \dfrac{5\pi}{3}$. (Watch this video to verify.) We can use the arc length formula $s = r_1\theta_1$ to find the distance that the small gear rotates.

$r_2 = 7.5$ cm
$r_1 = 3.6$ cm

$s = r_1\theta_1$	Write the arc length formula.
$= (3.6 \text{ cm})\left(\dfrac{5\pi}{3}\right)$	Substitute $r_1 = 3.6$ cm and $\theta_1 = \dfrac{5\pi}{3}$.
$= 6\pi \text{ cm}$	Multiply.

Because the gears are connected, the larger gear will also rotate a distance of 6π cm. Therefore, we can use the radius of the larger gear, $r_2 = 7.5$ cm, and the arc length, $s = 6\pi$ cm, to find the amount of rotation of the larger gear.

$$s = r_2\theta_2 \qquad \text{Write the arc length formula.}$$

$$6\pi \text{ cm} = (7.5 \text{ cm})(\theta_2) \qquad \text{Substitute } s = 6\pi \text{ cm and } r_2 = 7.5 \text{ cm.}$$

$$\frac{6\pi}{7.5} = \theta_2 \qquad \text{Divide both sides by 7.5 cm.}$$

$$\theta_2 = \frac{4\pi}{5} \qquad \text{Simplify and rewrite.}$$

The larger gear rotates $\theta_2 = \dfrac{4\pi}{5}$ radians which can be converted to $\theta_2 = 144°$. You may want to watch this video to see the solution to this example worked out in detail.

You Try It Work through this You Try It problem.

Work Exercises 7–14 in this textbook or in the My MathLab **Study Plan.**

OBJECTIVE 3 UNDERSTANDING ANGULAR VELOCITY AND LINEAR VELOCITY

Suppose that an object is moving along a circular path such as the rotor blade of a helicopter, children riding on a merry-go-round, or a tire on a moving bicycle. In the following discussion, we assume that all objects are moving along a circular path at a constant speed.

There are two different measures of the speed or velocity of an object traveling on a circular path. **Angular velocity** measures the rotation of the object over time and is usually measured in radians per unit of time. **Linear velocity** measures the distance that a particular point on the object travels over time and is usually measured in distance (such as miles or meters) per unit of time. Typically, the Greek letter ω (omega) is used to describe angular velocity and the letter v is used to describe linear velocity.

Angular Velocity

The angular velocity, ω, of an object moving along a circular path is the amount of rotation in radians of the object per unit of time and is given by $\omega = \dfrac{\theta}{t}$.

Linear Velocity

The linear velocity, v, of an object moving along a circular path is the distance traveled by the object, s, per unit of time, t, and is given by $v = \dfrac{s}{t}$. Note that $s = r\theta$, where θ is given in radians.

Example 6 Computing the Angular Velocity and the Linear Velocity of the Earth

Determine the angular velocity and the linear velocity of a point on the equator of the Earth. Assume that the radius of the Earth at the equator is 3963 miles.

Solution To determine the angular velocity of a point on the equator of the Earth, we first use the fact that the Earth makes one complete revolution every 24 hours. Therefore, any point on the surface of the Earth, not at a pole, rotates $\theta = 2\pi$ radians every 24 hours. Thus, a point on the equator of the Earth has an angular velocity of

$$\omega = \frac{\theta}{t} = \frac{2\pi \text{ rad}}{24 \text{ hr}} = \frac{\pi}{12} \frac{\text{rad}}{\text{hr}}.$$

To determine the linear velocity of a point on the equator of the Earth, we use the fact that an object at the equator travels a distance, s, equal to the circumference of the Earth, in 24 hours. Therefore, an object at the equator travels $s = 2\pi r = (2\pi)(3963 \text{ miles}) = 7926\pi$ miles every 24 hours. Thus, a point on the equator of the Earth has a linear velocity of $v = \frac{s}{t} = \frac{7926\pi \text{ miles}}{24 \text{ hr}} \approx 1038$ mph. (See Figure 17.)

Figure 17 The angular velocity of an object located on the equator of the Earth is $\omega = \frac{\pi}{12} \frac{\text{rad}}{\text{hr}}$. The linear velocity of the same object is $v \approx 1038$ mph.

Example 7 Computing Angular Velocity

The propeller of a small airplane is rotating 1500 revolutions per minute. Find the angular velocity of the propeller in radians per minute.

Solution The propeller rotates 1500 revolutions per minute. In one minute, the propeller rotates through an angle of $\theta = \frac{2\pi \text{ rad}}{\text{rev}} \cdot 1500 \text{ rev} = 3000\pi$ rad. We now use the formula $\omega = \frac{\theta}{t}$ with $\theta = 3000\pi$ rad and $t = 1$ min to get $\omega = \frac{3000\pi \text{ rad}}{1 \text{ min}} = 3000\pi \frac{\text{rad}}{\text{min}}$.

Example 8 Computing Linear Velocity

In 2008, the old Ferris wheel at the world-famous Santa Monica pier in Santa Monica, California, was replaced with a new solar-powered Ferris wheel that contains thousands of energy-efficient LED (light-emitting diode) lights that can illuminate into many different colors and designs. The wheel is approximately 85 feet in diameter and rotates 2.5 revolutions per minute. Find the linear velocity (in mph) of this new Ferris wheel at the outer edge. Round to two decimal places.

Solution The linear velocity is given by the formula $v = \dfrac{s}{t}$, where s is the distance traveled and t is time.

First, we will find the distance travelled, s. Recall that $s = r\theta$, where θ is the central angle and r is the radius. There are 2π radians in one complete rotation. The wheel rotates 2.5 revolutions per minute, so in 1 minute the wheel rotates through an angle of $\theta = (2.5)(2\pi) = 5\pi$ radians. We can now use one-half of the diameter, or $r = 42.5$ feet, to get the distance traveled in 1 minute.

$$s = r\theta \qquad \text{Write the formula for arc length.}$$

$$s = (42.5\ \text{ft})(5\pi) \qquad \text{Substitute } r = 42.5 \text{ ft and } \theta = 5\pi.$$

$$s = (212.5)\pi\ \text{ft} \qquad \text{Multiply.}$$

Now we find the linear velocity, v. A point on the outer edge of the Ferris wheel travels $(212.5)\pi$ feet in 2.5 revolutions (or in 1 minute). Therefore,

$$v = \dfrac{s}{t} \qquad \text{Write the formula for linear velocity.}$$

$$v = \dfrac{(212.5)\pi\ \text{ft}}{1\ \text{min}} \qquad \text{Substitute } s = (212.5)\pi \text{ ft and } t = 1 \text{ minute.}$$

To convert the linear velocity to miles per hour, we use the fact that there are 60 minutes in 1 hour and that there are 5280 feet in 1 mile.

$$v = \dfrac{(212.5)\pi\ \cancel{\text{ft}}}{1\ \cancel{\text{min}}} \cdot \dfrac{60\ \cancel{\text{min}}}{1\ \text{hr}} \cdot \dfrac{1\ \text{mile}}{5280\ \cancel{\text{ft}}} \approx 7.59\ \text{mph}$$

Therefore, the outer edge of the Ferris wheel has a linear velocity of approximately 7.59 miles per hour.

Because angular velocity and linear velocity are related, we can develop a formula to convert easily from one to the other.

$$v = \dfrac{s}{t} \qquad \text{Write the formula for linear velocity.}$$

$$v = \dfrac{r\theta}{t} \qquad \text{Use the arc length formula } s = r\theta \text{ and substitute } r\theta \text{ for } s.$$

$$v = r\dfrac{\theta}{t} \qquad \dfrac{r\theta}{t} = r\dfrac{\theta}{t}$$

$$v = r\omega \qquad \text{Use the formula for angular velocity } \omega = \dfrac{\theta}{t} \text{ and substitute } \omega \text{ for } \dfrac{\theta}{t}.$$

Relationship between Linear Velocity and Angular Velocity

If ω is the angular velocity of an object moving along a circular path of radius r, then the linear velocity, v, is given by $v = r\omega$.

To illustrate the relationship between linear velocity and angular velocity, let's revisit Example 6, where we computed the angular velocity and linear velocity of an object located on the equator of the Earth.

Object Located on the Equator of the Earth:

$$\omega = \frac{\pi}{12}\frac{rad}{hr}$$

$$v = r\omega = (3963 \text{ miles})\left(\frac{\pi}{12}\frac{rad}{hr}\right) \approx 1038 \text{ mph}^*$$

You can see that if we know the angular velocity and the radius of the circular object, then we can easily determine the linear velocity.

*An angle measured in radians can be thought of as the ratio of an arc length of a sector of a circle to the radius, or $\theta = \frac{s}{r}$. Because s and r have the same unit of measure, the ratios of those units cancel each other out. Thus, a radian has no unit of measure and is said to be "dimensionless." When multiplying an expression involving radians with another expression involving any unit of measure, the radians can always be dropped from the final expression.

Example 9 Computing Linear Velocity

Suppose that the propeller blade from Example 7 is 2.5 meters in diameter. Find the linear velocity in meters per minute for a point located on the tip of the propeller.

Solution Using the formula $v = r\omega$ with $r = \frac{2.5}{2} = 1.25$ meters and $\omega = 3000\pi\frac{rad}{min}$, we get

$$v = r\omega = (1.25 \text{ m})\left(3000\pi\frac{rad}{min}\right) = 3750\pi\frac{m}{min} \approx 11{,}781\frac{m}{min}.$$

You Try It Work through this You Try It problem.

Work Exercises 15–21 in this textbook or in the MyMathLab Study Plan.

1.2 Exercises

1. Find the area of the sector of a circle of radius 12 cm formed by a central angle of $\frac{1}{12}$ radian.

2. Find the area of the sector of a circle of radius 30 inches formed by a central angle of $\frac{\pi}{5}$ radians.

3. Find the area of the sector of a circle of radius 8 meters formed by a central angle of 60°.

4. A 16-inch-diameter pizza is cut into various sizes. What is the area of a piece that was cut with a central angle of 32°?

5. A sector of a circle has an area of 30 square feet. Find the central angle that forms the sector if the radius is 7 feet.

6. A farmer uses an irrigation sprinkler system that pivots about the center of a circular field. Determine the length of the sprinkler if an area of 2140 square feet is watered after a pivot of 50°.

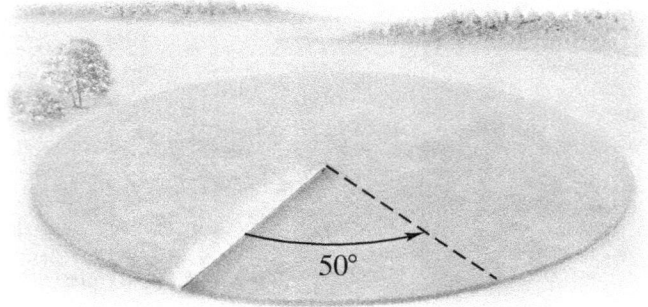

7. Find the length of the arc intercepted by a central angle of $\theta = \dfrac{\pi}{10}$ in a circle of radius $r = 30$ inches.

8. Find the length of the arc intercepted by a central angle of $\theta = \dfrac{4\pi}{3}$ in a circle of radius $r = 54$ centimeters.

9. Find the length of the arc intercepted by a central angle of $\theta = \dfrac{5\pi}{6}$ in a circle of radius $r = 31$ meters.

10. Find the length of the arc intercepted by a central angle of 305° in a circle of radius $r = 40$ kilometers.

11. Find the measure of the central angle (in radians) of a circle of radius $r = 25$ inches that intercepts an arc of length $s = 40$ inches.

12. On the planet Mercury, a 45° central angle intercepts an arc of approximately 379π miles. What is the radius of the planet mercury at the equator?

13. Find the central angle in degrees of an arc of 5000 km on the surface of Neptune if the radius of Neptune is 24,553 km.

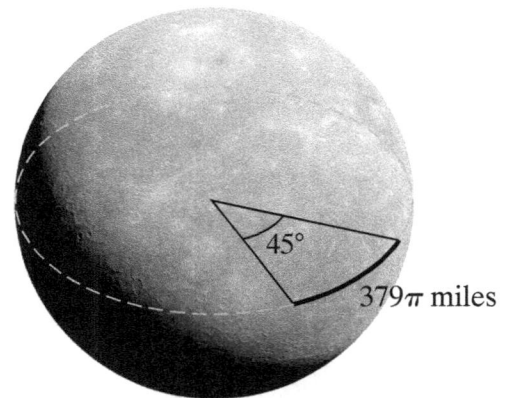

14. Two gears of a small mechanical device are connected so that the smaller gear turns the larger gear. If the smaller gear with a radius of 12.2 millimeters rotates 225°, how many degrees will the larger gear with a radius of 22.5 millimeters rotate?

15. Find the angular velocity in radians per minute of a microwave turntable that turns through an angle of 36° each second.

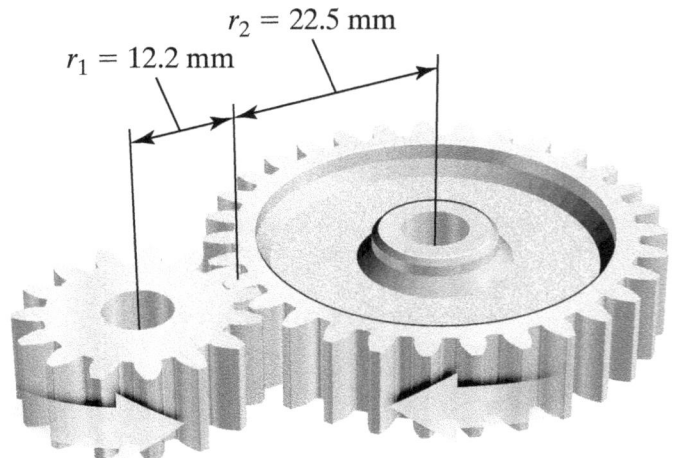

16. The use of wind-powered energy is becoming more and more prevalent around the world to produce clean energy. A windmill with a blade 40 feet long rotates at a rate of 50 revolutions per minute. Find the angular velocity of the blade.

17. What is the linear velocity of the tip of the windmill blade described in Exercise 16?

18. In 2009, the Singapore Flyer was the largest Ferris wheel in the world at approximately 75 meters in diameter. The wheel rotates once every 30 minutes. Find the linear speed of the Singapore Flyer at the outer edge of the wheel.

19. The blade of a circular saw rotates at a rate of 3000 revolutions per minute. What is the linear velocity in miles per hour of a point on the tip of the outer edge of a $7\frac{1}{4}$-inch-diameter blade?

20. Fishing line is being pulled from a circular fishing reel at a rate of 54 cm per second. Find the radius of the reel if it makes 160 revolutions per minute.

21. Suppose that two 3-blade windmills rotate 80 times per minute. What is the difference in linear velocity, in miles per hour, of the tips of the windmill blades if the larger windmill has 75-foot blades and the smaller windmill has 60-foot blades?

1.3 Triangles

THINGS TO KNOW

Before working through this section, be sure that you are familiar with the following concept:

VIDEO   ANIMATION   INTERACTIVE

You Try It

1. Converting between Degree Measure and Radian Measure (Section 1.1)

□   ⊘

OBJECTIVES

1 Classifying Triangles

2 Using the Pythagorean Theorem

OBJECTIVE 1   CLASSIFYING TRIANGLES

The word *trigonometry* comes from the Greek words *trigonon*, meaning "triangle," and *metron*, meaning "measure." In the first section of Chapter 1, we introduced angle measure. In this section, we introduce some fundamentals of triangles. You have undoubtedly worked with triangles in previous courses. Thus, much of this section may be review for you.

Every triangle has three angles and three sides. The sum of the measures of the three angles is equal to $180°$ or π radians. Two angles of a triangle are **congruent** if the measures of the two angles are the same. Similarly, two sides of a triangle are congruent if the lengths of the two sides are equal.

We can classify triangles based on the measures of their angles or the lengths of their sides. We start by classifying triangles based on their angle measures. We will use the notation $\sphericalangle$ to represent an angle. The notation $\sphericalangle A$ refers to angle A. A triangle with three acute angles is called an **acute triangle**. A triangle with one obtuse angle is called an **obtuse triangle**. A triangle with one right angle is called a **right triangle**. The side opposite the right angle of a right triangle is called the **hypotenuse** and the two other sides of a right triangle are called **legs**. Also, note that the hypotenuse and both legs can be referred to as "sides" but that the hypotenuse is never referred to as a "leg." Figure 18 illustrates the three classifications of triangles based on the measures of their angles.

Acute Triangle

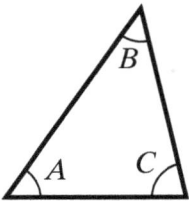

Obtuse Triangle

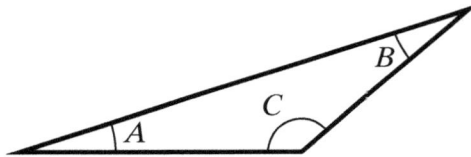

Right Triangle

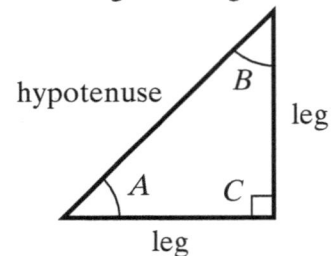

Acute triangles have three acute angles.
$\sphericalangle A < 90°$, $\sphericalangle B < 90°$, and $\sphericalangle C < 90°$

Obtuse triangles have one obtuse angle.
$\sphericalangle A < 90°$, $\sphericalangle B < 90°$, and $\sphericalangle C > 90°$

Right triangles have one right angle.
$\sphericalangle A < 90°$, $\sphericalangle B < 90°$, and $\sphericalangle C = 90°$

Figure 18
Classification of triangles based on their angle measures

We can also classify triangles based on their side lengths. We can represent that two sides of a triangle are congruent by drawing the same number of "tick marks" through the congruent sides. A triangle with no congruent sides is called a **scalene triangle**. See Figure 19a. A triangle with two congruent sides is called an **isosceles triangle**. See Figure 19b. The angles opposite the congruent sides of an isosceles triangle are also congruent. A special triangle with three congruent sides is called an **equilateral triangle**. Note that all three angles of an equilateral

triangle are congruent with a measure of $60°$ or $\dfrac{\pi}{3}$ radians. See Figure 19c.

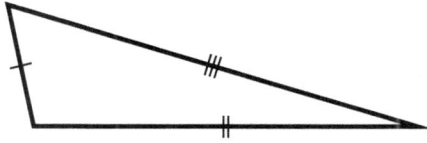

(a) Scalene triangles have no congruent sides. Each side is represented by a different number of "tick marks" that represents that the length of each side is different.

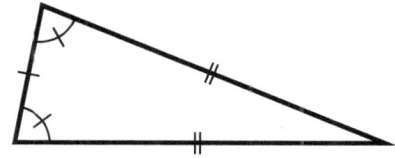

(b) Isosceles triangles have two congruent sides. The angles opposite the two congruent sides are congruent.

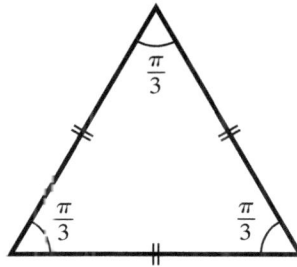

(c) Equilateral triangles have three congruent sides and three congruent angles. Each angle has a measure of 60° or $\frac{\pi}{3}$ radians.

Figure 19
The classification of triangles based on their side lengths

Example 1 Classifying a Triangle

Classify the given triangle as acute, obtuse, right, scalene, isosceles, or equilateral. State all that apply.

Solution The triangle is a right triangle because it has a right angle. The two legs of the triangle are congruent, as indicated by the single tick marks. Therefore, this is an isosceles right triangle.

My animation summary

⬚ The triangle in Example 1 is an isosceles right triangle. It is important to point out that every isosceles right triangle has two acute angles that have a measure of $\frac{\pi}{4}$ radians or 45°. Do you know why? Watch this animation for an explanation.

You Try It Work through this You Try It problem.

Work Exercises 1–3 in this textbook or in the MyMathLab Study Plan.

OBJECTIVE 2   USING THE PYTHAGOREAN THEOREM

The Pythagorean Theorem is one of the most widely recognized theorems in all of mathematics. The Pythagorean Theorem states a unique relationship between the lengths of the legs of a right triangle and the hypotenuse. This relationship does not hold for acute triangles or for obtuse triangles.

The Pythagorean Theorem

Given any right triangle, the sum of the squares of the lengths of the legs is equal to the square of the length of the hypotenuse.

In other words, if a and b are the lengths of the two legs and if c is the length of the hypotenuse, then $a^2 + b^2 = c^2$. See Figure 20.

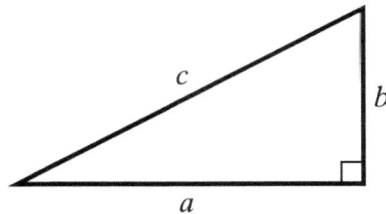

Figure 20 The Pythagorean Theorem: $a^2 + b^2 = c^2$

Example 2 Using the Pythagorean Theorem to Find the Lengths of the Missing Sides

Use the Pythagorean Theorem to find the length of the missing side of the given right triangle.

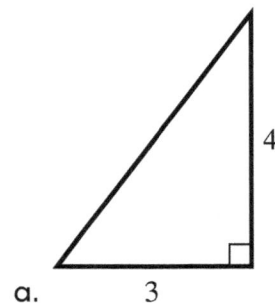

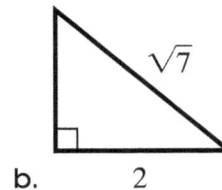

a.   3

b.   2

Solution

a. We are given the length of both legs. Using the equation $a^2 + b^2 = c^2$ with $a = 3$ and $b = 4$, we can solve for c.

$$a^2 + b^2 = c^2 \qquad \text{Write the equation representing the Pythagorean Theorem.}$$

$$(3)^2 + (4)^2 = c^2 \qquad \text{Substitute } a = 3 \text{ and } b = 4.$$

$$9 + 16 = c^2 \qquad \text{Simplify.}$$

$$25 = c^2 \qquad \text{Add.}$$

$$\pm 5 = c \qquad \text{Use the square root property.}$$

Because the length of the hypotenuse must be positive, the length is 5.

b. See if you can determine the length of the missing leg. When you think you are correct, view the solution.

Notice in Example 2a that all three sides of the right triangle have a length represented by an integer. This is very special, and these numbers are called Pythagorean triples. In addition to 3-4-5, other examples of Pythagorean triples are 5-12-13 and 8-15-17. Multiples of these numbers are also Pythagorean triples, such as 6-8-10, 9-12-15, 12-16-20, and so on. Being familiar with these Pythagorean triples can often save you time and effort in working trigonometric exercises, so you might want to become very familiar with them. You may want to view this interesting technique used for generating Pythagorean triples. ●

You Try It Work through this You Try It problem.

Work Exercises 4–8 in this textbook or in the MyMathLab Study Plan.

My video summary ⊙ **Example 3** Using the Pythagorean Theorem to Find the Lengths of the Missing Sides

A Major League baseball "diamond" is really a square. The distance between each consecutive base is 90 feet. What is the distance between home plate and second base? Round to two decimal places.

Solution Watch this video to see how to use the Pythagorean Theorem to determine that the distance between home plate and second base is approximately 127.28 feet.

90 feet

90 feet

You Try It Work through this You Try It problem.

Work Exercises 9–12 in this textbook or in the MyMathLab Study Plan.

OBJECTIVE 3   UNDERSTANDING SIMILAR TRIANGLES

Imagine drawing a triangle on a piece of paper and placing the paper in a copy machine. If you increase (or decrease) the size of the image to be copied, the resulting image will be a triangle of exactly the same shape as the original triangles but it will be a different size. Triangles that have the same shape but not necessarily the same size are called **similar triangles**. Triangles ABC and XYZ in Figure 21 are similar.

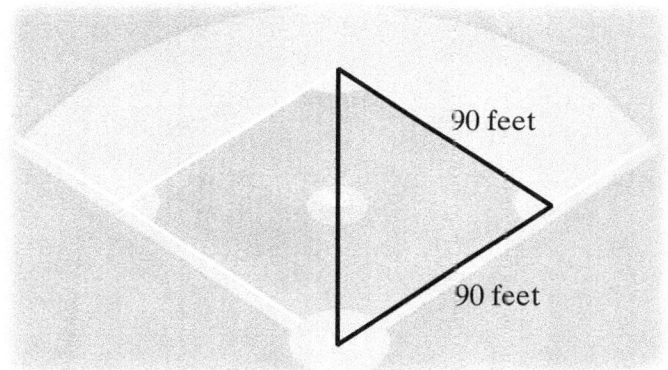

B

Y

A C X Z

Figure 21 Triangle ABC and triangle XYZ are similar. Similar triangles have the same shape but not necessarily the same size.

The measures of the corresponding angles of the triangles in Figure 21 are equal (as indicated by the small arcs with tick marks) but the lengths of the corresponding sides are obviously different. However, the ratio of the lengths of any two sides of triangle ABC is equal to the ratio of the lengths of the corresponding sides of triangle XYZ. For example, the ratio of side AB to side AC must be equal to the ratio of side XY to side XZ. This can be written as $\dfrac{AB}{AC} = \dfrac{XY}{XZ}$.

Properties of Similar Triangles

1. The corresponding angles have the same measure.

2. The ratio of the lengths of any two sides of one triangle is equal to the ratio of the lengths of the corresponding sides of the other triangle.

My video summary

⊙ **Example 4** Determining the Side Lengths of Similar Triangles

Triangles ABC and XYZ are similar. Find the lengths of the missing sides of triangle ABC.

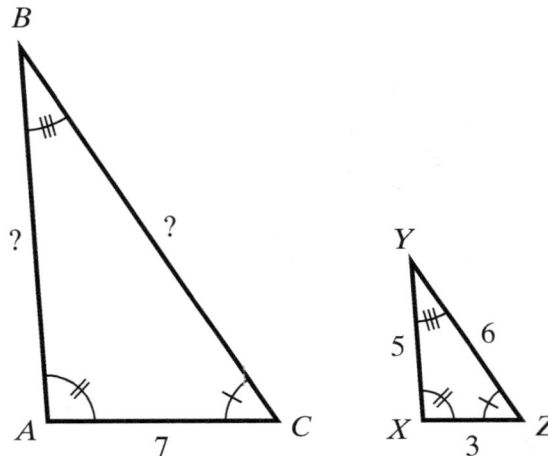

Solution We start by finding the length of side AB. Side AB corresponds to side XY. Also, side AC corresponds to side XZ. Therefore, we can set up the proportion $\dfrac{AB}{AC} = \dfrac{XY}{XZ}$.

$\dfrac{AB}{AC} = \dfrac{XY}{XZ}$ The lengths of corresponding sides of similar triangles are proportional.

$\dfrac{AB}{7} = \dfrac{5}{3}$ Substitute $AC = 7$, $XY = 5$, and $XZ = 3$.

$3AB = 35$ Multiply both sides by the LCD, which is $(3)(7) = 21$.

$AB = \dfrac{35}{3}$ Divide both sides by 3.

The length of side AB is $\dfrac{35}{3}$.

My video summary

⊙ To find the length of side BC, we can use the proportion $\dfrac{BC}{AC} = \dfrac{YZ}{XZ}$ to determine that the length of side BC is 14. Watch this video to see the entire solution to this example.

In Figure 22, we reconsider similar triangles ABC and XYZ from Example 4. Side AB corresponds to side XY, side BC corresponds to side YZ, and side AC corresponds to side XZ. Looking at the ratios of the lengths of these corresponding sides, we get $\dfrac{AB}{XY} = \dfrac{35/3}{5} = \dfrac{7}{3}$, $\dfrac{BC}{YZ} = \dfrac{14}{6} = \dfrac{7}{3}$, and $\dfrac{AC}{XZ} = \dfrac{7}{3}$. The number $\dfrac{7}{3}$ is called the **proportionality constant**. Note that the length of each side of triangle ABC is $\dfrac{7}{3}$ times the length of the corresponding side of triangle XYZ. Note also that the length of each side of triangle XYZ is $\dfrac{3}{7}$ times the length of the corresponding side of triangle ABC. For consistency, we will define the proportionality constant to always be greater than or equal to 1. Therefore, if two triangles are similar, then the length of a side of the larger triangle is k times the length of the corresponding side of the smaller triangle, where k is the proportionality constant. Also, the length of a side of the smaller triangle is $\dfrac{1}{k}$ times the length of the corresponding side of the larger triangle.

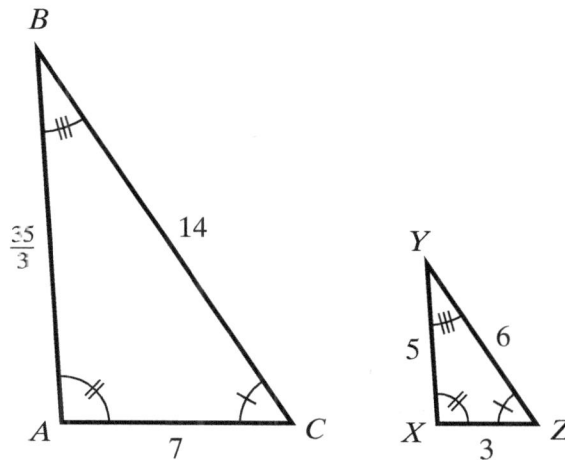

Figure 22 The lengths of the sides of triangle ABC are $k = \dfrac{7}{3}$ times the lengths of the corresponding sides of triangle XYZ.

Definition   Proportionality Constant of Similar Triangles

If two triangles are similar, there exists a constant $k \geq 1$ called the **proportionality constant of similar triangles** equal to the ratio of the lengths of corresponding sides. Given the similar triangles shown below, $k = \dfrac{a}{x} = \dfrac{b}{y} = \dfrac{c}{z}$, where $a \geq x$, $b \geq y$, and $c \geq z$.

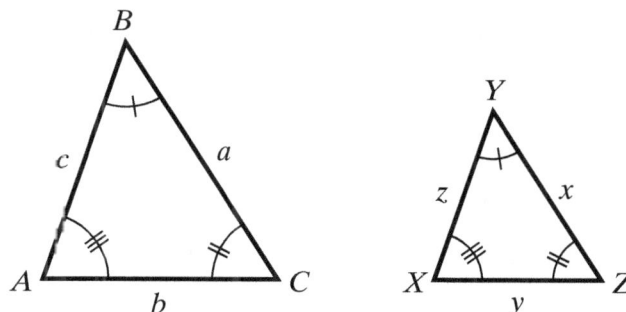

🔲 **Example 5** Determining the Proportionality Constant of Similar Triangles

The triangles below are similar. Find the proportionality constant. Then find the lengths of the missing sides.

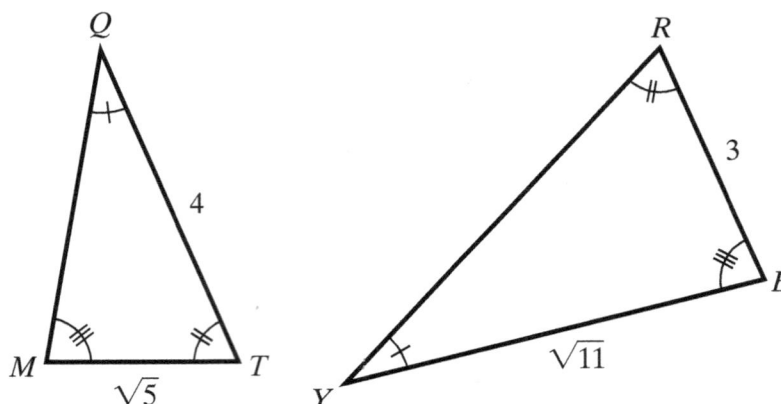

Solution Work through this animation to verify that $k = \dfrac{3}{\sqrt{5}}$ and that the lengths of missing sides YR and QM are $\dfrac{12}{\sqrt{5}}$ and $\dfrac{\sqrt{55}}{3}$, respectively.

You Try It Work through this You Try It problem.

Work Exercises 13–16 in this textbook or in the My MathLab Study Plan.

🔲 **Example 6** Determining the Side Lengths of Similar Right Triangles

The right triangles below are similar. Determine the lengths of the missing sides.

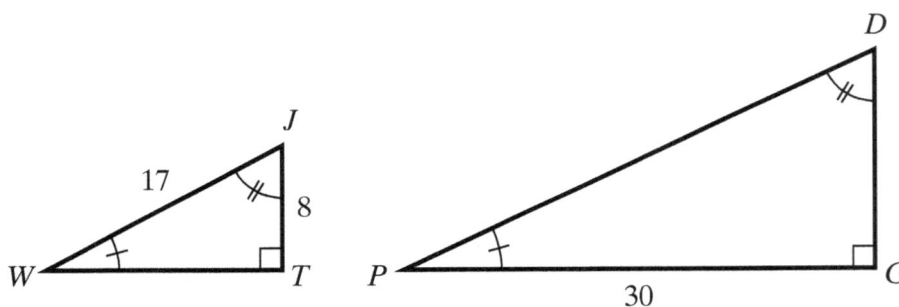

Solution Using the Pythagorean Theorem, we can determine the length of side WT. Once we have the length of side WT, we can find the proportionality constant, which will allow us to find the lengths of the missing sides of triangle PDG. Try to find the lengths of the missing sides yourself, and then watch this video to see if you are correct.

You Try It Work through this You Try It problem.

Work Exercises 17–20 in this textbook or in the My MathLab Study Plan.

OBJECTIVE 4   UNDERSTANDING THE SPECIAL RIGHT TRIANGLES

There are two "special" right triangles whose properties are worth memorizing. The first special right triangle has two acute angles of $\frac{\pi}{4}$ radians (45°).

The $\frac{\pi}{4}, \frac{\pi}{4}, \frac{\pi}{2}$ (45°, 45°, 90°) Right Triangle

Every isosceles right triangle has two acute angles that have a measure of $\frac{\pi}{4}$ radians (45°).

My animation summary

▶ Watch this animation for an explanation.

Because the triangle is isosceles, the length of the two legs must be congruent. If we let the length of a leg be a units, then we can use the Pythagorean Theorem to determine that the length of the hypotenuse is $a\sqrt{2}$ units. See Figure 23. The lengths of the sides of every $\frac{\pi}{4}, \frac{\pi}{4}, \frac{\pi}{2}$ right triangle has this relationship. For example, if we let $a = 1$, then the length of the hypotenuse is $\sqrt{2}$. See Figure 24. Watch this animation for a detailed explanation of the relationship between the lengths of the sides of the $\frac{\pi}{4}, \frac{\pi}{4}, \frac{\pi}{2}$ special right triangle.

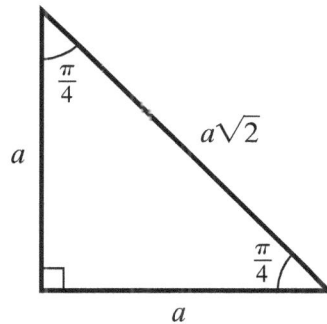

Figure 23 The length of the hypotenuse of the $\frac{\pi}{4}, \frac{\pi}{4}, \frac{\pi}{2}$ right triangle is $\sqrt{2}$ times the length of a leg.

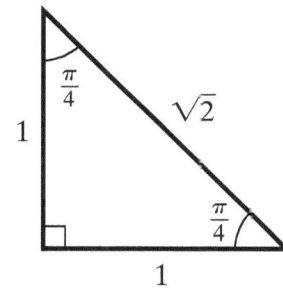

Figure 24 The $\frac{\pi}{4}, \frac{\pi}{4}, \frac{\pi}{2}$ right triangle with sides of lengths 1, 1, and $\sqrt{2}$

My animation summary

▶ **The $\frac{\pi}{6}, \frac{\pi}{3}, \frac{\pi}{2}$ (30°, 60°, 90°) Right Triangle**

The second special right triangle has acute angles of $\frac{\pi}{6}$ radians (30°) and $\frac{\pi}{3}$ radians (60°). We can construct this triangle by starting with an equilateral triangle whose angles all measure $\frac{\pi}{3}$ radians (60°). See Figure 25. We can then draw a perpendicular line segment that bisects one of the angles and one of the sides to create two $\frac{\pi}{6}, \frac{\pi}{3}, \frac{\pi}{2}$ right triangles. See Figure 26.

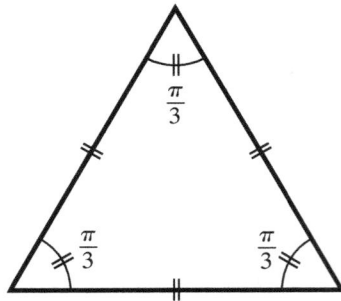

Figure 25
An equilateral triangle whose angles all measure $\frac{\pi}{3}$ radians and whose sides are all equal in length

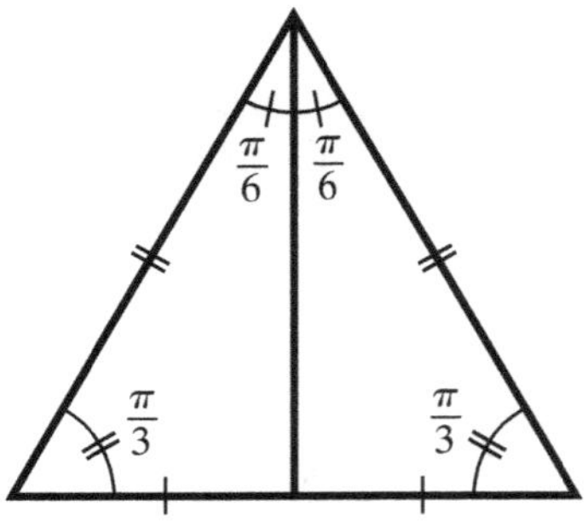

Figure 26
Bisecting an angle and a side of an equilateral triangle produces two $\frac{\pi}{6}, \frac{\pi}{3}, \frac{\pi}{2}$ right triangles.

Because we bisected one of the sides of the equilateral triangle to create two $\frac{\pi}{6}, \frac{\pi}{3}, \frac{\pi}{2}$ triangles, the length of the shortest side of a $\frac{\pi}{6}, \frac{\pi}{3}, \frac{\pi}{2}$ triangle must be exactly half of the length of the hypotenuse. (Or, the length of the hypotenuse must be twice as long as the length of the shortest side.) Therefore, if the length of the shortest leg (the leg opposite the $\frac{\pi}{6}$ angle) is a units, then the length of the hypotenuse is $2a$ units. We can use the Pythagorean Theorem to determine that the length of the other leg is $a\sqrt{3}$ units.

See Figure 27. If we let $a = 1$, then the length of the hypotenuse is 2 and the length of the side opposite the $\frac{\pi}{3}$ angle is $\sqrt{3}$. See Figure 28. Watch this animation for a complete explanation of the $\frac{\pi}{6}, \frac{\pi}{3}, \frac{\pi}{2}$ special right triangle.

You may be asking yourself, "Why are these triangles and angles so special?" Well, because of the relationships of the angles of the right triangles and the Pythagorean Theorem, we can easily determine the ratios of the sides of the triangles. This is not so simple with triangles having other angle measures. You will see in Section 1.4 how quickly we can work with these angles as opposed to angles of non-special triangles.

My animation summary

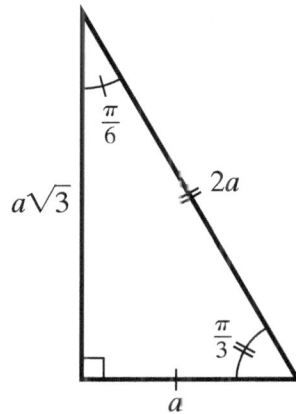

Figure 27 The length of the hypotenuse of the $\frac{\pi}{6}, \frac{\pi}{3}, \frac{\pi}{2}$ right triangle is 2 times the length of the shortest leg. The length of the longer leg is $\sqrt{3}$ times the length of the shorter leg.

Figure 28 The $\frac{\pi}{6}, \frac{\pi}{3}, \frac{\pi}{2}$ right triangle with sides of lengths 1, $\sqrt{3}$, and 2

My interactive video summary

⊙ **Example 7** Determining the Side Lengths of Special Triangles

Determine the lengths of the missing sides of each right triangle.

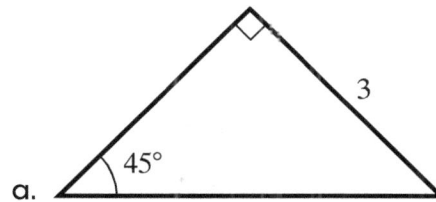

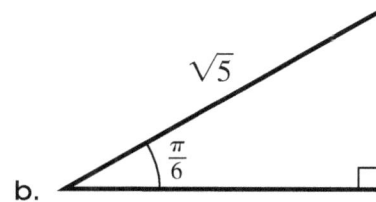

a.

b.

Solution

a. This is a $\frac{\pi}{4}, \frac{\pi}{4}, \frac{\pi}{2}$ right triangle similar to the $\frac{\pi}{4}, \frac{\pi}{4}, \frac{\pi}{2}$ right triangle having side

lengths of 1, 1, $\sqrt{2}$ The proportionality constant is $\frac{3}{1} = 3$. Therefore, the lengths

of the sides must be 3 times the lengths of the corresponding sides of the triangle with lengths 1, 1, $\sqrt{2}$. Thus, the missing lengths are 3 and $3\sqrt{2}$.

The lengths of the sides of the triangle on the right are 3 times the lengths of the corresponding sides of the similar triangle on the left.

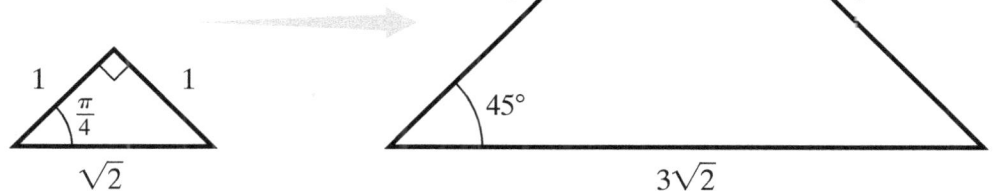

*My interactive
video summary*

⊙ **b.** Try finding the lengths of the missing sides of this triangle on your own. Work through this interactive video when you have finished to see if you are correct.

You Try It   Work through this You Try It problem.

Work Exercises 21–30 in this textbook or in the MyMathLab **Study Plan.**

OBJECTIVE 5   USING SIMILAR TRIANGLES TO SOLVE APPLIED PROBLEMS

Understanding similar triangles and the special right triangles can help us solve application problems involving triangles.

Example 8 Determining the Height of a Cell Tower

The shadow of a cell tower is 80 feet long. A boy 3 feet 9 inches tall is standing next to the tower. If the boy's shadow is 6 feet long, find the height of the cell tower.

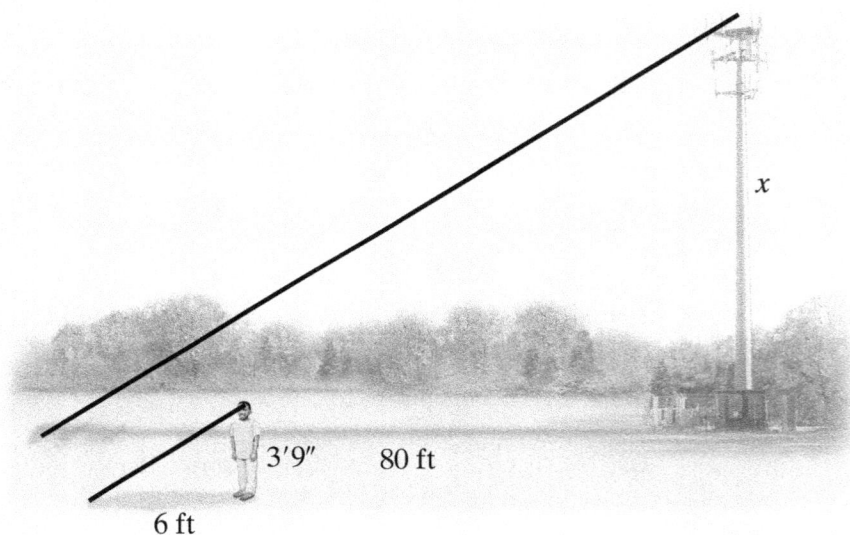

Solution   First, convert 9 inches to feet.

$$9 \text{ in} \cdot \frac{1 \text{ ft}}{12 \text{ in}} = \frac{9}{12} \text{ ft} = \frac{3}{4} \text{ ft} = .75 \text{ ft}$$

Therefore, the height of the boy is 3.75 feet (*not* 3.9 feet). We can now draw two similar right triangles representing this problem. Let x represent the height of the cell tower. We can then write a proportion to solve for x.

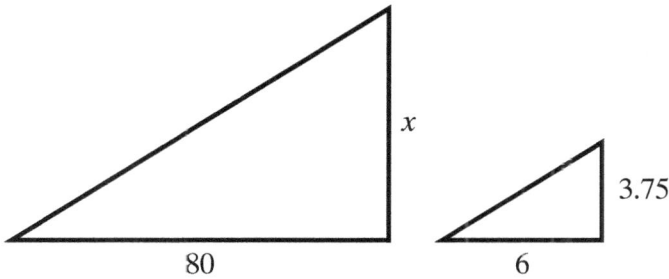

$$\frac{x}{3.75} = \frac{80}{6}$$ Write a proportion using the lengths of corresponding sides of similar triangles

$$6x = 300$$ Multiply both sides by the LCD, which is $(6)(3.75)$.

$$x = 50$$ Divide both sides by 6.

The height of the cell tower is 50 feet.

My video summary   ◉ **Example 9** Determining the Length of a Bridge

Two people are standing on opposite sides of a small river. One person is located at point Q, a distance of 20 feet from a bridge. The other person is standing on the southeast corner of the bridge at point P. The angle between the bridge and the line of sight from P to Q is 30°. Use this information to determine the length of the bridge and the distance between the two people. Round your answers to two decimal places as needed.

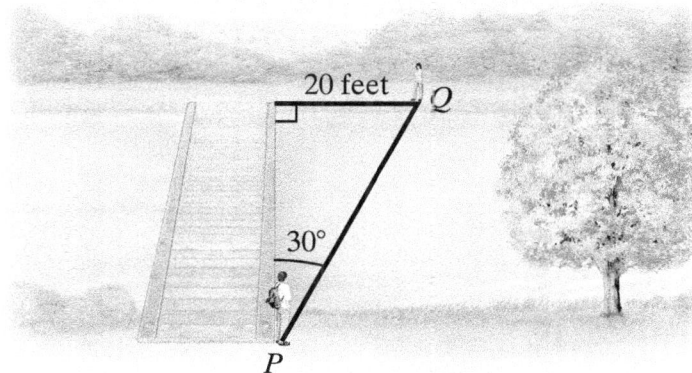

Solution The figure created by point Q, point P, and the bridge forms a 30°, 60°, 90° special right triangle that is similar to the 30°, 60°, 90° triangle with sides of lengths 1, $\sqrt{3}$, and 2. Try finding the length of the bridge on your own. Watch this video to see if you are correct.

You Try It Work through this You Try It problem.

Work Exercises 31–36 in this textbook or in the MyMathLab **Study Plan**.

1.3 Exercises

In Exercises 1–3, classify each triangle as acute, obtuse, right, scalene, isosceles, or equilateral. State all that apply.

1.
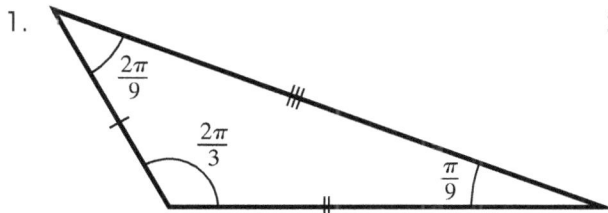

2.
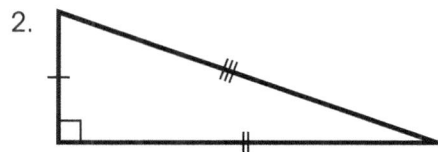

3.

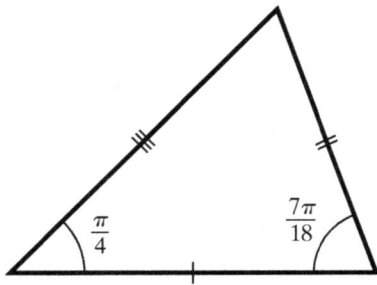

In Exercises 4–8, use the Pythagorean Theorem to find the missing side of each right triangle.

4.

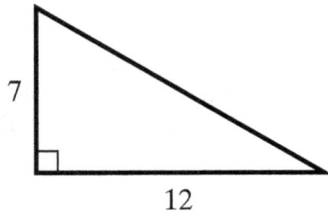

7

12

5.

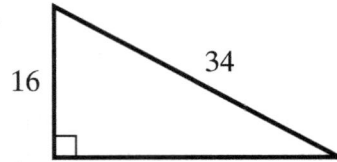

16

34

6.

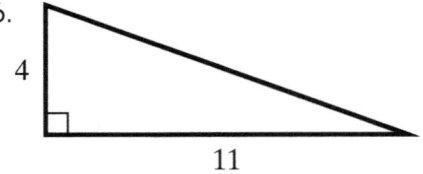

4

11

7.

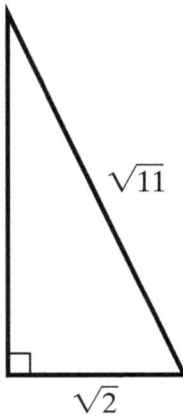

$\sqrt{11}$

$\sqrt{2}$

8.

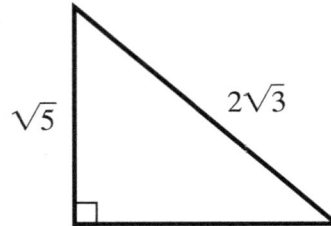

$\sqrt{5}$

$2\sqrt{3}$

9. An official National Collegiate Athletic Association (NCAA) women's softball "diamond" is really a square. The distance between each consecutive base is 60 feet. What is the distance between home plate and second base?

10. How far up the side of a building will a 9-meter ladder reach if the foot of the ladder is 2 meters from the base of the building?

11. The length of a rectangle is 1 inch less than twice the width. If the diagonal is 2 inches more than the length, find the dimensions of the rectangle.

12. The cross section of a camping tent forms a triangle. The slanted sides measure 7 feet and the base of the tent is 8 feet. What is the height of the tent?

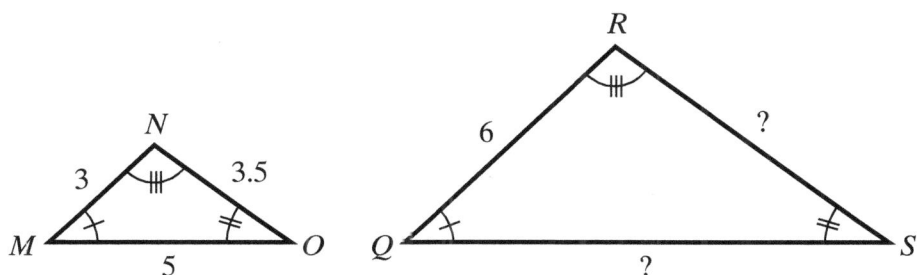

In Exercises 13–16, two similar triangles are given. Find the proportionality constant k such that the length of each side of the larger triangle is k times the length of the corresponding sides of the smaller triangle. Then find the lengths of the missing sides of each triangle.

13.

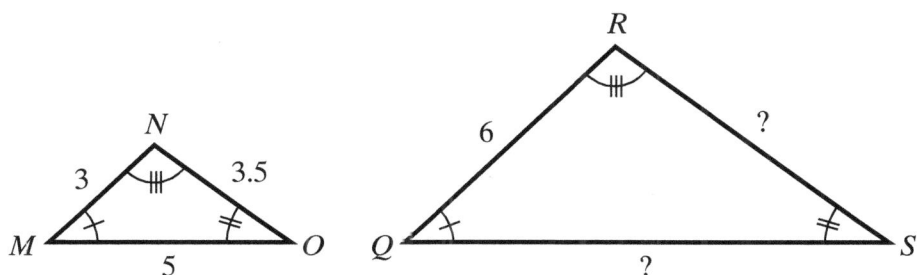

14.

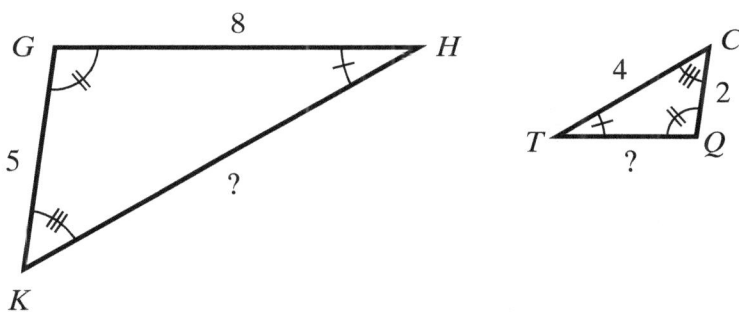

15.

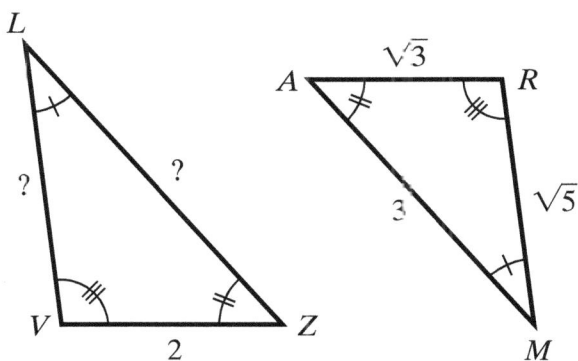

16.

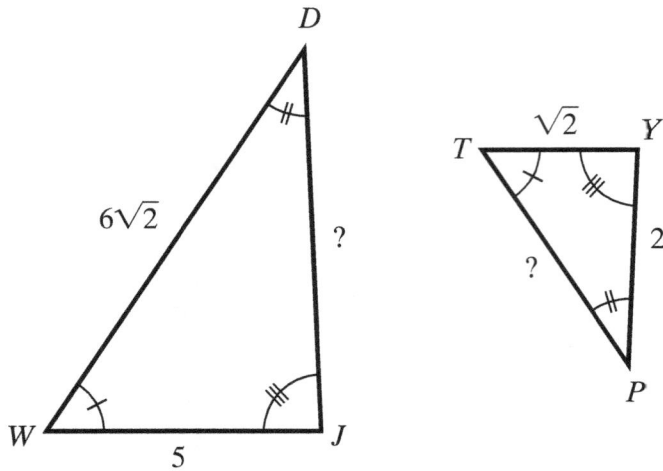

In Exercises 17–20, find the missing sides of the given similar right triangles.

17.

18.

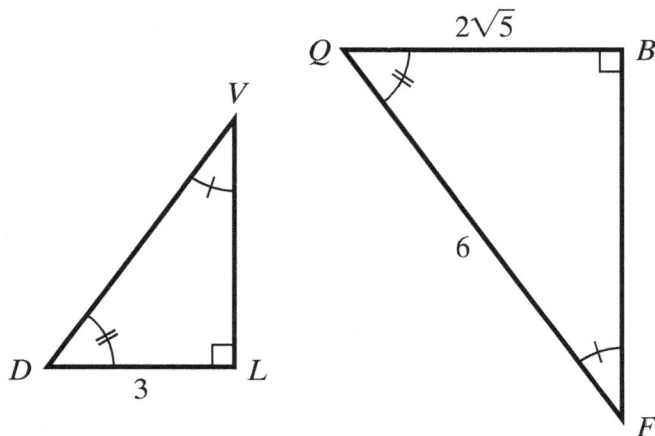

19.

20.

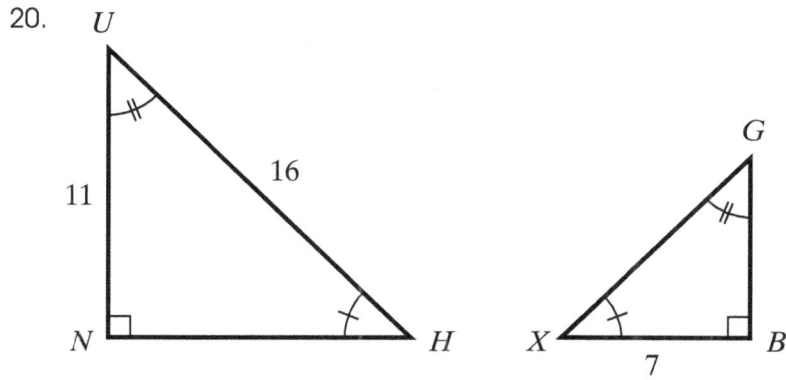

In Exercises 21–30, find the missing sides of each special right triangle.

21.

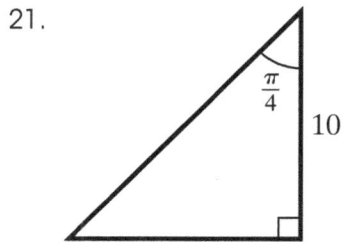

$\frac{\pi}{4}$

10

22.

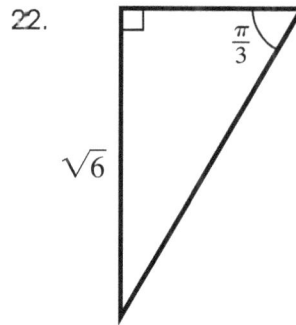

$\frac{\pi}{3}$

$\sqrt{6}$

23.

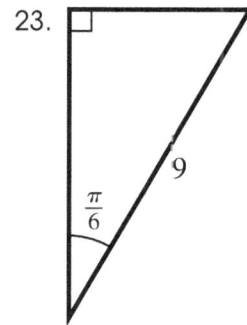

9

$\frac{\pi}{6}$

24.

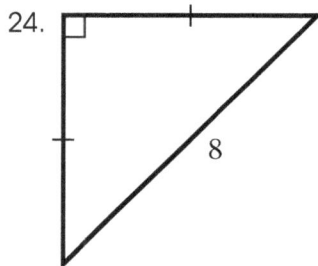

8

25.

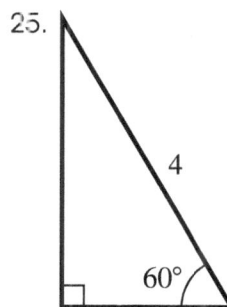

4

60°

26.

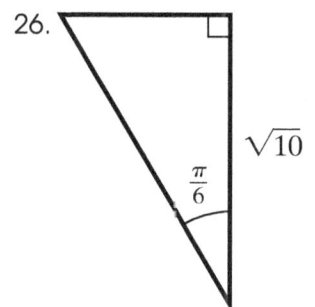

$\sqrt{10}$

$\frac{\pi}{6}$

27.

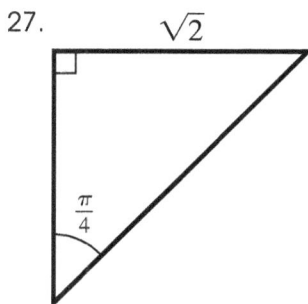

$\sqrt{2}$

$\frac{\pi}{4}$

28.

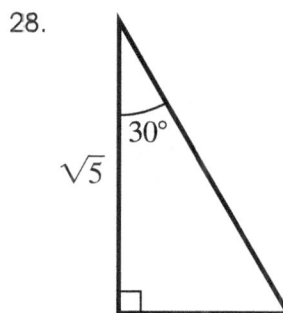

30°

$\sqrt{5}$

29.

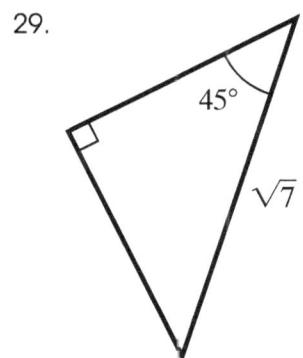

45°

$\sqrt{7}$

30.

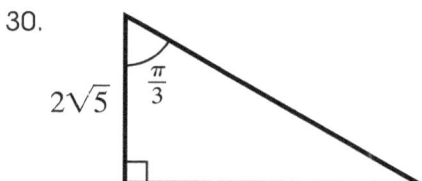

$2\sqrt{5}$

$\frac{\pi}{3}$

31. A 6-foot-tall man is standing next to a building. The man's shadow is 4 feet long. At the same time, the building casts a shadow that is 135 feet long. How tall is the building?

32. The length of the diagonal of a rectangular swimming pool shown in the figure is 40 feet. Determine the length and width of the pool.

33. Two ladders lean against a wall in such a way that the angle that they make with the ground is the same. The 15-foot ladder rests 8 feet up the wall. How much further up the wall will the 20-foot ladder reach?

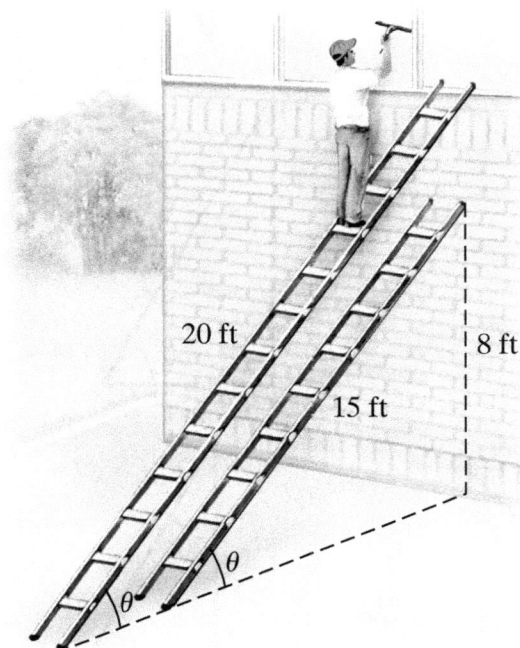

34. The **angle of elevation** is the angle from the horizontal looking up at an object. The angle of elevation of an airplane is $\frac{\pi}{4}$ radians, and the altitude of the plane is 12,000 feet. How far away is the plane "as the crow flies," which means through the air rather than on the ground?

35. Two people are standing on opposite sides of a small river. One person is located at point Q, a distance of 35 meters from a bridge. The other person is standing on the southeast corner of the bridge at point P. The angle between the bridge and the line of sight from P to Q is $\frac{\pi}{3}$ radians. Use this information to determine the length of the bridge and the distance between the two people. Round your answers to two decimal places as needed.

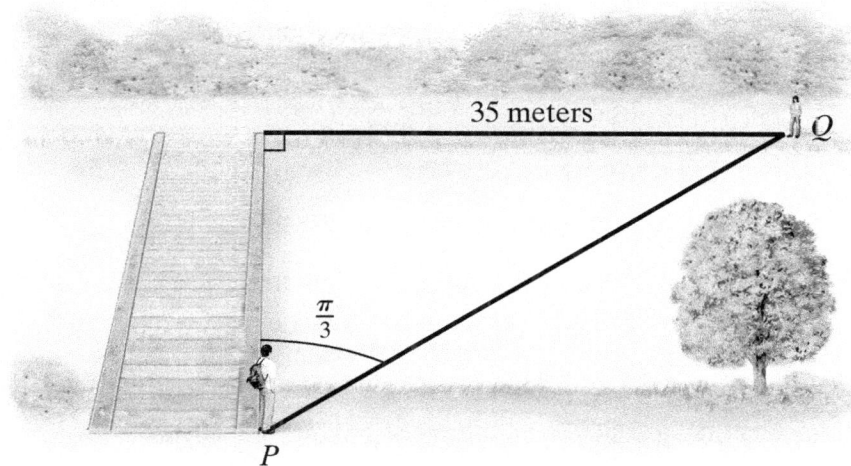

36. A team of forestry graduate students wants to determine the height of a giant sequoia tree using lasers. The laser is placed on top of a 2-foot tripod and shines a beam through the top of a 20-foot pole positioned 10 feet from the tripod and 140 feet from the approximate center of the base of the tree. Determine the height of the tree.

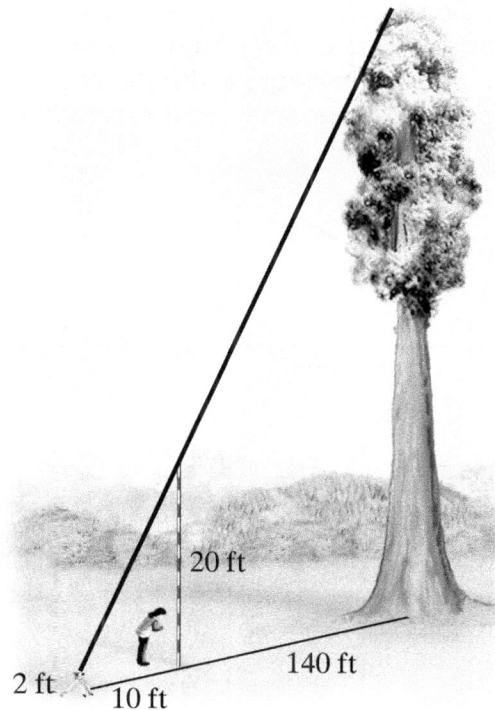

1.4 Right Triangle Trigonometry

THINGS TO KNOW

Before working through this section, be sure that you are familiar with the following concepts:

VIDEO   ANIMATION   INTERACTIVE

You Try It
1. Converting between Degree Measure and Radian Measure (Section 1.1)

You Try It
2. Understanding Similar Triangles (Section 1.3)

You Try It
3. Understanding the Special Right Triangles (Section 1.3)

OBJECTIVES

1 Understanding the Right Triangle Definitions of the Trigonometric Functions

2 Using the Special Right Triangles

3 Understanding the Fundamental Trigonometric Identities

4 Understanding Cofunctions

5 Evaluating Trigonometric Functions Using a Calculator

OBJECTIVE 1 UNDERSTANDING THE RIGHT TRIANGLE DEFINITIONS OF THE TRIGONOMETRIC FUNCTIONS

My video summary

Consider the right triangles in Figure 29 with the given **acute angle** θ. The side opposite the right angle is callec the **hypotenuse**. The hypotenuse is always the longest side of a right triangle. We label the two legs of the right triangle relative to the given angle θ. The leg opposite angle θ is appropriately named the **opposite side**. The other leg is called the **adjacent side**.

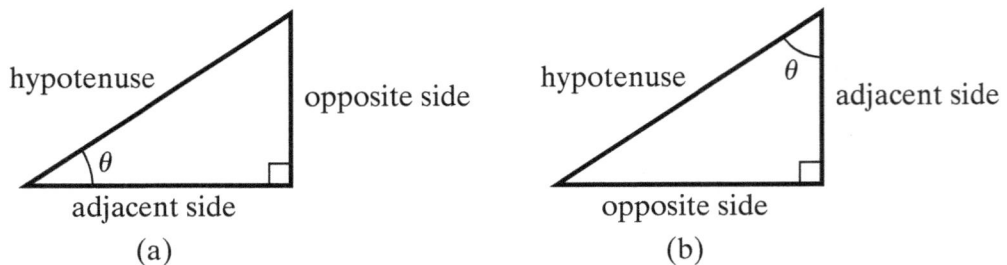

Figure 29 The sides of two right triangles relative to the given acute angle θ

We will be concerned about the *lengths* of each of these three sides. For simplicity, we will abbreviate the length of the hypotenuse as "*hyp*" The length of the opposite side will be referred to as "*opp*" and the length of the adjacent side will be denoted as "*adj*" See Figure 30.

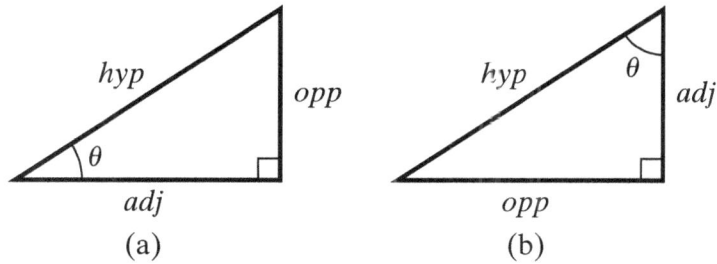

Figure 30 Two right triangles and a given acute angle θ with side lengths *hyp*, *opp*, and *adj*

(a)   (b)

Choosing side lengths two at a time, we can create a total of six ratios. These six ratios are $\dfrac{opp}{hyp}, \dfrac{adj}{hyp}, \dfrac{opp}{adj}, \dfrac{hyp}{opp}, \dfrac{hyp}{adj}$, and $\dfrac{adj}{opp}$. The value of each of these six ratios depends on the measure of the acute angle θ. Thus, the six ratios are functions of the variable θ and are called the **trigonometric functions of the acute angle** θ. For convenience, the six ratios have been given names. Historically, these six trigonometric functions have been named **sine** of theta, **cosine** of theta, **tangent** of theta, **cosecant** of theta, **secant** of theta, and **cotangent** of theta. The six functions are abbreviated as $\sin \theta$, $\cos \theta$, $\tan \theta$, $\csc \theta$, $\sec \theta$, and $\cot \theta$.

The Right Triangle Definitions of the Trigonometric Functions

Given a right triangle with acute angle θ and side lengths of *hyp*, *opp*, and *adj*, the six trigonometric functions of angle θ are defined as follows:

$$\sin \theta = \frac{opp}{hyp} \qquad \csc \theta = \frac{hyp}{opp}$$

$$\cos \theta = \frac{adj}{hyp} \qquad \sec \theta = \frac{hyp}{adj}$$

$$\tan \theta = \frac{opp}{adj} \qquad \cot \theta = \frac{adj}{opp}$$

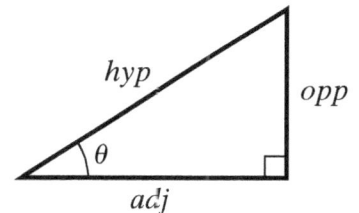

It is extremely important to memorize the six trigonometric functions. To memorize the ratios for $\sin \theta$, $\cos \theta$, and $\tan \theta$, you could use the following silly phrase:

"Some Old Horse Caught Another Horse Taking Oats Away"

$$\sin \theta = \frac{opp}{hyp} \qquad \cos \theta = \frac{adj}{hyp} \qquad \tan \theta = \frac{opp}{adj}$$

You might also use the acronym SOHCAHTOA, pronounced "So-Kah-Toe-Ah," to help you memorize these functions.

Once you have memorized the ratios for $\sin \theta$, $\cos \theta$, and $\tan \theta$, you can easily obtain the ratios of $\csc \theta$, $\sec \theta$, and $\cot \theta$. Note that the ratios for $\csc \theta$, $\sec \theta$, and $\cot \theta$ are simply the reciprocals of the ratios for $\sin \theta$, $\cos \theta$, and $\tan \theta$, respectively.

Recall from Section 1.3 that if two triangles are similar, then the ratio of the lengths of any two sides of one triangle is equal to the ratio of the lengths of the corresponding sides of the other triangle. Thus, the value of the six trigonometric functions for a specific acute angle θ will be exactly the same regardless of the size of the triangle. To illustrate this, consider the three similar triangles in Figure 31 below. Note that the values of θ and $\sin \theta$ are the same for each triangle even though the sides are of different length. This is because the ratio of any two sides of a right triangle is a function of the measure of the acute angle θ.

$$\sin \theta = \frac{12}{15} = \frac{4}{5} \qquad \sin \theta = \frac{10}{25/2} = 10 \cdot \frac{2}{25} = \frac{20}{25} = \frac{4}{5} \qquad \sin \theta = \frac{2}{2.5} = .8 = \frac{8}{10} = \frac{4}{5}$$

Figure 31 In each triangle, though the corresponding sides all have different lengths, the ratio of the length of the opposite side to the length of the hypotenuse is the same.

You should verify that the remaining five trigonometric functions of angle θ for each of the three similar right triangles are also equivalent. (To see the values of the six trigonometric functions for the three right triangles shown previously, consult this table.) It follows that the values of the six trigonometric functions of an acute angle θ depend only on the measure of θ, not on the size of the right triangle.

My interactive video summary

⊗ **Example 1** Evaluating the Trigonometric Functions Given a Right Triangle

Given the right triangle, evaluate the six trigonometric functions of the acute angle θ.

Solution We are given that $hyp = 5$ and $adj = 2$. We must first determine the length of the opposite side (opp) using the Pythagorean Theorem before we can evaluate the six trigonometric functions.

$$(adj)^2 + (opp)^2 = (hyp)^2 \qquad \text{Write an equation representing the Pythagorean Theorem.}$$

$$2^2 + (opp)^2 = 5^2 \qquad \text{Substitute } adj = 2 \text{ and } hyp = 5.$$

$$4 + (opp)^2 = 25 \qquad \text{Simplify.}$$

$$(opp)^2 = 21 \qquad \text{Subtract 4 from both sides.}$$

$$opp = \pm \sqrt{21} \qquad \text{Take the square root of both sides.}$$

Because opp represents the length of a side of a triangle, we use only the positive square root value. Therefore, $opp = \sqrt{21}$. We can now evaluate the six trigonometric functions of the acute angle θ.

$$\sin \theta = \frac{opp}{hyp} = \frac{\sqrt{21}}{5} \qquad \csc \theta = \frac{hyp}{opp} = \frac{5}{\sqrt{21}}$$

$$\cos \theta = \frac{adj}{hyp} = \frac{2}{5} \qquad \sec \theta = \frac{hyp}{adj} = \frac{5}{2}$$

$$\tan \theta = \frac{opp}{adj} = \frac{\sqrt{21}}{2} \qquad \cot \theta = \frac{adj}{opp} = \frac{2}{\sqrt{21}}$$

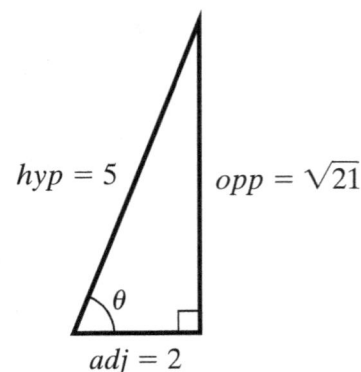

*My interactive
video summary*

⊗ You may wish to watch this interactive video to see each step of this solution. ●

Notice in Example 1 that for $\csc \theta$, we left the final answer as $\csc \theta = \dfrac{5}{\sqrt{21}}$. Answers of this form are perfectly acceptable and, in many situations, it is the preferred answer. Rationalizing the denominator is never going to be necessary in this text to get a correct answer. But if you would like to rationalize your answer, it will always be counted as correct as long as your work contains no errors. If you

My video summary ⊗ would like a quick review on how to rationalize the denominator, watch this video.

Though rationalizing is not required, simplifying will always be required. "Simplifying" means removing all perfect squares from any radicals. It also includes removing any common integer factors from the numerator and denominator. Common factors that appear as radicals should be removed, but it is not necessary to factor an integer into a product of radical factors in order to cancel one of the radical factors. Look at this table or see the following pages that demonstrate examples of simplification that will be required.

Examples of Simplifying Radical Expressions

Calculated Result	Simplification required	Correct simplified, unrationalized answer
$\dfrac{5}{\sqrt{21}}$	None	$\dfrac{5}{\sqrt{21}}$
$\dfrac{1}{\sqrt{2}}$	None	$\dfrac{1}{\sqrt{2}}$
$\dfrac{1}{\sqrt{24}}$	$\dfrac{1}{\sqrt{24}} = \dfrac{1}{\sqrt{4}\sqrt{6}} = \dfrac{1}{2\sqrt{6}}$	$\dfrac{1}{2\sqrt{6}}$
$\dfrac{3}{\sqrt{3}}$	None	$\dfrac{3}{\sqrt{3}}$
$\dfrac{3}{\sqrt{63}}$	$\dfrac{3}{\sqrt{63}} = \dfrac{3}{\sqrt{9}\sqrt{7}} = \dfrac{\cancel{3}}{\cancel{3}\sqrt{7}} = \dfrac{1}{\sqrt{7}}$	$\dfrac{1}{\sqrt{7}}$
$\dfrac{\sqrt{3}}{\sqrt{10}}$	None	$\dfrac{\sqrt{3}}{\sqrt{10}}$
$\dfrac{\sqrt{2}}{\sqrt{10}}$	$\dfrac{\sqrt{2}}{\sqrt{10}} = \dfrac{\cancel{\sqrt{2}}}{\cancel{\sqrt{2}}\sqrt{5}} = \dfrac{1}{\sqrt{5}}$	$\dfrac{1}{\sqrt{5}}$
$\dfrac{\sqrt{3}}{\sqrt{24}}$	$\dfrac{\sqrt{3}}{\sqrt{24}} = \dfrac{\cancel{\sqrt{3}}}{\cancel{\sqrt{3}}\sqrt{8}} = \dfrac{1}{\sqrt{8}} = \dfrac{1}{\sqrt{4}\sqrt{2}} = \dfrac{1}{2\sqrt{2}}$	$\dfrac{1}{2\sqrt{2}}$
$\dfrac{\sqrt{98}}{\sqrt{2}}$	$\dfrac{\sqrt{98}}{\sqrt{2}} = \dfrac{\cancel{\sqrt{2}}\sqrt{49}}{\cancel{\sqrt{2}}} = \sqrt{49} = 7$	7
$\dfrac{\sqrt{72}}{5}$	$\dfrac{\sqrt{72}}{5} = \dfrac{\sqrt{36}\sqrt{2}}{5} = \dfrac{6\sqrt{2}}{5}$	$\dfrac{6\sqrt{2}}{5}$

You Try It Work through this You Try It problem.

Work Exercises 1–5 in this textbook or in the MyMathLab Study Plan.

My video summary ⊙ **Example 2** Finding the Values of the Remaining Trigonometric Functions

If θ is an acute angle of a right triangle and if $\sin\theta = \dfrac{3}{4}$, then find the values of the remaining five trigonometric functions for angle θ.

Solution We know that $\sin\theta = \dfrac{opp}{hyp} = \dfrac{3}{4}$, where $opp = 3$ is the length of the side of a right triangle opposite an acute angle θ and $hyp = 4$ is the length of the hypotenuse. Using the Pythagorean Theorem, we find that the length of the adjacent side is $\sqrt{7}$.

Once we know the lengths of all three sides of the right triangle, we can determine the values of the remaining trigonometric functions. Try finding these values on your own. Once you have completed this work, watch this video to see if you are correct.

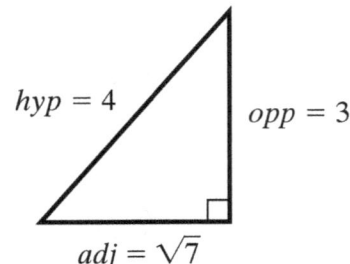

You Try It   Work through this You Try It problem.

Work Exercises 6–11 in this textbook or in the My Math Lab **Study Plan.**

OBJECTIVE 2   USING THE SPECIAL RIGHT TRIANGLES

My animation summary In Section 1.3, we introduced two special right triangles, the $\dfrac{\pi}{4}, \dfrac{\pi}{4}, \dfrac{\pi}{2}\ (45°, 45°, 90°)$ triangle and the $\dfrac{\pi}{6}, \dfrac{\pi}{3}, \dfrac{\pi}{2}\ (30°, 60°, 90°)$. You may want to watch this animation for a quick review of these two special right triangles. These two special triangles are seen in Figure 32.

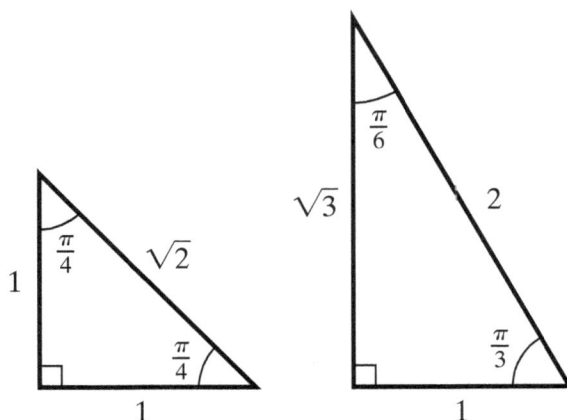

Figure 32 The special right triangles

Because we know the relationship between the sides of these two special right triangles, we can determine the exact values of the six trigonometric functions of θ by letting θ take on the values of $\dfrac{\pi}{6}, \dfrac{\pi}{4}$, and $\dfrac{\pi}{3}$. Table 1 shows the values of the sine, cosine, and tangent functions for these three acute angles. Notice that these values do not have to be rationalized.

That is, we don't have to write $\frac{1}{\sqrt{2}}$ as $\frac{\sqrt{2}}{2}$. It would be in your best interest to become very familiar with Table 1.

Table 1 The Trigonometric Functions for Acute Angles $\frac{\pi}{6}, \frac{\pi}{4}$, and $\frac{\pi}{3}$

θ	$\frac{\pi}{6}$ (30°)	$\frac{\pi}{4}$ (45°)	$\frac{\pi}{3}$ (60°)
$\sin \theta$	$\frac{1}{2}$	$\frac{1}{\sqrt{2}}$	$\frac{\sqrt{3}}{2}$
$\cos \theta$	$\frac{\sqrt{3}}{2}$	$\frac{1}{\sqrt{2}}$	$\frac{1}{2}$
$\tan \theta$	$\frac{1}{\sqrt{3}}$	1	$\sqrt{3}$

To find the values of $\csc \theta$, $\sec \theta$, and $\cot \theta$ for the angles in Table 1, we can use the fact that the ratios for $\csc \theta$, $\sec \theta$, and $\cot \theta$ are simply the reciprocals of the ratios for $\sin \theta$, $\cos \theta$, and $\tan \theta$, respectively. Try to determine the values of $\csc \theta$, $\sec \theta$, and $\cot \theta$ for each of the angles in Table 1. When finished, check your answers.

My video summary ⊙ **Example 3** Evaluating Trigonometric Functions Using the Special Right Triangles

Determine the value of $\csc \frac{\pi}{6} + \cot \frac{\pi}{4}$.

Solution Draw the two special right triangles and then find the value of $\csc \frac{\pi}{6}$ and $\cot \frac{\pi}{4}$.

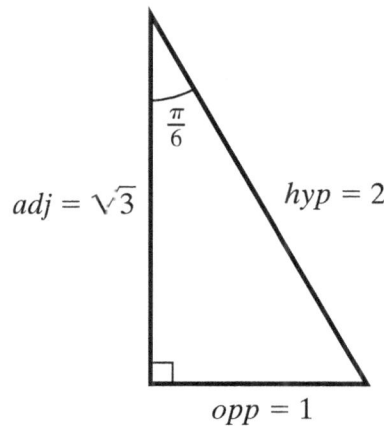

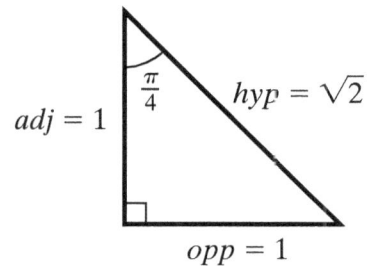

$$\csc \frac{\pi}{6} = \frac{hyp}{opp} = \frac{2}{1} = 2 \qquad \cot \frac{\pi}{4} = \frac{adj}{opp} = \frac{1}{1} = 1$$

Now, add the two expressions together.

$$\csc\frac{\pi}{6} + \cot\frac{\pi}{4} = 2 + 1 = 3$$

You Try It   Work through this You Try It problem.

Work Exercises 12–18 in this textbook or in the My Math Lab **Study Plan.**

Example 4 Determining the Measure of an Acute Angle

My video summary   ⊙ Determine the measure of the acute angle θ for which $\sec\theta = 2$.

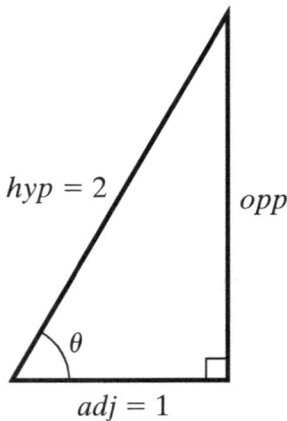

Solution By definition, the secant of an acute angle of a right triangle is equal to the ratio of the length of the hypotenuse to the length of the side adjacent to the angle. This ratio is given to be 2. Therefore, $\sec\theta = \dfrac{hyp}{adj} = \dfrac{2}{1}$. Note that we need to insert the 1 in the denominator of the whole number 2 to "see" the length of the adjacent side, *adj*. We can use this information to draw the following right triangle.

We can use the Pythagorean Theorem to determine that the length of the opposite side is $opp = \sqrt{3}$. We can see that this right triangle is a special $\dfrac{\pi}{6}, \dfrac{\pi}{3}, \dfrac{\pi}{2}$ triangle. The angle opposite the side of length $\sqrt{3}$ is always $\dfrac{\pi}{3}$. Therefore, $\theta = \dfrac{\pi}{3}$.

You may wish to watch this video to see a complete solution to this example.   ●

You Try It   Work through this You Try It problem.

Work Exercises 19–21 in this textbook or in the My Math Lab **Study Plan.**

OBJECTIVE 3   UNDERSTANDING THE FUNDAMENTAL TRIGONOMETRIC IDENTITIES

Now that you know the ratios of the lengths of the sides of a right triangle and their correspondingly named trigonometric functions of θ, it is time to recognize some of the relationships among these six trigonometric functions, called identities. Throughout this text, you will learn many trigonometric identities. Simply put, trigonometric identities are equalities involving trigonometric expressions that hold true for any angle θ for which all expressions are defined.

We start by introducing the **quotient identities**, the **reciprocal identities**, and the **Pythagorean identities.** We will derive these identities by referring to the right triangle with acute angle θ and sides of lengths *opp*, *adj*, and *hyp* seen in Figure 33. Therefore, for each identity stated in this section, it is assumed that θ is an acute angle.

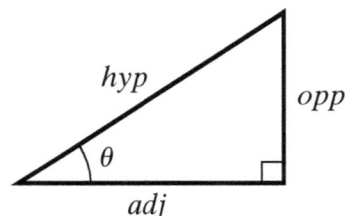

Figure 33 A right triangle and acute angle θ with side lengths *hyp*, *opp*, and *adj*

My video summary ◎ **THE QUOTIENT IDENTITIES**

Recall that $\sin\theta = \dfrac{opp}{hyp}$ and $\cos\theta = \dfrac{adj}{hyp}$. Therefore, the quotient of $\sin\theta$ and $\cos\theta$ is

$$\frac{\sin\theta}{\cos\theta} = \frac{\dfrac{opp}{hyp}}{\dfrac{adj}{hyp}} = \frac{opp}{\cancel{hyp}} \cdot \frac{\cancel{hyp}}{adj} = \frac{opp}{adj} = \tan\theta.$$

Similarly, the quotient of $\cos\theta$ and $\sin\theta$ is

$$\frac{\cos\theta}{\sin\theta} = \frac{\dfrac{adj}{hyp}}{\dfrac{opp}{hyp}} = \frac{adj}{\cancel{hyp}} \cdot \frac{\cancel{hyp}}{opp} = \frac{adj}{opp} = \cot\theta.$$

The two equations $\tan\theta = \dfrac{\sin\theta}{\cos\theta}$ and $\cot\theta = \dfrac{\cos\theta}{\sin\theta}$ are known as the quotient identities.

The Quotient Identities

$$\tan\theta = \frac{\sin\theta}{\cos\theta} \qquad \cot\theta = \frac{\cos\theta}{\sin\theta}$$

My video summary ◎ **THE RECIPROCAL IDENTITIES**

In algebra, the reciprocal of a real number a is defined as $\dfrac{1}{a}$. Similarly, the reciprocal of $\sin\theta$ is $\dfrac{1}{\sin\theta}$. If θ is the acute angle of the right triangle from Figure 33, then

$$\frac{1}{\sin\theta} = \frac{1}{\dfrac{opp}{hyp}} = \frac{hyp}{opp} = \csc\theta.$$

The identity $\csc\theta = \dfrac{1}{\sin\theta}$ is one of six reciprocal identities. Watch this video to see how to derive each reciprocal identity.

The Reciprocal Identities

$$\sin\theta = \frac{1}{\csc\theta} \qquad \csc\theta = \frac{1}{\sin\theta}$$

$$\cos\theta = \frac{1}{\sec\theta} \qquad \sec\theta = \frac{1}{\cos\theta}$$

$$\tan\theta = \frac{1}{\cot\theta} \qquad \cot\theta = \frac{1}{\tan\theta}$$

Example 5 Using the Quotient and Reciprocal Identities

Given that $\sin\theta = \dfrac{5}{7}$ and $\cos\theta = \dfrac{2\sqrt{6}}{7}$, find the values of the remaining four trigonometric functions using identities.

Solution We can use the quotient identity $\tan \theta = \dfrac{\sin \theta}{\cos \theta}$ to find the value of $\tan \theta$.

$$\tan \theta = \frac{\sin \theta}{\cos \theta} = \frac{\dfrac{5}{7}}{\dfrac{2\sqrt{6}}{7}} = \frac{5}{7} \cdot \frac{7}{2\sqrt{6}} = \frac{5}{2\sqrt{6}}$$

Now that we know the values of $\sin \theta$, $\cos \theta$, and $\tan \theta$, we can use the reciprocal identities to find the values of the remaining three trigonometric functions.

$$\csc \theta = \frac{1}{\sin \theta} = \frac{1}{\dfrac{5}{7}} = \frac{7}{5}$$

$$\sec \theta = \frac{1}{\cos \theta} = \frac{1}{\dfrac{2\sqrt{6}}{7}} = \frac{7}{2\sqrt{6}}$$

$$\cot \theta = \frac{1}{\tan \theta} = \frac{1}{\dfrac{5}{2\sqrt{6}}} = \frac{2\sqrt{6}}{5}$$

You Try It Work through this You Try It problem.

Work Exercises 22–23 in this textbook or in the MyMathLab Study Plan.

My video summary ⊙ **THE PYTHAGOREAN IDENTITIES**

Given the right triangle from Figure 33, we can use the Pythagorean Theorem to establish the equation $(opp)^2 + (adj)^2 = (hyp)^2$.

If we start with this equation and then divide both sides by $(hyp)^2$, we can establish the first of three Pythagorean identities.

$(opp)^2 + (adj)^2 = (hyp)^2$ Start with the equation representing the Pythagorean Theorem.

$\dfrac{(opp)^2}{(hyp)^2} + \dfrac{(adj)^2}{(hyp)^2} = \dfrac{(hyp)^2}{(hyp)^2}$ Divide both sides of the equation by $(hyp)^2$.

$\left(\dfrac{opp}{hyp}\right)^2 + \left(\dfrac{adj}{hyp}\right)^2 = 1$ Simplify.

$(\sin \theta)^2 + (\cos \theta)^2 = 1$ Substitute $\sin \theta = \dfrac{opp}{hyp}$ and $\cos \theta = \dfrac{adj}{hyp}$.

We will typically write $(\sin \theta)^2$ as $\sin^2 \theta$ and write $(\cos \theta)^2$ as $\cos^2 \theta$. Therefore, the first Pythagorean identity is $\sin^2 \theta + \cos^2 \theta = 1$. Watch this video to see how to derive the remaining two Pythagorean identities.

The Pythagorean Identities

$\sin^2 \theta + \cos^2 \theta = 1$ $1 + \tan^2 \theta = \sec^2 \theta$ $1 + \cot^2 \theta = \csc^2 \theta$

My interactive
video summary

⊙ Example 6 Using Identities to Find the Exact Value of a Trigonometric Expression

Use identities to find the exact value of each trigonometric expression.

a. $\tan 37° - \dfrac{\sin 37°}{\cos 37°}$

b. $\dfrac{1}{\cos^2 \dfrac{\pi}{9}} - \dfrac{1}{\cot^2 \dfrac{\pi}{9}}$

Solution

a. We can rewrite $\dfrac{\sin 37°}{\cos 37°}$ as $\tan 37°$ using the quotient identity $\tan \theta = \dfrac{\sin \theta}{\cos \theta}$.

Therefore, $\tan 37° - \dfrac{\sin 37°}{\cos 37°} = \tan 37° - \tan 37° = 0$.

b. Try using two reciprocal identities and a Pythagorean identity to find the exact value of this trigonometric expression. Watch this interactive video when you have finished to see if you are correct.

You Try It Work through this You Try It problem.

Work Exercises 24–28 in this textbook or in the MyMathLab Study Plan.

OBJECTIVE 4 UNDERSTANDING COFUNCTIONS

My video summary

⊙ Two positive angles are said to be **complementary** if the sum of their measures is $\dfrac{\pi}{2}$ radians (or 90°). The two acute angles of every right triangle are complementary. To see why, read this explanation. Thus, if θ is an acute angle of a right triangle, then the measure of the other acute angle must be $\dfrac{\pi}{2} - \theta$ (or $90° - \theta$ if θ is given in degrees). For example, if one acute angle of a right triangle is $\dfrac{\pi}{5}$, then the other acute angle is $\dfrac{\pi}{2} - \dfrac{\pi}{5} = \dfrac{5\pi}{10} - \dfrac{2\pi}{10} = \dfrac{3\pi}{10}$.

Consider the right triangle in Figure 34 with complementary acute angles of θ and $\dfrac{\pi}{2} - \theta$, and side lengths a, b, and c.

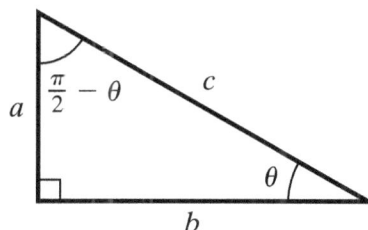

Figure 34 A right triangle with acute angles θ and $\dfrac{\pi}{2} - \theta$

Looking at Figure 34, we see that

$$\sin \theta = \frac{\text{length of side opposite } \theta}{\text{length of the hypotenuse}} = \frac{a}{c} \text{ and } \cos\left(\frac{\pi}{2} - \theta\right) = \frac{\text{length of side adjacent to } \left(\frac{\pi}{2} - \theta\right)}{\text{length of the hypotenuse}} = \frac{a}{c}.$$

Therefore, $\sin \theta = \cos\left(\frac{\pi}{2} - \theta\right)$. In words we say, "the sine of angle θ is equal to the cosine of the complement of angle θ." Trigonometric functions that have this relationship are called **cofunctions**. There are three pair of cofunctions: sine and cosine, tangent and cotangent, and secant and cosecant. We now state the cofunction identities.

Cofunction Identities

$$\sin \theta = \cos\left(\frac{\pi}{2} - \theta\right) \qquad \cos \theta = \sin\left(\frac{\pi}{2} - \theta\right)$$

$$\tan \theta = \cot\left(\frac{\pi}{2} - \theta\right) \qquad \cot \theta = \tan\left(\frac{\pi}{2} - \theta\right)$$

$$\sec \theta = \csc\left(\frac{\pi}{2} - \theta\right) \qquad \csc \theta = \sec\left(\frac{\pi}{2} - \theta\right)$$

If angle θ is given in degrees, replace $\frac{\pi}{2}$ with $90°$.

My interactive video summary

⊚ **Example 7** Using the Cofunction Identities to Rewrite or Evaluate Expressions

a. Rewrite the expression $\left(\cot\left(\frac{\pi}{2} - \theta\right)\right) \cos \theta$ as one of the six trigonometric functions of acute angle θ.

b. Determine the exact value of $\sec 55° \csc 35° - \tan 55° \cot 35°$.

Solution

a. Using cofunctions, we can rewrite $\cot\left(\frac{\pi}{2} - \theta\right)$ as $\tan \theta$ and then simplify.

$$\left(\cot\left(\frac{\pi}{2} - \theta\right)\right)\cos \theta \qquad \text{Start with the original expression.}$$

$$= \tan \theta \cos \theta \qquad \text{Use cofunctions to rewrite } \cot\left(\frac{\pi}{2} - \theta\right) \text{ as } \tan \theta.$$

$$= \frac{\sin \theta}{\cos \theta}\cos \theta \qquad \text{Use the quotient identity } \tan \theta = \frac{\sin \theta}{\cos \theta}.$$

Thus, the expression $\cot\left(\frac{\pi}{2} - \theta\right) \cos \theta$ is equivalent to $\sin \theta$.

b. Using cofunctions, we can rewrite sec 55° as csc (90° − 55°), which is equivalent to csc 35°. Similarly, we can rewrite tan 55° as cot (90° − 55°), which is equivalent to cot 35°. We now use these cofunction identities to simplify the original expression.

$$\sec 55° \csc 35° - \tan 55° \cot 35°$$ Start with the original expression.

$$= \csc 35° \csc 35° - \cot 35° \cot 35°$$ Substitute csc 35° for sec 55° and cot 35° for tan 55°.

$$= \csc^2 35° - \cot^2 35°$$ Simplify.

Note that the Pythagorean identity $1 + \cot^2 \theta = \csc^2 \theta$ is equivalent to $1 = \csc^2 \theta - \cot^2 \theta$. Therefore, $\csc^2 35° - \cot^2 35° = 1$. Hence, the exact value of the original expression, sec 55° csc 35° − tan 55° cot 35°, is 1.

My interactive video summary   ⊙ Watch this interactive video to see this solution worked out in detail.

You Try It Work through this You Try It problem.

Work Exercises 29–33 in this textbook or in the MyMathLab **Study Plan**.

OBJECTIVE 5   EVALUATING TRIGONOMETRIC FUNCTIONS USING A CALCULATOR

We now know how to find the exact values of trigonometric functions of the acute special angles using our knowledge of the special right triangles. But what about evaluating trigonometric functions of non-special angles? For this, we need to find decimal approximations of these values using a calculator. Take a few moments to locate the $\boxed{\sin}$ key, $\boxed{\cos}$ key, and $\boxed{\tan}$ key on your calculator. More importantly, make sure that you know how to set your calculator to degree mode or radian mode. Then try working through Example 8.

My video summary   ⊙ **Example 8** Approximate Trigonometric Expressions Using a Calculator

Evaluate each trigonometric expression using a calculator.
Round each answer to four decimal places.

a. $\sin \dfrac{8\pi}{7}$ **b.** $\cot 70°$ **c.** $\sec \dfrac{\pi}{5}$

Solution

a. First, make sure that your calculator is in radian mode.

$$\sin \frac{8\pi}{7} \approx -0.4339$$

b. First, make sure that your calculator is in degree mode. Since most calculators do not have a cotangent key, we will use the reciprocal identity $\cot \theta = \dfrac{1}{\tan \theta}$ and the $\boxed{\tan}$ key to evaluate the expression.

$$\cot 70° = \frac{1}{\tan 70°} \approx 0.3640$$

c. First, make sure that your calculator is in radian mode. Because most calculators do not have a secant key, we will use the reciprocal identity $\sec \theta = \dfrac{1}{\cos \theta}$ and the $\boxed{\cos}$ key to evaluate the expression.

$$\sec \frac{\pi}{5} = \frac{1}{\cos \dfrac{\pi}{5}} \approx 1.2361$$

My video summary ⊙ Watch this video to see how to enter these expressions into your calculator. ●

⚠ **It is extremely important to make sure that your calculator is set to the correct mode. In Example 8a, your calculator must be set to radian mode. Failing to set your calculator to the correct mode will yield erroneous answers, as illustrated below.**

Using Technology 📖

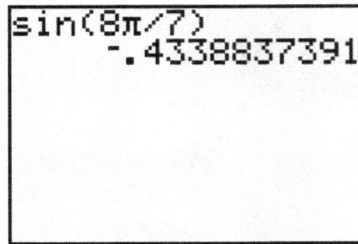

```
sin(8π/7)
        .0626231514
```

```
sin(8π/7)
       -.4338837391
```

The screenshot on the left shows the value of $\sin \dfrac{8\pi}{7}$ using a calculator set to degree mode. The screenshot on the right shows the correct value of $\sin \dfrac{8\pi}{7}$ using the same calculator set to radian mode. You can see the importance of setting your calculator to the correct mode!

You should quickly test yourself by clicking on the You Try It icon below to verify that you can evaluate trigonometric expressions using your calculator.

You Try It Work through this You Try It problem.

Work Exercises 34–37 in this textbook or in the MyMathLab Study Plan.

1.4 Exercises

Skill Check Exercises

For exercises SCE-1 through SCE-6, solve each equation.

SCE-1. $a^2 + (15)^2 = 17^2$ **SCE-2.** $a^2 + (\sqrt{3})^2 = 5^2$ **SCE-3.** $(24)^2 + b^2 = 25^2$

SCE-4. $(\sqrt{5})^2 + b^2 = 6^2$ **SCE-5.** $6^2 + 8^2 = c^2$ **SCE-6.** $2^2 + (\sqrt{7})^2 = c^2$

For exercises SCE-7 and SCE-8, simplify each expression by writing as a single term using a common denominator.

SCE-7. $\dfrac{\pi}{2} - \dfrac{\pi}{6}$

SCE-8. $\dfrac{\pi}{2} - \dfrac{5\pi}{12}$

In Exercises 1–5, a right triangle with acute angle θ is given. Evaluate the six trigonometric functions of the acute angle θ.

1.

2.

3.

4.

5.

6. If θ is an acute angle of a right triangle and if $\sin\theta = \dfrac{5}{13}$, then find the values of the remaining five trigonometric functions for angle θ.

7. If θ is an acute angle of a right triangle and if $\cos\theta = \dfrac{60}{61}$, then find the values of the remaining five trigonometric functions for angle θ.

8. If θ is an acute angle of a right triangle and if $\tan\theta = \dfrac{1}{2}$, then find the values of the remaining five trigonometric functions for angle θ.

9. If θ is an acute angle of a right triangle and if $\csc\theta = \sqrt{6}$, then find the values of the remaining five trigonometric functions for angle θ.

10. If θ is an acute angle of a right triangle and if $\sec\theta = \dfrac{7}{\sqrt{5}}$, then find the values of the remaining five trigonometric functions for angle θ.

11. If θ is an acute angle of a right triangle and if $\cot\theta = 3$, then find the values of the remaining five trigonometric functions for angle θ.

In Exercises 12–18, use special right triangles to evaluate each expression.

12. $\sin \dfrac{\pi}{6}$ **13.** $\sec 45°$ **14.** $\csc \dfrac{\pi}{3}$ **15.** $\csc 45° - \cot 30°$

16. $\sqrt{3} \sec \dfrac{\pi}{6} + \sqrt{2} \csc \dfrac{\pi}{4}$ **17.** $\sec^2 45° + \csc^2 60°$ **18.** $\dfrac{\sin^2 \dfrac{\pi}{3} + \cos^2 \dfrac{\pi}{6}}{\sec^2 \dfrac{\pi}{4}}$

19. Determine the measure of the acute angle θ for which $\csc \theta = 2$.

20. Determine the measure of the acute angle θ for which $\cot \theta = \sqrt{3}$.

21. Determine the measure of the acute angle θ for which $\sec \theta = \sqrt{2}$.

22. Given that $\sin \theta = \dfrac{7}{25}$ and $\cos \theta = \dfrac{24}{25}$, find the values of the remaining four trigonometric functions using identities.

23. Given that $\sin \theta = \dfrac{3}{\sqrt{11}}$ and $\cos \theta = \dfrac{\sqrt{2}}{\sqrt{11}}$, find the values of the remaining four trigonometric functions using identities.

In Exercises 24–26, use identities to find the exact value of each trigonometric expression. Assume that θ is an acute angle.

24. $\cot 19° - \dfrac{\cos 19°}{\sin 19°}$

25. $\dfrac{1}{\csc^2 \dfrac{\pi}{11}} + \dfrac{1}{\sec^2 \dfrac{\pi}{11}}$

26. $\tan \dfrac{5\pi}{12} \left(\dfrac{1}{\cos^2 \dfrac{5\pi}{12}} - \dfrac{1}{\cot^2 \dfrac{5\pi}{12}} \right) - \tan \dfrac{5\pi}{12}$

27. Use identities to rewrite the following expression as $\sin \theta$, $\cos \theta$, $\tan \theta$, $\csc \theta$, $\sec \theta$, or $\cot \theta$.

$$\sin \theta \sec^2 \theta \cot^2 \theta$$

28. Use identities to rewrite the following expression as $\sin \theta$, $\cos \theta$, $\tan \theta$, $\csc \theta$, $\sec \theta$, or $\cot \theta$.

$$\dfrac{\sin^2 \theta + \tan^2 \theta + \cos^2 \theta}{\sec \theta}$$

29. Find a cofunction equivalent to $\sin \dfrac{5\pi}{7}$.

30. Find a cofunction equivalent to $\cot 39°$.

31. Determine the exact value of $\tan \dfrac{\pi}{7} \cot \dfrac{5\pi}{14} - \sec \dfrac{\pi}{7} \csc \dfrac{5\pi}{14}$.

32. Rewrite the expression $\cos(90° - \theta)\sec\theta$ as one of the six trigonometric functions of acute angle θ.

33. Given that θ is an acute angle, rewrite the $\tan(\theta - 15°)$ as an equivalent value using its cofunction.

In Exercises 34–37, use a calculator to evaluate each trigonometric expression. Round your answer to four decimal places.

34. $\cos\dfrac{2\pi}{9}$ **35.** $\csc 81°$ **36.** $\sec\dfrac{4\pi}{13}$ **37.** $\sin 61°$

1.5 Trigonometric Functions of General Angles

THINGS TO KNOW

Before working through this section, be sure that you are familiar with the following concepts:

	VIDEO	ANIMATION	INTERACTIVE

You Try It 1. Understanding Radian Measure (Section 1.1) ⊙

You Try It 2. Converting between Degree Measure and Radian Measure (Section 1.1) ▭ ⊙

You Try It 3. Finding Coterminal Angles Using Radian Measure (Section 1.1)

You Try It 4. Understanding the Special Right Triangles (Section 1.3) ▭

You Try It 5. Understanding the Right Triangle Definitions of the Trigonometric Functions (Section 1.4) ⊙

OBJECTIVES

1 Understanding the Four Families of Special Angles

2 Understanding the Definitions of the Trigonometric Functions of General Angles

3 Finding the Values of the Trigonometric Functions of Quadrantal Angles

4 Understanding the Signs of the Trigonometric Functions

5 Determining Reference Angles

6 Evaluating Trigonometric Functions of Angles Belonging to the $\dfrac{\pi}{3}, \dfrac{\pi}{6},$ or $\dfrac{\pi}{4}$ Families

OBJECTIVE 1   UNDERSTANDING THE FOUR FAMILIES OF SPECIAL ANGLES

In Section 1.4, we defined the six trigonometric functions of an angle θ where θ was an acute angle of a right triangle. In this section, we will extend the definitions of the six trigonometric functions to include *all* angles for which each function is defined. We start by introducing four groups, or families, of special angles. These four families are known as **the quadrantal family, the $\frac{\pi}{3}$ family, the $\frac{\pi}{6}$ family, and the $\frac{\pi}{4}$ family.**

THE QUADRANTAL FAMILY OF ANGLES

My video summary

⊙ Recall that a quadrantal angle is an angle in standard position whose terminal side lies along either axis. Although there are infinitely many quadrantal angles, there are only four possible positions for the terminal side of the quadrantal angles. Any angle belonging to the quadrantal family must be coterminal with $0, \frac{\pi}{2}, \pi,$ or $\frac{3\pi}{2}$. See Figure 35. Watch this video for a detailed explanation of the quadrantal family of angles.

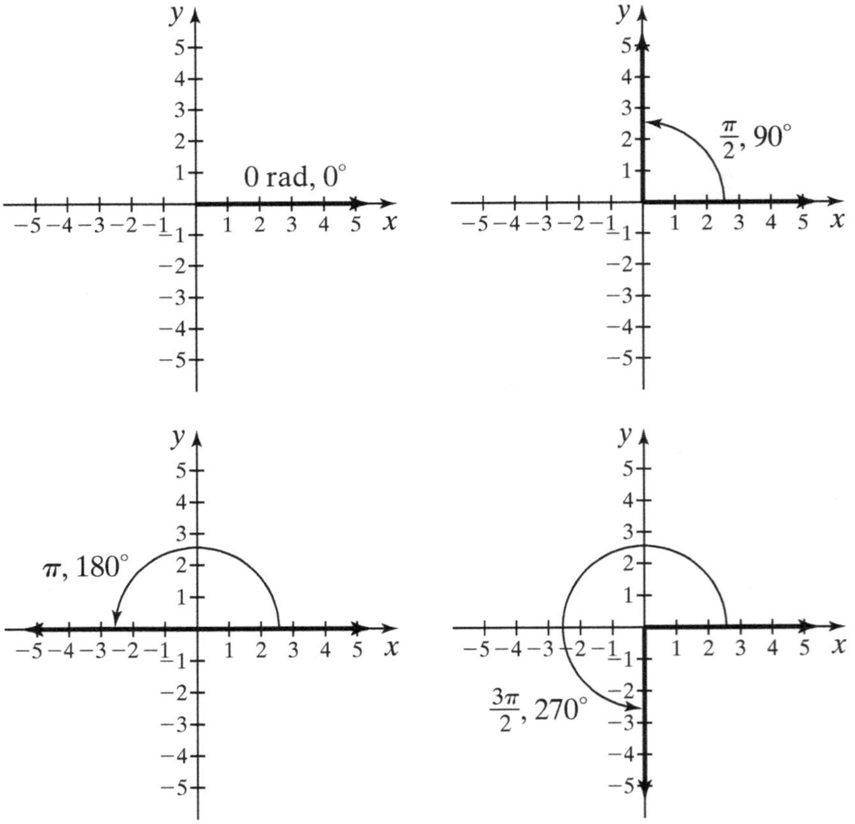

Figure 35 Angles belonging to the quadrantal family must be coterminal with $0, \frac{\pi}{2}, \pi,$ or $\frac{3\pi}{2}$.

My video summary ▶ **THE $\frac{\pi}{3}$ FAMILY OF ANGLES**

An angle in standard position belongs to the $\frac{\pi}{3}$ family if the angle is coterminal with $\frac{\pi}{3}, \frac{2\pi}{3}, \frac{4\pi}{3}$, or $\frac{5\pi}{3}$. See Figure 36. Angles belonging to the $\frac{\pi}{3}$ family must be an integer multiple of $\frac{\pi}{3}$ but must **not** be a quadrantal angle. For example, the angle $\theta = \frac{6\pi}{3}$ does not belong to the $\frac{\pi}{3}$ family because $\theta = \frac{6\pi}{3} = 2\pi$ and $\theta = 2\pi$ is a quadrantal angle. Watch this video for a detailed explanation of the $\frac{\pi}{3}$ family of angles.

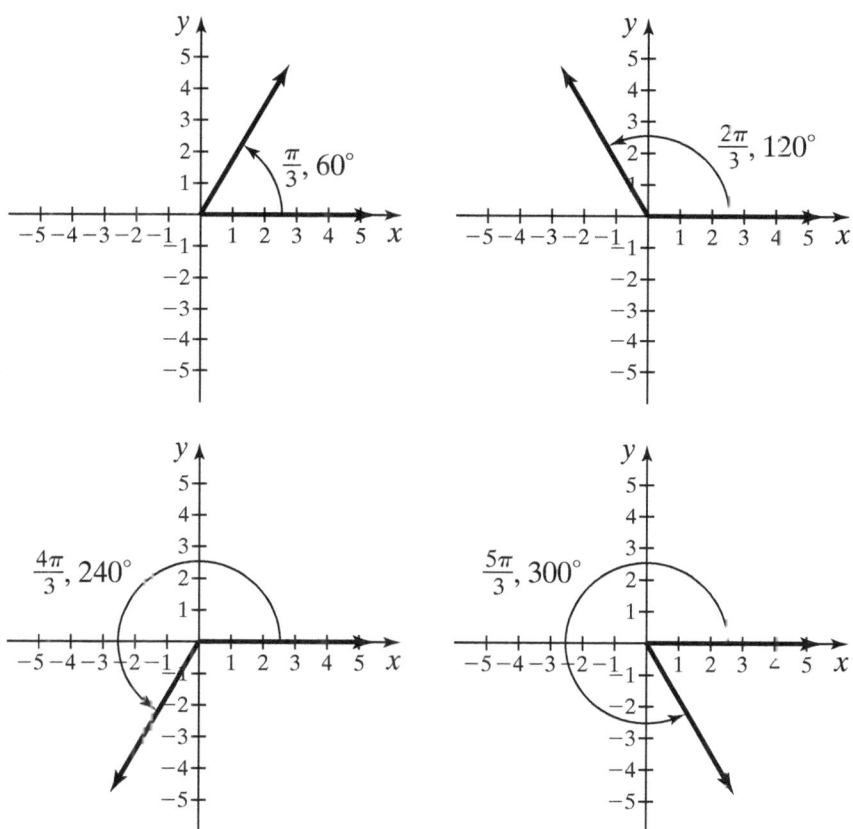

Figure 36 Angles belonging to the $\frac{\pi}{3}$ family must be coterminal with $\frac{\pi}{3}, \frac{2\pi}{3}, \frac{4\pi}{3}$, or $\frac{5\pi}{3}$.

My video summary ▶ **THE $\frac{\pi}{6}$ FAMILY OF ANGLES**

An angle in standard position belongs to the $\frac{\pi}{6}$ family if the angle is coterminal with $\frac{\pi}{6}, \frac{5\pi}{6}, \frac{7\pi}{6}$, or $\frac{11\pi}{6}$. See Figure 37. Angles belonging to the $\frac{\pi}{6}$ family must be an integer multiple of $\frac{\pi}{6}$ but must **not** belong to the quadrantal family or the $\frac{\pi}{3}$

family. For example, the angle $\theta = \dfrac{4\pi}{6}$ does not belong to the $\dfrac{\pi}{6}$ family because

$\theta = \dfrac{4\pi}{6} = \dfrac{2\pi}{3}$ and $\theta = \dfrac{2\pi}{3}$ belongs to the $\dfrac{\pi}{3}$ family of angles.

Watch this video for a detailed explanation of the $\dfrac{\pi}{6}$ family of angles.

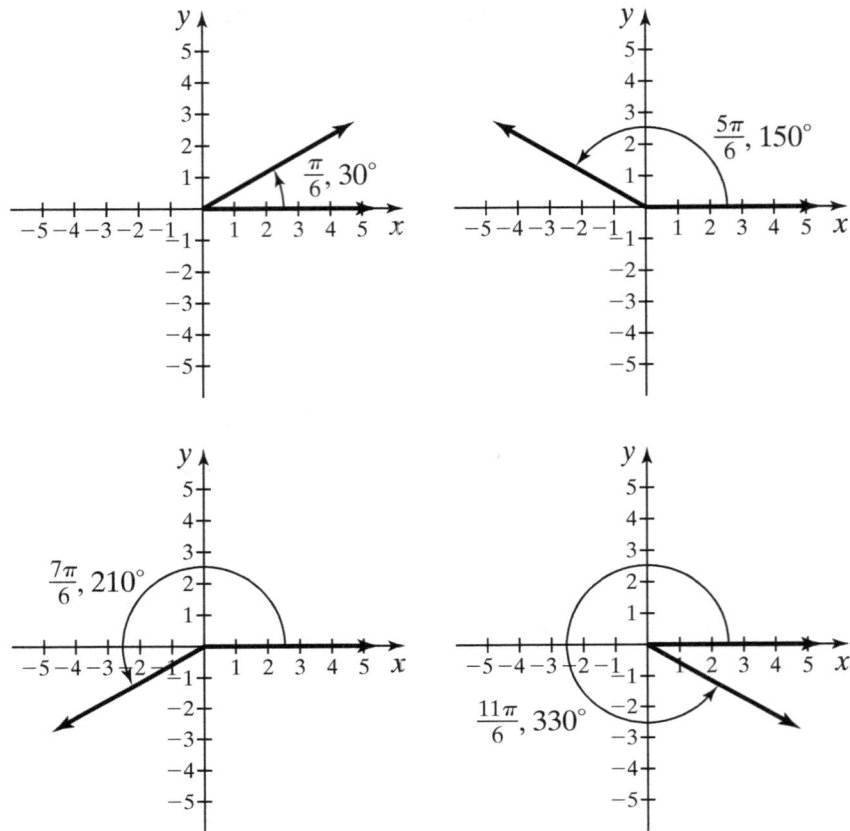

Figure 37 Angles belonging to the $\dfrac{\pi}{6}$ family must be coterminal

with $\dfrac{\pi}{6}, \dfrac{5\pi}{6}, \dfrac{7\pi}{6}$, or $\dfrac{11\pi}{6}$.

THE $\dfrac{\pi}{4}$ FAMILY OF ANGLES

My video summary ▷ An angle in standard position belongs to the $\dfrac{\pi}{4}$ family if the angle is coterminal

with $\dfrac{\pi}{4}, \dfrac{3\pi}{4}, \dfrac{5\pi}{4}$, or $\dfrac{7\pi}{4}$. See Figure 38. Angles belonging to the $\dfrac{\pi}{4}$ family must be an

integer multiple of $\dfrac{\pi}{4}$ but must **not** belong to the quadrantal family. For example,

the angle $\theta = \dfrac{10\pi}{4}$ does not belong to the $\dfrac{\pi}{4}$ family because $\theta = \dfrac{10\pi}{4} = \dfrac{5\pi}{2}$ and

$\theta = \dfrac{5\pi}{2}$ belongs to the quadrantal family. Watch this video for a detailed explana-

tion of the $\dfrac{\pi}{4}$ family of angles.

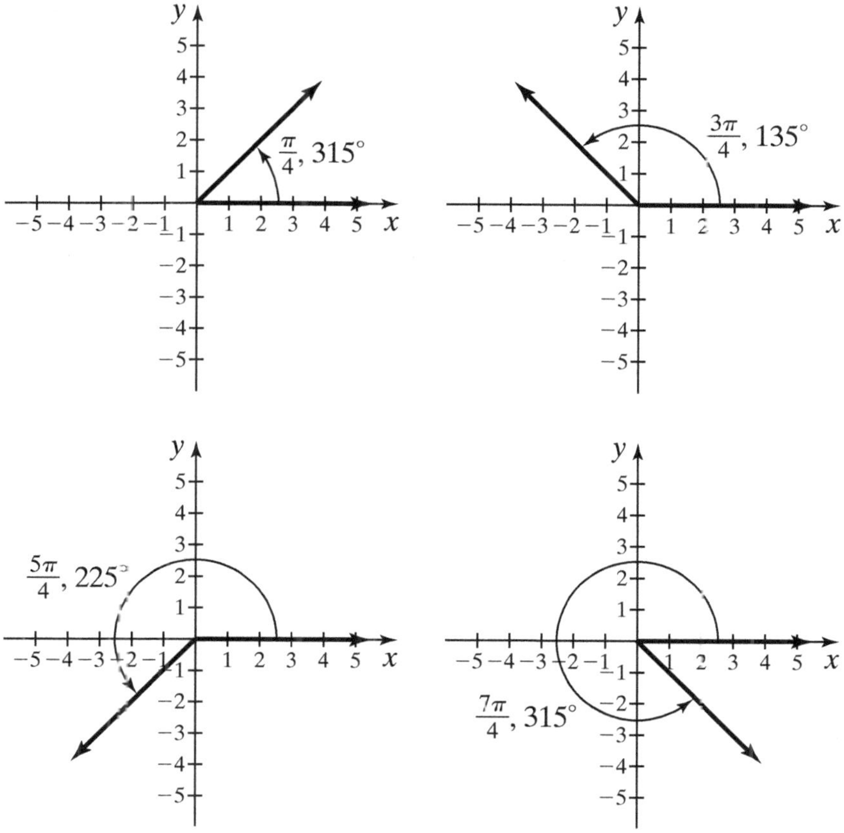

Figure 38 Angles belonging to the $\frac{\pi}{4}$ family must be coterminal

with $\frac{\pi}{4}, \frac{3\pi}{4}, \frac{5\pi}{4}$, or $\frac{7\pi}{4}$.

My interactive video summary

⊘ **Example 1** Working with Angles Belonging to the Families of Special Angles

Each of the given angles belongs to one of the four families of special angles. Determine the family of angles for which it belongs, sketch the angle, and then determine the angle of least nonnegative measure, θ_C, coterminal with the given angle.

a. $\theta = \dfrac{29\pi}{6}$ **b.** $\theta = \dfrac{14\pi}{2}$ **c.** $\theta = -\dfrac{18\pi}{4}$ **d.** $\theta = \dfrac{11\pi}{4}$

e. $\theta = \dfrac{14\pi}{6}$ **f.** $\theta = 420°$ **g.** $\theta = -495°$

Solution

a. The angle $\theta = \dfrac{29\pi}{6}$ is written in lowest terms and is 29 times $\dfrac{\pi}{6}$. Therefore,

$\theta = \dfrac{29\pi}{6}$ belongs to the $\dfrac{\pi}{6}$ family. Recall that the angle created from one

complete counterclockwise revolution has a measure of $2\pi = \dfrac{12\pi}{6}$ radians.

Two complete counterclockwise revolutions correspond to the angle $2(2\pi) = 2\left(\dfrac{12\pi}{6}\right) = \dfrac{24\pi}{6}$.

We can rewrite the angle

$$\theta = \dfrac{29\pi}{6} \text{ as } \theta = \dfrac{29\pi}{6} = \dfrac{24\pi}{6} + \dfrac{5\pi}{6} = 2(2\pi) + \dfrac{5\pi}{6}.$$

Therefore, we can sketch this angle by rotating the terminal side of the angle two complete counterclockwise revolutions plus an additional rotation of $\dfrac{5\pi}{6}$ radians.

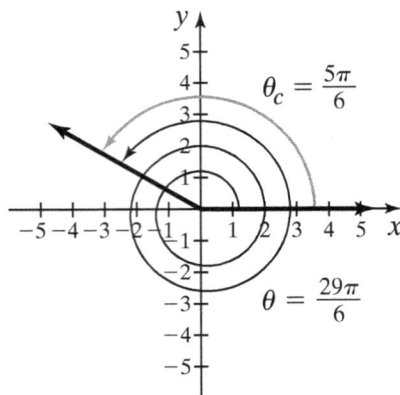

Figure 39

See Figure 39. The angle of least nonnegative measure coterminal with $\theta = \dfrac{29\pi}{6}$ is $\theta_C = \dfrac{5\pi}{6}$.

b. The angle $\theta = \dfrac{14\pi}{2}$ is not written in lowest terms. We can rewrite this angle in lowest terms as $\theta = \dfrac{14\pi}{2} = \dfrac{2 \cdot 7\pi}{2} = 7\pi$. Any angle that is an integer multiple of π belongs to the quadrantal family. Therefore, this angle belongs to the quadrantal family. The angle $\theta = 7\pi$ can be rewritten as $\theta = 3(2\pi) + \pi$. Therefore, this angle can be drawn by rotating the terminal side counterclockwise three complete revolutions plus an additional rotation of π radians. See Figure 40.

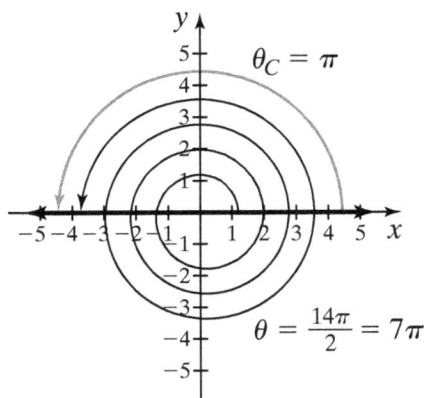

Figure 40

The terminal side of this angle lies along the negative x-axis. Therefore, the angle of least nonnegative measure coterminal with $\theta = 7\pi$ is $\theta_C = \pi$.

c. The angle $\theta = -\dfrac{18\pi}{4}$ is not written in lowest terms. We can rewrite this angle

in lowest terms as $\theta = -\dfrac{18\pi}{4} = -\dfrac{2 \cdot 9\pi}{2 \cdot 2} = -\dfrac{9\pi}{2}$. Any angle that is an inte-

ger multiple of $\dfrac{\pi}{2}$ belongs to the quadrantal family. Therefore, this angle

belongs to the quadrantal family. The angle $\theta = -\dfrac{9\pi}{2}$ can be rewritten as

$\theta = -\dfrac{8\pi}{2} - \dfrac{\pi}{2} = 2(-2\pi) - \dfrac{\pi}{2}$. Therefore, this angle can be drawn by rotating

the terminal side clockwise two complete revolutions plus an additional

rotation of $\dfrac{\pi}{2}$ radians in a clockwise direction. See Figure 41.

The terminal side of this angle lies along the negative y-axis. Therefore, the

angle of least nonnegative measure coterminal with $\theta = -\dfrac{9\pi}{2}$ is $\theta_C = \dfrac{3\pi}{2}$.

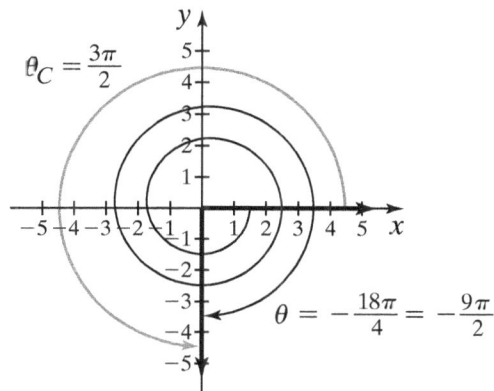

Figure 41

My interactive video summary

⊙ Try completing parts d–g on your own. Then watch this interactive video to see the remainder of this solution.

You Try It Work through this You Try It problem.

Work Exercises 1–12 in this textbook or in the MyMathLab Study Plan.

OBJECTIVE 2 UNDERSTANDING THE DEFINITIONS OF THE TRIGONOMETRIC FUNCTIONS OF GENERAL ANGLES

My animation summary

In Section 1.4, we presented the definitions of the trigonometric functions using the lengths of the sides and the hypotenuse of a right triangle. You may wish to review these definitions. The right triangle definitions of the trigonometric functions only applies to acute angles. We now want to define the trigonometric functions of *general angles*. The term *general angle* is used here to indicate that these angles are not restricted in size and can be either positive angles, negative angles, or zero.

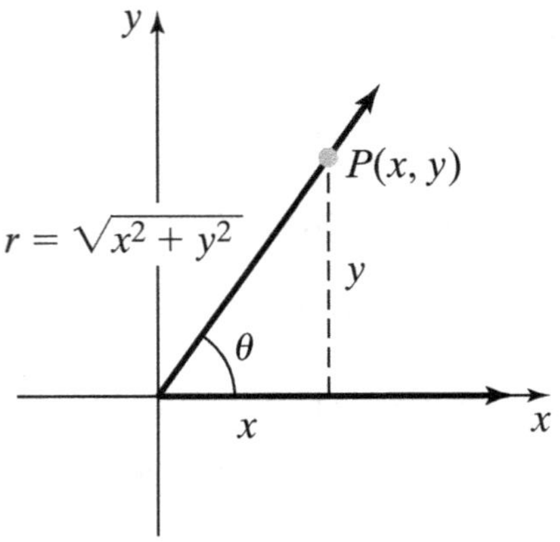

Figure 42 This illustrates an acute angle θ in Quadrant I.

In order to develop the definitions of trigonometric functions of general angles, let $P(x, y)$ be a point lying on the terminal side of an acute angle θ in standard position and let $r > 0$ represent the distance from the origin to point P. By the distance formula, $r = \sqrt{x^2 + y^2}$. We use the right triangle definitions of the trigonometric functions to write the trigonometric ratios in terms of x, y, and r. See Figure 42.

$$\sin \theta = \frac{opp}{hyp} = \frac{y}{r} \qquad \csc \theta = \frac{hyp}{opp} = \frac{r}{y}$$

$$\cos \theta = \frac{adj}{hyp} = \frac{x}{r} \qquad \sec \theta = \frac{hyp}{adj} = \frac{r}{x}$$

$$\tan \theta = \frac{opp}{adj} = \frac{y}{x} \qquad \cot \theta = \frac{adj}{opp} = \frac{x}{y}$$

We now use this representation to create a second set of definitions of trigonometric functions. These definitions will apply to general angles. Notice that the new definitions are consistent with our previous right triangle definitions.

The General Angle Definitions of the Trigonometric Functions

If $P(x, y)$ is a point on the terminal side of *any* angle θ in standard position and if $r = \sqrt{x^2 + y^2}$ is the distance from the origin to point P, then the six trigonometric functions of θ are defined as follows:

$$\sin \theta = \frac{y}{r} \qquad \csc \theta = \frac{r}{y}, y \neq 0$$

$$\cos \theta = \frac{x}{r} \qquad \sec \theta = \frac{r}{x}, x \neq 0$$

$$\tan \theta = \frac{y}{x}, x \neq 0 \qquad \cot \theta = \frac{x}{y}, y \neq 0$$

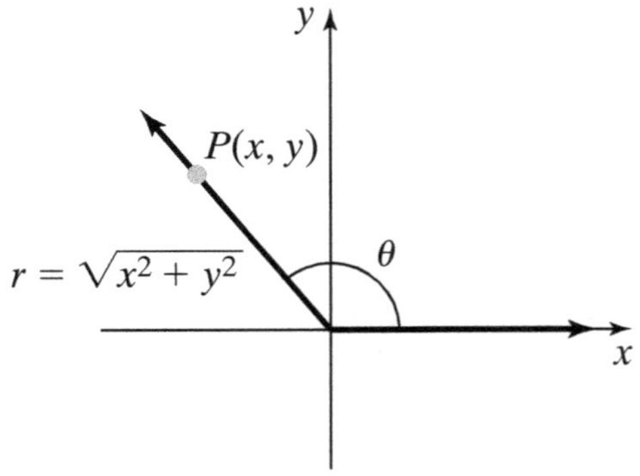

Because division by zero is undefined, it is important to note that four of the trigonometric functions will not be defined for certain angles. For example, $\tan \theta = \frac{y}{x}$ and $\sec \theta = \frac{r}{x}$ will be undefined when $x = 0$. A point having an x-coordinate of zero must lie along the y-axis. Therefore, the tangent and secant functions are undefined for any angle whose terminal side lies along the **y-axis**. Similarly, $\csc \theta = \frac{r}{y}$ and $\cot \theta = \frac{x}{y}$ are undefined when $y = 0$. A point having a y-coordinate of zero must lie along the x-axis. Thus, the cosecant and cotangent functions are undefined for any angle whose terminal side lies along the **x-axis**. We will explore this further when we discuss the trigonometric functions of quadrantal angles. The sine

$\left(\sin \theta = \dfrac{y}{r}\right)$ and cosine $\left(\cos \theta = \dfrac{x}{r}\right)$ functions are defined for any angle because

the value of r, which is in the denominator of each ratio, will never be zero.

My animation summary

You may want to watch this animation to see the development of the definitions of the trigonometric functions of general angles.

My video summary

Example 2 Finding the Values of the Six Trigonometric Functions for a General Angle

Suppose that the point $(-4, -6)$ is on the terminal side of an angle θ. Find the six trigonometric functions of θ.

Solution Always start by plotting the point, constructing the angle θ, and identifying the values of x, y, and r. We are given that $x = -4$ and $y = -6$. Using the equation $r = \sqrt{x^2 + y^2}$, we can find the value of r.

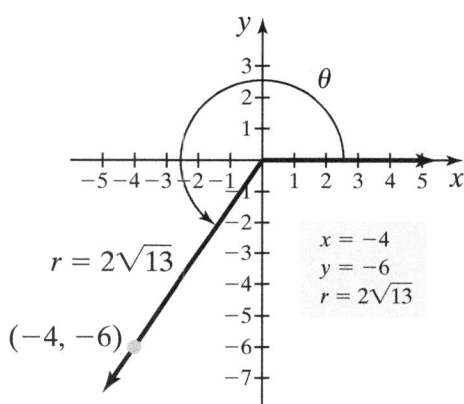

$$r = \sqrt{x^2 + y^2} \qquad \text{Write the equation representing the value of } r.$$

$$= \sqrt{(-4)^2 + (-6)^2} \qquad \text{Substitute } x = -4 \text{ and } y = -6.$$

$$= \sqrt{52} = 2\sqrt{13} \qquad \text{Simplify the radical.}$$

$r = 2\sqrt{13}$

$(-4, -6)$

$x = -4$
$y = -6$
$r = 2\sqrt{13}$

Now that we know the values of x, y, and r, we can use the general angle definitions of the trigonometric functions to find the values of $\sin \theta$, $\cos \theta$, and $\tan \theta$.

$$\sin \theta = \frac{y}{r} = \frac{-6}{2\sqrt{13}} = -\frac{3}{\sqrt{13}} \qquad \csc \theta = \frac{r}{y} = \frac{2\sqrt{13}}{-6} = -\frac{\sqrt{13}}{3}$$

$$\cos \theta = \frac{x}{r} = \frac{-4}{2\sqrt{13}} = -\frac{2}{\sqrt{13}} \qquad \sec \theta = \frac{r}{x} = \frac{2\sqrt{13}}{-4} = -\frac{\sqrt{13}}{2}$$

$$\tan \theta = \frac{y}{x} = \frac{-6}{-4} = \frac{3}{2} \qquad \cot \theta = \frac{x}{y} = \frac{-4}{-6} = \frac{2}{3}$$

Notice that you may leave a radical in the denominator as long as the radical is simplified and that all factors common to the numerator and denominator are removed. This includes common integer factors as well as radical factors. To see some examples of simplification that will be required, view this table. Watch this

My video summary

video to see a complete solution to this example.

You Try It Work through this You Try It problem.

Work Exercises 13–18 in this textbook or in the MyMathLab Study Plan.

OBJECTIVE 3 FINDING THE VALUES OF THE TRIGONOMETRIC FUNCTIONS OF QUADRANTAL ANGLES

If θ is an angle belonging to the quadrantal family, then the terminal side of the angle lies along an axis. Any point lying on the terminal side of an angle coterminal to 0 radian (0°) or coterminal to π radians (180°) lies along the x-axis and therefore has a y-coordinate of 0. Similarly, any point lying on the terminal side of an angle coterminal to $\frac{\pi}{2}$ radians (90°) or coterminal to $\frac{3\pi}{2}$ radians (270°) lies along the y-axis and has an x-coordinate of 0. To determine the values of the trigonometric functions of angles belonging to the quadrantal family, we can use *any* point lying on the terminal side and apply the definitions of the six trigonometric functions. See Figure 43.

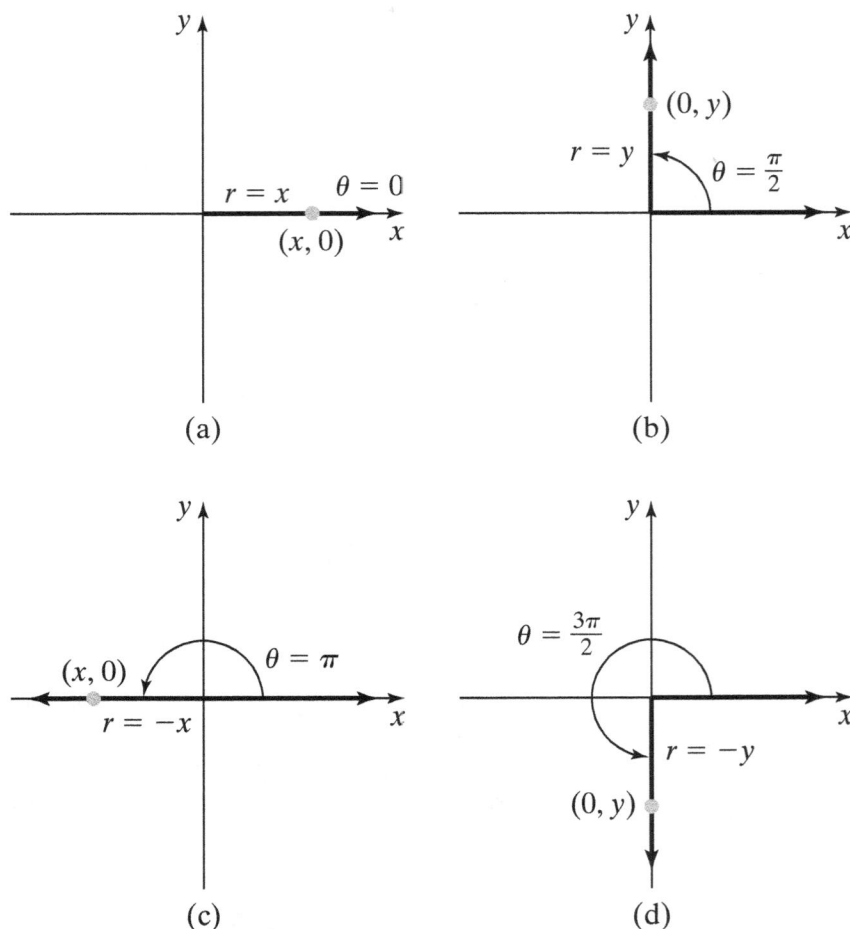

(a) (b) (c) (d)

Figure 43 Quadrantal angles with an arbitrary point located on the terminal side of each angle

Looking at Figure 43a, we can use x, $y = 0$, and $r = x$ to find the values of the six trigonometric functions for the angle $\theta = 0$ radian (or $\theta = 0°$).

$$\sin 0 = \frac{y}{r} = \frac{0}{x} = 0 \qquad \csc 0 = \frac{r}{y} = \frac{x}{0} \ (\text{undefined})$$

$$\cos 0 = \frac{x}{r} = \frac{x}{x} = 1 \qquad \sec 0 = \frac{r}{x} = \frac{x}{x} = 1$$

$$\tan 0 = \frac{y}{x} = \frac{0}{x} = 0 \qquad \cot 0 = \frac{x}{y} = \frac{x}{0} \ (\text{undefined})$$

We can see that the cosecant function and cotangent functions are undefined when $\theta = 0$. The cosecant and cotangent functions will be undefined for any quadrantal angle whose terminal side lies along the x-axis. Can you determine which functions will be undefined for quadrantal angles whose terminal side lies along the y-axis? Check your answer when finished. Use Figure 43 to evaluate the six trigonometric functions for $\theta = \frac{\pi}{2}, \theta = \pi$, and $\theta = \frac{3\pi}{2}$ on your own. Table 2 summarizes the values of the six trigonometric functions of quadrantal angles.

Table 2 Trigonometric Functions of Quadrantal Angles

θ	$\sin\theta$	$\cos\theta$	$\tan\theta$	$\csc\theta$	$\sec\theta$	$\cot\theta$
0	0	1	0	Undefined	1	Undefined
$\dfrac{\pi}{2}$	1	0	Undefined	1	Undefined	0
π	0	-1	0	Undefined	-1	Undefined
$\dfrac{3\pi}{2}$	-1	0	Undefined	-1	Undefined	0

You will find that, rather than trying to memorize this table, determining these values by drawing the graphs seen in **Figure 43** is much easier. Also, the values seen in Table 2 will become clear once you become familiar with the graphs of the trigonometric functions. The graphs of the trigonometric functions will be studied in depth in Chapter 2.

My interactive video summary

⊚ **Example 3** Finding the Values of the Six Trigonometric Functions of Quadrantal Angles

Without using a calculator, determine the value of the trigonometric function or state that the value is undefined.

a. $\cos(-11\pi)$ **b.** $\csc(-270°)$ **c.** $\tan\left(\dfrac{13\pi}{2}\right)$ **d.** $\sin(540°)$ **e.** $\cot\left(-\dfrac{7\pi}{2}\right)$

Solution Draw each angle and choose an arbitrary point (other than the origin) on the terminal side of the angle. Then use the general angle definitions of the trigonometric functions to find each value or state that the value is undefined. Once you have completed this, watch this interactive video to see if you are correct.

You Try It Work through this You Try It problem.

Work Exercises 19–30 in this textbook or in the MyMathLab Study Plan.

OBJECTIVE 4 UNDERSTANDING THE SIGNS OF THE TRIGONOMETRIC FUNCTIONS

My video summary

Suppose $0 < \theta < 2\pi$ and $\sin\theta = \frac{3}{5}$. What is the value of $\cos\theta$? By definition,

$\sin\theta = \frac{y}{r}$, so possible choices for y and r are $y = 3$ and $r = 5$. (We could have

picked $y = 6$ and $r = 10$ or $y = 9$ and $r = 15$ because $\frac{6}{10} = \frac{3}{5}$ and $\frac{9}{15} = \frac{3}{5}$.) Using

the equation $r = \sqrt{x^2 + y^2}$, we can solve for x to get $x = \pm 4$. (View these steps to

see how.) Looking at Figure 44, we see that there are two possible angles

(between 0 and 2π) for which $\sin\theta = \frac{3}{5}$. One possible angle appears in

Quadrant I with $x = 4$, $y = 3$, and $r = 5$. See Figure 44a. The other possible angle

appears in Quadrant II with $x = -4$, $y = 3$, and $r = 5$. See Figure 44b.

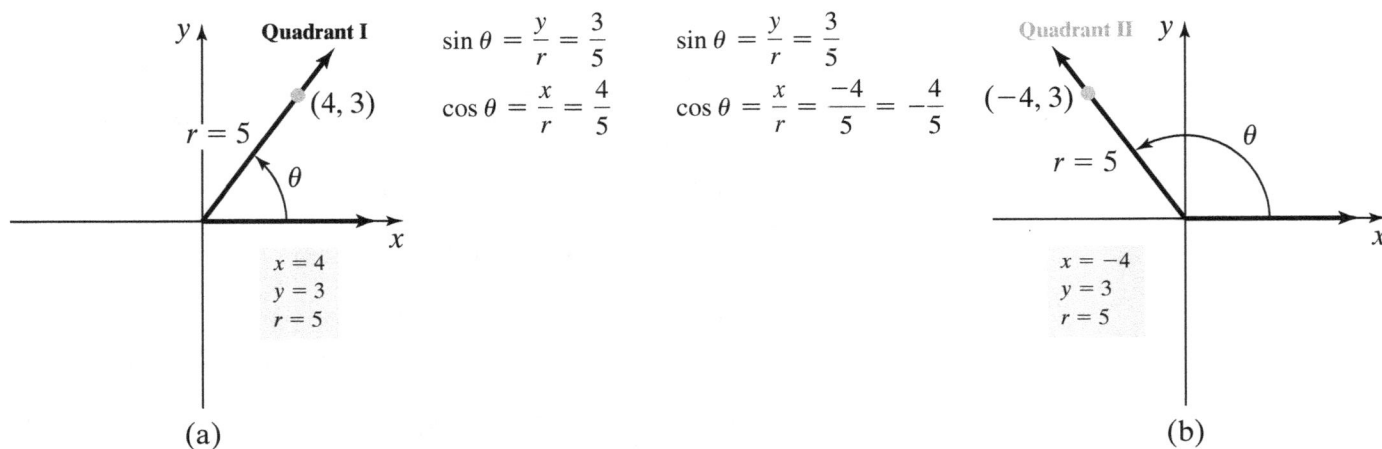

Quadrant I

$$\sin\theta = \frac{y}{r} = \frac{3}{5} \qquad \sin\theta = \frac{y}{r} = \frac{3}{5}$$

$$\cos\theta = \frac{x}{r} = \frac{4}{5} \qquad \cos\theta = \frac{x}{r} = \frac{-4}{5} = -\frac{4}{5}$$

(4, 3)

$r = 5$

θ

$x = 4$
$y = 3$
$r = 5$

(a)

Quadrant II

$(-4, 3)$

$r = 5$

θ

$x = -4$
$y = 3$
$r = 5$

(b)

Figure 44 There are two possible values of θ between 0 and 2π for which $\sin\theta = \frac{3}{5}$.

In Figure 44a, $\cos\theta = \frac{4}{5}$. In Figure 44b, $\cos\theta = -\frac{4}{5}$. Thus, it is impossible to determine

the value of $\cos\theta$ unless we know the quadrant in which the terminal side of the
angle lies. It is crucial to understand that the sign of each trigonometric function is
determined by the quadrant in which the terminal side of the angle lies. The value
of r is always positive. Thus, it follows that the sign of a trigonometric function
depends on the sign of the x- and y-coordinates of the ordered pair $P(x, y)$ lying
on the terminal side of the angle. The signs of x and y depends on the quadrant
in which the point is located. We now explore how to determine the sign of each
trigonometric function given a point on the terminal side of an angle in each of
the four quadrants.

QUADRANT I

In Quadrant I, $r > 0, x > 0$, and $y > 0$. Therefore, the ratios $\frac{y}{r}, \frac{x}{r}$, and $\frac{y}{x}$ (and their reciprocals) are all positive. Thus, for any angle θ whose terminal side lies in Quadrant I, $\sin \theta = \frac{y}{r} > 0, \cos \theta = \frac{x}{r} > 0,$ and $\tan \theta = \frac{y}{x} > 0$. Therefore, *all* six trigonometric functions are positive for *all* angles whose terminal side lies in Quadrant I. See Figure 45.

$$\sin \theta = \frac{y}{r} = \frac{+}{+} = +$$

$$\cos \theta = \frac{x}{r} = \frac{+}{+} = +$$

$$\tan \theta = \frac{y}{x} = \frac{+}{+} = +$$

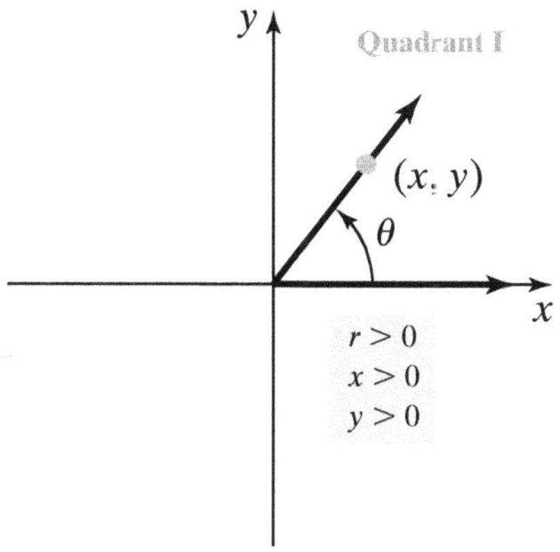

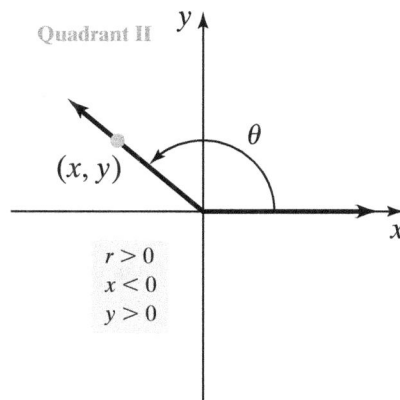

Figure 45 All trigonometric functions are positive for all angles in Quadrant I.

QUADRANT II

In Quadrant II, $r > 0, x < 0$, and $y > 0$. Therefore, $\frac{y}{r} > 0$ but $\frac{x}{r} < 0$ and $\frac{y}{x} < 0$. Thus, for any angle θ whose terminal side lies in Quadrant II, $\sin \theta = \frac{y}{r} > 0.$ $\cos \theta = \frac{x}{r} < 0,$

and $\tan \theta = \frac{y}{x} < 0$. We only need to remember that $\sin \theta$ (and its reciprocal $\csc \theta$) are positive for all angles with a terminal side lying in Quadrant II. All other trigonometric functions are negative. See Figure 46.

$$\sin \theta = \frac{y}{r} = \frac{+}{+} = +$$

$$\cos \theta = \frac{x}{r} = \frac{-}{+} = -$$

$$\tan \theta = \frac{y}{x} = \frac{+}{-} = -$$

Figure 46 The sine function is positive for all angles in Quadrant II.

QUADRANT III

In Quadrant III, $r > 0, x < 0$, and $y < 0$. Therefore, $\dfrac{y}{r} < 0$ and $\dfrac{x}{r} < 0$. The ratio $\dfrac{y}{x}$ is

positive because both x and y are negative values. Thus, for any angle θ whose

terminal side lies in Quadrant III, $\tan \theta = \dfrac{y}{x} > 0$, $\sin \theta = \dfrac{y}{r} < 0$, and $\cos \theta = \dfrac{x}{r} < 0$. We

need only remember that $\tan \theta$ (and its reciprocal $\cot \theta$) are positive for all angles whose terminal side lies in Quadrant III. All other trigonometric functions are negative. See Figure 47.

$$\sin \theta = \frac{y}{r} = \frac{-}{+} = -$$

$$\cos \theta = \frac{x}{r} = \frac{-}{+} = -$$

$$\tan \theta = \frac{y}{x} = \frac{-}{-} = +$$

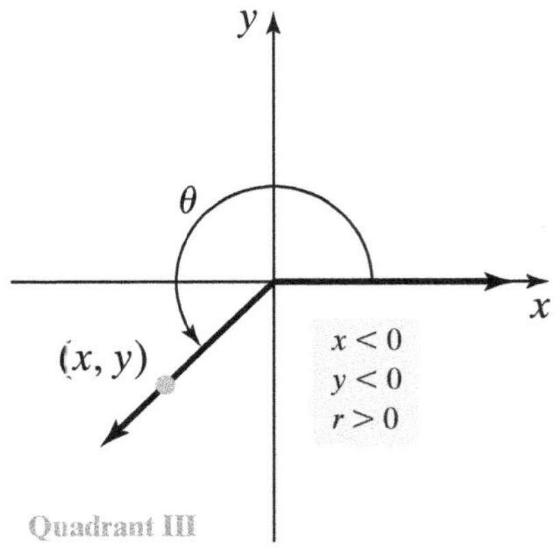

$x < 0$
$y < 0$
$r > 0$

Quadrant III

Figure 47 The tangent function is positive for all angles in Quadrant III.

QUADRANT IV

In Quadrant IV, $r > 0, x > 0$, and $y < 0$. Therefore, $\dfrac{x}{r} > 0$ but $\dfrac{y}{r} < 0$ and

$\dfrac{y}{x} < 0$. Thus, for any angle θ whose terminal side lies in Quadrant IV,

$\cos \theta = \dfrac{x}{r} > 0$, $\sin \theta = \dfrac{y}{r} < 0$, and $\tan \theta = \dfrac{y}{x} < 0$. We need to remember that only

$\cos \theta$ (and its reciprocal $\sec \theta$) are positive for all angles with a terminal side lying in Quadrant IV. All other trigonometric functions are negative. See Figure 48.

$$\sin \theta = \frac{y}{r} = \frac{-}{+} = -$$

$$\cos \theta = \frac{x}{r} = \frac{+}{+} = +$$

$$\tan \theta = \frac{y}{x} = \frac{-}{+} = -$$

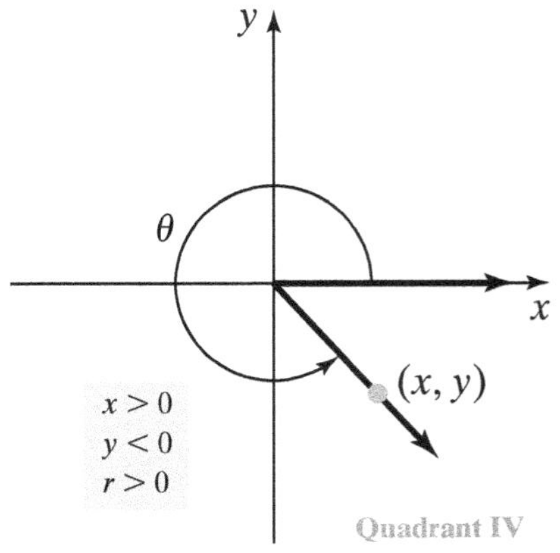

$x > 0$
$y < 0$
$r > 0$

Quadrant IV

Figure 48 The cosine function is positive for all angles in Quadrant IV.

We can memorize the signs of the trigonometric functions for angles whose terminal side lies in one of the four quadrants by remembering the acronym **ASTC**. These four letters stand for the following:

A: If the terminal side of an angle lies in **Quadrant I**, then **A**ll trigonometric functions are positive.

S: If the terminal side of an angle lies in **Quadrant II**, then the **S**ine function is positive.

T: If the terminal side of an angle lies in **Quadrant III**, then the **T**angent function is positive.

C: If the terminal side of an angle lies in **Quadrant IV**, the **C**osine function is positive.

An easy way to remember ASTC is to memorize the phrase "A Smart Trig Class" or the phrase "All Students Take Calculus." A summary of the signs of the trigonometric functions is shown in Figure 49.

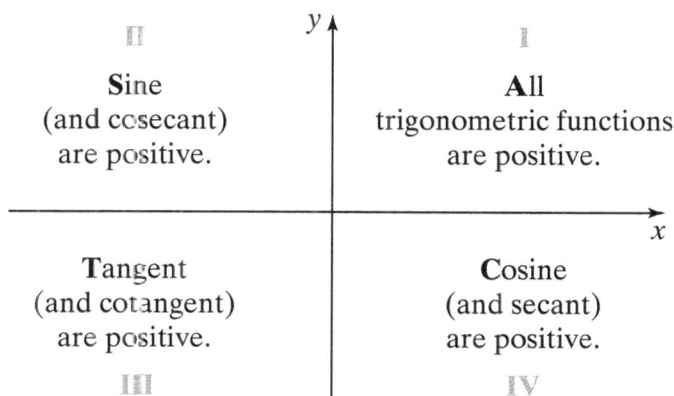

II

Sine
(and cosecant)
are positive.

y

I

All
trigonometric functions
are positive.

x

Tangent
(and cotangent)
are positive.

III

Cosine
(and secant)
are positive.

IV

Figure 49

The signs of the trigonometric functions $y = \sin x, y = \cos x,$ and $y = \tan x$ for angles having terminal sides in each of the four quadrants are displayed below in Figure 50.

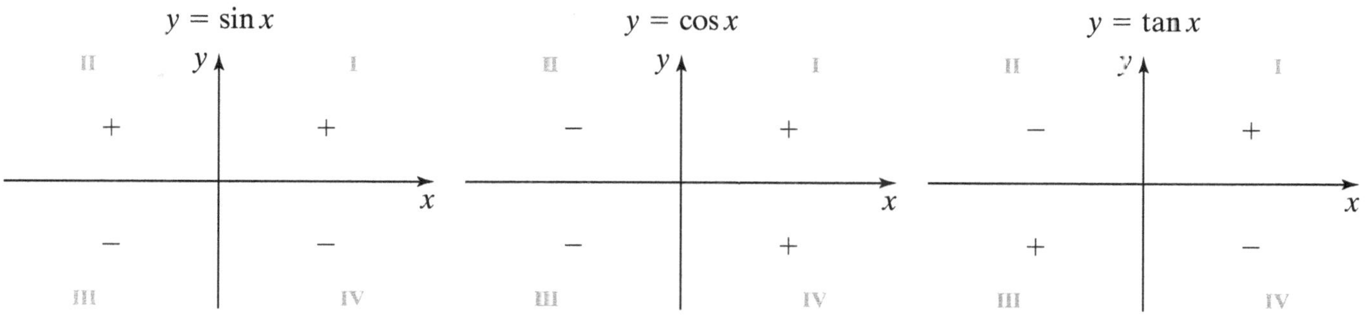

$y = \sin x$

II y I
+   +
                x
−   −
III   IV

The sine function is positive for angles with terminal sides lying in Quadrants I or II.

$y = \cos x$

II y I
−   +
                x
−   +
III   IV

The cosine function is positive for angles with terminal sides lying in Quadrants I or IV.

$y = \tan x$

II y I
−   +
                x
+   −
III   IV

The tangent function is positive for angles with terminal sides lying in Quadrants I or III.

Figure 50

My video summary   ⊙ **Example 4** Determining the Quadrant of an Angle and Evaluating a Trigonometric Function

Suppose θ is a positive angle in standard position such that $\sin \theta < 0$ and $\sec \theta > 0$.

a. Determine the quadrant in which the terminal side of angle θ lies.

b. Find the value of $\tan \theta$ if $\sec \theta = \sqrt{5}$.

Solution

a. The sine function is negative in Quadrant III and Quadrant IV. The secant function is the reciprocal of the cosine function which is positive in Quadrant I and Quadrant IV. The terminal side of angle θ must lie in the quadrant common to these two pairs. Therefore, the terminal side of angle θ must lie in Quadrant IV.

b. The terminal side of angle θ lies in Quadrant IV, so the value of y must be negative. Because $\sec \theta = \dfrac{\sqrt{5}}{1} = \dfrac{r}{x}$, we know that $r = \sqrt{5}$ and $x = 1$. We can use the equation $r = \sqrt{x^2 + y^2}$ to solve for y. Read these steps to verify that $y = -2$. We can now draw angle θ, which passes through the point $(1, -2)$. See Figure 51. By the definition of the tangent function, we get $\tan \theta = \dfrac{y}{x} = \dfrac{-2}{1} = -2.$

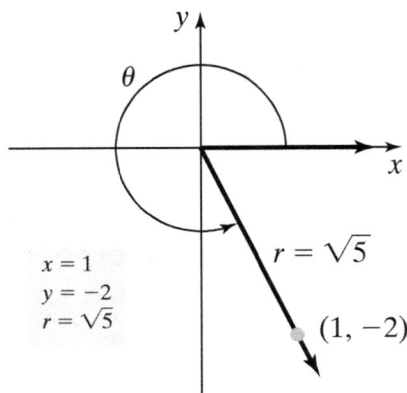

$x = 1$
$y = -2$
$r = \sqrt{5}$

$r = \sqrt{5}$

$(1, -2)$

Figure 51 Angle θ lies in Quadrant IV and passes through the point $(1, -2)$.

You Try It Work through this You Try It problem.

Work Exercises 31–40 in this textbook or in the MyMathLab Study Plan.

OBJECTIVE 5 DETERMINING REFERENCE ANGLES

Every angle θ (except those angles belonging to the quadrantal family) has a positive acute angle associated with it called the reference angle. The reference angle for θ is denoted as θ_R.

Definition   Reference Angle

The reference angle, θ_R, is the positive acute angle associated with a given angle θ. The reference angle is formed by the "nearest" x-axis and the terminal side of θ. Below are angles sketched in each of the four quadrants along with their corresponding reference angles.

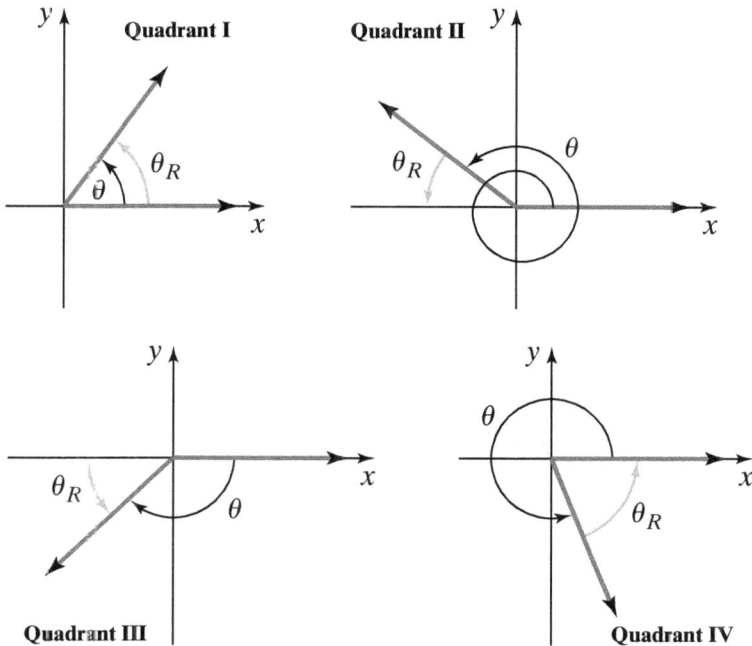

Quadrant I

y

θ_R

θ

x

Quadrant II

y

θ_R

θ

x

Quadrant III

y

θ_R

θ

x

Quadrant IV

y

θ

θ_R

x

You Try It

To find the measure of the reference angle, always start by drawing a picture. First, determine the angle of least nonnegative measure coterminal to θ, denoted as θ_C. You may want to review Section 1.1 to see how to find θ_C or try working through this exercise about determining θ_C. Note that if $0 < \theta < 2\pi$, then $\theta = \theta_C$. The measure of the reference angle θ_R depends on the quadrant in which the terminal side of θ_C lies. The four cases are outlined as follows.

Quadrant I

y

$\theta_C = \theta_R$

x

Quadrant II

y

$\theta_R = \pi - \theta_C$

θ_C

x

Case 1: If the terminal side of θ_C lies in Quadrant I, then $\theta_R = \theta_C$.

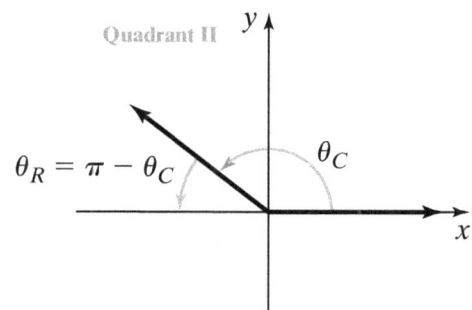

Case 2: If the terminal side of θ_C lies in Quadrant II, then $\theta_R = \pi - \theta_C$ (or $\theta_R = 180° - \theta_C$).

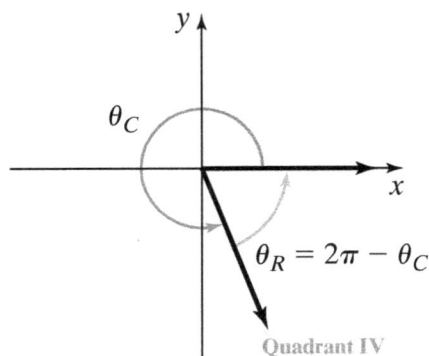

Case 3: If the terminal side of θ_C lies in Quadrant III, then $\theta_R = \theta_C - \pi$ (or $\theta_R = \theta_C - 180°$).

Case 4: If the terminal side of θ_C lies in Quadrant IV, then $\theta_R = 2\pi - \theta_C$ (or $\theta_R = 360° - \theta_C$).

My interactive video summary

⊙ **Example 5** Determining Reference Angles for Angles in Radians Belonging to the $\dfrac{\pi}{3}$, $\dfrac{\pi}{6}$, or $\dfrac{\pi}{4}$ Families

For each of the given angles, determine the reference angle.

a. $\theta = \dfrac{5\pi}{3}$ b. $\theta = \dfrac{11\pi}{4}$ c. $\theta = -\dfrac{25\pi}{6}$ d. $\theta = \dfrac{16\pi}{6}$

Solution

a. The angle $\theta = \dfrac{5\pi}{3}$ lies between 0 and 2π. Therefore, $\theta = \theta_C = \dfrac{5\pi}{3}$. The terminal

side of $\theta_C = \dfrac{5\pi}{3}$ lies in Quadrant IV. Therefore, $\theta_R = 2\pi - \theta_C$. Thus, the reference

angle is $\theta_R = 2\pi - \dfrac{5\pi}{3} = \dfrac{6\pi}{3} - \dfrac{5\pi}{3} = \dfrac{\pi}{3}$. See Figure 52.

b. The angle of least nonnegative measure coterminal with

$$\theta = \dfrac{11\pi}{4} \text{ is } \theta_C = \dfrac{11\pi}{4} - 2\pi = \dfrac{11\pi}{4} - \dfrac{8\pi}{4} = \dfrac{3\pi}{4}.$$

The terminal side of $\theta_C = \dfrac{3\pi}{4}$ lies in Quadrant II. Therefore, $\theta_R = \pi - \theta_C$. Thus,

$\theta_R = \pi - \dfrac{3\pi}{4} = \dfrac{4\pi}{4} - \dfrac{3\pi}{4} = \dfrac{\pi}{4}$. See Figure 53.

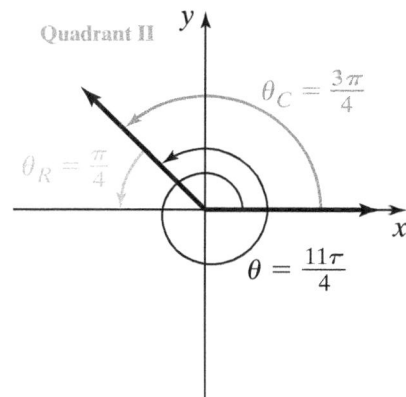

Figure 52

Figure 53

c. The angle of least nonnegative measure coterminal with $\theta = -\dfrac{25\pi}{6}$ is

$\theta_C = \dfrac{11\pi}{6}$. To see how to find θ_C, read this explanation. The terminal side of

$\theta_C = \dfrac{11\pi}{6}$ lies in Quadrant IV.

Therefore, $\theta_R = 2\pi - \theta_C$. Thus, $\theta_R = 2\pi - \theta_C = 2\pi - \dfrac{11\pi}{6} = \dfrac{12\pi}{6} - \dfrac{11\pi}{6} = \dfrac{\pi}{6}$.

See Figure 54.

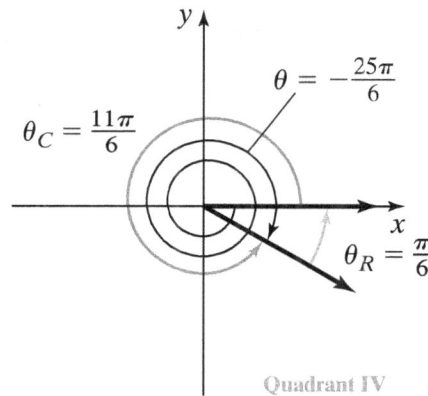

Figure 54

At this point, we can observe from Example 5a, 5b, and 5c that angles belonging

to the $\dfrac{\pi}{6}, \dfrac{\pi}{3}$, or $\dfrac{\pi}{4}$ families will always have a reference angle of $\dfrac{\pi}{6}, \dfrac{\pi}{3}$, or $\dfrac{\pi}{4}$,

respectively. However, when determining reference angles, make sure that the angle is written in lowest terms.

d. For the angle $\theta = \dfrac{16\pi}{6}$, do not assume that the reference angle is $\dfrac{\pi}{6}$. Note that

$\theta = \dfrac{16\pi}{6}$ can be reduced to $\theta = \dfrac{16\pi}{6} = \dfrac{2 \cdot 8\pi}{2 \cdot 3} = \dfrac{8\pi}{3}$. This is a special angle be-

longing to the $\dfrac{\pi}{3}$ family. Therefore, the reference angle is $\theta_R = \dfrac{\pi}{3}$.

My interactive video summary

⊙ You may want to work through this interactive video to see the solution to each part of Example 5.

We now know that if we are given an angle belonging to the $\frac{\pi}{6}, \frac{\pi}{3}$, or $\frac{\pi}{4}$ families, then the reference angle will be $\frac{\pi}{6}, \frac{\pi}{3}$, or $\frac{\pi}{4}$, respectively. But how do we find the reference angle of a given angle that does not belong to one of these families? In this situation, we must use the four cases outlined earlier to determine the reference angle.

My interactive video summary

⊙ **Example 6** Determining Reference Angles for Angles in Radians Not Belonging to the $\frac{\pi}{3}, \frac{\pi}{6}$, or $\frac{\pi}{4}$ Families

For each of the given angles, determine the reference angle.

a. $\theta = \dfrac{5\pi}{8}$ 　　　b. $\theta = \dfrac{22\pi}{9}$ 　　　c. $\theta = -\dfrac{5\pi}{7}$

Solution

a. The angle $\theta = \dfrac{5\pi}{8}$ lies between 0 and 2π. Therefore, $\theta = \theta_C = \dfrac{5\pi}{8}$. The terminal side of $\theta_C = \dfrac{5\pi}{8}$ lies in Quadrant II. Therefore, $\theta_R = \pi - \theta_C$. Thus, the reference angle is $\theta_R = \pi - \dfrac{5\pi}{8} = \dfrac{8\pi}{8} - \dfrac{5\pi}{8} = \dfrac{3\pi}{8}$. See Figure 55.

b. The angle of least nonnegative measure coterminal with $\theta = \dfrac{22\pi}{9}$ is

$$\theta_C = \dfrac{22\pi}{9} - 2\pi = \dfrac{22\pi}{9} - \dfrac{18\pi}{9} = \dfrac{4\pi}{9}.$$

Because $\dfrac{4\pi}{9} < \dfrac{\pi}{2}$, the terminal side of $\theta_C = \dfrac{4\pi}{9}$ lies in Quadrant I. Therefore, $\theta_R = \theta_C$. Thus, $\theta_R = \theta_C = \dfrac{4\pi}{9}$. See Figure 56.

Figure 55

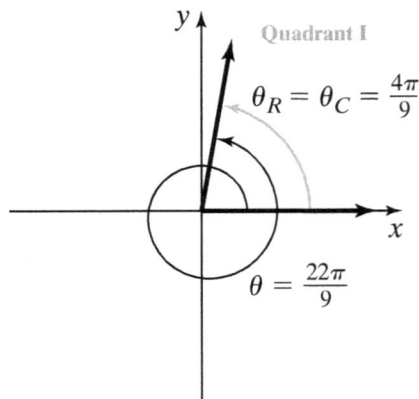

Figure 56

c. The angle of least nonnegative measure coterminal with

$$\theta = -\frac{5\pi}{7} \text{ is } \theta_C = -\frac{5\pi}{7} + 2\pi = -\frac{5\pi}{7} + \frac{14\pi}{7} = \frac{9\pi}{7}.$$

The terminal side of $\theta_C = \frac{9\pi}{7}$ lies in Quadrant III. Therefore, $\theta_R = \theta_C - \pi$.

Thus, $\theta_R = \theta_C - \pi = \frac{9\pi}{7} - \pi = \frac{9\pi}{7} - \frac{7\pi}{7} = \frac{2\pi}{7}$. See Figure 57.

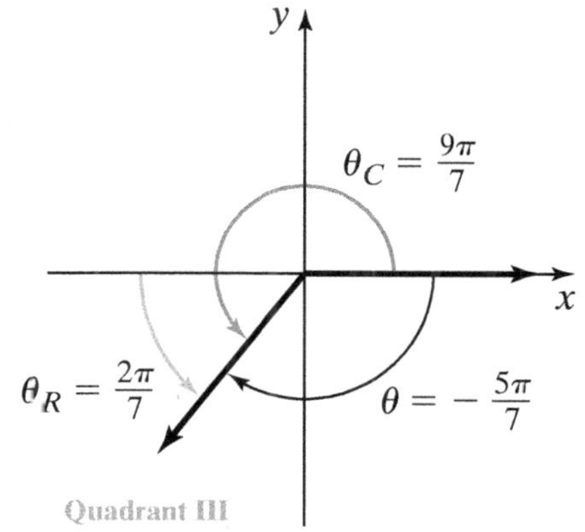

Figure 57

You may want to work through this interactive video to see the solution to each part of Example 6.

Let's now look at finding reference angles for angles given in degree measure.

My interactive video summary

⊘ **Example 7** Determining Reference Angles for Angles Given in Degrees

For each of the given angles, determine the reference angle.

a. $\theta = 225°$ b. $\theta = -233°$ c. $\theta = 510°$

Solution

a. The angle $\theta = 225°$ belongs to the $\frac{\pi}{4}$ (or 45°) family of angles because

$\theta = 225° = 5(45°)$.

Therefore, the reference angle is $\theta_R = 45°$. See Figure 58.

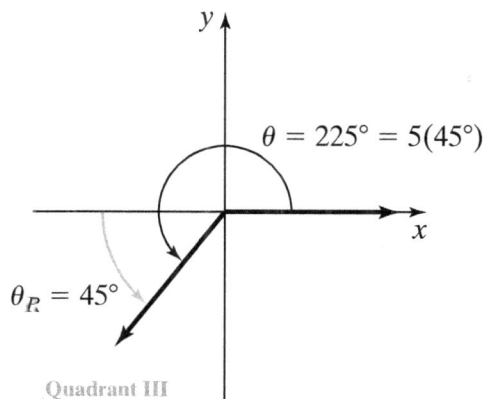

Figure 58

b. The angle $\theta = -233°$ does not belong to one of the special families of angles because $-233°$ is not an integer multiple of 30°, 45°, or 60°. The angle of least nonnegative measure coterminal with $\theta = -233°$ is $\theta_C = -233° + 360° = 127°$. The terminal side of angle $\theta_C = 127°$ lies in Quadrant II. Therefore, $\theta_R = 180° - \theta_C$. Thus, $\theta_R = 180° - \theta_C = 180° - 127° = 53°$. See Figure 59.

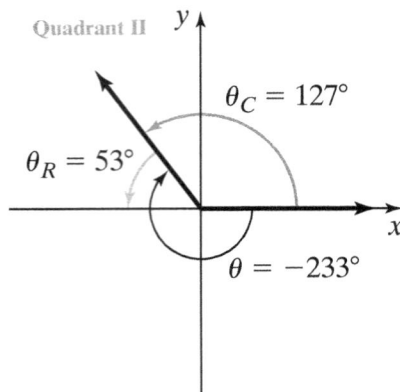

Figure 59

Try working part (c) on your own. Then work through this interactive video to see all solutions to Example 7.

You Try It Work through this You Try It problem.

Work Exercises 41–50 in this textbook or in the MyMathLab Study Plan.

OBJECTIVE 6 EVALUATING TRIGONOMETRIC FUNCTIONS OF ANGLES BELONGING TO THE $\dfrac{\pi}{3}, \dfrac{\pi}{6},$ OR $\dfrac{\pi}{4}$ FAMILIES

We are now ready to evaluate the trigonometric functions for any angle belonging to the $\dfrac{\pi}{3}, \dfrac{\pi}{6},$ or $\dfrac{\pi}{4}$ families. Before we continue, make sure that you have a good understanding of the two special right triangles discussed in Section 1.4. It is essential that you can use these special right triangles to evaluate all trigonometric functions for the acute angles of $\dfrac{\pi}{3}, \dfrac{\pi}{6},$ or $\dfrac{\pi}{4}$. At this point, you should be able to easily create Table 1 from Section 1.4. Click on one of the following three You Try It buttons to quickly test yourself before reading on.

You Try It Evaluate a trigonometric function when $\theta = \dfrac{\pi}{6}$ (or 30°).

You Try It Evaluate a trigonometric function when $\theta = \dfrac{\pi}{3}$ (or 60°).

You Try It Evaluate a trigonometric function when $\theta = \dfrac{\pi}{4}$ (or 45°).

Once you are comfortable evaluating trigonometric functions at angles of $\frac{\pi}{3}, \frac{\pi}{6}$, or $\frac{\pi}{4}$ using right triangles, you are ready to evaluate trigonometric functions for any angle belonging to the $\frac{\pi}{3}, \frac{\pi}{6}$, or $\frac{\pi}{4}$ families.

My video summary ⊚ **Example 8** Finding Trigonometric Function Values for an Angle Belonging to the $\frac{\pi}{4}$ Family

Find the values of the six trigonometric functions for $\theta = \frac{7\pi}{4}$.

Solution The angle $\theta = \frac{7\pi}{4}$ is located in Quadrant IV. The angle belongs to the $\frac{\pi}{4}$ family of angles. Thus, the reference angle is $\theta_R = \frac{\pi}{4}$. See Figure 60.

To find the values of the six trigonometric functions, we choose any point P on the terminal side of angle θ. The simplest point to choose is $P(1, -1)$, where $r = \sqrt{2}$. See Figure 61. To see why this point lies on the terminal side of angle theta, read this explanation.

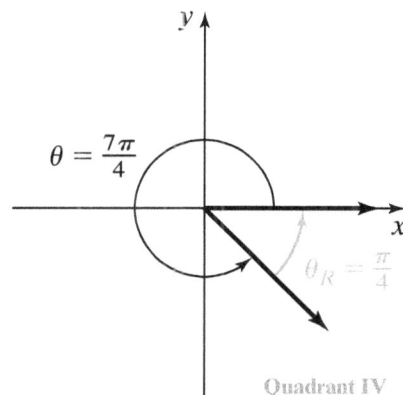

Figure 60 This illustrates the angle $\theta = \frac{7\pi}{4}$ and its reference angle $\theta_R = \frac{\pi}{4}$.

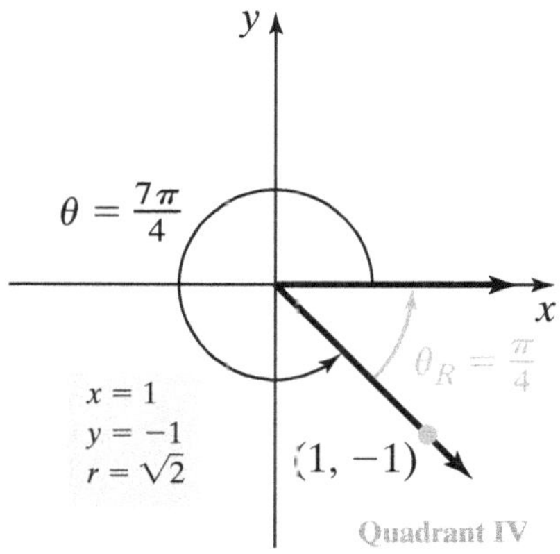

Figure 61

We can now use the general angle definitions to evaluate each trigonometric function.

$$\sin\left(\frac{7\pi}{4}\right) = \frac{y}{r} = \frac{-1}{\sqrt{2}} = -\frac{1}{\sqrt{2}} \qquad \csc\left(\frac{7\pi}{4}\right) = \frac{r}{y} = \frac{\sqrt{2}}{-1} = -\sqrt{2}$$

$$\cos\left(\frac{7\pi}{4}\right) = \frac{x}{r} = \frac{1}{\sqrt{2}} \qquad \sec\left(\frac{7\pi}{4}\right) = \frac{r}{x} = \frac{\sqrt{2}}{1} = \sqrt{2}$$

$$\tan\left(\frac{7\pi}{4}\right) = \frac{y}{x} = \frac{-1}{1} = -1 \qquad \cot\left(\frac{7\pi}{4}\right) = \frac{x}{y} = \frac{1}{-1} = -1$$

In Example 8, note that the absolute values of the trigonometric functions of $\theta = \dfrac{7\pi}{4}$ are the same as the absolute values of the trigonometric functions of the reference angle, $\theta_R = \dfrac{\pi}{4}$. In other words, the trigonometric functions of $\theta = \dfrac{7\pi}{4}$ are exactly the same as the trigonometric functions of $\theta_R = \dfrac{\pi}{4}$, except that they may have opposite signs.

$$\sin\left(\frac{7\pi}{4}\right) = -\frac{1}{\sqrt{2}} \xleftrightarrow{\text{opposite signs}} \frac{1}{\sqrt{2}} = \sin\left(\frac{\pi}{4}\right)$$

$$\cos\left(\frac{7\pi}{4}\right) = \frac{1}{\sqrt{2}} \xleftrightarrow{\text{both are positive}} \frac{1}{\sqrt{2}} = \cos\left(\frac{\pi}{4}\right)$$

$$\tan\left(\frac{7\pi}{4}\right) = -1 \xleftrightarrow{\text{opposite signs}} 1 = \tan\left(\frac{\pi}{4}\right)$$

$$\csc\left(\frac{7\pi}{4}\right) = -\sqrt{2} \xleftrightarrow{\text{opposite signs}} \sqrt{2} = \csc\left(\frac{\pi}{4}\right)$$

$$\sec\left(\frac{7\pi}{4}\right) = \sqrt{2} \xleftrightarrow{\text{both are positive}} \sqrt{2} = \sec\left(\frac{\pi}{4}\right)$$

$$\cot\left(\frac{7\pi}{4}\right) = -1 \xleftrightarrow{\text{opposite signs}} 1 = \cot\left(\frac{\pi}{4}\right)$$

Notice that only the cosine and secant functions are positive for both $\theta = \dfrac{7\pi}{4}$ and $\theta_R = \dfrac{\pi}{4}$. All other trigonometric functions of $\theta = \dfrac{7\pi}{4}$ are negative. We could have predicted this because we know by the acronym ASTC that only the **c**osine function and its reciprocal, the secant function, are positive for an angle whose terminal side lies in Quadrant IV. This example suggests a shortcut for evaluating the trigonometric functions for any angle belonging to the $\dfrac{\pi}{3}, \dfrac{\pi}{6},$ or $\dfrac{\pi}{4}$ families.

To evaluate a trigonometric function belonging to one of these families, we need only know the quadrant in which the terminal side of the angle lies and the reference angle. Thus, we can then use one of the special right triangles to evaluate the function. We now summarize this process.

Steps for Evaluating Trigonometric Functions of Angles Belonging to the $\dfrac{\pi}{3}, \dfrac{\pi}{6},$ or $\dfrac{\pi}{4}$ Families

Step 1. Draw the angle and determine the quadrant in which the terminal side of the angle lies.

Step 2. Determine if the sign of the function is positive or negative in that quadrant.

Step 3. Determine if the reference angle, θ_R, is $\dfrac{\pi}{3}, \dfrac{\pi}{6},$ or $\dfrac{\pi}{4}$.

Step 4. Use the appropriate special right triangle or your knowledge of Table 1 to determine the value of the trigonometric function.

---

*My interactive
video summary*

⊚ **Example 9** Evaluating Trigonometric Functions of Angles Belonging to the $\dfrac{\pi}{3}$, $\dfrac{\pi}{6}$, or $\dfrac{\pi}{4}$ Families

Find the exact value of each trigonometric expression without using a calculator.

a. $\sin\left(\dfrac{7\pi}{6}\right)$ **b.** $\cot\left(-\dfrac{22\pi}{3}\right)$ **c.** $\tan\left(\dfrac{11\pi}{4}\right)$

d. $\cos\left(\dfrac{11\pi}{3}\right)$ **e.** $\sec\left(\dfrac{5\pi}{6}\right)$ **f.** $\csc\left(-\dfrac{7\pi}{6}\right)$

Solution

a. Step 1. The angle $\theta = \dfrac{7\pi}{6}$ is sketched below with the terminal side of the angle located in Quadrant III.

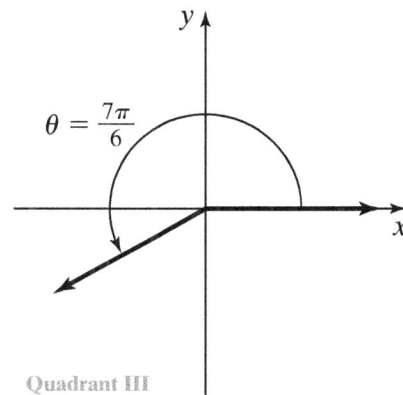

Quadrant III

Step 2. Because the terminal side of the angle lies in Quadrant III, we know that this quadrant corresponds with "T" from the acronym ASTC, meaning that *only* the tangent and its reciprocal, cotangent, are positive for angles lying in this quadrant. Thus, the value of $\sin\left(\dfrac{7\pi}{6}\right)$ must be negative.

Step 3. The angle belongs to the $\dfrac{\pi}{6}$ family so the reference angle is $\theta_R = \dfrac{\pi}{6}$.

Quadrant III

Step 4. We can use the **special $\frac{\pi}{6}, \frac{\pi}{3}, \frac{\pi}{2}$** right triangle or **Table 1** to get

$$\sin\left(\frac{\pi}{6}\right) = \frac{1}{2}.$$

Therefore, $\sin\left(\dfrac{7\pi}{6}\right) = \underbrace{\boxed{-}}_{\substack{\text{The sine}\\ \text{function is}\\ \text{negative in}\\ \text{Quadrant III.}}} \underbrace{\boxed{\tfrac{1}{2}}}_{\sin\left(\frac{\pi}{6}\right)=\frac{1}{2}} = -\frac{1}{2}.$

b. To evaluate $\cot\left(-\dfrac{22\pi}{3}\right)$, we follow the four-step process:

Step 1. The angle $\theta = -\dfrac{22\pi}{3}$ is sketched below with the terminal side of the angle located in Quadrant II.

$$\theta = -\frac{22\pi}{3}$$

Step 2. Because the terminal side of the angle lies in Quadrant II, we know that this quadrant corresponds with "S" from the acronym ASTC, meaning that *only* the sine and its reciprocal, cosecant, are positive for angles lying in this quadrant. Thus, the value of $\cot\left(-\dfrac{22\pi}{3}\right)$ must be negative.

Step 3. The angle $\theta = -\dfrac{22\pi}{3}$ belongs to the $\dfrac{\pi}{3}$ family so the reference angle is

$$\theta_R = \frac{\pi}{3}.$$

$$\theta_R = \frac{\pi}{3}$$

Step 4. We can use the **special** $\dfrac{\pi}{6}, \dfrac{\pi}{3}, \dfrac{\pi}{2}$ right triangle or **Table 1** to get

$$\cot\left(\dfrac{\pi}{3}\right) = \dfrac{1}{\sqrt{3}}.$$

Therefore, $\cot\left(-\dfrac{22\pi}{3}\right) = \underbrace{\boxed{-}}_{\substack{\text{The cotangent}\\\text{function is}\\\text{negative in}\\\text{Quadrant II.}}} \underbrace{\boxed{\dfrac{1}{\sqrt{3}}}}_{\cot\left(\frac{\pi}{3}\right)} = -\dfrac{1}{\sqrt{3}}.$

Watch this interactive video to see all solutions to Example 9.

You Try It Work through this You Try It problem.

Work Exercises 51–66 in this textbook or in the MyMathLab Study Plan.

1.5 Exercises

Skill Check Exercises

SCE-1. In what quadrant does the point $(12, -10)$ lie?

SCE-2. On which axis does the point $(-4, 0)$ lie?

1. List all angles written in degrees in the interval $[0°, 360°)$ that are members of the quadrantal family. List answers in order from smallest degree measure to largest degree measure.

2. List all angles written in radians in the interval $[0, 2\pi)$ that are members of the $\dfrac{\pi}{6}$ family. List answers in order from smallest radian measure to largest radian measure.

In Exercises 3–12, each angle belongs to one of the four families of special angles. Determine the family of angles for which it belongs, sketch the angle, and then determine the angle of least nonnegative measure, θ_C, coterminal with the given angle.

3. $\theta = \dfrac{5\pi}{2}$ 4. $\theta = 480°$ 5. $\theta = \dfrac{8\pi}{3}$ 6. $\theta = \dfrac{14\pi}{4}$ 7. $\theta = -315°$

8. $\theta = 6\pi$ 9. $\theta = -\dfrac{17\pi}{6}$ 10. $\theta = -420°$ 11. $\theta = -\dfrac{21\pi}{4}$ 12. $\theta = -\dfrac{22\pi}{6}$

In Exercises 13–18, a point lying on the terminal side of an angle θ is given. Find the exact value of the six trigonometric functions of θ.

13. $(4, 3)$ 14. $(-5, 12)$ 15. $(-2, -5)$

16. $(2, -8)$ 17. $\left(-\dfrac{1}{2}, \dfrac{\sqrt{3}}{2}\right)$ 18. $\left(-\dfrac{1}{\sqrt{2}}, -\dfrac{1}{\sqrt{2}}\right)$

In Exercises 19–30, without using a calculator, determine the value of each trigonometric function or state that the value is undefined.

SbS **19.** $\sin(8\pi)$ SbS **20.** $\cos(-270°)$ SbS **21.** $\tan\left(\dfrac{7\pi}{2}\right)$ SbS **22.** $\csc(-540°)$

SbS **23.** $\sec(-5\pi)$ SbS **24.** $\cot(180°)$ SbS **25.** $\sin(450°)$ SbS **26.** $\cos\left(-\dfrac{3\pi}{2}\right)$

SbS **27.** $\tan(-90°)$ SbS **28.** $\csc(9\pi)$ SbS **29.** $\sec(360°)$ SbS **30.** $\cot\left(\dfrac{13\pi}{2}\right)$

31. List the quadrants where the sine function has positive values.

32. List the quadrants where the secant function has positive values.

33. List the quadrants where the tangent function has negative values.

34. List the quadrants where the cosecant function has negative values.

35. If $\sin\theta < 0$ and $\cos\theta > 0$, determine the quadrant in which the terminal side of angle θ lies.

36. If $\tan\theta > 0$ and $\csc\theta > 0$, determine the quadrant in which the terminal side of angle θ lies.

37. If $\cos\theta > 0$ and $\tan\theta > 0$, determine the quadrant in which the terminal side of angle θ lies.

38. Find the value of $\csc\theta$ if $\cos\theta = -\dfrac{7}{25}$ and given that the terminal side of θ lies in Quadrant II.

39. Find the value of $\sin\theta$ if $\tan\theta = \dfrac{1}{4}$ and $\sec\theta < 0$.

40. Find the value of $\csc\theta$ if $\sec\theta = -\dfrac{5}{2}$ and $\tan\theta > 0$.

In Exercises 41–50, determine the reference angle for each of the given angles.

SbS **41.** $\theta = \dfrac{7\pi}{3}$ SbS **42.** $\theta = \dfrac{27\pi}{4}$ SbS **43.** $\theta = -\dfrac{37\pi}{3}$ SbS **44.** $\theta = -\dfrac{19\pi}{4}$ SbS **45.** $\theta = \dfrac{20\pi}{6}$

SbS **46.** $\theta = \dfrac{16\pi}{9}$ SbS **47.** $\theta = -\dfrac{32\pi}{7}$

48. $\theta = 570°$ SbS **49.** $\theta = -675°$ SbS **50.** $\theta = -341°$

SbS

In Exercises 51–66, find the exact value of each trigonometric expression without using a calculator.

SbS **51.** $\sin\left(\dfrac{5\pi}{6}\right)$ SbS **52.** $\cos\left(\dfrac{3\pi}{4}\right)$ SbS **53.** $\cot\left(\dfrac{4\pi}{3}\right)$ SbS **54.** $\csc\left(\dfrac{7\pi}{6}\right)$

SbS **55.** $\tan\left(-\dfrac{2\pi}{3}\right)$ SbS **56.** $\cos\left(-\dfrac{7\pi}{6}\right)$ SbS **57.** $\sec\left(-\dfrac{\pi}{4}\right)$ SbS **58.** $\cot\left(-\dfrac{5\pi}{4}\right)$

SbS **59.** $\sin\left(\dfrac{17\pi}{6}\right)$ SbS **60.** $\tan\left(\dfrac{11\pi}{3}\right)$ SbS **61.** $\sec\left(\dfrac{23\pi}{6}\right)$ SbS **62.** $\csc\left(\dfrac{10\pi}{3}\right)$

SbS **63.** $\cos\left(-\dfrac{15\pi}{4}\right)$ SbS **64.** $\sin\left(-\dfrac{17\pi}{6}\right)$ SbS **65.** $\cot\left(-\dfrac{11\pi}{3}\right)$ SbS **66.** $\sec\left(-\dfrac{10\pi}{3}\right)$

Brief Exercises

In Exercises 67–94, without using a calculator, determine the value of each trigonometric function or state that the value is undefined.

67. $\sin(8\pi)$ **68.** $\cos(-270°)$ **69.** $\tan\left(\dfrac{7\pi}{2}\right)$ **70.** $\csc(-540°)$ **71.** $\sec(-5\pi)$

72. $\cot(180°)$ **73.** $\sin(450°)$ **74.** $\cos\left(-\dfrac{3\pi}{2}\right)$ **75.** $\tan(-90°)$ **76.** $\csc(9\pi)$

77. $\sec(360°)$ **78.** $\cot\left(\dfrac{13\pi}{2}\right)$ **79.** $\sin\left(\dfrac{5\pi}{6}\right)$ **80.** $\cos\left(\dfrac{3\pi}{4}\right)$ **81.** $\cot\left(\dfrac{4\pi}{3}\right)$

82. $\csc\left(\dfrac{7\pi}{6}\right)$ **83.** $\tan\left(-\dfrac{2\pi}{3}\right)$ **84.** $\cos\left(-\dfrac{7\pi}{6}\right)$ **85.** $\sec\left(-\dfrac{\pi}{4}\right)$ **86.** $\cot\left(-\dfrac{5\pi}{4}\right)$

87. $\sin\left(\dfrac{17\pi}{6}\right)$ **88.** $\tan\left(\dfrac{11\pi}{3}\right)$ **89.** $\sec\left(\dfrac{23\pi}{6}\right)$ **90.** $\csc\left(\dfrac{10\pi}{3}\right)$ **91.** $\cos\left(-\dfrac{15\pi}{4}\right)$

92. $\sin\left(-\dfrac{17\pi}{6}\right)$ **93.** $\cot\left(-\dfrac{11\pi}{3}\right)$ **94.** $\sec\left(-\dfrac{10\pi}{3}\right)$

1.6 The Unit Circle

THINGS TO KNOW

Before working through this section, be sure that you are familiar with the following concepts:

VIDEO   ANIMATION   INTERACTIVE

You Try It   1. Converting between Degree Measure and Radian Measure (Section 1.1)

You Try It   2. Understanding the Special Right Triangles (Section 1.3)

You Try It   3. Understanding the Right Triangle Definitions of the Trigonometric Functions (Section 1.4)

You Try It   4. Understanding the Four Families of Special Angles (Section 1.5)

You Try It

5. Understanding the Definitions of the Trigonometric Functions of General Angles (Section 1.5)

You Try It

6. Evaluating Trigonometric Functions of Angles Belonging to the $\frac{\pi}{3}, \frac{\pi}{6}$, or $\frac{\pi}{4}$ Families (Section 1.5)

OBJECTIVES

1 Understanding the Definition of the Unit Circle

2 Using Symmetry to Determine Points on the Unit Circle

3 Understanding the Unit Circle Definitions of the Trigonometric Functions

4 Using the Unit Circle to Evaluate Trigonometric Functions at Increments of $\frac{\pi}{2}$

5 Using the Unit Circle to Evaluate Trigonometric Functions for Increments of $\frac{\pi}{6}, \frac{\pi}{4}$, and $\frac{\pi}{3}$

OBJECTIVE 1   UNDERSTANDING THE DEFINITION OF THE UNIT CIRCLE

My animation summary

Recall that the standard form of the equation of a circle with center (h, k) and radius r is given by $(x - h)^2 + (y - k)^2 = r^2$. A circle centered at the origin with a radius length of 1 unit is called the **unit circle**. The standard form of the equation of the unit circle is $x^2 + y^2 = 1$. Watch this animation for a further explanation of the unit circle.

The Unit Circle

A circle centered at the origin with a radius of 1 unit is called the **unit circle** whose equation is given by $x^2 + y^2 = 1$.

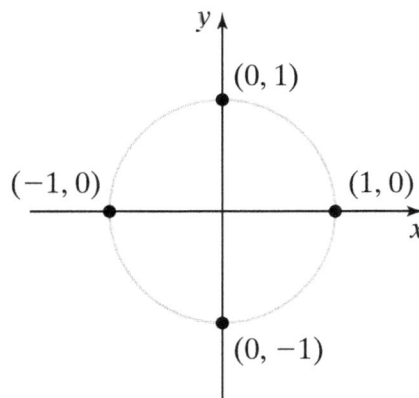

Note that the points $(1, 0)$, $(0, 1)$, $(-1, 0)$, and $(0, -1)$ all lie on the graph of the unit circle. These are the only four points that lie on the unit circle whose coordinates are integers. From the graph, you can see that the x- and y-coordinates of all other points that lie on the unit circle must be between -1 and 1 and cannot equal 0. A point (a, b) lies on the graph of the unit circle if and only if $a^2 + b^2 = 1$. In Example 1, one of the coordinates of a point that lies on the graph of the unit circle is given. Work through Example 1 to see if you can determine the value of the missing coordinate.

My interactive video summary

$\circledcirc$ **Example 1** Determine the Missing Coordinate of a Point That Lies on the Unit Circle

Determine the missing coordinate of a point that lies on the graph of the unit circle given the quadrant in which the point is located.

a. $\left(-\dfrac{1}{8}, y\right)$; Quadrant III   b. $\left(x, -\dfrac{\sqrt{3}}{2}\right)$; Quadrant IV   c. $\left(-\dfrac{1}{\sqrt{2}}, y\right)$; Quadrant II

Solution

a. We are given that $x = -\dfrac{1}{8}$. Substitute $x = -\dfrac{1}{8}$ into the equation of the unit circle and solve for y.

$$x^2 + y^2 = 1 \qquad \text{Write the equation of the unit circle.}$$

$$\left(-\dfrac{1}{8}\right)^2 + y^2 = 1 \qquad \text{Substitute } x = -\dfrac{1}{8}.$$

$$\dfrac{1}{64} + y^2 = 1 \qquad \text{Simplify.}$$

$$y^2 = 1 - \dfrac{1}{64} \qquad \text{Subtract } \dfrac{1}{64} \text{ from both sides.}$$

$$y^2 = \dfrac{63}{64} \qquad \text{Simplify.} \left(1 - \dfrac{1}{64} = \dfrac{64}{64} - \dfrac{1}{64} = \dfrac{63}{64}\right)$$

$$y = \pm\sqrt{\dfrac{63}{64}} = \pm\dfrac{3\sqrt{7}}{8} \qquad \text{Use the square root property and simplify.}$$

Because the y-coordinates of all points that lie in Quadrant III are *negative,* use $y = -\dfrac{3\sqrt{7}}{8}$.

Try finding the missing coordinates for parts (b) and (c) on your own. Then watch this interactive video to see if you are correct. $\bullet$

You Try It   Work through this You Try It problem.

Work Exercises 1–5 in this textbook or in the MyMathLab Study Plan.

OBJECTIVE 2   USING SYMMETRY TO DETERMINE POINTS THAT LIE ON THE UNIT CIRCLE

It is important to point out that the graph of the unit circle is symmetric about both axes and the origin. In Example 1b, we found that the point $\left(\dfrac{1}{2}, -\dfrac{\sqrt{3}}{2}\right)$ lies on the graph of the unit circle. Because the graph of the unit circle is symmetric about the *y*-axis, it follows that the point $\left(-\dfrac{1}{2}, -\dfrac{\sqrt{3}}{2}\right)$ also lies on the graph of the

unit circle. Now that we know two points lying on the unit circle located *below* the *x*-axis, we can use the fact that the unit circle is symmetric about the *x*-axis to find two more points lying on the unit circle *above* the *x*-axis. These two points are $\left(\dfrac{1}{2}, \dfrac{\sqrt{3}}{2}\right)$ and $\left(-\dfrac{1}{2}, \dfrac{\sqrt{3}}{2}\right)$. Note that we also could have used origin symmetry to locate these two points. In general, for any point (a, b) lying on the unit circle (not on an axis), the points $(-a, b), (a, -b)$, and $(-a, -b)$ must also lie on the unit circle. See Figure 62.

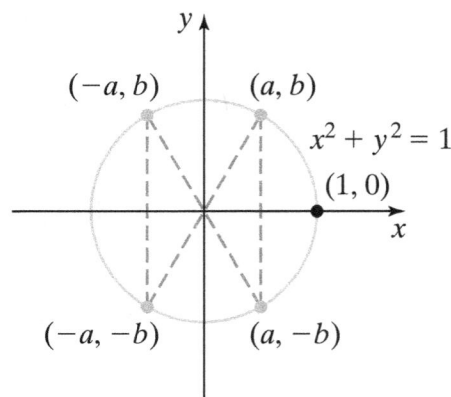

Figure 62 For any point (a, b) on the unit circle not lying on an axis, symmetry can be used to locate three more points.

My video summary ⊙ **Example 2** Finding Points Lying on the Graph of the Unit Circle

Verify that the point $\left(-\dfrac{1}{8}, -\dfrac{3\sqrt{7}}{8}\right)$ lies on the graph of the unit circle. Then use symmetry to find three other points that also lie on the graph of the unit circle.

Solution We can verify that the point $\left(-\dfrac{1}{8}, -\dfrac{3\sqrt{7}}{8}\right)$ lies on the graph of the unit circle by substituting $x = -\dfrac{1}{8}$ and $y = -\dfrac{3\sqrt{7}}{8}$ into the equation $x^2 + y^2 = 1$ to see if a true statement is produced. To see this verification, read these steps.

Because the unit circle has *y*-axis symmetry, we are guaranteed that the point $\left(\dfrac{1}{8}, -\dfrac{3\sqrt{7}}{8}\right)$ also lies on the graph of the unit circle. We now know two points on the graph of the unit circle that lie *below* the *x*-axis. We can use *x*-axis symmetry to find the corresponding points on the graph of the unit circle that lie *above* the *x*-axis. These two points are $\left(\dfrac{1}{8}, \dfrac{3\sqrt{7}}{8}\right)$ and $\left(-\dfrac{1}{8}, \dfrac{3\sqrt{7}}{8}\right)$. Watch this video to see this solution worked out in detail.

You Try It   Work through this You Try It problem.

Work Exercises 6–8 in this textbook or in the MyMathLab Study Plan.

OBJECTIVE 3   UNDERSTANDING THE UNIT CIRCLE DEFINITIONS OF THE TRIGONOMETRIC FUNCTIONS

My interactive video summary

So far in this text, we have seen two groups of definitions for the trigonometric functions. In Section 1.4, we saw the right triangle definitions of the trigonometric functions of acute angles. Then, in Section 1.5 we saw the definitions of the trigonometric functions of general angles.

We now turn our attention to the third set of definitions of the trigonometric functions that involve the unit circle. Suppose that t is the measure (in radians) of a central angle of a unit circle with a corresponding arc length, s. See Figure 63.

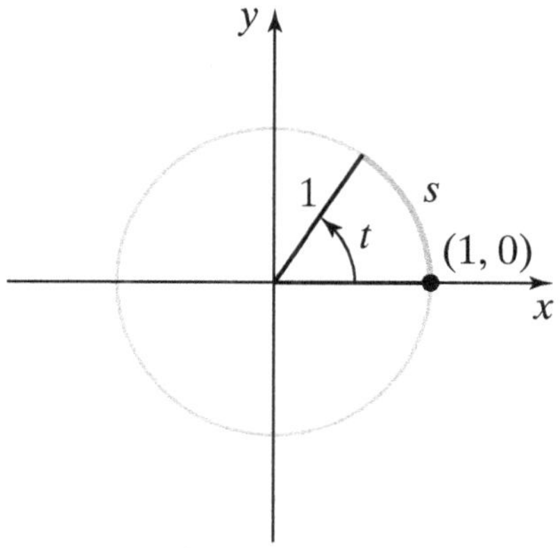

Figure 63 This illustrates the unit circle with a central angle of t radians and a corresponding arc of length s.

We can use the formula for the arc length of a sector of a circle, which we derived in Section 1.2, to find the arc length, s, of a unit circle that corresponds to a central angle of t radians.

$$s = r\theta \qquad \text{Write the formula for the arc length of a sector of a circle.}$$

$$s = 1 \cdot t \qquad \text{Substitute } r = 1 \text{ and } \theta = t.$$

$$s = t \qquad \text{Simplify.}$$

We see that the length of the intercepted arc seen in Figure 63 is $s = t$. **This means that the arc length of a sector of the *unit circle* is exactly equal to the measure of the central angle!** The arc length and the angle are represented by the same real number t.

We can now create the unit circle definitions of the trigonometric functions. To do this, let t be any real number and let $P(x, y)$ be the point on the unit circle that has an arc length of t units from the point $(1, 0)$. The measure of the central angle (in radians) is exactly the same as the arc length, t. If $t > 0$, then point P is obtained by rotating in a *counterclockwise* direction. See Figure 64a. If $t < 0$, then point P is obtained by rotating in a *clockwise* direction. See Figure 64b.

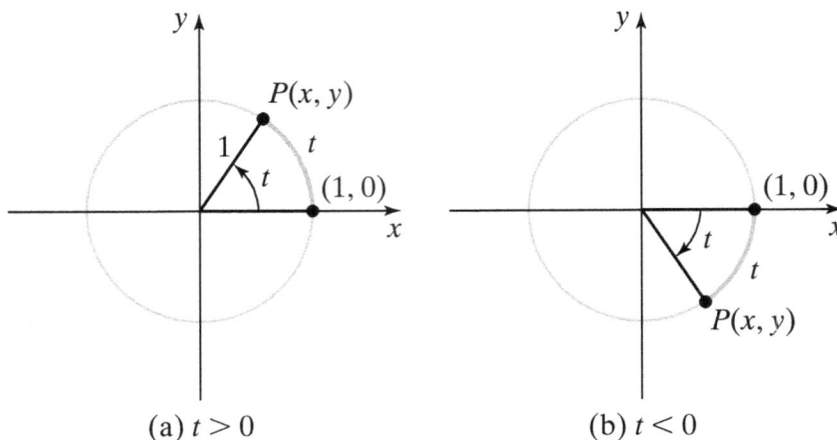

(a) $t > 0$ (b) $t < 0$ **Figure 64**

For the real number t and the corresponding point $P(x, y)$ lying on the graph of the unit circle, we define the cosine of t as the x-coordinate of P and the sine of t as the y-coordinate of P. Therefore, $\cos t = x$ and $\sin t = y$. This choice of x for the cosine of t and y for the sine of t is not made arbitrarily. If you remember the definitions of the trigonometric functions of general angles, the sine of an angle theta was defined as $\dfrac{y}{r}$ and the cosine of theta was defined as $\dfrac{x}{r}$. Therefore, it seems logical and consistent that we choose $\sin t$ to be y (which is $\dfrac{y}{r}$ when $r = 1$) and $\cos t$ to be x (which is $\dfrac{x}{r}$ when $r = 1$).

We now define all six of the trigonometric functions using the unit circle.

The Unit Circle Definitions of the Trigonometric Functions

For any real number t, if $P(x, y)$ is a point on the unit circle corresponding to t, then

$$\sin t = y \qquad\qquad \csc t = \frac{1}{y}, y \neq 0$$

$$\cos t = x \qquad\qquad \sec t = \frac{1}{x}, x \neq 0$$

$$\tan t = \frac{y}{x}, x \neq 0 \qquad \cot t = \frac{x}{y}, y \neq 0$$

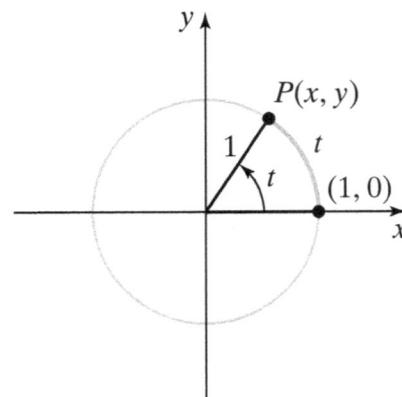

My video summary

⊘ Example 3 Evaluating Trigonometric Functions Using the Unit Circle

If $\left(-\frac{1}{4}, \frac{\sqrt{15}}{4}\right)$ is a point on the unit circle corresponding to a real number t, find the values of the six trigonometric functions of t.

Solution Draw the unit circle and plot the point $\left(-\frac{1}{4}, \frac{\sqrt{15}}{4}\right)$.

Using the unit circle definitions of *sine, cosine,* and *tangent,* we get

$$\sin t = y = \frac{\sqrt{15}}{4}, \quad \cos t = x = -\frac{1}{4}, \quad \text{and} \quad \tan t = \frac{y}{x} = \frac{\frac{\sqrt{15}}{4}}{-\frac{1}{4}} = \frac{\sqrt{15}}{4} \cdot \left(-\frac{4}{1}\right) = -\sqrt{15}$$

We can find the cosecant, secant, and cotangent of t using the reciprocals of the sine, cosine, and tangent functions. Thus,

$$\csc t = \frac{4}{\sqrt{15}}, \quad \cos t = -4, \quad \text{and} \quad \tan t = -\frac{1}{\sqrt{15}}.$$

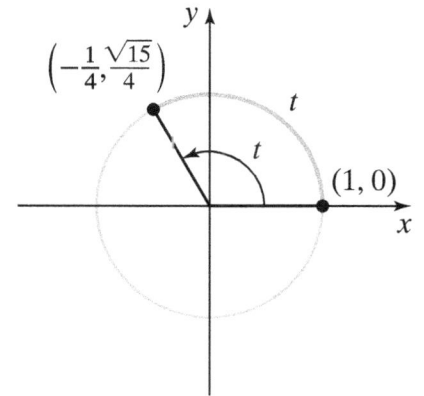

You Try It Work through this You Try It problem.

Work Exercises 9–11 in this textbook or in the My**Math**Lab **Study Plan.**

OBJECTIVE 4 USING THE UNIT CIRCLE TO EVALUATE TRIGONOMETRIC FUNCTIONS AT INCREMENTS OF $\frac{\pi}{2}$

My video summary

⊘ Understanding the relationship between certain values of t and specific points on the graph of the unit circle will help you quickly evaluate the trigonometric functions of these values. Undoubtedly, the easiest points on the unit circle to remember are the points $(1, 0)$, $(0, 1)$, $(-1, 0)$, and $(0, -1)$. The arc length from the point $(1, 0)$ to itself is 0 unit. Thus, the real number $t = 0$ corresponds to the point $(1, 0)$. See Figure 65a. To determine the values of t that correspond to the points $(0, 1)$, $(-1, 0)$, and $(0, -1)$, we use the fact that the circumference of a circle of radius, r, is $2\pi r$. Therefore, the circumference of the unit circle is $2\pi(1) = 2\pi$. The arc length from the point $(1, 0)$ to the point $(0, 1)$ is one-quarter of the circumference of the unit circle. Therefore, $t = \frac{1}{4} \cdot 2\pi = \frac{\pi}{2}$ corresponds to the point $(0, 1)$. See Figure 65b. The arc length from the point $(1, 0)$ to the point $(-1, 0)$ is one-half of the circumference of the unit circle. Thus, $t = \frac{1}{2} \cdot 2\pi = \pi$ corresponds to the point $(-1, 0)$. See Figure 65c. Finally, the arc length from the point $(1, 0)$ to the point $(0, -1)$ is three-quarters of the circumference of the unit circle. So, $t = \frac{3}{4} \cdot 2\pi = \frac{3\pi}{2}$ corresponds to the point $(0, -1)$. See Figure 65d.

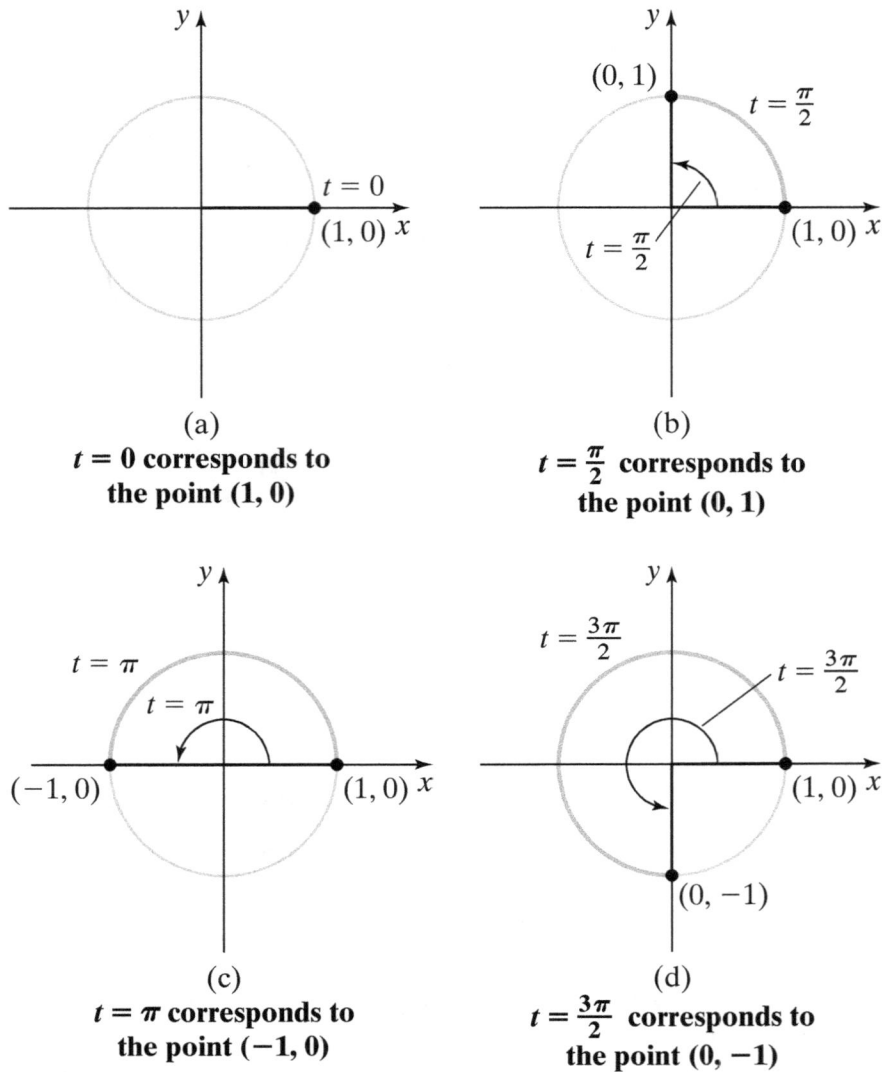

(a)

$t = 0$ corresponds to the point (1, 0)

(b)

$t = \dfrac{\pi}{2}$ corresponds to the point (0, 1)

(c)

$t = \pi$ corresponds to the point (−1, 0)

(d)

$t = \dfrac{3\pi}{2}$ corresponds to the point (0, −1)

Figure 65

We can now use the unit circle definitions to evaluate the trigonometric functions for values of t equal to 0, $\dfrac{\pi}{2}$, π, and $\dfrac{3\pi}{2}$. For example, using the fact that $\cos t = x$ and $\sin t = y$, we can use **Figure 66** to find the values of $\cos t$ and $\sin t$ for 0, $\dfrac{\pi}{2}$, π, and $\dfrac{3\pi}{2}$. Note in **Figure 66** that the numbers in blue represent the real number t and the x-coordinates of the corresponding points on the unit circle in green represent the value of $\cos t$. Similarly, the numbers in red represent the values of $\sin t$.

$$\cos 0 = 1$$
$$\sin 0 = 0$$

$$\cos \frac{\pi}{2} = 0$$
$$\sin \frac{\pi}{2} = 1$$

$$\cos \pi = -1$$
$$\sin \pi = 0$$

$$\cos \frac{3\pi}{2} = 0$$
$$\sin \frac{3\pi}{2} = -1$$

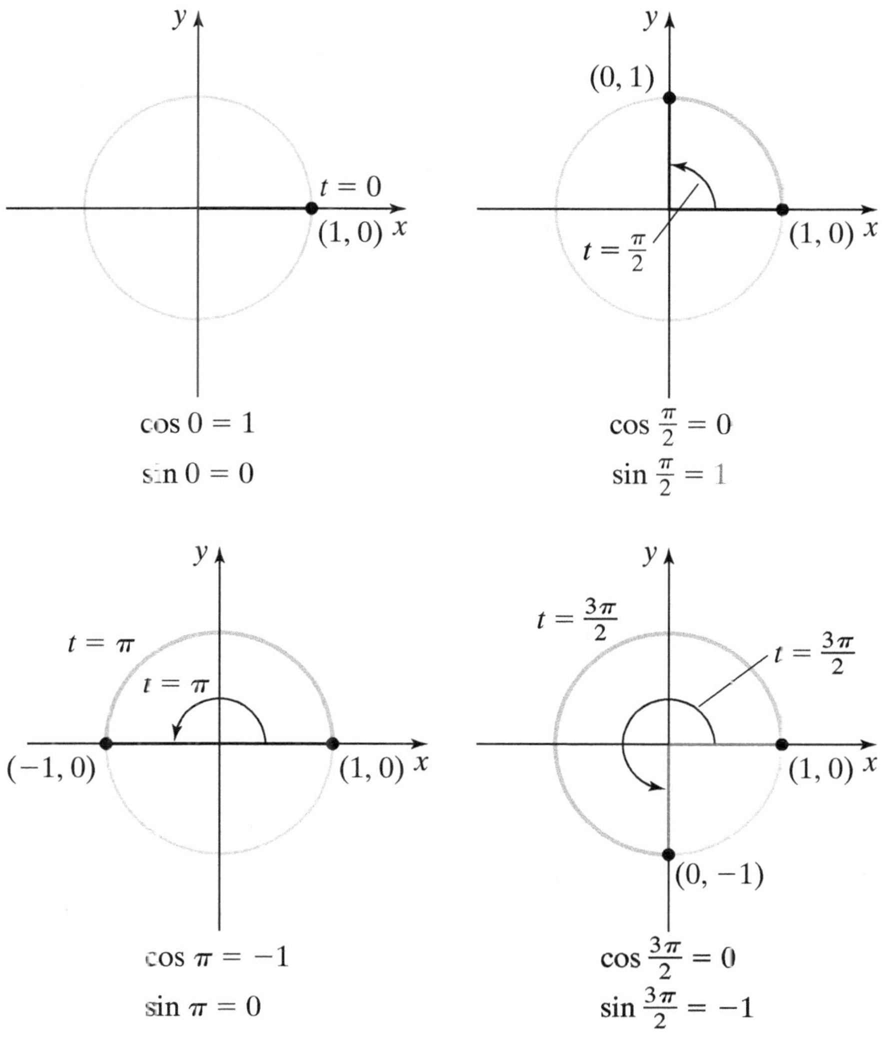

Figure 66 The values of $\cos t$ and $\sin t$ for $t = 0, \dfrac{\pi}{2}, \pi,$ and $\dfrac{3\pi}{2}$.

If we traverse the circle one complete revolution, an arc length of 2π, we are back at the point $(1, 0)$. Thus, $\sin 2\pi = 0$. We can continue around the circle again to obtain $\sin \dfrac{5\pi}{2} = 1, \sin 3\pi = 0, \sin \dfrac{7\pi}{2} = -1,$ and $\sin 4\pi = 0$. We can continue traversing the unit circle indefinitely. This is why the trigonometric functions are often referred to as "circular functions."

Because division by zero is undefined, the values of trigonometric functions may be undefined. For example, when $t = \dfrac{\pi}{2}$, the corresponding point on the unit circle is $(0, 1)$. Thus, $\tan \dfrac{\pi}{2} = \dfrac{y}{x} = \dfrac{1}{0}$, which is undefined. Table 3 shows the values of the six trigonometric functions for values of $0, \dfrac{\pi}{2}, \pi, \dfrac{3\pi}{2},$ and 2π. Use Figure 67 and watch this video for a complete explanation on how to fill out Table 3.

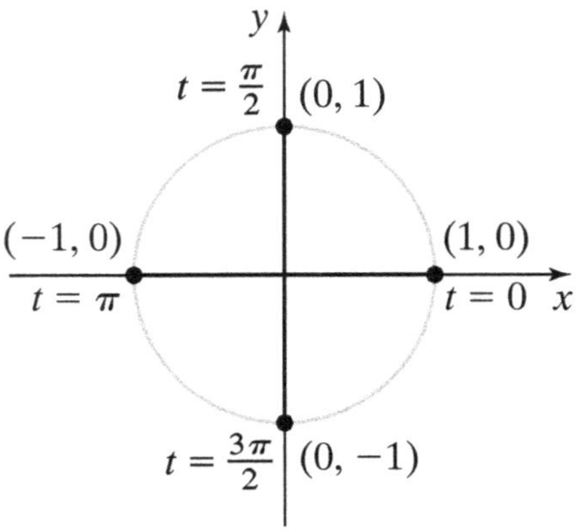

Figure 67

Table 3

t	$\sin t$	$\cos t$	$\tan t$	$\csc t$	$\sec t$	$\cot t$
$0, 2\pi$	0	1	0	Undefined	1	Undefined
$\dfrac{\pi}{2}$	1	0	Undefined	1	Undefined	0
π	0	-1	0	Undefined	-1	Undefined
$\dfrac{3\pi}{2}$	-1	0	Undefined	-1	Undefined	0

My interactive video summary

⊘ **Example 4** Evaluating Trigonometric Functions Using the Unit Circle

Use the unit circle to determine the value of each expression or state that it is undefined.

a. $\cos 11\pi$ **b.** $\tan \dfrac{5\pi}{2}$ **c.** $\csc\left(-\dfrac{3\pi}{2}\right)$

Solution

a. The number 11π is equal to $10\pi + \pi$, which is equivalent to the distance around the unit circle five times plus an additional half of a revolution. Thus, 11π corresponds to the point $(-1, 0)$ on the unit circle. Therefore, $\cos 11\pi = -1$. Work through parts (b) and (c) on your own. Then watch this **interactive video** to see if you are correct.

You Try It Work through this You Try It problem.

Work Exercises 12–19 in this textbook or in the MyMathLab Study Plan.

OBJECTIVE 5 USING THE UNIT CIRCLE TO EVALUATE TRIGONOMETRIC FUNCTIONS FOR INCREMENTS OF $\dfrac{\pi}{6}, \dfrac{\pi}{4}$, AND $\dfrac{\pi}{3}$

In Section 1.3, we introduced two special right triangles, the $\dfrac{\pi}{4}, \dfrac{\pi}{4}, \dfrac{\pi}{2}$ right triangle and the $\dfrac{\pi}{3}, \dfrac{\pi}{6}, \dfrac{\pi}{2}$ right triangle. We can use our knowledge of these two special right triangles to determine several more points that lie on the unit circle. If we look at the $\dfrac{\pi}{4}, \dfrac{\pi}{4}, \dfrac{\pi}{2}$ right triangle with side lengths of $1, 1, \sqrt{2}$ and divide the length of each side by $\sqrt{2}$, we can form a similar right triangle with side lengths of $\dfrac{1}{\sqrt{2}}, \dfrac{1}{\sqrt{2}}, 1$. See Figure 68.

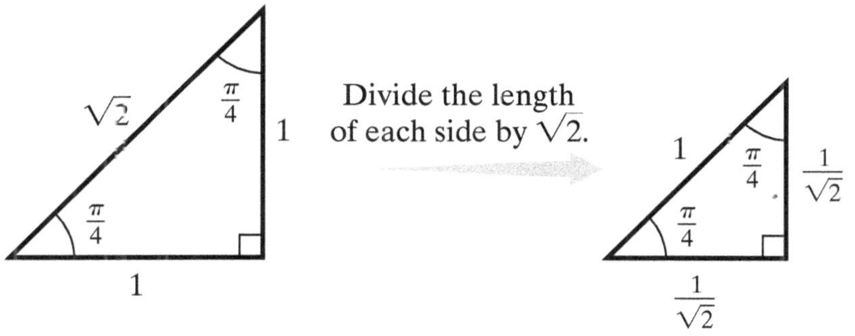

Figure 68

If we superimpose this triangle onto the unit circle, we see that the point $\left(\dfrac{1}{\sqrt{2}}, \dfrac{1}{\sqrt{2}}\right)$ lies on the unit circle. See Figure 69a. Similarly, we can use our knowledge of the $\dfrac{\pi}{3}, \dfrac{\pi}{6}, \dfrac{\pi}{2}$ right triangle to show that the points $\left(\dfrac{1}{2}, \dfrac{\sqrt{3}}{2}\right)$ and $\left(\dfrac{\sqrt{3}}{2}, \dfrac{1}{2}\right)$ also lie on the unit circle. See Figure 69b and Figure 69c. For a complete explanation, watch this interactive video.

My interactive video summary

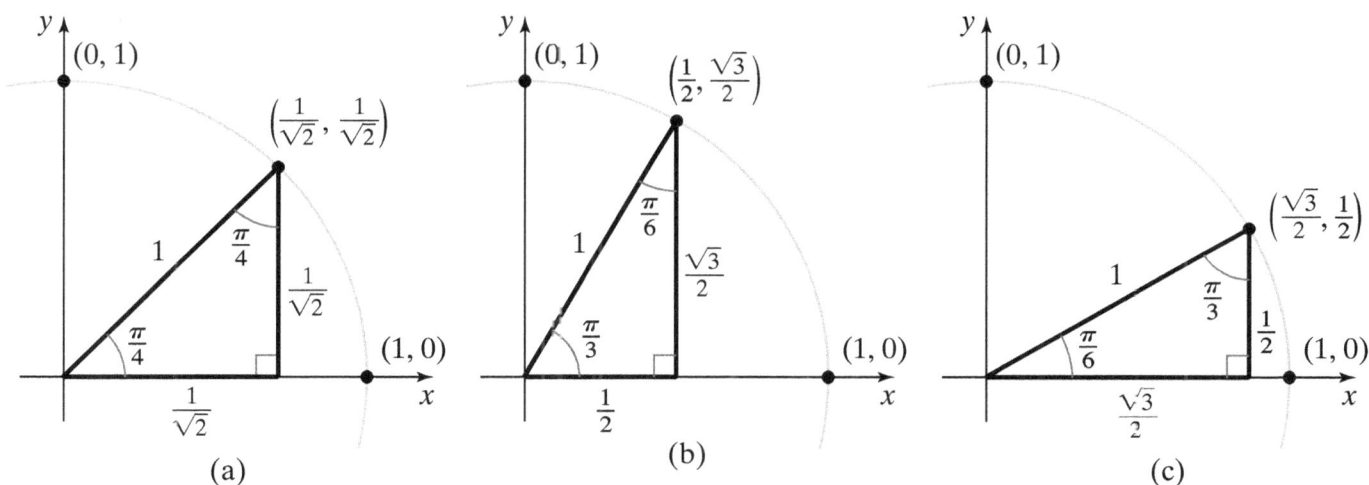

(a)

(b)

(c)

Figure 69

Now that we know that the points $\left(\dfrac{1}{\sqrt{2}}, \dfrac{1}{\sqrt{2}}\right), \left(\dfrac{1}{2}, \dfrac{\sqrt{3}}{2}\right)$, and $\left(\dfrac{\sqrt{3}}{2}, \dfrac{1}{2}\right)$ lie on the unit circle, we can use the symmetry properties of the unit circle to complete the unit circle, as shown in Figure 70. This unit circle contains the "special points" that correspond with "special values of t." For convenience, both the radian measure and degree measure is given for each corresponding point. You should become familiar with the unit circle shown in Figure 70.

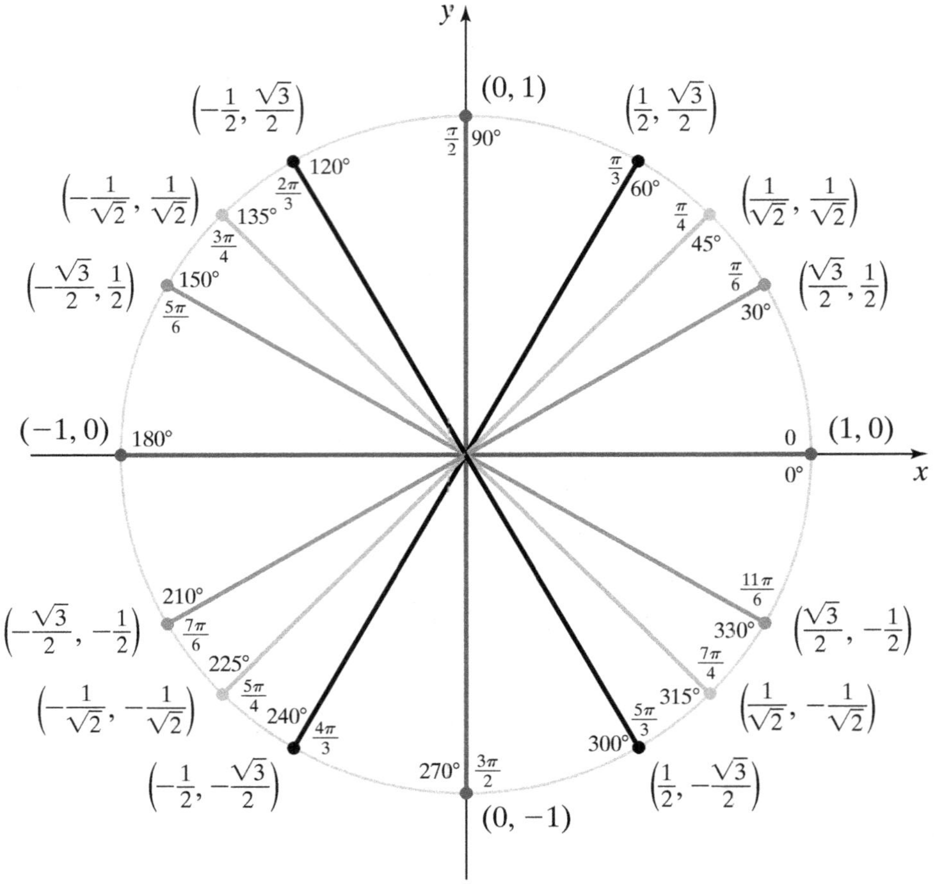

Figure 70
The Unit Circle

*My animation
summary*

Example 5 Evaluating Trigonometric Functions Using the Unit Circle

Use the unit circle to determine the following values.

a. $\tan\left(\dfrac{7\pi}{3}\right)$ b. $\sin\left(-\dfrac{3\pi}{4}\right)$ c. $\sec(480°)$ d. $\csc\left(-\dfrac{13\pi}{3}\right)$

Solution

a. We first need to identify the ordered pair on the unit circle that corresponds to $\dfrac{7\pi}{3}$. Watch this **animation** to see how to determine that this point is $\left(\dfrac{1}{2}, \dfrac{\sqrt{3}}{2}\right)$. Using the unit circle definition of the tangent function, we get

$$\tan\frac{7\pi}{3} = \frac{y}{x} = \frac{\dfrac{\sqrt{3}}{2}}{\dfrac{1}{2}} = \left(\frac{\sqrt{3}}{\cancel{2}}\right)(\cancel{2}) = \sqrt{3}.$$ Work through the entire **animation** to verify

the following:

b. $\sin\left(-\dfrac{3\pi}{4}\right) = -\dfrac{1}{\sqrt{2}}$

c. $\sec(480°) = -2$

d. $\csc\left(-\dfrac{13\pi}{3}\right) = -\dfrac{2}{\sqrt{3}}$

You Try It   Work through this You Try It problem.

Work Exercises 20–35 in this textbook or in the MyMathLab Study Plan.

1.6 Exercises

Skill Check Exercises

For exercises SCE-1 through SCE-6, solve each equation.

SCE-1. $x^2 + \left(\dfrac{4}{5}\right)^2 = 1$

SCE-2. $x^2 + \left(-\dfrac{\sqrt{3}}{2}\right)^2 = 1$

SCE-3. $\left(-\dfrac{3}{7}\right)^2 + y^2 = 1$

SCE-4. $\left(\dfrac{1}{\sqrt{2}}\right)^2 + y^2 = 1$

In Exercises 1–5, determine the missing coordinate of a point that lies on the graph of the unit circle, given the quadrant in which it is located.

1. $\left(\dfrac{2}{3}, \, y\right)$; Quadrant IV

2. $\left(-\dfrac{1}{5}, \, y\right)$; Quadrant II

3. $\left(x, \, -\dfrac{1}{7}\right)$; Quadrant III

4. $\left(x, \, \dfrac{2}{9}\right)$; Quadrant II

5. $\left(\dfrac{1}{3\sqrt{5}}, \, y\right)$; Quadrant IV

In Exercises 6–8, verify that the given point lies on the graph of the unit circle. Then use symmetry to find three other points that also lie on the graph of the unit circle.

6. $\left(\dfrac{1}{7}, \dfrac{4\sqrt{3}}{7}\right)$

7. $\left(-\dfrac{1}{2}, -\dfrac{\sqrt{3}}{2}\right)$

8. $\left(-\dfrac{4}{\sqrt{17}}, -\dfrac{1}{\sqrt{17}}\right)$

In Exercises 9–11, the given point lies on the graph of the unit circle and corresponds to a real number t. Find the values of the six trigonometric functions of t.

9. $\left(\dfrac{1}{7}, -\dfrac{4\sqrt{3}}{7}\right)$

10. $\left(-\dfrac{4}{\sqrt{17}}, -\dfrac{1}{\sqrt{17}}\right)$

11. $\left(-\dfrac{1}{2}, -\dfrac{\sqrt{3}}{2}\right)$

In Exercises 12–19, use the unit circle to determine the value of the given expression or state that it is undefined.

SbS 12. $\cos 4\pi$

SbS 13. $\cot\dfrac{7\pi}{2}$

SbS 14. $\tan\left(-\dfrac{11\pi}{2}\right)$

SbS 15. $\sec(-3\pi)$

SbS 16. $\csc 5\pi$

SbS 17. $\sin(-6\pi)$

SbS 18. $\cos\left(\dfrac{5\pi}{2}\right)$

SbS 19. $\csc\left(-\dfrac{17\pi}{2}\right)$

In Exercises 20–35, use the unit circle to determine the value of the given expression or state that it is undefined.

20. $\sin\left(\dfrac{5\pi}{6}\right)$ 21. $\cos\left(\dfrac{3\pi}{4}\right)$ 22. $\cot\left(\dfrac{4\pi}{3}\right)$ 23. $\csc(315°)$

24. $\tan\left(-\dfrac{2\pi}{3}\right)$ 25. $\cos\left(-\dfrac{7\pi}{6}\right)$ 26. $\sec\left(-\dfrac{\pi}{4}\right)$ 27. $\cot\left(-\dfrac{5\pi}{4}\right)$

28. $\sin\left(\dfrac{17\pi}{6}\right)$ 29. $\tan\left(\dfrac{11\pi}{3}\right)$ 30. $\sec\left(\dfrac{23\pi}{6}\right)$ 31. $\csc\left(\dfrac{10\pi}{3}\right)$

32. $\cos(-390°)$ 33. $\sin\left(-\dfrac{17\pi}{6}\right)$ 34. $\cot\left(-\dfrac{11\pi}{3}\right)$ 35. $\sec(-510°)$

Brief Exercises

In Exercises 36–59, use the unit circle to determine the value of the given expression or state that it is undefined.

36. $\cos 4\pi$ 37. $\cot\dfrac{7\pi}{2}$ 38. $\tan\left(-\dfrac{11\pi}{2}\right)$ 39. $\sec(-3\pi)$

40. $\csc 5\pi$ 41. $\sin(-6\pi)$ 42. $\cos\left(\dfrac{5\pi}{2}\right)$ 43. $\csc\left(-\dfrac{17\pi}{2}\right)$

44. $\sin\left(\dfrac{5\pi}{6}\right)$ 45. $\cos\left(\dfrac{3\pi}{4}\right)$ 46. $\cot\left(\dfrac{4\pi}{3}\right)$ 47. $\csc(315°)$

48. $\tan\left(-\dfrac{2\pi}{3}\right)$ 49. $\cos\left(-\dfrac{7\pi}{6}\right)$ 50. $\sec\left(-\dfrac{\pi}{4}\right)$ 51. $\cot\left(-\dfrac{5\pi}{4}\right)$

52. $\sin\left(\dfrac{17\pi}{6}\right)$ 53. $\tan\left(\dfrac{11\pi}{3}\right)$ 54. $\sec\left(\dfrac{23\pi}{6}\right)$ 55. $\csc\left(\dfrac{10\pi}{3}\right)$

56. $\cos(-390°)$ 57. $\sin\left(-\dfrac{17\pi}{6}\right)$ 58. $\cot\left(-\dfrac{11\pi}{3}\right)$ 59. $\sec(-510°)$

The Graphs of Trigonometric Functions

CHAPTER TWO CONTENTS

2.1 The Graphs of Sine and Cosine

THINGS TO KNOW

Before working through this section, be sure that you are familiar with the following concepts:

		VIDEO	ANIMATION	INTERACTIVE

You Try It 1. Using the Special Right Triangles (Section 1.4) ⊙

You Try It 2. Finding the Values of the Trigonometric Functions of Quadrantal Angles (Section 1.5) ⊙

You Try It 3. Evaluating Trigonometric Functions of Angles Belonging to the $\frac{\pi}{3}, \frac{\pi}{6}$, or $\frac{\pi}{4}$ Families (Section 1.5) ⊙

OBJECTIVES

1 Understanding the Graph of the Sine Function and Its Properties

2 Understanding the Graph of the Cosine Function and Its Properties

3 Sketching Graphs of the Form $y = A \sin x$ and $y = A \cos x$

4 Sketching Graphs of the Form $y = \sin(Bx)$ and $y = \cos(Bx)$

5 Sketching Graphs of the Form $y = A \sin(Bx)$ and $y = A \cos(Bx)$

6 Determine the Equation of a Function of the Form $y = A\sin(Bx)$ or $y = A\cos(Bx)$ Given Its Graph

Introduction to Section 2.1

Recall that in Chapter 1, we sketched an angle θ in standard position in the rectangular coordinate system. We used the ordered pair (x, y) to represent a point P lying on the terminal side of θ. From there, we were able to find the trigonometric values of that angle using the **general angle definitions of the trigonometric functions**. We now review these definitions.

The General Angle Definitions of the Trigonometric Functions

If $P(x, y)$ is a point on the terminal side of *any* angle θ in standard position and if $r = \sqrt{x^2 + y^2}$ is the distance from the origin to point P, then the six trigonometric functions of θ are defined as follows:

$$\sin \theta = \frac{y}{r} \qquad \csc \theta = \frac{r}{y}, \; y \neq 0$$

$$\cos \theta = \frac{x}{r} \qquad \sec \theta = \frac{r}{x}, \; x \neq 0$$

$$\tan \theta = \frac{y}{x}, \; x \neq 0 \qquad \cot \theta = \frac{x}{y}, \; y \neq 0$$

$$r = \sqrt{x^2 + y^2}$$

Now, in Chapter 2, we will use a rectangular coordinate system for a different purpose. We will plot points and label them (x, y), but the x-value (or **independent variable**) will represent an angle measured in radians and the y-value (or **dependent variable**) will represent a trigonometric function of the angle, either $y = \sin x$, $y = \cos x$, $y = \tan x$, $y = \csc x$, $y = \sec x$, or $y = \cot x$. For example, consider the function $y = \sin x$. When $x = \frac{\pi}{4}$, $y = \sin \frac{\pi}{4} = \frac{1}{\sqrt{2}}$. Thus, the ordered pair $\left(\frac{\pi}{4}, \frac{1}{\sqrt{2}}\right)$ represents a point that lies on the graph of $y = \sin x$.

Before we establish the graphs of $y = \sin x$ and $y = \cos x$, it is imperative that you can evaluate trigonometric functions at certain values. Take a moment to test yourself by clicking on any of the three You Try It exercises below.

You Try It Evaluating Trigonometric Functions for Angles Belonging to the Quadrantal Family

You Try It Using Special Right Triangles to Evaluate Trigonometric Functions

You Try It Evaluating Trigonometric Functions of Angles Belonging to the $\frac{\pi}{6}, \frac{\pi}{4}$, or $\frac{\pi}{3}$ Families

OBJECTIVE 1   UNDERSTANDING THE GRAPH OF THE SINE FUNCTION AND ITS PROPERTIES

My video summary   ⊙ We can sketch the graph of $y = \sin x$ by setting up a table (or tables) of values and plotting points. Take some time to fill in the four tables shown below. Table 1 includes values of x belonging to the quadrantal family. Table 2 includes values of x belonging to the $\dfrac{\pi}{6}$ family. Table 3 includes values of x belonging to the

$\dfrac{\pi}{4}$ family. Table 4 includes values of x belonging to the $\dfrac{\pi}{3}$ family. Once these four tables are complete, we can plot the corresponding ordered pairs to sketch the graph of $y = \sin x$ on the interval $[0, 2\pi]$. See Figure 1. Watch this video to see how to complete each of the four tables and to see how to sketch the graph of $y = \sin x$

Table 1 (View the completed Table 1.)

x	0	$\dfrac{\pi}{2}$	π	$\dfrac{3\pi}{2}$	2π
$y = \sin x$					

Table 2 (View the completed Table 2.)

x	$\dfrac{\pi}{6}$	$\dfrac{5\pi}{6}$	$\dfrac{7\pi}{6}$	$\dfrac{11\pi}{6}$
$y = \sin x$				

Table 3 (View the completed Table 3.)

x	$\dfrac{\pi}{4}$	$\dfrac{3\pi}{4}$	$\dfrac{5\pi}{4}$	$\dfrac{7\pi}{4}$
$y = \sin x$				

Table 4 (View the completed Table 4.)

x	$\dfrac{\pi}{3}$	$\dfrac{2\pi}{3}$	$\dfrac{4\pi}{3}$	$\dfrac{5\pi}{3}$
$y = \sin x$				

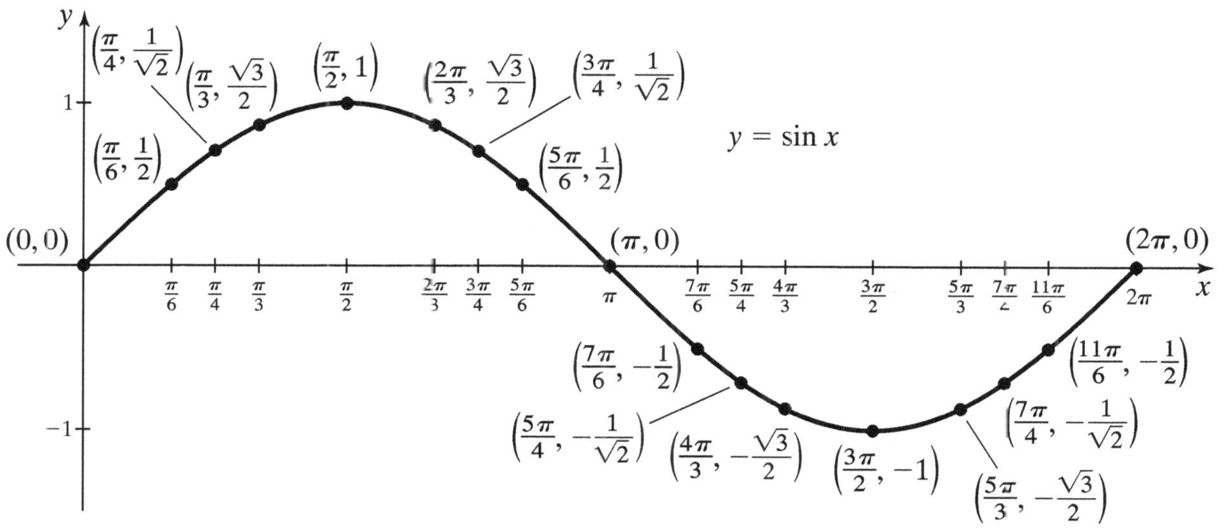

Figure 1 The graph of $y = \sin x$ on the interval $[0, 2\pi]$

The graph of $y = \sin x$ in Figure 1 is sketched on the interval $[0, 2\pi]$ and shows only one cycle of the graph. If we were to create another table of values choosing values of x between 2π and 4π (or between -2π and 0) to plot several more ordered pairs, we would see that the graph repeats itself. In fact, on every interval of length 2π, the graph of $y = \sin x$ repeats itself. We say that the graph of $y = \sin x$ is a **periodic function** with a period of 2π.

Definition   Periodic Function

A function f is said to be **periodic** if there is a positive number P such that $f(x + P) = f(x)$ for all x in the domain of f. The smallest number P for which f is periodic is called the **period** of f.

In Figure 2, we see three complete cycles of the graph of $y = \sin x$. As you can see in Figure 2, the graph of $y = \sin x$ repeats itself after every interval of length 2π. That is, for any value of x, $\sin x = \sin(x + 2\pi n)$, where n is any integer.

Note that the domain of $y = \sin x$ includes all real numbers, whereas the range includes all values between and including -1 and 1. We now list these and several other characteristics of the sine function. Watch this video for a detailed description of the characteristics of the sine function.

My video summary

◉ **Characteristics of the Sine Function**

The domain is $(-\infty, \infty)$.
The range is $[-1, 1]$.
The function is periodic with a period of $P = 2\pi$.
The y-intercept is 0.
The x-intercepts, or **zeros**, are of the form $n\pi$ where n is an integer.
The function is **odd**, which means $\sin(-x) = -\sin x$. The graph is symmetric about the origin.

The function obtains a **relative maximum** at $x = \dfrac{\pi}{2} + 2\pi n$, where n is an integer. The maximum value is 1.

The function obtains a **relative minimum** at $x = \dfrac{3\pi}{2} + 2\pi n$, where n is an integer. The minimum value is -1.

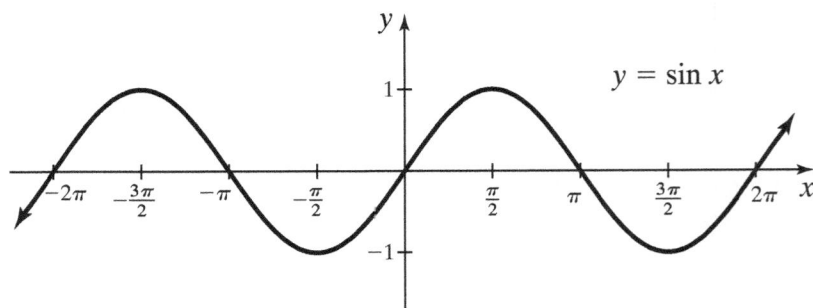

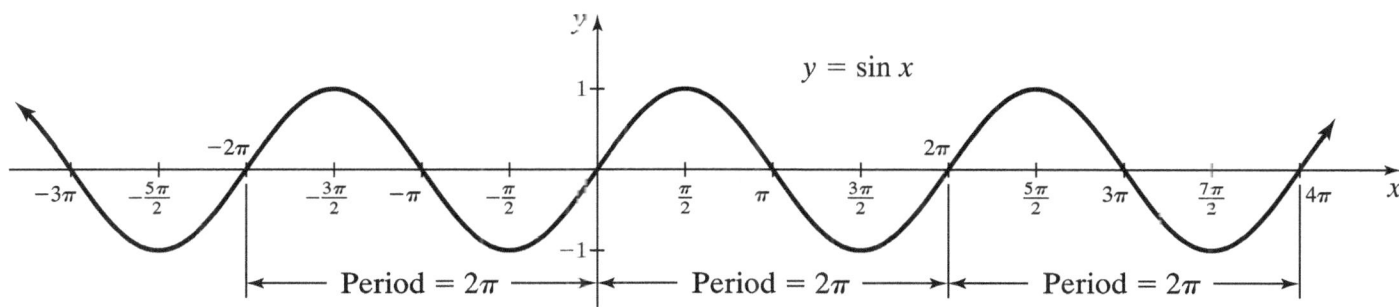

Figure 2 Three complete cycles of $y = \sin x$

Example 1 Understanding the Graph of $y = \sin x$

Using the graph of $y = \sin x$, list all values of x on the interval $\left[-3\pi, \dfrac{7\pi}{4}\right]$ that satisfy the ordered pair $(x, 0)$.

Solution First, sketch the graph of $y = \sin x$ on the interval $\left[-3\pi, \dfrac{7\pi}{4}\right]$.

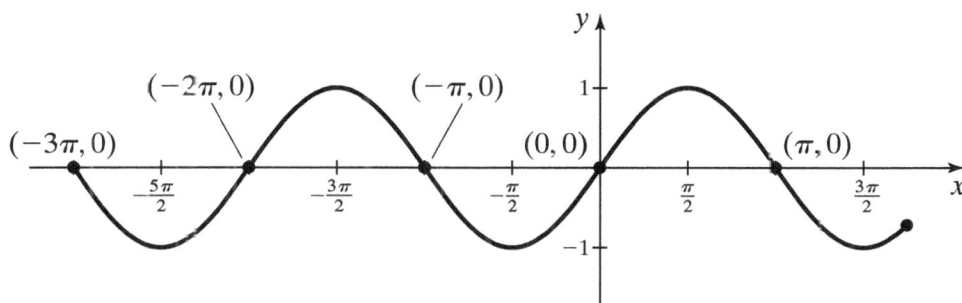

The values of x that satisfy the ordered pair $(x, 0)$ are precisely the x-intercepts of $y = \sin x$. The x-intercepts have the form $n\pi$, where n is an integer. We see from the graph that the x-intercepts on the interval $\left[-3\pi, \dfrac{7\pi}{4}\right]$ are $-3\pi, -2\pi, -\pi, 0$, and π.

You Try It Work through this You Try It problem.

Work Exercises 1–4 in this textbook or in the MyMathLab Study Plan.

Example 2 Using the Periodic Property of $y = \sin x$

Use the periodic property of $y = \sin x$ to determine which of the following expressions is equivalent to $\sin\left(\dfrac{23\pi}{6}\right)$.

i. $\sin\left(\dfrac{\pi}{6}\right)$ ii. $\sin\left(\dfrac{5\pi}{6}\right)$ iii. $\sin\left(\dfrac{11\pi}{6}\right)$ iv. $\sin\left(\dfrac{13\pi}{6}\right)$

Solution The function $y = \sin x$ has a period of 2π. This means that $\sin\left(\dfrac{23\pi}{6}\right) = \sin\left(\dfrac{23\pi}{6} + 2\pi n\right)$ for any integer n.

When $n = -1$, we get

$$\sin\left(\frac{23\pi}{6}\right) = \sin\left(\frac{23\pi}{6} + 2\pi(-1)\right) = \sin\left(\frac{23\pi}{6} - 2\pi\right) = \sin\left(\frac{23\pi}{6} - \frac{12\pi}{6}\right) = \sin\left(\frac{11\pi}{6}\right).$$

Thus, the answer is iii.

In the previous example, we could find infinitely many more expressions that are equivalent to $\sin\left(\dfrac{23\pi}{6}\right)$ using the periodic property of $y = \sin x$. Below are just a few.

For $n = 1$, $\sin\left(\dfrac{23\pi}{6}\right) = \sin\left(\dfrac{23\pi}{6} + 2\pi(1)\right) = \sin\left(\dfrac{23\pi}{6} + \dfrac{12\pi}{6}\right) = \sin\left(\dfrac{35\pi}{6}\right).$

For $n = -2$, $\sin\left(\dfrac{23\pi}{6}\right) = \sin\left(\dfrac{23\pi}{6} + 2\pi(-2)\right) = \sin\left(\dfrac{23\pi}{6} - 4\pi\right) = \sin\left(\dfrac{23\pi}{6} - \dfrac{24\pi}{6}\right) = \sin\left(-\dfrac{\pi}{6}\right).$

For $n = 2$, $\sin\left(\dfrac{23\pi}{6}\right) = \sin\left(\dfrac{23\pi}{6} + 2\pi(2)\right) = \sin\left(\dfrac{23\pi}{6} + 4\pi\right) = \sin\left(\dfrac{23\pi}{6} + \dfrac{24\pi}{6}\right) = \sin\left(\dfrac{47\pi}{6}\right).$

You Try It Work through this You Try It problem.

Work Exercises 5–7 in this textbook or in the MyMathLab Study Plan.

Example 3 Using the Odd Property of $y = \sin x$

Use the fact that $y = \sin x$ is an odd function to determine which of the following expressions is equivalent to $-\sin\left(\dfrac{9\pi}{16}\right)$.

i. $\sin\left(-\dfrac{9\pi}{16}\right)$ ii. $\sin\left(\dfrac{9\pi}{16}\right)$ iii. $-\sin\left(-\dfrac{9\pi}{16}\right)$

Solution Because $y = \sin x$ is an odd function, we know that $-\sin x = \sin(-x)$. Therefore, if $x = \dfrac{9\pi}{16}$, then $-\sin\left(\dfrac{9\pi}{16}\right) = \sin\left(-\dfrac{9\pi}{16}\right)$. Thus, the answer is i.

You Try It Work through this You Try It problem.

Work Exercise 8 in this textbook or in the MyMathLab Study Plan.

OBJECTIVE 2 UNDERSTANDING THE GRAPH OF THE COSINE FUNCTION AND ITS PROPERTIES

My video summary ⊙ To sketch the graph of $y = \cos x$, we once again create several tables of values and plot the corresponding ordered pairs. You should take some time and complete Table 5, Table 6, Table 7, and Table 8 shown below. Watch this video to see how to complete each of these four tables and to see how to sketch the graph of $y = \cos x$. One complete cycle of the graph of $y = \cos x$ is sketched in Figure 3.

Table 5 (View the completed Table 5.)

x	0	$\dfrac{\pi}{2}$	π	$\dfrac{3\pi}{2}$	2π
$y = \cos x$					

Table 6 (View the completed Table 6.)

x	$\dfrac{\pi}{6}$	$\dfrac{5\pi}{6}$	$\dfrac{7\pi}{6}$	$\dfrac{11\pi}{6}$
$y = \cos x$				

Table 7 (View the completed Table 7.)

x	$\dfrac{\pi}{4}$	$\dfrac{3\pi}{4}$	$\dfrac{5\pi}{4}$	$\dfrac{7\pi}{4}$
$y = \cos x$				

Table 8 (View the completed Table 8.)

x	$\dfrac{\pi}{3}$	$\dfrac{2\pi}{3}$	$\dfrac{4\pi}{3}$	$\dfrac{5\pi}{3}$
$y = \cos x$				

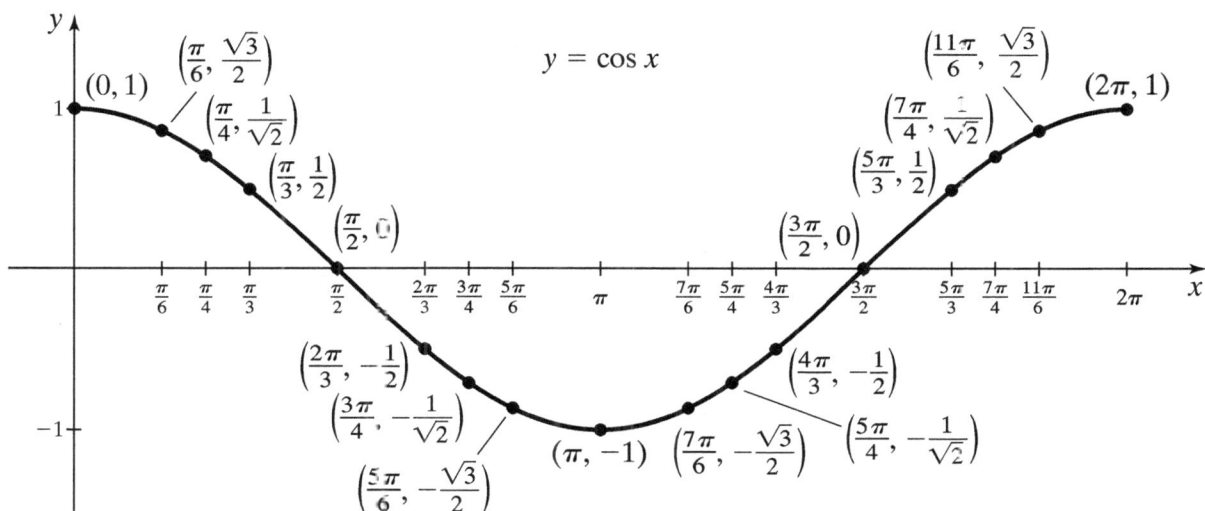

Figure 3 The graph of $y = \cos x$ on the interval $[0, 2\pi]$

Just like $y = \sin x$, the function $y = \cos x$ is periodic with a period of 2π. In Figure 4, we see three complete cycles of the graph of $y = \cos x$.

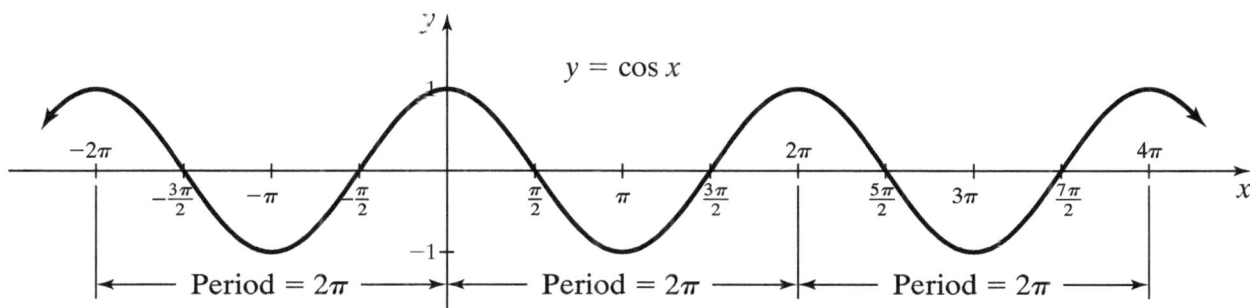

Figure 4 Three complete cycles of $y = \cos x$

As you can see in Figure 4, the domcin of $y = \cos x$ includes all real numbers, whereas the range includes all values between and including -1 and 1. We now list these and several other characteristics of the cosine function. Watch this video for a detailed description of the characteristics of the cosine function.

My video summary

⊘ Characteristics of the Cosine Function

The domain is $(-\infty, \infty)$.
The range is $[-1, 1]$.
The function is periodic with a period of $P = 2\pi$.
The y-intercept is 1.

The x-intercepts, or zeros, are of the form $(2n + 1) \cdot \dfrac{\pi}{2}$, where n is an integer.

The function is even, which means $\cos(-x) = \cos x$. The graph is symmetric about the y-axis.
The function obtains a relative maximum at $x = 2\pi n$, where n is an integer. The maximum value is 1.
The function obtains a relative minimum at $x = \pi + 2\pi n$, where n is an integer. The minimum value is -1.

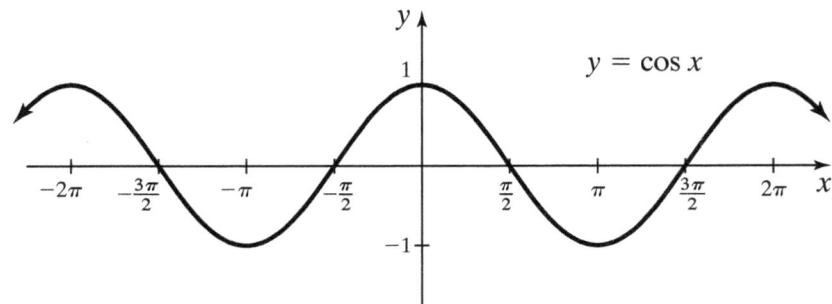

Example 4 Understanding the Graph of $y = \cos x$

Using the graph of $y = \cos x$, list all values of x on the interval $[-\pi, 2\pi]$ that satisfy the ordered pair $\left(x, \dfrac{1}{2}\right)$.

Solution We know that $\cos\left(\dfrac{\pi}{3}\right) = \dfrac{1}{2}$. Using the even property of $y = \cos x$, we also know that $\cos\left(-\dfrac{\pi}{3}\right) = \dfrac{1}{2}$. Using the periodic property of $y = \cos x$, we know that $\cos\left(-\dfrac{\pi}{3} + 2\pi\right) = \cos\left(\dfrac{5\pi}{3}\right) = \dfrac{1}{2}$. Therefore, the values of x that satisfy the ordered pair $\left(x, \dfrac{1}{2}\right)$ on the interval $[-\pi, 2\pi]$ are $-\dfrac{\pi}{3}, \dfrac{\pi}{3}$, and $\dfrac{5\pi}{3}$. You should verify that we have accurately found all x- values by sketching the graph of $y = \cos x$ on the interval $[-\pi, 2\pi]$.

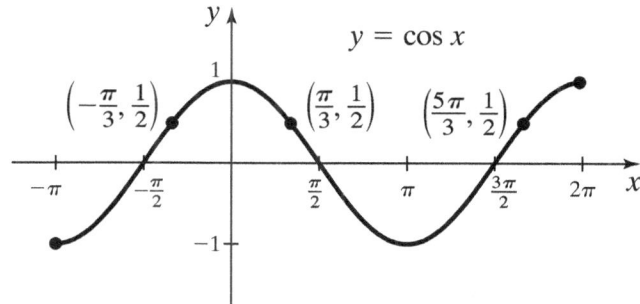

You Try It   Work through this You Try It problem.

Work Exercises 9–16 in this textbook or in the MyMathLab Study Plan.

Now that you have a basic understanding of the graphs of $y = \sin x$ and $y = \cos x$, we will introduce variations of these graphs. As we go forward, we will focus on sketching one cycle of each graph with the understanding that this cycle repeats itself indefinitely. We will focus on sketching five points on every graph. These five points evenly divide one cycle of the sine or cosine curve into fourths, or quarters. Therefore, we will call these five points **quarter points**. You should memorize the coordinates of these quarter points because they will help you sketch complicated sine and cosine curves. We now sketch one cycle of the sine and cosine curves using these five quarter points over the interval $[0, 2\pi]$.

The Five Quarter Points of $y = \sin x$

1. $(0, 0)$ 2. $\left(\dfrac{\pi}{2}, 1\right)$ 3. $(\pi, 0)$ 4. $\left(\dfrac{3\pi}{2}, -1\right)$

5. $(2\pi, 0)$

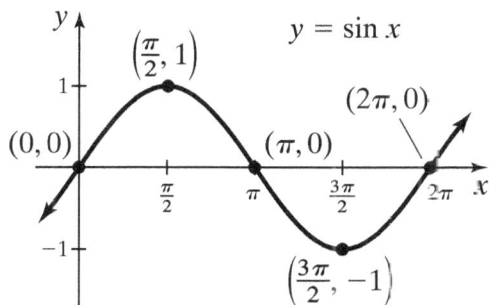

The Five Quarter Points of $y = \cos x$

1. $(0, 1)$ 2. $\left(\dfrac{\pi}{2}, 0\right)$ 3. $(\pi, -1)$ 4. $\left(\dfrac{3\pi}{2}, 0\right)$

5. $(2\pi, 1)$

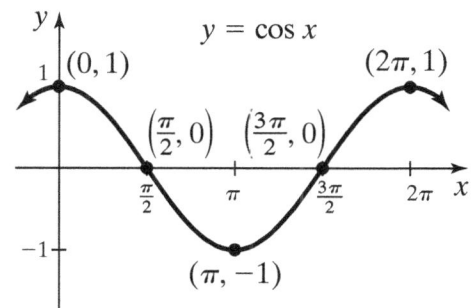

OBJECTIVE 3   SKETCHING GRAPHS OF THE FORM $y = A \sin x$ AND $y = A \cos x$

My video summary ⊙ There are four factors that affect the graph of a sine or cosine curve. The first is amplitude.

Definition   Amplitude

The **amplitude** of a sine or cosine curve is the measure of half the distance between the maximum and minimum values.

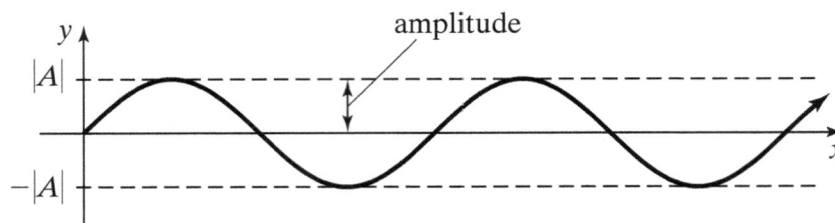

Trigonometric functions of the form $y = A \sin x$ and $y = A \cos x$ have an amplitude of $|A|$ and a range of $[-|A|, |A|]$.

Knowing the amplitude will help us to determine the maximum and minimum values of functions of the form $y = A \sin x$ or $y = A \cos x$. For example, the function $y = 2 \sin x$ has an amplitude of $|A| = |2| = 2$ and a range of $[-2, 2]$. Thus, the maximum value of $y = 2 \sin x$ is 2 and the minimum value is -2. The graphs of $y = \sin x$ and $y = 2 \sin x$ are shown on the same rectangular coordinate system in Figure 5.

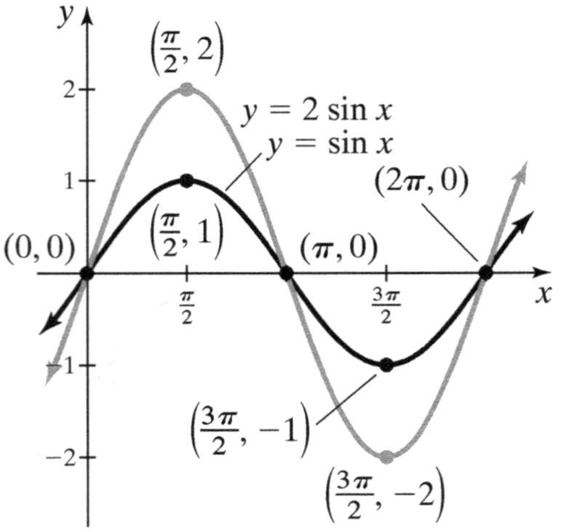

Figure 5 This illustrates one cycle of the graphs of $y = \sin x$ and $y = 2 \sin x$.

Using Technology 📱

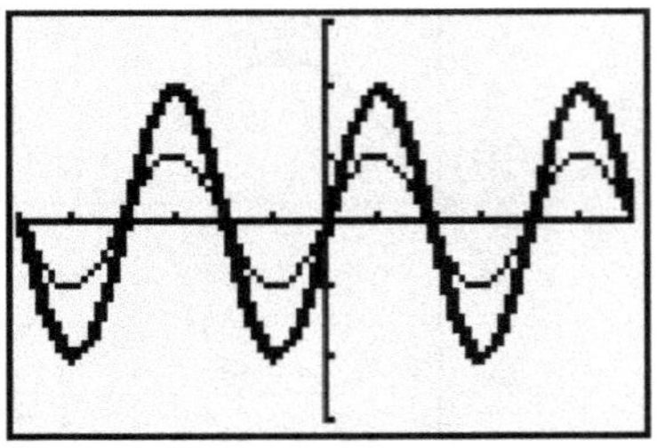

Figure 6 This illustrates several cycles of the graphs of $y = \sin x$ and $y = 2 \sin x$ using a graphing utility.

Notice in Figure 5 that the x-coordinates of all quarter points on each graph are the same. This is because the period of each function is 2π. We simply multiplied each y-coordinate of the quarter points of $y = \sin x$ by 2 to obtain the five quarter points used to sketch the graph of $y = 2\sin x$. Because the graph of $y = \sin x$ has three quarter points with y-coordinates of 0, multiplying those values by 2 causes no change in the values of the corresponding y-coordinates on the graph of $y = 2\sin x$.

My video summary

▷ **Example 5** Sketching a graph of the form $y = A\cos x$

Determine the amplitude and range of $y = -\dfrac{2}{3}\cos x$ and then sketch the graph.

Solution The amplitude of $y = -\dfrac{2}{3}\cos x$ is $A = \left|-\dfrac{2}{3}\right| = \dfrac{2}{3}$. Thus the range is $\left[-\dfrac{2}{3}, \dfrac{2}{3}\right]$.

To sketch the graph, we multiply the y-coordinates of the quarter points of $y = \cos x$ by $-\dfrac{2}{3}$ to obtain the quarter points for $y = -\dfrac{2}{3}\cos x$.

Quarter points for $y = \cos x$: $(0, 1)$, $\left(\dfrac{\pi}{2}, 0\right)$, $(\pi, -1)$, $\left(\dfrac{3\pi}{2}, 0\right)$, $(2\pi, 1)$

Quarter points for $y = -\dfrac{2}{3}\cos x$: $\left(0, -\dfrac{2}{3}\right)$, $\left(\dfrac{\pi}{2}, 0\right)$, $\left(\pi, \dfrac{2}{3}\right)$, $\left(\dfrac{3\pi}{2}, 0\right)$, $\left(2\pi, -\dfrac{2}{3}\right)$

One cycle of the graphs of $y = \cos x$ and $y = -\dfrac{2}{3}\cos x$ is sketched in Figure 7.

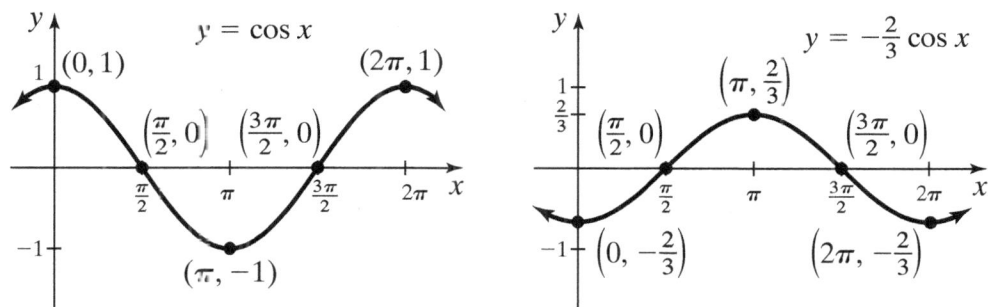

Figure 7 One cycle of the graphs of $y = \cos x$ and $y = -\dfrac{2}{3}\cos x$

In the previous example, we sketched the graph of $y = -\dfrac{2}{3}\cos x$ using amplitude and the five quarter points. Note that there is another way to sketch the graph of $y = -\dfrac{2}{3}\cos x$. You might recall from an algebra course that if $A > 1$, then the graph of a function $y = Af(x)$ can be obtained by **vertically stretching** the graph of the function $y = f(x)$ by a factor of A. If $0 < A < 1$, the graph of $y = Af(x)$ can be obtained by **vertically compressing** the graph of $y = Af(x)$ by a factor of A.

You may want to review the vertical stretch transformation or vertical compression transformation by watching the brief animations shown below. These topics are typically taught in an algebra course. (Note that in the animations below, a lower case "a" is used to describe the constant. In this section, a capital "A" is used to describe the constant.)

My animation summary

▣ To review the concept of a vertical stretch, watch this animation.

▣ To review the concept of a vertical compression, watch this animation.

▣ Also, the graph of $y = -f(x)$ can be obtained by reflecting the graph of $y = f(x)$ about the x-axis. To review the concept of a reflection about the x-axis, watch this animation.

Therefore, using transformations, we can sketch the graph of $y = -\dfrac{2}{3}\cos x$ by first vertically compressing the graph of $y = \cos x$ by a factor of $\dfrac{2}{3}$ to obtain the graph of $y = \dfrac{2}{3}\cos x$. Then we can reflect the graph of $y = \dfrac{2}{3}\cos x$ about the x-axis, producing the graph of $y = -\dfrac{2}{3}\cos x$. To see how to sketch the graph of $y = -\dfrac{2}{3}\cos x$ using transformations, read these steps.

You Try It Work through this You Try It problem.

Work Exercises 17–26 in this textbook or in the MyMathLab Study Plan.

OBJECTIVE 4 SKETCHING GRAPHS OF THE FORM $y = \sin(Bx)$ AND $y = \cos(Bx)$

The second factor that affects a sine or cosine curve is a change in period. Functions of the form $y = \sin(Bx)$ and $y = \cos(Bx)$ have a period other than 2π when $B \neq 1$. To determine the period of $y = \sin(Bx)$ and $y = \cos(Bx)$, first recall that the period of $y = \sin x$ and $y = \cos x$ is 2π. That is, the graphs of $y = \sin x$ and $y = \cos x$ complete one cycle for $0 \leq x \leq 2\pi$. For now, suppose that $B > 0$. Then the graph of $y = \sin(Bx)$ and $y = \cos(Bx)$ will complete one cycle for $0 \leq Bx \leq 2\pi$. We can solve this three-part inequality to determine the period of $y = \sin(Bx)$ and $y = \cos(Bx)$.

$0 \leq Bx \leq 2\pi$ The graph of $y = \sin(Bx)$ and $y = \cos(Bx)$ will complete one cycle for values of x satisfying this inequality.

$\dfrac{0}{B} \leq \dfrac{Bx}{B} \leq \dfrac{2\pi}{B}$ Divide each part by B. The direction of the inequality remains the same because $B > 0$.

$0 \leq x \leq \dfrac{2\pi}{B}$ Simplify.

The graph of the functions $y = \sin(Bx)$ and $y = \cos(Bx)$ will complete one cycle for $0 \leq x \leq \dfrac{2\pi}{B}$, which means that the period, P, of $y = \sin(Bx)$ and $y = \cos(Bx)$ is

$$P = \dfrac{2\pi}{B}.$$

Determining the Period of $y = \sin(Bx)$ and $y = \cos(Bx)$

The **period** of a sine or cosine curve is equal to $P = \dfrac{2\pi}{B}$, where $B > 0$.

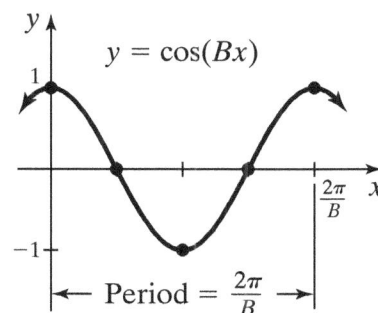

For $B > 0$, the graphs of $y = \sin(Bx)$ and $y = \cos(Bx)$ can be obtained by subdividing the interval $\left[0, \dfrac{2\pi}{B}\right]$ into four equal subintervals of length $\dfrac{2\pi}{B} \div 4$ and by plotting five quarter points. The y-coordinate of each quarter point of $y = \sin(Bx)$ and $y = \cos(Bx)$ has the same value as the y-coordinate of each quarter point of $y = \sin x$ and by $y = \cos x$, respectively.

My interactive video summary

$\circledcirc$ **Example 6** Sketching graphs of the form $y = \sin(Bx)$ and $y = \cos(Bx)$

Determine the period and sketch the graph of each function.

a. $y = \sin(2x)$ **b.** $y = \cos\left(\dfrac{1}{2}x\right)$ **c.** $y = \sin(\pi x)$

Solution

a. The period of $y = \sin(2x)$ is $P = \dfrac{2\pi}{B} = \dfrac{2\pi}{2} = \pi$. Next, we divide the interval $[0, \pi]$ into four equal subintervals of length $\pi \div 4 = \dfrac{\pi}{4}$ by starting with 0 and adding $\dfrac{\pi}{4}$ to the x-coordinate of each successive quarter point. Thus, the x-coordinates of the five quarter points are $0, \dfrac{\pi}{4}, \dfrac{\pi}{2}, \dfrac{3\pi}{4}$, and π. The y-coordinates of the five quarter points of $y = \sin(2x)$ will be the same as the y-coordinates of the five quarter points of $y = \sin x$.

Quarter points for $y = \sin x$: $(0,0), \quad \left(\dfrac{\pi}{2}, 1\right), \quad (\pi, 0), \quad \left(\dfrac{3\pi}{2}, -1\right), \quad (2\pi, 0)$

Quarter points for $y = \sin(2x)$: $(0,0), \quad \left(\dfrac{\pi}{4}, 1\right), \quad \left(\dfrac{\pi}{2}, 0\right), \quad \left(\dfrac{3\pi}{4}, -1\right), \quad (\pi, 0)$

One cycle of the graphs of $y = \sin x$ and $y = \sin(2x)$ is sketched in Figure 8. Watch this interactive video to see how to sketch the graph of $y = \sin(2x)$.

Figure 8 The graphs of $y = \sin x$ and $y = \sin(2x)$

b. The period of $y = \cos\left(\dfrac{1}{2}x\right)$ is $P = \dfrac{2\pi}{B} = \dfrac{2\pi}{\dfrac{1}{2}} = 2\pi \cdot \dfrac{2}{1} = 4\pi$. Next, we divide the

interval $[0, 4\pi]$ into four equal subintervals of length $4\pi \div 4 = \pi$ by starting with 0 and adding π to the x-coordinate of each successive quarter point. The x-coordinates of the five quarter points are $0, \pi, 2\pi, 3\pi$, and 4π. The y-coordinates of the five quarter points of $y = \cos\left(\dfrac{1}{2}x\right)$ will be the same as the y-coordinates of the five quarter points of $y = \cos x$.

Quarter points for $y = \cos x$: $(0, 1)$, $\left(\dfrac{\pi}{2}, 0\right)$, $(\pi, -1)$, $\left(\dfrac{3\pi}{2}, 0\right)$, $(2\pi, 1)$

Quarter points for $y = \cos\left(\dfrac{1}{2}x\right)$: $(0, 1)$, $(\pi, 0)$, $(2\pi, -1)$, $(3\pi, 0)$, $(4\pi, 1)$

My interactive video summary

▶ Watch this interactive video to see how to sketch the graph of

$y = \cos\left(\dfrac{1}{2}x\right)$ as seen below in Figures 9 and 10.

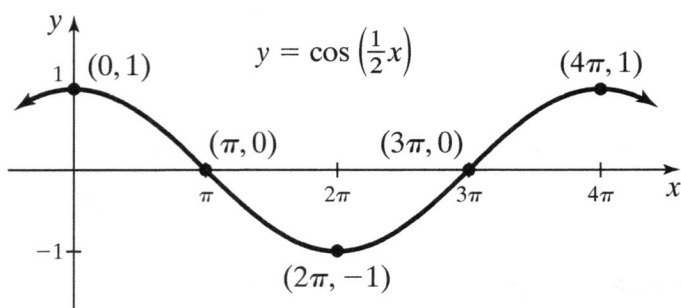

Figure 9 This illustrates one cycle of the graph of $y = \cos\left(\dfrac{1}{2}x\right)$.

Using Technology

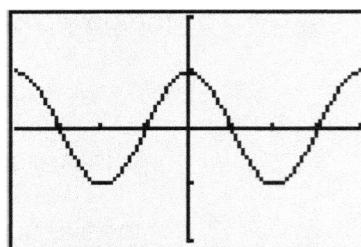

Figure 10 This illustrates two cycles of the graph of $y = \cos\left(\dfrac{1}{2}x\right)$ using a graphing utility.

c. The period of $y = \sin(\pi x)$ is $P = \dfrac{2\pi}{B} = \dfrac{2\pi}{\pi} = 2$. Next, we divide the interval $[0, 2]$

into four equal subintervals of length $\dfrac{2}{4} = \dfrac{1}{2}$ by starting with 0 and adding $\dfrac{1}{2}$ to

the x-coordinate of each successive quarter point. Thus, the x-coordinates of the five quarter points are $0, \frac{1}{2}, 1, \frac{3}{2},$ and 2. The y-coordinates of the five quarter points of $y = \sin(\pi x)$ will be the same as the y-coordinates of the five quarter points of $y = \sin x$. The graph of $y = \sin(\pi x)$ can be seen in Figures 11 and 12.

Quarter points for $y = \sin x$: $(0,0)$, $\left(\frac{\pi}{2}, 1\right)$, $(\pi, 0)$, $\left(\frac{3\pi}{2}, -1\right)$, $(2\pi, 0)$

Quarter points for $y = \sin(\pi x)$: $(0,0)$, $\left(\frac{1}{2}, 1\right)$, $(1, 0)$, $\left(\frac{3}{2}, -1\right)$, $(2, 0)$

My interactive video summary

⊙ Watch this interactive video see the complete solution to this example.

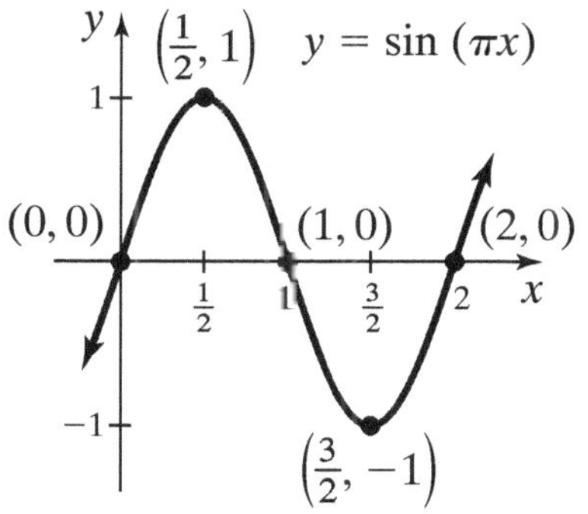

Using Technology

Figure 11 This illustrates one cycle of the graph of $y = \sin(\pi x)$.

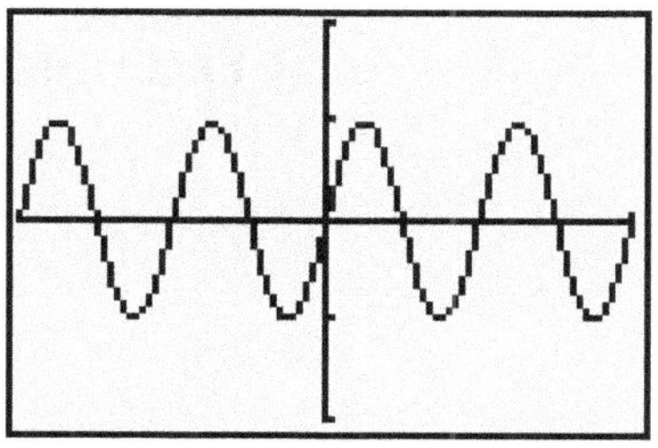

Figure 12 This illustrates several cycles of the graph of $y = \sin(\pi x)$ using a graphing utility.

It is noteworthy that we could have used transformation techniques to sketch the three graphs from Example 6. In Example 6a we sketched a function of the form $y = \sin(Bx)$, where $B = 2$. Because $B > 1$, the graph of a function $y = f(Bx)$ can be obtained by **horizontally compressing** the graph of the function $y = f(x)$ by dividing each x-coordinate of $y = f(x)$ by B.

Similarly, if $0 < B < 1$, the graph of $y = f(Bx)$ can be obtained by **horizontally stretching** the graph of $y = f(x)$ by a factor of B.

You may want to review the horizontal stretch transformation or horizontal compression transformation. These topics are typically first introduced in an algebra course. Click on the animations below to review this concept. (Note that in the animations below, a lowercase "a" is used to describe the constant. In this section, we are going to use a capital "B" to describe this constant.)

My animation summary

▢ To review the concept of a horizontal compression, watch this animation.

▢ To review the concept of a horizontal stretch, watch this animation.

To see how to sketch each of the functions from Example 6 using transformations, click on the Show Graph link located under the following three functions.

$$y = \sin(2x) \qquad y = \cos\left(\frac{1}{2}x\right) \qquad y = \sin(\pi x)$$

Show Graph   Show Graph   Show Graph

Now that we know how to sketch the graphs of $y = \sin(Bx)$ and $y = \cos(Bx)$, where $B > 0$, we can use the even and odd properties of cosine and sine to sketch the graphs of $y = \sin(-Bx)$ and $y = \cos(-Bx)$.

Recall that $y = \sin x$ is an odd function. This means that $\sin (-x) = -\sin x$ for all x in the domain of $y = \sin x$. It follows that

$$\sin (-Bx) = -\sin (Bx).$$

Similarly, $y = \cos x$ is an even function. This means that $\cos (-x) = \cos x$ for all x in the domain of $y = \cos x$. It follows that

$$\cos (-Bx) = \cos (Bx).$$

Therefore, when $B > 0$, we can sketch the graph of $y = \sin (-Bx)$ by simply sketching the graph of $y = -\sin (Bx)$. This graph can be obtained by reflecting the graph of $y = \sin (Bx)$ about the x-axis. The graph of $y = \cos (-Bx)$ is the exact same graph as the graph of $y = \cos (Bx)$. The period of $y = \sin (-Bx)$ and $y = \cos (-Bx)$ is the same as the period of $y = \sin (Bx)$ and $y = \cos (Bx)$, which is $P = \dfrac{2\pi}{B}$.

My interactive video summary

⊚ **Example 7** Sketching graphs of the form $y = \sin (-Bx)$ and $y = \cos (-Bx)$

Determine the period and sketch the graph of each function.

a. $y = \sin (-2x)$ **b.** $y = \cos \left(-\dfrac{1}{2}x\right)$ **c.** $y = \sin (-\pi x)$

Solution

a. We use the odd property of the sine function to rewrite $y = \sin (-2x)$ as $y = -\sin (2x)$. We have already sketched the graph of $y = \sin (2x)$ in Example 6a. See Figure 8. We did this by finding the period of $y = \sin (2x)$ to be $P = \dfrac{2\pi}{B} = \dfrac{2\pi}{2} = \pi$ and then dividing the interval $[0, \pi]$ into four equal subintervals of length $\pi \div 4 = \dfrac{\pi}{4}$. This produced x-coordinates of the five quarter points of $y = \sin (2x)$, namely $0, \dfrac{\pi}{4}, \dfrac{\pi}{2}, \dfrac{3\pi}{4}$, and π.

To sketch the graph of $y = -\sin (2x)$ using the graph of $y = \sin (2x)$, we can multiply the y-coordinates of the quarter points of $y = \sin (2x)$ by -1 to obtain the quarter points for $y = -\sin (2x)$ or we can simply reflect the graph of $y = \sin (2x)$ about the x-axis. See Figure 13.

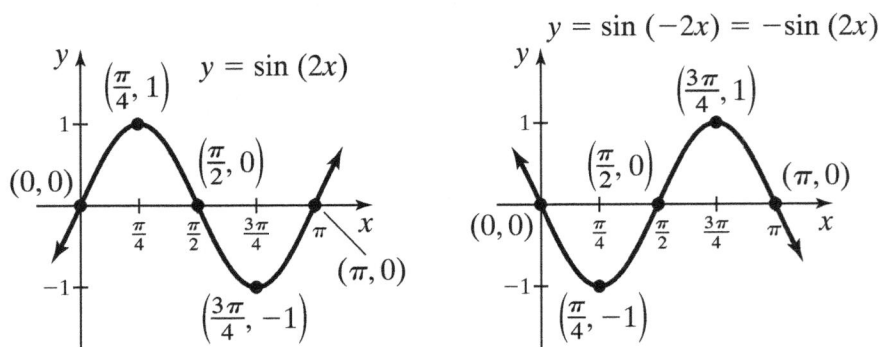

Figure 13 The graphs of $y = \sin (2x)$ and $y = \sin (-2x)$

b. Because the cosine function is even, the graph of $y = \cos\left(-\frac{1}{2}x\right)$ is the same as the graph of $y = \cos\left(\frac{1}{2}x\right)$, which we have already sketched in Example 6. See Figure 9.

We sketch the graph of $y = \cos\left(-\frac{1}{2}x\right) = \cos\left(\frac{1}{2}x\right)$ again in Figure 14.

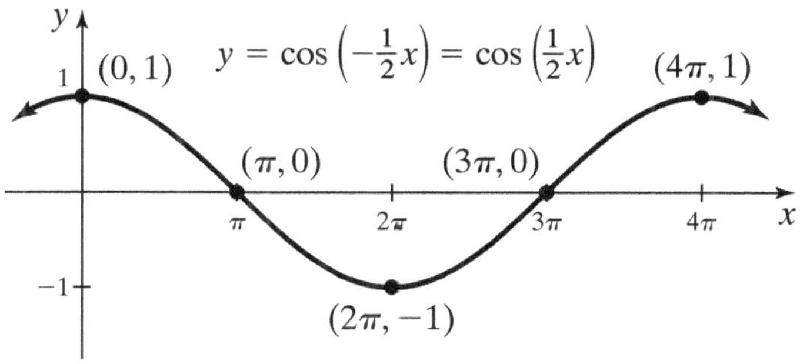

Figure 14 One cycle of the graph of $y = \cos\left(-\frac{1}{2}x\right) = \cos\left(\frac{1}{2}x\right)$

c. Try sketching the graph of $y = \sin(-\pi x)$ on your own. Watch this interactive video when you have finished to see if you are correct.

We could have used transformation techniques to sketch the graphs of the two functions from Example 7.

My animation summary

▢ Recall that the graph of $y = f(-x)$ can be obtained by reflecting the graph of $y = f(x)$ about the y-axis. To review this transformation, watch this short animation.

In Example 7 we could have sketched the graphs of $y = \sin(-2x)$, $y = \cos\left(-\frac{1}{2}x\right)$, and $y = \sin(-\pi x)$ by first using a horizontal compression (or stretch) followed by a reflection about the y-axis to sketch both of these graphs. To see how to sketch these two functions using transformations, click on the Show Graph links below.

$y = \sin(-2x)$	$y = \cos\left(-\frac{1}{2}x\right)$	$y = \sin(-\pi x)$
Show Graph	Show Graph	Show Graph

You Try It Work through this You Try It problem.

Work Exercises 27–36 in this textbook or in the MyMathLab Study Plan.

OBJECTIVE 5 SKETCHING GRAPHS OF THE FORM $y = A \sin(Bx)$ AND $y = A \cos(Bx)$

We are now ready to combine the graphing techniques learned in the previous two objectives to sketch the graphs of functions of the form $y = A \sin(Bx)$ and $y = A \cos(Bx)$. Below is a six-step process that will help you sketch functions of this form.

Steps for Sketching Functions of the Form $y = A \sin (Bx)$
and $y = A \cos (Bx)$

Step 1. If $B < 0$, use the even and odd properties of the sine and cosine function to rewrite the function in an equivalent form such that $B > 0$.

We now use this new form to determine A and B.

Step 2. Determine the amplitude and range. The amplitude is $|A|$. The range is $[-|A|, |A|]$.

Step 3. Determine the period. The period is $P = \dfrac{2\pi}{B}$.

Step 4. An interval for one complete cycle is $\left[0, \dfrac{2\pi}{B}\right]$. Subdivide this interval into four equal subintervals of length $\dfrac{2\pi}{B} \div 4$ by starting with 0 and adding $\left(\dfrac{2\pi}{B} \div 4\right)$ to the x-coordinate of each successive quarter point.

Step 5. Multiply the y-coordinates of the quarter points of $y = \sin x$ or $y = \cos x$ by A to determine the y-coordinates of the corresponding quarter points for the new graph.

Step 6. Connect the quarter points to obtain one complete cycle.

My interactive video summary

⊙ **Example 8** Sketching graphs of the form $y = A \sin (Bx)$ and $y = A \cos (Bx)$

Use the six-step process outlined in this section to sketch each graph.

a. $y = 3 \sin (4x)$ **b.** $y = -2 \cos \left(\dfrac{1}{3}x\right)$ **c.** $y = -6 \sin \left(-\dfrac{\pi x}{2}\right)$

Solution

a. Step 1. For the function $y = 3 \sin (4x)$, we see that $B = 4 > 0$. Therefore, we do not need to use the odd property of sine to rewrite the function, and we continue to step 2.

Step 2. The amplitude of $y = 3 \sin (4x)$ is $|A| = |3| = 3$. Therefore, the range is $[-3, 3]$. The maximum value is 3 and the minimum value is -3.

Step 3. The period is $P = \dfrac{2\pi}{B} = \dfrac{2\pi}{4} = \dfrac{\pi}{2}$.

Step 4. An interval for one complete cycle of the graph is $\left[0, \dfrac{\pi}{2}\right]$. We divide this interval into four equal subintervals of length $\dfrac{\pi}{2} \div 4 = \dfrac{\pi}{2} \cdot \dfrac{1}{4} = \dfrac{\pi}{8}$ by starting with 0 and adding $\dfrac{\pi}{8}$ to the x-coordinate of each successive quarter point. Thus, the x-coordinates of the five quarter points are $0, \dfrac{\pi}{8}, \dfrac{\pi}{4}, \dfrac{3\pi}{8}$, and $\dfrac{\pi}{2}$.

Step 5. Multiply the y-coordinate of each of the five quarter points of $y = \sin x$ by 3 to obtain the y-coordinates of the corresponding quarter points of $y = 3 \sin(4x)$. The y-coordinates of the five quarter points are $0, 3, 0, -3$, and 0.

Step 6. From the information gathered from the previous steps, we get the following.

Quarter points for $y = \sin x$: $(0, 0), \quad \left(\dfrac{\pi}{2}, 1\right), \quad (\pi, 0), \quad \left(\dfrac{3\pi}{2}, -1\right), \quad (2\pi, 0)$

Quarter points for $y = 3 \sin(4x)$: $(0, 0), \quad \left(\dfrac{\pi}{8}, 3\right), \quad \left(\dfrac{\pi}{4}, 0\right), \quad \left(\dfrac{3\pi}{8}, -3\right), \quad \left(\dfrac{\pi}{2}, 0\right)$

We connect these quarter points with a smooth curve to complete the sketch of $y = 3 \sin(4x)$. See Figures 15 and 16.

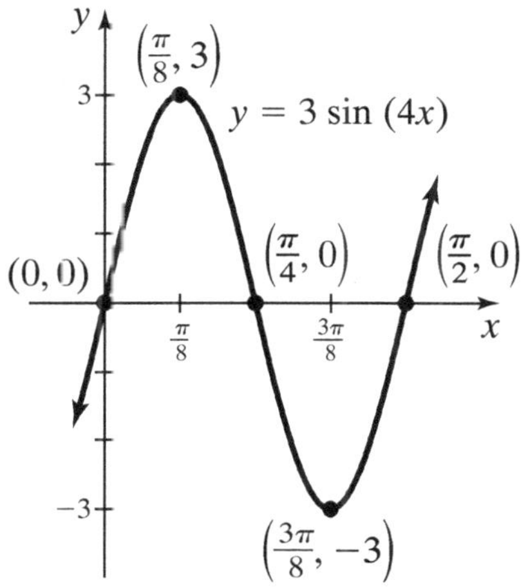

Using Technology

Figure 15 This illustrates one cycle of the graph of $y = 3 \sin(4x)$.

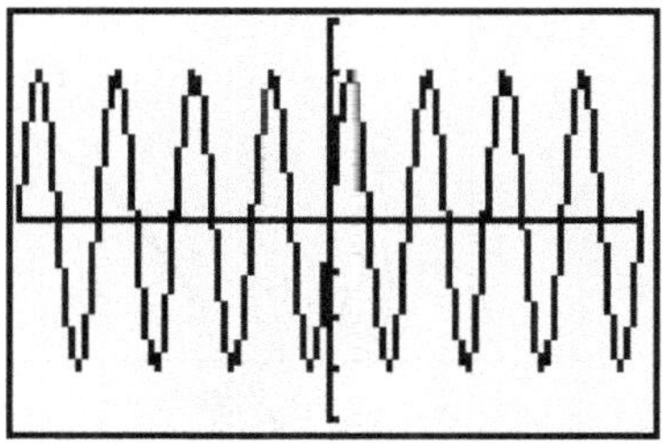

Figure 16 This illustrates several cycles of the graph of $y = 3 \sin(4x)$ using a graphing utility.

My interactive video summary

You may want to work through this interactive video to see how to sketch the graph of $y = 3 \sin(4x)$.

b. Step 1. For the function $y = -2 \cos\left(\dfrac{1}{3}x\right)$, we see that $B = \dfrac{1}{3} > 0$. Therefore, we do not need to use the even property of cosine to rewrite the function, and we continue to step 2.

Step 2. The amplitude of $y = -2 \cos\left(\dfrac{1}{3}x\right)$ is $|A| = |-2| = 2$. Therefore, the range is $[-2, 2]$. The maximum value is 2 and the minimum value is -2.

Step 3. The period is $P = \dfrac{2\pi}{B} = \dfrac{2\pi}{\dfrac{1}{3}} = 2\pi \cdot 3 = 6\pi$.

Step 4. An interval for one complete cycle of the graph is $[0, 6\pi]$. We divide this interval into four equal subintervals of length $6\pi \div 4 = \dfrac{6\pi}{4} = \dfrac{3\pi}{2}$ by starting with 0 and adding $\dfrac{3\pi}{2}$ to the x-coordinate of each successive quarter point. Thus, the x-coordinates of the five quarter points are 0, $\dfrac{3\pi}{2}$, 3π, $\dfrac{9\pi}{2}$, and 6π.

Step 5. Multiply the y-coordinate of each of the five quarter points of $y = \cos x$ by -2 to obtain the y-coordinates of the corresponding quarter points of $y = -2\cos\left(\dfrac{1}{3}x\right)$. The y-coordinates of the five quarter points are $-2,\ 0,\ 2,\ 0,$ and -2.

Step 6. From the information gathered from the previous steps, we get the following.

Quarter points for $y = \cos x$: $\quad (0, 1),\quad \left(\dfrac{\pi}{2}, 0\right),\quad (\pi, -1),\quad \left(\dfrac{3\pi}{2}, 0\right),\quad (2\pi, 1)$

Quarter points for $y = -2\cos\left(\dfrac{1}{3}x\right)$: $\quad (0, -2),\quad \left(\dfrac{3\pi}{2}, 0\right),\quad (3\pi, 2),\quad \left(\dfrac{9\pi}{2}, 0\right),\quad (6\pi, -2)$

We connect these quarter points with a smooth curve to complete the sketch of $y = -2\cos\left(\dfrac{1}{3}x\right)$. See Figures 17 and 18.

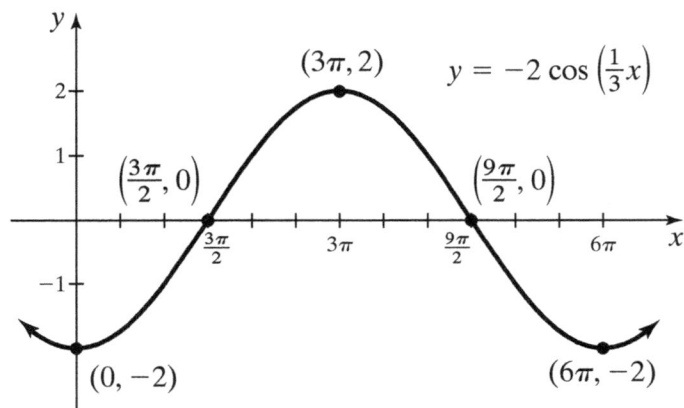

Figure 17 This illustrates one cycle of the graph of $y = -2\cos\left(\dfrac{1}{3}x\right)$.

Using Technology

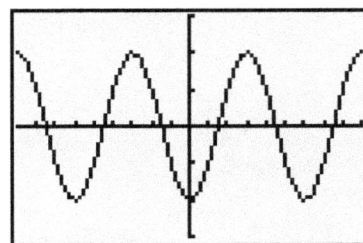

Figure 18 This illustrates several cycles of the graph of $y = -2\cos\left(\dfrac{1}{3}x\right)$ using a graphing utility.

My interactive video summary

⊙ You may want to work through this interactive video to see how to sketch the graph of $y = -2\cos\left(\dfrac{1}{3}x\right)$.

c. Step 1. For the function $y = -6\sin\left(-\dfrac{\pi x}{2}\right)$, we see that $B = -\dfrac{\pi}{2} < 0$. Therefore, we use the odd property of sine to rewrite the function using a positive value of B.

$y = -6\sin\left(-\dfrac{\pi x}{2}\right)$ \quad Write the original function.

$y = 6\sin\left(\dfrac{\pi x}{2}\right)$ \quad Use the odd property of $y = \sin x$: $\sin(-x) = -\sin x$.

We now continue through the steps using the function $y = 6 \sin\left(\dfrac{\pi x}{2}\right)$ with $A = 6$ and $B = \dfrac{\pi}{2}$.

Step 2. The amplitude of $y = 6 \sin\left(\dfrac{\pi x}{2}\right)$ is $|A| = |6| = 6$. Therefore, the range is $[-6, 6]$. The maximum value is 6 and the minimum value is -6.

Step 3. The period of $y = 6 \sin\left(\dfrac{\pi x}{2}\right)$ is $P = \dfrac{2\pi}{B} = 2\pi \div \dfrac{\pi}{2} = 2\pi \cdot \dfrac{2}{\pi} = 4$.

Step 4. An interval for one complete cycle of the graph is $[0, 4]$. We divide this interval into four equal subintervals of length $\dfrac{4}{4} = 1$ by starting with 0 and adding 1 to the x-coordinate of each successive quarter point. Thus, the x-coordinates of the five quarter points are 0, 1, 2, 3, and 4.

Step 5. Multiply the y-coordinates of the five quarter points of $y = \sin x$ by 6 to obtain the y-coordinates of the corresponding quarter points of $y = 6 \sin\left(\dfrac{\pi x}{2}\right)$. The y-coordinates of the five quarter points are 0, 6, 0, -6, and 0.

Step 6. From the information gathered from the previous steps, we get the following.

Quarter points for $y = \sin x$: $(0, 0)$, $\left(\dfrac{\pi}{2}, 1\right)$, $(\pi, 0)$, $\left(\dfrac{3\pi}{2}, -1\right)$, $(2\pi, 0)$

Quarter points for $y = 6 \sin\left(\dfrac{\pi x}{2}\right)$: $(0, 0)$, $(1, 6)$, $(2, 0)$, $(3, -6)$, $(4, 0)$

We connect these quarter points with a smooth curve to complete the sketch of $y = 6 \sin\left(\dfrac{\pi x}{2}\right)$. See Figures 19 and 20.

$$y = -6 \sin\left(-\dfrac{\pi x}{2}\right) \text{ or } y = 6 \sin\left(\dfrac{\pi x}{2}\right)$$

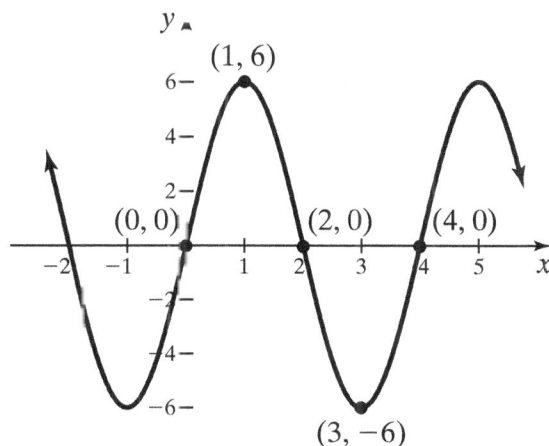

Figure 19 This illustrates the graph of
$y = -6 \sin\left(-\dfrac{\pi x}{2}\right)$ or $y = 6 \sin\left(\dfrac{\pi x}{2}\right)$.

Using Technology

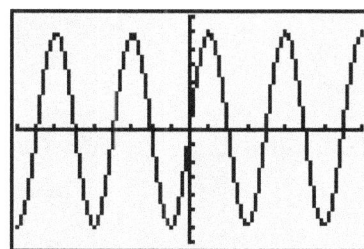

Figure 20 This illustrates several cycles of the graph of
$y = -6 \sin\left(-\dfrac{\pi x}{2}\right)$
or $y = 6 \sin\left(\dfrac{\pi x}{2}\right)$ using a graphing utility.

My interactive video summary

▶ You may want to work through this interactive video to see how to sketch the graph of $y = -6 \sin\left(-\dfrac{\pi x}{2}\right)$.

You Try It Work through this You Try It problem.

Work Exercises 37–52 in this textbook or in the MyMathLab Study Plan.

OBJECTIVE 6 DETERMINE THE EQUATION OF A FUNCTION OF THE FORM $y = A \sin (Bx)$ OR $y = A \cos (Bx)$ GIVEN ITS GRAPH

We now know how to sketch the graphs of $y = A \sin (Bx)$ and $y = A \cos (Bx)$. Suppose that we are given a graph whose function is given by $y = A \sin (Bx)$ or $y = A \cos (Bx)$ for $B > 0$. (We will assume that the value of B is positive. Otherwise, there could be more than one correct answer because of the even nature of the cosine function.) To determine the proper function we must determine three things:

1. Is the given graph a representation of a function of the form $y = A \sin (Bx)$ or $y = A \cos (Bx)$?

2. What is the value of B?

3. What is the value of A?

Therefore, we can establish the following three steps for determining the proper function.

Steps for Determining the Equation of a Function of the Form
$y = A \sin (Bx)$ or $y = A \cos (Bx)$ Given the Graph

Step 1. Determine whether the given graph is a representation of a function of the form $y = A \sin (Bx)$ or $y = A \cos (Bx)$. Choose $y = A \sin (Bx)$ if the graph passes through the origin or choose $y = A \cos (Bx)$ if the graph does not pass through the origin.*

Step 2. Determine the period $P = \dfrac{2\pi}{B}$ then use the period to determine the value of $B > 0$.

Step 3. Use the information from the previous two steps then use one of the given points on the graph to solve for A.

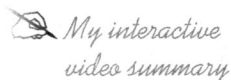

My interactive video summary

▶ **Example 9** Determining the Equation of a Function Given the Graph that it Represents

One cycle of the graphs of three trigonometric functions of the from $y = A \sin (Bx)$ or $y = A \cos (Bx)$ for $B > 0$ are given below. Determine the equation of the function represented by each graph.

---

*As long as we restrict the choices of the equations to $y = A \sin (Bx)$ and $y = A \cos (Bx)$ where $B > 0$, this will be true. Without these restrictions, there would be more than one correct equation represented by the given graph.

a.

$\left(\frac{3\pi}{2}, 2\right)$

$\left(\frac{\pi}{2}, -2\right)$

b.

$\left(\frac{3\pi}{2}, 4\right)$

$\left(\frac{9\pi}{2}, -4\right)$

c.

$\left(\frac{\pi}{5}, \frac{4}{3}\right)$

$\left(0, -\frac{4}{3}\right)$

$\left(\frac{2\pi}{5}, -\frac{4}{3}\right)$

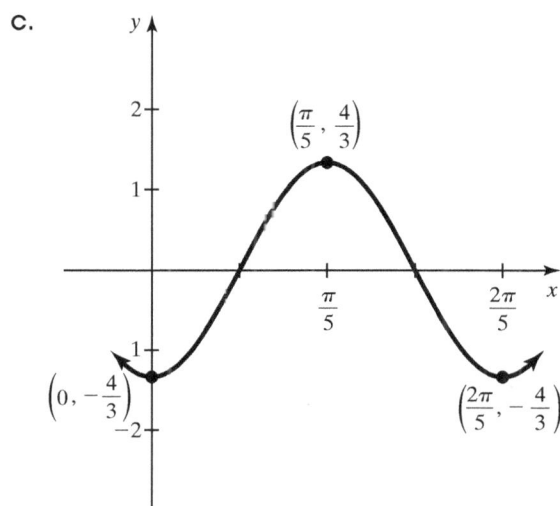

Solution Work through the interactive video to verify that the function whose graph is given in part a) is $y = -2\sin x$.

Work through the interactive video to verify that the function whose graph is given in part b) is $y = 4\sin\left(\frac{1}{3}x\right)$.

c. **Step 1.** The function whose graph is given in part (c) does **not** pass through the origin. Therefore, a possible representation of this graph is a function of the form $y = A\cos(Bx)$.

Step 2. The graph completes one cycle on the interval $\left[0, \frac{2\pi}{5}\right]$. Thus, the period is $P = \frac{2\pi}{5} = \frac{2\pi}{B}$. Therefore, $B = 5$.

Step 3. The function is of the form $y = A\cos(5x)$ and the graph passes through the point $\left(0, -\frac{4}{3}\right)$.

Therefore when $x = 0$, the value of y is $-\dfrac{4}{3}$. We can use this information to determine the value of A.

$$y = A \cos (5x)$$ Start with the equation $y = A \cos (5x)$.

$$-\frac{4}{3} = A \cos \left(5(0)\right)$$ Substitute $x = 0$ and $y = -\dfrac{4}{3}$.

$$-\frac{4}{3} = A \cos 0$$ Simplify.

$$-\frac{4}{3} = A$$ Evaluate $\cos 0 = 1$.

Therefore, $A = -\dfrac{4}{3}$ and the function whose graph is represented in part (c) is

$$y = -\frac{4}{3} \cos (5x).$$

You Try It Work through this You Try It problem.

Work Exercises 53–64 in this textbook or in the MyMathLab **Study Plan**.

2.1 Exercises

Skill Check Exercises

In Exercises SCE-1 and SCE-2, perform the indicated operations and simplify.

SCE-1. $\dfrac{2\pi}{5} \div 4$ SCE-2. $\dfrac{2\pi}{\dfrac{\pi}{3}}$

1. Sketch the graph of $y = \sin x$, and identify the properties that apply.

   The function is an even function.

   The function is an odd function.

   The function is increasing on the interval $\left(-\pi, -\dfrac{\pi}{2}\right)$.

   The function is decreasing on the interval $\left(\pi, \dfrac{3\pi}{2}\right)$.

   The domain is $(-\infty, \infty)$.

   The range is $(-1, 1)$.

   The range is $[-1, 1]$.

   The y-intercept is 0.

   The y-intercept is 1.

The zeros are of the form $(2n + 1) \cdot \dfrac{\pi}{2}$, where n is any integer.

The zeros are of the form $n\pi$, where n is any integer.

The function obtains a relative maximum at $x = \dfrac{\pi}{2} + 2\pi n$, where n is an integer.

The function obtains a relative maximum at $x = 2\pi n$, where n is an integer.

The function obtains a relative minimum at $x = \dfrac{3\pi}{2} + 2\pi n$, where n is an integer.

The function obtains a relative minimum at $x = \pi n$, where n is an integer.

2. Using the graph of $y = \sin x$, list all values of x on the interval $\left[-2\pi, \dfrac{5\pi}{2} \right]$ that satisfy the ordered pair $\left(x, \dfrac{1}{2} \right)$.

3. Using the graph of $y = \sin x$, list all values of x on the interval $\left[-\dfrac{11\pi}{4}, \dfrac{\pi}{2} \right]$ that satisfy the ordered pair $\left(x, \dfrac{1}{\sqrt{2}} \right)$.

4. Using the graph of $y = \sin x$, list all values of x on the interval $\left[-\dfrac{7\pi}{2}, \dfrac{4\pi}{3} \right]$ that satisfy the ordered pair $\left(x, -\dfrac{\sqrt{3}}{2} \right)$.

5. Use the periodic property of $y = \sin x$ to determine which of the following expressions is equivalent to $\sin\left(\dfrac{7\pi}{3} \right)$.

i. $\sin\left(-\dfrac{\pi}{3} \right)$ ii. $\sin\left(\dfrac{4\pi}{3} \right)$ iii. $\sin\left(\dfrac{\pi}{3} \right)$ iv. $\sin\left(\dfrac{5\pi}{3} \right)$

6. Use the periodic property of $y = \sin x$ to determine which of the following expressions is equivalent to $\sin\left(-\dfrac{5\pi}{4} \right)$.

i. $\sin\left(-\dfrac{13\pi}{4} \right)$ ii. $\sin\left(-\dfrac{\pi}{4} \right)$ iii. $\sin\left(\dfrac{5\pi}{4} \right)$ iv. $\sin\left(-\dfrac{9\pi}{4} \right)$

7. Use the periodic property of $y = \sin x$ to determine which of the following expressions is equivalent to $\sin\left(-\dfrac{25\pi}{6} \right)$.

i. $\sin\left(\dfrac{\pi}{6} \right)$ ii. $\sin\left(\dfrac{23\pi}{6} \right)$ iii. $\sin\left(\dfrac{29\pi}{6} \right)$ iv. $\sin\left(-\dfrac{31\pi}{6} \right)$

8. Use the fact that $y = \sin x$ is an odd function to determine which of the following expressions is equivalent to $-\sin\left(\dfrac{9\pi}{16}\right)$.

i. $\sin\left(-\dfrac{9\pi}{16}\right)$ **ii.** $\sin\left(\dfrac{9\pi}{16}\right)$ **iii.** $-\sin\left(-\dfrac{9\pi}{16}\right)$ **iv.** $-\sin\left(-\dfrac{\pi}{16}\right)$

9. Sketch the graph of $y = \cos x$, and identify the properties that apply.

The function is an even function.

The function is an odd function.

The function is increasing on the interval $\left(-\pi, -\dfrac{\pi}{2}\right)$.

The function is decreasing on the interval $\left(\pi, \dfrac{3\pi}{2}\right)$.

The domain is $(-\infty, \infty)$.

The range is $(-1, 1)$.

The range is $[-1, 1]$.

The y-intercept is 0.

The y-intercept is 1.

The zeros are of the form $(2n + 1) \cdot \dfrac{\pi}{2}$, where n is any integer.

The zeros are of the form $n\pi$, where n is any integer.

The function obtains a relative maximum at $x = \dfrac{\pi}{2} + 2\pi n$, where n is an integer.

The function obtains a relative maximum at $x = 2\pi n$, where n is an integer.

The function obtains a relative minimum at $x = \dfrac{3\pi}{2} + 2\pi n$, where n is an integer.

The function obtains a relative minimum at $x = \pi n$, where n is an integer.

10. Using the graph of $y = \cos x$, list all values of x on the interval $\left[-3\pi, \dfrac{5\pi}{4}\right]$ that satisfy the ordered pair $(x, 0)$.

11. Using the graph of $y = \cos x$, list all values of x on the interval $\left[-\dfrac{11\pi}{4}, \dfrac{\pi}{2}\right]$ that satisfy the ordered pair $\left(x, \dfrac{1}{\sqrt{2}}\right)$.

12. Using the graph of $y = \cos x$, list all values of x on the interval $\left[-\dfrac{7\pi}{2}, \dfrac{4\pi}{3}\right]$ that satisfy the ordered pair $\left(x, -\dfrac{\sqrt{3}}{2}\right)$.

13. Use the periodic property of $y = \cos x$ to determine which of the following expressions is equivalent to $\cos\left(\dfrac{25\pi}{6}\right)$.

i. $\cos\left(\dfrac{5\pi}{6}\right)$ ii. $\cos\left(\dfrac{7\pi}{6}\right)$ iii. $\cos\left(\dfrac{13\pi}{6}\right)$ iv. $\cos\left(-\dfrac{7\pi}{6}\right)$

14. Use the periodic property of $y = \cos x$ to determine which of the following expressions is equivalent to $\cos\left(-\dfrac{4\pi}{3}\right)$.

i. $\cos\left(\dfrac{2\pi}{3}\right)$ ii. $\cos\left(-\dfrac{\pi}{3}\right)$ iii. $\cos\left(\dfrac{7\pi}{3}\right)$ iv. $\cos\left(\dfrac{5\pi}{3}\right)$

15. Use the periodic property of $y = \cos x$ to determine which of the following expressions is equivalent to $\cos\left(-\dfrac{17\pi}{4}\right)$.

i. $\cos\left(\dfrac{5\pi}{4}\right)$ ii. $\cos\left(-\dfrac{9\pi}{4}\right)$ iii. $\cos\left(\dfrac{3\pi}{4}\right)$ iv. $\cos\left(-\dfrac{11\pi}{4}\right)$

16. Use the fact that $y = \cos x$ is an even function to determine which of the following expressions is equivalent to $-\cos\left(\dfrac{7\pi}{13}\right)$.

i. $-\cos\left(-\dfrac{7\pi}{13}\right)$ ii. $\cos\left(-\dfrac{7\pi}{13}\right)$ iii. $\cos\left(\dfrac{7\pi}{13}\right)$ iv. $\cos\left(-\dfrac{\pi}{13}\right)$

In Exercises 17–26, determine the amplitude and range of each function and sketch its graph.

17. $y = 3 \sin x$ **18.** $y = 4 \cos x$ **19.** $y = \dfrac{1}{3}\sin x$ **20.** $y = \dfrac{1}{4}\cos x$

21. $y = \dfrac{3}{2}\sin x$ **22.** $y = -2 \sin x$ **23.** $y = -5 \cos x$ **24.** $y = -\dfrac{3}{4}\sin x$

25. $y = -\dfrac{1}{2}\cos x$ **26.** $y = -\dfrac{5}{4}\cos x$

In Exercises 27–36, determine the period and sketch the graph of each function.

27. $y = \sin(3x)$ **28.** $y = \cos\left(\dfrac{1}{3}x\right)$ **29.** $y = \sin\left(\dfrac{3}{2}x\right)$ **30.** $y = \cos(3\pi x)$

31. $y = \sin\left(\dfrac{\pi}{3}x\right)$ **32.** $y = \cos(-4x)$ **33.** $y = \sin\left(-\dfrac{1}{4}x\right)$ **34.** $y = \cos\left(-\dfrac{5}{2}x\right)$

35. $y = \sin(-3\pi x)$ **36.** $y = \cos\left(-\dfrac{\pi}{2}x\right)$

In Exercises 37–52, determine the amplitude, range, and period of each function and sketch the graph.

SbS 37. $y = 2 \sin (3x)$

SbS 38. $y = -\dfrac{5}{3} \cos (4x)$

SbS 39. $y = \dfrac{3}{4} \sin (-2x)$

SbS 40. $y = -4 \cos (-3x)$

SbS 41. $y = 5 \cos \left(\dfrac{1}{2} x\right)$

SbS 42. $y = \dfrac{1}{2} \sin \left(-\dfrac{1}{4} x\right)$

SbS 43. $y = -\cos \left(-\dfrac{1}{5} x\right)$

SbS 44. $y = 6 \sin \left(\dfrac{3}{2} x\right)$

SbS 45. $y = 2 \sin \left(-\dfrac{4}{3} x\right)$

SbS 46. $y = -2 \cos \left(\dfrac{5}{4} x\right)$

SbS 47. $y = -\sin \left(-\dfrac{5}{2} x\right)$

SbS 48. $y = 3 \cos \left(\dfrac{\pi}{2} x\right)$

SbS 49. $y = -4 \sin \left(-\dfrac{\pi}{3} x\right)$

SbS 50. $y = \dfrac{5}{4} \cos (\pi x)$

SbS 51. $y = -\dfrac{1}{2} \sin (-\pi x)$

SbS 52. $y = -6 \cos \left(-\dfrac{2\pi}{3} x\right)$

In Exercises 53–62, one cycle of the graph a trigonometric function of the from $y = A \sin (Bx)$ or $y = A \cos (Bx)$ for $B > 0$ is given. Determine the equation of the function represented by each graph.

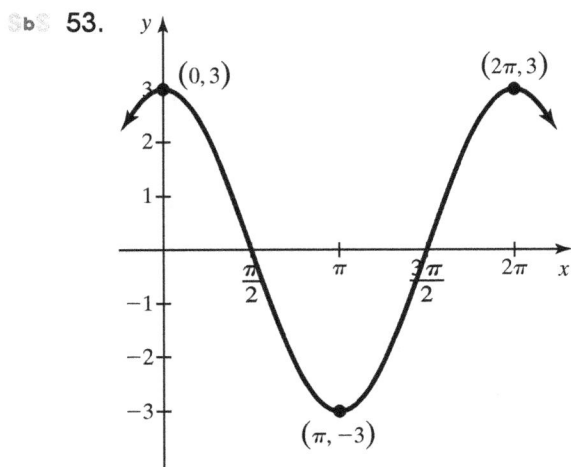

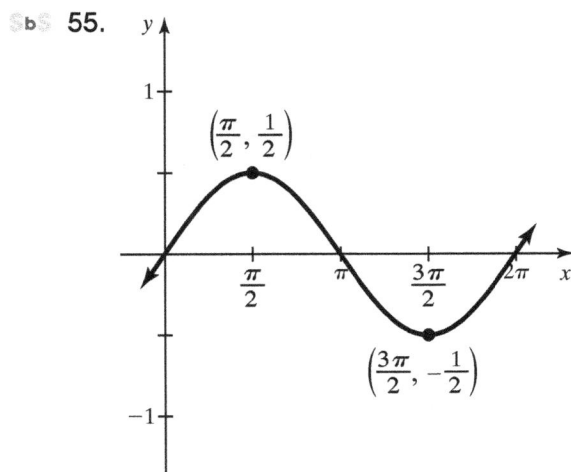

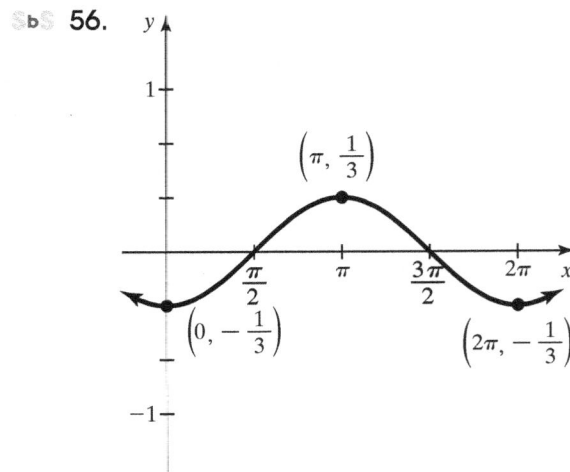

SbS 53.

SbS 54.

SbS 55.

SbS 56.

57.

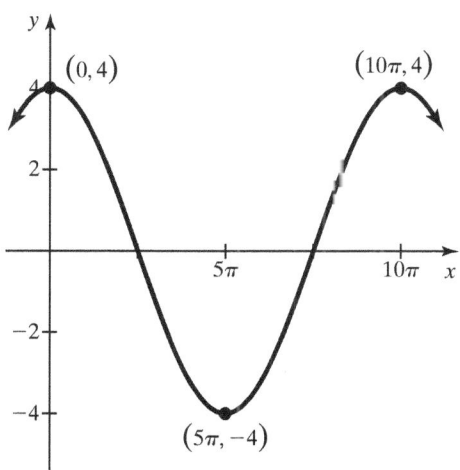

Points shown: $(0, 1)$, $(\pi, 1)$, $\left(\dfrac{\pi}{2}, -1\right)$

58.

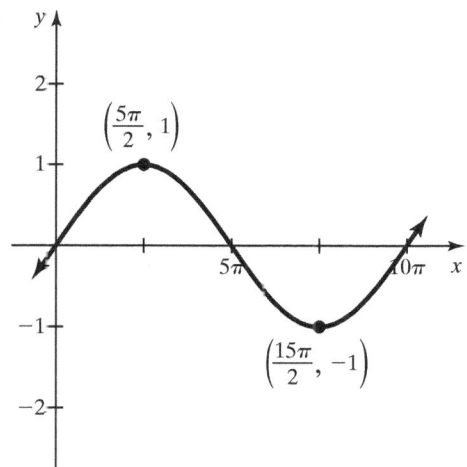

Points shown: $\left(\dfrac{5\pi}{2}, 1\right)$, $\left(\dfrac{15\pi}{2}, -1\right)$

59.

Points shown: $\left(\dfrac{\pi}{8}, 3\right)$, $\left(\dfrac{3\pi}{8}, -3\right)$

60.

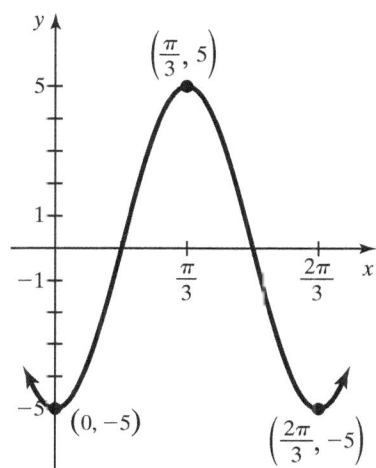

Points shown: $\left(\dfrac{\pi}{3}, 5\right)$, $(0, -5)$, $\left(\dfrac{2\pi}{3}, -5\right)$

61.

Points shown: $(0, 4)$, $(10\pi, 4)$, $(5\pi, -4)$

62.

Points shown: $(6\pi, 2)$, $(2\pi, -2)$

Brief Exercises

In Exercises 63–65, determine the amplitude of each function.

63. $y = -3 \sin x$ **64.** $y = \cos(-3x)$ **65.** $y = -2 \cos\left(\dfrac{5}{4}x\right)$

In Exercises 66–68, determine the range of each function.

66. $y = -3 \sin x$ **67.** $y = \cos(-3x)$ **68.** $y = -2 \cos\left(\dfrac{5}{4}x\right)$

In Exercises 69–71, determine the period of each function.

69. $y = -3 \sin x$ **70.** $y = \cos(-3x)$ **71.** $y = -2 \cos\left(\dfrac{5}{4}x\right)$

In Exercises 72–74, determine the coordinates for the five quarter points of each function that correspond to the five quarter points of $y = \sin x$ or $y = \cos x$.

72. $y = -3 \sin x$ **73.** $y = \cos(-3x)$ **74.** $y = -2 \cos\left(\dfrac{5}{4}x\right)$

In Exercises 75–90, sketch the graph.

75. $y = 2 \sin(3x)$ **76.** $y = -\dfrac{5}{3}\cos(4x)$ **77.** $y = \dfrac{3}{4}\sin(-2x)$ **78.** $y = -4 \cos(-3x)$

79. $y = 5 \cos\left(\dfrac{1}{2}x\right)$ **80.** $y = \dfrac{1}{2}\sin\left(-\dfrac{1}{4}x\right)$ **81.** $y = -\cos\left(-\dfrac{1}{5}x\right)$ **82.** $y = 6 \sin\left(\dfrac{3}{2}x\right)$

83. $y = 2 \sin\left(-\dfrac{4}{3}x\right)$ **84.** $y = -2 \cos\left(\dfrac{5}{4}x\right)$ **85.** $y = -\sin\left(-\dfrac{5}{2}x\right)$ **86.** $y = 3 \cos\left(\dfrac{\pi}{2}x\right)$

87. $y = -4 \sin\left(-\dfrac{\pi}{3}x\right)$ **88.** $y = \dfrac{5}{4}\cos(\pi x)$ **89.** $y = -\dfrac{1}{2}\sin(-\pi x)$ **90.** $y = -6 \cos\left(-\dfrac{2\pi}{3}x\right)$

In Exercises 91–100, one cycle of the graph a trigonometric function of the from $y = A \sin(Bx)$ or $y = A \cos(Bx)$ for $B > 0$ is given. Determine the equation of the function represented by each graph.

91.

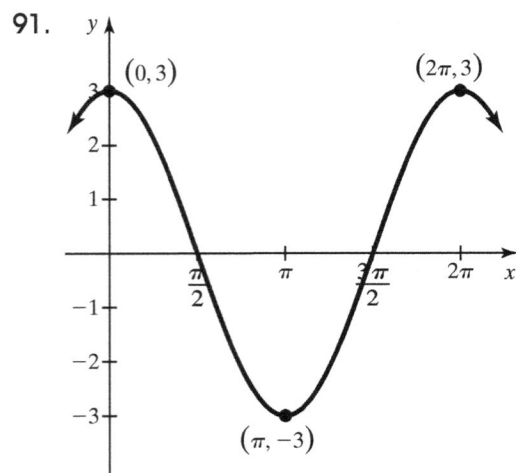

92.

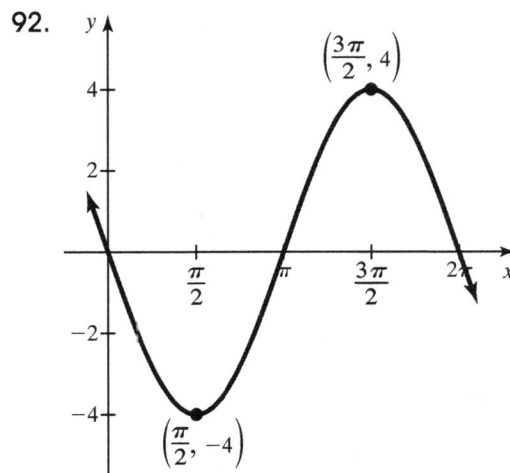

93.

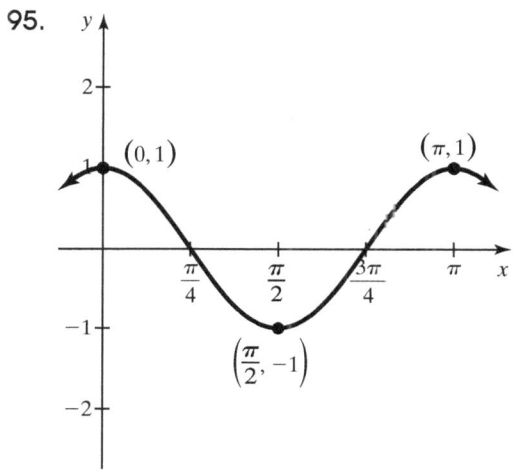

94.

95.

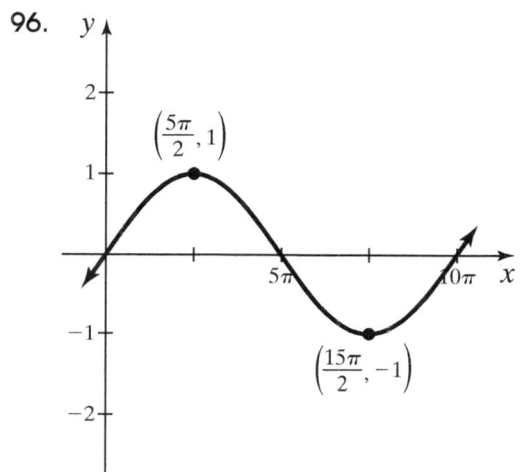

96.

97.

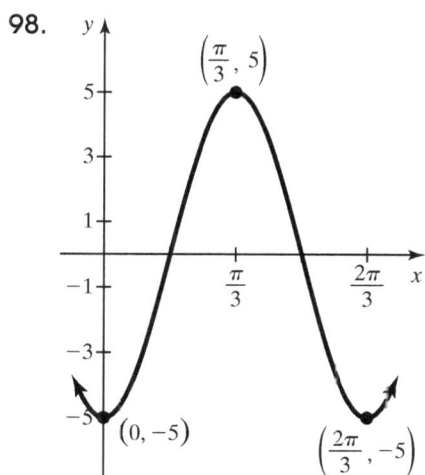

98.

99.

100.

2.2 More on the Graphs of Sine and Cosine

THINGS TO KNOW

Before working through this section, be sure that you are familiar with the following concepts:

VIDEO   ANIMATION   INTERACTIVE

You Try It   1. Sketching Graphs of the Form $y = A \sin x$ and $y = A \cos x$ (Section 2.1)

You Try It   2. Sketching Graphs of the Form $y = \sin (Bx)$ and $y = \cos (Bx)$ (Section 2.1)

You Try It   3. Sketching Graphs of the Form $y = A \sin (Bx)$ and $y = A \cos (Bx)$ (Section 2.1)

OBJECTIVES

1 Sketching Graphs of the Form $y = \sin (x - C)$ and $y = \cos (x - C)$

2 Sketching Graphs of the Form $y = A \sin (Bx - C)$ and $y = A \cos (Bx - C)$

3 Sketching Graphs of the Form $y = A \sin (Bx - C) + D$ and $y = A \cos (Bx - C) + D$

4 Determine the Equation of a Function of the Form $y = A \sin (Bx - C) + D$ or $y = A \cos (Bx - C) + D$ Given Its Graph

Introduction to Section 2.2

In Section 2.1, we mentioned that there are four factors that affect the graph of a sine or cosine curve. The first factor was **amplitude** and the second factor was **period**. The next two are **phase shift** and **vertical shift**. Before we learn about these last two factors that can affect a sine or cosine curve, take a moment and work through the following You Try It exercises to make sure that you have a complete understanding of amplitude and period. Do this before starting Objective 1.

You Try It   Sketching A Graph of the Form $y = A \sin x$

You Try It   Sketching A Graph of the Form $y = A \cos x$

You Try It   Sketching A Graph of the Form $y = \sin (Bx)$

You Try It   Sketching a Graph of the Form $y = \cos (Bx)$

You Try It   Sketching a Graph of the Form $y = A \sin (Bx)$

You Try It   Sketching a Graph of the Form $y = A \cos (Bx)$

OBJECTIVE 1   SKETCHING GRAPHS OF THE FORM $y = \sin (x - C)$
AND $y = \cos (x - C)$

My animation summary

The third factor that can affect the graph of a sine or cosine curve is known as **phase shift**. In this objective, for functions of the form $y = A \sin (Bx - C)$ and $y = A \cos (Bx - C)$, we consider only the case where $A = 1$ and $B = 1$. We will look at functions of the form $y = A \sin (Bx - C)$ and $y = A \cos (Bx - C)$ for any nonzero values of A and B in Objective 2. Before we define phase shift, it is worthy to point out that if C is a positive constant, then the graph of a function of the form $y = f (x - C)$ can be obtained by horizontally shifting the graph of $y = f (x)$ to the right C units. The horizontal shift transformation is typically first introduced in an algebra course. Watch this animation, which illustrates a horizontal shift to the right.

My animation summary

Similarly, if C is a positive constant, the graph of $y = f (x + C)$ can be obtained by horizontally shifting the graph of $y = f (x)$ to the *left* C units. Watch this animation that illustrates a horizontal shift to the left.

It follows that if $C > 0$, the graph of $y = \sin (x - C)$ can be obtained by horizontally shifting each quarter point of $y = \sin x$ to the *right* C units and the graph of $y = \sin (x + C)$ can be obtained by horizontally shifting each quarter point of $y = \sin x$ to the *left* C units. If $C > 0$ the graph of $y = \cos (x - C)$ can be obtained by horizontally shifting each quarter point of $y = \cos x$ to the *right* C units and the graph of $y = \cos (x + C)$ can be obtained by horizontally shifting each quarter point of $y = \cos x$ to the *left* C units. See Figure 21.

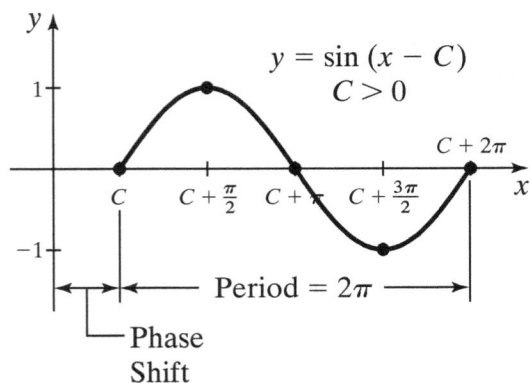

Figure 21

As you can see in Figure 21, the amplitude for each graph is $|A| = |1| = 1$ and the range is $[-1, 1]$. The period of each graph is $P = \dfrac{2\pi}{B} = \dfrac{2\pi}{1} = 2\pi$. Notice that the increments between the x-coordinates of each quarter point of all graphs remain unchanged at $\dfrac{2\pi}{4} = \dfrac{\pi}{2}$. Only the x-coordinates of each quarter point of the graph of $y = \sin(x - C)$ and $y = \cos(x - C)$ are different from the x-coordinates of the quarter points of one cycle of the graph of $y = \sin x$ and $y = \cos x$, respectively. The x-coordinates of the quarter points of $y = \sin(x - C)$ and $y = \cos(x - C)$ depend on the constant C. In general, the number $\dfrac{C}{B}$ is known

as the **phase shift**. Because $B = 1$, we call the number $\dfrac{C}{B} = \dfrac{C}{1} = C$ the **phase shift** for functions of the form $y = \sin(x - C)$ or $y = \cos(x - C)$.

The Graphs of $y = \sin(x - C)$ or $y = \cos(x - C)$

The amplitude is 1 (because $A = 1$) and the range is $[-1, 1]$.

The period is $P = 2\pi \left(\text{Because } B = 1 \text{ and } P = \dfrac{2\pi}{B} \right)$.

The phase shift is $C \left(\text{Because } B = 1 \text{ and the phase shift is } \dfrac{C}{B} \right)$.

The x-coordinates of the quarter points are $C, C + \dfrac{\pi}{2}, C + \pi, C + \dfrac{3\pi}{2}$, and $C + 2\pi$.

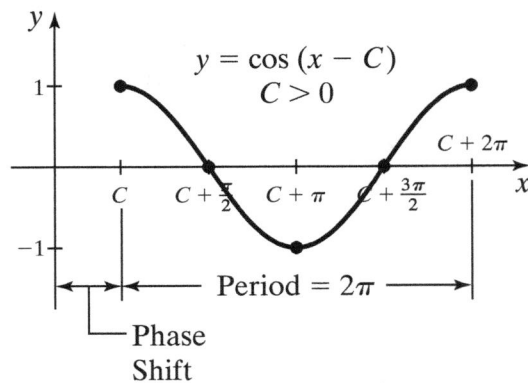

⊙ **Example 1** Sketching graphs of the form $y = \sin(x - C)$ and $y = \cos(x - C)$

Determine the phase shift and sketch the graph of each function.

a. $y = \cos(x - \pi)$ **b.** $y = \sin\left(x + \dfrac{\pi}{2}\right)$

Solution

a. The amplitude of $y = \cos(x - \pi)$ is 1 and the range is $[-1, 1]$. The period is $P = 2\pi$. The phase shift is $C = \pi$. We know that the x-coordinate of the first quarter point of $y = \cos x$ is $x = 0$. Because the phase shift is π, the x-coordinate of the first quarter point of $y = \cos(x - \pi)$ is $x = 0 + \pi = \pi$. The x-coordinate of the last quarter point is equal to the x-coordinate of the first quarter point plus the period, which is $x = \pi + 2\pi = 3\pi$. Next, we divide the interval $[\pi, 3\pi]$ into four equal subintervals of length $P \div 4 = 2\pi \div 4 = \dfrac{\pi}{2}$ by starting with π and adding $\dfrac{\pi}{2}$ to the x-coordinate of each successive quarter point. Thus, the x-coordinates of the five quarter points are π, $\dfrac{3\pi}{2}$, 2π, $\dfrac{5\pi}{2}$, and 3π. The y-coordinates of the five quarter points of $y = \cos(x - \pi)$ will be exactly the same as the y-coordinates of the five quarter points of $y = \cos x$. We now state the quarter points of $y = \cos x$ and $y = \cos(x - \pi)$. The graphs of these functions can be seen in Figure 22.

Quarter points for $y = \cos x$: $(0, 1)$, $\left(\dfrac{\pi}{2}, 0\right)$, $(\pi, -1)$, $\left(\dfrac{3\pi}{2}, 0\right)$, $(2\pi, 1)$

Quarter points for $y = \cos(x - \pi)$: $(\pi, 1)$, $\left(\dfrac{3\pi}{2}, 0\right)$, $(2\pi, -1)$, $\left(\dfrac{5\pi}{2}, 0\right)$, $(3\pi, 1)$

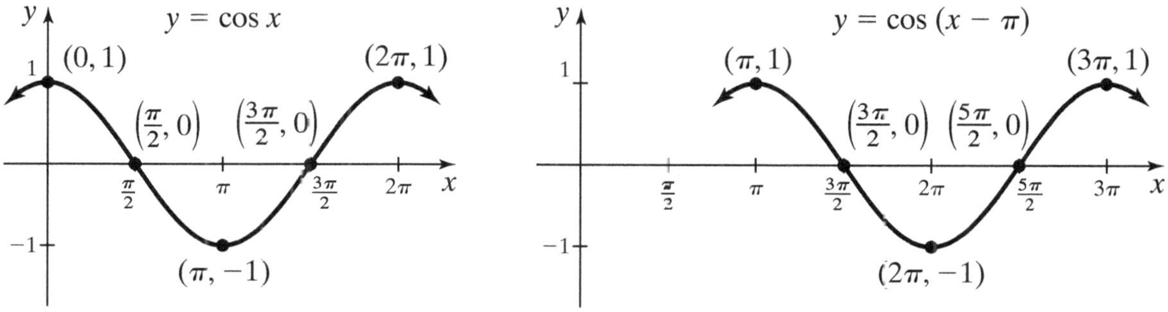

Figure 22 One cycle of the graphs of $y = \cos x$ and $y = \cos(x - \pi)$

b. The amplitude of $y = \sin\left(x + \dfrac{\pi}{2}\right)$ is 1 and the range is $[-1, 1]$. The period is $P = 2\pi$. The phase shift is $C = -\dfrac{\pi}{2}$. We know that the x-coordinate of the first quarter point of $y = \sin x$ is $x = 0$. Because the phase shift is $-\dfrac{\pi}{2}$, the x-coordinate of the first quarter point of $y = \sin\left(x + \dfrac{\pi}{2}\right)$ is $x = 0 - \dfrac{\pi}{2} = -\dfrac{\pi}{2}$. The x-coordinate of the last quarter point is equal to the x-coordinate of the first quarter point plus the period, which is $x = -\dfrac{\pi}{2} + 2\pi = \dfrac{3\pi}{2}$. Next, we divide

the interval $\left[-\dfrac{\pi}{2}, \dfrac{3\pi}{2}\right]$ into four equal subintervals of length $P \div 4 = 2\pi \div 4 = \dfrac{\pi}{2}$ by

starting with $-\dfrac{\pi}{2}$ and adding $\dfrac{\pi}{2}$ to the x-coordinate of each successive quarter

point. Thus, the x-coordinates of the five quarter points are $-\dfrac{\pi}{2}, 0, \dfrac{\pi}{2}, \pi$, and $\dfrac{3\pi}{2}$.

The y-coordinates of the five quarter points of $y = \sin\left(x + \dfrac{\pi}{2}\right)$ will be exactly

the same as the y-coordinates of the five quarter points of $y = \sin x$. We now

state the quarter points of $y = \sin x$ and $y = \sin\left(x + \dfrac{\pi}{2}\right)$. The graphs of these

functions can be seen in Figure 23.

Quarter points for $y = \sin x$: $\qquad$ $(0, 0)$, $\qquad$ $\left(\dfrac{\pi}{2}, 1\right)$, $\quad$ $(\pi, 0)$, $\quad$ $\left(\dfrac{3\pi}{2}, -1\right)$, $\quad$ $(2\pi, 0)$

Quarter points for $y = \sin\left(x + \dfrac{\pi}{2}\right)$: $\left(-\dfrac{\pi}{2}, 0\right)$, $\quad$ $(0, 1)$, $\quad$ $\left(\dfrac{\pi}{2}, 0\right)$, $\quad$ $(\pi, -1)$, $\quad$ $\left(\dfrac{3\pi}{2}, 0\right)$

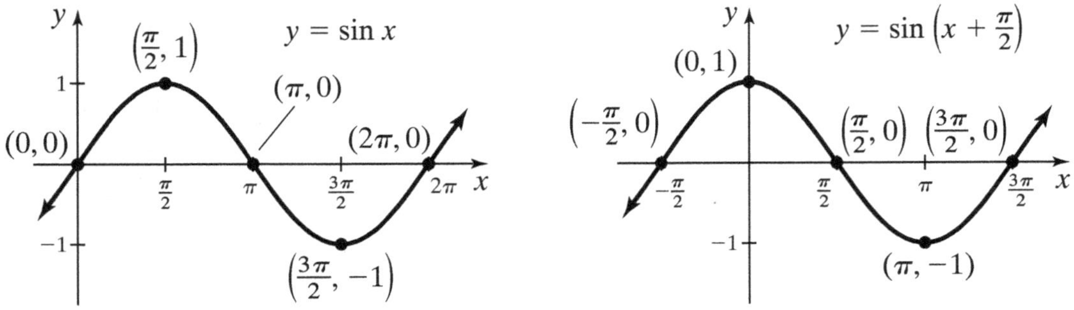

Figure 23 One cycle of the graphs of $y = \sin x$ and $y = \sin\left(x + \dfrac{\pi}{2}\right)$

Functions of the form $y = \sin(x - C)$ and $y = \cos(x - C)$ (such as $y = \cos(x - \pi)$

and $y = \sin\left(x + \dfrac{\pi}{2}\right)$ from Example 1) are fairly easy to sketch using transformations.

Using transformations, we can sketch the graph of $y = \cos(x - \pi)$ by horizontally
shifting the graph of $y = \cos x$ to the right π units. Similarly, we can sketch the graph

of $y = \sin\left(x + \dfrac{\pi}{2}\right)$ by horizontally shifting the graph of $y = \sin\left(x + \dfrac{\pi}{2}\right)$ to the left $\dfrac{\pi}{2}$

units. To see how to sketch each of the functions from Example 1 using transformations, click on the Show Graph link located under the following two functions.

$$y = \cos(x - \pi) \qquad y = \sin\left(x + \dfrac{\pi}{2}\right)$$

Show Graph $\qquad\qquad$ Show Graph

If we take a closer look at the graph of $y = \sin\left(x + \dfrac{\pi}{2}\right)$, we see that it is identical to

the graph of $y = \cos x$. See Figure 24. Therefore, we can establish the trigonometric

identity, $\sin\left(x + \dfrac{\pi}{2}\right) = \cos x$.

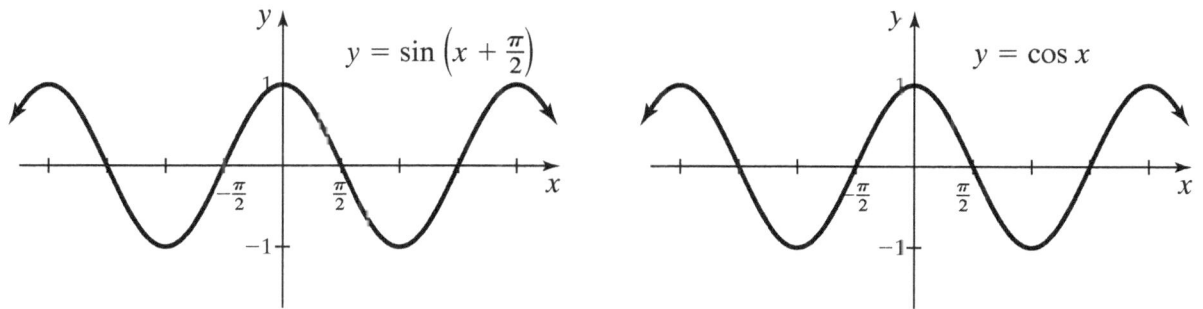

$$y = \sin\left(x + \frac{\pi}{2}\right)$$

$$y = \cos x$$

Figure 24 The graph of $y = \sin\left(x + \dfrac{\pi}{2}\right)$ is the same as the graph of $y = \cos x$.

A trigonometric identity is an equality involving trigonometric functions that are true for every single value of the given variable for which both sides of the equality are defined. We have been exposed to a few trigonometric identities earlier in the text such as the reciprocal identities and the Pythagorean identities. Trigonometric identities will be discussed extensively in Chapter 3.

If we sketch the graph of $y = \cos\left(x - \dfrac{\pi}{2}\right)$, we see that this graph is exactly the same as the graph of $y = \sin x$. This leads to the following two trigonometric identities.

Relationship between Graphs of Sine and Cosine Functions

$$\sin\left(x + \frac{\pi}{2}\right) = \cos x \qquad\qquad \cos\left(x - \frac{\pi}{2}\right) = \sin x$$

You Try It   Work through this You Try It problem.

Work Exercises 1–5 in this textbook or in the MyMathLab Study Plan.

OBJECTIVE 2   SKETCHING GRAPHS OF THE FORM $y = A \sin(Bx - C)$ AND $y = A \cos(Bx - C)$

We now want to discuss trigonometric functions of the form $y = A \sin(Bx - C)$ and $y = A \cos(Bx - C)$. The amplitude is $|A|$ and the range is $[-|A|, |A|]$. At first glance, it may appear that these graphs have a phase shift of C units. This is *not* true unless $B = 1$! To determine the period and phase shift of functions of this form, we use the fact that the period of $y = A \sin x$ and $y = A \cos x$ is $P = 2\pi$. That is, the graphs of $y = A \sin x$ and $y = A \cos x$ complete one cycle for $0 \le x \le 2\pi$. For now, suppose that $B > 0$. Then the graph of $y = A \sin(Bx - C)$ and $y = A \cos(Bx - C)$ will complete one cycle for $0 \le Bx - C \le 2\pi$. We can solve this three-part inequality to determine the period and phase shift of $y = A \sin(Bx - C)$ and $y = A \cos(Bx - C)$.

$0 \le Bx - C \le 2\pi$ The graph of $y = A \sin(Bx - C)$ and $y = A \cos(Bx - C)$ will complete one cycle for values of x satisfying this inequality.

$C \le Bx \le 2\pi + C$ Add C to each part of the inequality.

$\dfrac{C}{B} \le x \le \dfrac{2\pi}{B} + \dfrac{C}{B}$ Divide each part by B and simplify. The direction of the inequality remains the same because $B > 0$.

The graph of the functions $y = A \sin (Bx - C)$ and $y = A \cos (Bx - C)$ will therefore complete one cycle on the interval $\left[\dfrac{C}{B}, \dfrac{C}{B} + \dfrac{2\pi}{B} \right]$. The length of this interval is $\left(\dfrac{2\pi}{B} + \dfrac{C}{B} \right) - \left(\dfrac{C}{B} \right) = \dfrac{2\pi}{B}$. Thus, the period is $P = \dfrac{2\pi}{B}$ and the phase shift is $\dfrac{C}{B}$.

Note that if $B < 0$, then we will use the fact that cosine is an **even function** and that sine is an **odd function** to rewrite the functions using a *positive* value of B. Thus, functions of the form $y = A \sin (Bx - C)$ and $y = A \cos (Bx - C)$ can *always* be rewritten in an equivalent factored form in which $B > 0$. For example, consider the function $y = 3 \sin (\pi - 2x)$.

$y = 3 \sin (\pi - 2x)$	Write the original function.
$y = 3 \sin (-2x + \pi)$	Reorder inside the parentheses.
$y = 3 \sin (-(2x - \pi))$	Factor out a negative within the parentheses.
$y = -3 \sin (2x - \pi)$	Use the odd property of $y = \sin x$: $\sin (-x) = -\sin x$.

As you can see, we have rewritten the original function into an equivalent form such that $B > 0$. When sketching trigonometric functions, we will always rewrite the functions using a positive value of B.

When $B \neq 1$, we may choose to factor out B from the **argument** of the function so that the function's characteristics are more apparent.

$y = -3 \sin (2x - \pi)$	Start with the form of the function where $B > 0$.
$y = -3 \sin \left(2x - \dfrac{2\pi}{2} \right)$	Rewrite the second term within the parentheses so that $B = 2$ is in the numerator.
$y = -3 \sin \left(2 \left(x - \dfrac{\pi}{2} \right) \right)$	Factor out 2.

Amplitude: $|A| = |-3| = 3$ **Period:** $P = \dfrac{2\pi}{B} = \dfrac{2\pi}{2} = \pi$ **Phase Shift:** $\dfrac{C}{B} = \dfrac{\pi}{2}$

We will sketch the graph of $y = 3 \sin (\pi - 2x)$ in Example 2c. We now list a seven-step process for sketching the graphs of functions of the form $y = A \sin (Bx - C)$ and $y = A \cos (Bx - C)$.

Steps for Sketching Functions of the Form $y = A \sin (Bx - C)$ and $y = A \cos (Bx - C)$

Step 1. If $B < 0$, rewrite the function in an equivalent form such that $B > 0$. Use the odd property of the sine function or the even property of the cosine function.

It is often helpful to factor out B when $B \neq 1$ such that

$$y = A \sin (Bx - C) = A \sin \left(B \left(x - \frac{C}{B} \right) \right)$$

or

$$y = A \cos (Bx - C) = A \cos \left(B \left(x - \frac{C}{B} \right) \right).$$

In the factored form, the amplitude, period, and phase shift are more apparent.

Step 2. The amplitude is $|A|$. The range is $[-|A|, |A|]$.

Step 3. The period is $P = \dfrac{2\pi}{B}$.

Step 4. The phase shift is $\dfrac{C}{B}$.

Step 5. The x-coordinate of the first quarter point is $\dfrac{C}{B}$. The x-coordinate of the last quarter point is $\dfrac{C}{B} + P$. An interval for one complete cycle is $\left[\dfrac{C}{B}, \dfrac{C}{B} + P\right]$. Subdivide this interval into four equal subintervals of length $P \div 4$ by starting with $\dfrac{C}{B}$ and adding $(P \div 4)$ to the x-coordinate of each successive quarter point.

Step 6. Multiply the y-coordinates of the quarter points of $y = \sin x$ or $y = \cos x$ by A to determine the y-coordinates of the corresponding quarter points for $y = A \sin(Bx - C)$ and $y = A \cos(Bx - C)$.

Step 7. Connect the quarter points to obtain one complete cycle.

My interactive video summary

⊙ **Example 2** Sketching graphs of the form $y = A \sin(Bx - C)$ and $y = A \cos(Bx - C)$

Sketch the graph of each function.

a. $y = 3 \sin(2x - \pi)$ **b.** $y = -2 \cos\left(3x + \dfrac{\pi}{2}\right)$

c. $y = 3 \sin(\pi - 2x)$ **d.** $y = -2 \cos\left(-3x + \dfrac{\pi}{2}\right)$

Solution

a. Step 1. We see that $B = 2 > 0$. Therefore, there is no need to use the odd property of sine to rewrite the function.

We can factor this function as $y = 3 \sin(2x - \pi) = 3 \sin\left(2\left(x - \dfrac{\pi}{2}\right)\right)$ to determine more easily the function's characteristics. Note that $A = 3, B = 2$, and $\dfrac{C}{B} = \dfrac{\pi}{2}$.

Step 2. The amplitude is $|A| = |3| = 3$. Thus, the range is $[-3, 3]$.

Step 3. The period is $P = \dfrac{2\pi}{B} = \dfrac{2\pi}{2} = \pi$.

Step 4. The phase shift is $\dfrac{C}{B} = \dfrac{\pi}{2}$.

Step 5. The x-coordinate of the first quarter point is $x = \dfrac{C}{B} = \dfrac{\pi}{2}$. The x-coordinate of the last quarter point is equal to the x-coordinate of the first quarter point plus the period, which is $x = \dfrac{\pi}{2} + \pi = \dfrac{3\pi}{2}$. Next, we divide the interval $\left[\dfrac{\pi}{2}, \dfrac{3\pi}{2}\right]$ into four equal subintervals of length $P \div 4 = \pi \div 4 = \dfrac{\pi}{4}$ by starting with $\dfrac{\pi}{2}$ and adding $\dfrac{\pi}{4}$ to the x-coordinate of each successive quarter point. Thus, the x-coordinates of the five quarter points are $\dfrac{\pi}{2}, \dfrac{3\pi}{4}, \pi, \dfrac{5\pi}{4},$ and $\dfrac{3\pi}{2}$.

Step 6. The y-coordinates of the five quarter points will be $A = 3$ times the corresponding y-coordinates of the five quarter points of $y = \sin x$. We now state the quarter points of $y = \sin x$ and $y = 3 \sin (2x - \pi) = 3 \sin\left(2\left(x - \dfrac{\pi}{2}\right)\right)$.

Quarter points for $y = \sin x$:
$\quad (0, 0), \quad \left(\dfrac{\pi}{2}, 1\right), \quad (\pi, 0), \quad \left(\dfrac{3\pi}{2}, -1\right), \quad (2\pi, 0)$

$\quad\quad (3) \cdot 0 = 0 \mid (3) \cdot 1 = 3 \mid (3) \cdot 0 = 0 \mid (3) \cdot (-1) = -3 \mid (3) \cdot 0 = 0$

Quarter points for $y = 3 \sin (2x - \pi)$:

$= 3 \sin\left(2\left(x - \dfrac{\pi}{2}\right)\right) \quad \left(\dfrac{\pi}{2}, 0\right), \quad \left(\dfrac{3\pi}{4}, 3\right), \quad (\pi, 0), \quad \left(\dfrac{5\pi}{4}, -3\right), \quad \left(\dfrac{3\pi}{2}, 0\right)$

Step 7. We connect these quarter points with a smooth curve to complete one cycle of the graph of $y = 3 \sin (2x - \pi)$. See Figure 25. Several cycles of the graph can be seen in Figure 26.

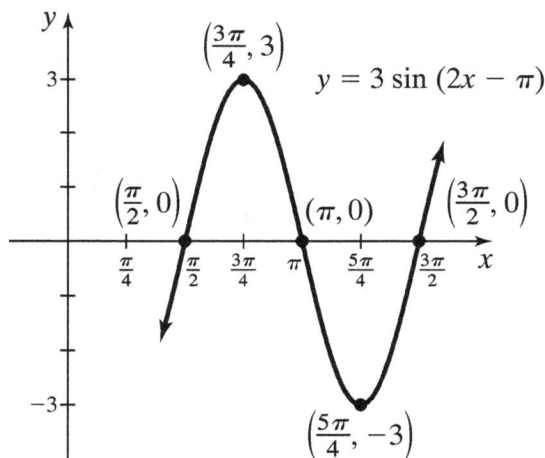

Figure 25 One complete cycle of the graph of $y = 3 \sin (2x - \pi)$

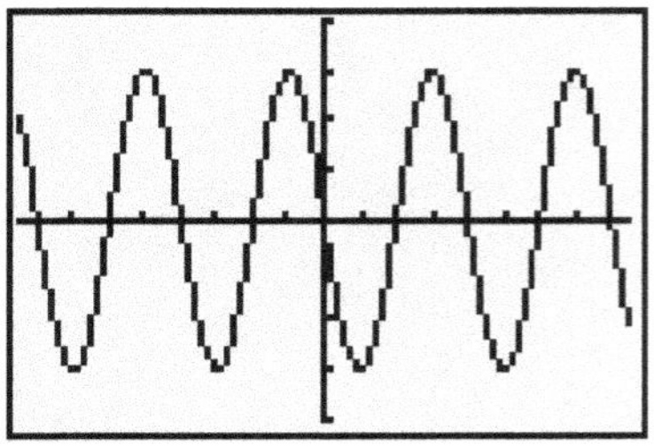

Figure 26 Several cycles of the graph of $y = 3 \sin (2x - \pi)$ using a graphing utility

My interactive video summary

To see how to sketch the function $y = 3 \sin (2x - \pi)$, watch this interactive video.

b. Step 1. For the function $y = -2\cos\left(3x + \dfrac{\pi}{2}\right)$, we see that $B = 3 > 0$. Therefore, there is no need to use the even property of cosine to rewrite the function.

We can factor this function as $y = -2\cos\left(3x + \dfrac{\pi}{2}\right) =$
$-2\cos\left(3\left(x - \left(-\dfrac{\pi}{6}\right)\right)\right)$ to more easily determine the function's characteristics. Note that $A = -2$, $B = 3$, and $\dfrac{C}{B} = -\dfrac{\pi}{6}$.

Step 2. The amplitude is $|A| = |-2| = 2$. Thus, the range is $[-2, 2]$.

Step 3. The period is $P = \dfrac{2\pi}{B} = \dfrac{2\pi}{3}$.

Step 4. The phase shift is $\dfrac{C}{B} = -\dfrac{\pi}{6}$.

Step 5. The x-coordinate of the first quarter point is $x = \dfrac{C}{B} = -\dfrac{\pi}{6}$. The x-coordinate of the last quarter point is equal to the x-coordinate of the first quarter point plus the period, which is $x = -\dfrac{\pi}{6} + \dfrac{2\pi}{3} = -\dfrac{\pi}{6} + \dfrac{4\pi}{6} = \dfrac{3\pi}{6} = \dfrac{\pi}{2}$.

Next, we divide the interval $\left[-\dfrac{\pi}{6}, \dfrac{\pi}{2}\right]$ into four equal subintervals of length $P \div 4 = \dfrac{2\pi}{3} \div 4 = \dfrac{2\pi}{3} \cdot \dfrac{1}{4} = \dfrac{2\pi}{12} = \dfrac{\pi}{6}$ by starting with $-\dfrac{\pi}{6}$ and adding $\dfrac{\pi}{6}$ to the x-coordinate of each successive quarter point. Thus, the x-coordinates of the five quarter points are $-\dfrac{\pi}{6}, 0, \dfrac{\pi}{6}, \dfrac{\pi}{3}$, and $\dfrac{\pi}{2}$.

Step 6. The y-coordinates of the five quarter points will be $A = -2$ times the corresponding y-coordinates of the five quarter points of $y = \cos x$. We now state the quarter points of $y = \cos x$ and $y = -2\cos\left(3x + \dfrac{\pi}{2}\right) = -2\cos\left(3\left(x - \left(-\dfrac{\pi}{6}\right)\right)\right)$.

Quarter points for $y = \cos x$:

$(0, 1),\qquad \left(\dfrac{\pi}{2}, 0\right),\qquad (\pi, -1),\qquad \left(\dfrac{3\pi}{2}, 0\right),\qquad (2\pi, 1)$

$(-2)\cdot 1 = -2 \quad | \quad (-2)\cdot 0 = 0 \quad | \quad (-2)\cdot(-1) = 2 \quad | \quad (-2)\cdot 0 = 0 \quad | \quad (-2)\cdot 1 = -2$

Quarter points for $y = -2\cos\left(3x + \dfrac{\pi}{2}\right)$
$= -2\cos\left(3\left(x - \left(-\dfrac{\pi}{6}\right)\right)\right)$:

$\left(-\dfrac{\pi}{6}, -2\right),\qquad (0, 0),\qquad \left(\dfrac{\pi}{6}, 2\right),\qquad \left(\dfrac{\pi}{3}, 0\right),\qquad \left(\dfrac{\pi}{2}, -2\right)$

Step 7. The graph of $y = -2\cos\left(3x + \dfrac{\pi}{2}\right)$ can be seen in Figure 27 and Figure 28.

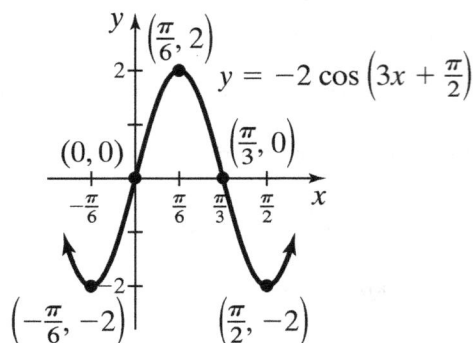

Figure 27 One complete cycle of the graph of $y = -2\cos\left(3x + \dfrac{\pi}{2}\right)$

Using Technology 📱

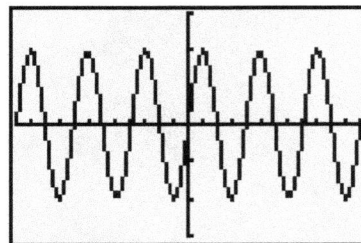

Figure 28 Several cycles of the graph of $y = -2\cos\left(3x + \dfrac{\pi}{2}\right)$ using a graphing utility

My interactive video summary

▶ To see how to sketch the function $y = -2\cos\left(3x + \dfrac{\pi}{2}\right)$, watch this interactive video.

c. Step 1. For the function $y = 3\sin(\pi - 2x)$, we see that $B = -2 < 0$. Therefore, we use the odd property of the sine function to rewrite the function as an equivalent function using a positive value of B.

$y = 3\sin(\pi - 2x)$	Write the original function.
$y = 3\sin(-2x + \pi)$	Reorder inside the parentheses.
$y = 3\sin(-(2x - \pi))$	Factor out a negative within the parentheses.
$y = -3\sin(2x - \pi)$	Use the odd property of $y = \sin x$: $\sin(-x) = -\sin x$.

We can now factor this function as $y = -3\sin(2x - \pi) = -3\sin\left(2\left(x - \dfrac{\pi}{2}\right)\right)$ to determine more easily the function's characteristics.

Note that $A = -3, B = 2$, and $\dfrac{C}{B} = \dfrac{\pi}{2}$.

Step 2. The amplitude is $|A| = |-3| = 3$. Thus, the range is $[-3, 3]$.

Step 3. The period is $P = \dfrac{2\pi}{B} = \dfrac{2\pi}{2} = \pi$.

Step 4. The phase shift is $\dfrac{C}{B} = \dfrac{\pi}{2}$.

Step 5. The x-coordinate of the first quarter point is $x = \dfrac{C}{B} = \dfrac{\pi}{2}$. The x-coordinate of the last quarter point is equal to the x-coordinate of the first quarter point plus the period, which is $x = \dfrac{\pi}{2} + \pi = \dfrac{\pi}{2} + \dfrac{2\pi}{2} = \dfrac{3\pi}{2}$. Next, we divide the interval $\left[\dfrac{\pi}{2}, \dfrac{3\pi}{2}\right]$ into four equal subintervals of length $P \div 4 = \pi \div 4 = \dfrac{\pi}{4}$ by starting with $\dfrac{\pi}{2}$ and adding $\dfrac{\pi}{4}$ to the x-coordinate of each successive quarter point. Thus, the x-coordinates of the five quarter points are $\dfrac{\pi}{2}, \dfrac{3\pi}{4}, \pi, \dfrac{5\pi}{4}$, and $\dfrac{3\pi}{2}$.

Step 6. The y-coordinates of the five quarter points of $y = -3 \sin\left(2\left(x - \dfrac{\pi}{2}\right)\right)$

will be $A = -3$ times the corresponding y-coordinates of the five quarter points of $y = \sin x$. We now state the quarter points of $y = \sin x$ and

$$y = 3 \sin\left(\pi - 2x\right) = -3 \sin\left(2x - \pi\right) = -3 \sin\left(2\left(x - \dfrac{\pi}{2}\right)\right).$$

Quarter points for $y = \sin x$: $(0, 0)$, $\left(\dfrac{\pi}{2}, 1\right)$, $(\pi, 0)$, $\left(\dfrac{3\pi}{2}, -1\right)$, $(2\pi, 0)$

$(-3) \cdot 0 = 0$ $(-3) \cdot 1 = -3$ $(-3) \cdot 0 = 0$ $(-3) \cdot (-1) = 3$ $(-3) \cdot 0 = 0$

Quarter points for $y = 3 \sin\left(\pi - 2x\right)$

$= -3 \sin\left(2x - \pi\right)$

$= -3 \sin\left(2\left(x - \dfrac{\pi}{2}\right)\right)$: $\left(\dfrac{\pi}{2}, 0\right)$, $\left(\dfrac{3\pi}{4}, -3\right)$, $(\pi, 0)$, $\left(\dfrac{5\pi}{4}, 3\right)$, $\left(\dfrac{3\pi}{2}, 0\right)$

Step 7. We can now connect the quarter points with a smooth curve to create one complete cycle of the graph of $y = 3 \sin\left(\pi - 2x\right)$. See Figure 29. Several cycles of the graph can be seen in Figure 30.

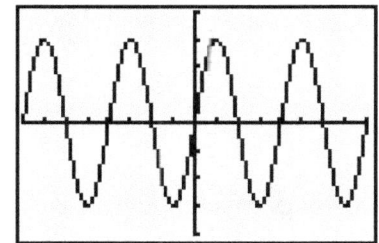

Figure 29 One complete cycle of the graph of $y = 3 \sin\left(\pi - 2x\right)$

Figure 30 Several cycles of the graph of $y = 3 \sin\left(\pi - 2x\right)$ using a graphing utility

My interactive video summary

To see how to sketch the function $y = 3 \sin\left(\pi - 2x\right)$, watch this interactive video.

d. Step 1. For the function $y = -2 \cos\left(-3x + \dfrac{\pi}{2}\right)$, we see that $B = -3 < 0$.

Therefore, we use the even property of the cosine function to rewrite the function as an equivalent function using a positive value of B.

$$y = -2 \cos\left(-3x + \dfrac{\pi}{2}\right) \qquad \text{Write the original function.}$$

$$y = -2 \cos\left(-\left(3x - \dfrac{\pi}{2}\right)\right) \qquad \text{Factor out the negative within the parentheses.}$$

$$y = -2 \cos\left(3x - \dfrac{\pi}{2}\right) \qquad \begin{array}{l} \text{Use the even property of} \\ y = \cos x \colon \cos\left(-x\right) = \cos x. \end{array}$$

We can now factor this function as $y = -2\cos\left(3x - \dfrac{\pi}{2}\right) = -2\cos\left(3\left(x - \dfrac{\pi}{6}\right)\right)$ to more easily determine the function's characteristics. Note that $A = -2$, $B = 3$, and $\dfrac{C}{B} = \dfrac{\pi}{6}$.

Step 2. The amplitude is $|A| = |-2| = 2$. Thus, the range is $[-2, 2]$.

Step 3. The period is $P = \dfrac{2\pi}{B} = \dfrac{2\pi}{3}$.

Step 4. The phase shift is $\dfrac{C}{B} = \dfrac{\pi}{6}$.

Step 5. The x-coordinate of the first quarter point is $x = \dfrac{C}{B} = \dfrac{\pi}{6}$. The x-coordinate of the last quarter point is equal to the x-coordinate of the first quarter point plus the period, which is $x = \dfrac{\pi}{6} + \dfrac{2\pi}{3} = \dfrac{\pi}{6} + \dfrac{4\pi}{6} = \dfrac{5\pi}{6}$. Next, we divide the interval $\left[\dfrac{\pi}{6}, \dfrac{5\pi}{6}\right]$ into four equal subintervals of length $P \div 4 = \dfrac{2\pi}{3} \div 4 = \dfrac{2\pi}{3} \cdot \dfrac{1}{4} = \dfrac{2\pi}{12} = \dfrac{\pi}{6}$. We do this by starting with $\dfrac{\pi}{6}$ and adding $\dfrac{\pi}{6}$ to the x-coordinate of each successive quarter point. Thus, the x-coordinates of the five quarter points are $\dfrac{\pi}{6}, \dfrac{\pi}{3}, \dfrac{\pi}{2}, \dfrac{2\pi}{3}$, and $\dfrac{5\pi}{6}$.

Step 6. The y-coordinates of the five quarter points will be $A = -2$ times the corresponding y-coordinates of the five quarter points of $y = \cos x$. We now state the quarter points of $y = \cos x$ and $y = -2\cos\left(-3x + \dfrac{\pi}{2}\right) = -2\cos\left(3\left(x - \dfrac{\pi}{6}\right)\right)$.

Quarter points for $y = \cos x$:

$(0, 1), \qquad \left(\dfrac{\pi}{2}, 0\right), \qquad (\pi, -1), \qquad \left(\dfrac{3\pi}{2}, 0\right), \qquad (2\pi, 1)$

$\begin{array}{ccccc} (-2)\cdot 1 = -2 & (-2)\cdot 0 = 0 & (-2)\cdot -1 = 2 & (-2)\cdot 0 = 0 & (-2)\cdot 1 = -2 \end{array}$

Quarter points for $y = -2\cos\left(-3x + \dfrac{\pi}{2}\right)$

$= -2\cos\left(3\left(x - \dfrac{\pi}{6}\right)\right)$: $\quad \left(\dfrac{\pi}{6}, -2\right), \quad \left(\dfrac{\pi}{3}, 0\right), \quad \left(\dfrac{\pi}{2}, 2\right), \quad \left(\dfrac{2\pi}{3}, 0\right), \quad \left(\dfrac{5\pi}{6}, -2\right)$

Step 7. The graph of $y = -2\cos\left(-3x + \dfrac{\pi}{2}\right)$ can be seen in Figures 31 and 32.

$$y = -2 \cos\left(-3x + \frac{\pi}{2}\right)$$

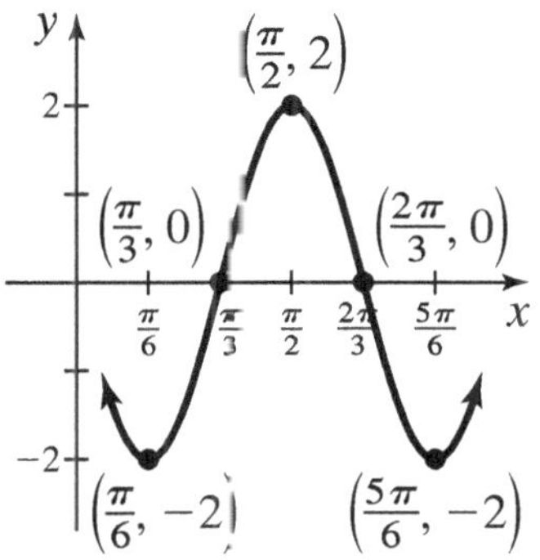

Figure 31 One complete cycle of the graph of
$$y = -2 \cos\left(-3x + \frac{\pi}{2}\right)$$

Using Technology

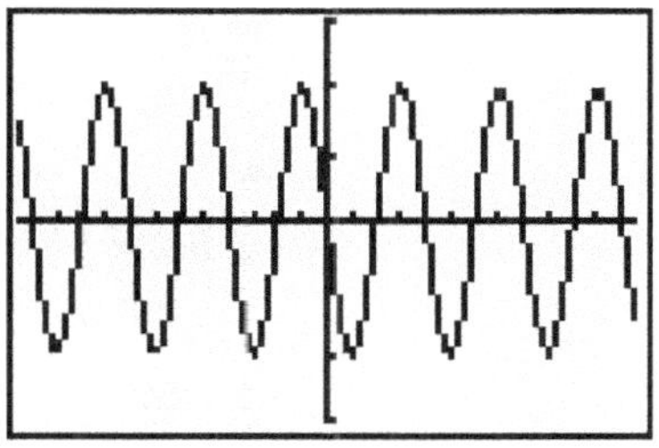

Figure 32 Several cycles of the graph of $y = -2 \cos\left(-3x + \frac{\pi}{2}\right)$ using a graphing utility

My interactive video summary

▶ To see how to sketch the function $y = -2 \cos\left(-3x + \frac{\pi}{2}\right)$, watch this interactive video.

It is possible to use transformations to sketch functions of the form $y = A \sin(Bx - C)$ and $y = A \cos(Bx - C)$ as seen in Example 2. Such functions require several transformations. When sketching functions that require multiple transformations, we will always use the following order of transformations.

1. Horizontal shifts

2. Horizontal stretches/compressions

3. Reflection about the y-axis

4. Vertical stretches/compressions

5. Reflection about the x-axis

6. Vertical shifts

Functions of the form $y = A \sin(Bx - C)$ and $y = A \cos(Bx - C)$ never require a vertical shift. However, we will encounter functions requiring vertical shifts in Example 3. To see how to sketch each of the functions from Example 2 using transformations, click on the Show Graph link below the following four functions.

$y = 3 \sin(2x - \pi)$ $y = -2 \cos\left(3x + \frac{\pi}{2}\right)$ $y = 3 \sin(-2x - \pi)$ $y = -2 \cos\left(-3x + \frac{\pi}{2}\right)$

Show Graph   Show Graph   Show Graph   Show Graph

You Try It Work through this You Try It problem.

Work Exercises 6–15 in this textbook or in the MyMathLab Study Plan.

OBJECTIVE 3 SKETCHING GRAPHS OF THE FORM $y = A \sin(Bx - C) + D$ AND $y = A \cos(Bx - C) + D$

Recall that when we add a nonzero constant to a function $y = f(x)$, we get $y = f(x) + D$. The graph of $y = f(x) + D$ can be obtained by adding D to the corresponding y-coordinates of each ordered pair lying on the graph of $y = f(x)$.

Therefore, the graph of $y = A \sin (Bx - C) + D$ and $y = A \cos (Bx - C) + D$ can be obtained by adding D to the y-coordinate of each quarter point of the graphs of $y = A \sin (Bx - C)$ and $y = A \cos (Bx - C)$, respectively. Because the range of $y = A \sin (Bx - C)$ and $y = A \cos (Bx - C)$ is $[-|A|, |A|]$, the range of functions of the form $y = A \sin (Bx - C) + D$ and $y = A \cos (Bx - C) + D$ is $[-|A| + D, |A| + D]$. We now summarize a process for sketching functions of the form $y = A \sin (Bx - C) + D$ and $y = A \cos (Bx - C) + D$.

Steps for Sketching Functions of the Form $y = A \sin (Bx - C) + D$ and $y = A \cos (Bx - C) + D$

Step 1. If $B < 0$, rewrite the function in an equivalent form such that $B > 0$. Use the odd property of the sine function or the even property of the cosine function.

It is often helpful to factor out B when $B \neq 1$ such that

$$y = A \sin (Bx - C) + D = A \sin \left(B\left(x - \frac{C}{B} \right) \right) + D \text{ or}$$

$$y = A \cos (Bx - C) + D = A \cos \left(B\left(x - \frac{C}{B} \right) \right) + D. \text{ In the factored}$$

form, the amplitude, period, and phase shift are more apparent.

Step 2. The amplitude is $|A|$. The range is $[-|A| + D, |A| + D]$.

Step 3. The period is $P = \dfrac{2\pi}{B}$.

Step 4. The phase shift is $\dfrac{C}{B}$.

Step 5. The x-coordinate of the first quarter point is $\dfrac{C}{B}$. The x-coordinate of the last quarter point is $\dfrac{C}{B} + P$. An interval for one complete cycle is $\left[\dfrac{C}{B}, \dfrac{C}{B} + P \right]$. Subdivide this interval into four equal subintervals of length $P \div 4$ by starting with $\dfrac{C}{B}$ and adding $(P \div 4)$ to the x-coordinate of each successive quarter point.

Step 6. Multiply the y-coordinates of the quarter points of $y = \sin x$ or $y = \cos x$ by A and then add D to determine the y-coordinates of the corresponding quarter points for $y = A \sin (Bx - C) + D$ and $y = A \cos (Bx - C) + D$.

Step 7. Connect the quarter points to obtain one complete cycle.

My interactive video summary

⊙ **Example 3** Sketching graphs of the form $y = A \sin (Bx - C) + D$ and $y = A \cos (Bx - C) + D$

Sketch the graph of each function.

a. $y = 3 \sin \left(2x - \dfrac{\pi}{2} \right) - 1$

b. $y = 4 - \cos (-\pi x + 2)$

Solution

a. **Step 1.** For the function $y = 3 \sin \left(2x - \dfrac{\pi}{2} \right) - 1$, we see that $B = 2 > 0$.

Therefore, there is no need to use the odd property of sine to rewrite the function. We can now factor this function as

$$y = 3 \sin \left(2x - \frac{\pi}{2} \right) - 1 = 3 \sin \left(2\left(x - \frac{\pi}{4} \right) \right) - 1 \text{ to determine more}$$

easily the function's characteristics. Note that $A = 3, B = 2, \dfrac{C}{B} = \dfrac{\pi}{4}$, and $D = -1$.

Step 2. The amplitude is $|A| = |3| = 3$.
The range is $[-|A| + D, |A| + D] = [-3 + (-1), 3 + (-1)] = [-4, 2]$.

Step 3. The period is $P = \dfrac{2\pi}{B} = \dfrac{2\pi}{2} = \pi$.

Step 4. The phase shift is $\dfrac{C}{B} = \dfrac{\pi}{4}$.

Step 5. The x-coordinate of the first quarter point is $x = \dfrac{C}{B} = \dfrac{\pi}{4}$. The x-coordinate of the last quarter point is equal to the x-coordinate of the first quarter point plus the period, which is $x = \dfrac{\pi}{4} + \pi = \dfrac{5\pi}{4}$. An interval for one complete cycle of the graph is $\left[\dfrac{\pi}{4}, \dfrac{5\pi}{4} \right]$. We divide this interval into four equal subintervals of length $\dfrac{P}{4} = \dfrac{\pi}{4}$ by starting with $\dfrac{\pi}{4}$ and adding $\dfrac{\pi}{4}$ to the x-coordinate of each successive quarter point. Thus, the x-coordinates of the five quarter points are $\dfrac{\pi}{4}, \dfrac{\pi}{2}, \dfrac{3\pi}{4}, \pi$, and $\dfrac{5\pi}{4}$.

Step 6. Multiply each y-coordinate of the quarter points of $y = \sin x$ by $A = 3$ and then add $D = -1$ to obtain the quarter points of

$$y = 3 \sin \left(2x - \frac{\pi}{2} \right) - 1 = 3 \sin \left(2\left(x - \frac{\pi}{4} \right) \right) - 1.$$

Quarter points for $y = \sin x$:

$(0, 0), \quad \left(\dfrac{\pi}{2}, 1 \right), \quad (\pi, 0), \quad \left(\dfrac{3\pi}{2}, -1 \right), \quad (2\pi, 0)$

$(3)\ 0 + (-1) = -1 \quad (3) \cdot 1 + (-1) = 2 \quad (3) \cdot 0 + (-1) = -1 \quad (3) \cdot (-1) + (-1) = -4 \quad (3) \cdot 0 + (-1) = -1$

Quarter points for $y = 3 \sin \left(2x - \dfrac{\pi}{2} \right) - 1$

$= 3 \sin \left(2\left(x - \dfrac{\pi}{4} \right) \right) - 1: \quad \left(\dfrac{\pi}{4}, -1 \right), \quad \left(\dfrac{\pi}{2}, 2 \right), \quad \left(\dfrac{3\pi}{4}, -1 \right), \quad (\pi, -4), \quad \left(\dfrac{5\pi}{4}, -1 \right)$

Step 7. We connect these quarter points with a smooth curve to complete one cycle of the graph of $y = 3 \sin \left(2x - \dfrac{\pi}{2} \right) - 1$. See Figures 33 and 34.

$$y = 3 \sin \left(2x - \frac{\pi}{2}\right) - 1$$

Figure 33 One complete cycle of the graph of $y = 3 \sin \left(2x - \dfrac{\pi}{2}\right) - 1$

Using Technology

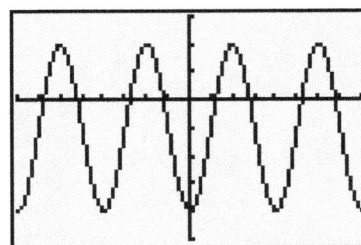

Figure 34 Several cycles of the graph of $y = 3 \sin \left(2x - \dfrac{\pi}{2}\right) - 1$ using a graphing utility

My interactive video summary

▶ To see how to sketch the graph of $y = 3 \sin \left(2x - \dfrac{\pi}{2}\right) - 1$, watch this interactive video.

b. Step 1. For the function $y = 4 - \cos(-\pi x + 2)$, we see that $B = -\pi < 0$. Therefore, we use the even property of the cosine function to rewrite the function as an equivalent function using a positive value of B.

$y = 4 - \cos(-\pi x + 2)$	Write the original function.
$y = -\cos(-\pi x + 2) + 4$	Reorder the terms.
$y = -\cos(-(\pi x - 2)) + 4$	Factor out the negative inside the parentheses.
$y = -\cos(\pi x - 2) + 4$	Use the even property of $y = \cos x$: $\cos(-x) = \cos x$.

We now factor this function as $y = -\cos(\pi x - 2) + 4 = -\cos\left(\pi\left(x - \dfrac{2}{\pi}\right)\right) + 4$

to determine more easily the function's characteristics.

Note that $A = -1$, $B = \pi$, $\dfrac{C}{B} = \dfrac{2}{\pi}$, and $D = 4$.

Step 2. The amplitude is $|A| = |-1| = 1$.
The range is $[-|A| + D, |A| + D] = [-1 + 4, 1 + 4] = [3, 5]$.

Step 3. The period is $P = \dfrac{2\pi}{B} = \dfrac{2\pi}{\pi} = 2$.

Step 4. The phase shift is $\dfrac{C}{B} = \dfrac{2}{\pi}$.

Step 5. We know that the x-coordinate of the first quarter point is $x = \dfrac{C}{B} = \dfrac{2}{\pi}$.
The x-coordinate of the last quarter point is equal to the x-coordinate

of the first quarter point plus the period, which is $x = \dfrac{2}{\pi} + 2 = \dfrac{2}{\pi} + \dfrac{2\pi}{\pi} = \dfrac{2 + 2\pi}{\pi}$. An interval for one complete cycle of the graph is $\left[\dfrac{2}{\pi}, \dfrac{2 + 2\pi}{\pi}\right]$.

We divide this interval into four equal subintervals of length $\dfrac{P}{4} = \dfrac{2}{4} = \dfrac{1}{2}$ by starting with $\dfrac{2}{\pi}$ and adding $\dfrac{1}{2}$ to the x-coordinate of each successive quarter point. Thus, the x-coordinates of the five quarter points are

$$\dfrac{2}{\pi}, \ \dfrac{2}{\pi} + \dfrac{1}{2}, \ \dfrac{2}{\pi} + 1, \ \dfrac{2}{\pi} + \dfrac{3}{2}, \ \text{and} \ \dfrac{2 + 2\pi}{\pi}.$$

These x-coordinates can be simplified and rewritten as

$$\dfrac{2}{\pi}, \ \dfrac{4 + \pi}{2\pi}, \ \dfrac{2 + \pi}{\pi}, \ \dfrac{4 + 3\pi}{2\pi}, \ \text{and} \ \dfrac{2 + 2\pi}{\pi}.$$

Step 6. Multiply each y-coordinate of the quarter points of $y = \cos x$ by $A = -1$ and then add $D = 4$ to obtained the corresponding y-coordinates of the quarter points of $y = 4 - \cos(-\pi x + 2) = -\cos\left(\pi\left(x - \dfrac{2}{\pi}\right)\right) + 4$.

Quarter points for $y = \cos x$: $\quad (0, 1), \quad \left(\dfrac{\pi}{2}, 0\right), \quad (\pi, -1), \quad \left(\dfrac{3\pi}{2}, 0\right), \quad (2\pi, 1)$

$$(-1)\cdot 1 + 4 = 3 \mid (-1)\cdot 0 + 4 = 4 \mid (-1)\cdot(-1) + 4 = 5 \mid (-1)\cdot 0 + 4 = 4 \mid (-1)\cdot 1 + 4 = 3$$

Quarter points for $y = 4 - \cos(-\pi x + 2)$

$= -\cos\left(\pi\left(x - \dfrac{2}{\pi}\right)\right) + 4{:}\quad \left(\dfrac{2}{\pi}, 3\right), \quad \left(\dfrac{4 + \pi}{2\pi}, 4\right), \quad \left(\dfrac{2 + \pi}{\pi}, 5\right), \quad \left(\dfrac{4 + 3\pi}{2\pi}, 4\right), \quad \left(\dfrac{2 + 2\pi}{\pi}, 3\right)$

Step 7. We connect these quarter points with a smooth curve to complete one cycle of the graph of $y = 4 - \cos(-\pi x + 2)$. See Figures 35 and 36.

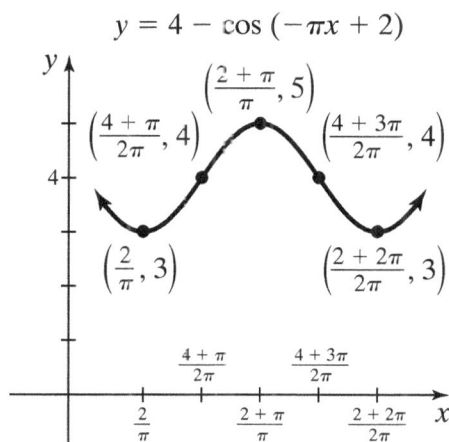

Figure 35 One complete cycle of the graph of $y = 4 - \cos(-\pi x + 2)$

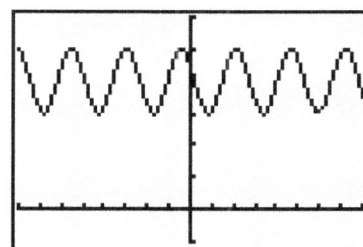

Using Technology

Figure 36 Several cycles of the graph of $y = 4 - \cos(-\pi x + 2)$ using a graphing utility.

My interactive video summary

▶ To see how to sketch the graph of $y = 4 - \cos(-\pi x + 2)$, watch this interactive video.

We can sketch the graphs seen in Example 3 using transformations if we carefully follow the order of transformations. Try sketching the functions

$y = 3\sin\left(2x - \dfrac{\pi}{2}\right) - 1$ and $y = 4 - \cos\left(-\pi x + 2\right)$ using transformations and

then click on the Show Graph links below to see if you are correct.

$$y = 3\sin\left(2x - \dfrac{\pi}{2}\right) - 1 \qquad\qquad y = 4 - \cos\left(-\pi x + 2\right)$$

Show Graph   Show Graph

You Try It   Work through this You Try It problem.

Work Exercises 16–25 in this textbook or in the My MathLab **Study Plan.**

OBJECTIVE 4   DETERMINE THE EQUATION OF A FUNCTION OF THE FORM
$y = A\sin(Bx - C) + D$ OR $y = A\cos(Bx - C) + D$ GIVEN ITS GRAPH

Suppose that you were asked to determine the equation of a function that describes the graph seen in Figure 37 where the function is of the form $y = A\sin(Bx - C) + D$ or $y = A\cos(Bx - C) + D$.

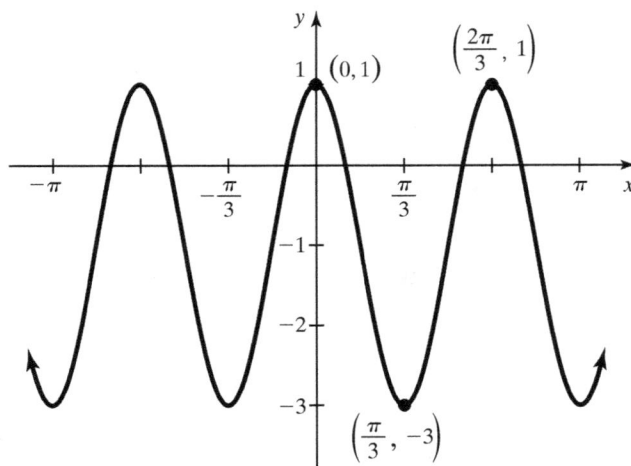

Figure 37

Furthermore, suppose that you were asked to choose a function that describes the graph seen in Figure 37 where the function is chosen from the five functions listed below.

$$y = 2\cos(3x) - 1$$

$$y = 2\sin\left(3x + \dfrac{\pi}{2}\right) - 1$$

$$y = -2\cos(3x + \pi) - 1$$

$$y = -2\sin\left(3x - \dfrac{\pi}{2}\right) - 1$$

$$y = 2\cos(-3x) - 1$$

Which of the five functions above would you choose to accurately describe the graph displayed in Figure 37? The answer is **all five functions describe the graph** seen in Figure 37! You might want to take a few minutes and indeed verify that all of these functions describe the same graph.

Therefore, given the graph of a sine curve or cosine curve, it is impossible to determine a unique function that describes the graph unless certain assumptions are made. In this objective, when given the graph of a function of the form $y = A \sin(Bx - C) + D$ or $y = A \cos(Bx - C) + D$, we will first state whether the given graph is a sine curve or a cosine curve and we will assume that $B > 0$. We will also always label the five quarter points of one cycle of the given graph to correspond with the five quarter points of the graph of $y = \sin x$ or $y = \cos x$ over the interval $[0, 2\pi]$. If these assumptions are made, then we can determine a unique function that describes the given graph.

We can follow the six steps outlined below to determine the function that describes a given sine or cosine curve.

Steps for Determining an Equation of a Function of the Form
$y = A \sin(Bx - C) + D$ **or** $y = A \cos(Bx - C) + D$ **Given the Graph**

Step 1. Subtract the x-coordinate of the first quarter point from the x-coordinate of the fifth quarter point to determine the period.

Step 2. Use the equation $P = \dfrac{2\pi}{B}$ to determine the value of $B > 0$.

Step 3. The x-coordinate of the first quarter point represents the phase shift, $\dfrac{C}{B}$. Use this information to determine the value of C.

Step 4. The amplitude is $|A| = \dfrac{|b - a|}{2}$ where $[a, b]$ is the range of the given graph.

Step 5. If the given graph is a sine curve, then $A = |A|$ if the graph is *increasing* from the first quarter point to the second quarter point. Otherwise, $A = -|A|$.

If the given graph is a cosine curve, then $A = |A|$ if the graph is *decreasing* from the first quarter point to the second quarter point. Otherwise, $A = -|A|$.

Step 6. Choose the first quarter point (x_1, y_1) that lies on the given graph and the corresponding first quarter point of the graph of $y = \sin x$ or $y = \cos x$. Note that the first quarter point of $y = \sin x$ is $(0, 0)$ and the first quarter point of $y = \cos x$ is $(0, 1)$.

The relationship between y_1, the y-coordinate of the first quarter point of the given graph, and the y-coordinate of the first quarter point of $y = \sin x$ or $y = \cos x$ is given by the following:

If the graph is a sine curve, then $\quad y_1 = (0)A + D$.

If the graph is a cosine curve, then $y_1 = (1)A + D$.

Use this information to determine the value of D.

My interactive video summary

⊙ Example 4 Determining an Equation of a Function Given the Graph that it Represents

a. The graph of a function of the form $y = A\cos(Bx - C) + D$ where $B > 0$ is given below. The five quarter points of one cycle of the graph are labeled. These five quarter points on the graph correspond to the five quarter points of the graph of $y = \cos x$ over the interval $[0, 2\pi]$. Determine the specific function that is represented by the given graph based on the association of the labeled quarter points and the quarter points of the graph of $y = \cos x$ over the interval $[0, 2\pi]$.

$$y = A\cos(Bx - C) + D$$

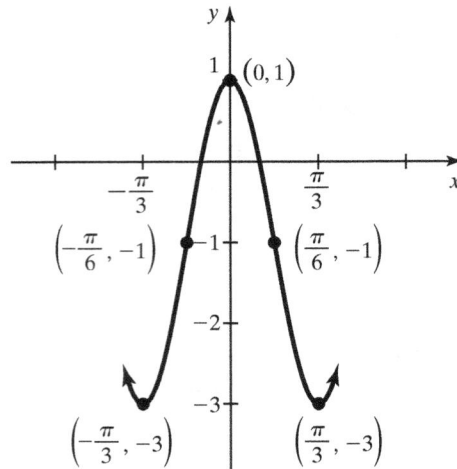

b. The graph of a function of the form $y = A\sin(Bx - C) + D$ where $B > 0$ is given below. The five quarter points of one cycle of the graph are labeled. These five quarter points on the graph correspond to the five quarter points of the graph of $y = \sin x$ over the interval $[0, 2\pi]$. Determine the specific function that is represented by the given graph based on the association of the labeled quarter points and the quarter points of the graph of $y = \sin x$ over the interval $[0, 2\pi]$.

$$y = A\sin(Bx - C) + D$$

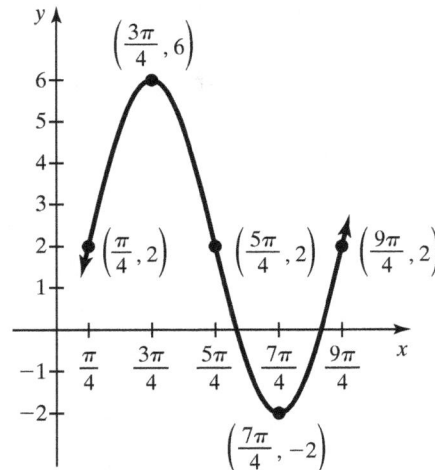

Solution

a. The given graph is shown on the following page. Before completing Step 1, we compare the coordinates of the five quarter points of the graph of $y = \cos x$ over the interval $[0, 2\pi]$ with the five quarter points of the given graph.

$$y = A\cos(Bx - C) + D$$

	First	Second	Third	Fourth	Fifth
Quarter Points of $y = \cos x$:	$(0, 1)$	$\left(\dfrac{\pi}{2}, 0\right)$	$(\pi, -1)$	$\left(\dfrac{3\pi}{2}, 0\right)$	$(2\pi, 1).$
Quarter Points of the Given Graph:	$\left(-\dfrac{\pi}{3}, -3\right)$	$\left(-\dfrac{\pi}{6}, -1\right)$	$(0, 1)$	$\left(\dfrac{\pi}{6}, -1\right)$	$\left(\dfrac{\pi}{3}, -3\right)$

Step 1. Looking at the graph we see that the x-coordinate of the first quarter point is $-\dfrac{\pi}{3}$ and the x-coordinate of the fifth quarter point is $\dfrac{\pi}{3}$.

Therefore, the period is $\dfrac{\pi}{3} - \left(-\dfrac{\pi}{3}\right) = \dfrac{\pi}{3} + \dfrac{\pi}{3} = \dfrac{2\pi}{3}.$

Step 2. The period is $P = \dfrac{2\pi}{B} = \dfrac{2\pi}{3}$. Therefore, $B = 3.$

Step 3. The x-coordinate of the first quarter point is $-\dfrac{\pi}{3}$. Therefore, the phase shift is $\dfrac{C}{B} = -\dfrac{\pi}{3}$. From Step 2, we know that $B = 3$. Thus, $\dfrac{C}{B} = -\dfrac{\pi}{3}$ so $C = -\pi.$

Step 4. The range of this graph is $[-3, 1]$. Therefore, the amplitude is
$$|A| = \frac{|1 - (-3)|}{2} = \frac{|4|}{2} = 2.$$

Step 5. The given graph is a cosine curve that *increases* between the first quarter point and the second quarter point. Therefore, the value of A must be negative. Hence, $A = -|A| = -2.$

Step 6. The first quarter point of the given graph is $(x_1, y_1) = \left(-\dfrac{\pi}{3}, -3\right)$. The first quarter point of the graph of $y = \cos x$ is $(0, 1)$. Therefore, we can use the equation $y_1 = (1)A + D$ to solve for D.

$y_1 = (1)A + D$	Start with the equation describing the y-coordinate of the first quarter point of the given graph.
$-3 = (1)(-2) + D$	Substitute $y_1 = -3$. and $A = -2.$

Solve this equation for D to get $D = -1.$

We see that $A = -2$, $B = 3$, $C = -\pi$, and $D = -1$. Therefore, substituting these values into the function of the form $y = A\cos(Bx - C) + D$, we see that the function which describes the given graph using the association with the quarter points of the function $y = \cos x$ over the interval $[0, 2\pi]$ is $y = -2\cos(3x - (-\pi)) - 1$ or $y = -2\cos(3x + \pi) - 1$. You should sketch the graph of $y = -2\cos(3x + \pi) - 1$ to make sure that it matches the given graph.

b. Work through this interactive video to verify that the function of the form $y = A\sin(Bx - C) + D$ whose graph is shown on the right is $y = 4\sin\left(x - \dfrac{\pi}{4}\right) + 2$.

$$y = A\sin(Bx - C) + D$$

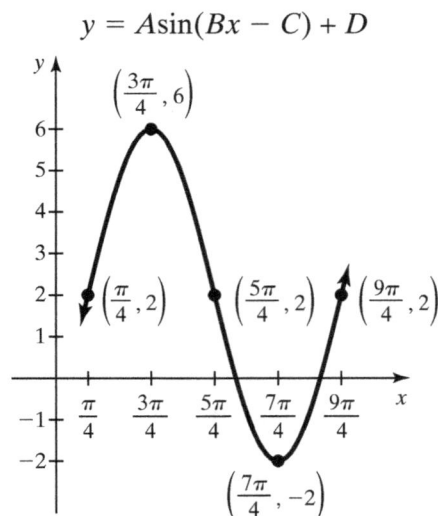

You Try It Work through this You Try It problem.

Work Exercises 26–33 in this textbook or in the MyMathLab Study Plan.

2.2 Exercises

Skill Check Exercises

In Exercises SCE-1 and SCE-3, solve the inequality. Write your answer in interval notation form.

SCE-1. $0 \le 2x - \pi \le 2\pi$ **SCE-2.** $0 \le 3x - \dfrac{\pi}{6} \le 2\pi$ **SCE-3.** $0 \le \pi x - 3 \le 2\pi$

In Exercises SCE-4 and SCE-5, perform the indicated operations and simplify.

SCE-4. $\left(\dfrac{3\pi}{2} - \dfrac{\pi}{2}\right) \div 4$ **SCE-5.** $\left(\dfrac{3\pi}{4} - \dfrac{\pi}{4}\right) \div 4$

In Exercises SCE-6 through SCE-9, an algebraic expression is given. On the second line, complete the factorization by correctly filling in the parentheses. The factored expression on the second line should be equivalent to the expression on the first line.

SCE-6. $(4x - \pi)$ **SCE-7.** $\left(5x - \dfrac{\pi}{3}\right)$ **SCE-8.** $(-3x + \pi)$

     $= 4(\quad\quad)$ $= 5(\quad\quad)$ $= -3(\quad\quad)$

SCE-9. $\left(-4x + \dfrac{\pi}{3}\right)$ **SCE-10.** $(\pi - 2x)$ **SCE-11.** $(\pi x - 3)$

     $= -4(\quad\quad)$ $= -2(\quad\quad)$ $= \pi(\quad\quad)$

In Exercises 1–5, determine the amplitude, range, period, and phase shift and then sketch the graph.

1. $y = \sin(x - \pi)$

2. $y = \cos\left(x + \dfrac{\pi}{2}\right)$

3. $y = \sin\left(x - \dfrac{\pi}{6}\right)$

4. $y = \sin\left(x + \dfrac{2\pi}{3}\right)$

5. $y = \cos(x - 2)$

In Exercises 6–15, determine the amplitude, range, period, and phase shift and then sketch the graph of each function.

6. $y = \cos(3x + \pi)$

7. $y = -\sin(2x - \pi)$

8. $y = 3\cos(4x - \pi)$

9. $y = -2\sin(6x + \pi)$

10. $y = 4\cos(-2x - \pi)$

11. $y = 3\sin\left(-2x - \dfrac{\pi}{2}\right)$

12. $y = -2\sin\left(3x - \dfrac{\pi}{2}\right)$

13. $y = -4\cos\left(-x - \dfrac{\pi}{3}\right)$

14. $y = -\cos(\pi - 2x)$

15. $y = 4\sin(\pi x - 3)$

In Exercises 16–25, determine the amplitude, range, period, and phase shift and then sketch the graph of each function.

16. $y = 2\cos(3x + \pi) + 1$

17. $y = -\sin(2x - \pi) - 2$

18. $y = 3\cos(4x - \pi) - 1$

19. $y = 1 - 2\sin(6x + \pi)$

20. $y = 4\cos(-2x - \pi) + 3$

21. $y = -3\sin\left(-2x + \dfrac{\pi}{4}\right) - 4$

22. $y = 5 - 2\sin\left(3x - \dfrac{\pi}{2}\right)$

23. $y = -4\cos\left(-x - \dfrac{\pi}{3}\right) + 3$

24. $y = -\cos(\pi - 2x) + 3$

25. $y = 4\sin(\pi x - 3) - 2$

In Exercises 26–33, the graph is of a function of the form $y = A\cos(Bx - C) + D$ or $y = A\sin(Bx - C) + D$ where $B > 0$ is given. The five quarter points of one cycle of the graph are labeled. These five quarter points on the graph correspond to the five quarter points of the graph of either $y = \cos x$ or $y = \sin x$ over the interval $[0, 2\pi]$. Determine the equation of the specific function that is represented by the given graph based on the association of the labeled quarter points and the quarter points of the graph of either $y = \sin x$ or $y = \cos x$ over the interval $[0, 2\pi]$.

26.
$$y = A\sin(Bx - C) + D, \ B > 0$$

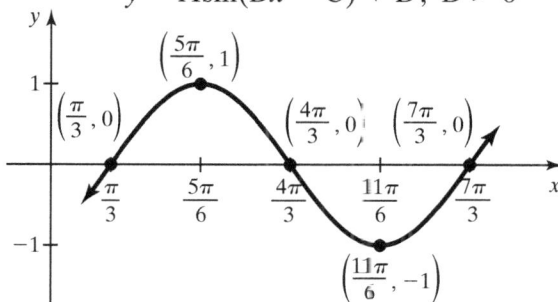

SbS **27.**

$$y = A\cos(Bx - C) + D, B > 0$$

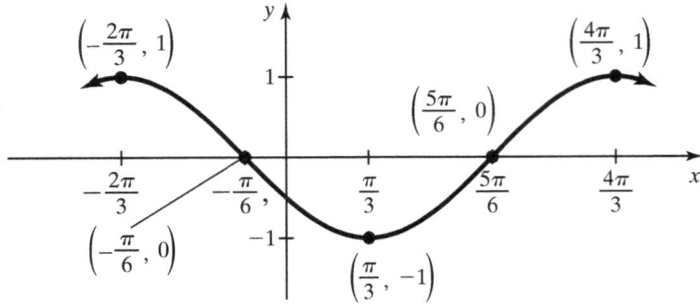

SbS **28.**

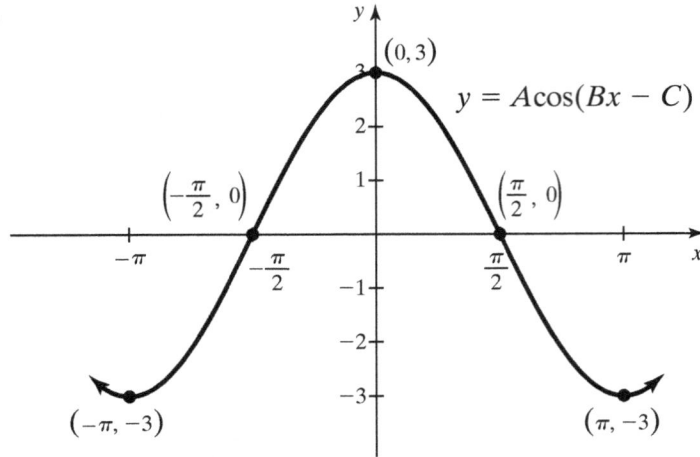

$$y = A\cos(Bx - C) + D, B > 0$$

SbS **29.** $y = A\sin(Bx - C) + D, B > 0$

SbS **30.** $y = A\sin(Bx - C) + D, B > 0$

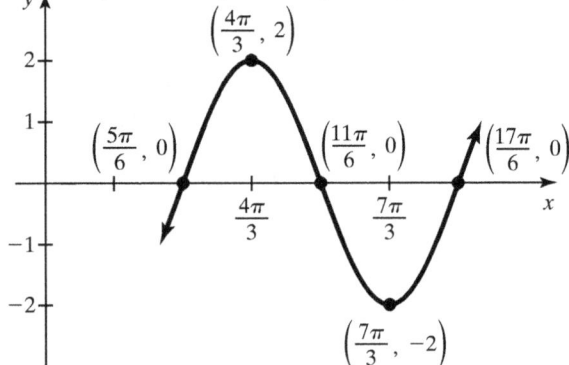

2-56   **Chapter 2**   The Graphs of Trigonometric Functions

31.

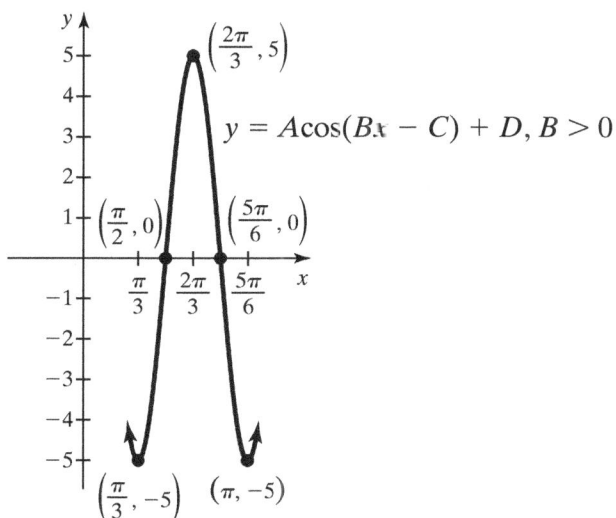

$y = A\cos(Bx - C) + D, B > 0$

Points shown: $\left(\frac{2\pi}{3}, 5\right)$, $\left(\frac{\pi}{2}, 0\right)$, $\left(\frac{5\pi}{6}, 0\right)$, $\left(\frac{\pi}{3}, -5\right)$, $(\pi, -5)$

32. $y = A\cos(Bx - C) + D, B > 0$

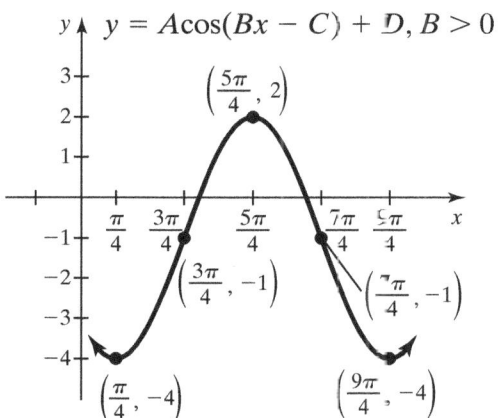

Points shown: $\left(\frac{5\pi}{4}, 2\right)$, $\left(\frac{3\pi}{4}, -1\right)$, $\left(\frac{7\pi}{4}, -1\right)$, $\left(\frac{\pi}{4}, -4\right)$, $\left(\frac{9\pi}{4}, -4\right)$

33.

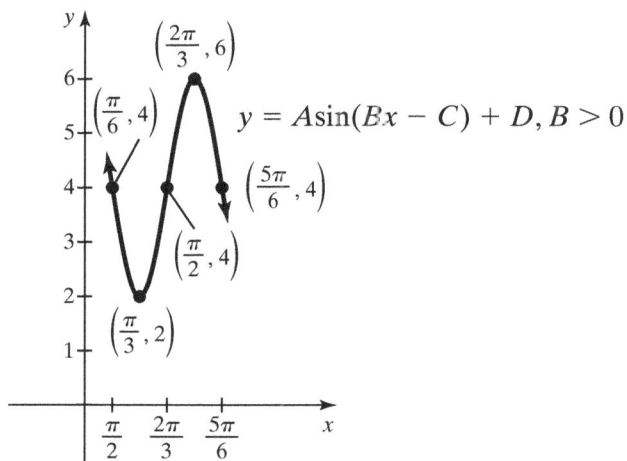

$y = A\sin(Bx - C) + D, B > 0$

Points shown: $\left(\frac{2\pi}{3}, 6\right)$, $\left(\frac{\pi}{6}, 4\right)$, $\left(\frac{5\pi}{6}, 4\right)$, $\left(\frac{\pi}{2}, 4\right)$, $\left(\frac{\pi}{3}, 2\right)$

Brief Exercises

In Exercises 34–36, determine the amplitude of each function.

34. $y = \sin\left(x - \frac{\pi}{6}\right)$ **35.** $y = -3\cos(2x + \pi)$ **36.** $y = 2\sin\left(-x + \frac{\pi}{3}\right) - 2$

In Exercises 37–39, determine the range of each function.

37. $y = \sin\left(x - \dfrac{\pi}{6}\right)$ **38.** $y = -3\cos(2x + \pi)$ **39.** $y = 2\sin\left(-x + \dfrac{\pi}{3}\right) - 2$

In Exercises 40–42, determine the period of each function.

40. $y = \sin\left(x - \dfrac{\pi}{6}\right)$ **41.** $y = -3\cos(2x + \pi)$ **42.** $y = 2\sin\left(-x + \dfrac{\pi}{3}\right) - 2$

In Exercises 43–45, determine the phase shift of each function.

43. $y = \sin\left(x - \dfrac{\pi}{6}\right)$ **44.** $y = -3\cos(2x + \pi)$ **45.** $y = 2\sin\left(-x + \dfrac{\pi}{3}\right) - 2$

In Exercises 46–48, determine the coordinates for the five quarter points of each function that correspond to the five quarter points of $y = \sin x$ or $y = \cos x$.

46. $y = \sin\left(x - \dfrac{\pi}{6}\right)$ **47.** $y = -3\cos(2x + \pi)$ **48.** $y = 2\sin\left(-x + \dfrac{\pi}{3}\right) - 2$

In Exercises 49–72, sketch the graph of each function.

49. $y = \sin(x - \pi)$ **50.** $y = \cos\left(x + \dfrac{\pi}{2}\right)$ **51.** $y = \sin\left(x - \dfrac{\pi}{6}\right)$

52. $y = \sin\left(x + \dfrac{2\pi}{3}\right)$ **53.** $y = \cos(x - 2)$ **54.** $y = \cos(3x + \pi)$

55. $y = -\sin(2x - \pi)$ **56.** $y = 3\cos(4x - \pi)$ **57.** $y = -2\sin(6x + \pi)$

58. $y = 4\cos(-2x - \pi)$ **59.** $y = 3\sin\left(-2x - \dfrac{\pi}{2}\right)$ **60.** $y = -2\sin\left(3x - \dfrac{\pi}{2}\right)$

61. $y = -4\cos\left(-x - \dfrac{\pi}{3}\right)$ **62.** $y = -\cos(\pi - 2x)$ **63.** $y = 4\sin(\pi x - 3)$

64. $y = 2\cos(3x + \pi) + 1$ **65.** $y = -\sin(2x - \pi) - 2$ **66.** $y = 3\cos(4x - \pi) - 1$

67. $y = 1 - 2\sin(6x + \pi)$ **68.** $y = 4\cos(-2x - \pi) + 3$ **69.** $y = -3\sin\left(-2x + \dfrac{\pi}{4}\right) - 4$

70. $y = 5 - 2\sin\left(3x - \dfrac{\pi}{2}\right)$ **71.** $y = -4\cos\left(-x - \dfrac{\pi}{3}\right) + 3$ **72.** $y = -\cos(\pi - 2x) + 3$

In Exercises 73–80, the graph of a function of the form $y = A \cos(Bx - C) + D$ or $y = A \sin(Bx - C) + D$ where $B > 0$ is given. The five quarter points of one cycle of the graph are labeled. These five quarter points on the graph correspond to the five quarter points of the graph of either $y = \cos x$ or $y = \sin x$ over the interval $[0, 2\pi]$. Determine the equation of the specific function that is represented by the given graph based on the association of the labeled quarter points and the quarter points of the graph of either $y = \sin x$ or $y = \cos x$ over the interval $[0, 2\pi]$.

73.

$$y = A\sin(Bx - C) + D, \ B > 0$$

74.

$$y = A\cos(Bx - C) + D, \ B > 0$$

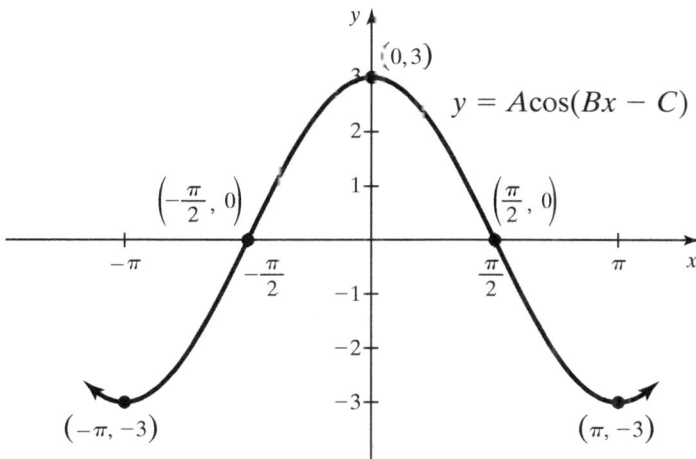

75.

$$y = A\cos(Bx - C) + D, \ B > 0$$

76.

$y = A\sin(Bx - C) + D, B > 0$

$\left(\frac{4\pi}{3}, 4\right)$

$\left(-\frac{\pi}{6}, 0\right)$ $\left(\frac{5\pi}{6}, 0\right)$ $\left(\frac{11\pi}{6}, 0\right)$

$\left(\frac{\pi}{3}, -4\right)$

77.

$y = A\sin(Bx - C) + D, B > 0$

$\left(\frac{4\pi}{3}, 2\right)$

$\left(\frac{5\pi}{6}, 0\right)$ $\left(\frac{11\pi}{6}, 0\right)$ $\left(\frac{17\pi}{6}, 0\right)$

$\left(\frac{7\pi}{3}, -2\right)$

78.

$\left(\frac{2\pi}{3}, 5\right)$

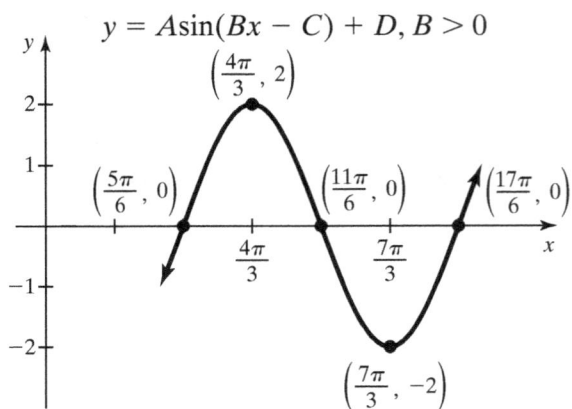

$y = A\cos(Bx - C) + D, B > 0$

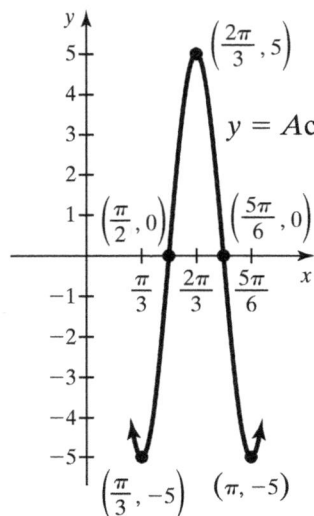

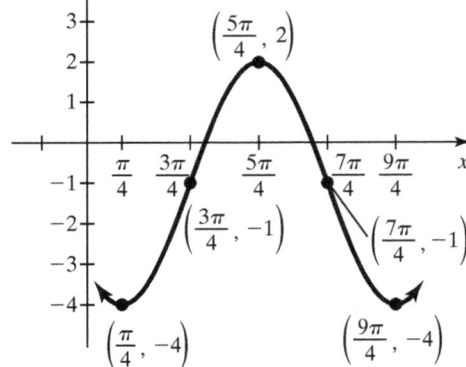

$\left(\frac{\pi}{2}, 0\right)$ $\left(\frac{5\pi}{6}, 0\right)$

$\left(\frac{\pi}{3}, -5\right)$ $(\pi, -5)$

79.

$y = A\cos(Bx - C) + D, B > 0$

$\left(\frac{5\pi}{4}, 2\right)$

$\left(\frac{3\pi}{4}, -1\right)$ $\left(\frac{7\pi}{4}, -1\right)$

$\left(\frac{\pi}{4}, -4\right)$ $\left(\frac{9\pi}{4}, -4\right)$

80.

$\left(\frac{2\pi}{3}, 6\right)$

$\left(\frac{\pi}{6}, 4\right)$

$y = A\sin(Bx - C) + D, B > 0$

$\left(\frac{5\pi}{6}, 4\right)$

$\left(\frac{\pi}{2}, 4\right)$

$\left(\frac{\pi}{3}, 2\right)$

2.3 The Graphs of the Tangent, Cotangent, Cosecant, and Secant Functions

THINGS TO KNOW

Before working through this section, be sure that you are familiar with the following concepts:

		VIDEO	ANIMATION	INTERACTIVE

You Try It   1. Using the Special Right Triangles (Section 1.4) ⊘ VIDEO

You Try It   2. Finding the Values of the Trigonometric Functions of Quadrantal Angles (Section 1.5) ⊘ VIDEO

You Try It   3. Evaluating Trigonometric Functions of Angles Belonging to the $\frac{\pi}{3}$, $\frac{\pi}{6}$, or $\frac{\pi}{4}$ Families (Section 1.5) ⊘ INTERACTIVE

You Try It   4. Sketching Graphs of the Form $y = A \sin (Bx)$ and $y = A \cos (Bx)$ (Section 2.1) ⊘ INTERACTIVE

You Try It   5. Sketching Graphs of the Form $y = A \sin (Bx - C) + D$ and $y = A \cos (Bx - C) + D$ (Section 2.2) ⊘ INTERACTIVE

OBJECTIVES

1 Understanding the Graph of the Tangent Function and Its Properties
2 Sketching Graphs of the Form $y = A \tan (Bx - C) + D$
3 Understanding the Graph of the Cotangent Function and Its Properties
4 Sketching Graphs of the Form $y = A \cot (Bx - C) + D$
5 Determine the Equation of a Function of the Form $y = A \tan (Bx - C) + D$ or $y = A \cot (Bx - C) + D$ Given Its Graph
6 Understanding the Graph of the Cosecant and Secant Functions and Their Properties
7 Sketching Graphs of the Form $y = A \csc (Bx - C) + D$ and $y = A \sec (Bx - C) + D$

OBJECTIVE 1   UNDERSTANDING THE GRAPH OF THE TANGENT FUNCTION AND ITS PROPERTIES

My video summary   ⊘ In Sections 2.1 and 2.2, we sketched the sine and cosine functions, as well as variations of them. In this section, we will sketch the tangent, cotangent, secant, and cosecant functions, as well as variations of them. We start by introducing the graph of $y = \tan x$. Watch this video to see how the graph of the tangent function is established.

Unlike the functions $y = \sin x$ and $y = \cos x$, the domain of $y = \tan x$ is *not* all real numbers. To determine the domain of $y = \tan x$, we start by using a **quotient identity** to rewrite $y = \tan x$ as $y = \dfrac{\sin x}{\cos x}$. Because division by zero is never allowed, we see that the tangent function is not defined when the denominator, $\cos x$, is equal to zero. The **zeros of $y = \cos x$** are of the form $(2n+1) \cdot \dfrac{\pi}{2}$, where n is an integer. Therefore, the domain of $y = \tan x$ is all real numbers *except* values of x of the form $(2n+1) \cdot \dfrac{\pi}{2}$, where n is an integer. Thus, the tangent function is undefined for values of x belonging to the set

$$\left\{ \ldots, -\frac{5\pi}{2}, -\frac{3\pi}{2}, -\frac{\pi}{2}, \frac{\pi}{2}, \frac{3\pi}{2}, \frac{5\pi}{2}, \ldots \right\}.$$

We will see that the graph approaches a **vertical asymptote** at these values.

We can determine the x-intercepts of the tangent function by observing that $y = \tan x = \dfrac{\sin x}{\cos x}$ will equal zero only when the numerator, $\sin x$, is equal to zero. Therefore, the x-intercepts, or zeros, of $y = \tan x = \dfrac{\sin x}{\cos x}$ will be precisely the zeros of $y = \sin x$. Thus, the x-intercepts of $y = \tan x$ are of the form $n\pi$, where n is an integer. Therefore, the x-intercepts belong to the set $\{\ldots, -3\pi, -2\pi, \pi, 0, \pi, 2\pi, 3\pi, \ldots\}$.

We begin by creating a table of values and plotting points to sketch one cycle of the graph of $y = \tan x$. We start by choosing $x = 0, x = \dfrac{\pi}{6}, x = \dfrac{\pi}{4}$, and $x = \dfrac{\pi}{3}$. See Table 9 and Figure 38.

Table 9

x	$y = \tan x$
0	0
$\dfrac{\pi}{6}$	$\dfrac{1}{\sqrt{3}} \approx 0.577$
$\dfrac{\pi}{4}$	1
$\dfrac{\pi}{3}$	$\sqrt{3} \approx 1.732$

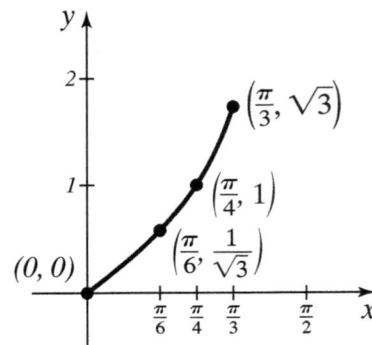

Figure 38 A portion of the graph $y = \tan x$

If we continue to choose values of x closer and closer to $x = \dfrac{\pi}{2}$, we see that the values of $y = \tan x$ increase without bound and the graph approaches the vertical asymptote $x = \dfrac{\pi}{2}$. See Table 10 and Figure 39.

Table 10

x	$y = \tan x$
$\dfrac{4\pi}{10}$	≈ 3.078
$\dfrac{49\pi}{100}$	≈ 31.821
$\dfrac{499\pi}{1000}$	≈ 318.309
$\dfrac{4999\pi}{10{,}000}$	≈ 3183.099
$\dfrac{5000\pi}{10{,}000} = \dfrac{\pi}{2}$	Undefined

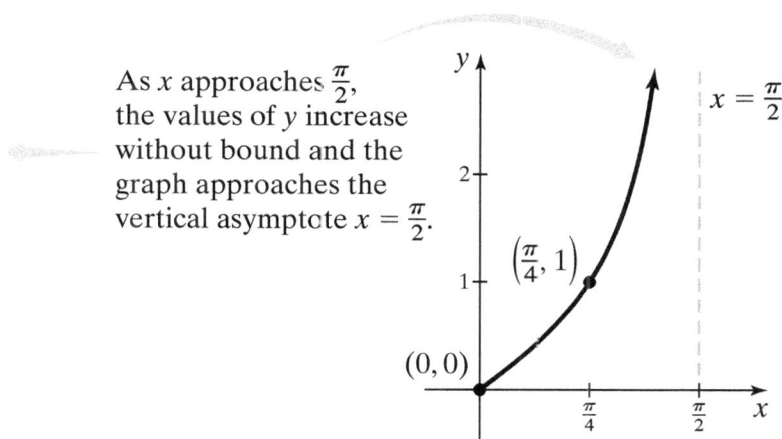

As x approaches $\frac{\pi}{2}$, the values of y increase without bound and the graph approaches the vertical asymptote $x = \frac{\pi}{2}$.

Figure 39 The graph of $y = \tan x$ on the interval $\left[0, \dfrac{\pi}{2}\right)$

The tangent function is an odd function, which means that the graph is symmetric about the origin. Therefore, we can reflect the portion of the graph seen in Figure 39 about the origin to get one complete cycle of the graph of $y = \tan x$ on the interval $\left(-\dfrac{\pi}{2}, \dfrac{\pi}{2}\right)$. See Figure 40.

Figure 40 One cycle of the graph of $y = \tan x$ on the interval $\left(-\dfrac{\pi}{2}, \dfrac{\pi}{2}\right)$

We will call the graph seen in Figure 40 the **principal cycle** of the graph of $y = \tan x$. Notice that the graph of $y = \tan x$ completes this principal cycle on the interval $\left(-\dfrac{\pi}{2}, \dfrac{\pi}{2}\right)$. The length of this interval is π units. Therefore, the tangent function is periodic with a period of $P = \pi$. Each cycle, or period, of the graph lies between consecutive values of x for which the function is undefined. Also note that each cycle of the graph of $y = \tan x$ is **one-to-one**. We can obtain a complete graph of $y = \tan x$ by repeating the graph seen in Figure 40 infinitely many times in either direction.

For each cycle of the graph of $y = \tan x$, there are three special points that will help us sketch the graph. The first point is called the **center point**. The center point is located at the center of the graph of each cycle. The center point of the principal cycle of the graph of $y = \tan x$ is located at the origin. Note that when we sketch the graphs of variations of the tangent function in Objective 2, this center point will not necessarily be located at the origin.

The two other special points are called **halfway points**. One halfway point is located halfway between the x-coordinate of the center point and the vertical asymptote to the left of the center point and has a y-coordinate of -1. The other halfway point is located halfway between the x-coordinate of the center point and the vertical asymptote to the right of the center point and has a y-coordinate of 1. See Figure 41.

We now state the characteristics of the tangent function.

Characteristics of the Tangent Function

The domain is $\left\{ x \mid x \neq (2n+1) \cdot \dfrac{\pi}{2}, \text{ where } n \text{ is an integer} \right\}$.

The range is $(-\infty, \infty)$.

The function is periodic with a period of $P = \pi$. The principal cycle of the graph occurs on the interval $\left(-\dfrac{\pi}{2}, \dfrac{\pi}{2} \right)$.

The function has infinitely many vertical asymptotes with equations $x = (2n+1) \cdot \dfrac{\pi}{2}$, where n is an integer.

The y-intercept is 0.

For each cycle there is one center point. The x-coordinates of the center points are also the x-intercepts, or zeros, and are of the form $n\pi$, where n is an integer.

For each cycle there are two halfway points. The halfway point to the left of the x-intercept has a y-coordinate of -1. The halfway point to the right of the x-intercept has a y-coordinate of 1.

The function is odd, which means $\tan(-x) = -\tan x$. The graph is symmetric about the origin.

The graph of each cycle of $y = \tan x$ is one-to-one.

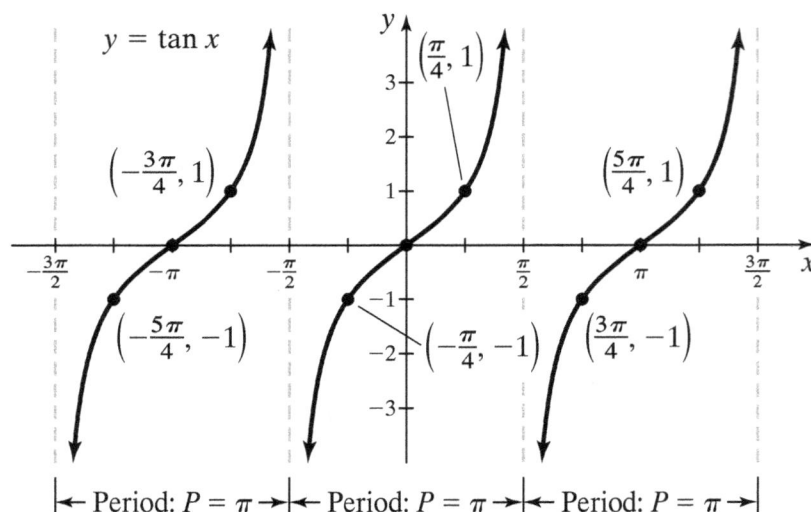

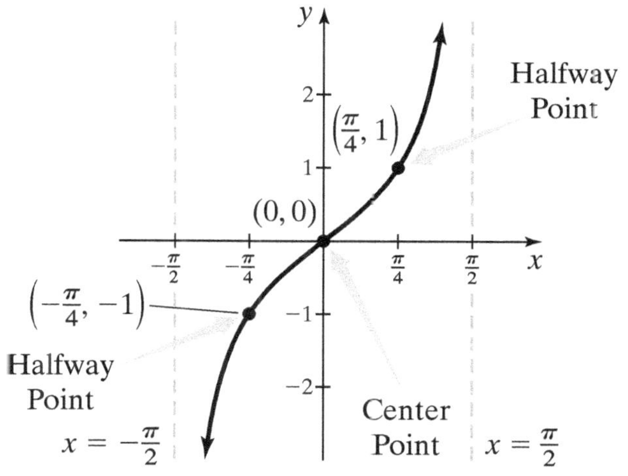

Figure 41 The principal cycle of the graph of $y = \tan x$ with the center point and the two halfway points labeled

My video summary ⊙ **Example 1** Understanding the Graph of $y = \tan x$

List all halfway points of $y = \tan x$ on the interval $\left[-\pi, \dfrac{5\pi}{2}\right]$ that have a y-coordinate of -1.

Solution From the graph of $y = \tan x$ on the previous page, or using the fact that the graph of each cycle of the graph of $y = \tan x$ is one-to-one, we know that there is exactly one point from each cycle that has a y-coordinate of -1. The only halfway point of the principal cycle of $y = \tan x$ with a y-coordinate of -1 is $\left(-\dfrac{\pi}{4}, -1\right)$.

To find all other halfway points with a y-coordinate of -1, we use the fact that the tangent function is periodic with a period of $P = \pi$. Therefore, every halfway point with a y-coordinate of -1 must be a distance of plus or minus π units horizontally from the previous halfway point with a y-coordinate of -1.

We can express all halfway points with a y-coordinate of -1 as $\left(-\dfrac{\pi}{4} + k\pi, -1\right)$, where k is an integer. We now choose integer values of k, making sure that the resulting x-coordinate is in the interval $\left[-\pi, \dfrac{5\pi}{2}\right]$.

$k = -1$: $\left(-\dfrac{\pi}{4} + (-1)\pi, -1\right) \rightarrow \left(\cancel{-\dfrac{5\pi}{4}, -1}\right)$ $\left(x = -\dfrac{5\pi}{4} \text{ is not in the interval } \left[-\pi, \dfrac{5\pi}{2}\right]\right)$

$k = 1$: $\left(-\dfrac{\pi}{4} + (1)\pi, -1\right) \rightarrow \left(\dfrac{3\pi}{4}, -1\right)$

$k = 2$: $\left(-\dfrac{\pi}{4} + (2)\pi, -1\right) \rightarrow \left(\dfrac{7\pi}{4}, -1\right)$

$k = 3$: $\left(-\dfrac{\pi}{4} + (3)\pi, -1\right) \rightarrow \left(\cancel{\dfrac{11\pi}{4}, -1}\right)$ $\left(x = \dfrac{11\pi}{4} \text{ is not in the interval } \left[-\pi, \dfrac{5\pi}{2}\right]\right)$

See Figure 42.

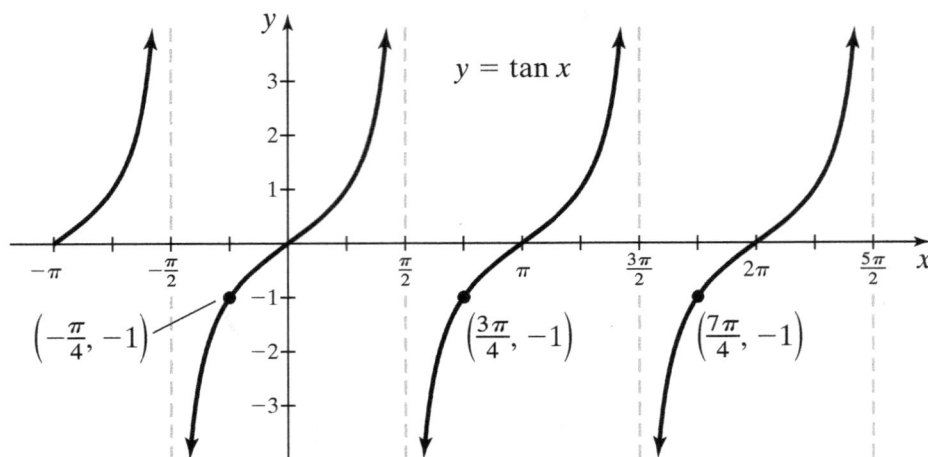

Figure 42 All halfway points of $y = \tan x$ on the interval $\left[-\pi, \dfrac{5\pi}{2}\right]$ that have a y-coordinate of -1

My video summary ▷ The three halfway points of $y = \tan x$ that have a y-coordinate of -1 on the interval $\left[-\pi, \dfrac{5\pi}{2}\right]$ are

$$\left(-\frac{\pi}{4}, -1\right), \left(\frac{3\pi}{4}, -1\right), \text{ and } \left(\frac{7\pi}{4}, -1\right).$$

You may want to watch this video to see this solution worked out in detail.

You Try It Work through this You Try It problem.

Work Exercises 1–5 in this textbook or in the MyMathLab Study Plan.

OBJECTIVE 2 SKETCHING FUNCTIONS OF THE FORM $y = A \tan (Bx - C) + D$

We will now sketch variations of the tangent function. These functions have the form $y = A \tan (Bx - C) + D$. Just like the graph of $y = \tan x$, the graph of $y = A \tan (Bx - C) + D$ has infinitely many cycles. We will focus on sketching the principal cycle. The principal cycle of $y = A \tan (Bx - C) + D$ corresponds to the principal cycle of $y = \tan x$ and has two vertical asymptotes, one center point, and two halfway points. To sketch the principal cycle of functions of the form $y = A \tan (Bx - C) + D$, we follow the process outlined below.

Steps for Sketching Functions of the Form $y = A \tan (Bx - C) + D$

Step 1. If $B < 0$, rewrite the function in an equivalent form such that $B > 0$. Use the odd property of the tangent function.

We now use this new form to determine $A, B, C,$ and D.

Step 2. Determine the interval and the equations of the vertical asymptotes of the principal cycle. The interval for the principal cycle can be found by solving the inequality $-\dfrac{\pi}{2} < Bx - C < \dfrac{\pi}{2}$.

The vertical asymptotes of the principal cycle occur at the endpoints of the interval of the principal cycle.

Step 3. The period is $P = \dfrac{\pi}{B}$.

Step 4. Determine the center point of the principal cycle of $y = A \tan (Bx - C) + D$. The x-coordinate of the center point is located midway between the vertical asymptotes of the principal cycle. The y-coordinate of the center point is D. Note that when $D = 0$, the x-coordinate of the center point is the x-intercept.

Step 5. Determine the coordinates of the two halfway points of the principal cycle of $y = A \tan (Bx - C) + D$. Each x-coordinate of a halfway point is located halfway between the x-coordinate of the center point and a vertical asymptote. The y-coordinates of these points are A times the y-coordinate of the corresponding halfway point of $y = \tan x$ plus D.

Step 6. Sketch the vertical asymptotes, plot the center point, and plot the two halfway points. Connect these points with a smooth curve. Complete the sketch, showing appropriate behavior of the graph as it approaches each asymptote.

My interactive video summary

⊙ **Example 2** Sketching graphs of the form $y = A \tan (Bx - C) + D$

For each function, determine the interval for the principal cycle. Then for the principal cycle, determine the equations of the vertical asymptotes, the coordinates of the center point, and the coordinates of the halfway points. Sketch the graph.

a. $y = \tan\left(x - \dfrac{\pi}{6}\right)$ **b.** $y = 4\tan(\pi - 2x) + 3$ **c.** $y = \dfrac{1}{2}\tan(3x) - 1$

Solution

a. Step 1. For the function $y = \tan\left(x - \dfrac{\pi}{6}\right)$, we see that $B > 0$.

Therefore, there is no need to rewrite the function and we can skip **step 1**.

Note that $A = 1$, $B = 1$, $C = \dfrac{\pi}{6}$, and $D = 0$.

Step 2. The interval for the principal cycle can be found by solving the inequality $-\dfrac{\pi}{2} < x - \dfrac{\pi}{6} < \dfrac{\pi}{2}$.

$-\dfrac{\pi}{2} < x - \dfrac{\pi}{6} < \dfrac{\pi}{2}$ Write the inequality.

$-\dfrac{\pi}{2} + \dfrac{\pi}{6} < x < \dfrac{\pi}{2} + \dfrac{\pi}{6}$ Add $\dfrac{\pi}{6}$ to all parts of the inequality.

$-\dfrac{3\pi}{6} + \dfrac{\pi}{6} < x < \dfrac{3\pi}{6} + \dfrac{\pi}{6}$ Get a common denominator.

$-\dfrac{2\pi}{6} < x < \dfrac{4\pi}{6}$ Combine terms.

$-\dfrac{\pi}{3} < x < \dfrac{2\pi}{3}$ Simplify.

The principal cycle occurs on the interval $\left(-\dfrac{\pi}{3}, \dfrac{2\pi}{3}\right)$. Therefore, the equations of the vertical asymptotes of the principal cycle are $x = -\dfrac{\pi}{3}$ and $x = \dfrac{2\pi}{3}$.

Step 3. The period is $P = \dfrac{\pi}{B} = \dfrac{\pi}{1} = \pi$.

Step 4. The x-coordinate of the center point of $y = \tan\left(x - \dfrac{\pi}{6}\right)$ is located midway between the two vertical asymptotes. Therefore, the x-coordinate of the center point is $x = \dfrac{-\dfrac{\pi}{3} + \dfrac{2\pi}{3}}{2} = \dfrac{\dfrac{\pi}{3}}{2} = \dfrac{\pi}{3} \cdot \dfrac{1}{2} = \dfrac{\pi}{6}$. The y-coordinate of the center point is $D = 0$. Thus, the coordinates of the center point are $\left(\dfrac{\pi}{6}, 0\right)$.

Step 5. The x-coordinate of the halfway point located to the *left* of the center point is $-\dfrac{\pi}{12}$ and the x-coordinate of the halfway point located to the *right* of the center point is $\dfrac{5\pi}{12}$. To see how to obtain these values, read these steps. The y-coordinates of these two halfway points will be $A = 1$ times the corresponding y-coordinate of the halfway point of the principal cycle of $y = \tan x$ plus $D = 0$.

Halfway points of the principal cycle of $y = \tan x$: $\left(-\dfrac{\pi}{4}, -1\right)$, $\left(\dfrac{\pi}{4}, 1\right)$

$$1 \cdot (-1) + 0 = -1 \qquad 1 \cdot 1 + 0 = 1$$

Halfway points of the principal cycle of $y = \tan\left(x - \dfrac{\pi}{6}\right)$: $\left(-\dfrac{\pi}{12}, -1\right)$, $\left(\dfrac{5\pi}{12}, 1\right)$

My interactive video summary

⊙ **Step 6.** We now sketch the vertical asymptotes,

$$x = -\dfrac{\pi}{3} \text{ and } x = \dfrac{2\pi}{3};$$

plot the center point $\left(\dfrac{\pi}{6}, 0\right)$; and plot the two halfway points,

$$\left(-\dfrac{\pi}{12}, -1\right) \text{ and } \left(\dfrac{5\pi}{12}, 1\right).$$

We connect these three special points with a smooth curve showing the appropriate behavior of the graph as it approaches each asymptote. Watch this interactive video to see how to sketch the graph of $y = \tan\left(x - \dfrac{\pi}{6}\right)$.

The graph of $y = \tan\left(x - \dfrac{\pi}{6}\right)$ can be seen in Figures 43 and 44.

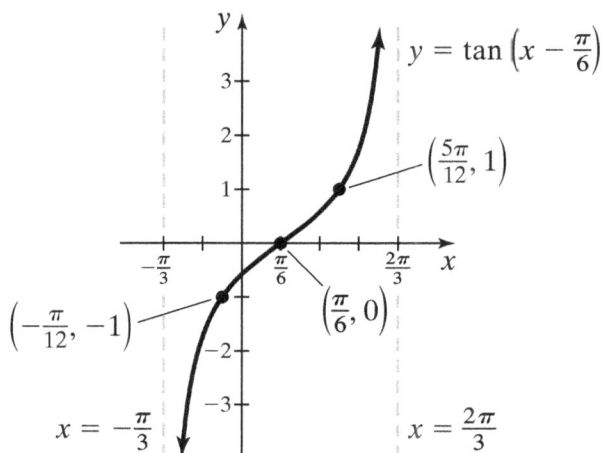

Figure 43 The principal cycle of the graph of

$$y = \tan\left(x - \frac{\pi}{6}\right)$$

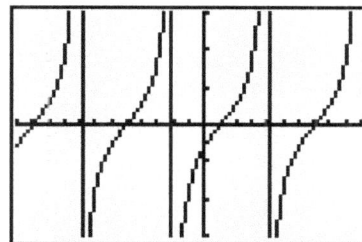

Using Technology

Figure 44 Several cycles of the graph of

$$y = \tan\left(x - \frac{\pi}{6}\right) \text{ using a graphing utility}$$

b. Step 1. For the function $y = 4\tan(\pi - 2x) + 3$, we see that $B = -2$, which is less than zero. Therefore, we will rewrite the function as an equivalent function using a positive value of B.

$y = 4\tan(\pi - 2x) + 3$	Write the original function.
$y = 4\tan(-2x + \pi) + 3$	Reorder the terms.
$y = 4\tan(-(2x - \pi)) + 3$	Factor out a negative inside the parentheses.
$y = -4\tan(2x - \pi) + 3$	Use the odd property of $y = \tan x$: $\tan(-x) = -\tan x$.

We see that the function $y = 4\tan(\pi - 2x) + 3$ is equivalent to $y = -4\tan(2x - \pi) + 3$. We now continue using the function $y = -4\tan(2x - \pi) + 3$ with $A = -4$, $B = 2$, $C = \pi$, and $D = 3$.

Step 2. The interval for the principal cycle can be found by solving the inequality $-\frac{\pi}{2} < 2x - \pi < \frac{\pi}{2}$.

$-\dfrac{\pi}{2} < 2x - \pi < \dfrac{\pi}{2}$	Write the inequality.
$-\dfrac{\pi}{2} + \pi < 2x < \dfrac{\pi}{2} + \pi$	Add π to all parts of the inequality.
$-\dfrac{\pi}{2} - \dfrac{2\pi}{2} < 2x < \dfrac{\pi}{2} + \dfrac{2\pi}{2}$	Get a common denominator.
$\dfrac{\pi}{2} < 2x < \dfrac{3\pi}{2}$	Simplify.
$\dfrac{\pi}{4} < x < \dfrac{3\pi}{4}$	Divide all parts of the inequality by 2.

The principal cycle occurs on the interval $\left(\dfrac{\pi}{4}, \dfrac{3\pi}{4}\right)$. Thus, the equations of the vertical asymptotes of the principal cycle are $x = \dfrac{\pi}{4}$ and $x = \dfrac{3\pi}{4}$.

Step 3. The period is $P = \dfrac{\pi}{B} = \dfrac{\pi}{2}$.

Step 4. The x-coordinate of the center point of $y = -4\tan(2x - \pi) + 3$ is located midway between the two vertical asymptotes. Thus, the x-coordinate of the center point is

$$x = \frac{\dfrac{\pi}{4} + \dfrac{3\pi}{4}}{2} = \frac{\dfrac{4\pi}{4}}{2} = \frac{\pi}{2}.$$

The y-coordinate of the center point is $D = 3$. Therefore, the coordinates of the center point of

$$y = -4\tan(2x - \pi) + 3 \text{ are } \left(\frac{\pi}{2}, 3\right).$$

Step 5. The x-coordinates of the two halfway points are $x = \dfrac{3\pi}{8}$ and $x = \dfrac{5\pi}{8}$.

To see how to obtain these values, read these steps. The y-coordinates of these two halfway points will be $A = -4$ times the corresponding y-coordinate of the halfway point of the principal cycle of $y = \tan x$ plus $D = 3$.

Halfway points of the principal cycle of $y = \tan x$: $\left(-\dfrac{\pi}{4}, -1\right)$ $\left(\dfrac{\pi}{4}, 1\right)$

$(-4)\cdot(-1) + 3 = 7 \quad (-4)\cdot(1) + 3 = -1$

Halfway points of the principal cycle of $y = -4\tan(2x - \pi) + 3$: $\left(\dfrac{3\pi}{8}, 7\right)$ $\left(\dfrac{5\pi}{8}, -1\right)$

My interactive video summary

Step 6. We now sketch the vertical asymptotes,

$$x = \frac{\pi}{4} \text{ and } x = \frac{3\pi}{4};$$

plot the center point $\left(\dfrac{\pi}{2}, 3\right)$; and plot the halfway points $\left(\dfrac{3\pi}{8}, 7\right)$ and $\left(\dfrac{5\pi}{8}, -1\right)$. We connect these three special points with a smooth curve showing the appropriate behavior of the graph as it approaches each asymptote. Watch this interactive video to see how to sketch the graph of $y = 4\tan(\pi - 2x) + 3$. The graph of $y = 4\tan(\pi - 2x) + 3$ can be seen in Figures 45 and 46.

$y = 4\tan(\pi - 2x) + 3$

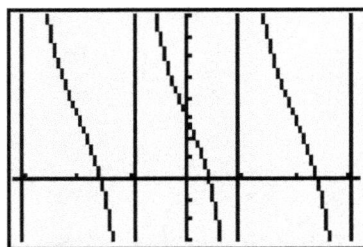

Figure 45 The principal cycle of the graph of $y = 4\tan(\pi - 2x) + 3$

Using Technology

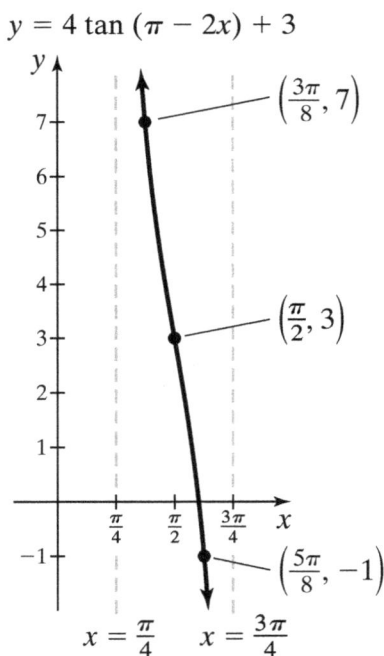

Figure 46 Several cycles of the graph of $y = 4\tan(\pi - 2x) + 3$ using a graphing utility

My interactive
video summary

> **c.** See if you can sketch the graph of $y = \dfrac{1}{2}\tan(3x) - 1$ on your own. Watch
this interactive video to see if you are correct.

We could have used transformations to sketch the functions from Example 2.
Click on the Show Graph link under the following functions to see how to sketch
the function using transformations.

$$y = \tan\left(x - \frac{\pi}{6}\right) \qquad y = 4\tan(\pi - 2x) + 3 \qquad y = \frac{1}{2}\tan(3x) - 1$$

Show Graph   Show Graph   Show Graph

You Try It   Work through this You Try It problem.

Work Exercises 6–19 in this textbook or in the MyMathLab Study Plan.

OBJECTIVE 3   UNDERSTANDING THE GRAPH OF THE COTANGENT FUNCTION
AND ITS PROPERTIES

My video summary

> We can establish the graph of $y = \cot x = \dfrac{\cos x}{\sin x}$ in much the same way
as we did with the tangent function. Watch this video to see how to establish
the principal cycle of the graph of $y = \cot x$ seen in Figure 42. The period of
$y = \cot x$ is $P = \pi$ and the graph of the principal cycle of the cotangent function occurs on the interval $(0, \pi)$. The equations of the vertical asymptotes of
the principal cycle are $x = 0$ (the y-axis) and $x = \pi$. The center point of the
principal cycle is $\left(\dfrac{\pi}{2}, 0\right)$ and the two halfway points of the principal cycle
are $\left(\dfrac{\pi}{4}, 1\right)$ and $\left(\dfrac{3\pi}{4}, -1\right)$. Also note that the graph of each cycle of $y = \cot x$
is one-to-one.

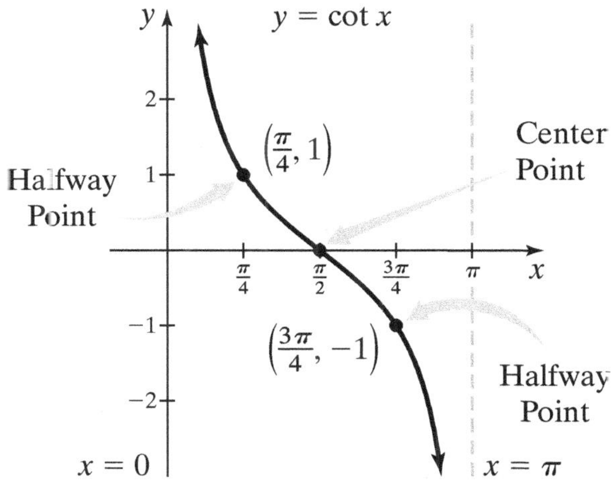

Figure 47 The principal cycle of the graph
of $y = \cot x$ on the interval $(0, \pi)$
with the center point and the two
halfway points labeled

We now state the characteristics of the cotangent function.

Characteristics of the Cotangent Function

The domain is $\{x \mid x \neq n\pi, \text{ where } n \text{ is an integer}\}$.
The range is $(-\infty, \infty)$.
The function is periodic with a period of $P = \pi$.
The principal cycle of the graph occurs on the interval $(0, \pi)$.

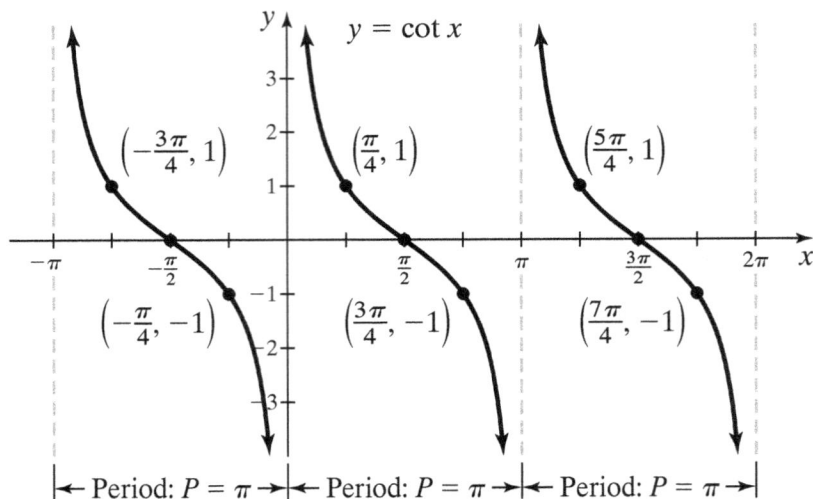

The function has infinitely many vertical asymptotes with equations $x = n\pi$, where n is an integer.
For each cycle, there is one center point. The x-coordinates of the center points are also the x-intercepts, or zeros, and are of the form $(2n + 1) \cdot \dfrac{\pi}{2}$, where n is an integer.
For each cycle, there are two halfway points. The halfway point to the left of the x-intercept has a y-coordinate of 1. The halfway point to the right of the x-intercept has a y-coordinate of -1.
The function is odd, which means $\cot(-x) = -\cot x$. The graph is symmetric about the origin.
The graph of one cycle is one-to-one.

My video summary ⓥ **Example 3** Understanding the Graph of $y = \cot x$

List all points on the graph of $y = \cot x$ on the interval $[-2\pi, 2\pi]$ that have a y-coordinate of $-\sqrt{3}$.

Solution From the graph of $y = \cot x$ and using the property that each cycle of the graph of $y = \cot x$ is one-to-one, we know that there is exactly one point from each cycle that has a y-coordinate of $-\sqrt{3}$. Using our knowledge of right triangle trigonometry, we know that $\cot\left(\dfrac{\pi}{6}\right) = \sqrt{3}$. To see why this is true, read this explanation.

Therefore, the point $\left(\dfrac{\pi}{6}, \sqrt{3}\right)$ lies on the graph of $y = \cot x$. Using the fact that the cotangent function is odd, we know that the point $\left(-\dfrac{\pi}{6}, -\sqrt{3}\right)$ also lies on the graph. This point lies on the given interval $[-2\pi, 2\pi]$. To find all other points with a y-coordinate of $-\sqrt{3}$, we use the fact that the cotangent function is periodic with a period of $P = \pi$. Therefore, every point with a y-coordinate of $-\sqrt{3}$ must be of the form $\left(-\dfrac{\pi}{6} + k\pi, -\sqrt{3}\right)$, where k is an integer. We now choose negative integer values of k and positive integer values of k, making sure that the resulting x-coordinate is in the interval $[-2\pi, 2\pi]$.

$k = -2$: $\left(-\dfrac{\pi}{6} + (-2)\pi, -\sqrt{3}\right) \to \cancel{\left(-\dfrac{13\pi}{6}, -\sqrt{3}\right)}$ $\left(x = -\dfrac{13\pi}{6} \text{ is not in the interval } [-2\pi, 2\pi]\right)$

$k = -1$: $\left(-\dfrac{\pi}{6} + (-1)\pi, -\sqrt{3}\right) \to \left(-\dfrac{7\pi}{6}, -\sqrt{3}\right)$

$k = 1$: $\left(-\dfrac{\pi}{6} + (1)\pi, -\sqrt{3}\right) \to \left(\dfrac{5\pi}{6}, -\sqrt{3}\right)$

$k = 2$: $\left(-\dfrac{\pi}{6} + (2)\pi, -\sqrt{3}\right) \to \left(\dfrac{11\pi}{6}, -\sqrt{3}\right)$

$k = 3$: $\left(-\dfrac{\pi}{6} + (3)\pi, -\sqrt{3}\right) \to \cancel{\left(\dfrac{17\pi}{6}, -\sqrt{3}\right)}$ $\left(x = \dfrac{17\pi}{6} \text{ is not in the interval } [-2\pi, 2\pi]\right)$

See Figure 48.

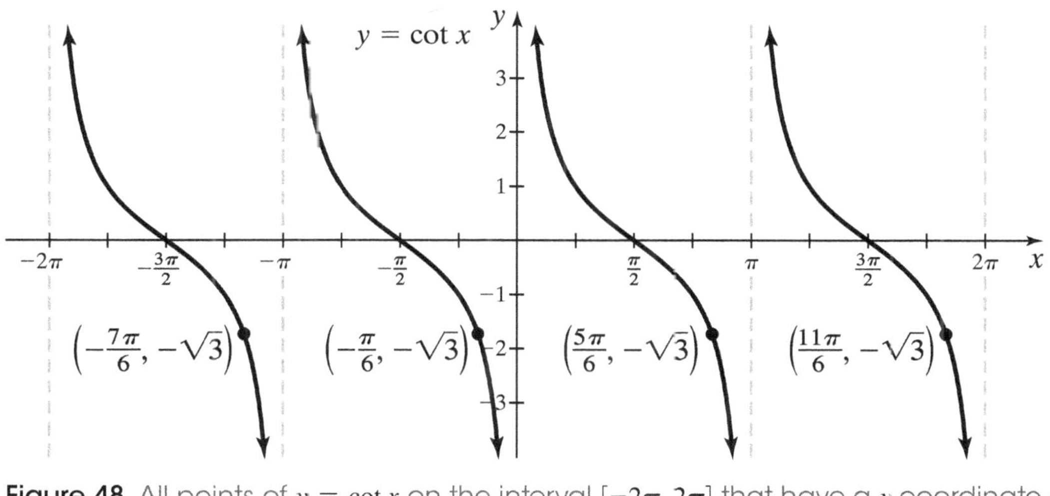

Figure 48 All points of $y = \cot x$ on the interval $[-2\pi, 2\pi]$ that have a y-coordinate of $-\sqrt{3}$

My video summary ▶ The four points of $y = \cot x$ on the interval $[-2\pi, 2\pi]$ that have a y-coordinate of $-\sqrt{3}$ are

$$\left(-\dfrac{7\pi}{6}, -\sqrt{3}\right), \left(-\dfrac{\pi}{6}, -\sqrt{3}\right), \left(\dfrac{5\pi}{6}, -\sqrt{3}\right) \text{ and } \left(\dfrac{11\pi}{6}, -\sqrt{3}\right).$$

You may want to watch this video to see this solution worked out in detail.

You Try It Work through this You Try It problem.

Work Exercises 20–24 in this textbook or in the MyMathLab Study Plan.

OBJECTIVE 4   SKETCHING FUNCTIONS OF THE FORM $y = A \cot (Bx - C) + D$

To sketch the principal cycle of functions of the form $y = A \cot (Bx - C) + D$, we follow the process outlined below.

Steps for Sketching Functions of the Form $y = A \cot (Bx - C) + D$

Step 1. If $B < 0$, rewrite the function in an equivalent form such that $B > 0$. Use the odd property of the cotangent function.

We now use this new form to determine A, B, C, and D.

Step 2. Determine the interval and the equations of the vertical asymptotes of the principal cycle. The interval for the principal cycle can be found by solving the inequality $0 < Bx - C < \pi$. The vertical asymptotes of the principal cycle occur at the endpoints of the interval of the principal cycle.

Step 3. The period is $P = \dfrac{\pi}{B}$.

Step 4. Determine the center point of the principal cycle of $y = A \cot (Bx - C) + D$. The x-coordinate of the center point is located midway between the vertical asymptotes of the principal cycle. The y-coordinate of the center point is D. Note that when $D = 0$ the x-coordinate of the center point is the x-intercept.

Step 5. Determine the coordinates of the two halfway points of the principal cycle of $y = A \cot (Bx - C) + D$ Each x-coordinate of a halfway point is located halfway between the x-coordinate of the center point and a vertical asymptote. The y-coordinates of these points are A times the y-coordinate of the corresponding halfway point of $y = \cot x$ plus D.

Step 6. Sketch the vertical asymptotes, plot the center point, and plot the two halfway points. Connect these points with a smooth curve. Complete the sketch showing appropriate behavior of the graph as it approaches each asymptote.

⬙ *My interactive video summary*

⊙ **Example 4** Sketching graphs of the form $y = A \cot (Bx - C) + D$

For each function, determine the interval for the principal cycle. Then for the principal cycle, determine the equations of the vertical asymptotes, the coordinates of the center point, and the coordinates of the halfway points. Sketch the graph.

a. $y = \cot (2x + \pi) + 1$ **b.** $y = -3 \cot \left(x - \dfrac{\pi}{4} \right)$

Solution

a. Step 1. For the function $y = \cot (2x + \pi) + 1$, we see that $B > 0$. Therefore, there is no need to rewrite the function, and we can skip **step 1**. Note that $A = 1$, $B = 2$, $C = -\pi$, and $D = 1$.

    Step 2. The interval for the principal cycle can be found by solving the inequality $0 < 2x + \pi < \pi$.

$$0 < 2x + \pi < \pi \qquad \text{Write the inequality.}$$

$$0 - \pi < 2x < \pi - \pi \qquad \text{Subtract } \pi \text{ from all parts of the inequality.}$$

$$-\pi < 2x < 0 \qquad \text{Simplify.}$$

$$-\frac{\pi}{2} < x < 0 \qquad \text{Divide all parts of the inequality by 2.}$$

The principal cycle occurs on the interval $\left(-\frac{\pi}{2}, 0\right)$. The equations of the vertical asymptotes of the principal cycle are $x = -\frac{\pi}{2}$ and $x = 0$.

Step 3. The period is $P = \frac{\pi}{B} = \frac{\pi}{2}$.

Step 4. The x-coordinate of the center point of $y = \cot(2x + \pi) + 1$ is located midway between the two vertical asymptotes. Therefore, the

x-coordinate of the center point is $x = \dfrac{-\frac{\pi}{2} + 0}{2} = -\frac{\pi}{2} \div 2 = -\frac{\pi}{2} \cdot \frac{1}{2} = -\frac{\pi}{4}$.

The y-coordinate of the center point is $D = 1$. Thus, the coordinates of the center point are $\left(-\frac{\pi}{4}, 1\right)$.

Step 5 There are two halfway points in the interval $\left(-\frac{\pi}{2}, 0\right)$. The x-coordinate of the halfway point located to the *left* of the center point is $-\frac{3\pi}{8}$ and the x-coordinate of the halfway point located to the *right* of the center point is $-\frac{\pi}{8}$. To see how to obtain these values, read these steps. Now, list the y-coordinates of the two halfway points of the principal cycle of $y = \cot x$. Next, multiply these values by $A = 1$ and add $D = 1$ to get the y-coordinates of the halfway points.

Halfway points of the principal cycle of $y = \cot x$: $\left(\frac{\pi}{4}, 1\right) \quad \left(\frac{3\pi}{4}, -1\right)$

$$(1)\cdot(1) + 1 = 2 \quad\Big|\quad (1)\cdot(-1) + 1 = 0$$

Halfway points of the principal cycle of $y = \cot(2x + \pi)$: $\left(-\frac{3\pi}{8}, 2\right) \quad \left(-\frac{\pi}{8}, 0\right)$

My interactive video summary

Step 6. We now sketch the vertical asymptotes, $x = -\frac{\pi}{2}$ and $x = 0$; plot the center point $\left(-\frac{\pi}{4}, 1\right)$; and plot the halfway points $\left(-\frac{3\pi}{8}, 2\right)$ and $\left(-\frac{\pi}{8}, 0\right)$. We connect these points with a smooth curve showing the appropriate behavior of the graph as it approaches each asymptote. Watch this interactive video to see how to sketch the graph of $y = \cot(2x + \pi) + 1$. The graph of $y = \cot(2x + \pi) + 1$ can be seen in Figures 49 and 50.

$$y = \cot (2x + \pi) + 1$$

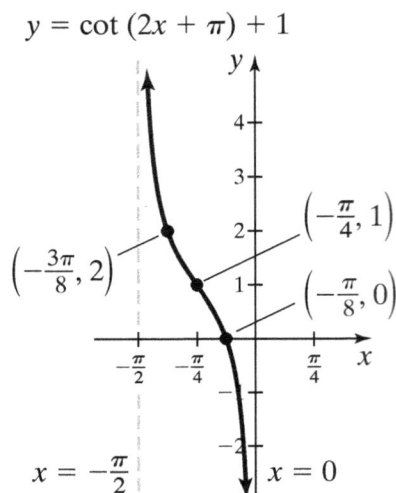

Figure 49 The principal cycle
of the graph of
$y = \cot (2x + \pi) + 1$

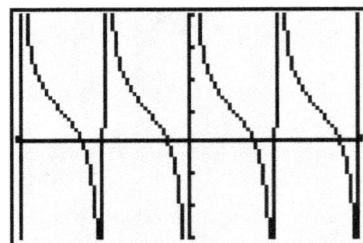

Using Technology

Figure 50 Several cycles of the graph of
$y = \cot (2x + \pi) + 1$ using a
graphing utility

*My interactive
video summary*

⊙ **b.** See if you can sketch the graph of $y = -3 \cot \left(x - \dfrac{\pi}{4} \right)$ on your own. To see
how to sketch this graph, watch this interactive video.

We could have used transformations to sketch the functions from Example 4.
Click on the Show Graph link under the following functions to see how to
sketch the function using transformations.

$$y = \cot (2x + \pi) + 1 \qquad y = -3 \cot \left(x - \dfrac{\pi}{4} \right)$$

Show Graph   Show Graph

You Try It Work through this You Try It problem.

Work Exercises 25–38 in this textbook or in the My MathLab **Study Plan.**

OBJECTIVE 5 DETERMINE THE EQUATION OF A FUNCTION OF THE FORM
$y = A \tan (Bx - C) + D$ OR $y = A \cot (Bx - C) + D$ GIVEN ITS GRAPH

We now know how to sketch the graphs of $y = A \tan (Bx - C) + D$ and
$y = A \cot (Bx - C) + D$. Suppose that we are given the principal cycle of the
graph whose function is given by $y = A \tan (Bx - C) + D$ or $y = A \cot (Bx - C) + D$
for $B > 0$. (We will assume that the value of B is positive. Otherwise, there could be
more than one correct answer because of the odd nature of the tangent and
cotangent functions.) To determine the proper function we must first identify the
following five characteristics of the graph of the given principal cycle.

1. $x = a$, the equation of the left-most vertical asymptote

2. $x = b$, the equation of the right-most vertical asymptote

3. (x_1, y_1), the coordinates of the left-most halfway point

4. (x_2, y_2), the coordinates of the center point

5. (x_3, y_3), the coordinates of the right-most halfway point

We will use these five characteristics to help us to determine the values of A, B, C and D. Figure 51 on the following page illustrates the principal cycle of a possible tangent or cotangent curve having the five characteristics above.

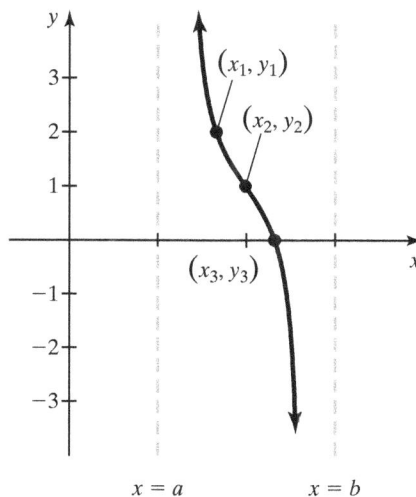

Figure 51 The principal cycle of the graph of a tangent or cotangent function of the form $y = A \tan (Bx - C) + D$ or $y = A \cot (Bx - C) + D$.

Note that the period of the graph seen in Figure 51 must be $b - a$, the distance between the right-most vertical asymptote and the left-most vertical asymptote. Using the fact that the period of a tangent curve or cotangent curve is $P = \dfrac{\pi}{B}$, we can easily determine the value of B.

Recall that the interval for the principal cycle of a tangent curve is $-\dfrac{\pi}{2} < Bx - C < \dfrac{\pi}{2}$ whereas the interval for the principal cycle of a cotangent curve is $0 < Bx - C < \pi$. If we solve these two inequalities, we get

$$\underbrace{\frac{-\dfrac{\pi}{2} + C}{B}}_{a} < x < \underbrace{\frac{\dfrac{\pi}{2} + C}{B}}_{b} \quad \text{and} \quad \underbrace{\frac{C}{B}}_{a} < x < \underbrace{\frac{\pi + C}{B}}_{b} \text{ respectively.}$$

Therefore, if we are given the principal cycle of the graph of a tangent function or a cotangent function whose vertical asymptotes are $x = a$, $x = b$ where $a < b$, then we can establish the following equations:

If the graph is a tangent curve, then $\dfrac{-\dfrac{\pi}{2} + C}{B} = a$ and $\dfrac{\dfrac{\pi}{2} + C}{B} = b$.

If the graph is a cotangent curve, then $\dfrac{C}{B} = a$ and $\dfrac{\pi + C}{B} = b$.

These equations can be used to determine the value of C. We can determine the values of A and D by comparing the given half-way points and the center point with the known half-way points and center point of the graph of $y = \tan x$ or $y = \cot x$. We can follow the five steps outlined on the following page to determine the equation of the function that describes a given tangent or cotangent curve.

Steps for Determining the Equation of a Function of the Form $y = A \tan (Bx - C) + D$ or $y = A \cot (Bx - C) + D$ Given the Graph

Step 1. Use the two vertical asymptotes of the given principal cycle of the graph to determine the period. If $x = a$ and $x = b$ where $a < b$ are the equations of the vertical asymptotes of the principal cycle, then the period is $P = b - a$.

Step 2. Use the equation $P = \dfrac{\pi}{B}$ to determine the value of $B > 0$.

Step 3. If the given graph is a tangent curve, then use the equation $\dfrac{-\dfrac{\pi}{2} + C}{B} = a$

or $\dfrac{\dfrac{\pi}{2} + C}{B} = b$ to determine the value of C.

If the given graph is a cotangent curve, then use the equation $\dfrac{C}{B} = a$ or

$\dfrac{\pi + C}{B} = b$ to determine the value of C.

Step 4. The y-coordinate of the center point represents the value of D.

Step 5. Identify the halfway points (x_1, y_1) and (x_3, y_3) that lie on the given graph.

If the given graph is a **tangent curve**, then use the relationship between (x_3, y_3), the coordinates of the **right-most** halfway point of the given graph, and $\left(\dfrac{\pi}{4}, 1\right)$, the **right-most** halfway point of the principal cycle of the graph of $y = \tan x$ to get the equation $y_3 = A + D$. Use this information to determine the value of A.

If the given graph is a **cotangent curve**, then use the relationship between (x_1, y_1), the coordinates of the **left-most** halfway point of the given graph, and $\left(\dfrac{\pi}{4}, 1\right)$, the **left-most** halfway point of the principal cycle of the graph of $y = \cot x$ to get the equation $y_1 = A + D$. Use this information to determine the value of A.

My interactive video summary

▶ **Example 5** Determining the Equation of a Function Given the Graph that it Represents

The principal cycle of the graphs of two trigonometric functions of the from $y = A \tan (Bx - C) + D$ or $y = A \cot (Bx - C) + D$ for $B > 0$ are given below. Determine the equation of the function represented by each graph.

a. $y = A\tan(Bx - C) + D$

b. $y = A\cot(Bx - C) + D$

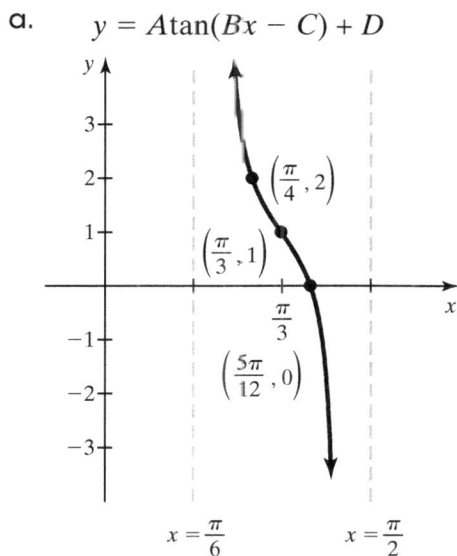

$\left(\dfrac{\pi}{4}, 2\right)$

$\left(\dfrac{\pi}{3}, 1\right)$

$\left(\dfrac{5\pi}{12}, 0\right)$

$x = \dfrac{\pi}{6}$ $x = \dfrac{\pi}{2}$

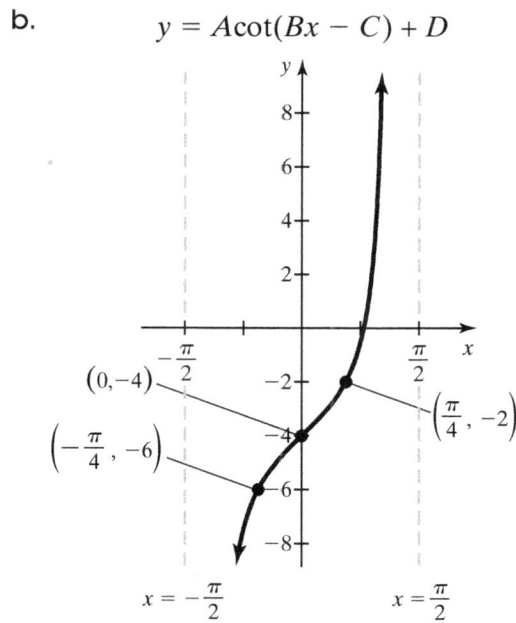

$(0, -4)$

$\left(-\dfrac{\pi}{4}, -6\right)$

$\left(\dfrac{\pi}{4}, -2\right)$

$x = -\dfrac{\pi}{2}$ $x = \dfrac{\pi}{2}$

Solution

a. First, use the given graph to identify the following.

1. The equation of the left-most vertical asymptote is $x = \dfrac{\pi}{6}$.

2. The equation of the right-most vertical asymptote is $x = \dfrac{\pi}{2}$.

3. The coordinates of the left-most halfway point are $(x_1, y_1) = \left(\dfrac{\pi}{4}, 2\right)$.

4. The coordinates of the center point are $(x_2, y_2) = \left(\dfrac{\pi}{3}, 1\right)$.

5. The coordinates of the right-most halfway point are $(x_3, y_3) = \left(\dfrac{5\pi}{12}, 0\right)$.

Step 1. The period is the distance between the two vertical asymptotes of the principal cycle of the given graph. Thus, the period is

$$P = b - a = \frac{\pi}{2} - \frac{\pi}{6} = \frac{\pi}{3}.$$

Step 2. The period is $P = \dfrac{\pi}{B} = \dfrac{\pi}{3}$. Therefore, $B = 3$.

Step 3. The given graph is a tangent curve. Therefore, we can use the equation

$$\frac{-\dfrac{\pi}{2} + C}{B} = a \text{ or } \frac{\dfrac{\pi}{2} + C}{B} = b \text{ to determine the value of } C.$$

$$\frac{\dfrac{\pi}{2} + C}{B} = b \qquad \text{Start with one of the equations.}$$

$$\frac{\dfrac{\pi}{2} + C}{3} = \frac{\pi}{2} \qquad \text{Substitute } B = 3 \text{ and } b = \frac{\pi}{2}.$$

2.3 The Graphs of the Tangent, Cotangent, Cosecant, and Secant Functions   2-79

$$\frac{\pi}{2} + C = \frac{3\pi}{2} \qquad \text{Multiply both sides of the equation by 3.}$$

$$C = \pi \qquad \text{Solve for } C.$$

Step 4. The y-coordinate of the center point is 1. Therefore, $D = 1$.

Step 5. Since the given graph is a tangent curve, we identify the right-most halfway point whose coordinates are $(x_3, y_3) = \left(\dfrac{5\pi}{12}, 0\right)$ and use the equation $y_3 = A + D$ to determine the value of A.

$$y_3 = A + D \qquad \text{Write the equation relating } y_3, A, \text{ and } D.$$

$$0 = A + 1 \qquad \text{Substitute } y_3 = 0 \text{ and } D = 1.$$

$$A = -1 \qquad \text{Solve for } A.$$

We see that $A = -1, B = 3, C = \pi$, and $D = 1$. Therefore, substituting these values into the function of the form $y = A \tan (Bx - C) + D$, we see that the function which describes the principal cycle of the given graph is $y = -\tan (3x - \pi) + 1$. You should sketch the graph of $y = -\tan (3x - \pi) + 1$ on your own to make sure that it matches the given graph. The graph of several cycles of the function $y = -\tan (3x - \pi) + 1$ is shown in Figure 52.

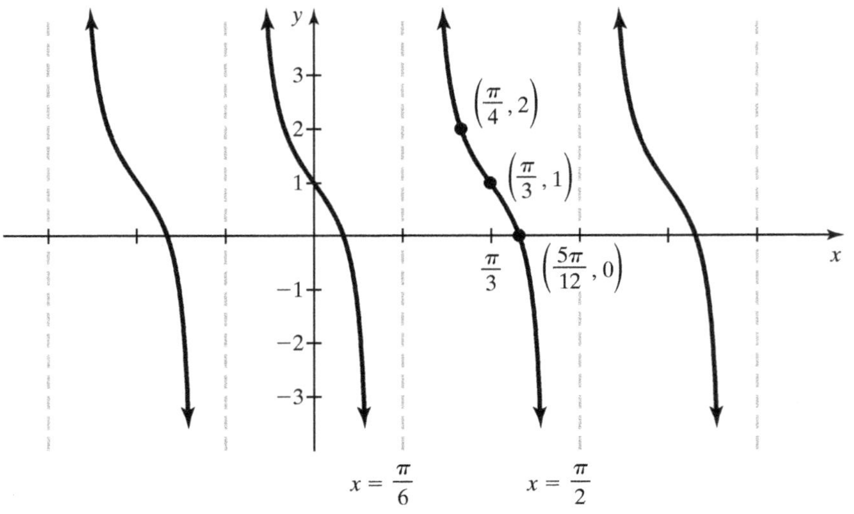

Figure 52 Several cycles of the graph of
$$y = -\tan (3x - \pi) + 1.$$

b. Work through this interactive video to verify that the function of the form $y = A \cot (Bx - C) + D$ whose graph is shown below is
$$y = -2 \cot \left(x + \frac{\pi}{2}\right) - 4.$$

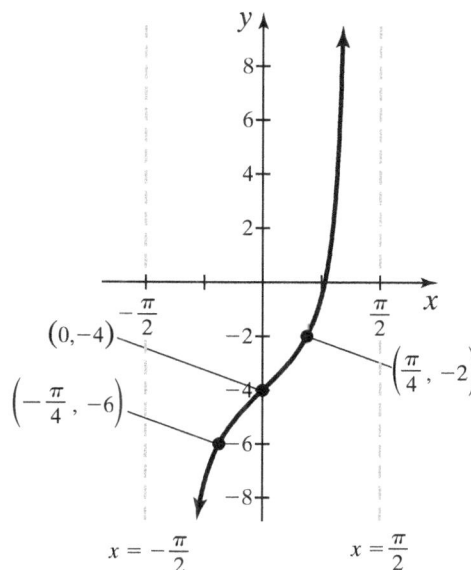

You Try It   Work through this You Try It problem.

Work Exercises 39–44 in this textbook or in the MyMathLab Study Plan.

OBJECTIVE 6   UNDERSTANDING THE GRAPHS OF THE COSECANT AND SECANT FUNCTIONS AND THEIR PROPERTIES

My interactive video summary

To sketch the graph of the cosecant function, we use a reciprocal identity to rewrite $y = \csc x$ as $y = \dfrac{1}{\sin x}$ and use our knowledge of the graph of the sine function to help sketch the graph of the cosecant function. Note that if x is any value such that $\sin x \neq 0$, then the value of $y = \csc x$ will be the reciprocal of the value of $\sin x$. If $\sin x = 0$, then the cosecant function is undefined. Therefore, the domain of the cosecant function is all real numbers except values of x for which x is a zero of $y = \sin x$. Thus, the domain is all real numbers except integer multiples of π. The graph of $y = \csc x$ has infinitely many vertical asymptotes of the form $x = n\pi$, where n is an integer. Recall that the graph of $y = \sin x$ obtains a relative maximum value of 1 at $x = \dfrac{\pi}{2} + 2\pi n$, where n is an integer. Because of the reciprocal relationship between $y = \sin x$ and $y = \csc x$, these values will be relative *minimums* of the graph of $y = \csc x$. Likewise, the relative minimum values of $y = \sin x$ are -1, which occur at $x = \dfrac{3\pi}{2} + 2\pi n$, where n is an integer. These values will be relative *maximums* of the graph of $y = \csc x$. Watch this interactive video to see how to establish the graph of $y = \csc x$ and its characteristics.

Characteristics of the Cosecant Function

The domain is $\{x \mid x \neq n\pi, \text{ where } n \text{ is an integer}\}$.

The range is $(-\infty, -1] \cup [1, \infty)$.

The function is periodic with a period of $P = 2\pi$.

The function has infinitely many vertical asymptotes with equations $x = n\pi$, where n is an integer.

The function obtains a relative maximum at $x = \dfrac{3\pi}{2} + 2\pi n$, where n is an integer. The relative maximum value is -1.

The function obtains a relative minimum at $x = \dfrac{\pi}{2} + 2\pi n$, where n is an integer. The relative minimum value is 1.

The function is odd, which means $\csc(-x) = -\csc x$. The graph is symmetric about the origin.

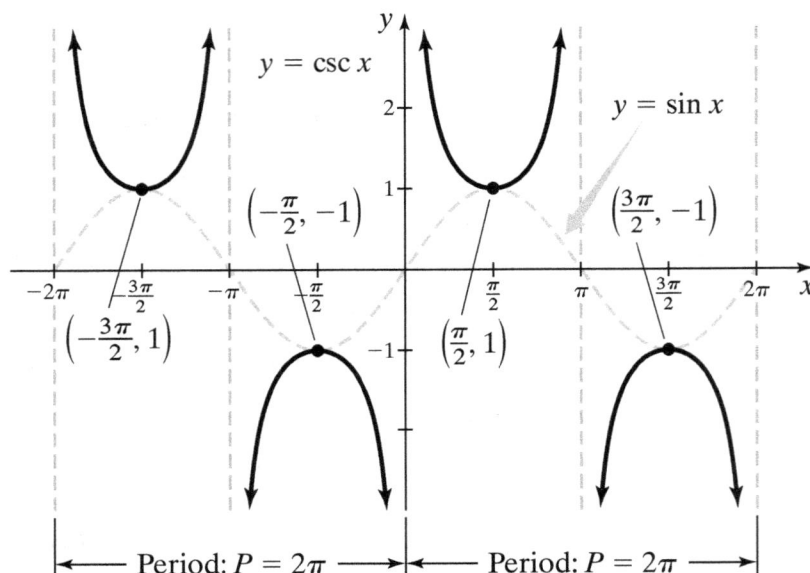

My interactive video summary

We can sketch the graph of $y = \sec x$ in much the same way as we did with the graph of $y = \csc x$. Because we can use a reciprocal identity to rewrite $y = \sec x$ as $y = \dfrac{1}{\cos x}$, we can use our knowledge of the graph of the cosine function to help sketch the graph of the secant function. Watch this interactive video to see how to establish the graph of $y = \sec x$ and its characteristics that are shown on the following page.

Characteristics of the Secant Function

The domain is $\left\{ x \mid x \neq (2n + 1) \cdot \dfrac{\pi}{2} \text{, where } n \text{ is an integer} \right\}$.

The range is $(-\infty, -1] \cup [1, \infty)$.

The function is periodic with a period of $P = 2\pi$.

The function has infinitely many vertical asymptotes with equations $x = (2n + 1) \cdot \dfrac{\pi}{2}$, where n is an integer.

The function obtains a relative maximum at $x = (2n + 1)\pi$, where n is an integer. The relative maximum value is -1.

The function obtains a relative minimum at $2\pi n$, where n is an integer. The relative minimum value is 1.

The function is even, which means $\sec(-x) = \sec x$. The graph is symmetric about the y-axis.

You Try It   Work through this You Try It problem.

Work Exercises 45–46 in this textbook or in the MyMathLab Study Plan.

OBJECTIVE 7   SKETCHING FUNCTIONS OF THE FORM $y = A \csc (Bx - C) + D$ AND $y = A \sec (Bx - C) + D$

To sketch functions of the form $y = A \csc (Bx - C) + D$ and $y = A \sec (Bx - C) + D$, we must first sketch the graph of the corresponding reciprocal function. The corresponding reciprocal function of $y = A \csc (Bx - C) + D$ is $y = A \sin (Bx - C) + D$ and the corresponding reciprocal function of $y = A \sec (Bx - C) + D$ is $y = A \cos (Bx - C) + D$. We sketch the corresponding reciprocal functions by following the steps that were outlined in Section 2.2. Before going on, make sure that you can sketch functions of the form $y = A \sin (Bx - C) + D$ and $y = A \cos (Bx - C) + D$. Click on the You Try It icon below to see if you understand how to sketch such functions.

You Try It

An example of sketching functions of the form
$y = A \sin (Bx - C) + D$ and $y = A \cos (Bx - C) + D$

To sketch functions of the form $y = A \csc (Bx - C) + D$ and $y = A \sec (Bx - C) + D$, we follow the steps outlined below.

Steps for Sketching Functions of the Form $y = A \csc (Bx - C) + D$ and $y = A \sec (Bx - C) + D$

Step 1. Lightly sketch at least two cycles of the corresponding reciprocal function using the process outlined in Section 2.2. If $D \neq 0$, lightly sketch two reciprocal functions, one with $D = 0$ and one with $D \neq 0$.

Step 2. Sketch the vertical asymptotes. The vertical asymptotes will correspond to the x-intercepts of the reciprocal function $y = A \sin (Bx - C)$ or $y = A \cos (Bx - C)$.

Step 3. Plot all maximum and minimum points on the graph of $y = A \sin (Bx - C) + D$ or $y = A \cos (Bx - C) + D$.

Step 4. Draw smooth curves through each point from step 3, making sure to approach the vertical asymptotes.

My interactive video summary

⊚ **Example 5** Sketching graphs of the form $y = A \csc (Bx - C) + D$ and $y = A \sec (Bx - C) + D$

Determine the equations of the vertical asymptotes and all relative maximum and relative minimum points of two cycles of each function and then sketch its graph.

a. $y = -2 \csc (x + \pi)$

b. $y = -\csc (\pi x) - 2$

c. $y = 3 \sec (\pi - x)$

d. $y = \sec \left(2x + \dfrac{\pi}{2} \right) + 1$

Solution The solution to parts a and d are shown below. If you would like to see the solution to parts b and c, watch this interactive video.

a. Step 1. Sketch the corresponding reciprocal function, $y = -2 \sin (x + \pi)$. Two complete cycles of $y = -2 \sin (x + \pi)$ are sketched in Figure 53.

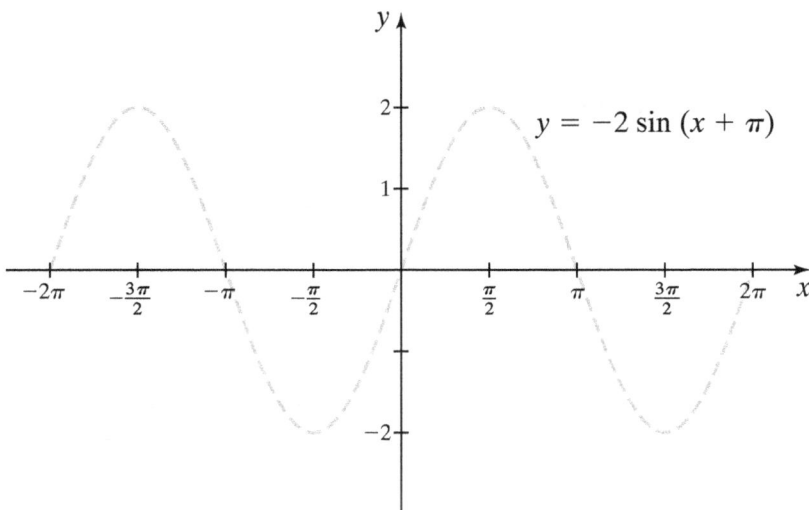

$y = -2 \sin (x + \pi)$

Figure 53 Two complete cycles of the graph of $y = -2 \sin (x + \pi)$

Step 2. As you can see in Figure 53, the zeros of $y = -2 \sin (x + \pi)$ on the interval $[-2\pi, 2\pi]$ are $-2\pi, -\pi, 0, \pi$, and 2π. Therefore, the equations of the vertical asymptotes of the graph of $y = -2 \csc (x + \pi)$ are $x = -2\pi, x = -\pi, x = 0, x = \pi$, and $x = 2\pi$.

Step 3. The relative maximum points of the graph of $y = -2 \csc (x + \pi)$ are $\left(-\dfrac{\pi}{2}, -2\right)$ and $\left(\dfrac{3\pi}{2}, -2\right)$. The relative minimum points are $\left(-\dfrac{3\pi}{2}, 2\right)$ and $\left(\dfrac{\pi}{2}, 2\right)$.

Step 4. We draw smooth curves through the points from step 3, making sure to approach the vertical asymptotes. The graph of $y = -2 \csc (x + \pi)$ can be seen in Figure 54 and Figure 55.

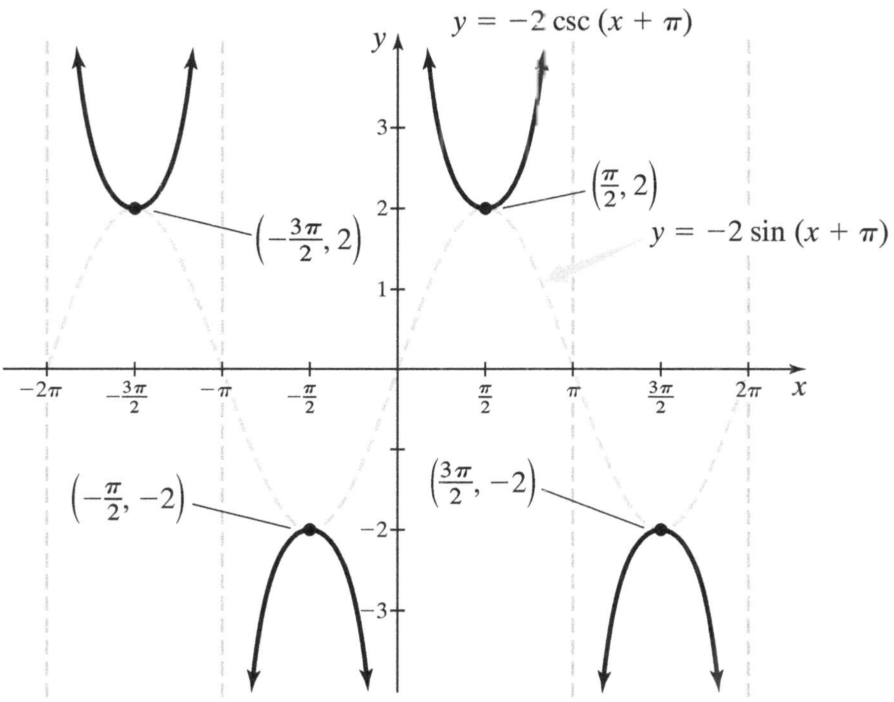

Figure 54 Two cycles of the graph of $y = -2 \csc (x + \pi)$

Using Technology

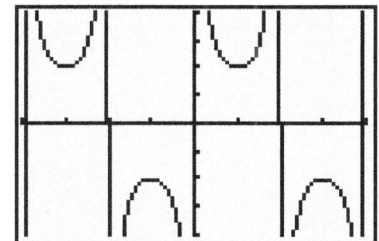

Figure 55 Two cycles of the graph of $y = -2 \csc (x + \pi)$ using a graphing utility

b. Step 1. Sketch the corresponding reciprocal functions,

$$y = \cos\left(2x + \frac{\pi}{2}\right) \text{ and } y = \cos\left(2x + \frac{\pi}{2}\right) + 1.$$

Two complete cycles of $y = \cos\left(2x + \dfrac{\pi}{2}\right)$ and $y = \cos\left(2x + \dfrac{\pi}{2}\right) + 1$ are sketched in Figure 56.

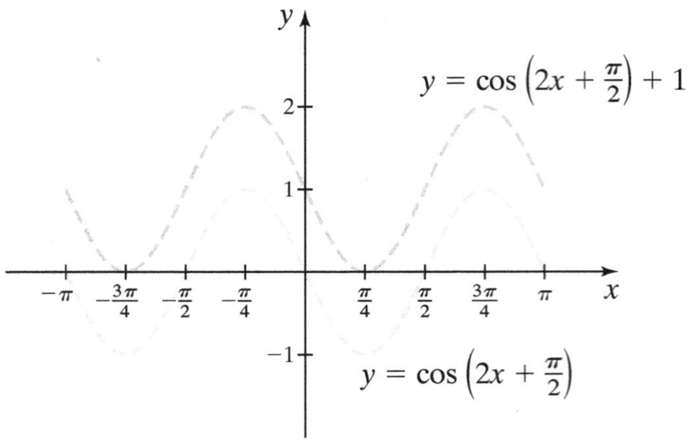

$$y = \cos\left(2x + \frac{\pi}{2}\right) + 1$$

$$y = \cos\left(2x + \frac{\pi}{2}\right)$$

Figure 56 Two cycles of the graphs of $y = \cos\left(2x + \frac{\pi}{2}\right)$ and $y = \cos\left(2x + \frac{\pi}{2}\right) + 1$

Step 2. As you can see in Figure 56, the zeros of

$y = \cos\left(2x + \frac{\pi}{2}\right)$ on the interval $[-\pi, \pi]$ are $-\pi, -\frac{\pi}{2}, 0, \frac{\pi}{2},$ and π.

Therefore, the equations of the vertical asymptotes of the graph of $y = \sec\left(2x + \frac{\pi}{2}\right) + 1$ are

$$x = -\pi, x = -\frac{\pi}{2}, x = 0 \, x = \frac{\pi}{2}, \text{ and } x = \pi.$$

Step 3. The relative maximum and relative minimum points of the graph of $y = \sec\left(2x + \frac{\pi}{2}\right) + 1$ on the interval $[-\pi, \pi]$ are

$$\left(-\frac{3\pi}{4}, 0\right), \left(-\frac{\pi}{4}, 2\right), \left(\frac{\pi}{4}, 0\right), \text{ and } \left(\frac{3\pi}{4}, 2\right).$$

Figure 57

Two cycles of the graphs

of $y = \sec\left(2x + \frac{\pi}{2}\right) + 1$

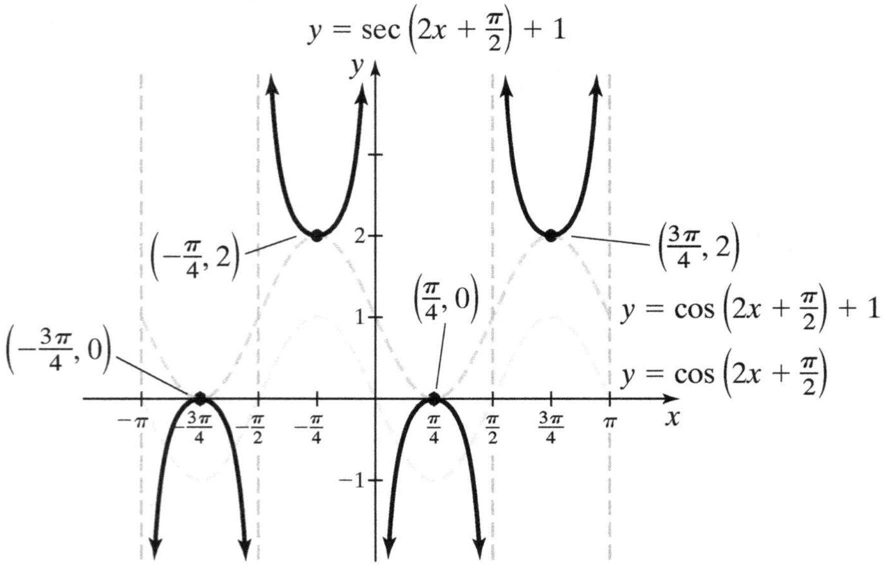

Step 4. We draw a smooth curve through the points from step 3, making sure to approach the vertical asymptotes. The graph of $y = \sec\left(2x + \frac{\pi}{2}\right) + 1$ can be seen in Figures 57 and 58.

$$y = \sec\left(2x + \frac{\pi}{2}\right) + 1$$

$\left(-\frac{\pi}{4}, 2\right)$

$\left(\frac{3\pi}{4}, 2\right)$

$\left(\frac{\pi}{4}, 0\right)$

$y = \cos\left(2x + \frac{\pi}{2}\right) + 1$

$\left(-\frac{3\pi}{4}, 0\right)$

$y = \cos\left(2x + \frac{\pi}{2}\right)$

Using Technology

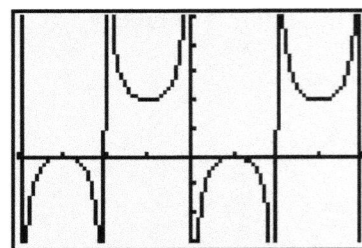

Figure 58 Two cycles of the graph of $y = \sec\left(2x + \frac{\pi}{2}\right) + 1$ using a graphing utility

To see how to sketch the functions from Example 5 using transformations, click on the Show Graph link under each function below.

$$y = -2\csc(x + \pi) \qquad y = \csc(\pi x) - 2 \qquad y = 3\sec(\pi - x) \qquad y = \sec\left(2x + \frac{\pi}{2}\right) + 1$$

Show Graph   Show Graph   Show Graph   Show Graph

You Try It   Work through this You Try It problem.

Work Exercises 47–56 in this textbook or in the MyMathLab Study Plan.

2.3 Exercises

Skill Check Exercises

In Exercises SCE-1 through SCE-4, solve the inequality. Write your answer in interval notation form.

SCE-1. $-\dfrac{\pi}{2} < 2x - \pi < \dfrac{\pi}{2}$

SCE-2. $0 < 3x + \pi < \pi$

SCE-3. $-\dfrac{\pi}{2} < \dfrac{1}{3}x < \dfrac{\pi}{2}$

SCE-4. $0 < \dfrac{1}{5}x < \pi$

1. Sketch the graph of $y = \tan x$, and then identify as many of the following properties that apply.

   The function is an even function.

   The function is an odd function.

   The domain is all real numbers except odd integer multiples of $\dfrac{\pi}{2}$.

   The range is $(-\infty, \infty)$.

   The function has a period of $P = 2\pi$.

   The zeros are of the form $\dfrac{n\pi}{2}$, where n is an odd integer.

   Every halfway point has a y-coordinate of -1 or 1.

   The interval for the graph of the principal cycle is $\left(-\dfrac{\pi}{2}, \dfrac{\pi}{2}\right)$.

   The y-intercept is 1.

   The domain is all real numbers except integer multiples of π.

   The y-intercept is 0.

   Every halfway point has an x-coordinate of -1 or 1.

   The range is all real numbers except odd integer multiples of $\dfrac{\pi}{2}$.

   The function has a period of $P = \pi$.

   The zeros are of the form $n\pi$, where n is an integer.

   The interval for the graph of the principal cycle is $(0, \pi)$.

2. List all points on the graph of $y = \tan x$ on the interval $\left[-2\pi, \dfrac{5\pi}{2}\right]$ that have a y-coordinate of $\sqrt{3}$.

3. List all points on the graph of $y = \tan x$ on the interval $\left[-3\pi, \dfrac{\pi}{2}\right]$ that have a y-coordinate of $-\dfrac{1}{\sqrt{3}}$.

4. List all halfway points of $y = \tan x$ on the interval $[-2\pi, \pi]$.

5. Use the periodic property of $y = \tan x$ to determine which of the following expressions is equivalent to $\tan\left(\dfrac{4\pi}{3}\right)$.

   i. $\tan\left(\dfrac{2\pi}{3}\right)$ **ii.** $\tan\left(-\dfrac{\pi}{3}\right)$ **iii.** $\tan\left(-\dfrac{2\pi}{3}\right)$ **iv.** $\tan\left(\dfrac{5\pi}{3}\right)$

For Exercises 6–19, given each function, determine the interval for the principal cycle. Determine the period. Then for the principal cycle, determine the equations of the vertical asymptotes, the coordinates of the center point, and the coordinates of the halfway points. Sketch the graph.

SbS **6.** $y = \tan\left(x + \dfrac{\pi}{4}\right)$ SbS **7.** $y = 3\tan\left(x - \dfrac{\pi}{6}\right)$ SbS **8.** $y = -2\tan\left(x - \dfrac{\pi}{4}\right)$

SbS **9.** $y = \tan(3x + \pi)$ SbS **10.** $y = \tan(2x)$ SbS **11.** $y = \tan\left(\dfrac{1}{2}x\right)$

SbS **12.** $y = 2\tan(-3x)$ SbS **13.** $y = \tan\left(\dfrac{1}{2}x + \pi\right)$ SbS **14.** $y = \tan(\pi - 2x)$

SbS **15.** $y = 3\tan(2x + \pi) - 1$ SbS **16.** $y = 4\tan(-2x + \pi) + 1$ SbS **17.** $y = \dfrac{1}{2}\tan(-3x) + 4$

SbS **18.** $y = 1 - \tan\left(x - \dfrac{\pi}{2}\right)$ SbS **19.** $y = -3\tan(\pi - 2x) - 1$

20. Sketch the graph of $y = \cot x$, and then identify as many of the following properties that apply.

The function is an even function.

The function is an odd function.

The domain is all real numbers except odd integer multiples of $\dfrac{\pi}{2}$.

The range is $(-\infty, \infty)$.

The function has a period of $P = 2\pi$.

The zeros are of the form $\dfrac{n\pi}{2}$, where n is an odd integer.

Every halfway point has a y-coordinate of -1 or 1.

The interval for the graph of the principal cycle is $\left(-\dfrac{\pi}{2}, \dfrac{\pi}{2}\right)$.

The y-intercept is 1.

The domain is all real numbers except integer multiples of π.

The y-intercept is 0.

Every halfway point has an x-coordinate of -1 or 1.

The range is all real numbers except odd integer multiples of $\dfrac{\pi}{2}$.

The function has a period of $P = \pi$.

The zeros are of the form $n\pi$, where n is an integer.

The interval for the graph of the principal cycle is $(0, \pi)$.

21. List all points on the graph of $y = \cot x$ on the interval $\left[-2\pi, \dfrac{5\pi}{2}\right]$ that have a y-coordinate of $\sqrt{3}$.

22. List all points on the graph of $y = \cot x$ on the interval $\left[-3\pi, \dfrac{\pi}{2}\right]$ that have a y-coordinate of $-\dfrac{1}{\sqrt{3}}$.

23. List all halfway points of $y = \cot x$ on the interval $[-2\pi, \pi]$.

24. Use the periodic property of $y = \cot x$ to determine which of the following expressions is equivalent to $\cot\left(\dfrac{13\pi}{6}\right)$.

    i. $\cot\left(-\dfrac{5\pi}{6}\right)$ **ii.** $\cot\left(-\dfrac{\pi}{6}\right)$ **iii.** $\cot\left(\dfrac{11\pi}{6}\right)$ **iv.** $\cot\left(\dfrac{5\pi}{6}\right)$

For Exercises 25–38, given each function, determine the interval for the principal cycle. Determine the period. Then for the principal cycle, determine the equations of the vertical asymptotes, the coordinates of the center point, and the coordinates of the halfway points. Sketch the graph.

25. $y = \cot\left(x + \dfrac{\pi}{4}\right)$ **26.** $y = 3\cot\left(x - \dfrac{\pi}{2}\right)$ **27.** $y = -2\cot\left(x - \dfrac{\pi}{3}\right)$

28. $y = \cot(3x + \pi)$ **29.** $y = \cot(2x)$ **30.** $y = \cot\left(\dfrac{1}{3}x\right)$

31. $y = 2\cot(-3x)$ **32.** $y = -3\cot(-2x + \pi)$ **33.** $y = \cot(\pi - 2x)$

34. $y = \cot\left(\dfrac{1}{2}x + \pi\right)$ **35.** $y = 3\cot(2x + \pi) - 1$ **36.** $y = \dfrac{1}{2}\cot(-3x) + 4$

37. $y = 1 - \cot\left(x - \dfrac{\pi}{2}\right)$ **38.** $y = -3\cot(\pi - 2x) - 1$

In Exercises 39–44, the principal cycle of the graph of a trigonometric function of the from $y = A\tan(Bx - C) + D$ or $y = A\cot(Bx - C) + D$ for $B > 0$ is given. Determine the equation of the function represented by each graph.

39. $y = A\tan(Bx - C) + D$

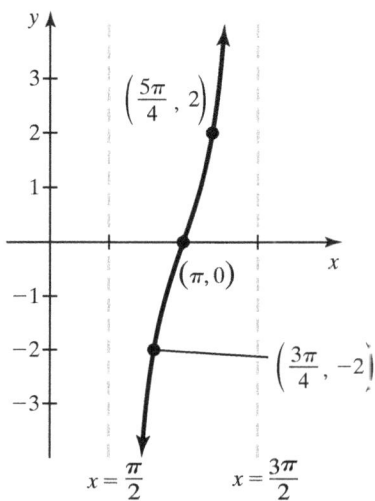

40. $y = A\cot(Bx - C) + D$

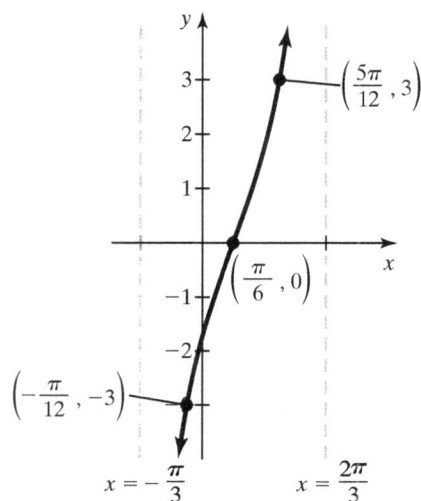

41. $y = A\tan(Bx - C) + D$

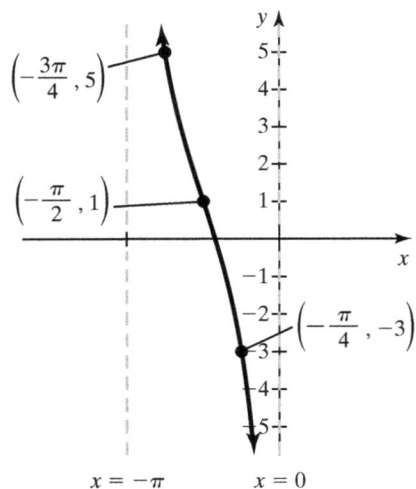

42. $y = A\cot(Bx - C) + D$

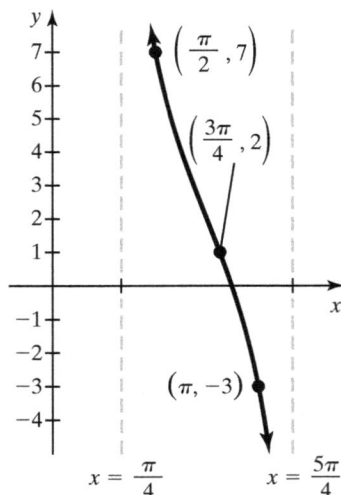

43. $y = A\cot(Bx - C) + D$

44. $y = A\tan(Bx - C) + D$

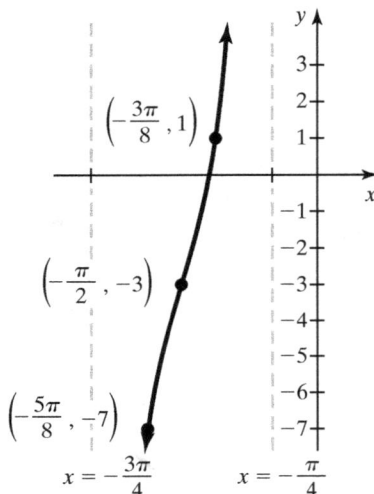

45. Sketch the graph of $y = \csc x$, and then identify as many of the following properties that apply.

The function is an even function.

The function is an odd function.

The domain is all real numbers except integer multiples of π.

The range is $(-\infty, \infty)$.

The function has a period of $P = 2\pi$.

The equations of the vertical asymptotes are of the form $x = n\pi$, where n is an integer.

The function obtains a relative maximum at $x = -\dfrac{\pi}{2} + 2\pi n$, where n is an integer.

The domain is all real numbers except odd integer multiples of $\dfrac{\pi}{2}$.

The range is $(-\infty, -1] \cup [1, \infty)$.

The function obtains a relative maximum at $x = \pi n$, where n is an odd integer.

The equations of the vertical asymptotes are of the form $x = \dfrac{n\pi}{2}$, where n is an odd integer.

The function has a period of $P = \pi$.

46. Sketch the graph of $y = \sec x$, and then identify as many of the following properties that apply.

The function is an even function.

The function is an odd function.

The domain is all real numbers except integer multiples of π.

The range is $(-\infty, \infty)$.

The function has a period of $P = 2\pi$.

The equations of the vertical asymptotes are of the form $x = n\pi$, where n is an integer.

The function obtains a relative maximum at $x = -\dfrac{\pi}{2} + 2\pi n$, where n is an integer.

The domain is all real numbers except odd integer multiples of $\dfrac{\pi}{2}$.

The range is $(-\infty, -1] \cup [1, \infty)$.

The function obtains a relative maximum at $x = \pi n$, where n is an odd integer.

The equations of the vertical asymptotes are of the form $x = \dfrac{n\pi}{2}$, where n is an odd integer.

The function has a period of $P = \pi$.

In Exercises 47–56, determine the equations of the vertical asymptotes and all relative maximum and relative minimum points of two cycles of each function and then sketch its graph.

47. $y = 2 \sec (x - \pi)$

48. $y = -2 \sec (x - \pi)$

49. $y = \csc \left(x - \dfrac{\pi}{4} \right)$

50. $y = -\csc \left(x - \dfrac{\pi}{4} \right)$

51. $y = 2 \sec (3x + \pi)$

52. $y = -3 \csc (2x + \pi)$

53. $y = 4 \sec (2x - \pi) + 3$

54. $y = -3 \csc \left(x - \dfrac{\pi}{4} \right) - 4$

55. $y = 5 - 2 \sec (x - \pi)$

56. $y = \csc (\pi - 2x) + 3$

Brief Exercises

In Exercises 57–61, determine the interval of the principal cycle of each function.

57. $y = 3 \tan \left(x - \dfrac{\pi}{6} \right)$

58. $y = \cot (2x)$

59. $y = \tan (3x + \pi)$

60. $y = \cot (\pi - 2x)$

61. $y = -3 \tan (-2x + \pi) + 1$

In Exercises 62–66, determine the period of each function.

62. $y = 3 \tan \left(x - \dfrac{\pi}{6} \right)$

63. $y = \cot (2x)$

64. $y = \tan (3x + \pi)$

65. $y = \cot (\pi - 2x)$

66. $y = -3 \tan (-2x + \pi) + 1$

In Exercises 67–71, determine the equations of the vertical asymptotes of the principal cycle of each function.

67. $y = 3 \tan \left(x - \dfrac{\pi}{6} \right)$

68. $y = \cot (2x)$

69. $y = \tan (3x + \pi)$

70. $y = \cot (\pi - 2x)$

71. $y = -3 \tan (-2x + \pi) + 1$

In Exercises 72–76, determine the coordinates of the center point of the principal cycle of each function.

72. $y = 3 \tan \left(x - \dfrac{\pi}{6} \right)$ **73.** $y = \cot (2x)$ **74.** $y = \tan (3x + \pi)$

75. $y = \cot (\pi - 2x)$ **76.** $y = -3 \tan (-2x + \pi) + 1$

In Exercises 77–81, determine the coordinates of the two halfway points of the principal cycle of each function.

77. $y = 3 \tan \left(x - \dfrac{\pi}{6} \right)$ **78.** $y = \cot (2x)$ **79.** $y = \tan (3x + \pi)$

80. $y = \cot (\pi - 2x)$ **81.** $y = -3 \tan (-2x + \pi) + 1$

In Exercises 82–87, the principal cycle of the graph of a trigonometric function of the from $y = A \tan (Bx - C) + D$ or $y = A \cot (Bx - C) + D$ for $B > 0$ is given. Determine the equation of the function represented by each graph.

82. $y = A\tan(Bx - C) + D$

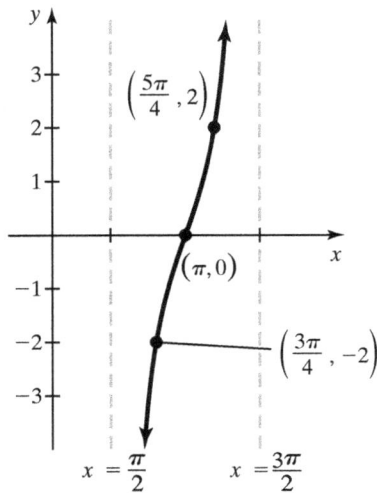

83. $y = A\cot(Bx - C) + D$

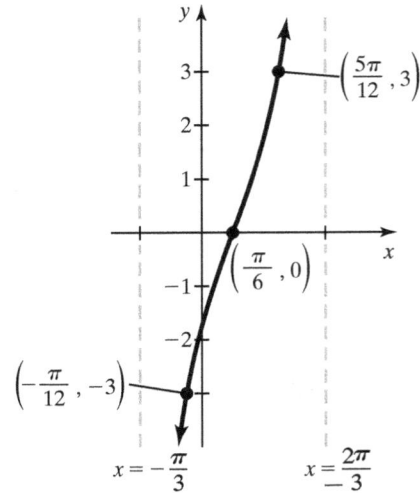

84. $y = A\tan(Bx - C) + D$

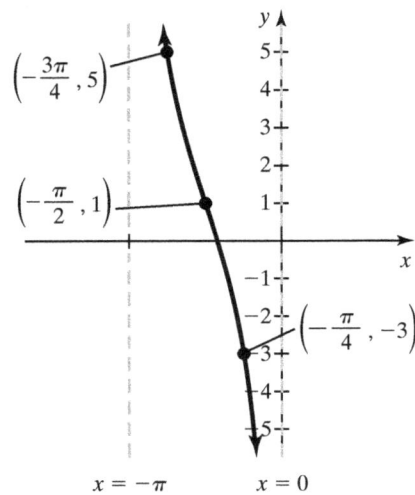

85. $y = A\cot(Bx - C) + D$

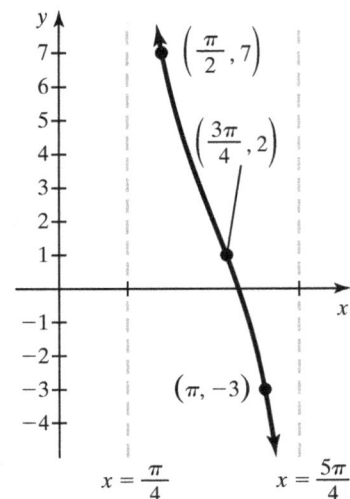

86. $y = A\cot(Bx - C) + D$

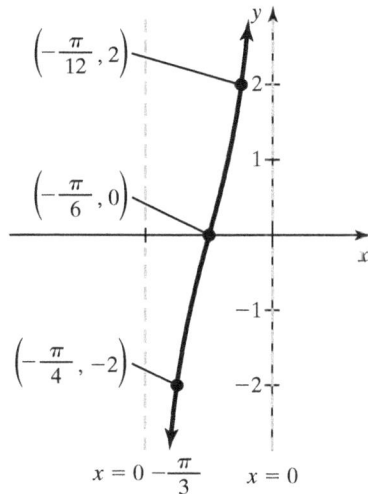

87. $y = A\tan(Bx - C) + D$

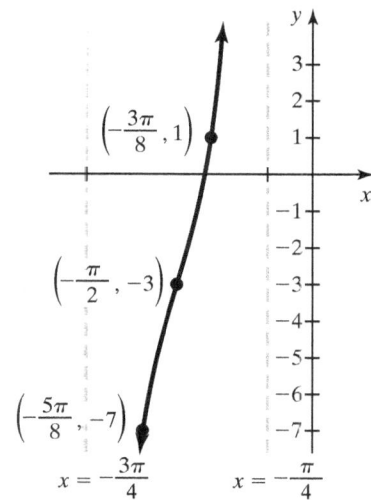

88. Determine the equations of the four vertical asymptotes of the graph of $y = 2\sec(x - \pi)$ on the interval $[-\pi, 3\pi]$.

89. Determine the equations of the four vertical asymptotes of the graph of $y = 4\csc(2x + \pi) + 1$ on the interval $\left[-\dfrac{\pi}{2}, \dfrac{3\pi}{2}\right]$.

90. Determine the coordinates of the three relative minimum points of the graph of $y = 2\sec(x - \pi)$ on the interval $[-\pi, 3\pi]$.

91. Determine the coordinates of the two relative minimum points of the graph of $y = 4\csc(2x + \pi) + 1$ on the interval $\left[-\dfrac{\pi}{2}, \dfrac{3\pi}{2}\right]$.

92. Determine the coordinates of the two relative maximum points of the graph of $y = 2\sec(x - \pi)$ on the interval $[-\pi, 3\pi]$.

93. Determine the coordinates of the two relative maximum points of the graph of $y = 4\csc(2x + \pi) + 1$ on the interval $\left[-\dfrac{\pi}{2}, \dfrac{3\pi}{2}\right]$.

In Exercises 94–109, sketch each function.

94. $y = \tan\left(x + \dfrac{\pi}{4}\right)$ **95.** $y = \tan(2x)$ **96.** $y = \tan(3x + \pi)$ **97.** $y = 3\tan(2x + \pi) - 1$

98. $y = \cot\left(x - \dfrac{\pi}{3}\right)$ **99.** $y = \cot(3x)$ **100.** $y = \cot(2x - \pi)$ **101.** $y = 3\cot(2x + \pi) - 1$

102. $y = 2\tan(-3x)$ **103.** $y = 2\cot(-3x)$ **104.** $y = -3\tan(\pi - 2x) - 1$

105. $y = -3\cot(\pi - 2x) - 1$ **106.** $y = \csc\left(x - \dfrac{\pi}{4}\right)$ **107.** $y = 2\sec(x - \pi)$

108. $y = 4\sec(2x - \pi) + 3$ **109.** $y = \csc(\pi - 2x) + 3$

2.4 Inverse Trigonometric Functions I

THINGS TO KNOW

Before working through this section, be sure that you are familiar with the following concepts:

		VIDEO	ANIMATION	INTERACTIVE
You Try It	1. Determining Whether a Function Is One-to-One Using the Horizontal Line Test	⊙	▭	
You Try It	2. Understanding the Definition of an Inverse Function	⊙		
You Try It	3. Sketching the Graph of an Inverse Function		▭	
You Try It	4. Understanding the Special Right Triangles (Section 1.3)		▭	
You Try It	5. Understanding the Right Triangle Definitions of the Trigonometric Functions (Section 1.4)	⊙		
You Try It	6. Understanding the Signs of the Trigonometric Functions (Section 1.5)	⊙		
You Try It	7. Determining Reference Angles (Section 1.5)			⊙
You Try It	8. Evaluating Trigonometric Functions of Angles Belonging to the $\frac{\pi}{3}, \frac{\pi}{6},$ or $\frac{\pi}{4}$ Families (Section 1.5)			⊙
You Try It	9. Understanding the Graph of the Sine Function and Its Properties (Section 2.1)	⊙		
You Try It	10. Understanding the Graph of the Cosine Function and Its Properties (Section 2.1)	⊙		
You Try It	11. Understanding the Graph of the Tangent Function and Its Properties (Section 2.3)			

OBJECTIVES

1 Understanding and Finding the Exact and Approximate Values of the Inverse Sine Function

2 Understanding and Finding the Exact and Approximate Values of the Inverse Cosine Function

3 Understanding and Finding the Exact and Approximate Values of the Inverse Tangent Function

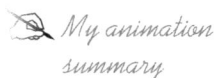

....

Introduction to Section 2.4

My animation summary

Recall that if a graph of a function passes the horizontal line test, then the graph represents a one-to-one function. *Only* one-to-one functions have inverse functions. However, it is possible to restrict the domain of a function that is not one-to-one to produce a one-to-one function. For example, the function $f(x) = x^2 + 1$ is not one-to-one because its graph does not pass the horizontal line test. However, if we restrict the domain of $f(x) = x^2 + 1$ to the interval $(-\infty, 0]$, then the graph of f on this restricted domain passes the horizontal line test, is one-to-one, and thus has an inverse function. Watch this animation to see how to sketch this one-to-one function and its inverse function.

OBJECTIVE 1 UNDERSTANDING AND FINDING THE EXACT AND APPROXIMATE VALUES OF THE INVERSE SINE FUNCTION

In this section, we want to focus our attention on inverse trigonometric functions. We first review the sine function, $y = \sin x$. Think of the independent variable, x, as an angle given in radians. Furthermore, as you read through this section, every time you encounter an angle, try to visualize the quadrant or axis in which the terminal side of the angle lies. For example, the angle $\frac{\pi}{4}$ has a terminal side that lies in Quadrant I, the angle π has a terminal side that lies on the negative x-axis, the angle $\frac{11\pi}{6}$ has a terminal side that lies in Quadrant IV, and so on. Figure 59 illustrates the graph of $y = \sin x$. We label each subinterval of length $\frac{\pi}{2}$ with the Roman numerals I, II, III, or IV to illustrate that the angle located in that interval has a terminal side that lies in the listed quadrant.

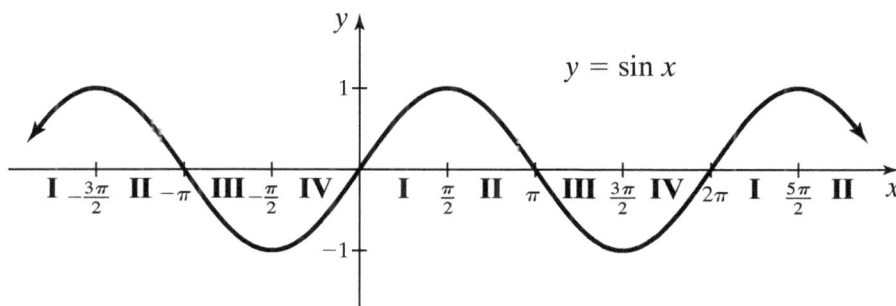

Figure 59 The Roman numerals under each subinterval indicate that any angle located within the interval has a terminal side that lies in that quadrant.

My animation summary

The domain of $y = \sin x$ is $(-\infty, \infty)$. The sine function is clearly not one-to-one because the graph fails the horizontal line test. See Figure 60. However, perhaps we can restrict the domain of $y = \sin x$ by limiting it to a closed interval to produce a graph that is one-to-one. But what interval do we choose? There are actually infinitely many ways to restrict the domain of $y = \sin x$ to produce a graph that is one-to-one. Traditionally, it has been agreed upon to use the interval $-\dfrac{\pi}{2} \le x \le \dfrac{\pi}{2}$ as the restricted domain. Note that this interval shows the entire range of $[-1, 1]$. See Figure 61. This graph is a one-to-one function and therefore has an inverse function. Watch the first part of this **animation** for further explanation.

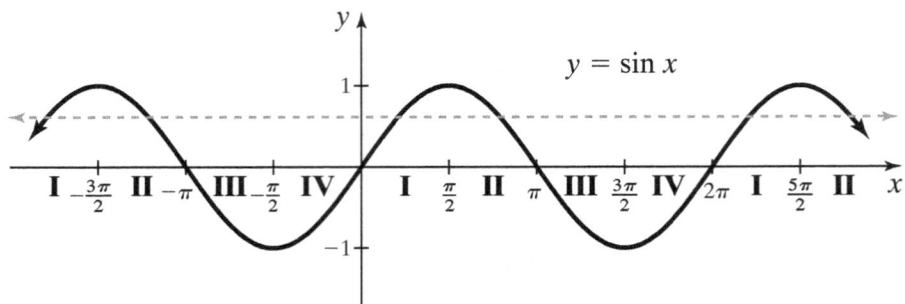

Figure 60 The graph of the function $y = \sin x$ fails the horizontal line test and is therefore not a one-to-one function. This function does not have an inverse function.

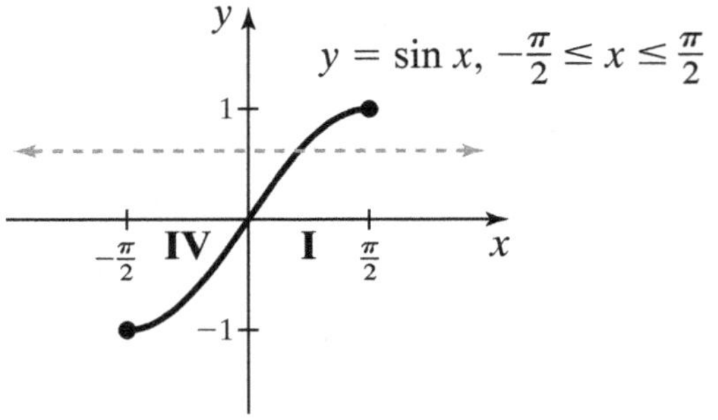

Figure 61 The graph of the function $y = \sin x$ on the restricted domain $-\dfrac{\pi}{2} \le x \le \dfrac{\pi}{2}$ passes the horizontal line test and is therefore a one-to-one function. This function has an inverse function.

My animation summary

We now want to establish the graph of the inverse sine function. For any **one-to-one** function f, the graph of the inverse function, f^{-1}, can be obtained by reversing the coordinates of each of the ordered pairs lying on the graph of f. In other words, for any point (a, b) lying on the graph of f, the point (b, a) must lie on the graph of f^{-1}.

Therefore, starting with the graph of the restricted sine function, we can obtain the graph of $y = \sin^{-1} x$ by reversing the coordinates of the ordered pairs that lie on the graph of the restricted sine function. We then plot these ordered pairs and connect them with a smooth curve to obtain the graph of $y = \sin^{-1} x$. This function is called the **inverse sine function**. Be sure to work through this **animation** to see exactly how to obtain the graphs shown in Figures 62 and 63.

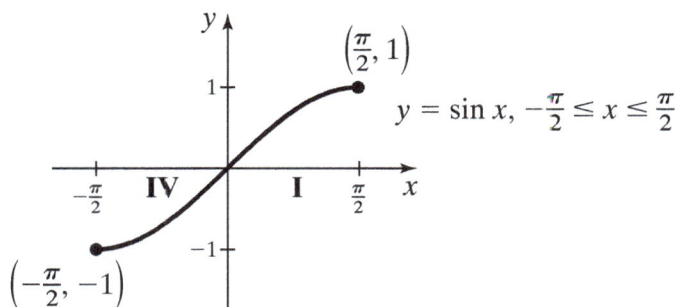

Figure 62 The graph of $y = \sin x$ on the restricted domain $-\dfrac{\pi}{2} \le x \le \dfrac{\pi}{2}$ with range $-1 \le y \le 1$

If $0 < x < 1$, then the value of $\sin^{-1} x$ is an angle whose terminal side lies in Quadrant I.

If $-1 < x < 0$, then the value of $\sin^{-1} x$ is an angle whose terminal side lies in Quadrant IV.

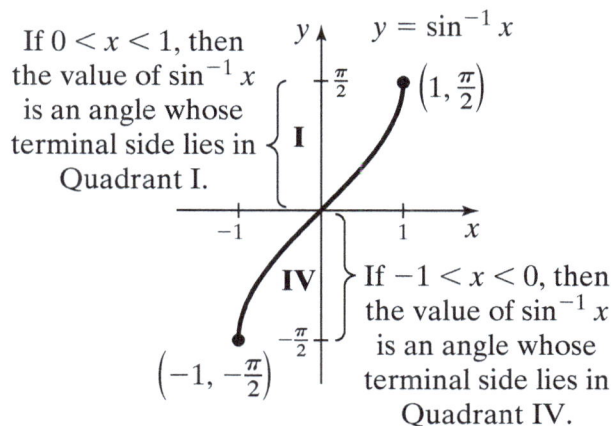

Figure 63 The domain of $y = \sin^{-1}(x)$ is $-1 \le x \le 1$ and the range is $-\dfrac{\pi}{2} \le y \le \dfrac{\pi}{2}$.

Definition   Inverse Sine Function

The **inverse sine function**, denoted as $y = \sin^{-1} x$, is the inverse of $y = \sin x$, $-\dfrac{\pi}{2} \le x \le \dfrac{\pi}{2}$.

The domain of $y = \sin^{-1} x$ is $-1 \le x \le 1$ and the range is $-\dfrac{\pi}{2} \le y \le \dfrac{\pi}{2}$.

(Note that an alternative notation for $\sin^{-1} x$ is $\arcsin x$.)

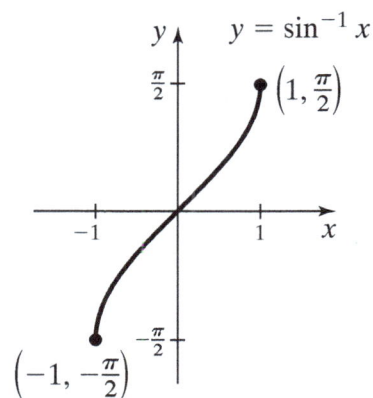

⚠ **Do not confuse the notation** $\sin^{-1} x$ **with** $(\sin x)^{-1} = \dfrac{1}{\sin x} = \csc x$.

The negative 1 is not an exponent! Thus, $\sin^{-1} x \ne \dfrac{1}{\sin x}$.

We are often interested in determining the exact value of $\sin^{-1} x$. Before proceeding, you must have a complete understanding of all of the prerequisite topics stated in the Things to Know pages located at the beginning of this section. To determine the exact value of $\sin^{-1} x$, we will think of the expression $\sin^{-1} x$ as the angle on the interval $\left[-\dfrac{\pi}{2}, \dfrac{\pi}{2} \right]$ whose sine is equal to x. For this reason, we will use our conventional angle notation, θ, to represent the value of $\sin^{-1} x$. If $\theta = \sin^{-1} x$, then $-\dfrac{\pi}{2} \le \theta \le \dfrac{\pi}{2}$.

Therefore, the terminal side of angle θ must lie in Quadrant I, in Quadrant IV, on the positive x-axis, or the positive y-axis, or on the negative y-axis. See Figure 64.

Figure 64 If $\theta = \sin^{-1} x$, then $-\dfrac{\pi}{2} \le \theta \le \dfrac{\pi}{2}$ and the terminal side of angle θ lies in Quadrant I, in Quadrant IV, on the positive x-axis, on the positive y-axis, or on the negative y-axis.

To determine the exact value of $\sin^{-1} x$, we use the four-step process outlined on the following page.

Steps for Determining the Exact Value of $\sin^{-1}x$

Step 1. If x is in the interval $[-1, 1]$, then the value of $\sin^{-1}x$ must be an angle in the interval $\left[-\dfrac{\pi}{2}, \dfrac{\pi}{2} \right]$.

Step 2. Let $\sin^{-1}x = \theta$ such that $\sin \theta = x$.

Step 3. If $\sin \theta = 0$, then $\theta = 0$ and the terminal side of angle θ lies on the positive x-axis.

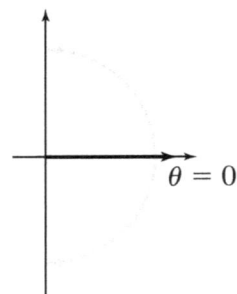

If $\sin \theta > 0$, then $0 < \theta \le \dfrac{\pi}{2}$ and the terminal side of angle θ lies in Quadrant I or on the positive y-axis.

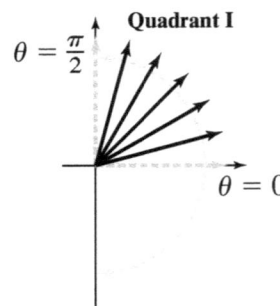

If $\sin \theta < 0$, then $-\dfrac{\pi}{2} \le \theta < 0$ and the terminal side of angle θ lies in Quadrant IV or on the negative y-axis.

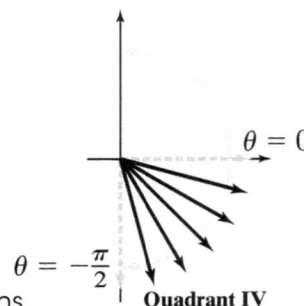

Step 4. Use your knowledge of the two special right triangles, the graphs of the trigonometric functions, Table 1 from Section 1.4, or Table 2 from Section 1.5 to determine the angle in the correct quadrant whose sine is x.

⊗ **Example 1** Determining the Exact Value of an Inverse Sine Expression

Determine the exact value of each expression.

a. $\sin^{-1}\left(\frac{1}{2}\right)$ **b.** $\sin^{-1}\left(-\frac{\sqrt{3}}{2}\right)$

Solution

a. Step 1. Because $x = \frac{1}{2}$ is in the interval $[-1, 1]$, the value of $\sin^{-1}\left(\frac{1}{2}\right)$ must be an angle in the interval $\left[-\frac{\pi}{2}, \frac{\pi}{2}\right]$.

Step 2. Let $\sin^{-1}\left(\frac{1}{2}\right) = \theta$ such that $\sin\theta = \frac{1}{2}$.

Step 3. Because $\sin\theta = \frac{1}{2} > 0$, it follows that $0 < \theta \le \frac{\pi}{2}$. Therefore, we are looking for an angle whose terminal side lies in Quadrant I or on the positive y-axis. See Figure 65.

Step 4. Using our knowledge of the special $\frac{\pi}{6}, \frac{\pi}{3}, \frac{\pi}{2}$ right triangle or Table 1, we know that $\sin\frac{\pi}{6} = \frac{1}{2}$. The angle $\theta = \frac{\pi}{6}$ is the **only** angle for which $0 < \theta \le \frac{\pi}{2}$ whose terminal side lies in Quadrant I or on the positive y-axis such that $\sin\frac{\pi}{6} = \frac{1}{2}$. See Figure 66. Therefore,

$$\sin^{-1}\left(\frac{1}{2}\right) = \frac{\pi}{6}.$$

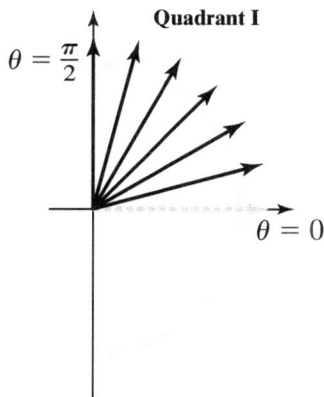

Figure 65 The angle $\theta = \sin^{-1}\left(\frac{1}{2}\right)$ must have a terminal side that lies in Quadrant I or on the positive y-axis.

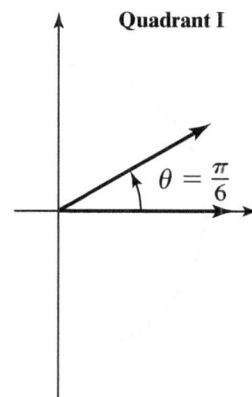

Figure 66 The angle $\theta = \frac{\pi}{6}$ is the only angle in the interval $0 < \theta \le \frac{\pi}{2}$ whose terminal side lies in Quadrant I or on the positive y-axis such that $\sin\theta = \frac{1}{2}$.

⊗ Work through this interactive video to see each step of this solution.

b. To find the value of $\sin^{-1}\left(-\dfrac{\sqrt{3}}{2}\right)$, we follow the four-step process.

Step 1. Because $x = -\dfrac{\sqrt{3}}{2}$ is in the interval $[-1, 1]$, the value of $\sin^{-1}\left(-\dfrac{\sqrt{3}}{2}\right)$ must be an angle in the interval $\left[-\dfrac{\pi}{2}, \dfrac{\pi}{2}\right]$.

Step 2. Let $\sin^{-1}\left(-\dfrac{\sqrt{3}}{2}\right) = \theta$ such that $\sin\theta = -\dfrac{\sqrt{3}}{2}$.

Step 3. Because $\sin\theta = -\dfrac{\sqrt{3}}{2} < 0$, it follows that $-\dfrac{\pi}{2} \leq \theta < 0$. Therefore, we are looking for an angle whose terminal side lies in Quadrant IV or on the negative y-axis. See Figure 67.

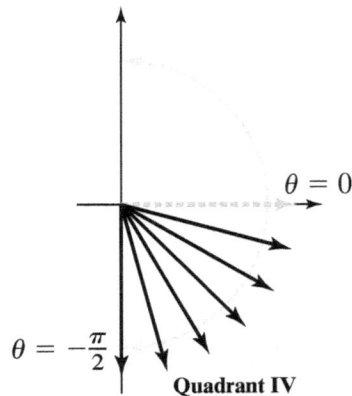

Figure 67 The angle $\theta = \sin^{-1}\left(-\dfrac{\sqrt{3}}{2}\right)$ must have a terminal side that lies in Quadrant IV or on the negative y-axis.

Step 4. Using our knowledge of the special $\dfrac{\pi}{6}, \dfrac{\pi}{3}, \dfrac{\pi}{2}$ right triangle or Table 1, we know that $\sin\dfrac{\pi}{3} = \dfrac{\sqrt{3}}{2}$. However, $\theta \neq \dfrac{\pi}{3}$ because the terminal side of this angle does **not** lie in Quadrant IV or on the negative y-axis. We must identify the angle θ that satisfies the following three conditions:

The reference angle is $\theta_R = \dfrac{\pi}{3}$.

The terminal side lies in Quadrant IV or on the negative y-axis.

$-\dfrac{\pi}{2} \leq \theta < 0$.

The only such angle is $\theta = -\dfrac{\pi}{3}$. See Figure 68. Therefore, $\sin^{-1}\left(-\dfrac{\sqrt{3}}{2}\right) = -\dfrac{\pi}{3}$.

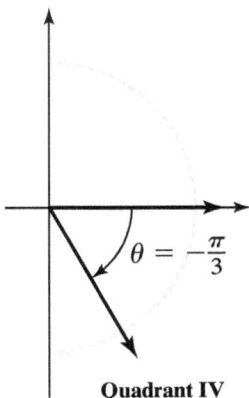

Figure 68 The angle $\theta = -\dfrac{\pi}{3}$ is the only angle in the interval $-\dfrac{\pi}{2} \leq \theta < 0$ whose terminal side lies in Quadrant IV or on the negative y-axis such that $\sin\theta = -\dfrac{\sqrt{3}}{2}$.

My interactive video summary

⊙ Work through this interactive video to see each step of this solution.

You Try It Work through this You Try It problem.

Work Exercises 1–3 in this textbook or in the MyMathLab Study Plan.

The steps for determining the exact value of $\sin^{-1}x$ will not apply when the argument, x, is not a recognized value associated with special right triangles or quadrantal angles. In this situation, the use of a calculator will be necessary to find an approximate value of $\sin^{-1}x$.

Example 2 Finding the Approximate Value of an Inverse Sine Expression

Use a calculator to approximate each value, or state that the value does not exist.

a. $\sin^{-1}(.7)$ **b.** $\sin^{-1}(-.95)$ **c.** $\sin^{-1}(3)$

Solution

a. Because $x = .7$ is in the interval $[-1, 1]$, the value of $\sin^{-1}(.7)$ must be an angle in the interval $\left[-\dfrac{\pi}{2}, \dfrac{\pi}{2}\right]$. Before using a calculator to approximate this value, it is good practice to think about what type of answer we can expect. If we let $\sin^{-1}(.7) = \theta$, then $\sin\theta = .7$. Because $\sin\theta = .7 > 0$, then $0 < \theta \leq \dfrac{\pi}{2}$, where $\dfrac{\pi}{2} \approx 1.5708$. Therefore, the approximate value of $\sin^{-1}(.7)$ must be between 0 and 1.5708.

Using a calculator in radian mode, type [2nd] [sin] and then .7. We find that $\sin^{-1}(.7) \approx 0.7754$ radian. The value of $\theta \approx 0.7754$ seems reasonable because $0 < 0.7754 \leq \dfrac{\pi}{2}$.

b. Because $x = -.95$ is in the interval $[-1, 1]$, the value of $\sin^{-1}(-.95)$ must be an angle in the interval $\left[-\dfrac{\pi}{2}, \dfrac{\pi}{2}\right]$. If $\sin^{-1}(-.95) = \theta$, then $\sin\theta = -.95 < 0$.

Thus, $-\dfrac{\pi}{2} \leq \theta < 0$. Therefore, the approximate value of $\sin^{-1}(-.95)$ must be negative but greater than or equal to $-\dfrac{\pi}{2} \approx -1.5708$.

Using a calculator in radian mode, type [2nd] [sin] and then $-.95$. We find that $\sin^{-1}(-.95) \approx -1.2532$ radians, which is certainly in the range $-\dfrac{\pi}{2} \leq \theta < 0$.

c. Looking at the expression, $\sin^{-1}(3)$, we see that 3 does not belong to the interval $-1 \leq x \leq 1$. Therefore, 3 is not in the domain of $y = \sin^{-1}x$. Thus, $\sin^{-1}(3)$ does not exist. Also, if $\theta = \sin^{-1}(3)$, then $\sin\theta = 3$, which is impossible because the maximum value of the sine function is 1.

Figure 69 illustrates what happens when we input $\sin^{-1}(3)$ into a calculator.

Using Technology

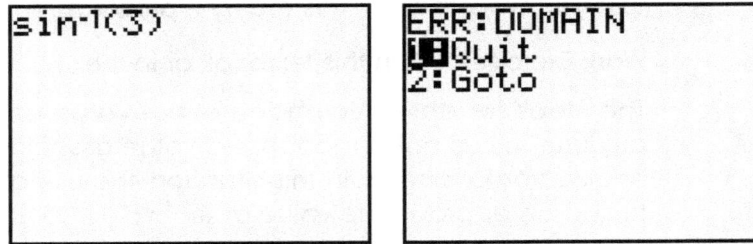

Figure 69 An input of $\sin^{-1}(3)$ using a graphing utility results in a domain error.

You Try It   Work through this You Try It problem.

Work Exercises 4–6 in this textbook or in the MyMathLab Study Plan.

OBJECTIVE 2   UNDERSTANDING AND FINDING THE EXACT AND APPROXIMATE VALUES OF THE INVERSE COSINE FUNCTION

We establish the inverse cosine function in much the same way as we did with the inverse sine function. First, the cosine function is defined for all real numbers and is not one-to-one because the graph does not pass the horizontal line test. See Figure 70. Note that we label each subinterval of length $\dfrac{\pi}{2}$ with the Roman numerals I, II, III, or IV to illustrate that the angle located in that interval has a terminal side that lies in the listed quadrant.

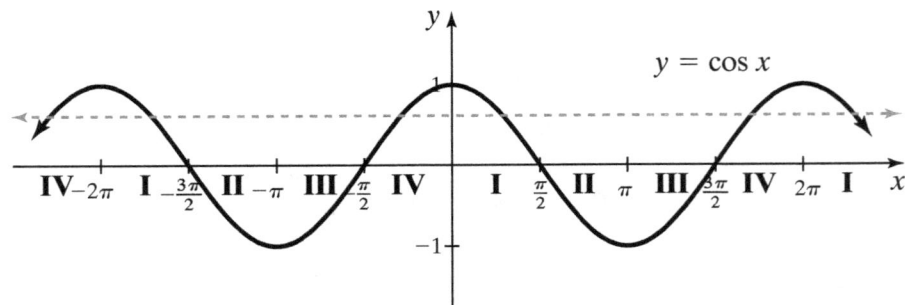

Figure 70 The graph of the function $y = \cos x$ fails the horizontal line test and is therefore not a one-to-one function. This function does not have an inverse function.

My animation summary

As with the sine function, we can restrict the domain of the cosine function to produce a graph that passes the horizontal line test and hence is one-to-one. Although there are infinitely many ways to restrict the domain of the cosine function to produce a graph that is one-to-one, most mathematicians have agreed on the interval $0 \le x \le \pi$ as the common restricted interval. The graph of the restricted cosine function on the interval $0 \le x \le \pi$ passes the horizontal line test, is one-to-one, and hence has an inverse function. See Figure 71. Work through the first part of this animation for further explanation.

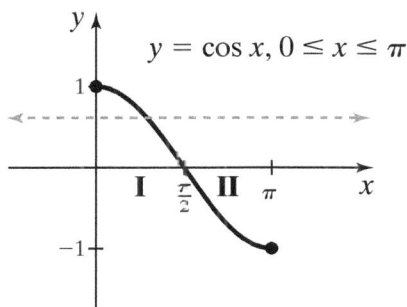

$y = \cos x, 0 \le x \le \pi$

Figure 71 The graph of the function $y = \cos x$ on the restricted domain $0 \le x \le \pi$ passes the horizontal line test and is therefore a one-to-one function. This function has an inverse function.

My animation summary

To determine the graph of the inverse of the restricted cosine function, we can reverse the coordinates of the ordered pairs, plot the points, and connect with a smooth curve. See Figures 72 and Figure 73. The function seen in Figure 72 is called the **inverse cosine function.** Work through the second part of this animation to see exactly how to obtain the graph of the inverse cosine function.

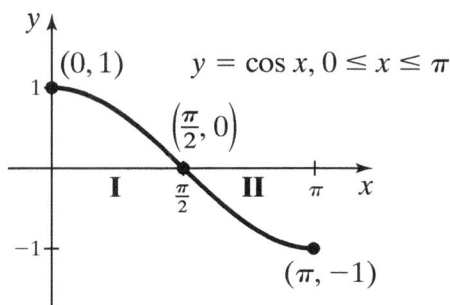

$y = \cos x, 0 \le x \le \pi$ Start with the restricted cosine function and reverse the coordinates of each ordered pair to produce the graph of the inverse cosine function.

Figure 72 The graph of $y = \cos x$ on the restricted domain $0 \le x \le \pi$ with range $-1 \le y \le 1$

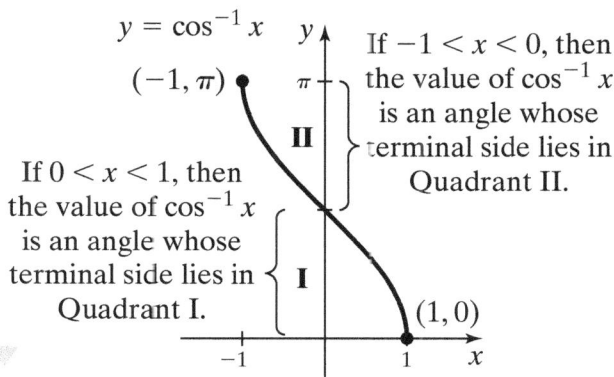

$y = \cos^{-1} x$

If $0 < x < 1$, then the value of $\cos^{-1} x$ is an angle whose terminal side lies in Quadrant I.

If $-1 < x < 0$, then the value of $\cos^{-1} x$ is an angle whose terminal side lies in Quadrant II.

Figure 73 The domain of $y = \cos^{-1} x$ is $-1 \le x \le 1$ and the range is $0 \le y \le \pi$

Definition   Inverse Cosine Function

The **inverse cosine function,** denoted as $y = \cos^{-1} x$, is the inverse of $y = \cos x, 0 \le x \le \pi$.

The domain of $y = \cos^{-1} x$ is $-1 \le x \le 1$ and the range is $0 \le y \le \pi$.

(Note that an alternative notation for $\cos^{-1} x$ is arccos x.)

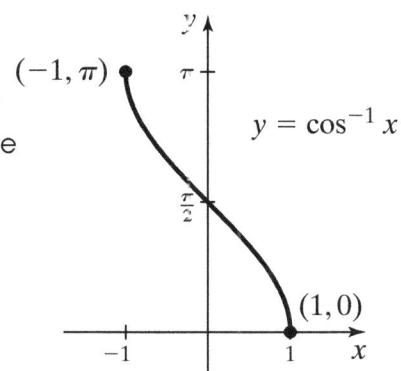

$y = \cos^{-1} x$

We will think of the expression $\cos^{-1} x$ as the angle on the interval $[0, \pi]$ whose cosine is equal to x. It is important to stress that the interval $[0, \pi]$ is quite different from the interval used to describe the range of the inverse sine function. Recall that the range of the inverse sine function is $\left[-\dfrac{\pi}{2}, \dfrac{\pi}{2} \right]$. As mentioned earlier when we described the inverse sine function, it is customary to use the angle notation, θ, to represent the value of $\cos^{-1} x$. Therefore, if $\theta = \cos^{-1} x$, then $0 \leq \theta \leq \pi$. Thus, the terminal side of angle θ must lie in Quadrant I, in Quadrant II, on the positive y-axis, on the positive x-axis, or on the negative x-axis. See Figure 74.

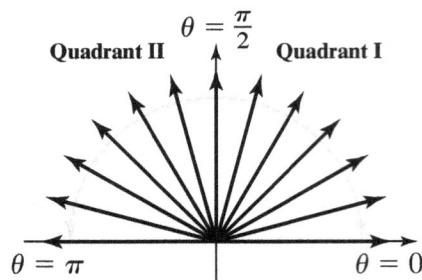

Figure 74 If $\theta = \cos^{-1} x$, then $0 \leq \theta \leq \pi$ and the terminal side of angle θ lies in Quadrant I, in Quadrant II, on the positive y-axis, on the positive x-axis, or on the negative x-axis.

To determine the exact value of an inverse cosine expression, we follow the four-step process outlined on the following page.

Steps for Determining the Exact Value of $\cos^{-1} x$

Step 1. If x is in the interval $[-1, 1]$, then the value of $\cos^{-1} x$ must be an angle in the interval $[0, \pi]$.

Step 2. Let $\cos^{-1} x = \theta$ such that $\cos \theta = x$.

Step 3. If $\cos \theta = 0$, then $\theta = \dfrac{\pi}{2}$ and the terminal side of angle θ lies on the positive y-axis.

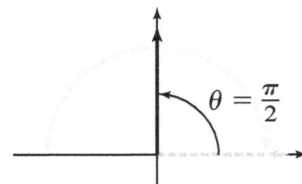

If $\cos \theta > 0$, then $0 \leq \theta < \dfrac{\pi}{2}$ and the terminal side of angle θ lies in Quadrant I or on the positive x-axis.

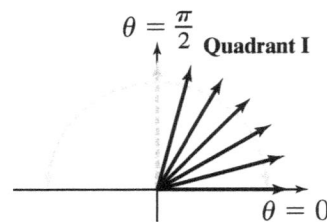

If $\cos \theta < 0$, then $\dfrac{\pi}{2} < \theta \leq \pi$ and the terminal side of angle θ lies in Quadrant II or on the negative x-axis.

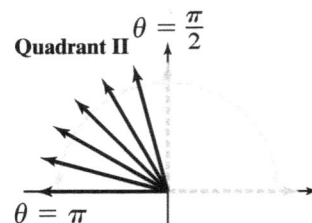

Step 4. Use your knowledge of the two special right triangles, the graphs of the trigonometric functions, Table 1 from Section 1.4, or Table 2 from Section 1.5 to determine the angle in the correct quadrant whose cosine is x.

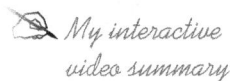

My interactive video summary

▶ **Example 3** Finding the Exact Value of an Inverse Cosine Expression

Determine the exact value of each expression.

a. $\cos^{-1}(1)$ **b.** $\cos^{-1}\left(-\dfrac{1}{\sqrt{2}}\right)$

Solution

a. Step 1. Because $x = 1$ is in the interval $[-1, 1]$, the value of $\cos^{-1}(1)$ must be an angle in the interval $[0, \pi]$.

Step 2. Let $\cos^{-1}(1) = \theta$ such that $\cos \theta = 1$.

Step 3. Because $\cos \theta = 1 > 0$, it follows that $0 \le \theta < \dfrac{\pi}{2}$. Therefore, we are looking for an angle whose terminal side lies in Quadrant I or on the positive x-axis. See Figure 75.

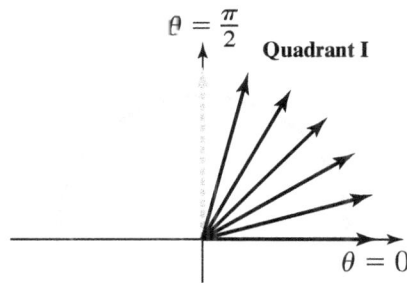

Figure 75 The angle $\theta = \cos^{-1}(1)$ must have a terminal side that lies in Quadrant I or on the positive x-axis.

Step 4. From the graph of $y = \cos \theta$ or from Section 1.5 Table 2, we know that if $\theta = 0$, then $\cos 0 = 1$. The angle $\theta = 0$ is the *only* angle for which $0 \le \theta < \dfrac{\pi}{2}$ whose terminal side lies in Quadrant I or on the positive x-axis such that $\cos 0 = 1$. See Figure 76. Therefore, $\cos^{-1}(1) = 0$.

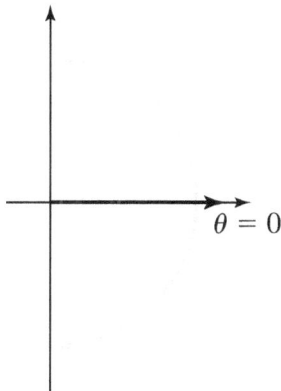

Figure 76 The angle $\theta = 0$ is the only angle in the interval $0 \le \theta < \dfrac{\pi}{2}$ whose terminal side lies in Quadrant I or on the positive x-axis such that $\cos \theta = 1$.

My interactive video summary

⊙ Work through this interactive video to see each step of this solution.

b. To find the value of $\cos^{-1}\left(-\dfrac{1}{\sqrt{2}}\right)$, we follow the four-step process.

Step 1. Because $x = -\dfrac{1}{\sqrt{2}}$ is in the interval $[-1, 1]$, the value of $\cos^{-1}\left(-\dfrac{1}{\sqrt{2}}\right)$ must be an angle in the interval $[0, \pi]$.

Step 2. Let $\cos^{-1}\left(-\dfrac{1}{\sqrt{2}}\right) = \theta$ such that $\cos \theta = -\dfrac{1}{\sqrt{2}}$.

Step 3. Because $\cos\theta = -\dfrac{1}{\sqrt{2}} < 0$, it follows that $\dfrac{\pi}{2} < \theta \le \pi$. Thus, we are looking for an angle whose terminal side lies in Quadrant II or on the negative x-axis. See Figure 77.

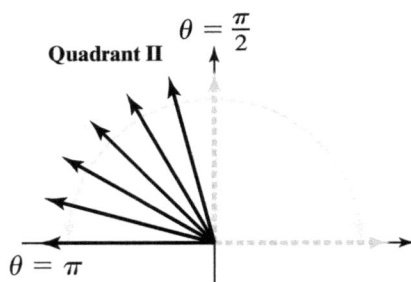

Figure 77 The angle $\theta = \cos^{-1}\left(-\dfrac{1}{\sqrt{2}}\right)$ must have a terminal side that lies in Quadrant II or on the negative x-axis.

Step 4. Using our knowledge of the special $\dfrac{\pi}{4}, \dfrac{\pi}{4}, \dfrac{\pi}{2}$ right triangle or Table 1 from Section 1.4, we know that $\cos\dfrac{\pi}{4} = \dfrac{1}{\sqrt{2}}$. However, $\theta \ne \dfrac{\pi}{4}$ because the terminal side of this angle does **not** lie in Quadrant II or on the negative x-axis. We must identify the angle θ that satisfies the following three conditions:

The reference angle is $\theta_R = \dfrac{\pi}{4}$.

The terminal side lies in Quadrant II or on the negative x-axis.

$\dfrac{\pi}{2} < \theta \le \pi$.

The only such angle is $\theta = \dfrac{3\pi}{4}$. See Figure 78. Therefore, $\cos^{-1}\left(-\dfrac{1}{\sqrt{2}}\right) = \dfrac{3\pi}{4}$.

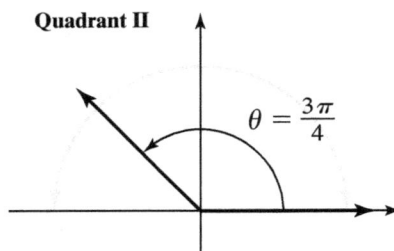

Figure 78 The angle $\theta = \dfrac{3\pi}{4}$ is the only angle in the interval $\dfrac{\pi}{2} < \theta \le \pi$ whose terminal side lies in Quadrant II or on the negative x-axis such that $\cos\theta = -\dfrac{1}{\sqrt{2}}$.

My interactive video summary

⊙ Work through this **interactive video** to see each step of this solution.

You Try It Work through this You Try It problem.

Work Exercises 7–9 in this textbook or in the MyMathLab Study Plan.

Example 4 Finding the Approximate Value of an Inverse Cosine Expression

Use a calculator to approximate each value, or state that the value does not exist.

a. $\cos^{-1}(1.5)$ **b.** $\cos^{-1}(-.25)$

Solution

a. Note that 1.5 is not in the interval $[-1, 1]$. Therefore, 1.5 is not in the domain of $y = \cos^{-1} x$. Also, if $\theta = \cos^{-1}(1.5)$, then $\cos\theta = 1.5$, which is not possible. Thus, $\cos^{-1}(1.5)$ does not exist.

b. Because $x = -.25$ is in the interval $[-1, 1]$, the value of $\cos^{-1}(-.25)$ must be an angle in the interval $[0, \pi]$. If $\cos^{-1}(-.25) = \theta$, then $\cos\theta = -.25 < 0$. Thus, $\frac{\pi}{2} < \theta \le \pi$. Therefore, the approximate value of $\cos^{-1}(-.25)$ must be a decimal value greater than $\frac{\pi}{2} \approx 1.5708$ but less than or equal to $\pi \approx 3.1416$. Using a calculator in radian mode, type $\boxed{\text{2nd}}$ $\boxed{\text{cos}}$ and then $-.25$. We find that $\cos^{-1}(-.25) \approx 1.8235$ radians. This value represents an angle in Quadrant II between $\frac{\pi}{2}$ and π, as predicted.

You Try It Work through this You Try It problem.

Work Exercises 10–12 in this textbook or in the MyMathLab Study Plan.

OBJECTIVE 3 UNDERSTANDING AND FINDING THE EXACT AND APPROXIMATE VALUES OF THE INVERSE TANGENT FUNCTION

Recall that the function $y = \tan x$ is defined for all values except odd integer multiples of $\frac{\pi}{2}$. The tangent function on its entire domain is not one-to-one because the the graph does not pass the horizontal line test. See Figure 79. However, we established in Section 2.3 that the principal cycle of the tangent function is one-to-one. See Figure 80. Once again, in Figure 79 and 80, we label each subinterval of length $\frac{\pi}{2}$ with the Roman numerals I, II, III, or IV to illustrate that the angle located in that interval has a terminal side that lies in the listed quadrant.

My animation summary

Although there are infinitely many ways to restrict the domain of the tangent function to produce a graph that is one-to-one, it is customary to use the principal cycle of the tangent function, the portion of the graph restricted to the domain $-\frac{\pi}{2} < x < \frac{\pi}{2}$. Because this restricted tangent function is one-to-one, it must have an

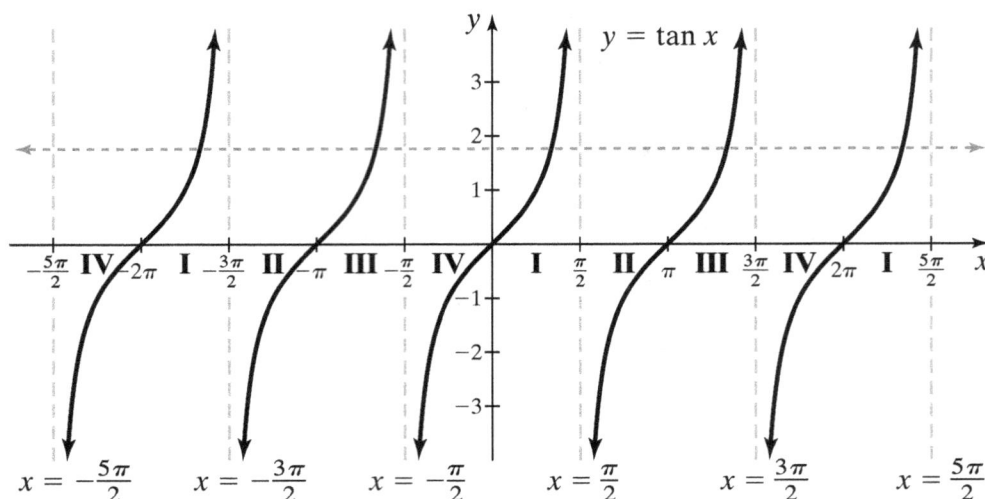

Figure 79 The tangent function fails the horizontal line test and is not one-to-one. This function does not have an inverse function.

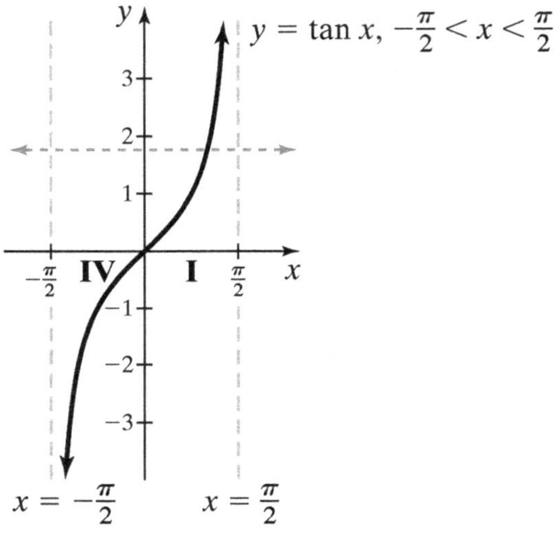

Figure 80 The principal cycle of the tangent function on the interval $-\dfrac{\pi}{2} < x < \dfrac{\pi}{2}$ passes the horizontal line test and is one-to-one. This function has an inverse function.

inverse function. To determine the graph of the inverse of the restricted tangent function, we can reverse the coordinates of the ordered pairs, plot the points, and connect with a smooth curve. See Figures 81 and 82. The inverse of the restricted tangent function is called the **inverse tangent function** and is denoted as $y = \tan^{-1} x$. Notice that the domain of the restricted tangent function is $\left(-\dfrac{\pi}{2}, \dfrac{\pi}{2}\right)$. This is precisely the range of $y = \tan^{-1} x$. Likewise, the range of the restricted tangent function is $(-\infty, \infty)$. This is exactly the domain of $y = \tan^{-1} x$.

The restricted tangent function has two vertical asymptotes at $x = -\dfrac{\pi}{2}$ and $x = \dfrac{\pi}{2}$. The inverse tangent function has two horizontal asymptotes at $y = -\dfrac{\pi}{2}$ and $y = \dfrac{\pi}{2}$. Work through this animation to see how to obtain the inverse tangent function.

Figure 81 The principal cycle of the graph of $y = \tan x$ on the restricted domain

$$-\frac{\pi}{2} < x < \frac{\pi}{2} \text{ with range } -\infty < y < \infty$$

If $x > 0$, then the value of $\tan^{-1} x$ is an angle whose terminal side lies in Quadrant I.

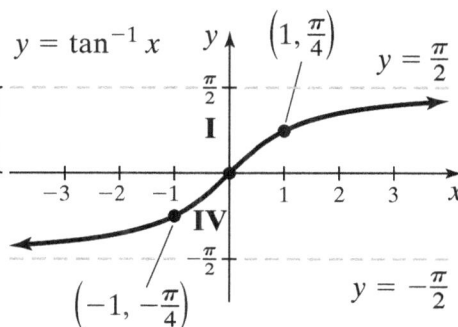

If $x < 0$, then the value of $\tan^{-1} x$ is an angle whose terminal side lies in Quadrant IV.

Figure 82 The domain of $y = \tan^{-1} x$ is $-\infty < x < \infty$ and the range

is $-\frac{\pi}{2} < y < \frac{\pi}{2}$

Definition Inverse Tangent Function

The **inverse tangent function**, denoted as $y = \tan^{-1} x$, is the inverse of $y = \tan x$,

$$-\frac{\pi}{2} < x < \frac{\pi}{2}.$$

The domain of $y = \tan^{-1} x$ is all real numbers and the range is $-\frac{\pi}{2} < y < \frac{\pi}{2}$.

(Note that an alternative notation for $\tan^{-1} x$ is arctan x.)

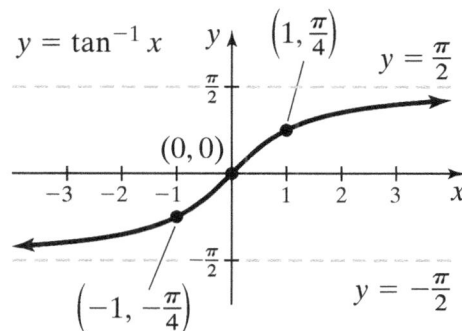

We will think of the expression $\tan^{-1} x$ as the angle on the interval $\left(-\frac{\pi}{2}, \frac{\pi}{2}\right)$ whose tangent is equal to x. For this reason, we continue to use our conventional angle notation, θ, to represent the value of $\tan^{-1} x$. If $\theta = \tan^{-1} x$, then $-\frac{\pi}{2} < \theta < \frac{\pi}{2}$.

Therefore, the terminal side of angle θ must lie in Quadrant I, in Quadrant IV, or on the positive x-axis. See Figure 83.

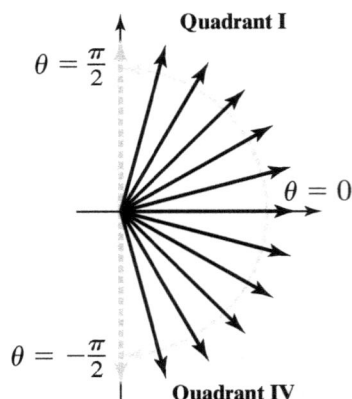

Figure 83 If $\theta = \tan^{-1} x$, then $-\dfrac{\pi}{2} < \theta < \dfrac{\pi}{2}$ and the terminal side of angle θ lies in Quadrant I, in Quadrant IV, or on the positive x-axis.

To determine the exact value of an inverse tangent expression, we follow the four-step process below.

Steps for Determining the Exact Value of $\tan^{-1} x$

Step 1. The value of $\tan^{-1} x$ must be an angle in the interval $\left(-\dfrac{\pi}{2}, \dfrac{\pi}{2} \right)$.

Step 2. Let $\tan^{-1} x = \theta$ such that $\tan \theta = x$.

Step 3. If $\tan \theta = 0$, then $\theta = 0$ and the terminal side of angle θ lies on the positive x-axis.

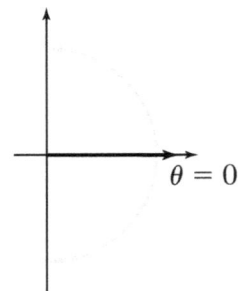

If $\tan \theta > 0$, then $0 < \theta < \dfrac{\pi}{2}$ and the terminal side of angle θ lies in Quadrant I.

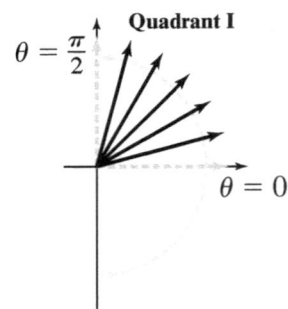

If $\tan \theta < 0$, then $-\dfrac{\pi}{2} < \theta < 0$ and the terminal side of angle θ lies in Quadrant IV.

Step 4. Use your knowledge of the two special right triangles, the graphs of the trigonometric functions, Table 1 from Section 1.4, or Table 2 from Section 1.5 to determine the angle in the correct quadrant whose tangent is x.

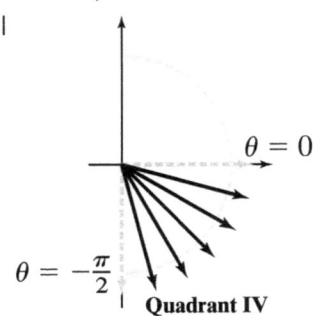

My interactive video summary

▷ **Example 5** Finding the Exact Value of an Inverse Tangent Expression

Determine the exact value of each expression.

a. $\tan^{-1}\left(\dfrac{1}{\sqrt{3}}\right)$ **b.** $\tan^{-1}\left(-\dfrac{1}{\sqrt{3}}\right)$

Solution

a. Step 1. The value of $\tan^{-1}\left(\dfrac{1}{\sqrt{3}}\right)$ must be an angle in the interval $\left(-\dfrac{\pi}{2}, \dfrac{\pi}{2}\right)$.

    Step 2. Let $\tan^{-1}\left(\dfrac{1}{\sqrt{3}}\right) = \theta$ such that $\tan\theta = \dfrac{1}{\sqrt{3}}$.

    Step 3. Because $\tan\theta = \dfrac{1}{\sqrt{3}} > 0$, then $0 < \theta < \dfrac{\pi}{2}$. Thus, we are looking for an angle whose terminal side lies in Quadrant I. See Figure 84.

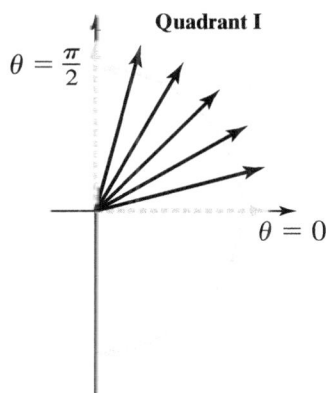

Figure 84 The angle $\theta = \tan^{-1}\left(\dfrac{1}{\sqrt{3}}\right)$ must have a terminal side that lies in Quadrant I.

    Step 4. Using our knowledge of the special $\dfrac{\pi}{6}, \dfrac{\pi}{3}, \dfrac{\pi}{2}$ right triangle or Table 1, we know that $\tan\dfrac{\pi}{6} = \dfrac{1}{\sqrt{3}}$. The angle $\theta = \dfrac{\pi}{6}$ is the **only** angle for which $0 < \theta < \dfrac{\pi}{2}$ whose terminal side lies in Quadrant I such that $\tan\theta = \dfrac{1}{\sqrt{3}}$. See Figure 85. Therefore, $\tan^{-1}\left(\dfrac{1}{\sqrt{3}}\right) = \dfrac{\pi}{6}$.

My interactive video summary

▷ Work through this interactive video to see each step of this solution.

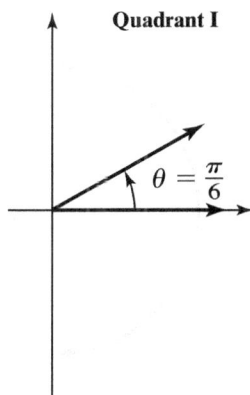

Figure 85 The angle $\theta = \dfrac{\pi}{6}$ is the only angle in the interval $0 < \theta < \dfrac{\pi}{2}$ whose terminal side lies in Quadrant I such that $\tan\theta = \dfrac{1}{\sqrt{3}}$.

b. To find the value of $\tan^{-1}\left(-\dfrac{1}{\sqrt{3}}\right)$, we follow the four-step process.

Step 1. The value of $\tan^{-1}\left(-\dfrac{1}{\sqrt{3}}\right)$ must be an angle in the interval $\left(-\dfrac{\pi}{2}, \dfrac{\pi}{2}\right)$.

Step 2. Let $\tan^{-1}\left(-\dfrac{1}{\sqrt{3}}\right) = \theta$ such that $\tan\theta = -\dfrac{1}{\sqrt{3}}$.

Step 3. Because $\tan\theta = -\dfrac{1}{\sqrt{3}} < 0$, then $-\dfrac{\pi}{2} < \theta < 0$. Thus, we are looking for an angle whose terminal side lies in Quadrant IV. See Figure 86.

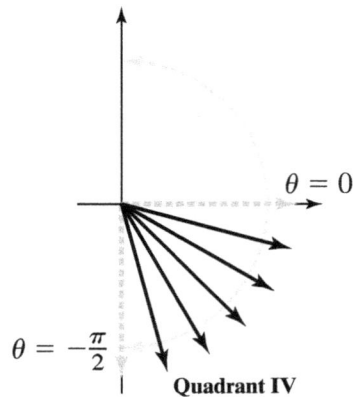

Figure 86 The angle $\theta = \tan^{-1}\left(-\dfrac{1}{\sqrt{3}}\right)$ must have a terminal side that lies in Quadrant IV.

Step 4. Using our knowledge of the special $\dfrac{\pi}{6}, \dfrac{\pi}{3}, \dfrac{\pi}{2}$ right triangle or Table 1, we know that $\tan\dfrac{\pi}{6} = \dfrac{1}{\sqrt{3}}$. Therefore, we must identify the angle θ that satisfies the following three conditions:

The reference angle is $\theta_R = \dfrac{\pi}{6}$.

The terminal side lies in Quadrant IV.

$-\dfrac{\pi}{2} < \theta < 0$.

The only such angle is $\theta = -\dfrac{\pi}{6}$. See Figure 87. Therefore, $\tan^{-1}\left(-\dfrac{1}{\sqrt{3}}\right) = -\dfrac{\pi}{6}$.

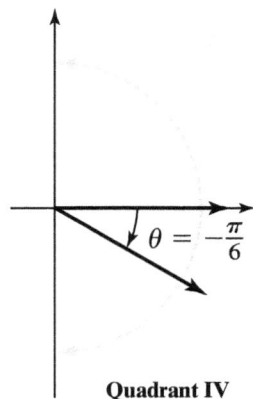

Figure 87 The angle $\theta = -\dfrac{\pi}{6}$ is the only angle in the interval $-\dfrac{\pi}{2} < \theta < 0$ whose terminal side lies in Quadrant IV such that $\tan\theta = -\dfrac{1}{\sqrt{3}}$.

My interactive video summary

⊚ Work through this interactive video to see each step of this solution.

You Try It Work through this You Try It problem.

Work Exercises 13–15 in this textbook or in the MyMathLab **Study Plan.**

Example 6 Finding the Approximate Value of an Inverse Tangent Expression

Use a calculator to approximate the value of $\tan^{-1}(20)$, or state that the value does not exist.

Solution As always, it is good practice to think about the type of result that we can expect when using a calculator. If $\theta = \tan^{-1}(20)$, then $\tan \theta = 20 > 0$. Thus, θ is an angle whose terminal side is located in Quadrant I. If we think of the graph of $y = \tan^{-1} x$, we know that as the values of x get large, then the value of y approaches the horizontal asymptote, $y = \dfrac{\pi}{2}$. See Figure 88. Therefore, we can expect the value of $\tan^{-1}(20)$ to be an angle "close to" $\dfrac{\pi}{2} \approx 1.5708$.

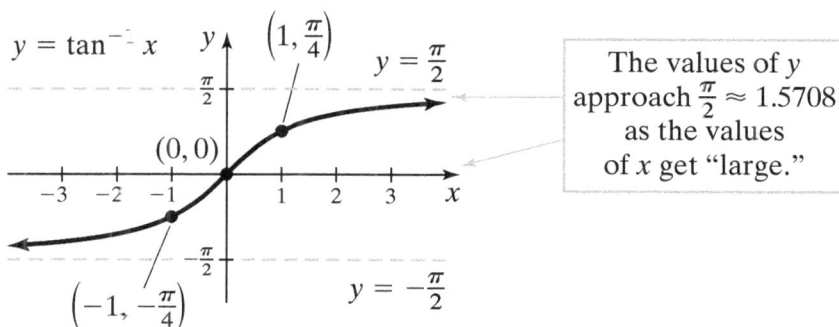

Figure 88 The values of y approach $\dfrac{\pi}{2} \approx 1.5708$ as the values of x get "large."

Using a calculator in radian mode, type [2nd] [tan] and then 20. We find that $\tan^{-1}(20) \approx 1.521$ radians. Notice that this value is "close to" $\dfrac{\pi}{2} \approx 1.5708$, as expected.

Note Because the domain of $y = \tan^{-1} x$ is all real numbers, every real number input value will produce a real number output value, unlike with Example 2c or Example 4a, when we were using a calculator to evaluate an inverse sine expression and inverse cosine expression.

You Try It Work through this You Try It problem.

Work Exercises 16–17 in this textbook or in the MyMathLab **Study Plan.**

2.4 Exercises

In Exercises 1–3, determine the exact value of each expression.

SbS 1. $\sin^{-1}\left(\dfrac{\sqrt{3}}{2}\right)$ SbS 2. $\sin^{-1}\left(-\dfrac{1}{\sqrt{2}}\right)$ SbS 3. $\sin^{-1}(1)$

In Exercises 4–6, use a calculator to approximate each value, or state that the value does not exist.

SbS 4. $\sin^{-1}(.33)$ SbS 5. $\sin^{-1}(-.87)$ SbS 6. $\sin^{-1}(-2.5)$

In Exercises 7–9, determine the exact value of each expression.

SbS 7. $\cos^{-1}\left(\dfrac{\sqrt{3}}{2}\right)$ SbS 8. $\cos^{-1}\left(-\dfrac{1}{\sqrt{2}}\right)$ SbS 9. $\cos^{-1}(1)$

In Exercises 10–12, use a calculator to approximate each value, or state that the value does not exist.

SbS 10. $\cos^{-1}(.35)$ SbS 11. $\cos^{-1}(-.92)$ SbS 12. $\cos^{-1}(3.1)$

In Exercises 13–15, determine the exact value of each expression.

SbS 13. $\tan^{-1}\left(\dfrac{1}{\sqrt{3}}\right)$ SbS 14. $\tan^{-1}(-\sqrt{3})$ SbS 15. $\tan^{-1}(1)$

In Exercises 16–17, use a calculator to approximate each value, or state that the value does not exist.

SbS 16. $\tan^{-1}(.76)$ SbS 17. $\tan^{-1}(5.4)$

Brief Exercises

18. The value of $\sin^{-1} x$ represents an angle in what interval?

19. The value of $\cos^{-1} x$ represents an angle in what interval?

20. The value of $\tan^{-1} x$ represents an angle in what interval?

21. If $\theta = \sin^{-1}\left(\dfrac{\sqrt{3}}{2}\right)$, then determine the quadrant in which or the axis on which the terminal side of θ lies.

22. If $\theta = \sin^{-1}\left(-\dfrac{1}{\sqrt{2}}\right)$, then determine the quadrant in which or the axis on which the terminal side of θ lies.

23. If $\theta = \sin^{-1}(1)$, then determine the quadrant in which or the axis on which the terminal side of θ lies.

24. If $\theta = \cos^{-1}\left(\dfrac{\sqrt{3}}{2}\right)$, then determine the quadrant in which or the axis on which the terminal side of θ lies.

25. If $\theta = \cos^{-1}\left(-\dfrac{1}{\sqrt{2}}\right)$, then determine the quadrant in which or the axis on which the terminal side of θ lies.

26. If $\theta = \cos^{-1}(1)$, then determine the quadrant in which or the axis on which the terminal side of θ lies.

27. If $\theta = \tan^{-1}\left(\dfrac{1}{\sqrt{3}}\right)$, then determine the quadrant in which or the axis on which the terminal side of θ lies.

28. If $\theta = \tan^{-1}(-\sqrt{3})$, then determine the quadrant in which or the axis on which the terminal side of θ lies.

29. If $\theta = \tan^{-1}(1)$, then determine the quadrant in which or the axis on which the terminal side of θ lies.

In Exercises 30–38, determine the exact value of the given expression.

30. $\sin^{-1}\left(\dfrac{\sqrt{3}}{2}\right)$

31. $\sin^{-1}\left(-\dfrac{1}{\sqrt{2}}\right)$

32. $\sin^{-1}(1)$

33. $\cos^{-1}\left(\dfrac{\sqrt{3}}{2}\right)$

34. $\cos^{-1}\left(-\dfrac{1}{\sqrt{2}}\right)$

35. $\cos^{-1}(1)$

36. $\tan^{-1}\left(\dfrac{1}{\sqrt{3}}\right)$

37. $\tan^{-1}(-\sqrt{3})$

38. $\tan^{-1}(1)$

2.5 Inverse Trigonometric Functions II

THINGS TO KNOW

Before working through this section, be sure that you are familiar with the following concepts:

| | VIDEO | ANIMATION | INTERACTIVE |

You Try It
1. Understanding the Composition Cancellation Equations — VIDEO, INTERACTIVE

You Try It
2. Understanding the Special Right Triangles (Section 1.3) — ANIMATION

You Try It
3. Understanding the Right Triangle Definitions of the Trigonometric Functions (Section 1.4) — VIDEO

You Try It
4. Understanding the Signs of the Trigonometric Functions (Section 1.5) — VIDEO

You Try It
5. Determining Reference Angles (Section 1.5) — INTERACTIVE

You Try It 6. Evaluating Trigonometric Functions of Angles Belonging to the $\frac{\pi}{3}, \frac{\pi}{6}$, or $\frac{\pi}{4}$ Families (Section 1.5)

You Try It 7. Understanding the Inverse Sine Function (Section 2.4)

You Try It 8. Understanding the Inverse Cosine Function (Section 2.4)

You Try It 9. Understanding the Inverse Tangent Function (Section 2.4)

OBJECTIVES

1 Evaluating Composite Functions Involving Inverse Trigonometric Functions of the Form $f \circ f^{-1}$ and $f^{-1} \circ f$

2 Evaluating Composite Functions Involving Inverse Trigonometric Functions of the Form $f \circ g^{-1}$ and $f^{-1} \circ g$

3 Understanding the Inverse Cosecant, Inverse Secant, and Inverse Cotangent Functions

4 Writing Trigonometric Expressions as Algebraic Expressions

Introduction to Section 2.5

Before reading this section, it is crucial that you understand the three inverse trigonometric functions that were discussed in Section 2.4. These functions are $y = \sin^{-1}x$, $y = \cos^{-1}x$, and $y = \tan^{-1}x$. Below is a quick overview of each of these functions.

The inverse sine function, $y = \sin^{-1}x$

Domain: $-1 \leq x \leq 1$ **Range:** $-\frac{\pi}{2} \leq y \leq \frac{\pi}{2}$

The range of the inverse sine function represents an angle whose terminal side lies in Quadrant I, Quadrant IV, on the positive x-axis, on the positive y-axis, or on the negative y-axis.

The inverse cosine function, $y = \cos^{-1} x$

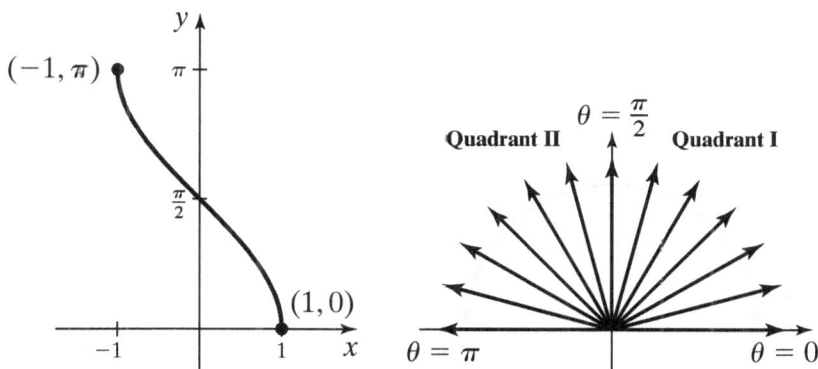

Domain: $-1 \le x \le 1$ **Range:** $0 \le y \le \pi$

The range of the inverse cosine function represents an angle whose terminal side lies in Quadrant I, Quadrant II, on the positive y-axis, on the positive x-axis, or on the negative x-axis.

The inverse tangent function, $y = \tan^{-1} x$

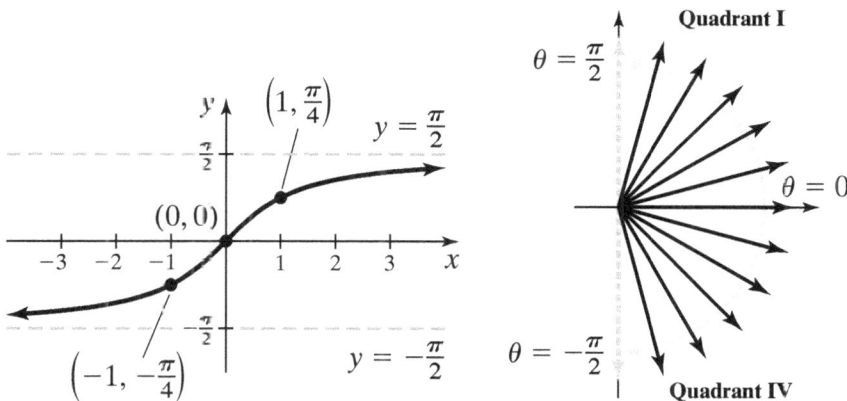

Domain: All real numbers   **Range:** $-\dfrac{\pi}{2} < y < \dfrac{\pi}{2}$

The range of the inverse tangent function represents an angle whose terminal side lies in Quadrant I, Quadrant IV, or on the positive x-axis.

OBJECTIVE 1 EVALUATING COMPOSITE FUNCTIONS INVOLVING INVERSE TRIGONOMETRIC FUNCTIONS OF THE FORM $f \circ f^{-1}$ AND $f^{-1} \circ f$

For any one-to-one function f and its inverse function f^{-1}, the following two composition cancellation equations are satisfied:

$f(f^{-1}(x)) = x$ for all x in the domain of f^{-1} and $f^{-1}(f(x)) = x$ for all x in the domain of f

Therefore, if we encounter a trigonometric expression of the form $(f \circ f^{-1})(x)$ or $(f^{-1} \circ f)(x)$, we can use the cancellation equations *only* if we verify that x is in the domain of the "inner" function. This gives the following cancellation equations for compositions of inverse trigonometric functions.

Cancellation Equations for Compositions of Inverse Trigonometric Functions

Cancellation Equations for the Restricted Sine Function and Its Inverse

$$\sin(\sin^{-1} x) = x \text{ for all } x \text{ in the interval } [-1, 1]$$

$$\sin^{-1}(\sin \theta) = \theta \text{ for all } \theta \text{ in the interval } \left[-\frac{\pi}{2}, \frac{\pi}{2}\right]$$

Cancellation Equations for the Restricted Cosine Function and Its Inverse

$$\cos(\cos^{-1} x) = x \text{ for all } x \text{ in the interval } [-1, 1]$$

$$\cos^{-1}(\cos \theta) = \theta \text{ for all } \theta \text{ in the interval } [0, \pi]$$

Cancellation Equations for the Restricted Tangent Function and Its Inverse

$$\tan(\tan^{-1} x) = x \text{ for all } x \text{ in the interval } (-\infty, \infty)$$

$$\tan^{-1}(\tan \theta) = \theta \text{ for all } \theta \text{ in the interval } \left(-\frac{\pi}{2}, \frac{\pi}{2}\right)$$

My interactive video summary

⊙ **Example 1** Finding the Exact Value of a Composite Expression of the Form $f \circ f^{-1}$

Find the exact value of each expression or state that it does not exist.

a. $\sin\left(\sin^{-1}\frac{1}{2}\right)$ **b.** $\cos\left(\cos^{-1}\frac{3}{2}\right)$ **c.** $\tan(\tan^{-1}(8.2))$ **d.** $\sin(\sin^{-1}(1.3))$

Solution

a. Here, we have an expression of the form $\sin(\sin^{-1} x)$. The value of $x = \frac{1}{2}$ is in the interval $[-1, 1]$. Thus we can use the cancellation equation, $\sin(\sin^{-1} x) = x$. Therefore, $\sin\left(\sin^{-1}\frac{1}{2}\right) = \frac{1}{2}$.

We do not have to use a cancellation equation to evaluate the expression $\sin\left(\sin^{-1}\frac{1}{2}\right)$, as shown in the following alternate method for evaluating $\sin\left(\sin^{-1}\frac{1}{2}\right)$:

$$\sin\left(\sin^{-1}\frac{1}{2}\right) \qquad \text{Write the original expression.}$$

$$= \sin\left(\frac{\pi}{6}\right) \qquad \text{Evaluate } \sin^{-1}\frac{1}{2} \text{ using the four-step process for determining the exact value of } \mathbf{sin^{-1}}\textbf{\textit{x}}.$$

$$= \frac{1}{2} \qquad \text{Evaluate } \sin\left(\frac{\pi}{6}\right) \text{ using a special right triangle or Table 1 from Section 1.4.}$$

b. To evaluate the expression $\cos\left(\cos^{-1}\frac{3}{2}\right)$, we cannot use a cancellation equation because $x = \frac{3}{2}$ is *not* in the domain of the inverse cosine function.

In fact, because $x = \frac{3}{2}$ is not in the domain of the inverse cosine function, the expression $\cos^{-1}\frac{3}{2}$ does not exist. $\Big($There is no angle whose cosine is equal to $\frac{3}{2}$.$\Big)$ Therefore, $\cos\left(\cos^{-1}\frac{3}{2}\right)$ is undefined.

⚠ **Do not get into the habit of using a calculator to evaluate the composition of trigonometric expressions because it is possible to get false results. Displays of two different graphing devices are shown on the following page. The screenshot on the far right of Figure 89 displays an "incorrect" result because this particular calculator is able to handle imaginary and complex numbers.**

Using Technology 🔲

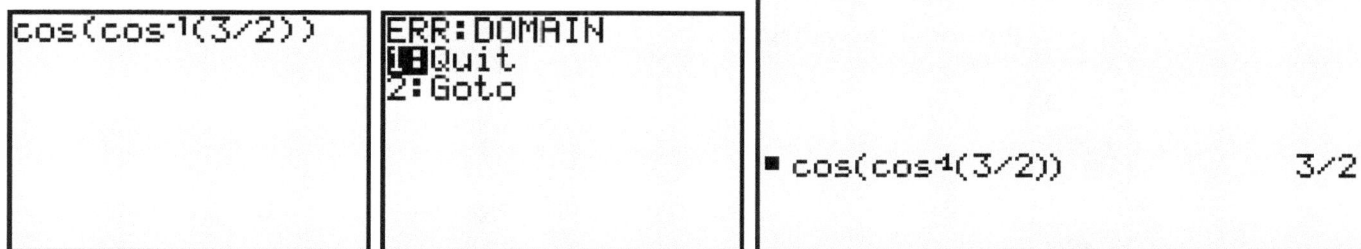

```
cos(cos-1(3/2))
```

```
ERR:DOMAIN
1▪Quit
2:Goto
```

```
F1▼▾  F2▼   F3▼  F4▼  F5
▾f─ Algebra Calc Other PrgmIO




■ cos(cos⁻¹(3/2))            3/2

```

Figure 89 An input of $\cos\left(\cos^{-1}\left(\frac{3}{2}\right)\right)$ using a graphing utility results in a domain error on some calculators but may give a false result on others.

My interactive video summary

▷ Try working parts c and d on your own. When you have completed your work, view the solution, or watch this interactive video to see the worked-out solution. ●

You Try It Work through this You Try It problem.

Work Exercises 1–9 in this textbook or in the MyMathLab Study Plan.

My interactive video summary

▷ **Example 2** Finding the Exact Value of a Composite Expression of the Form $f^{-1} \circ f$

Find the exact value of each expression or state that it does not exist.

a. $\sin^{-1}\left(\sin\dfrac{\pi}{6}\right)$ b. $\cos^{-1}\left(\cos\dfrac{2\pi}{3}\right)$ c. $\sin^{-1}\left(\sin\dfrac{4\pi}{3}\right)$ d. $\tan^{-1}\left(\tan\dfrac{7\pi}{10}\right)$

Solution

a. The terminal side of $\theta = \dfrac{\pi}{6}$ lies in Quadrant I and is thus is in the domain of the restricted sine function, $\left[-\dfrac{\pi}{2}, \dfrac{\pi}{2}\right]$. Therefore, we can use the cancellation equation $\sin^{-1}(\sin\theta) = \theta$. It follows that, $\sin^{-1}\left(\sin\dfrac{\pi}{6}\right) = \dfrac{\pi}{6}$.

b. The terminal side of $\theta = \dfrac{2\pi}{3}$ lies in Quadrant II and is thus in the domain of the restricted cosine function, $[0, \pi]$. Therefore, we can use the cancellation equation $\cos^{-1}(\cos\theta) = \theta$. It follows that $\cos^{-1}\left(\cos\dfrac{2\pi}{3}\right) = \dfrac{2\pi}{3}$.

c. To evaluate the expression $\sin^{-1}\left(\sin\dfrac{4\pi}{3}\right)$, we first note that the terminal side of $\theta = \dfrac{4\pi}{3}$ lies in Quadrant III and is **not** in the interval $\left[-\dfrac{\pi}{2}, \dfrac{\pi}{2}\right]$. Therefore, we cannot directly use a cancellation equation. However, we can first evaluate the "inner expression" $\sin\dfrac{4\pi}{3}$.

$\sin^{-1}\left(\sin\dfrac{4\pi}{3}\right)$ Write the original expression.

$= \sin^{-1}\left(-\dfrac{\sqrt{3}}{2}\right)$ Evaluate the inner expression, $\sin\dfrac{4\pi}{3}$, using the steps for evaluating trigonometric functions of general angles.

$= -\dfrac{\pi}{3}$ Evaluate the inverse sine expression using the four-step process for determining the exact value of $\sin^{-1}x$.

Therefore, $\sin^{-1}\left(\sin\dfrac{4\pi}{3}\right) = -\dfrac{\pi}{3}$.

My interactive video summary

▷ Work through this interactive video to see every step of this solution.

d. To evaluate the expression $\tan^{-1}\left(\tan\dfrac{7\pi}{10}\right)$, we first note that $\dfrac{7\pi}{10}$ is **not** in the interval $\left(-\dfrac{\pi}{2},\dfrac{\pi}{2}\right)$. We can first try to evaluate the "inner expression" $\tan\dfrac{7\pi}{10}$. However, $\dfrac{7\pi}{10}$ does not belong to one of the four families of special angles. Therefore, we cannot determine the *exact* value of $\tan\dfrac{7\pi}{10}$. So we must develop an alternate strategy to evaluate $\tan^{-1}\left(\tan\dfrac{7\pi}{10}\right)$. Let $\tan^{-1}\left(\tan\dfrac{7\pi}{10}\right)=\theta$. Because the value of the inverse tangent function is an angle that must be in the interval $\left(-\dfrac{\pi}{2},\dfrac{\pi}{2}\right)$, we are looking for an angle whose terminal side lies in Quadrant I or Quadrant IV.

The terminal side of $\dfrac{7\pi}{10}$ lies in Quadrant II and has a reference angle of $\dfrac{3\pi}{10}$. The tangent function is negative in Quadrant II. The tangent function is also negative in Quadrant IV. The tangent of any angle in Quadrant IV having a reference angle of $\theta_R=\dfrac{3\pi}{10}$ will be equivalent to the value of $\tan\dfrac{7\pi}{10}$. The only angle θ in the interval $\left(-\dfrac{\pi}{2},\dfrac{\pi}{2}\right)$ whose terminal side lies in Quadrant IV and has a reference angle of $\theta_R=\dfrac{3\pi}{10}$ is $\theta=-\dfrac{3\pi}{10}$. Therefore, $\tan\dfrac{7\pi}{10}=\tan\left(-\dfrac{3\pi}{10}\right)$. See Figure 90.

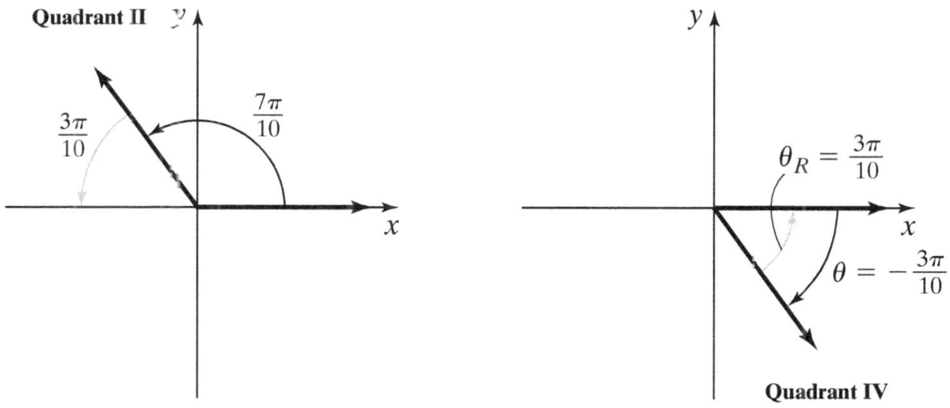

Figure 90 The angles $\dfrac{7\pi}{10}$ and $-\dfrac{3\pi}{10}$ both have a reference angle of $\dfrac{3\pi}{10}$. Since the tangent function is negative in Quadrants II and IV, it follows that $\tan\dfrac{7\pi}{10}=\tan\left(-\dfrac{3\pi}{10}\right)$.

Therefore, we can rewrite the expression $\tan^{-1}\left(\tan\dfrac{7\pi}{10}\right)$ as $\tan^{-1}\left(\tan\left(-\dfrac{3\pi}{10}\right)\right)$. Because $\theta=-\dfrac{3\pi}{10}$ is in the interval $\left(-\dfrac{\pi}{2},\dfrac{\pi}{2}\right)$, we can use the cancellation equation $\tan^{-1}(\tan \theta) = \theta$. It follows that $\tan^{-1}\left(\tan \dfrac{7\pi}{10}\right) = \tan^{-1}\left(\tan \left(-\dfrac{3\pi}{10}\right)\right) = -\dfrac{3\pi}{10}$.

My interactive video summary

▶ Work through this interactive video to see this solution worked out in its entirety.

You Try It Work through this You Try It problem.

Work Exercises 10–27 in this textbook or in the MyMathLab Study Plan.

OBJECTIVE 2   EVALUATING COMPOSITE FUNCTIONS INVOLVING INVERSE TRIGONOMETRIC FUNCTIONS OF THE FORM $f \circ g^{-1}$ AND $f^{-1} \circ g$

When we encounter composite functions involving inverse trigonometric functions that are not of the form $f \circ f^{-1}$ or $f^{-1} \circ f$, as in Examples 1 and 2, then the cancellation equations for inverse trigonometric functions *cannot* be directly used. To evaluate composite functions of the form $f \circ g^{-1}$ or $f^{-1} \circ g$, where f and g are trigonometric functions and f^{-1} and g^{-1} are inverse trigonometric functions, first try to evaluate the "inner function" and then evaluate the "outer function."

My interactive video summary

▶ **Example 3** Finding the Exact Value of a Composite Expression of the Form $f \circ g^{-1}$

Find the exact value of each expression or state that it does not exist.

a. $\cos\left(\tan^{-1}\sqrt{3}\right)$ **b.** $\csc\left(\cos^{-1}\left(-\dfrac{\sqrt{3}}{2}\right)\right)$ **c.** $\sec\left(\sin^{-1}\left(-\dfrac{\sqrt{5}}{8}\right)\right)$

Solution

a. To evaluate the expression $\cos\left(\tan^{-1}\sqrt{3}\right)$, first evaluate the "inner expression" $\tan^{-1}\sqrt{3}$, and then evaluate the "outer expression."

$\cos\left(\tan^{-1}\sqrt{3}\right)$ Write the original expression.

$= \cos\left(\dfrac{\pi}{3}\right)$ Evaluate the inverse tangent expression using the four-step process for determining the exact value of **$\tan^{-1}x$.**

$= \dfrac{1}{2}$ Evaluate $\cos\left(\dfrac{\pi}{3}\right)$ using a special right triangle or Table 1 from Section 1.4.

My interactive video summary

▶ **b.** Try to evaluate $\csc\left(\cos^{-1}\left(-\dfrac{\sqrt{3}}{2}\right)\right)$ on your own. Check your answer or watch this interactive video to see the solution worked out in detail.

c. To evaluate $\sec\left(\sin^{-1}\left(-\dfrac{\sqrt{5}}{8}\right)\right)$, let $\sin^{-1}\left(-\dfrac{\sqrt{5}}{8}\right) = \theta$, where $-\dfrac{\pi}{2} \leq \theta \leq \dfrac{\pi}{2}$ and $\sin \theta = -\dfrac{\sqrt{5}}{8}$. Therefore, we wish to evaluate the expression

$\sec\left(\sin^{-1}\left(-\dfrac{\sqrt{5}}{8}\right)\right) = \sec \theta$. The terminal side of θ lies in Quadrant IV because

$\sin \theta = -\dfrac{\sqrt{5}}{8} < 0$. By the definition of the sine function for general angles, we know that $\sin \theta = \dfrac{y}{r} = -\dfrac{\sqrt{5}}{8}$. Thus, the terminal side of angle θ lies in Quadrant IV and passes through the point having a y-coordinate of $-\sqrt{5}$, a distance of 8 units from the origin. See Figure 91.

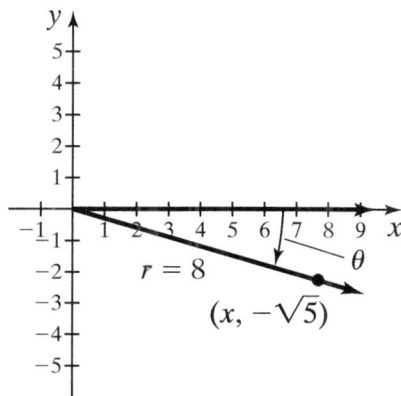

Figure 91 The terminal side of angle θ lies in Quadrant IV and passes through the point $(x, -\sqrt{5})$, a distance of 8 units from the origin.

To find the value of x, we use the formula $r = \sqrt{x^2 + y^2}$ to get $x = \pm\sqrt{59}$.

Because the point $(x, -\sqrt{5})$ lies in Quadrant IV, we know that x must be **positive**. Therefore, $x = \sqrt{59}$.

Using the general angle definition of the secant function, we get

$$\sec\left(\sin^{-1}\left(-\frac{\sqrt{5}}{8}\right)\right) = \sec \theta = \frac{r}{x} = \frac{8}{\sqrt{59}}.$$

My interactive video summary

⊙ Watch this interactive video to see this solution worked out in detail.

You Try It   Work through this You Try It problem.

Work Exercises 28–39 in this textbook or in the MyMathLab Study Plan.

My interactive video summary

⊙ **Example 4** Finding the Exact Value of a Composite Expression of the Form $f^{-1} \circ g$

Find the exact value of each expression or state that it does not exist.

a. $\sin^{-1}\left(\cos\left(-\dfrac{2\pi}{3}\right)\right)$ **b.** $\cos^{-1}\left(\sin\dfrac{\pi}{7}\right)$

Solution

a. To evaluate the expression $\sin^{-1}\left(\cos\left(-\dfrac{2\pi}{3}\right)\right)$, first evaluate the "inner expression" $\cos\left(-\dfrac{2\pi}{3}\right)$, and then evaluate the "outer expression."

$$\sin^{-1}\left(\cos\left(-\frac{2\pi}{3}\right)\right)$$ Write the original expression.

$$= \sin^{-1}\left(-\frac{1}{2}\right)$$ Evaluate the inner expression, $\cos\left(-\frac{2\pi}{3}\right)$ using the steps for evaluating trigonometric functions of general angles.

$$= -\frac{\pi}{6}$$ Evaluate the inverse sine expression using the four-step process for determining the exact value of $\sin^{-1}x$.

Work through this interactive video to see every step of this solution.

b. We cannot determine the exact value of the "inner expression" of $\cos^{-1}\left(\sin\frac{\pi}{7}\right)$ because the angle $\frac{\pi}{7}$ does not belong to one of the four families of special angles. However, it is possible to rewrite the expression $\sin\frac{\pi}{7}$ as an equivalent expression involving cosine using a cofunction identity.

$$\cos^{-1}\left(\sin\frac{\pi}{7}\right)$$ Write the original expression.

$$= \cos^{-1}\left(\cos\left(\frac{\pi}{2} - \frac{\pi}{7}\right)\right)$$ Use the cofunction identity $\sin\theta = \cos\left(\frac{\pi}{2} - \theta\right)$.

$$= \cos^{-1}\left(\cos\frac{5\pi}{14}\right)$$ Simplify.

Therefore, we see that we can rewrite the expression $\cos^{-1}\left(\sin\frac{\pi}{7}\right)$ as $\cos^{-1}\left(\cos\frac{5\pi}{14}\right)$. Because $\frac{5\pi}{14}$ is on the interval $[0, \pi]$, we can use the cancellation equation $\cos^{-1}(\cos\theta) = \theta$. It follows that $\cos^{-1}\left(\sin\frac{\pi}{7}\right) = \cos^{-1}\left(\cos\frac{5\pi}{14}\right) = \frac{5\pi}{14}$.

My interactive video summary

⊙ Work through this interactive video to this example worked out in detail. ●

You Try It Work through this You Try It problem.

Work Exercises 40–47 in this textbook or in the MyMathLab Study Plan.

OBJECTIVE 3 UNDERSTANDING THE INVERSE COSECANT, INVERSE SECANT, AND INVERSE COTANGENT FUNCTIONS

The remaining three inverse trigonometric functions are $y = \csc^{-1}x$, $y = \sec^{-1}x$, and $y = \cot^{-1}x$. Their graphs and definitions are shown below. Click on the video icon next to each definition to see how each graph is obtained.

My video summary

> **Definition**   Inverse Cosecant Function

The **inverse cosecant function**, denoted as $y = \csc^{-1} x$, is the inverse of
$y = \csc x, \left[-\dfrac{\pi}{2}, 0 \right) \cup \left(0, \dfrac{\pi}{2} \right]$.

The domain of $y = \csc^{-1} x$ is $(-\infty, -1] \cup [1, \infty)$ and the range is $\left[-\dfrac{\pi}{2}, 0 \right) \cup \left(0, \dfrac{\pi}{2} \right]$.

$y = \csc^{-1} x$

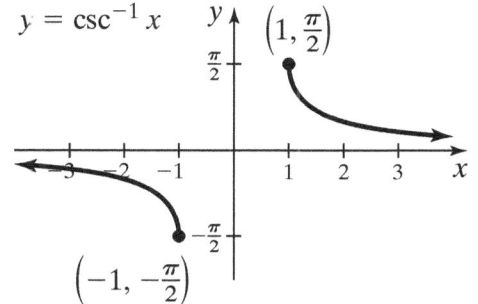

My video summary

> **Definition**   Inverse Secant Function

The **inverse secant function**, denoted as $y = \sec^{-1} x$, is the inverse of
$y = \sec x, \left[0, \dfrac{\pi}{2} \right) \cup \left(\dfrac{\pi}{2}, \pi \right]$.

The domain of $y = \sec^{-1} x$ is $(-\infty, -1] \cup [1, \infty)$ and the range is $\left[0, \dfrac{\pi}{2} \right) \cup \left(\dfrac{\pi}{2}, \pi \right]$.

$y = \sec^{-1} x$

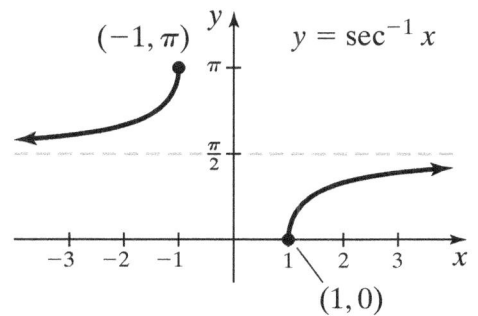

My video summary

> **Definition**   Inverse Cotangent Function*

The **inverse cotangent function**, denoted as $y = \cot^{-1} x$, is the inverse of $y = \cot x, \left(-\dfrac{\pi}{2}, 0 \right) \cup \left(0, \dfrac{\pi}{2} \right]$.

The domain of $y = \cot^{-1} x$ is $(-\infty, \infty)$ and the range is $\left(-\dfrac{\pi}{2}, 0 \right) \cup \left(0, \dfrac{\pi}{2} \right]$.

$y = \cot^{-1} x$

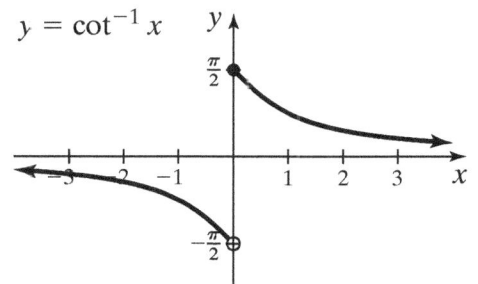

*Some texts define the inverse cotangent function as the inverse of $y = \cot x, 0 < x < \pi$.

Example 5 Finding the Exact Value of an Inverse Secant Expression

Find the exact value of $\sec^{-1}(-\sqrt{2})$ or state that it does not exist.

Solution We follow steps similar to the steps used for determining the exact value of $\cos^{-1} x$.

Step 1. Let $\theta = \sec^{-1}(-\sqrt{2})$.

Step 2. Rewrite $\theta = \sec^{-1}(-\sqrt{2})$ as $\sec \theta = -\sqrt{2}$ (or $\cos \theta = -\dfrac{1}{\sqrt{2}}$), where

$$0 \le \theta < \frac{\pi}{2} \text{ or } \frac{\pi}{2} < \theta \le \pi.$$

Step 3. Because $\sec\theta = -\sqrt{2} < 0$ (or $\cos\theta = -\dfrac{1}{\sqrt{2}} < 0$), we are looking for an angle whose terminal side lies in Quadrant II such that $\dfrac{\pi}{2} < \theta \le \pi$.

Step 4. Using our knowledge of the special $\dfrac{\pi}{4}, \dfrac{\pi}{4}, \dfrac{\pi}{2}$ right triangle or Table 1 from Section 1.4, we know that $\sec\dfrac{\pi}{4} = \sqrt{2}$ (or $\cos\dfrac{\pi}{4} = \dfrac{1}{\sqrt{2}}$.) However, $\theta \ne \dfrac{\pi}{4}$ because the terminal side of this angle does **not** lie in Quadrant II. We must identify the angle that satisfies the following three conditions:

- The reference angle is $\theta_R = \dfrac{\pi}{4}$.
- The terminal side lies in Quadrant II.
- $\dfrac{\pi}{2} < \theta \le \pi$.

The only such angle is $\theta = \dfrac{3\pi}{4}$. Therefore, $\sec^{-1}\left(-\sqrt{2}\right) = \dfrac{3\pi}{4}$.

You Try It Work through this You Try It problem.

Work Exercises 48–50 in this textbook or in the My MathLab **Study Plan.**

Most calculators do not have inverse cosecant, inverse secant, or inverse cotangent keys. To use a calculator, we need to rewrite the given inverse cosecant, inverse secant, or inverse cotangent expression as an expression involving the inverse sine, inverse cosine, or inverse tangent, respectively. We illustrate this by rewriting $y = \sec^{-1} x$ as a function involving inverse cosine.

$y = \sec^{-1} x$ Start with the inverse secant function.

$\sec y = x$ $y = \sec^{-1} x$ means that $x = \sec y$.

$\dfrac{1}{\cos y} = x$ Use the reciprocal identity $\sec\theta = \dfrac{1}{\cos\theta}$.

$1 = x\cos y$ Multiply both sides by $\cos y$.

$\dfrac{1}{x} = \cos y$ Divide both sides by x.

$\cos^{-1}\left(\dfrac{1}{x}\right) = y$ Rewrite as an expression involving inverse cosine.

Therefore, $y = \sec^{-1} x = \cos^{-1}\left(\dfrac{1}{x}\right)$, except where either expression is undefined.

Example 6 illustrates how a calculator can be used to approximate an inverse cosecant, inverse secant, or inverse cotangent expression.

Example 6 Finding the Approximate Value of an Inverse Trigonometric Expression

Use a calculator to approximate each value or state that the value does not exist.

a. $\sec^{-1}(5)$ **b.** $\cot^{-1}(-10)$ **c.** $\csc^{-1}(0.4)$

Solution

a. We first note that 5 is in the interval $(-\infty, -1] \cup [1, \infty)$. Now, rewrite the expression $\sec^{-1}(5)$ as $\cos^{-1}\left(\frac{1}{5}\right)$. Using a calculator in radian mode, type $\boxed{2^{nd}}$ $\boxed{\cos}$ and then $\frac{1}{5}$. We find that $\sec^{-1}(5) = \cos^{-1}\left(\frac{1}{5}\right) \approx 1.3694$ radians.

b. We know that $\cot^{-1}(-10)$ exists because the domain of the inverse cotangent function is all real numbers. Now, rewrite the expression $\cot^{-1}(-10)$ as $\tan^{-1}\left(-\frac{1}{10}\right)$. Using a calculator in radian mode, type $\boxed{2^{nd}}$ $\boxed{\tan}$ and then $-\frac{1}{10}$. We find that $\cot^{-1}(-10) = \tan^{-1}\left(-\frac{1}{10}\right) \approx -0.9967$ radian.

c. The domain of the cosecant function is $(-\infty, -1] \cup [1, \infty)$. Because 0.4 is not on this interval, this means that the value of $\csc^{-1}(0.4)$ does not exist.

You Try It Work through this You Try It problem.

Work Exercises 51–53 in this textbook or in the MyMathLab Study Plan.

OBJECTIVE 4 WRITING TRIGONOMETRIC EXPRESSIONS AS ALGEBRAIC EXPRESSIONS

Given an expression involving inverse trigonometric functions, it is often useful (especially in calculus) to rewrite the expression as an algebraic expression involving the given variable, as in Example 7.

My video summary

⊙ Example 7 Rewrite a Trigonometric Expression as an Algebraic Expression

Rewrite the trigonometric expression $\sin(\tan^{-1}u)$ as an algebraic expression involving the variable u. Assume that $\tan^{-1}u$ represents an angle whose terminal side is located in Quadrant I.

Solution Let $\theta = \tan^{-1}u$. Then θ is the angle in Quadrant I whose tangent is u or $\frac{u}{1}$. Hence, $\tan\theta = \frac{u}{1}$. Because $u > 1$, we know that the terminal side of θ lies in Quadrant I. We can construct a right triangle with acute angle θ with sides of lengths u and 1. See Figure 92.

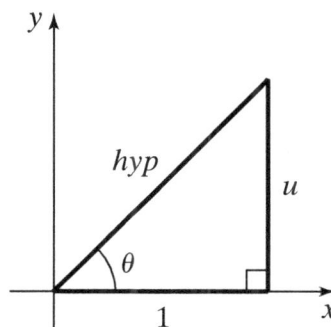

Figure 92

By the Pythagorean Theorem, we can find the length of the hypotenuse, *hyp*, in terms of *u*.

$$a^2 + b^2 = c^2 \qquad \text{Write the Pythagorean Theorem.}$$
$$1^2 + u^2 = (hyp)^2 \qquad \text{Substitute } a = 1, b = u, \text{ and } c = hyp.$$
$$1 + u^2 = (hyp)^2 \qquad \text{Simplify.}$$
$$\sqrt{1 + u^2} = hyp \qquad \begin{array}{l}\text{Take square roots of both sides.}\\ \text{(Ignore the negative square root.)}\end{array}$$

Thus, the length of the hypotenuse is $\sqrt{1 + u^2}$. See Figure 93.

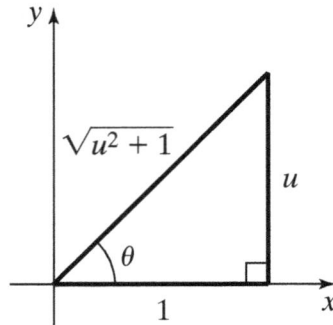

Figure 93

Therefore, $\sin(\tan^{-1} u) = \sin \theta = \dfrac{opp}{hyp} = \dfrac{u}{\sqrt{u^2 + 1}}$.

You Try It   Work through this You Try It problem.

Work Exercises 54–58 in this textbook or in the MyMathLab Study Plan.

2.5 Exercises

In Exercises 1–50, find the exact value of each expression or state that it does not exist.

SbS 1. $\sin\left(\sin^{-1} \dfrac{1}{\sqrt{2}}\right)$

SbS 2. $\cos\left(\cos^{-1} \dfrac{\sqrt{3}}{2}\right)$

SbS 3. $\tan\left(\tan^{-1}\left(\dfrac{1}{\sqrt{3}}\right)\right)$

SbS 4. $\sin\left(\sin^{-1}\left(-\dfrac{\sqrt{3}}{2}\right)\right)$

SbS 5. $\cos\left(\cos^{-1}\left(-\dfrac{1}{2}\right)\right)$

SbS 6. $\tan\left(\tan^{-1}(-\sqrt{3})\right)$

SbS 7. $\sin\left(\sin^{-1} \dfrac{7}{9}\right)$

SbS 8. $\cos\left(\cos^{-1}\left(-\dfrac{8}{5}\right)\right)$

SbS 9. $\tan(\tan^{-1} 35.4)$

SbS 10. $\sin^{-1}\left(\sin \dfrac{\pi}{7}\right)$

SbS 11. $\cos^{-1}\left(\cos \dfrac{\pi}{5}\right)$

SbS 12. $\tan^{-1}\left(\tan \dfrac{\pi}{3}\right)$

SbS 13. $\sin^{-1}\left(\sin\left(-\dfrac{\pi}{4}\right)\right)$

SbS 14. $\cos^{-1}\left(\cos \dfrac{3\pi}{4}\right)$

SbS 15. $\tan^{-1}\left(\tan \dfrac{7\pi}{6}\right)$

SbS 16. $\sin^{-1}\left(\sin \dfrac{14\pi}{3}\right)$

SbS 17. $\cos^{-1}\left(\cos \dfrac{9\pi}{4}\right)$

SbS 18. $\sin^{-1}\left(\sin\left(-\dfrac{7\pi}{2}\right)\right)$

19. $\cos^{-1}(\cos 22\pi)$

20. $\tan^{-1}\left(\tan \dfrac{8\pi}{3}\right)$

21. $\sin^{-1}\left(\sin \dfrac{13\pi}{4}\right)$

22. $\cos^{-1}\left(\cos \dfrac{19\pi}{6}\right)$

23. $\tan^{-1}\left(\tan \dfrac{12\pi}{5}\right)$

24. $\sin^{-1}\left(\sin \dfrac{15\pi}{7}\right)$

25. $\cos^{-1}\left(\cos \dfrac{17\pi}{9}\right)$

26. $\sin^{-1}\left(\sin \dfrac{13\pi}{12}\right)$

27. $\cos^{-1}\left(\cos \dfrac{13\pi}{10}\right)$

28. $\sin\left(\tan^{-1}\sqrt{3}\right)$

29. $\tan\left(\cos^{-1}(0)\right)$

30. $\cos\left(\sin^{-1}\left(\dfrac{\sqrt{3}}{2}\right)\right)$

31. $\sec\left(\sin^{-1}\left(\dfrac{1}{2}\right)\right)$

32. $\csc\left(\tan^{-1}\left(\dfrac{1}{\sqrt{3}}\right)\right)$

33. $\cos\left(\tan^{-1}(-1)\right)$

34. $\sin\left(\cos^{-1}\left(-\dfrac{1}{2}\right)\right)$

35. $\tan\left(\sin^{-1}\left(-\dfrac{1}{\sqrt{2}}\right)\right)$

36. $\sec\left(\cos^{-1}\left(\dfrac{4}{7}\right)\right)$

37. $\csc\left(\tan^{-1}\left(\dfrac{7}{5}\right)\right)$

38. $\tan\left(\sin^{-1}\left(\dfrac{\sqrt{3}}{4}\right)\right)$

39. $\csc\left(\cos^{-1}\left(-\dfrac{2}{\sqrt{11}}\right)\right)$

40. $\sin^{-1}\left(\cos\left(\dfrac{\pi}{3}\right)\right)$

41. $\cos^{-1}\left(\tan\left(-\dfrac{5\pi}{6}\right)\right)$

42. $\tan^{-1}\left(\sin\left(-\dfrac{7\pi}{4}\right)\right)$

43. $\sin^{-1}\left(\tan\left(\dfrac{11\pi}{3}\right)\right)$

44. $\cos^{-1}\left(\sin\left(\dfrac{5\pi}{4}\right)\right)$

45. $\tan^{-1}\left(\cos\left(\dfrac{-19\pi}{6}\right)\right)$

46. $\sin^{-1}\left(\cos\left(\dfrac{\pi}{11}\right)\right)$

47. $\cos^{-1}\left(\sin\left(\dfrac{19\pi}{9}\right)\right)$

48. $\sec^{-1}\left(-\dfrac{2}{\sqrt{3}}\right)$

49. $\csc^{-1}(2)$

50. $\cot^{-1}\left(-\dfrac{1}{\sqrt{3}}\right)$

In Exercises 51–53, use a calculator to approximate each value or state that the value does not exist.

51. $\cot^{-1}(15)$

52. $\sec^{-1}(-.92)$

53. $\csc^{-1}(3.1)$

In Exercises 54–58, rewrite each trigonometric expression as an algebraic expression involving the variable u. Assume that $u > 0$ and that the value of the "inner" inverse trigonometric expression represents an angle θ such that $0 < \theta < \dfrac{\pi}{2}$.

54. $\cos\left(\sin^{-1}u\right)$

55. $\tan\left(\cos^{-1}2u\right)$

56. $\cos\left(\sin^{-1}\dfrac{3}{u}\right)$

57. $\cot\left(\tan^{-1}\dfrac{u}{\sqrt{11}}\right)$

58. $\sec\left(\sin^{-1}\dfrac{u}{\sqrt{u^2 + 121}}\right)$

Brief Exercises

In Exercises 59–105, find the exact value of each expression or state that it does not exist.

59. $\sin\left(\sin^{-1}\dfrac{1}{\sqrt{2}}\right)$

60. $\cos\left(\cos^{-1}\dfrac{\sqrt{3}}{2}\right)$

61. $\tan\left(\tan^{-1}\left(\dfrac{1}{\sqrt{3}}\right)\right)$

62. $\sin\left(\sin^{-1}\left(-\dfrac{\sqrt{3}}{2}\right)\right)$

63. $\cos\left(\cos^{-1}\left(-\dfrac{1}{2}\right)\right)$

64. $\tan\left(\tan^{-1}(-\sqrt{3})\right)$

65. $\sin\left(\sin^{-1}\dfrac{7}{9}\right)$

66. $\cos\left(\cos^{-1}\left(-\dfrac{8}{5}\right)\right)$

67. $\tan\left(\tan^{-1}35.4\right)$

68. $\sin^{-1}\left(\sin\dfrac{\pi}{7}\right)$

69. $\cos^{-1}\left(\cos\dfrac{\pi}{5}\right)$

70. $\tan^{-1}\left(\tan\dfrac{\pi}{3}\right)$

71. $\sin^{-1}\left(\sin\left(-\dfrac{\pi}{4}\right)\right)$

72. $\cos^{-1}\left(\cos\dfrac{3\pi}{4}\right)$

73. $\tan^{-1}\left(\tan\dfrac{7\pi}{6}\right)$

74. $\sin^{-1}\left(\sin\dfrac{14\pi}{3}\right)$

75. $\cos^{-1}\left(\cos\dfrac{9\pi}{4}\right)$

76. $\sin^{-1}\left(\sin\left(-\dfrac{7\pi}{2}\right)\right)$

77. $\cos^{-1}(\cos 22\pi)$

78. $\tan^{-1}\left(\tan\dfrac{8\pi}{3}\right)$

79. $\sin^{-1}\left(\sin\dfrac{13\pi}{4}\right)$

80. $\cos^{-1}\left(\cos\dfrac{19\pi}{6}\right)$

81. $\tan^{-1}\left(\tan\dfrac{12\pi}{5}\right)$

82. $\sin^{-1}\left(\sin\dfrac{15\pi}{7}\right)$

83. $\cos^{-1}\left(\cos\dfrac{17\pi}{9}\right)$

84. $\sin^{-1}\left(\sin\dfrac{13\pi}{12}\right)$

85. $\cos^{-1}\left(\cos\dfrac{13\pi}{10}\right)$

86. $\sin\left(\tan^{-1}\sqrt{3}\right)$

87. $\tan\left(\cos^{-1}(0)\right)$

88. $\cos\left(\sin^{-1}\left(\dfrac{\sqrt{3}}{2}\right)\right)$

89. $\sec\left(\sin^{-1}\left(\dfrac{1}{2}\right)\right)$

90. $\csc\left(\tan^{-1}\left(\dfrac{1}{\sqrt{3}}\right)\right)$

91. $\cos\left(\tan^{-1}(-1)\right)$

92. $\sin\left(\cos^{-1}\left(-\dfrac{1}{2}\right)\right)$

93. $\tan\left(\sin^{-1}\left(-\dfrac{1}{\sqrt{2}}\right)\right)$

94. $\sec\left(\cos^{-1}\left(\dfrac{4}{7}\right)\right)$

95. $\csc\left(\tan^{-1}\left(\dfrac{7}{5}\right)\right)$

96. $\tan\left(\sin^{-1}\left(\dfrac{\sqrt{3}}{4}\right)\right)$

97. $\csc\left(\cos^{-1}\left(-\dfrac{2}{\sqrt{11}}\right)\right)$

98. $\sin^{-1}\left(\cos\left(\dfrac{\pi}{3}\right)\right)$

99. $\cos^{-1}\left(\tan\left(-\dfrac{5\pi}{6}\right)\right)$

100. $\tan^{-1}\left(\sin\left(-\dfrac{7\pi}{4}\right)\right)$

101. $\sin^{-1}\left(\tan\left(\dfrac{11\pi}{3}\right)\right)$

102. $\cos^{-1}\left(\sin\left(\dfrac{5\pi}{4}\right)\right)$

103. $\tan^{-1}\left(\cos\left(\dfrac{-19\pi}{6}\right)\right)$

104. $\sin^{-1}\left(\cos\left(\dfrac{\pi}{11}\right)\right)$

105. $\cos^{-1}\left(\sin\left(\dfrac{19\pi}{9}\right)\right)$

Trigonometric Identities

CHAPTER THREE CONTENTS

3.1 Trigonometric Identities

THINGS TO KNOW

Before working through this section, be sure that you are familiar with the following concepts:

VIDEO   ANIMATION   INTERACTIVE

You Try It 1. Understanding the Quotient Identities for Acute Angles (Section 1.4) ⊘

You Try It 2. Understanding the Reciprocal Identities for Acute Angles (Section 1.4) ⊘

You Try It 3. Understanding the Pythagorean Identities for Acute Angles (Section 1.4) ⊘

OBJECTIVES

1 Reviewing the Fundamental Identities

2 Substituting Known Identities to Verify an Identity

3 Changing to Sines and Cosines to Verify an Identity

4 Factoring to Verify an Identity

5 Separating a Single Quotient into Multiple Quotients to Verify an Identity

6 Combining Fractional Expressions to Verify an Identity

7 Multiplying by Conjugates to Verify an Identity

8 Summarizing the Techniques for Verifying Identities

OBJECTIVE 1   REVIEWING THE FUNDAMENTAL IDENTITIES

In Section 1.4, we established some fundamental trigonometric identities that were valid for all acute angles. We used right triangles to prove many of these identities. We now review these fundamental identities (the quotient identities, the reciprocal identities, and the Pythagorean identities), which are true for *all* values of θ for which each trigonometric expression is defined.

The Quotient Identities

$$\tan \theta = \frac{\sin \theta}{\cos \theta} \quad \cot \theta = \frac{\cos \theta}{\sin \theta}$$

The Reciprocal Identities

$$\sin \theta = \frac{1}{\csc \theta} \quad \csc \theta = \frac{1}{\sin \theta}$$

$$\cos \theta = \frac{1}{\sec \theta} \quad \sec \theta = \frac{1}{\cos \theta}$$

$$\tan \theta = \frac{1}{\cot \theta} \quad \cot \theta = \frac{1}{\tan \theta}$$

The Pythagorean Identities

$$\sin^2\theta + \cos^2\theta = 1$$

$$1 + \tan^2\theta = \sec^2\theta$$

$$1 + \cot^2\theta = \csc^2\theta$$

Note that there are several variations of the Pythagorean identities such as $\sin^2\theta = 1 - \cos^2\theta$ that are obtained by a simple algebraic manipulation of one of the three listed identities. You should become very familiar with the various forms of the Pythagorean identities.

We have also explored the graphs and properties of each of the trigonometric functions in depth in Chapter 2. Recall that one of the properties of the sine, tangent, cosecant, and cotangent functions is that each of these functions is an odd function. In contrast, the cosine and secant functions are even functions. These facts give rise to the following **odd and even properties.**

The Odd Properties

$$\sin (-\theta) = -\sin \theta \quad \tan (-\theta) = -\tan \theta$$

$$\csc (-\theta) = -\csc \theta \quad \cot (-\theta) = -\cot \theta$$

The Even Properties

$$\cos (-\theta) = \cos \theta$$

$$\sec (-\theta) = \sec \theta$$

You Try It   Work through this You Try It problem.

Work Exercises 1–7 in this textbook or in the MyMathLab Study Plan.

OBJECTIVE 2   SUBSTITUTING KNOWN IDENTITIES TO VERIFY AN IDENTITY

We can use the fundamental identities to determine other trigonometric identities. When given a trigonometric identity, it is important to remember that the identity is valid for *all* values of the independent variable for which both sides of the identity are defined. In this section, we will explore several different techniques that can be used to verify identities. Verifying trigonometric identities is the process of showing that the expression on one side of the identity can be simplified to be the exact expression that appears on the other side. We typically start with what appears to be the more complicated side of the identity and then attempt to use known identities to transform that side into the trigonometric expression that appears on the other side of the identity. There is no one correct way to verify identities. In fact, there can be several ways to verify an identity. We start with a technique that first involves substituting one or more known trigonometric identities and then simplifying. Be especially aware of squared trigonometric expressions because they may indicate the use of a Pythagorean identity.

My interactive
video summary

⊚ **Example 1** Verifying an Identity

Verify each identity.

a. $\tan x \cot x = 1$

b. $\sec^2 3x + \cot^2 3x - \tan^2 3x = \csc^2 3x$

c. $(5 \sin y + 2 \cos y)^2 + (5 \cos y - 2 \sin y)^2 = 29$

Solution

a. The left-hand side of the identity is more complicated. Note that we can first use a reciprocal identity to rewrite $\cot x$ and then simplify.

$$\tan x \cot x \overset{?}{=} 1 \qquad \text{Write the original identity.}$$

$$\tan x \cdot \frac{1}{\tan x} \overset{?}{=} 1 \qquad \text{Use the quotient identity } \cot x = \frac{1}{\tan x}.$$

$$\cancel{\tan x} \cdot \frac{1}{\cancel{\tan x}} \overset{?}{=} 1 \qquad \text{Cancel the common factors.}$$

$$1 = 1 \qquad \text{The left-hand side is identical to the right-hand side.}$$

My interactive
video summary

⊚ The left-hand side is now identical to the right-hand side. Therefore, the identity is verified. You may wish to work through this interactive video to see each step of this solution. Note that if we multiply any trigonometric function by its reciprocal function, then the result will always be 1. To see these products worked out, read these steps.

b. The left-hand side appears to be more difficult than the right-hand side. We see that the left-hand side contains three squared trigonometric expressions. This is an indication that one or more Pythagorean identities might be used.

$$\sec^2 3x + \cot^2 3x - \tan^2 3x \overset{?}{=} \csc^2 3x \qquad \text{Write the original identity.}$$

$$\cot^2 3x + (\sec^2 3x - \tan^2 3x) \overset{?}{=} \csc^2 3x \qquad \text{Rearrange the terms.}$$

$$\cot^2 3x + 1 \overset{?}{=} \csc^2 3x \qquad \text{Use the Pythagorean identity } \sec^2\theta - \tan^2\theta = 1.$$

$$\csc^2 3x = \csc^2 3x \qquad \text{Use the Pythagorean identity } \cot^2\theta + 1 = \csc^2\theta.$$

My interactive
video summary

⊚ The left-hand side is now identical to the right-hand side. Therefore, the identity is verified. You may wish to work through this interactive video to see each step of this solution.

⚠ **In the previous example, note that we treated the argument, $3x$, just as if it were a single symbol like θ. Never separate an argument:**

$$\sec^2 3x = \sec^2(3x) \text{ but } \sec^2 3x \neq \sec^2(3) \cdot x$$

c. It does not appear that we can directly substitute a known identity. However, squaring the two expressions will produce squared trigonometric expressions that may allow us to use a Pythagorean identity.

$$(5 \sin y + 2 \cos y)^2 + (5 \cos y - 2 \sin y)^2 \overset{?}{=} 29 \qquad \text{Write the original identity.}$$

$$29 \sin^2 y + 29 \cos^2 y \overset{?}{=} 29 \qquad \text{Square each expression on the left-hand side. To see this squaring process, read these steps.}$$

$$29(\sin^2 y + \cos^2 y) \overset{?}{=} 29 \qquad \text{Factor out the common factor of 29.}$$

$$29(1) \overset{?}{=} 29 \qquad \text{Use the Pythagorean identity } \sin^2\theta + \cos^2\theta = 1.$$

$$29 = 29 \qquad \text{Multiply.}$$

The left-hand side is now identical to the right-hand side. Therefore, the identity is verified. You may wish to work through this interactive video to see each step of this solution.

You Try It Work through this You Try It problem.

Work Exercises 8–12 in this textbook or in the MyMathLab Study Plan.

OBJECTIVE 3 CHANGING TO SINES AND COSINES TO VERIFY AN IDENTITY

When we have simple forms of multiple trigonometric functions on both sides of the identity, it is often efficient first to rewrite all of the functions on one side of the identity in terms of sines and cosines and then simplify.

Example 2 Verifying an Identity

Verify each identity.

a. $\sin^2 t = \tan t \cot t - \cos^2 t$

b. $\dfrac{\sec \theta \csc \theta}{\cot \theta} = \sec^2\theta$

c. $\dfrac{\cos(-\theta)}{\sec \theta} + \sin(-\theta) \csc \theta = -\sin^2\theta$

Solution

a. The right-hand side of the identity $\sin^2 t = \tan t \cot t - \cos^2 t$ looks like the more complicated side. Therefore, we will rewrite the right-hand side in terms of sines and cosines.

$$\sin^2 t \overset{?}{=} \tan t \cot t - \cos^2 t \qquad \text{Write the original identity.}$$

$$\sin^2 t \overset{?}{=} \frac{\sin t}{\cos t} \cdot \frac{\cos t}{\sin t} - \cos^2 t \qquad \text{Use the quotient identities } \tan t = \frac{\sin t}{\cos t} \text{ and } \cot t = \frac{\cos t}{\sin t}.$$

$$\sin^2 t \overset{?}{=} \frac{\cancel{\sin t}}{\cancel{\cos t}} \cdot \frac{\cancel{\cos t}}{\cancel{\sin t}} - \cos^2 t \qquad \text{Cancel common factors.}$$

$$\sin^2 t \overset{?}{=} 1 - \cos^2 t \qquad \text{Simplify.}$$

$$\sin^2 t = \sin^2 t \qquad \text{Use the Pythagorean identity } 1 - \cos^2\theta = \sin^2\theta.$$

⊗ The right-hand side is now identical to the left-hand side. Therefore, the identity is verified. Note that we could have used an alternate method to verify this identity without changing the right-hand side to sines and cosines. To see an alternate solution, read these **steps** or work through this **interactive video**.

b. It appears that the trigonometric expression on the left-hand side of the identity $\dfrac{\sec \theta \csc \theta}{\cot \theta} = \sec^2 \theta$ is more complicated than the expression on the right-hand side. Therefore, we will start by rewriting the left-hand side in terms of sines and cosines.

$$\dfrac{\sec \theta \csc \theta}{\cot \theta} \overset{?}{=} \sec^2 \theta \qquad \text{Write the original identity.}$$

$$\dfrac{\dfrac{1}{\cos \theta} \cdot \dfrac{1}{\sin \theta}}{\dfrac{\cos \theta}{\sin \theta}} \overset{?}{=} \sec^2 \theta \qquad \text{Use the reciprocal identities } \sec \theta = \dfrac{1}{\cos \theta} \text{ and } \csc \theta = \dfrac{1}{\sin \theta} \text{ and the quotient identity } \cot \theta = \dfrac{\cos \theta}{\sin \theta}.$$

$$\dfrac{\dfrac{1}{\sin \theta \cos \theta}}{\dfrac{\cos \theta}{\sin \theta}} \overset{?}{=} \sec^2 \theta \qquad \text{Multiply the expressions in the numerator.}$$

$$\dfrac{1}{\sin \theta \cos \theta} \div \dfrac{\cos \theta}{\sin \theta} \overset{?}{=} \sec^2 \theta \qquad \text{Rewrite as the division of two expressions.}$$

$$\dfrac{1}{\sin \theta \cos \theta} \cdot \dfrac{\sin \theta}{\cos \theta} \overset{?}{=} \sec^2 \theta \qquad \text{Multiply the first expression by the reciprocal of the second expression.}$$

$$\dfrac{1}{\cancel{\sin \theta} \cos \theta} \cdot \dfrac{\cancel{\sin \theta}}{\cos \theta} \overset{?}{=} \sec^2 \theta \qquad \text{Cancel the common factors.}$$

$$\dfrac{1}{\cos^2 \theta} \overset{?}{=} \sec^2 \theta \qquad \text{Multiply.}$$

$$\left(\dfrac{1}{\cos \theta}\right)^2 \overset{?}{=} \sec^2 \theta \qquad \text{Rewrite } \dfrac{1}{\cos^2 \theta} \text{ as } \left(\dfrac{1}{\cos \theta}\right)^2.$$

$$(\sec \theta)^2 \overset{?}{=} \sec^2 \theta \qquad \text{Use the reciprocal identity } \dfrac{1}{\cos \theta} = \sec \theta.$$

$$\sec^2 \theta = \sec^2 \theta \qquad \text{Rewrite } (\sec \theta)^2 \text{ as } \sec^2 \theta.$$

⊗ The left-hand side is now identical to the right-hand side. Therefore, the identity is verified. Work through this **interactive video** to see each step of this solution.

c. Try to verify the identity $\dfrac{\cos(-\theta)}{\sec \theta} + \sin(-\theta) \csc \theta = -\sin^2 \theta$ on your own.

To see the solution, read these **steps** or work through part c of this interactive video.

You Try It   Work through this You Try It problem.

Work Exercises 13–18 in this textbook or in the MyMathLab **Study Plan**.

OBJECTIVE 4 FACTORING TO VERIFY AN IDENTITY

In the solution to Example 1c, we factored out a common factor of 29 to help us verify the identity. We may encounter identities where it may be necessary to use a more complicated factoring technique. We now quickly review some factoring techniques.

You should also be very comfortable factoring trinomials of the form $ax^2 + bx + c$. Watch this video to review how to factor trinomials that have a leading coefficient equal to 1. You might also want to watch this video to review how to factor trinomials that have a leading coefficient not equal to 1.

It is also worthwhile to remember the following special factoring formulas. These formulas hold true for all algebraic and trigonometric expressions a and b.

Difference of Two Squares $a^2 - b^2 = (a + b)(a - b)$

Perfect Square Formulas $a^2 + 2ab + b^2 = (a + b)^2$ and $a^2 - 2ab + b^2 = (a - b)^2$

Sum of Two Cubes $a^3 + b^3 = (a + b)(a^2 - ab + b^2)$

Difference of Two Cubes $a^3 - b^3 = (a - b)(a^2 + ab + b^2)$

Example 3 Verifying a Trigonometric Identity by Factoring

Verify each trigonometric identity.

a. $\sin x - \cos^2 x \sin x = \sin^3 x$

b. $\dfrac{\tan^3 \alpha - 1}{\tan \alpha - 1} = \sec^2 \alpha + \tan \alpha$

c. $\dfrac{8 \sin^2 \theta - 2 \sin \theta - 3}{1 + 2 \sin \theta} = 4 \sin \theta - 3$

Solution

a. Note that both terms of the left-hand side of the identity share a common factor of $\sin x$. Therefore, we start by first factoring out $\sin x$ from both terms on the left-hand side and then substituting with a known identity as we first did in Objective 2.

$$\sin x - \cos^2 x \sin x \overset{?}{=} \sin^3 x \qquad \text{Write the original trigonometric identity.}$$

$$\sin x \, (1 - \cos^2 x) \overset{?}{=} \sin^3 x \qquad \text{Factor out the common factor of } \sin x.$$

$$\sin x \, (\sin^2 x) \overset{?}{=} \sin^3 x \qquad \text{Use the Pythagorean identity } 1 - \cos^2 \theta = \sin^2 \theta.$$

$$\sin^3 x = \sin^3 x \qquad \text{Multiply.}$$

My interactive video summary

⊙ The left-hand side is now identical to the right-hand side. Therefore, the identity is verified. Watch this interactive video to see each step of this verification process.

b. The numerator of the left-hand expression is a difference of two cubes of the form $a^3 - b^3$, where $a = \tan \alpha$ and $b = 1$. Start by factoring the difference of two cubes and then look for a known identity to substitute and simplify.

$$\frac{\tan^3 \alpha - 1}{\tan \alpha - 1} \overset{?}{=} \sec^2 \alpha + \tan \alpha \qquad \text{Write the original trigonometric identity.}$$

$$\frac{(\tan \alpha - 1)(\tan^2\alpha + \tan \alpha + 1)}{\tan \alpha - 1} \stackrel{?}{=} \sec^2\alpha + \tan \alpha$$

Use the difference of two cubes formula $a^3 - b^3 = (a - b)(a^2 + ab - b^2)$ with $a = \tan \alpha$ and $b = 1$.

$$\frac{(\cancel{\tan \alpha - 1})(\tan^2\alpha + \tan \alpha + 1)}{\cancel{\tan \alpha - 1}} \stackrel{?}{=} \sec^2\alpha + \tan \alpha$$

Cancel common factors.

$$(1 + \tan^2 \alpha) + \tan \alpha \stackrel{?}{=} \sec^2\alpha + \tan \alpha$$

Rearrange the terms.

$$\sec^2 \alpha + \tan \alpha = \sec^2 \alpha + \tan \alpha$$

Use the Pythagorean identity $1 + \tan^2\theta = \sec^2\theta$.

My interactive video summary

⊘ The left-hand side is now identical to the right-hand side. Therefore, the identity is verified. Watch this interactive video to see each step of this verification process.

c. Try verifying the identity $\dfrac{8 \sin^2\theta - 2 \sin \theta - 3}{1 + 2 \sin \theta} = 4 \sin \theta - 3$ on your own. Try factoring the numerator on the left-hand side by treating it as a trinomial. Check your work by watching this interactive video.

You Try It Work through this You Try It problem.

Work Exercises 19–23 in this textbook or in the MyMath Lab **Study Plan**.

OBJECTIVE 5   SEPARATING A SINGLE QUOTIENT INTO MULTIPLE QUOTIENTS TO VERIFY AN IDENTITY

When one side of a trigonometric identity is a quotient of the form $\dfrac{A + B}{C}$, where C is a single trigonometric expression, then it is often advantageous to begin by separating the quotient into multiple quotients. We do this using the algebraic property $\dfrac{A + B}{C} = \dfrac{A}{C} + \dfrac{B}{C}$.

My video summary ⊘ **Example 4** Verifying a Trigonometric Identity by Separating a Single Quotient into Multiple Quotients

Verify the trigonometric identity $\dfrac{\sin \alpha + \cos \alpha}{\cos \alpha} - \dfrac{\sin \alpha + \cos \alpha}{\sin \alpha} = \tan \alpha - \cot \alpha$.

Solution The left-hand side appears to be more complex than the right-hand side. Note that the denominator of each quotient on the left-hand side of the identity contains one term. We can therefore separate each quotient into two separate quotients, substitute, and then simplify.

$$\frac{\sin \alpha + \cos \alpha}{\cos \alpha} - \frac{\sin \alpha + \cos \alpha}{\sin \alpha} \stackrel{?}{=} \tan \alpha - \cot \alpha$$

Write the original trigonometric identity.

$$\frac{\sin \alpha}{\cos \alpha} + \frac{\cos \alpha}{\cos \alpha} - \frac{\sin \alpha}{\sin \alpha} - \frac{\cos \alpha}{\sin \alpha} \stackrel{?}{=} \tan \alpha - \cot \alpha$$

Separate each single quotient into two quotients.

$$\tan \alpha + 1 - 1 - \cot \alpha \stackrel{?}{=} \tan \alpha - \cot \alpha$$

Use two quotient identities and rewrite the quotients $\dfrac{\cos \alpha}{\cos \alpha}$ and $\dfrac{\sin \alpha}{\sin \alpha}$ as 1.

$$\tan \alpha - \cot \alpha = \tan \alpha - \cot \alpha$$

Combine like terms.

The left-hand side is now identical to the right-hand side. Therefore, the identity is verified. Watch this **video** to see each step of this verification process.

You Try It Work through this You Try It problem.

Work Exercises 24–26 in this textbook or in the MyMathLab **Study Plan**.

OBJECTIVE 6 COMBINING FRACTIONAL EXPRESSIONS TO VERIFY AN IDENTITY

When two or more fractional expressions appear on one side of an identity and the other side contains only one term, it may be useful to begin by combining all fractions into a single quotient using a common denominator, substituting if necessary, and then simplifying.

My video summary ⊘ **Example 5** Verifying a Trigonometric Identity by Combining Fractional Expressions

Verify the trigonometric identity $\dfrac{1 - \csc\theta}{\cot\theta} - \dfrac{\cot\theta}{1 - \csc\theta} = 2\tan\theta$.

Solution The left-hand side looks more complicated than the right-hand side. So we will focus on transforming the left-hand side.

$$\frac{1 - \csc\theta}{\cot\theta} - \frac{\cot\theta}{1 - \csc\theta} \overset{?}{=} 2\tan\theta \qquad \text{Write the original trigonometric identity.}$$

$$\frac{(1 - \csc\theta)(1 - \csc\theta) - (\cot\theta)(\cot\theta)}{(\cot\theta)(1 - \csc\theta)} \overset{?}{=} 2\tan\theta \qquad \begin{array}{l}\text{Combine the two expressions to write} \\ \text{as a single quotient using the LCD} \\ (\cot\theta)(1 - \csc\theta).\end{array}$$

$$\frac{1 - 2\csc\theta + \csc^2\theta - \cot^2\theta}{(\cot\theta)(1 - \csc\theta)} \overset{?}{=} 2\tan\theta \qquad \text{Multiply.}$$

$$\frac{1 - 2\csc\theta + 1}{(\cot\theta)(1 - \csc\theta)} \overset{?}{=} 2\tan\theta \qquad \begin{array}{l}\text{Use the Pythagorean identity} \\ \csc^2\theta - \cot^2\theta = 1.\end{array}$$

$$\frac{2 - 2\csc\theta}{(\cot\theta)(1 - \csc\theta)} \overset{?}{=} 2\tan\theta \qquad \text{Combine like terms.}$$

$$\frac{2(1 - \csc\theta)}{(\cot\theta)(1 - \csc\theta)} \overset{?}{=} 2\tan\theta \qquad \text{Factor out the common factor of 2.}$$

$$\frac{2\cancel{(1 - \csc\theta)}}{(\cot\theta)\cancel{(1 - \csc\theta)}} \overset{?}{=} 2\tan\theta \qquad \text{Cancel common factors.}$$

$$\frac{2}{\cot\theta} \overset{?}{=} 2\tan\theta \qquad \text{Simplify.}$$

$$2\tan\theta = 2\tan\theta \qquad \text{Use the reciprocal identity } \frac{1}{\cot\theta} = \tan\theta.$$

The left-hand side is now identical to the right-hand side. Therefore, the identity is verified. Watch this **video** to see each step of this verification process.

You Try It Work through this You Try It problem.

Work Exercises 27–30 in this textbook or in the MyMathLab Study Plan.

OBJECTIVE 7 MULTIPLYING BY CONJUGATES TO VERIFY IDENTITIES

Given an expression of the form $A + B$, we define its conjugate as the expression $A - B$. When verifying an identity, if the numerator or denominator of one of the expressions is of the form $A + B$, try first multiplying the numerator and denominator of the expression by $\dfrac{A - B}{A - B} = 1$. We illustrate this technique in Example 6.

My video summary

⊙ **Example 6** Verifying a Trigonometric Identity by Multiplying the Numerator and Denominator by a Conjugate

Verify the trigonometric identity $\dfrac{\sin \theta}{\csc \theta + 1} = \dfrac{1 - \sin \theta}{\cot^2 \theta}$.

Solution Neither side of the identity looks more complicated than the other. We could try to separate the right-hand side into two terms. However, this is not an efficient first step because the left-hand side does not contain two separate terms. So, multiply the left-hand side by the conjugate of its denominator.

$$\frac{\sin \theta}{\csc \theta + 1} \overset{?}{=} \frac{1 - \sin \theta}{\cot^2 \theta} \qquad \text{Write the original trigonometric identity.}$$

$$\frac{\sin \theta}{\csc \theta + 1} \cdot \frac{\csc \theta - 1}{\csc \theta - 1} \overset{?}{=} \frac{1 - \sin \theta}{\cot^2 \theta} \qquad \text{Multiply the numerator and denominator by } 1 = \frac{\csc \theta - 1}{\csc \theta - 1}.$$

$$\frac{\sin \theta \csc \theta - \sin \theta}{\csc^2 \theta - 1} \overset{?}{=} \frac{1 - \sin \theta}{\cot^2 \theta} \qquad \text{Multiply the terms in the numerator and denominator.}$$

$$\frac{\sin \theta \cdot \dfrac{1}{\sin \theta} - \sin \theta}{\cot^2 \theta} \overset{?}{=} \frac{1 - \sin \theta}{\cot^2 \theta} \qquad \text{Use the reciprocal identity } \csc \theta = \frac{1}{\sin \theta} \text{ and the Pythagorean identity } \csc^2 \theta - 1 = \cot^2 \theta.$$

$$\frac{\cancel{\sin \theta} \cdot \dfrac{1}{\cancel{\sin \theta}} - \sin \theta}{\cot^2 \theta} \overset{?}{=} \frac{1 - \sin \theta}{\cot^2 \theta} \qquad \text{Cancel common factors.}$$

$$\frac{1 - \sin \theta}{\cot^2 \theta} = \frac{1 - \sin \theta}{\cot^2 \theta}$$

The left-hand side is now identical to the right-hand side. Therefore, the identity is verified. Watch this video to see each step of this verification process. ●

You Try It Work through this You Try It problem.

Work Exercises 31–32 in this textbook or in the MyMathLab Study Plan.

OBJECTIVE 8 SUMMARIZING THE TECHNIQUES FOR VERIFYING IDENTITIES

As stated earlier, there is no one correct way to verify an identity. This fact makes it sometimes difficult to know how to begin. But it seems that often there are one or two initial techniques that will allow for the most efficient verification process, and the key is to recognize those before you begin.

We have seen six different techniques that can be used as the initial step for verifying trigonometric identities, and we have discussed how to recognize when to use each technique. We have also seen that often other techniques along with simplification may be needed in subsequent steps to complete the verification process.

We now state some generic guidelines for verifying trigonometric identities, and we summarize the six techniques learned in this section along with tips on how to recognize when each should be used.

A Summary for Verifying Trigonometric Identities

First, start with what appears to be the more difficult side of the given identity. Try to transform this side so that it eventually matches identically to the other side. Don't hesitate to start over and work with the other side of the identity if you are having trouble. Try using one of the following techniques to begin and then use others if necessary to continue the verification process.

1. Look for ways to use known identities such as the reciprocal identities, quotient identities, and even/odd properties. If the identity includes a squared trigonometric expression, try using a variation of a Pythagorean identity.

2. Try rewriting each trigonometric expression in terms of sines and cosines.

3. Factor out a greatest common factor and use algebraic factoring techniques such as factoring the difference of two squares or the sum/difference of two cubes.

4. If a single term appears in the denominator of a quotient, try separating the quotient into two or more quotients:

$$\frac{A+B}{C} = \frac{A}{C} + \frac{B}{C}$$

5. If there are two or more fractional expressions, try combining the expressions using a common denominator:

$$\frac{A}{B} + \frac{C}{D} = \frac{AD+BC}{BD}$$

6. If the numerator or denominator of one or more quotients contains an expression of the form $A + B$, try multiplying the numerator and denominator by its conjugate $A - B$.

$$\frac{C}{A+B} = \frac{C}{A+B} \cdot \frac{A-B}{A-B} \quad \text{or} \quad \frac{A+B}{C} = \frac{A+B}{C} \cdot \frac{A-B}{A-B}$$

My interactive video summary

⊘ **Example 7** Verifying Trigonometric Identities

Verify each trigonometric identity.

a. $\dfrac{\sin^2 t + 6\sin t + 9}{\sin t + 3} = \dfrac{3\csc t + 1}{\csc t}$

b. $\dfrac{2\csc\theta}{\sec\theta} + \dfrac{\cos\theta}{\sin\theta} = 3\cot\theta$

c. $\dfrac{1 - \sin\theta}{\cos\theta} + \dfrac{\cos\theta}{1 - \sin\theta} = 2\sec\theta$

Solution

a. It appears that the left-hand side is more complicated than the right-hand side. It also appears that we can factor the numerator on the left-hand side.

$\dfrac{\sin^2 t + 6\sin t + 9}{\sin t + 3} \overset{?}{=} \dfrac{3\csc t + 1}{\csc t}$ Write the original trigonometric identity.

$\dfrac{\cancel{(\sin t + 3)}(\sin t + 3)}{\cancel{\sin t + 3}} \overset{?}{=} \dfrac{3\csc t + 1}{\csc t}$ Factor the numerator and cancel common factors.

$\sin t + 3 \overset{?}{=} \dfrac{3\csc t + 1}{\csc t}$ Simplify.

The left-hand side is now more simplified but it is not yet identical to the right-hand side. The denominator on the right-hand side is $\csc t$. Therefore, we will use a reciprocal identity to rewrite $\sin t$ as $\dfrac{1}{\csc t}$. We can then apply the technique of combining fractional expressions.

$\dfrac{1}{\csc t} + 3 \overset{?}{=} \dfrac{3\csc t + 1}{\csc t}$ Use the reciprocal identity $\sin t = \dfrac{1}{\csc t}$.

$\dfrac{1 + 3\csc t}{\csc t} \overset{?}{=} \dfrac{3\csc t + 1}{\csc t}$ Combine using the LCD $\csc t$.

$\dfrac{3\csc t + 1}{\csc t} = \dfrac{3\csc t + 1}{\csc t}$ Rearrange the terms.

The left-hand side is now identical to the right-hand side. As you can see, we used two different techniques (factoring and combining fractional expressions) to verify this identity.

My interactive video summary

⊘ Try to verify the identities for parts b and c on your own. Use the summary for verifying trigonometric identities to assist you. Work through this interactive video when you are through to see the verification process. Note that this interactive video illustrates two different ways to verify the identities in parts b and c.   ●

You Try It Work through this You Try It problem.

Work Exercises 33–43 in this textbook or in the My‌‌Math‌‌Lab **Study Plan.**

3.1 Exercises

1. Complete the quotient identity: $\tan \theta = \dfrac{\sin \theta}{\boxed{}}$.

2. Complete the quotient identity: $\dfrac{\cos \theta}{\sin \theta} = \boxed{}$.

3. Complete the reciprocal identity: $\csc \theta = \dfrac{1}{\boxed{}}$.

4. Complete the reciprocal identity: $\boxed{} = \dfrac{1}{\sec \theta}$.

5. Complete the Pythagorean identity: $\boxed{} + \cos^2 \theta = 1$.

6. Complete the Pythagorean identity: $\sec^2 \theta - \boxed{} = \boxed{}$.

7. Use fundamental identities to complete the identity: $1 - \dfrac{1}{\cos^2 \theta} = \boxed{}$.

In Exercises 8–43, verify each identity.

8. $1 + \cot^2(-\theta) = \csc^2 \theta$

9. $\dfrac{\cot^2 \beta + 1}{\csc \beta} = \csc \beta$

10. $\tan^2 4x + \csc^2 4x - \cot^2 4x = \sec^2 4x$

11. $(3 \cos \theta - 4 \sin \theta)^2 + (4 \cos \theta + 3 \sin \theta)^2 = 25$

12. $7 \sin^2 \theta + 4 \cos^2 \theta = 4 + 3 \sin^2 \theta$

13. $\cos \theta \csc \theta = \cot \theta$

14. $\tan(-x) \cos x = -\sin x$

15. $\cos \theta \tan \theta \csc \theta = 1$

16. $\sec \theta - \cos \theta = \sin \theta \tan \theta$

17. $\csc t \sin t - \sin^2 t = \cos^2 t$

18. $\dfrac{\csc(\theta)}{\sin \theta} - \cos(-\theta) \sec(-\theta) = \cot^2 \theta$

19. $\sec x + \tan^2 x \sec x = \sec^3 x$

20. $\dfrac{\sin^2 \beta - \cos^2 \beta}{\sin \beta - \cos \beta} = \sin \beta + \cos \beta$

21. $\dfrac{\cot^3 \theta + 1}{\cot \theta + 1} = \csc^2 \theta - \cot \theta$

22. $\dfrac{6 \csc^2 \theta - 7 \csc \theta - 3}{1 + 3 \csc \theta} = 2 \csc \theta - 3$

23. $\sin^4 \theta - \cos^4 \theta = 2 \sin^2 \theta - 1$

24. $\dfrac{1 + 2 \sec \theta}{\sec \theta} = 2 + \cos \theta$

25. $\dfrac{\cot \theta + 1}{\csc \theta} = \sin \theta + \cos \theta$

26. $\dfrac{\tan \alpha + \cot \alpha}{\tan \alpha} - \dfrac{\cot \alpha + \tan \alpha}{\cot \alpha} = \cot^2 \alpha - \tan^2 \alpha$

27. $\dfrac{\sin x + \tan x + 1}{\cos x} = \sec x + \tan x + \sin x \sec^2 x$

28. $\dfrac{1 - \cos \theta}{\sin \theta} + \dfrac{\sin \theta}{1 - \cos \theta} = 2 \csc \theta$

29. $\dfrac{\sin t}{1 + \cos t} + \cot t = \csc t$

30. $\dfrac{\sec \beta}{\sin \beta} - \dfrac{\sin \beta}{\sec \beta} = \dfrac{\tan^2 \beta + \cos^2 \beta}{\tan \beta}$

31. $\dfrac{\sin \theta}{1 - \cos \theta} = \dfrac{1 + \cos \theta}{\sin \theta}$

32. $\dfrac{\sec t - 1}{\tan t} = \dfrac{\tan t}{\sec t + 1}$

33. $\cot^2 3x + \sec^2 3x - \tan^2 3x = \csc^2 3x$

34. $1 + \cot^2(-\theta) = \csc^2 \theta$

35. $\tan(-x) \cos x = -\sin x$

36. $\dfrac{\csc \theta + 1}{\cot \theta} = \sec \theta + \tan \theta$

37. $\dfrac{1 - \sec \theta}{\tan \theta} - \dfrac{\tan \theta}{1 - \sec \theta} = 2 \cot \theta$

38. $\dfrac{\csc\theta}{\sec\theta} + \dfrac{4\cos\theta}{\sin\theta} = 5\cot\theta$

39. $\dfrac{\cos^2 t + 3\cos t - 10}{\cos t + 5} = \dfrac{1 - 2\sec t}{\sec t}$

40. $1 + \dfrac{1 - \cot^2 x}{1 + \cot^2 x} = 2\sin^2 x$

41. $\sec^4\theta - \tan^4\theta = 2\sec^2\theta - 1$

42. $\dfrac{\tan(-\theta)}{\cot(-\theta)} - \sin\theta\csc(-\theta) = \sec^2\theta$

43. $\dfrac{\cos^3\theta + \sin^3\theta}{\cos\theta + \sin\theta} = 1 - \cos\theta\sin\theta$

3.2 The Sum and Difference Formulas

THINGS TO KNOW

Before working through this section, be sure that you are familiar with the following concepts:

| | VIDEO | ANIMATION | INTERACTIVE |

You Try It
1. Understanding Cofunctions (Section 1.4) ⊙

You Try It
2. Evaluating Trigonometric Functions of Angles Belonging to the $\dfrac{\pi}{3}, \dfrac{\pi}{6}$, or $\dfrac{\pi}{4}$ Families (Section 1.5) ⊙

You Try It
3. Finding the Exact and Approximate Values of an Inverse Sine Expression (Section 2.4) ⊙

You Try It
4. Finding the Exact and Approximate Values of an Inverse Cosine Expression (Section 2.4) ⊙

OBJECTIVES

1 Understanding the Sum and Difference Formulas for the Cosine Function

2 Understanding the Sum and Difference Formulas for the Sine Function

3 Understanding the Sum and Difference Formulas for the Tangent Function

4 Using the Sum and Difference Formulas to Verify Identities

5 Using the Sum and Difference Formulas to Evaluate Expressions Involving Inverse Trigonometric Functions

OBJECTIVE 1   UNDERSTANDING THE SUM AND DIFFERENCE FORMULAS FOR THE COSINE FUNCTION

In the previous section, we verified trigonometric identities using a wide variety of techniques. In this section we introduce several new trigonometric identities. Unlike the identities in the previous section, where our identities contained only one variable, many of the identities in this section contain two variables. In this section, we will focus on formulas involving the sum and difference of two angles. We begin by introducing the sum and difference formulas for the cosine function.

The Sum and Difference Formulas for the Cosine Function

$$\cos(\alpha + \beta) = \cos\alpha\cos\beta - \sin\alpha\sin\beta$$

Cosine of the Sum of Two Angles Formula

$$\cos(\alpha - \beta) = \cos\alpha\cos\beta + \sin\alpha\sin\beta$$

Cosine of the Difference of Two Angles Formula

My animation summary

These sum and difference formulas are true for *all* real numbers α and β, although we typically think of α and β as angles. We start by proving the cosine of the difference of two angles formula. To prove this formula, we will assume that α and β are angles such that $0 < \beta < \alpha < 2\pi$. It is a very good exercise to work carefully through the animation proving this formula. Before you work through the animation proof, be prepared to draw two angles in standard position on a piece of paper. You will also need to remember the distance formula. As you work through the animation, you will need to use the pause button to draw and label your angles and points. Also, don't be afraid to rewind when necessary!

The cosine of the sum of two angles formula can be derived using the cosine of the difference of two angles along with the even and odd properties. To see how this is done, read this proof.

My interactive video summary

⊚ **Example 1** Finding the Exact Value of a Trigonometric Expression Involving Cosine

Find the exact value of each trigonometric expression without the use of a calculator.

a. $\cos\left(\dfrac{2\pi}{3} + \dfrac{3\pi}{4}\right)$ **b.** $\cos(225° - 150°)$

Solution

a. To evaluate the expression $\cos\left(\dfrac{2\pi}{3} + \dfrac{3\pi}{4}\right)$, we can use the cosine of the sum of two angles formula.

$$\cos\left(\frac{2\pi}{3} + \frac{3\pi}{4}\right)$$

Write the original expression.

$$= \cos\frac{2\pi}{3}\cos\frac{3\pi}{4} - \sin\frac{2\pi}{3}\sin\frac{3\pi}{4}$$

Use the cosine of the sum of two angles formula.

$$= \left(-\frac{1}{2}\right)\left(-\frac{1}{\sqrt{2}}\right) - \left(\frac{\sqrt{3}}{2}\right)\left(\frac{1}{\sqrt{2}}\right)$$

Evaluate each trigonometric function. (See how to evaluate each function by watching the interactive video.)

$$= \frac{1}{2\sqrt{2}} - \frac{\sqrt{3}}{2\sqrt{2}} = \frac{1 - \sqrt{3}}{2\sqrt{2}}$$

Simplify.

b. To evaluate the expression $\cos(225° - 150°)$, we can use the cosine of the difference of two angles formula. Try to find the exact value of this expression on your own. When you're finished, check your solution or work through this interactive video to see each step worked out in detail.

You Try It Work through this You Try It problem.

Work Exercises 1–8 in this textbook or in the MyMathLab Study Plan.

My interactive video summary

⊙ **Example 2** Finding the Exact Value of a Trigonometric Expression Involving Cosine

Find the exact value of each trigonometric expression without the use of a calculator.

a. $\cos\left(\dfrac{7\pi}{12}\right)\cos\left(\dfrac{5\pi}{12}\right) + \sin\left(\dfrac{7\pi}{12}\right)\sin\left(\dfrac{5\pi}{12}\right)$ **b.** $\cos\left(\dfrac{7\pi}{12}\right)$ **c.** $\cos(-75°)$

Solution

a. To evaluate the expression $\cos\left(\dfrac{7\pi}{12}\right)\cos\left(\dfrac{5\pi}{12}\right) + \sin\left(\dfrac{7\pi}{12}\right)\sin\left(\dfrac{5\pi}{12}\right)$, we recognize that this expression represents one side of the cosine of the difference of two angles formula with $\alpha = \dfrac{7\pi}{12}$ and $\beta = \dfrac{5\pi}{12}$. Therefore,

$$\cos\left(\frac{7\pi}{12}\right)\cos\left(\frac{5\pi}{12}\right) + \sin\left(\frac{7\pi}{12}\right)\sin\left(\frac{5\pi}{12}\right) = \cos\left(\frac{7\pi}{12} - \frac{5\pi}{12}\right) = \cos\left(\frac{2\pi}{12}\right)$$

$$= \cos\frac{\pi}{6} = \frac{\sqrt{3}}{2}.$$

b. To evaluate the expression $\cos\left(\dfrac{7\pi}{12}\right)$, we first need to write the angle $\dfrac{7\pi}{12}$ as the sum or difference of angles whose cosine is known. We know the exact value of $\cos\left(\dfrac{k\pi}{6}\right)$, $\cos\left(\dfrac{k\pi}{4}\right)$, $\cos\left(\dfrac{k\pi}{3}\right)$, and $\cos\left(\dfrac{k\pi}{2}\right)$, where k is an integer.

So, perhaps we can rewrite $\dfrac{7\pi}{12}$ as the sum or difference of angles of the form $\dfrac{k\pi}{6}, \dfrac{k\pi}{4}, \dfrac{k\pi}{3}$, and $\dfrac{k\pi}{2}$.

Note that $\dfrac{7\pi}{12} = \dfrac{3\pi}{12} + \dfrac{4\pi}{12} = \dfrac{\pi}{4} + \dfrac{\pi}{3}$. Therefore, using the cosine of the sum of two angles formula, we get

$$\cos\left(\frac{7\pi}{12}\right) = \cos\left(\frac{\pi}{4} + \frac{\pi}{3}\right) = \cos\frac{\pi}{4}\cos\frac{\pi}{3} - \sin\frac{\pi}{4}\sin\frac{\pi}{3}.$$

My interactive video summary

⊙ Try evaluating this expression on your own. When you're finished, check your answer or watch this interactive video to see the video solution.

c. To evaluate the expression $\cos(-75°)$, we first need to rewrite $-75°$ as the sum or difference of angles whose cosine is known. We know that we can determine the cosine of any angle that is an integer multiple of $30°, 45°, 60°$, or $90°$. So, we try the sum or difference of integer multiples of these angles until we find a combination that works. Note that $3 \cdot (45°) = 135°$ and $-75° = 60° - 135°$. Therefore, using the cosine of the difference of two angles formula, we get

$$\cos(-75°) = \cos(60° - 135°) = \cos 60° \cos(135°) + \sin 60° \sin(135°).$$

My interactive video summary

⊙ Try evaluating this expression on your own. When you're finished, check your answer or watch this interactive video to see the video solution.

You Try It Work through this You Try It problem.

Work Exercises 9–13 in this textbook or in the MyMathLab **Study Plan.**

My video summary ⊙ **Example 3** Finding the Exact Value of a Trigonometric Expression

Suppose that the terminal side of angle α lies in Quadrant IV and the terminal side of angle β lies in Quadrant III. If $\cos \alpha = \dfrac{4}{7}$ and $\sin \beta = -\dfrac{8}{13}$, find the exact value of $\cos(\alpha + \beta)$.

Solution To evaluate the expression $\cos(\alpha + \beta)$ using the cosine of the sum of two angles formula, we must first determine the value of $\sin \alpha$ and $\cos \beta$.

To find the value of $\sin \alpha$, sketch an angle α with the terminal side of α lying in Quadrant IV. Because $\cos \alpha = \dfrac{4}{7} = \dfrac{x}{r}$, we can choose the point $(4, y)$ lying on the terminal side of α, a distance of $r = 7$ units from the origin. See Figure 1.

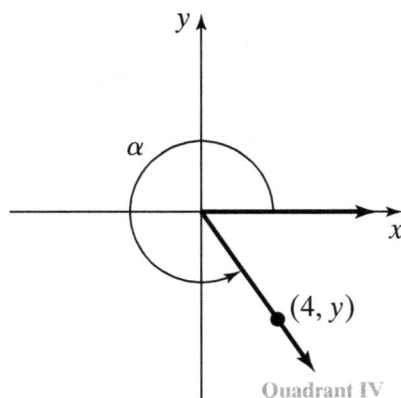

Figure 1 Angle α and the point $(4, y)$ located a distance of $r = 7$ units from the origin

Because the point $(4, y)$ lies in Quadrant IV, we know that the value of y must be negative. Using the relationship $x^2 + y^2 = r^2$, we can determine that $y = -\sqrt{33}$. To see how we obtain this y-coordinate, read through these steps. Therefore,

$$\sin \alpha = \frac{y}{r} = -\frac{\sqrt{33}}{7}.$$

We can determine the value of $\cos \beta$ in a similar fashion. Read through these steps to see how we determine that $\cos \beta = -\dfrac{\sqrt{105}}{13}$. Now that we know the values of $\sin \alpha$, $\cos \alpha$, $\sin \beta$, and $\cos \beta$, we can evaluate $\cos (\alpha + \beta)$.

$$\cos (\alpha + \beta) = \cos \alpha \cos \beta - \sin \alpha \sin \beta \qquad \text{Write the cosine of the sum of two angles formula.}$$

$$= \left(\frac{4}{7}\right)\left(-\frac{\sqrt{105}}{13}\right) - \left(-\frac{\sqrt{33}}{7}\right)\left(-\frac{8}{13}\right) \qquad \text{Substitute the appropriate values.}$$

$$= \frac{-4\sqrt{105} - 8\sqrt{33}}{91} \qquad \text{Simplify.}$$

You Try It   Work through this You Try It problem.

Work Exercises 14 and 15 in this textbook or in the MyMathLab Study Plan.

OBJECTIVE 2   UNDERSTANDING THE SUM AND DIFFERENCE FORMULAS FOR THE SINE FUNCTION

Before we introduce the sum and difference formulas for the sine function, it is important to revisit the cofunction identities because these identities are necessary to prove the sine of the sum of two angles formula. In Section 1.4 we proved that the cofunction identities were true for acute angles. Using formulas from this section, we can verify that these cofunction identities are true for *all* angles. We now restate the cofunction identities for sine and cosine.

Cofunction Identities for Sine and Cosine

$$\cos \left(\frac{\pi}{2} - \theta\right) = \sin \theta \qquad \sin \left(\frac{\pi}{2} - \theta\right) = \cos \theta$$

We can now introduce the sum and difference formulas for the sine function.

The Sum and Difference Formulas for the Sine Function

$\sin (\alpha + \beta) = \sin \alpha \cos \beta + \cos \alpha \sin \beta$ Sine of the Sum of Two Angles Formula

$\sin (\alpha - \beta) = \sin \alpha \cos \beta - \cos \alpha \sin \beta$ Sine of the Difference of Two Angle Formula

---

My video summary ⊙ To derive the sine of the sum of two angles formula, we use a cofunction identity and the cosine of the difference of two angles formula. To see the derivation of this formula, watch this video.

My video summary ⊙ We derive the sine of the difference of two angles formula by rewriting the left-hand side as $\sin (\alpha + (-\beta))$ and then use the sine of the sum of two angles formula and even and odd properties of cosine and sine. To see the derivation of this formula, watch this video.

My interactive video summary

⊘ Example 4 Finding the Exact Value of a Trigonometric Expression Involving Sine

Find the exact value of each trigonometric expression without the use of a calculator.

a. $\sin\left(-\dfrac{\pi}{3} + \dfrac{5\pi}{4}\right)$ **b.** $\sin(12°)\cos(78°) + \cos(12°)\sin(78°)$ **c.** $\sin(15°)$

Solution

a. To evaluate the expression $\sin\left(-\dfrac{\pi}{3} + \dfrac{5\pi}{4}\right)$, we can use the sine of the sum of two angles formula.

$\sin\left(-\dfrac{\pi}{3} + \dfrac{5\pi}{4}\right)$ Write the original expression.

$= \sin\left(-\dfrac{\pi}{3}\right)\cos\left(\dfrac{5\pi}{4}\right) + \cos\left(-\dfrac{\pi}{3}\right)\sin\left(\dfrac{5\pi}{4}\right)$ Use the sine of the sum of two angles formula.

$= \left(-\dfrac{\sqrt{3}}{2}\right)\left(-\dfrac{1}{\sqrt{2}}\right) + \left(\dfrac{1}{2}\right)\left(-\dfrac{1}{\sqrt{2}}\right)$ Evaluate each trigonometric function. (See how to evaluate each function by watching the interactive video.)

$= \dfrac{\sqrt{3}}{2\sqrt{2}} - \dfrac{1}{2\sqrt{2}} = \dfrac{\sqrt{3}-1}{2\sqrt{2}}$ Simplify.

Note that we could have rearranged the order of the angles and rewritten the expression $\sin\left(-\dfrac{\pi}{3} + \dfrac{5\pi}{4}\right)$ as $\sin\left(\dfrac{5\pi}{4} - \dfrac{\pi}{3}\right)$ and then used the sine of the difference of two angles formula to simplify the expression. To see this alternate method worked out in detail, read these steps.

b. To evaluate the expression $\sin(12°)\cos(78°) + \cos(12°)\sin(78°)$, we recognize that this expression represents one side of the sine of the sum of two angles formula with $\alpha = 12°$ and $\beta = 78°$. Therefore,

$$\sin(12°)\cos(78°) + \cos(12°)\sin(78°) = \sin(12° + 78°) = \sin 90° = 1.$$

c. To evaluate the expression $\sin(15°)$, we first need to write the angle $15°$ as the sum or difference of angles whose sine is known. Note that $15° = 60° - 45°$. Therefore, using the sine of the difference of two angles formula, we get

$$\sin(15°) = \sin(60° - 45°) = \sin 60° \cos 45° - \cos 60° \sin 45°.$$

My interactive video summary

⊘ Try evaluating this expression on your own. When you're finished, check your answer or watch this interactive video to see the video solution.

You Try It   Work through this You Try It problem.

Work Exercises 16–28 in this textbook or in the MyMathLab Study Plan.

My video summary   **⊘ Example 5** Finding the Exact Value of a Trigonometric Expression

Suppose that α is an angle such that $\tan \alpha = \dfrac{5}{7}$ and $\cos \alpha < 0$. Also, suppose that β is an angle such that $\sec \beta = -\dfrac{4}{3}$ and $\csc \beta > 0$. Find the exact value of $\sin(\alpha + \beta)$.

Solution To evaluate the expression $\sin(\alpha + \beta)$ using the sine of the sum of two angles formula, we must first determine the value of $\sin \alpha$, $\cos \alpha$, $\sin \beta$, and $\cos \beta$.

To find the value of $\sin \alpha$ and $\cos \alpha$, we must first determine the quadrant in which the terminal side of angle α lies. Because $\tan \alpha > 0$, we know that the terminal side of α could lie in either Quadrant I or Quadrant III. Also, because $\cos \alpha < 0$, we know that the terminal side of α could lie in Quadrant II or Quadrant III. Therefore, angle α lies in Quadrant III.

Because $\tan \alpha = \dfrac{5}{7} = \dfrac{y}{x}$, we can choose the point $(-7, -5)$ that lies on the terminal side of α. Using the relationship $x^2 + y^2 = r^2$, we can determine that $r = \sqrt{74}$. To see exactly how we obtained this value of r, read these steps. See Figure 2.

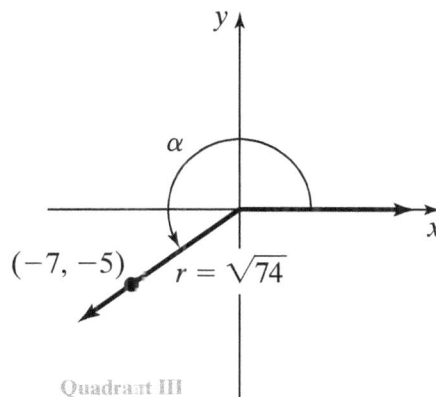

Figure 2 Angle α and the point $(-7, -5)$, which is located a distance of $r = \sqrt{74}$ units from the origin

My video summary ⊙ Thus, $\sin \alpha = \dfrac{y}{r} = \dfrac{-5}{\sqrt{74}}$ and $\cos \alpha = \dfrac{x}{r} = \dfrac{-7}{\sqrt{74}}$. Watch this video to see that

$\sin \beta = \dfrac{\sqrt{7}}{4}$ and $\cos \beta = -\dfrac{3}{4}$. Now that we know the values of $\sin \alpha$, $\cos \alpha$, $\sin \beta$, and $\cos \beta$, we can evaluate $\sin(\alpha + \beta)$.

$$\sin(\alpha + \beta) = \sin \alpha \cos \beta + \cos \alpha \sin \beta \qquad \text{Write the sine of the sum of two angles formula.}$$

$$= \left(\frac{-5}{\sqrt{74}}\right)\left(-\frac{3}{4}\right) + \left(\frac{-7}{\sqrt{74}}\right)\left(\frac{\sqrt{7}}{4}\right) \qquad \text{Substitute the appropriate values.}$$

$$= \frac{15 - 7\sqrt{7}}{4\sqrt{74}} \qquad \text{Simplify.}$$

You Try It Work through this You Try It problem.

Work Exercises 29–30 in this textbook or in the MyMathLab Study Plan.

OBJECTIVE 3 UNDERSTANDING THE SUM AND DIFFERENCE FORMULAS FOR THE TANGENT FUNCTION

We now introduce the final sum and difference formulas, which are the sum and difference formulas for the tangent function.

The Sum and Difference Formulas for the Tangent Function

$$\tan(\alpha + \beta) = \frac{\tan\alpha + \tan\beta}{1 - \tan\alpha\tan\beta}$$ Tangent of the Sum of Two Angles Formula

$$\tan(\alpha - \beta) = \frac{\tan\alpha - \tan\beta}{1 + \tan\alpha\tan\beta}$$ Tangent of the Difference of Two Angles Formula

---

My video summary ⊙ To derive the tangent of the sum of two angles formula, we use the identity

$$\tan(\alpha+\beta) = \frac{\sin(\alpha+\beta)}{\cos(\alpha+\beta)}$$ and then use the sum of two angles formulas for sine and cosine. To see the derivation of the tangent of the sum of two angles formula, watch this video.

My video summary ⊙ To derive the tangent of the difference of two angles formula, we use the odd property of the tangent function and the tangent of the sum of two angles formula. To see the derivation of the tangent of the difference of two angles formula, watch this video.

My interactive video summary

⊙ **Example 6 Finding the Exact Value of a Trigonometric Expression**

Find the exact value of each trigonometric expression without the use of a calculator.

a. $\tan\left(\dfrac{5\pi}{6} + \dfrac{3\pi}{4}\right)$ **b.** $\tan\left(\dfrac{\pi}{12}\right)$

Solution

a. To evaluate the expression $\tan\left(\dfrac{5\pi}{6} + \dfrac{3\pi}{4}\right)$, we can use the tangent of the sum of two angles formula.

$$\tan\left(\frac{5\pi}{6} + \frac{3\pi}{4}\right)$$ Write the original expression.

$$= \frac{\tan\left(\dfrac{5\pi}{6}\right) + \tan\left(\dfrac{3\pi}{4}\right)}{1 - \tan\left(\dfrac{5\pi}{6}\right)\tan\left(\dfrac{3\pi}{4}\right)}$$ Use the tangent of the sum of two angles formula.

$$= \frac{-\dfrac{1}{\sqrt{3}} + (-1)}{1 - \left(-\dfrac{1}{\sqrt{3}}\right)(-1)}$$ Evaluate each trigonometric function. (See how to evaluate each function by watching the interactive video.)

$$= \frac{-\dfrac{1}{\sqrt{3}} - 1}{1 - \dfrac{1}{\sqrt{3}}}$$ Simplify.

$$= \dfrac{\dfrac{-1 - \sqrt{3}}{\sqrt{3}}}{\dfrac{\sqrt{3} - 1}{\sqrt{3}}}$$

Combine terms in the numerator and denominator.

$$= \dfrac{-1 - \sqrt{3}}{\sqrt{3}} \div \dfrac{\sqrt{3} - 1}{\sqrt{3}}$$

Rewrite as the division of two expressions.

$$= \dfrac{-1 - \sqrt{3}}{\sqrt{3}} \cdot \dfrac{\sqrt{3}}{\sqrt{3} - 1}$$

Multiply the first expression by the reciprocal of the second expression.

$$= \dfrac{-1 - \sqrt{3}}{\cancel{\sqrt{3}}} \cdot \dfrac{\cancel{\sqrt{3}}}{\sqrt{3} - 1}$$

Cancel common factors.

$$= \dfrac{-1 - \sqrt{3}}{\sqrt{3} - 1}$$

Simplify.

Recall that rationalizing the denominator is never required in this text. Therefore, the expression $\dfrac{-1 - \sqrt{3}}{\sqrt{3} - 1}$ is a perfectly acceptable answer. However, if we do rationalize the denominator, we see that an equivalent answer is $-2 - \sqrt{3}$. To see how we obtain this, read these steps. Thus,

$$\tan\left(\dfrac{5\pi}{6} + \dfrac{3\pi}{4}\right) = \dfrac{-1 - \sqrt{3}}{\sqrt{3} - 1} \text{ or } \tan\left(\dfrac{5\pi}{6} + \dfrac{3\pi}{4}\right) = -2 - \sqrt{3}.$$

b. To evaluate the expression $\tan\left(\dfrac{\pi}{12}\right)$, we first need to write the angle $\dfrac{\pi}{12}$ as the sum or difference of angles whose tangent is known. Note that $\dfrac{\pi}{12} = \dfrac{4\pi}{12} - \dfrac{3\pi}{12}$ $= \dfrac{\pi}{3} - \dfrac{\pi}{4}$. Therefore, using the tangent of the difference of two angles formula, we get

$$\tan\left(\dfrac{\pi}{12}\right) = \tan\left(\dfrac{\pi}{3} - \dfrac{\pi}{4}\right) = \dfrac{\tan\left(\dfrac{\pi}{3}\right) - \tan\left(\dfrac{\pi}{4}\right)}{1 + \tan\left(\dfrac{\pi}{3}\right)\tan\left(\dfrac{\pi}{4}\right)}.$$

My interactive video summary

⊙ Try evaluating this expression on your own. When you're finished, check your answer or watch this interactive video to see the video solution. ●

You Try It Work through this You Try It problem.

Work Exercises 31–45 in this textbook or in the MyMathLab Study Plan.

OBJECTIVE 4   USING THE SUM AND DIFFERENCE FORMULAS TO VERIFY IDENTITIES

Recall that verifying a trigonometric identity is the process of showing that the expression on one side of the identity can be simplified to be the exact expression that appears on the other side. We typically start with what appears to be the more complicated side of the identity and then attempt to use known identities to transform that side into the trigonometric expression that appears on the other side of the identity. However, if one side of the identity includes an expression of the form $\cos(\alpha \pm \beta)$, $\sin(\alpha \pm \beta)$, or $\tan(\alpha \pm \beta)$, then first substitute one of the sum or difference formulas discussed in this section. Next, use the strategies we developed in Section 3.1 for verifying identities. To review these strategies, read this summary of techniques that were introduced in Section 3.1.

Example 7 Verifying a Trigonometric Identity

Verify the trigonometric identity $\sin(2\theta) = 2\sin\theta\cos\theta$.

Solution We start by writing the left-hand side as $\sin(\theta + \theta)$ and then using the sine of the sum of two angles formula.

$$\sin(2\theta) \overset{?}{=} 2\sin\theta\cos\theta \qquad \text{Write the original trigonometric identity.}$$

$$\sin(\theta + \theta) \overset{?}{=} 2\sin\theta\cos\theta \qquad \text{Write } 2\theta \text{ as } \theta + \theta.$$

$$\sin\theta\cos\theta + \cos\theta\sin\theta \overset{?}{=} 2\sin\theta\cos\theta \qquad \text{Use the sine of the sum of two angles formula.}$$

$$2\sin\theta\cos\theta = 2\sin\theta\cos\theta \qquad \text{Combine like terms.}$$

The left-hand side is now identical to the right-hand side. Therefore, the identity is verified. The identity $\sin(2\theta) = 2\sin\theta\cos\theta$ is called a **double angle identity** and will be discussed in Section 3.3.

My video summary ⊙ **Example 8** Verifying a Trigonometric Identity

Verify the trigonometric identity $\csc(\alpha - \beta) = \dfrac{\sin\alpha\cos\beta + \cos\alpha\sin\beta}{\sin^2\alpha - \sin^2\beta}$.

Solution Although the right-hand side of the identity may look more complicated, we will start with the left-hand side because we can first use a reciprocal identity to rewrite $\csc(\alpha - \beta)$ as $\dfrac{1}{\sin(\alpha - \beta)}$. Then we can use the sine of the difference of two angles formula to rewrite the denominator.

$$\csc(\alpha - \beta) \overset{?}{=} \frac{\sin\alpha\cos\beta + \cos\alpha\sin\beta}{\sin^2\alpha - \sin^2\beta} \qquad \text{Write the original trigonometric identity.}$$

$$\frac{1}{\sin(\alpha - \beta)} \overset{?}{=} \frac{\sin\alpha\cos\beta + \cos\alpha\sin\beta}{\sin^2\alpha - \sin^2\beta} \qquad \text{Use the reciprocal identity } \csc(\alpha - \beta) = \frac{\Rightarrow}{\sin(\alpha - \beta)}.$$

$$\frac{1}{\sin\alpha\cos\beta - \cos\alpha\sin\beta} \overset{?}{=} \frac{\sin\alpha\cos\beta + \cos\alpha\sin\beta}{\sin^2\alpha - \sin^2\beta} \qquad \text{Use the sine of the difference of two angles formula.}$$

$$\frac{1}{\sin\alpha\cos\beta - \cos\alpha\sin\beta} \cdot \frac{\sin\alpha\cos\beta + \cos\alpha\sin\beta}{\sin\alpha\cos\beta + \cos\alpha\sin\beta} \overset{?}{=} \frac{\sin\alpha\cos\beta + \cos\alpha\sin\beta}{\sin^2\alpha - \sin^2\beta} \qquad \text{Multiply the numerator and denominator by the conjugate of the denominator.}$$

$$\frac{\sin\alpha\cos\beta + \cos\alpha\sin\beta}{\sin^2\alpha\cos^2\beta - \cos^2\alpha\sin^2\beta} \overset{?}{=} \frac{\sin\alpha\cos\beta + \cos\alpha\sin\beta}{\sin^2\alpha - \sin^2\beta} \qquad \text{Multiply the terms in the numerator and denominator.}$$

$$\frac{\sin\alpha\cos\beta + \cos\alpha\sin\beta}{\sin^2\alpha(1 - \sin^2\beta) - (1 - \sin^2\alpha)\sin^2\beta} \overset{?}{=} \frac{\sin\alpha\cos\beta + \cos\alpha\sin\beta}{\sin^2\alpha - \sin^2\beta} \qquad \text{Substitute } 1 - \sin\beta \text{ for } \cos^2\beta$$

$$\frac{\sin \alpha \cos \beta + \cos \alpha \sin \beta}{\sin^2 \alpha - \sin^2 \alpha \sin^2 \beta - \sin^2 \beta + \sin^2 \alpha \sin^2 \beta} \stackrel{?}{=} \frac{\sin \alpha \cos \beta + \cos \alpha \sin \beta}{\sin^2 \alpha - \sin^2 \beta} \qquad \text{Multiply.}$$

$$\frac{\sin \alpha \cos \beta + \cos \alpha \sin \beta}{\sin^2 \alpha - \sin^2 \beta} = \frac{\sin \alpha \cos \beta + \cos \alpha \sin \beta}{\sin^2 \alpha - \sin^2 \beta} \qquad \text{Combine like terms.}$$

My video summary ⊙ The left-hand side is now identical to the right-hand side. Therefore, the identity is verified. Watch this video to see each step of this verification process. ●

You Try It Work through this You Try It problem.

Work Exercises 46–52 in this textbook or in the MyMathLab **Study Plan**.

OBJECTIVE 5 USING THE SUM AND DIFFERENCE FORMULAS TO EVALUATE EXPRESSIONS INVOLVING INVERSE TRIGONOMETRIC FUNCTIONS

The sum and difference formulas discussed in this section may be helpful when we encounter trigonometric expressions involving one or more inverse trigonometric functions. Before working through the next example, you may want to review the inverse trigonometric functions that were first introduced in Section 2.4. Click on one of the following links for a quick review of these functions.

A Quick Review of Inverse Trigonometric Functions

$$y = \sin^{-1} x \qquad y = \cos^{-1} x \qquad y = \tan^{-1} x$$

My video summary ⊙ **Example 9** Finding the Exact Value of a Trigonometric Expression Involving Inverse Trigonometric Expressions

Find the exact value of the expression $\cos\left(\sin^{-1}\left(\frac{1}{5}\right) + \cos^{-1}\left(-\frac{3}{4}\right)\right)$ without using a calculator.

Solution If we let $\sin^{-1}\left(\frac{1}{5}\right) = \alpha$ and $\cos^{-1}\left(-\frac{3}{4}\right) = \beta$, then we are trying to find the value of the cosine of the sum of α and β.

Remember that if $y = \sin^{-1} x$, then $\sin y = x$, where $-\frac{\pi}{2} \le y \le \frac{\pi}{2}$. This means that the terminal side of angle y must lie in Quadrant I or Quadrant IV. Because $\sin^{-1}\left(\frac{1}{5}\right) = \alpha$, then $\sin \alpha = \frac{1}{5}$. Because $\sin \alpha = \frac{1}{5} > 0$, we know that α is an angle whose terminal side lies in Quadrant I. Therefore, $\cos \alpha > 0$ because *all* trigonometric functions are positive for angles whose terminal side lies in Quadrant I. To find $\cos \alpha$, we use the Pythagorean identity $\sin^2 \alpha + \cos^2 \alpha = 1$ or $\cos \alpha = \sqrt{1 - \sin^2 \alpha}$. Read these steps to see how to obtain $\cos \alpha = \frac{2\sqrt{6}}{5}$.

Similarly, if $y = \cos^{-1} x$, then $\cos y = x$, where $0 \le y \le \pi$. This means that the terminal side of angle y must lie in Quadrant I or Quadrant II. Because $\cos^{-1}\left(-\frac{3}{4}\right) = \beta$, then $\cos \beta = -\dfrac{3}{4}$. Because $\cos \beta = -\dfrac{3}{4} < 0$, we know that β is an angle whose terminal side lies in Quadrant II. Therefore, $\sin \alpha > 0$ because the sine function is positive for all angles whose terminal side lies in Quadrant II. We can use the identity $\sin^2 \beta + \cos^2 \beta = 1$ or $\sin \beta = \sqrt{1 - \cos^2 \beta}$ to get $\sin \beta = \dfrac{\sqrt{7}}{4}$. To see how we obtained this, read these steps.

By the cosine of the sum of two angles formula we can find the exact value of $\cos\left(\sin^{-1}\left(\dfrac{1}{5}\right) + \cos^{-1}\left(-\dfrac{3}{4}\right)\right)$.

$\cos\left(\sin^{-1}\left(\dfrac{1}{5}\right) + \cos^{-1}\left(-\dfrac{3}{4}\right)\right)$ Write the original expression.

$= \cos(\alpha + \beta)$ Write as the cosine of the sum of two angles.

$= \cos\alpha\cos\beta - \sin\alpha\sin\beta$ Use the cosine of the sum of two angles formula.

$= \left(\dfrac{2\sqrt{6}}{5}\right)\left(-\dfrac{3}{4}\right) - \left(\dfrac{1}{5}\right)\left(\dfrac{\sqrt{7}}{4}\right)$ Substitute the appropriate values.

$= \dfrac{-6\sqrt{6} - \sqrt{7}}{20}$ Simplify.

You Try It Work through this You Try It problem.

Work Exercises 52–55 in this textbook or in the MyMathLab Study Plan.

3.2 Exercises

Skill Check Exercises

For exercises SCE-1 through SCE-3 simplify each expression.

SCE-1. $\left(\dfrac{1}{\sqrt{2}} \cdot \dfrac{1}{2} + \dfrac{1}{\sqrt{2}} \cdot \dfrac{\sqrt{3}}{2}\right)$ SCE-2. $\left(\dfrac{-5}{\sqrt{61}}\right)\left(\dfrac{-4}{\sqrt{41}}\right) + \left(\dfrac{-6}{\sqrt{61}}\right)\left(\dfrac{5}{\sqrt{41}}\right)$ SCE-3. $\dfrac{(-1) + (-\sqrt{3})}{1 - (-1)(-\sqrt{3})}$

In Exercises 1–13, find the exact value of each trigonometric expression without the use of a calculator.

1. $\cos\left(\dfrac{\pi}{3} + \dfrac{\pi}{4}\right)$ 2. $\cos(30° + 45°)$ 3. $\cos\left(\dfrac{\pi}{4} - \dfrac{\pi}{6}\right)$ 4. $\cos(60° - 45°)$

5. $\cos\left(\dfrac{7\pi}{6} + \dfrac{5\pi}{4}\right)$ 6. $\cos(210° + 135°)$ 7. $\cos\left(\dfrac{5\pi}{4} - \dfrac{7\pi}{6}\right)$ 8. $\cos(120° - 225°)$

9. $\cos\left(\dfrac{11\pi}{12}\right)\cos\left(\dfrac{7\pi}{12}\right) + \sin\left(\dfrac{11\pi}{12}\right)\sin\left(\dfrac{7\pi}{12}\right)$ 10. $\cos(88°)\cos(58°) + \sin(88°)\sin(58°)$

11. $\cos\left(\dfrac{7\pi}{12}\right)$ 12. $\cos(75°)$ 13. $\sec\left(-\dfrac{\pi}{12}\right)$

14. Suppose that the terminal side of angle α lies in Quadrant I and the terminal side of angle β lies in Quadrant IV. If $\sin \alpha = \dfrac{8}{17}$ and $\cos \beta = \dfrac{3}{\sqrt{11}}$, find the exact value of $\cos(\alpha + \beta)$.

15. Suppose that the terminal side of angle α lies in Quadrant II and the terminal side of angle β lies in Quadrant I. If $\tan \alpha = -\dfrac{8}{15}$ and $\cos \beta = \dfrac{3}{4}$, find the exact value of $\cos(\alpha - \beta)$.

In Exercises 16–28, find the exact value of each trigonometric expression without the use of a calculator.

16. $\sin\left(\dfrac{\pi}{6} + \dfrac{\pi}{4}\right)$ 17. $\sin(30° + 45°)$ 18. $\sin\left(\dfrac{\pi}{4} - \dfrac{\pi}{6}\right)$ 19. $\sin(60° - 45°)$

20. $\sin\left(\dfrac{7\pi}{6} + \dfrac{5\pi}{4}\right)$ 21. $\sin(210° + 135°)$ 22. $\sin\left(\dfrac{5\pi}{4} - \dfrac{7\pi}{6}\right)$ 23. $\sin(120° - 225°)$

24. $\sin\left(\dfrac{\pi}{12}\right)\cos\left(\dfrac{2\pi}{3}\right) + \cos\left(\dfrac{\pi}{12}\right)\sin\left(\dfrac{2\pi}{3}\right)$ 25. $\sin(25°)\cos(5°) + \cos(25°)\sin(5°)$

26. $\sin\left(\dfrac{11\pi}{12}\right)$ 27. $\sin(105°)$ 28. $\csc\left(-\dfrac{\pi}{12}\right)$

29. Suppose that the terminal side of angle α lies in Quadrant I and the terminal side of angle β lies in Quadrant IV. If $\sin \alpha = \dfrac{8}{17}$ and $\cos \beta = \dfrac{3}{\sqrt{11}}$, find the exact value of $\sin(\alpha + \beta)$.

30. Suppose that α is an angle such that $\tan \alpha = \dfrac{4}{9}$ and $\sin \alpha < 0$. Also, suppose that β is an angle such that $\cot \beta = -\dfrac{2}{7}$ and $\sec \beta > 0$. Find the exact value of $\sin(\alpha + \beta)$.

In Exercises 31–43, find the exact value of each trigonometric expression without the use of a calculator.

31. $\tan\left(\dfrac{\pi}{6} + \dfrac{\pi}{4}\right)$ 32. $\tan(60° + 45°)$ 33. $\tan\left(\dfrac{\pi}{3} - \dfrac{\pi}{4}\right)$

34. $\tan(45° - 60°)$ 35. $\tan\left(\dfrac{5\pi}{3} + \dfrac{7\pi}{4}\right)$ 36. $\tan(240° + 315°)$

37. $\tan\left(\dfrac{7\pi}{6} - \dfrac{9\pi}{4}\right)$ 38. $\tan(225° - 300°)$ 39. $\dfrac{\tan 40° + \tan 20°}{1 - \tan 40° \tan 20°}$

40. $\dfrac{\tan\left(\dfrac{4\pi}{5}\right) - \tan\left(\dfrac{2\pi}{15}\right)}{1 + \tan\left(\dfrac{4\pi}{5}\right)\tan\left(\dfrac{2\pi}{15}\right)}$ 41. $\tan(15°)$ 42. $\tan\left(\dfrac{17\pi}{12}\right)$

43. $\cot\left(-\dfrac{5\pi}{12}\right)$

44. Suppose that the terminal side of angle α lies in Quadrant II and the terminal side of angle β lies in Quadrant III. If $\sin \alpha = \dfrac{1}{4}$ and $\cos \beta = -\dfrac{3}{5}$, find the exact value of $\tan(\alpha + \beta)$.

45. Suppose that the terminal side of angle α lies in Quadrant IV and the terminal side of angle β lies in Quadrant III. If $\cos \alpha = \dfrac{5}{\sqrt{61}}$ and $\sin \beta = -\dfrac{7}{\sqrt{91}}$, find the exact value of

For Exercises 46–51, verify each identity.

46. $\sin\left(\dfrac{\pi}{2} + \theta\right) = \cos\theta$

47. $\cos(\alpha + \beta) + \cos(\alpha - \beta) = 2\cos\alpha\cos\beta$

48. $\dfrac{\cos(\alpha + \beta)}{\sin\alpha\cos\beta} = \cot\alpha - \tan\beta$

49. $\dfrac{\cos(\alpha - \beta)}{\cos(\alpha + \beta)} = \dfrac{1 + \tan\alpha\tan\beta}{1 - \tan\alpha\tan\beta}$

50. $\tan(\alpha - \beta) = \dfrac{\cot\beta - \cot\alpha}{\cot\alpha\cot\beta + 1}$

51. $\csc(\alpha - \beta) = \dfrac{\csc\alpha\csc\beta}{\cot\beta - \cot\alpha}$

For Exercises 52–55, find the exact value of each expression without the use of a calculator.

52. $\sin\left(\tan^{-1}1 + \cos^{-1}\dfrac{\sqrt{3}}{2}\right)$

53. $\cos\left(\sin^{-1}\left(\dfrac{5}{13}\right) + \cos^{-1}\left(-\dfrac{12}{13}\right)\right)$

54. $\sin\left(\sin^{-1}\left(\dfrac{1}{6}\right) - \cos^{-1}\left(-\dfrac{3}{7}\right)\right)$

55. $\tan\left(\sin^{-1}\left(\dfrac{1}{\sqrt{2}}\right) + \cos^{-1}\left(-\dfrac{1}{2}\right)\right)$

3.3 The Double-Angle and Half-Angle Formulas

THINGS TO KNOW

Before working through this section, be sure that you are familiar with the following concepts:

VIDEO   ANIMATION   INTERACTIVE

You Try It

1. Evaluating Trigonometric Functions of Angles Belonging to the $\dfrac{\pi}{3}, \dfrac{\pi}{6},$ or $\dfrac{\pi}{4}$ Families (Section 1.5) ⊚

You Try It

2. Finding the Exact and Approximate Values of an Inverse Sine Expression (Section 2.4) ⊚

You Try It

3. Finding the Exact and Approximate Values of an Inverse Cosine Expression (Section 2.4) ⊚

You Try It

4. Finding the Exact and Approximate Values of an Inverse Tangent Expression (Section 2.4) ⊚

You Try It

5. Understanding the Sum and Difference Formulas for the Sine Function (Section 3.2) ⊚

You Try It

6. Understanding the Sum and Difference Formulas for the Cosine Function (Section 3.2) ⊚

OBJECTIVES

1 Understanding the Double-Angle Formulas

2 Understanding the Power Reduction Formulas

3 Understanding the Half-Angle Formulas

4 Using the Double-Angle, Power Reduction, and Half-Angle Formulas to Verify Identities

5 Using the Double-Angle and Half-Angle Formulas to Evaluate Expressions Involving Inverse Trigonometric Functions

---

OBJECTIVE 1 UNDERSTANDING THE DOUBLE-ANGLE FORMULAS

In Section 3.2 we introduced the sum and difference formulas for sine, cosine, and tangent. We can use these formulas to derive the following **double-angle formulas**.

Double-Angle Formulas

$$\sin 2\theta = 2 \sin \theta \cos \theta \qquad \cos 2\theta = \cos^2 \theta - \sin^2 \theta \qquad \tan 2\theta = \frac{2 \tan \theta}{1 - \tan^2 \theta}$$

My interactive video summary

⊘ Work through this interactive video to see how to use the sum formulas to derive each of the three double-angle formulas. Using a form of the Pythagorean identity $\sin^2 \theta + \cos^2 \theta = 1$, we can rewrite the formula for $\cos 2\theta$ in two additional ways.

$\cos 2\theta = \cos^2 \theta - \sin^2 \theta$	Write the double-angle formula for cosine.
$= (1 - \sin^2 \theta) - \sin^2 \theta$	Use the identity $\cos^2 \theta = 1 - \sin^2 \theta$.
$= 1 - 2 \sin^2 \theta$	Simplify.
$\cos 2\theta = \cos^2 \theta - \sin^2 \theta$	Write the double-angle formula for cosine.
$= \cos^2 \theta - (1 - \cos^2 \theta)$	Use the identity $\sin^2 \theta = 1 - \cos^2 \theta$.
$= \cos^2 \theta - 1 + \cos^2 \theta$	Distribute.
$= 2 \cos^2 \theta - 1$	Simplify.

We now state all three forms of the double-angle formula for cosine.

The Double-Angle Formulas for Cosine

$$\cos 2\theta = \cos^2 \theta - \sin^2 \theta$$
$$\cos 2\theta = 1 - 2 \sin^2 \theta$$
$$\cos 2\theta = 2 \cos^2 \theta - 1$$

My interactive video summary

⊘ **Example 1** Evaluating Trigonometric Expressions Using Double-Angle Formulas

Rewrite each expression as the sine, cosine, or tangent of a double angle. Then evaluate the expression without using a calculator.

a. $\cos^2\left(\dfrac{11\pi}{12}\right) - \sin^2\left(\dfrac{11\pi}{12}\right)$ **b.** $2\sin 67.5° \cos 67.5°$

c. $\dfrac{2\tan\left(-\dfrac{\pi}{8}\right)}{1 - \tan^2\left(-\dfrac{\pi}{8}\right)}$ **d.** $2\cos^2 105° - 1$

Solution

a. To evaluate the expression $\cos^2\left(\dfrac{11\pi}{12}\right) - \sin^2\left(\dfrac{11\pi}{12}\right)$, we should recognize that this expression represents the right-hand side of the double-angle formula $\cos 2\theta = \cos^2\theta - \sin^2\theta$, where $\theta = \dfrac{11\pi}{12}$.

$\cos^2\left(\dfrac{11\pi}{12}\right) - \sin^2\left(\dfrac{11\pi}{12}\right)$ Write the original expression.

$= \cos\left(2 \cdot \dfrac{11\pi}{12}\right)$ Use the double-angle formula $\cos 2\theta = \cos^2\theta - \sin^2\theta$.

$= \cos\left(\dfrac{11\pi}{6}\right)$ Simplify.

$= \dfrac{\sqrt{3}}{2}$ Evaluate the trigonometric function.

Therefore, $\cos^2\left(\dfrac{11\pi}{12}\right) - \sin^2\left(\dfrac{11\pi}{12}\right) = \dfrac{\sqrt{3}}{2}$.

Try working parts b–d on your own using one of the double-angle formulas. When finished, check your solutions or work through this interactive video to see each step worked out in detail.

You Try It   Work through this You Try It problem.

Work Exercises 1–10 in this textbook or in the My MathLab **Study Plan.**

My interactive video summary

⊙ **Example 2** Evaluating Trigonometric Expressions Using Double-Angle Formulas

Suppose that the terminal side of angle θ lies in Quadrant II such that $\sin\theta = \dfrac{5}{7}$.

Find the values of $\sin 2\theta$, $\cos 2\theta$, and $\tan 2\theta$.

Solution We start by sketching an angle whose terminal side lies in Quadrant II such that $\sin\theta = \dfrac{5}{7} = \dfrac{y}{r}$. See Figure 3.

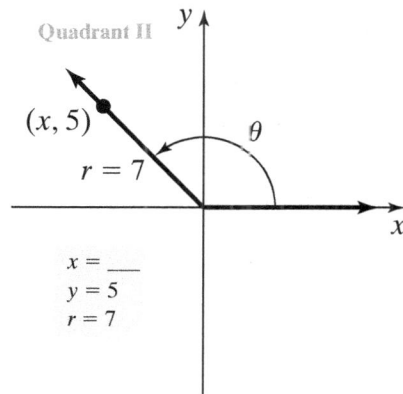

Figure 3 Angle θ and the point $(x, 5)$ located in Quadrant II, a distance of $r = 7$ units from the origin

Because the point $(x, 5)$ lies in Quadrant II, we know that the value of x must be negative. We can use the relationship $x^2 + y^2 = r^2$ to determine the value of x. Read through these steps to see how we determine that $x = -2\sqrt{6}$. See Figure 4.

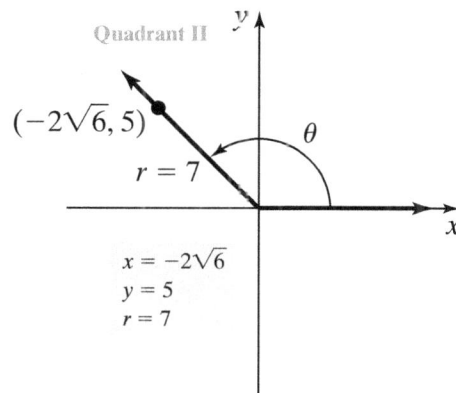

Figure 4 Angle θ lies in Quadrant II and passes through the point $(-2\sqrt{6}, 5)$.

We now summarize the information gathered thus far about angle θ.

$$\sin \theta = \frac{5}{7} \qquad \cos \theta = \frac{x}{r} = -\frac{2\sqrt{6}}{7} \qquad \tan \theta = \frac{y}{x} = \frac{5}{-2\sqrt{6}} = -\frac{5}{2\sqrt{6}}$$

Use the double-angle formulas to determine the values of $\sin 2\theta$, $\cos 2\theta$, and $\tan 2\theta$.

$\sin 2\theta = 2 \sin \theta \cos \theta$ Write the double-angle formula for sine.

$\qquad = 2\left(\dfrac{5}{7}\right)\left(-\dfrac{2\sqrt{6}}{7}\right)$ Substitute $\sin \theta = \dfrac{5}{7}$ and $\cos \theta = -\dfrac{2\sqrt{6}}{7}$.

$\qquad = -\dfrac{20\sqrt{6}}{49}$ Multiply.

$\cos 2\theta = \cos^2 \theta - \sin^2 \theta$ Write a double-angle formula for cosine.

$\qquad = \left(-\dfrac{2\sqrt{6}}{7}\right)^2 - \left(\dfrac{5}{7}\right)^2$ Substitute $\sin \theta = \dfrac{5}{7}$ and $\cos \theta = -\dfrac{2\sqrt{6}}{7}$.

$\qquad = \dfrac{24}{49} - \dfrac{25}{49}$ Square both terms.

$\qquad = -\dfrac{1}{49}$ Subtract.

3.3 The Double-Angle and Half-Angle Formulas   3-29

My interactive video summary

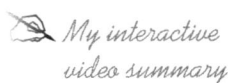

⊙ Therefore, $\sin 2\theta = -\dfrac{20\sqrt{6}}{49}$ and $\cos 2\theta = -\dfrac{1}{49}$. Try to determine the value of $\tan 2\theta$ on your own. When you're finished, check your answer or watch this interactive video to see the solution worked out in detail.

You Try It   Work through this You Try It problem.

Work Exercises 11–14 in this textbook or in the MyMathLab Study Plan.

Example 3 Evaluating Trigonometric Expressions Using Double-Angle Formulas

If $\tan 2\theta = -\dfrac{24}{7}$ for $\dfrac{3\pi}{2} < 2\theta < 2\pi$, then find the values of $\sin\theta$, $\cos\theta$, and $\tan\theta$.

Solution We start by sketching angle 2θ located in Quadrant IV between $\dfrac{3\pi}{2}$ and 2π such that $\tan 2\theta = \dfrac{y}{x} = -\dfrac{24}{7}$. We can use the relationship $x^2 + y^2 = r^2$ to determine the value of r. Read these steps to see how we determine that $r = 25$. See Figure 5.

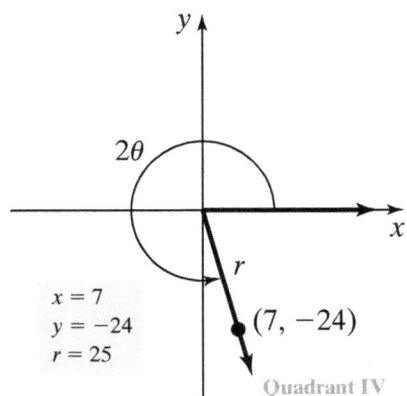

$x = 7$
$y = -24$
$r = 25$

$(7, -24)$

Quadrant IV

Figure 5 This illustrates angle 2θ and the point $(7, -24)$ located in Quadrant IV, a distance of $r = 25$ from the origin.

We now summarize the information gathered thus far about angle 2θ.

$$\sin 2\theta = \frac{y}{r} = -\frac{24}{25} \qquad \cos 2\theta = \frac{x}{r} = \frac{7}{25} \qquad \tan 2\theta = -\frac{24}{7}$$

Before we determine the value of $\sin\theta$, $\cos\theta$, and $\tan\theta$, we need to determine the quadrant in which the terminal side of θ lies. Dividing each part of this inequality $\dfrac{3\pi}{2} < 2\theta < 2\pi$ by 2, we get $\dfrac{3\pi}{4} < \theta < \pi$. Thus, the terminal side of angle θ lies in Quadrant II.

We can determine the value of $\sin\theta$ by using a double-angle formula. We choose $\cos 2\theta = 1 - 2\sin^2\theta$ because this formula contains only $\cos 2\theta$ and $\sin\theta$ and allows us to solve easily for $\sin\theta$.

$$\cos 2\theta = 1 - 2\sin^2\theta \qquad \text{Start with a double-angle formula for cosine that involves } \sin\theta.$$

$$\frac{7}{25} = 1 - 2\sin^2\theta \qquad \text{Substitute } \cos 2\theta = \frac{7}{25}.$$

$$\frac{7}{25} - 1 = -2\sin^2\theta \qquad \text{Subtract 1 from both sides.}$$

$$\frac{7}{25} - \frac{25}{25} = -2\sin^2\theta \qquad \text{Use 25 as the LCD.}$$

$$-\frac{18}{25} = -2\sin^2\theta \qquad \text{Subtract.}$$

$$\frac{9}{25} = \sin^2\theta \qquad \text{Divide both sides by } -2.$$

$$\pm\frac{3}{5} = \sin\theta \qquad \text{Take square roots of both sides.}$$

My video summary ◯ Because the terminal side of θ lies in Quadrant II, we know that $\sin\theta$ must be positive. Therefore, $\sin\theta = \frac{3}{5}$. We can now determine the values of $\cos\theta$ and $\tan\theta$.

Watch this video to verify that $\cos\theta = -\frac{4}{5}$ and $\tan\theta = -\frac{3}{4}$.

You Try It Work through this You Try It problem.

Work Exercises 15–18 in this textbook or in the MyMathLab Study Plan.

OBJECTIVE 2 UNDERSTANDING THE POWER REDUCTION FORMULAS

In calculus, when encountering expressions that involve trigonometric functions that are raised to a power, it is often useful to rewrite the expression so that it involves only trigonometric functions that are raised to a power of 1. We can rewrite *even* powers of trigonometric functions using the following **power reduction formulas.** The first two power reduction formulas are alternate forms of the double-angle formulas for cosine, and the third formula is the quotient of the first two.

The Power Reduction Formulas

$$\sin^2\theta = \frac{1 - \cos 2\theta}{2} \qquad \cos^2\theta = \frac{1 + \cos 2\theta}{2} \qquad \tan^2\theta = \frac{1 - \cos 2\theta}{1 + \cos 2\theta}$$

My interactive video summary ◯ Work through this interactive video to see how we can use the double-angle formulas to establish the three power reduction formulas.

My video summary ◯ **Example 4** Reducing the Power of a Trigonometric Function Using a Power Reduction Formula

Rewrite the function $f(x) = 6\sin^4 x$ as an equivalent function containing only cosine terms raised to a power of 1.

Solution Start by rewriting $6 \sin^4 x$ as $6 (\sin^2 x)^2$, and then make the substitution $\sin^2 x = \dfrac{1 - \cos 2x}{2}$.

$$6 \sin^4 x = 6 (\sin^2 x)^2 \qquad \text{Write } \sin^4 x \text{ as } (\sin^2 x)^2.$$

$$= 6 \left(\frac{1 - \cos 2x}{2} \right)^2 \qquad \begin{array}{l} \text{Use the power reduction formula} \\ \sin^2 x = \dfrac{1 - \cos 2x}{2}. \end{array}$$

$$= 6 \left(\frac{1 - 2 \cos 2x + \cos^2 2x}{4} \right) \qquad \begin{array}{l} \text{Square the numerator and} \\ \text{denominator.} \end{array}$$

$$= \frac{3}{2} (1 - 2 \cos 2x + \cos^2 2x) \qquad \text{Write } \dfrac{6}{4} \text{ as } \dfrac{3}{2}.$$

$$= \frac{3}{2} - 3 \cos 2x + \frac{3}{2} \cos^2 2x \qquad \text{Distribute.}$$

$$= \frac{3}{2} - 3 \cos 2x + \frac{3}{2} \left(\frac{1 + \cos 2 (2x)}{2} \right) \qquad \begin{array}{l} \text{Use the power reduction formula} \\ \cos^2 \theta = \dfrac{1 + \cos 2\theta}{2}, \text{ where } \theta = 2x. \end{array}$$

$$= \frac{3}{2} - 3 \cos 2x + \frac{3}{4} (1 + \cos 4x) \qquad \text{Simplify.}$$

$$= \frac{3}{2} - 3 \cos 2x + \frac{3}{4} + \frac{3}{4} \cos 4x \qquad \text{Distribute.}$$

$$= \frac{9}{4} - 3 \cos 2x + \frac{3}{4} \cos 4x \qquad \text{Combine like terms.}$$

Therefore, we can rewrite $f(x) = 6 \sin^4 x$ as $f(x) = \dfrac{9}{4} - 3 \cos 2x + \dfrac{3}{4} \cos 4x.$

You Try It Work through this You Try It problem.

Work Exercises 19–23 in this textbook or in the MyMathLab Study Plan.

OBJECTIVE 3 UNDERSTANDING THE HALF-ANGLE FORMULAS

Starting with the power reduction formula $\sin^2 \theta = \dfrac{1 - \cos 2\theta}{2}$, we can take the square root of both sides of the equation to get $\sin \theta = \sqrt{\dfrac{1 - \cos 2\theta}{2}}$ or $\sin \theta = -\sqrt{\dfrac{1 - \cos 2\theta}{2}}$. The choice of which root to use depends on the quadrant in which the terminal side of θ lies Similarly, we can determine that $\cos \theta = \sqrt{\dfrac{1 + \cos 2\theta}{2}}$ or $\cos \theta = -\sqrt{\dfrac{1 + \cos 2\theta}{2}}$. Using the substitution $\theta = \dfrac{\alpha}{2}$, we obtain the following **half-angle formulas** for sine and cosine.

Note that the choice of which form of $\sin \left(\dfrac{\alpha}{2} \right)$ and $\cos \left(\dfrac{\alpha}{2} \right)$ to use depends on

The Half-Angle Formulas for Sine and Cosine

$$\sin\left(\frac{\alpha}{2}\right) = \sqrt{\frac{1-\cos\alpha}{2}} \qquad \text{for } \frac{\alpha}{2} \text{ in Quadrant I or Quadrant II.}$$

$$\sin\left(\frac{\alpha}{2}\right) = -\sqrt{\frac{1-\cos\alpha}{2}} \qquad \text{for } \frac{\alpha}{2} \text{ in Quadrant III or Quadrant IV.}$$

$$\cos\left(\frac{\alpha}{2}\right) = \sqrt{\frac{1+\cos\alpha}{2}} \qquad \text{for } \frac{\alpha}{2} \text{ in Quadrant I or Quadrant IV.}$$

$$\cos\left(\frac{\alpha}{2}\right) = -\sqrt{\frac{1+\cos\alpha}{2}} \qquad \text{for } \frac{\alpha}{2} \text{ in Quadrant II or Quadrant III.}$$

the quadrant in which the terminal side of angle $\frac{\alpha}{2}$ lies. We use the acronym ASTC first discussed in Section 1.5 to determine the signs of trigonometric functions for angles lying in each of the four quadrants. Refer to this figure for a quick review of the signs of the trigonometric functions.

My interactive video summary

To derive the half-angle formula for tangent, we start with the power reduction formula $\tan^2\theta = \frac{1-\cos 2\theta}{1+\cos 2\theta}$ and take the square root of both sides to get $\tan\theta = \sqrt{\frac{1-\cos 2\theta}{1+\cos 2\theta}}$ or $\tan\theta = -\sqrt{\frac{1-\cos 2\theta}{1+\cos 2\theta}}$. Again, the choice of which root to use depends on the quadrant in which the terminal side of θ lies. Using the substitution $\theta = \frac{\alpha}{2}$, we obtain the half-angle formulas $\tan\left(\frac{\alpha}{2}\right) = \sqrt{\frac{1-\cos\alpha}{1+\cos\alpha}}$ or $\tan\left(\frac{\alpha}{2}\right) = -\sqrt{\frac{1-\cos\alpha}{1+\cos\alpha}}$. We can derive two other formulas for $\tan\left(\frac{\alpha}{2}\right)$ that do not include any radicals and can be used regardless of the quadrant in which the terminal side of $\frac{\alpha}{2}$ lies.

Watch this interactive video to see how to derive the half-angle formulas for tangent listed on the following page.

The Half-Angle Formulas for Tangent

$$\tan\left(\frac{\alpha}{2}\right) = \sqrt{\frac{1-\cos\alpha}{1+\cos\alpha}} \qquad \text{for } \frac{\alpha}{2} \text{ in Quadrant I or Quadrant III.}$$

$$\tan\left(\frac{\alpha}{2}\right) = -\sqrt{\frac{1-\cos\alpha}{1+\cos\alpha}} \qquad \text{for } \frac{\alpha}{2} \text{ in Quadrant II or Quadrant IV.}$$

$$\tan\left(\frac{\alpha}{2}\right) = \frac{1-\cos\alpha}{\sin\alpha} \qquad \text{for } \frac{\alpha}{2} \text{ in any quadrant.}$$

$$\tan\left(\frac{\alpha}{2}\right) = \frac{\sin\alpha}{1+\cos\alpha} \qquad \text{for } \frac{\alpha}{2} \text{ in any quadrant.}$$

Example 5 Evaluating Trigonometric Expressions Using Half-Angle Formulas

Use a half-angle formula to evaluate each expression without using a calculator.

a. $\sin\left(-\dfrac{7\pi}{8}\right)$ **b.** $\cos(-15°)$ **c.** $\tan\left(\dfrac{11\pi}{12}\right)$

Solution

a. To evaluate the expression $\sin\left(-\dfrac{7\pi}{8}\right)$ using a half-angle formula, we first identify $\dfrac{\alpha}{2}$ and α.

$$\dfrac{\alpha}{2} = -\dfrac{7\pi}{8} \qquad \text{Let } \dfrac{\alpha}{2} \text{ represent the given angle.}$$

$$2 \cdot \dfrac{\alpha}{2} = 2 \cdot \left(-\dfrac{7\pi}{8}\right) \qquad \text{Multiply by 2.}$$

$$\alpha = -\dfrac{7\pi}{4} \qquad \text{Simplify.}$$

The terminal side of $\dfrac{\alpha}{2} = -\dfrac{7\pi}{8}$ lies in Quadrant III. See Figure 6.

The sine function is negative for all angles lying in Quadrant III, which indicates that $\sin\left(-\dfrac{7\pi}{8}\right)$ is a negative value. Thus, we must use the half-angle formula $\sin\left(\dfrac{\alpha}{2}\right) = -\sqrt{\dfrac{1-\cos\alpha}{2}}$ with $\dfrac{\alpha}{2} = -\dfrac{7\pi}{8}$ and $\alpha = -\dfrac{7\pi}{4}$.

$$\sin\left(\dfrac{\alpha}{2}\right) = -\sqrt{\dfrac{1-\cos\alpha}{2}} \qquad \begin{array}{l}\text{Write the half-angle formula for sine using}\\ \text{the negative square root.}\end{array}$$

$$\sin\left(-\dfrac{7\pi}{8}\right) = -\sqrt{\dfrac{1-\cos\left(-\dfrac{7\pi}{4}\right)}{2}} \qquad \text{Substitute } \dfrac{\alpha}{2} = -\dfrac{7\pi}{8} \text{ and } \alpha = -\dfrac{7\pi}{4}.$$

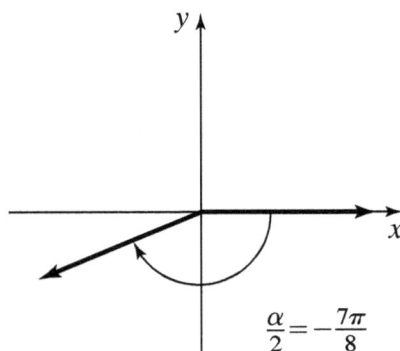

$$\dfrac{\alpha}{2} = -\dfrac{7\pi}{8}$$

Quadrant III

Figure 6

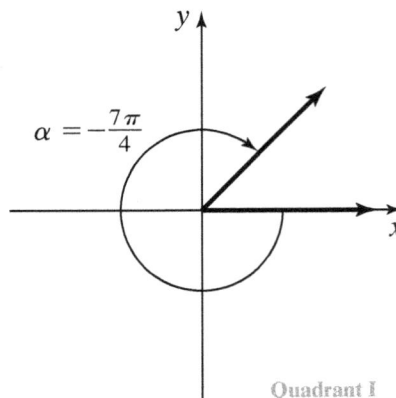

$$\alpha = -\dfrac{7\pi}{4}$$

Quadrant I

Figure 7

The terminal side of $\alpha = -\dfrac{7\pi}{4}$ lies in Quadrant I. See Figure 7. The cosine of any angle whose terminal side lies in Quadrant I is *positive*. When we evaluate the expression $\cos\left(-\dfrac{7\pi}{4}\right)$, we get $\cos\left(-\dfrac{7\pi}{4}\right) = \dfrac{1}{\sqrt{2}}$.

$$\sin\left(-\frac{7\pi}{8}\right) = -\sqrt{\frac{1 - \cos\left(-\dfrac{7\pi}{4}\right)}{2}}$$

$$= -\sqrt{\frac{1 - \dfrac{1}{\sqrt{2}}}{2}} \qquad \text{Evaluate: } \cos\left(-\frac{7\pi}{4}\right) = \frac{1}{\sqrt{2}}.$$

$$= -\sqrt{\frac{\dfrac{\sqrt{2} - 1}{\sqrt{2}}}{2}} \qquad \text{Combine the expressions in the numerator.}$$

$$= -\sqrt{\frac{\sqrt{2} - 1}{2\sqrt{2}}} \qquad \text{Simplify: } -\sqrt{\frac{\dfrac{\sqrt{2} - 1}{\sqrt{2}}}{2}} = -\sqrt{\frac{\sqrt{2} - 1}{\sqrt{2}} \div 2}$$

$$= -\sqrt{\frac{\sqrt{2} - 1}{\sqrt{2}} \cdot \frac{1}{2}} = -\sqrt{\frac{\sqrt{2} - 1}{2\sqrt{2}}}.$$

My interactive video summary

⊛ We see that $\sin\left(-\dfrac{7\pi}{8}\right) = -\sqrt{\dfrac{\sqrt{2} - 1}{2\sqrt{2}}}$. Watch this interactive video to see every step of this solution. Note that if we choose to rationalize the expression under the radical (which is not required), we can obtain an equivalent solution of $\sin\left(-\dfrac{7\pi}{8}\right) = -\dfrac{\sqrt{2 - \sqrt{2}}}{2}$.

My interactive video summary

⊛ **b.** Try to evaluate the expression $\cos(-15°)$ on your own. Watch this interactive video to check your work.

c. To evaluate the expression $\tan\left(\dfrac{11\pi}{12}\right)$ using a half-angle formula, we first identify $\dfrac{\alpha}{2}$ and α.

$$\frac{\alpha}{2} = \frac{11\pi}{12} \qquad \text{Let } \frac{\alpha}{2} \text{ represent the given angle.}$$

$$2 \cdot \frac{\alpha}{2} = 2 \cdot \left(\frac{11\pi}{12}\right) \qquad \text{Multiply by 2.}$$

$$\alpha = \frac{11\pi}{6} \qquad \text{Simplify.}$$

We can use any of the three half-angle formulas for tangent. For this example, we will use the formula $\tan\left(\dfrac{\alpha}{2}\right) = \dfrac{\sin\alpha}{1 + \cos\alpha}$

$$\tan\left(\frac{11\pi}{12}\right) = \frac{\sin\left(\dfrac{11\pi}{6}\right)}{1 + \cos\left(\dfrac{11\pi}{6}\right)} \qquad \begin{array}{l}\text{Write the original expression then} \\ \text{use a half-angle formula for } \tan\left(\dfrac{x}{2}\right).\end{array}$$

$$= \frac{-\frac{1}{2}}{1 + \frac{\sqrt{3}}{2}}$$ Evaluate: $\sin\left(\frac{11\pi}{6}\right) = -\frac{1}{2}$ and $\cos\left(\frac{11\pi}{6}\right) = \frac{\sqrt{3}}{2}$.

$$= \frac{-\frac{1}{2}}{\frac{2 + \sqrt{3}}{2}}$$ Combine the expressions in the denominator.

$$= -\frac{1}{2} \div \frac{2 + \sqrt{3}}{2}$$ Rewrite as the division of two expressions.

$$= -\frac{1}{2} \cdot \frac{2}{2 + \sqrt{3}}$$ Multiply the first expression by the reciprocal of the second expression.

$$= -\frac{1}{\cancel{2}} \cdot \frac{\cancel{2}}{2 + \sqrt{3}}$$ Cancel common factors.

$$= -\frac{1}{2 + \sqrt{3}}$$ Simplify.

My interactive video summary

⊙ Therefore, $\tan\left(\frac{11\pi}{12}\right) = -\frac{1}{2 + \sqrt{3}}$. You may wish to work through this interactive video to see each step of this solution. Note that if we choose to rationalize this answer (which is not required), we can obtain an equivalent solution $\tan\left(\frac{11\pi}{12}\right) = \sqrt{3} - 2$.

You Try It Work through this You Try It problem.

Work Exercises 24–39 in this textbook or in the MyMathLab Study Plan.

My interactive video summary

⊙ **Example 6** Evaluating Trigonometric Expressions Using Half-Angle Formulas

Suppose that $\csc\alpha = \frac{8}{3}$ such that $\frac{\pi}{2} < \alpha < \pi$.

Find the values of $\sin\left(\frac{\alpha}{2}\right)$, $\cos\left(\frac{\alpha}{2}\right)$, and $\tan\left(\frac{\alpha}{2}\right)$.

Solution To determine the values of $\sin\left(\frac{\alpha}{2}\right)$, $\cos\left(\frac{\alpha}{2}\right)$, and $\tan\left(\frac{\alpha}{2}\right)$, we must first find the values of $\sin\alpha$ and $\cos\alpha$. The value of $\sin\alpha$ is the reciprocal of $\csc\alpha = \frac{8}{3}$.

Thus, $\sin\alpha = \frac{3}{8}$. To determine the value of $\cos\alpha$, we start by sketching an angle whose terminal side lies in Quadrant II and that illustrates $\csc\alpha = \frac{8}{3} = \frac{r}{y}$. See Figure 8.

The point $(x, 3)$ lies in Quadrant II; thus we know that the value of x must be negative. We can use the relationship $x^2 + y^2 = r^2$ to determine the value of x. Read these steps to see how we determine that $x = -\sqrt{55}$. See Figure 9.

(eText Screens 3.3-1–3.3-40)

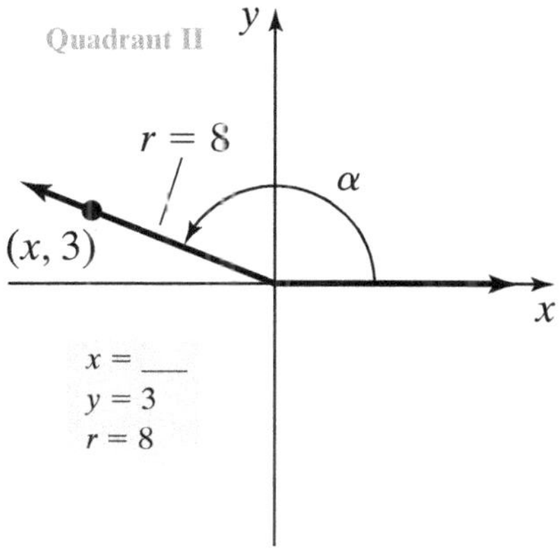

Figure 8 This illustrates angle α and the point $(x, 3)$, located in Quadrant II a distance of $r = 8$ units from the origin.

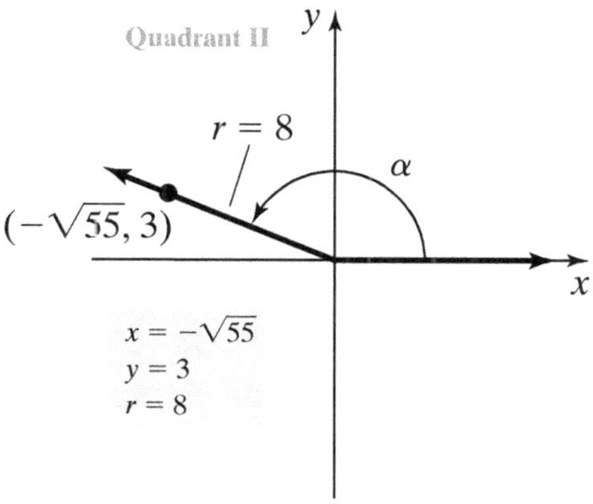

Figure 9 This illustrates angle α and the point $\left(-\sqrt{55}, 3\right)$ located in Quadrant II, a distance of $r = 8$ from the origin.

We now summarize the information gathered thus far about angle α.

$$\sin\alpha = \frac{y}{r} = \frac{3}{8} \qquad \cos\alpha = \frac{x}{r} = -\frac{\sqrt{55}}{8}$$

Now we consider the angle $\frac{\alpha}{2}$. We were given that $\frac{\pi}{2} < \alpha < \pi$.

Dividing each part of this inequality by 2, we get

$$\frac{\pi}{4} < \frac{\alpha}{2} < \frac{\pi}{2}.$$

My interactive video summary

This indicates that the terminal side of $\frac{\alpha}{2}$ lies in Quadrant I. Therefore, the values of $\sin\left(\frac{\alpha}{2}\right)$, $\cos\left(\frac{\alpha}{2}\right)$, and $\tan\left(\frac{\alpha}{2}\right)$ must all be positive because all trigonometric functions are positive for any angle whose terminal side lies in Qadrant I. We can now use the formulas

$$\sin\left(\frac{\alpha}{2}\right) = \sqrt{\frac{1-\cos\alpha}{2}},\ \cos\left(\frac{\alpha}{2}\right) = \sqrt{\frac{1+\cos\alpha}{2}},\ \text{and } \tan\left(\frac{\alpha}{2}\right) = \frac{\sin\alpha}{1+\cos\alpha}$$

with $\sin\alpha = \frac{3}{8}$ and $\cos\alpha = \frac{-\sqrt{55}}{8}$ to determine the values of each half-angle. Work through this interactive video to verify that

$$\sin\left(\frac{\alpha}{2}\right) = \frac{\sqrt{8+\sqrt{55}}}{4},\ \cos\left(\frac{\alpha}{2}\right) = \frac{\sqrt{8-\sqrt{55}}}{4},\ \text{and } \tan\left(\frac{\alpha}{2}\right) = \frac{3}{8-\sqrt{55}}.$$

You Try It Work through this You Try It problem.

Work Exercises 40–45 in this textbook or in the MyMathLab Study Plan.

OBJECTIVE 4 USING THE DOUBLE-ANGLE, POWER REDUCTION, AND HALF-ANGLE FORMULAS TO VERIFY IDENTITIES

We can use formulas discussed in this section, along with any identity discussed so far in this text, to verify trigonometric identities. As always, we will verify an identity by trying to transform one side of the identity into the exact trigonometric expression that appears on the opposite side. We typically start with what appears to be the more complicated side of the identity and then attempt to use known identities to transform that side into the trigonometric expression that appears on the other side of the identity. However, if one side of the identity includes a trigonometric expression involving 2θ or $\dfrac{\theta}{2}$, then first substitute one of the formulas discussed in this section. Next, use the strategies we developed in Section 3.1 for verifying identities. To review these strategies, read this summary of techniques that were introduced in Section 3.1. We now summarize the formulas discussed in this section.

Double-Angle Formulas

$$\sin 2\theta = 2 \sin \theta \cos \theta$$

$$\cos 2\theta = \cos^2\theta - \sin^2\theta$$

$$\cos 2\theta = 1 - 2 \sin^2\theta$$

$$\cos 2\theta = 2 \cos^2\theta - 1$$

$$\tan 2\theta = \frac{2 \tan \theta}{1 - \tan^2\theta}$$

Power Reduction Formulas

$$\sin^2 \theta = \frac{1 - \cos 2\theta}{2}$$

$$\cos^2 \theta = \frac{1 + \cos 2\theta}{2}$$

$$\tan^2 \theta = \frac{1 - \cos 2\theta}{1 + \cos 2\theta}$$

Half-Angle Formulas

$$\sin\left(\frac{\alpha}{2}\right) = \sqrt{\frac{1 - \cos \alpha}{2}} \qquad \text{for } \frac{\alpha}{2} \text{ in Quadrant I or Quadrant II.}$$

$$\sin\left(\frac{\alpha}{2}\right) = -\sqrt{\frac{1 - \cos \alpha}{2}} \qquad \text{for } \frac{\alpha}{2} \text{ in Quadrant III or Quadrant IV.}$$

$$\cos\left(\frac{\alpha}{2}\right) = \sqrt{\frac{1 + \cos \alpha}{2}} \qquad \text{for } \frac{\alpha}{2} \text{ in Quadrant I or Quadrant IV.}$$

$$\cos\left(\frac{\alpha}{2}\right) = -\sqrt{\frac{1 + \cos \alpha}{2}} \qquad \text{for } \frac{\alpha}{2} \text{ in Quadrant II or Quadrant III.}$$

$$\tan\left(\frac{\alpha}{2}\right) = \sqrt{\frac{1 - \cos \alpha}{1 + \cos \alpha}} \qquad \text{for } \frac{\alpha}{2} \text{ in Quadrant I or Quadrant III.}$$

$$\tan\left(\frac{\alpha}{2}\right) = -\sqrt{\frac{1 - \cos \alpha}{1 + \cos \alpha}} \qquad \text{for } \frac{\alpha}{2} \text{ in Quadrant II or Quadrant IV.}$$

$$\tan\left(\frac{\alpha}{2}\right) = \frac{1 - \cos \alpha}{\sin \alpha} \qquad \text{for } \frac{\alpha}{2} \text{ in any quadrant.}$$

$$\tan\left(\frac{\alpha}{2}\right) = \frac{\sin \alpha}{1 + \cos \alpha} \qquad \text{for } \frac{\alpha}{2} \text{ in any quadrant.}$$

Example 7 Verifying a Trigonometric Identity

Verify the trigonometric identity $\dfrac{\cos (2\theta)}{1 + \sin (2\theta)} = \dfrac{\cos \theta - \sin \theta}{\cos \theta + \sin \theta}$.

Solution We start by rewriting the left-hand side using the double-angle formulas for sine and cosine.

$$\frac{\cos(2\theta)}{1 + \sin(2\theta)} \overset{?}{=} \frac{\cos\theta - \sin\theta}{\cos\theta + \sin\theta}$$

Write the original trigonometric identity.

$$\frac{\cos^2\theta - \sin^2\theta}{1 + 2\sin\theta\cos\theta} \overset{?}{=} \frac{\cos\theta - \sin\theta}{\cos\theta + \sin\theta}$$

Choose $\cos 2\theta = \sin^2\theta - \cos^2\theta$ because it is a good match for the right-hand side. Use the double-angle formula for sine.

$$\frac{\cos^2\theta - \sin^2\theta}{\sin^2\theta + \cos^2\theta + 2\sin\theta\cos\theta} \overset{?}{=} \frac{\cos\theta - \sin\theta}{\cos\theta + \sin\theta}$$

Use the Pythagorean identity $1 = \sin^2\theta + \cos^2\theta$ to eliminate the 1.

$$\frac{\cos^2\theta - \sin^2\theta}{\cos^2\theta + 2\sin\theta\cos\theta + \sin^2\theta} \overset{?}{=} \frac{\cos\theta - \sin\theta}{\cos\theta + \sin\theta}$$

Rearrange the terms in the denominator.

$$\frac{(\cos\theta + \sin\theta)(\cos\theta - \sin\theta)}{(\cos\theta + \sin\theta)(\cos\theta + \sin\theta)} \overset{?}{=} \frac{\cos\theta - \sin\theta}{\cos\theta + \sin\theta}$$

Factor the numerator and denominator.

$$\frac{\cancel{(\cos\theta + \sin\theta)}(\cos\theta - \sin\theta)}{\cancel{(\cos\theta + \sin\theta)}(\cos\theta + \sin\theta)} \overset{?}{=} \frac{\cos\theta - \sin\theta}{\cos\theta + \sin\theta}$$

Cancel common factors.

$$\frac{\cos\theta - \sin\theta}{\cos\theta + \sin\theta} \overset{?}{=} \frac{\cos\theta - \sin\theta}{\cos\theta + \sin\theta}$$

Simplify.

The left-hand side is now identical to the right-hand side. Therefore, the identity is verified.

You Try It Work through this You Try It problem.

Work Exercises 46–52 in this textbook or in the MyMathLab **Study Plan**.

OBJECTIVE 5   USING THE DOUBLE-ANGLE AND HALF-ANGLE FORMULAS TO EVALUATE EXPRESSIONS INVOLVING INVERSE TRIGONOMETRIC FUNCTIONS

The double-angle and half-angle formulas discussed in this section may be helpful when we encounter trigonometric expressions involving one or more inverse trigonometric functions. Before working through the next example, you should review the inverse trigonometric functions that were first introduced in Section 2.4. Click on one of the following links for a quick review of these functions.

A Quick Review of Inverse Trigonometric Functions

$$y = \sin^{-1}x \qquad y = \cos^{-1}x \qquad y = \tan^{-1}x$$

My video summary

Example 8 Finding the Exact Value of a Trigonometric Expression Involving Inverse Trigonometric Expressions

Find the exact value of the expression $\cos\left(\frac{1}{2}\sin^{-1}\left(-\frac{3}{11}\right)\right)$ without the use of a calculator.

Solution If we let $\sin^{-1}\left(-\frac{3}{11}\right) = \alpha$, then we are trying to find the value of $\cos\left(\frac{\alpha}{2}\right)$.

Because $\sin^{-1}\left(-\frac{3}{11}\right) = \alpha$, then $\sin\alpha = -\frac{3}{11}$ and $-\frac{\pi}{2} \le \alpha \le \frac{\pi}{2}$. Thus the terminal side of α lies in either Quadrant I or Quadrant IV. Because $\sin\alpha < 0$, we know that α must lie in Quadrant IV. To find the value of $\cos\alpha$, we use the Pythagorean identity $\sin^2\alpha + \cos^2\alpha = 1$ or $\cos\alpha = \sqrt{1 - \sin^2\alpha}$. Read these steps to see how to obtain $\cos\alpha = \dfrac{4\sqrt{7}}{11}$.

Because the terminal side of the angle lies in Quadrant IV, we know that $-\dfrac{\pi}{2} < \alpha < 0$. Dividing each part of this inequality by 2, we get $-\dfrac{\pi}{4} < \dfrac{\alpha}{2} < 0$.

Therefore, the terminal side of $\dfrac{\alpha}{2}$ also lies in Quadrant IV. Thus, $\cos\left(\dfrac{\alpha}{2}\right) > 0$ because the cosine function is positive for all angles whose terminal side lies in Quadrant IV.

We therefore use the **positive form** of the half-angle formula for cosine to find the exact value of $\cos\left(\dfrac{1}{2}\sin^{-1}\left(-\dfrac{3}{11}\right)\right)$.

$\cos\left(\dfrac{1}{2}\sin^{-1}\left(-\dfrac{3}{11}\right)\right)$ Write the original expression.

$= \cos\left(\dfrac{\alpha}{2}\right)$ Write as the cosine of a half-angle, where $\alpha = \sin^{-1}\left(-\dfrac{3}{11}\right)$.

$= \sqrt{\dfrac{1 + \cos\alpha}{2}}$ Use the positive form of the half-angle formula for cosine.

$= \sqrt{\dfrac{1 + \dfrac{4\sqrt{7}}{11}}{2}}$ Substitute $\cos\alpha = \dfrac{4\sqrt{7}}{11}$.

$= \sqrt{\dfrac{\dfrac{11 + 4\sqrt{7}}{11}}{2}}$ Combine the expressions in the denominator.

$= \sqrt{\dfrac{11 + 4\sqrt{7}}{22}}$ Simplify.

Therefore, $\cos\left(\dfrac{1}{2}\sin^{-1}\left(-\dfrac{3}{11}\right)\right) = \sqrt{\dfrac{11 + 4\sqrt{7}}{22}}$.

You Try It   Work through this You Try It problem.

Work Exercises 53–72 in this textbook or in the MyMathLab Study Plan.

3.3 Exercises

Skill Check Exercises

In Exercises SCE-1 through SCE-5 simplify each expression.

SCE-1. $\sqrt{\dfrac{1 - \dfrac{1}{\sqrt{2}}}{2}}$ SCE-2. $\sqrt{\dfrac{1 - \left(-\dfrac{1}{\sqrt{2}}\right)}{2}}$ SCE-3. $-\sqrt{\dfrac{1 - \dfrac{5}{11}}{2}}$

SCE-4. $\sqrt{\dfrac{1 - \dfrac{5}{23}}{1 + \dfrac{5}{23}}}$ **SCE-5.** $\sqrt{\dfrac{1 - \left(-\dfrac{1}{\sqrt{2}}\right)}{1 + \left(-\dfrac{1}{\sqrt{2}}\right)}}$

In Exercises 1–10, rewrite each expression as the sine, cosine, or tangent of a double-angle. Then find the exact value of the trigonometric expression without the use of a calculator.

1. $2 \sin\left(\dfrac{\pi}{8}\right) \cos\left(\dfrac{\pi}{8}\right)$ **2.** $2 \cos^2 75° - 1$ **3.** $\dfrac{2 \tan\left(\dfrac{7\pi}{8}\right)}{1 - \tan^2\left(\dfrac{7\pi}{8}\right)}$

4. $\cos^2\left(-\dfrac{5\pi}{12}\right) - \sin^2\left(-\dfrac{5\pi}{12}\right)$ **5.** $2 \sin 202.5° \cos 202.5°$ **6.** $1 - 2 \sin^2\left(\dfrac{-5\pi}{8}\right)$

7. $\dfrac{2 \tan(-67.5°)}{1 - \tan^2(-67.5°)}$ **8.** $\cos^2 105° - \sin^2 105°$ **9.** $1 - 2 \sin^2(-157.5°)$

10. $2 \cos^2\left(\dfrac{-9\pi}{8}\right) - 1$

In Exercises 11–14, use the given information to determine the values of $\sin 2\theta$, $\cos 2\theta$, and $\tan 2\theta$.

11. $\sin \theta = \dfrac{2}{13}$; The terminal side of θ lies in Quadrant II.

12. $\cos \theta = \dfrac{21}{29}$; The terminal side of θ lies in Quadrant IV.

13. $\sin \theta = -\dfrac{8}{17}$; The terminal side of θ lies in Quadrant III.

14. $\cot \theta = 10$; The terminal side of θ lies in Quadrant III.

In Exercises 15–18, use the given information to determine the values of $\sin \theta$, $\cos \theta$, and $\tan \theta$.

15. $\sin 2\theta = \dfrac{5}{13}$; $\dfrac{\pi}{2} < 2\theta < \pi$ **16.** $\cos 2\theta = \dfrac{3}{5}$; $\dfrac{3\pi}{2} < 2\theta < 2\pi$

17. $\tan 2\theta = \dfrac{2}{7}$; $\pi < 2\theta < \dfrac{3\pi}{2}$ **18.** $\csc 2\theta = -\dfrac{7}{5}$; $\dfrac{3\pi}{2} < 2\theta < 2\pi$

In Exercises 19–23, rewrite the given function as an equivalent function containing only cosine terms raised to a power of 1.

19. $f(x) = -3 \sin^2 x$ **20.** $f(x) = 7 \cos^2 x$ **21.** $f(x) = 2 \sin^4 x$

22. $f(x) = -3 \cos^4 x$ **23.** $f(x) = -2 \sin^2 x \cos^2 x$

In Exercises 24–31, use a half-angle formula to evaluate each expression without using a calculator. **Note to instructors:** The final answer to these exercises is **not** fully simplified.

24. $\sin\left(-\dfrac{3\pi}{8}\right)$ **25.** $\cos\left(-22.5°\right)$ **26.** $\tan 157.5°$ **27.** $\sin 67.5°$

28. $\cos\left(\dfrac{\pi}{12}\right)$ **29.** $\tan\left(\dfrac{5\pi}{12}\right)$ **30.** $\csc\left(\dfrac{9\pi}{8}\right)$ **31.** $\sec\left(\dfrac{13\pi}{12}\right)$

In Exercises 32–39, use a half-angle formula to evaluate each expression without using a calculator. **Note to instructors:** The final answer to these exercises is completely simplified.

32. $\sin\left(-\dfrac{3\pi}{8}\right)$ **33.** $\cos\left(-22.5°\right)$ **34.** $\tan 157.5°$ **35.** $\sin 67.5°$

36. $\cos\left(\dfrac{\pi}{12}\right)$ **37.** $\tan\left(\dfrac{5\pi}{12}\right)$ **38.** $\csc\left(\dfrac{9\pi}{8}\right)$ **39.** $\sec\left(\dfrac{13\pi}{12}\right)$

In Exercises 40–45, use the given information to determine the values of $\sin\left(\dfrac{\alpha}{2}\right)$, $\cos\left(\dfrac{\alpha}{2}\right)$, and $\tan\left(\dfrac{\alpha}{2}\right)$.

40. $\sin\alpha = \dfrac{12}{13}; \dfrac{\pi}{2} < \alpha < \pi$ **41.** $\cos\alpha = -\dfrac{21}{29}; \dfrac{\pi}{2} < \alpha < \pi$

42. $\tan\alpha = \dfrac{12}{5}; \pi < \alpha < \dfrac{3\pi}{2}$ **43.** $\sec\alpha = -\dfrac{5}{4}; \pi < \alpha < \dfrac{3\pi}{2}$

44. $\cot\alpha = \dfrac{1}{3}; \pi < \alpha < \dfrac{3\pi}{2}$ **45.** $\cos\alpha = \dfrac{5}{11}; \dfrac{3\pi}{2} < \alpha < 2\pi$

In Exercises 46–52, verify each identity.

46. $\dfrac{\tan^2\theta - 1}{1 + \tan^2\theta} = -\cos 2\theta$ **47.** $\cot 2\theta = \dfrac{1}{2}\sec\theta\csc\theta - \tan\theta$ **48.** $\sec 2\theta = \dfrac{\csc^2\theta}{\csc^2\theta - 2}$

49. $\tan\theta = \dfrac{1 - \cos 2\theta}{\sin 2\theta}$ **50.** $\sin^2\dfrac{\theta}{2} = \dfrac{\tan\theta - \sin\theta}{2\tan\theta}$ **51.** $\tan\dfrac{\theta}{2} = \dfrac{1}{\csc\theta + \cot\theta}$

52. $\cot\dfrac{\theta}{2} - \cot\theta = \csc\theta$

In Exercises 53–56, use a double angle formula to find the exact value of each expression without using a calculator.

53. $\sin\left(2\cos^{-1}\dfrac{\sqrt{3}}{2}\right)$ **54.** $\cos\left(2\sin^{-1}\left(-\dfrac{1}{\sqrt{2}}\right)\right)$

55. $\sin\left(2\tan^{-1}\dfrac{5}{2}\right)$ **56.** $\cos\left(2\sin^{-1}\left(-\dfrac{5}{9}\right)\right)$

In Exercises 57–64, use a half-angle formula to evaluate each expression without using a calculator. **Note to instructors:** The final answer to these exercises is **not** simplified.

57. $\sin\left(\frac{1}{2}\cos^{-1}\left(-\frac{1}{2}\right)\right)$ **58.** $\cos\left(\frac{1}{2}\cos^{-1}\left(-\frac{1}{\sqrt{2}}\right)\right)$ **59.** $\sin\left(\frac{1}{2}\cos^{-1}\left(-\frac{3}{7}\right)\right)$

60. $\cos\left(\frac{1}{2}\cos^{-1}\left(\frac{5}{11}\right)\right)$ **61.** $\sin\left(\frac{1}{2}\tan^{-1}(-1)\right)$ **62.** $\cos\left(\frac{1}{2}\tan^{-1}\left(\frac{1}{\sqrt{3}}\right)\right)$

63. $\sin\left(\frac{1}{2}\tan^{-1}\left(\frac{5}{7}\right)\right)$ **64.** $\cos\left(\frac{1}{2}\tan^{-1}(-5)\right)$

In Exercises 65–72, use a half-angle formula to evaluate each expression without using a calculator. **Note to instructors:** The final answer to these exercises is completely simplified.

65. $\sin\left(\frac{1}{2}\cos^{-1}\left(-\frac{1}{2}\right)\right)$ **66.** $\cos\left(\frac{1}{2}\cos^{-1}\left(-\frac{1}{\sqrt{2}}\right)\right)$ **67.** $\sin\left(\frac{1}{2}\cos^{-1}\left(-\frac{3}{7}\right)\right)$

68. $\cos\left(\frac{1}{2}\cos^{-1}\left(\frac{5}{11}\right)\right)$ **69.** $\sin\left(\frac{1}{2}\tan^{-1}(-1)\right)$ **70.** $\cos\left(\frac{1}{2}\tan^{-1}\left(\frac{1}{\sqrt{3}}\right)\right)$

71. $\sin\left(\frac{1}{2}\tan^{-1}\left(\frac{5}{7}\right)\right)$ **72.** $\cos\left(\frac{1}{2}\tan^{-1}(-5)\right)$

Brief Exercises

In Exercises 73–80, use a half-angle formula to evaluate each expression without using a calculator. **Note to instructors:** The final answer to these exercises is **not** completely simplified.

73. $\sin\left(-\frac{3\pi}{8}\right)$ **74.** $\cos(-22.5°)$ **75.** $\tan 157.5°$ **76.** $\sin 67.5°$

77. $\cos\left(\frac{\pi}{12}\right)$ **78.** $\tan\left(\frac{5\pi}{12}\right)$ **79.** $\csc\left(\frac{9\pi}{8}\right)$ **80.** $\sec\left(\frac{13\pi}{12}\right)$

In Exercises 81–88, use a half-angle formula to evaluate each expression without using a calculator. **Note to instructors:** The final answer to these exercises is completely simplified.

81. $\sin\left(-\frac{3\pi}{8}\right)$ **82.** $\cos(-22.5°)$ **83.** $\tan 157.5°$ **84.** $\sin 67.5°$

85. $\cos\left(\frac{\pi}{12}\right)$ **86.** $\tan\left(\frac{5\pi}{12}\right)$ **87.** $\csc\left(\frac{9\pi}{8}\right)$ **88.** $\sec\left(\frac{13\pi}{12}\right)$

In Exercises 89–96, use a half-angle formula to evaluate each expression without using a calculator. **Note to instructors:** The final answer to these exercises is not completely simplified.

89. $\sin\left(\frac{1}{2}\cos^{-1}\left(-\frac{1}{2}\right)\right)$ **90.** $\cos\left(\frac{1}{2}\cos^{-1}\left(-\frac{1}{\sqrt{2}}\right)\right)$ **91.** $\sin\left(\frac{1}{2}\cos^{-1}\left(-\frac{3}{7}\right)\right)$

92. $\cos\left(\dfrac{1}{2}\cos^{-1}\left(\dfrac{5}{11}\right)\right)$ **93.** $\sin\left(\dfrac{1}{2}\tan^{-1}(-1)\right)$ **94.** $\cos\left(\dfrac{1}{2}\tan^{-1}\left(-\dfrac{1}{\sqrt{3}}\right)\right)$

95. $\sin\left(\dfrac{1}{2}\tan^{-1}\left(\dfrac{5}{7}\right)\right)$ **96.** $\cos\left(\dfrac{1}{2}\tan^{-1}(-5)\right)$

In Exercises 97–104, use a half-angle formula to evaluate each expression without using a calculator. **Note to instructors:** The final answer to these exercises is completely simplified.

97. $\sin\left(\dfrac{1}{2}\cos^{-1}\left(-\dfrac{1}{2}\right)\right)$ **98.** $\cos\left(\dfrac{1}{2}\cos^{-1}\left(-\dfrac{1}{\sqrt{2}}\right)\right)$ **99.** $\sin\left(\dfrac{1}{2}\cos^{-1}\left(-\dfrac{3}{7}\right)\right)$

100. $\cos\left(\dfrac{1}{2}\cos^{-1}\left(\dfrac{5}{11}\right)\right)$ **101.** $\sin\left(\dfrac{1}{2}\tan^{-1}(-1)\right)$ **102.** $\cos\left(\dfrac{1}{2}\tan^{-1}\left(-\dfrac{1}{\sqrt{3}}\right)\right)$

103. $\sin\left(\dfrac{1}{2}\tan^{-1}\left(\dfrac{5}{7}\right)\right)$ **104.** $\cos\left(\dfrac{1}{2}\tan^{-1}(-5)\right)$

3.4 The Product-to-Sum and Sum-to-Product Formulas

THINGS TO KNOW

Before working through this section, be sure that you are familiar with the following concepts:

VIDEO   ANIMATION   INTERACTIVE

You Try It

1. Evaluating Trigonometric Functions of Angles Belonging to the $\dfrac{\pi}{3}, \dfrac{\pi}{6},$ or $\dfrac{\pi}{4}$ Families (Section 1.5)

You Try It

2. Understanding the Sum and Difference Formulas for the Sine Function (Section 3.2)

You Try It

3. Understanding the Sum and Difference Formulas for the Cosine Function (Section 3.2)

OBJECTIVES

1 Understanding the Product-to-Sum Formulas

2 Understanding the Sum-to-Product Formulas

3 Using the Product-to-Sum and Sum-to-Product Formulas to Verify Identities

Introduction to Section 3.4

In this section, we will learn formulas that will allow us to convert the product of trigonometric functions into the sum or difference of trigonometric functions and vice versa. We start by introducing the **product-to-sum formulas.**

OBJECTIVE 1   UNDERSTANDING THE PRODUCT-TO-SUM FORMULAS

There are four product-to-sum formulas. That is, there are four different formulas that can be used to change the product of two trigonometric expressions into the sum or difference of two trigonometric expressions. These formulas are especially useful in calculus. Although the formulas appear to be difficult to memorize, they are actually fairly easy to derive provided that you can recall the sum and difference formulas for sine and cosine.

Product-to-Sum Formulas

$$\sin \alpha \sin \beta = \frac{1}{2}\left[\cos(\alpha - \beta) - \cos(\alpha + \beta)\right]$$

$$\cos \alpha \cos \beta = \frac{1}{2}\left[\cos(\alpha - \beta) + \cos(\alpha + \beta)\right]$$

$$\sin \alpha \cos \beta = \frac{1}{2}\left[\sin(\alpha + \beta) + \sin(\alpha - \beta)\right]$$

$$\cos \alpha \sin \beta = \frac{1}{2}\left[\sin(\alpha + \beta) - \sin(\alpha - \beta)\right]$$

My interactive video summary

⊘ Work through this interactive video to see how to use the sum and difference formulas for sine and cosine to derive each of the four product-to-sum formulas.

My interactive video summary

⊘ **Example 1** Writing the Product of Two Functions as a Sum or Difference of Two Functions

Write each product as a sum or difference containing only sines or cosines.

a. $\sin 4\theta \sin 2\theta$ **b.** $\cos\left(\dfrac{19\theta}{2}\right)\sin\left(\dfrac{\theta}{2}\right)$ **c.** $\cos 11\theta \cos 5\theta$ **d.** $\sin 6\theta \cos 3\theta$

Solution

a. To write the product $\sin(4\theta)\sin(2\theta)$ as a sum or difference of sines and cosines, we must first identify the correct product-to-sum formula. Because we are given the product of two sine expressions, we will use the formula

$$\sin \alpha \sin \beta = \frac{1}{2}\left[\cos(\alpha - \beta) - \cos(\alpha + \beta)\right] \text{ with } \alpha = 4\theta \text{ and } \beta = 2\theta.$$

$\sin \alpha \sin \beta = \dfrac{1}{2}\left[\cos(\alpha - \beta) - \cos(\alpha + \beta)\right]$ Write the appropriate product-to-sum formula.

$\sin(4\theta)\sin(2\theta) = \dfrac{1}{2}\left[\cos(4\theta - 2\theta) - \cos(4\theta + 2\theta)\right]$ Substitute $\alpha = 4\theta$ and $\beta = 2\theta$.

$= \dfrac{1}{2}\left[\cos(2\theta) - \cos(6\theta)\right]$ Simplify the arguments.

$= \dfrac{1}{2}\cos(2\theta) - \dfrac{1}{2}\cos(6\theta)$ Distribute.

b. To write the product $\cos\left(\dfrac{19\theta}{2}\right)\sin\left(\dfrac{\theta}{2}\right)$ as a sum or difference of sines or cosines, we must first identify the correct product-to-sum formula. Because we are given the product of a cosine expression and a sine expression, we will use the formula $\cos\alpha\sin\beta = \dfrac{1}{2}\left[\sin(\alpha+\beta) - \sin(\alpha-\beta)\right]$ with $\alpha = \dfrac{19\theta}{2}$ and $\beta = \dfrac{\theta}{2}$.

$\cos\alpha\sin\beta = \dfrac{1}{2}\left[\sin(\alpha+\beta) - \sin(\alpha-\beta)\right]$ — Write the appropriate product-to-sum formula.

$\cos\left(\dfrac{19\theta}{2}\right)\sin\left(\dfrac{\theta}{2}\right) = \dfrac{1}{2}\left[\sin\left(\dfrac{19\theta}{2} + \dfrac{\theta}{2}\right) - \sin\left(\dfrac{19\theta}{2} - \dfrac{\theta}{2}\right)\right]$ — Substitute $\alpha = \dfrac{19\theta}{2}$ and $\beta = \dfrac{\theta}{2}$.

$= \dfrac{1}{2}\left[\sin 10\theta - \sin 9\theta\right]$ — Simplify the arguments.

$= \dfrac{1}{2}\sin 10\theta - \dfrac{1}{2}\sin 9\theta$ — Distribute.

My interactive video summary

▶ Try to determine the solution to part c and part d on your own by first choosing the appropriate product-to-sum formula. When you're finished, check your solutions, or watch this interactive video to see the solutions to all four examples. ●

You Try It Work through this You Try It problem.

Work Exercises 1–4 in this textbook or in the MyMathLab Study Plan.

My video summary

▶ **Example 2 Evaluating the Product of a Trigonometric Expression**

Determine the exact value of the expression $\sin\left(\dfrac{3\pi}{8}\right)\cos\left(\dfrac{\pi}{8}\right)$ without the use of a calculator.

Solution To evaluate the expression $\sin\left(\dfrac{3\pi}{8}\right)\cos\left(\dfrac{\pi}{8}\right)$, first use the product-to-sum formula $\sin\alpha\cos\beta = \dfrac{1}{2}\left[\sin(\alpha+\beta) + \sin(\alpha-\beta)\right]$ with $\alpha = \dfrac{3\pi}{8}$ and $\beta = \dfrac{\pi}{8}$.

$\sin\left(\dfrac{3\pi}{8}\right)\cos\left(\dfrac{\pi}{8}\right)$ — Write the original expression.

$= \dfrac{1}{2}\left[\sin\left(\dfrac{3\pi}{8} + \dfrac{\pi}{8}\right) + \sin\left(\dfrac{3\pi}{8} - \dfrac{\pi}{8}\right)\right]$ — Use the formula $\sin\alpha\cos\beta = \dfrac{1}{2}[\sin(\alpha+\beta) + \sin(\alpha-\beta)]$ with $\alpha = \dfrac{3\pi}{8}$ and $\beta = \dfrac{\pi}{8}$.

$= \dfrac{1}{2}\left[\sin\dfrac{\pi}{2} + \sin\dfrac{\pi}{4}\right]$ — Simplify the arguments.

$= \dfrac{1}{2}\left[1 + \dfrac{1}{\sqrt{2}}\right]$ — Substitute $\sin\dfrac{\pi}{2} = 1$ and $\sin\dfrac{\pi}{4} = \dfrac{1}{\sqrt{2}}$.

$= \dfrac{1}{2}\left[\dfrac{\sqrt{2}+1}{\sqrt{2}}\right] = \dfrac{\sqrt{2}+1}{2\sqrt{2}}$ — Rewrite using a common denominator and multiply.

Therefore, $\sin\left(\dfrac{3\pi}{8}\right)\cos\left(\dfrac{\pi}{8}\right) = \dfrac{\sqrt{2}+1}{2\sqrt{2}}$. Watch this video to see every step of this solution. Note that the expression $\dfrac{\sqrt{2}+1}{2\sqrt{2}}$ can be written as $\dfrac{2+\sqrt{2}}{4}$ if we choose to rationalize the denominator.

You Try It Work through this You Try It problem.

Work Exercises 5–8 in this textbook or in the MyMathLab Study Plan.

OBJECTIVE 2 UNDERSTANDING THE SUM-TO-PRODUCT FORMULAS

We can use the four product-to-sum formulas to verify each of the four sum-to-product formulas. These formulas are particularly useful when solving trigonometric equations. (See Section 3.5, Example 4d.)

Sum-to-Product Formulas

$$\sin\alpha + \sin\beta = 2\sin\left(\frac{\alpha+\beta}{2}\right)\cos\left(\frac{\alpha-\beta}{2}\right)$$

$$\sin\alpha - \sin\beta = 2\sin\left(\frac{\alpha-\beta}{2}\right)\cos\left(\frac{\alpha+\beta}{2}\right)$$

$$\cos\alpha + \cos\beta = 2\cos\left(\frac{\alpha+\beta}{2}\right)\cos\left(\frac{\alpha-\beta}{2}\right)$$

$$\cos\alpha - \cos\beta = -2\sin\left(\frac{\alpha+\beta}{2}\right)\sin\left(\frac{\alpha-\beta}{2}\right)$$

My interactive video summary

⊙ Work through this interactive video to see how to use the product-to-sum formulas to verify each of the sum-to-product formulas.

My video summary

⊙ **Example 3** Writing the Sum or Difference of Two Functions as a Product of Two Functions

Write each sum or difference as a product of sines and/or cosines.

a. $\sin 5\theta - \sin 3\theta$ 　　　　　　　　**b.** $\cos\left(\dfrac{3\theta}{2}\right) - \cos\left(\dfrac{17\theta}{2}\right)$

Solution

a. To write the sum $\sin 5\theta + \sin 3\theta$ as a product of sines and/or cosines, we must first identify the correct sum-to-product formula. Because we are given the sum of two sine expressions, we will use the formula

$$\sin\alpha + \sin\beta = 2\sin\left(\frac{\alpha+\beta}{2}\right)\cos\left(\frac{\alpha-\beta}{2}\right) \text{ with } \alpha = 5\theta \text{ and } \beta = 3\theta.$$

$$\sin\alpha + \sin\beta = 2\sin\left(\frac{\alpha + \beta}{2}\right)\cos\left(\frac{\alpha - \beta}{2}\right)$$

Write the appropriate product-to-sum formula.

$$\sin 5\theta + \sin 3\theta = 2\sin\left(\frac{5\theta + 3\theta}{2}\right)\cos\left(\frac{5\theta - 3\theta}{2}\right)$$

Substitute $\alpha = 5\theta$ and $\beta = 3\theta$.

$$= 2\sin 4\theta \cos\theta$$

Simplify the arguments.

Watch this interactive video to see each step of this solution.

b. To write the difference $\cos\left(\dfrac{3\theta}{2}\right) - \cos\left(\dfrac{17\theta}{2}\right)$ as a product of sines and/or cosines, we must first identify the correct sum-to-product formula. Because we are given the difference of two cosine expressions, we will use the formula $\cos\alpha - \cos\beta = -2\sin\left(\dfrac{\alpha + \beta}{2}\right)\sin\left(\dfrac{\alpha - \beta}{2}\right)$ with $\alpha = \dfrac{3\theta}{2}$ and $\beta = \dfrac{17\theta}{2}$.

$$\cos\alpha - \cos\beta = -2\sin\left(\frac{\alpha + \beta}{2}\right)\sin\left(\frac{\alpha - \beta}{2}\right)$$

Write the appropriate product-to-sum formula.

$$\cos\left(\frac{3\theta}{2}\right) - \cos\left(\frac{17\theta}{2}\right) = -2\sin\left(\frac{\frac{3\theta}{2} + \frac{17\theta}{2}}{2}\right)\sin\left(\frac{\frac{3\theta}{2} - \frac{17\theta}{2}}{2}\right)$$

Substitute $\alpha = \dfrac{3\theta}{2}$ and $\beta = \dfrac{17\theta}{2}$.

$$= -2\sin\left(\frac{\frac{20\theta}{2}}{2}\right)\sin\left(\frac{-\frac{14\theta}{2}}{2}\right)$$

Combine the expressions in the numerator of each argument.

$$= -2\sin 5\theta \sin\left(\frac{-7\theta}{2}\right)$$

Simplify each argument.

My interactive video summary

⊚ Watch this interactive video to see each step of this solution.

You Try It Work through this You Try It problem.

Work Exercises 9–12 in this textbook or in the MyMathLab Study Plan.

My interactive video summary

⊚ **Example 4** Evaluating the Difference of Two Trigonometric Expressions

Determine the exact value of the expression $\sin\left(\dfrac{\pi}{12}\right) - \sin\left(\dfrac{17\pi}{12}\right)$ without the use of a calculator.

Solution To evaluate the expression $\sin\left(\dfrac{\pi}{12}\right) - \sin\left(\dfrac{17\pi}{12}\right)$, first use the sum-to-product formula $\sin\alpha - \sin\beta = 2\sin\left(\dfrac{\alpha - \beta}{2}\right)\cos\left(\dfrac{\alpha + \beta}{2}\right)$ with $\alpha = \dfrac{\pi}{12}$ and $\beta = \dfrac{17\pi}{12}$.

$$\sin\left(\frac{\pi}{12}\right) - \sin\left(\frac{17\pi}{12}\right)$$ Write the original expression.

$$= 2\sin\left(\frac{\dfrac{\pi}{12} - \dfrac{17\pi}{12}}{2}\right)\cos\left(\frac{\dfrac{\pi}{12} + \dfrac{17\pi}{12}}{2}\right)$$ Use the formula $\sin\alpha - \sin\beta = 2\sin\left(\dfrac{\alpha - \beta}{2}\right)$ $\cos\left(\dfrac{\alpha + \beta}{2}\right)$ with $\alpha = \dfrac{\pi}{12}$ and $\beta = \dfrac{17\pi}{12}$.

Watch this interactive video to see how this expression simplifies to $2\sin\left(-\dfrac{2\pi}{3}\right)\cos\left(\dfrac{3\pi}{4}\right)$. Using our knowledge of the trigonometric functions of general angles, we can determine the values of $\sin\left(-\dfrac{2\pi}{3}\right)$ and $\cos\left(\dfrac{3\pi}{4}\right)$.

$$2\sin\left(-\frac{2\pi}{3}\right)\cos\left(\frac{3\pi}{4}\right)$$ Write the new expression that represents $\sin\left(\dfrac{\pi}{12}\right) - \sin\left(\dfrac{17\pi}{12}\right)$.

$$= 2\left(-\frac{\sqrt{3}}{2}\right)\left(-\frac{1}{\sqrt{2}}\right)$$ Substitute $\sin\left(-\dfrac{2\pi}{3}\right) = -\dfrac{\sqrt{3}}{2}$ and $\cos\left(\dfrac{3\pi}{4}\right) = -\dfrac{1}{\sqrt{2}}$.

$$= \frac{\sqrt{3}}{\sqrt{2}}$$ Multiply.

My interactive video summary   ⊙ Therefore, the exact value of $\sin\left(\dfrac{\pi}{12}\right) - \sin\left(\dfrac{17\pi}{12}\right)$ is $\dfrac{\sqrt{3}}{\sqrt{2}}$. This expression can also be written as $\sqrt{\dfrac{3}{2}}$ or $\dfrac{\sqrt{6}}{2}$. Watch this interactive video to see each step of this solution.   ●

You Try It   Work through this You Try It problem.

Work Exercises 13–16 in this textbook or in the MyMathLab Study Plan.

OBJECTIVE 3   USING THE PRODUCT-TO-SUM AND SUM-TO-PRODUCT FORMULAS TO VERIFY IDENTITIES

We can use formulas discussed in this section along with any identity discussed so far in this text to verify trigonometric identities. As always, we will verify an identity by trying to transform one side of the identity into the exact trigonometric expression that appears on the opposite side. We typically start with what appears to be the more complicated side of the identity and then attempt to use known identities to transform that side into the trigonometric expression that appears on the other side of the identity. However, if one side of the identity includes a trigonometric expression involving the sum or difference of sine and cosine, then first substitute the appropriate sum-to-product formula. Likewise, if one side of the identity includes the product of sines and/or cosines, then first substitute the appropriate product-to-sum formula. Next, use the strategies we developed in Section 3.1 for verifying identities. To review these strategies, read this summary of techniques that were introduced in Section 3.1.

Example 5 Verifying a Trigonometric Identity

Verify the trigonometric identity $\dfrac{\cos\theta + \cos 3\theta}{2\cos 2\theta} = \cos\theta$.

Solution We start by rewriting the numerator of the left-hand side using a sum-to-product formula.

$$\frac{\cos\theta + \cos 3\theta}{2\cos 2\theta} \overset{?}{=} \cos\theta \qquad \text{Write the original trigonometric identity.}$$

$$\frac{2\cos\left(\dfrac{\theta+3\theta}{2}\right)\cos\left(\dfrac{\theta-3\theta}{2}\right)}{2\cos 2\theta} \overset{?}{=} \cos\theta$$

Use the sum-to-product formula
$$\cos\alpha + \cos\beta = 2\cos\left(\frac{\alpha+\beta}{2}\right)\cos\left(\frac{\alpha-\beta}{2}\right)$$
with $\alpha = \theta$ and $\beta = 3\theta$.

$$\frac{2\cos 2\theta \cos(-\theta)}{2\cos 2\theta} \overset{?}{=} \cos\theta \qquad \text{Simplify each argument.}$$

$$\frac{2\cancel{\cos 2\theta}\cos(-\theta)}{2\cancel{\cos 2\theta}} \overset{?}{=} \cos\theta \qquad \text{Cancel common factors.}$$

$$\cos(-\theta) \overset{?}{=} \cos\theta \qquad \text{Simplify.}$$

$$\cos\theta \overset{?}{=} \cos\theta \qquad \text{Use the even property of cosine:}$$
$$\cos(-\theta) = \cos\theta.$$

The left-hand side is now identical to the right-hand side. Therefore, the identity is verified.

You Try It Work through this You Try It problem.

Work Exercises 17–21 in this textbook or in the MyMathLab Study Plan.

3.4 Exercises

In Exercises 1–4, write each product as the sum or difference containing only sines or cosines.

1. $\sin(8\theta)\sin(4\theta)$ **2.** $\sin(7\theta)\cos(3\theta)$ **3.** $\cos(\theta)\cos(5\theta)$ **4.** $\sin\left(\dfrac{5\theta}{2}\right)\cos\left(\dfrac{\theta}{2}\right)$

In Exercises 5–8, determine the exact value of each expression without the use of a calculator.

5. $\sin\left(\dfrac{5\pi}{24}\right)\cos\left(\dfrac{\pi}{24}\right)$ **6.** $(\sin 75°)(\sin 15°)$ **7.** $\cos\left(\dfrac{5\pi}{8}\right)\cos\left(\dfrac{3\pi}{8}\right)$ **8.** $(\cos 67.5°)(\sin 22.5°)$

In Exercises 9–12, write each sum or difference as a product containing only sines and/or cosines.

9. $\sin(9\theta) - \sin(7\theta)$ **10.** $\cos(5\theta) + \cos(3\theta)$

11. $\sin(11\theta) + \sin(9\theta)$ **12.** $\cos\left(\dfrac{5\theta}{2}\right) - \cos\left(\dfrac{11\theta}{2}\right)$

In Exercises 13–16, determine the exact value of each expression without the use of a calculator.

13. $\sin\left(\dfrac{5\pi}{12}\right) - \sin\left(\dfrac{11\pi}{12}\right)$ **14.** $\sin 225° + \sin 135°$

15. $\cos\left(\dfrac{19\pi}{12}\right) - \cos\left(\dfrac{\pi}{12}\right)$ **16.** $\cos 225° + \cos 135°$

In Exercises 17–21, verify each identity.

17. $\dfrac{\sin \theta - \sin 3\theta}{\cos \theta + \cos 3\theta} = -\tan \theta$

18. $\dfrac{\cos \theta + \cos 3\theta}{\sin \theta + \sin 3\theta} = \cot 2\theta$

19. $\dfrac{\sin x + \sin y}{\cos x + \cos y} = \tan \dfrac{x + y}{2}$

20. $\dfrac{\cos x + \cos y}{\cos x - \cos y} = -\cot \dfrac{x + y}{2} \cot \dfrac{x - y}{2}$

21. $\dfrac{\sin (6\theta) + \sin (8\theta)}{\sin (6\theta) - \sin (8\theta)} = -\dfrac{\tan (7\theta)}{\tan \theta}$

3.5 Trigonometric Equations

THINGS TO KNOW

Before working through this section, be sure that you are familiar with the following concepts:

| | | VIDEO | ANIMATION | INTERACTIVE |

You Try It   1. Solving Equations That Are Quadratic in Form — ⊙ (INTERACTIVE)

You Try It   2. Evaluating Trigonometric Functions of Angles Belonging to the $\dfrac{\pi}{3}, \dfrac{\pi}{6}$, or $\dfrac{\pi}{4}$ Families (Section 1.5) — ⊙ (INTERACTIVE)

You Try It   3. Finding the Exact and Approximate Values of an Inverse Sine Expression (Section 2.4) — ⊙ (INTERACTIVE)

You Try It   4. Finding the Exact and Approximate Values of an Inverse Cosine Expression (Section 2.4) — ⊙ (INTERACTIVE)

OBJECTIVES

1 Solving Trigonometric Equations That Are Linear in Form

2 Solving Trigonometric Equations That Are Quadratic in Form

3 Solving Trigonometric Equations Using Identities

4 Solving Other Types of Trigonometric Equations

5 Solving Trigonometric Equations Using a Calculator

Introduction to Section 3.5

Some trigonometric equations are true for *all* values of the variable for which each trigonometric function is defined. We call these **identities**. We have verified many identities throughout this chapter. Other trigonometric equations are only true for *specific* values of the variable (or no values at all). These are called **conditional trigonometric equations** or simply trigonometric equations.

Because of the periodic nature of trigonometric functions, we will often find that trigonometric equations have infinitely many solutions. We obviously cannot make a list of the infinitely many solutions. Therefore, we will describe these solutions using a formula (or formulas.) The most concise formula(s) describing the infinite set of solutions to a trigonometric equation is called the **general solution(s)**. We will also be interested in finding the **specific solution(s)** on a given restricted interval. We will typically try to find the solutions on the interval $[0, 2\pi)$.

My animation summary

▶ To motivate our discussion of trigonometric equations, consider the equation $\sin \theta = b$ for $-1 \le b \le 1$. In this equation, we are trying to determine all values θ for which the sine of θ is equal to the constant b. We can get a visual representation of the solutions to this equation by sketching the graphs $y = \sin \theta$ and $y = b$ on the same coordinate plane. The solutions to the equation $\sin \theta = b$ will be the first coordinate of all points of intersection of the two graphs. Notice in Figure 10 that the graph of $y = b$ for $-1 \le b \le 1$ will intersect the graph of $y = \sin \theta$ an infinite number of times. Thus, if $-1 \le b \le 1$, then the equation $\sin \theta = b$ has infinitely many solutions. Note that for $b > 1$ or for $b < -1$ the equation $\sin \theta = b$ has no solution because the line $y = b$ would not intersect the graph of $y = \sin \theta$. Work through this animation to see this visual representation of the solutions to this equation.

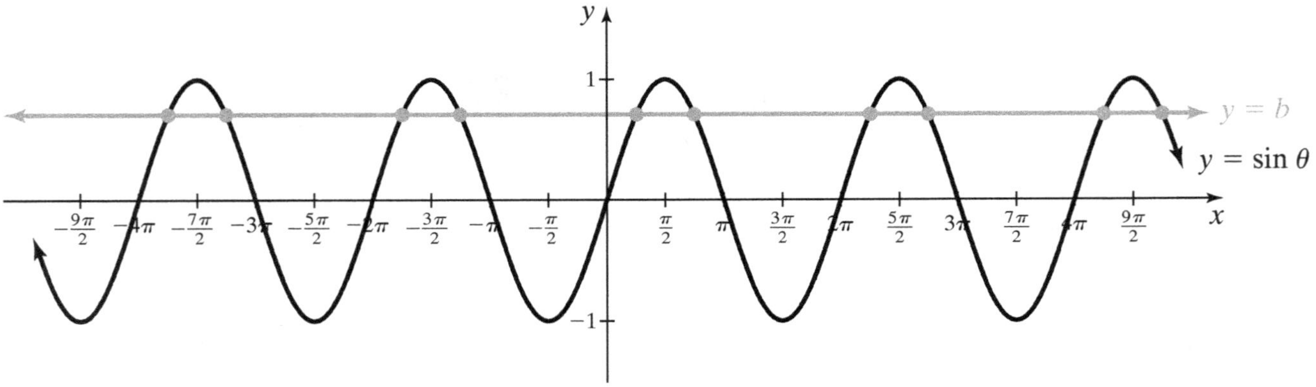

Figure 10 The graphs of $y = \sin \theta$ and $y = b$ for $-1 \le b \le 1$ intersect infinitely many times. Thus, the equation $\sin \theta = b$ has infinitely many solutions for $-1 \le b \le 1$.

My animation summary

▶ If we choose to restrict θ to angles on the interval $[0, 2\pi)$, then we see that there are exactly two points of intersection of $\sin \theta = b$ and $y = b$. See Figure 11. Thus, the equation $\sin \theta = b$ has exactly two solutions on this interval. This is **not** always the case. It is possible to have many more solutions on a restricted interval or have no solutions at all.

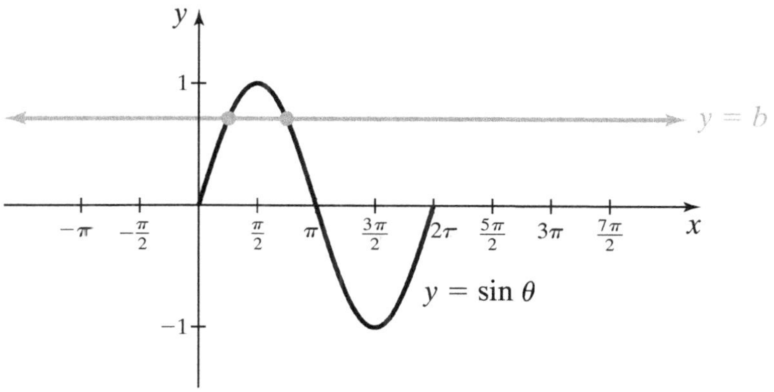

Figure 11 The graphs of $y = \sin\theta$ and $y = b$ for $-1 \le b \le 1$ intersect exactly twice on the interval $[0, 2\pi)$. Thus, the equation $\sin\theta = b$ has two solutions on the restricted interval $[0, 2\pi)$ for $-1 \le b \le 1$.

Throughout this section we will learn a variety of techniques used to solve trigonometric equations. First we will look at trigonometric equations that are linear in form.

OBJECTIVE 1 SOLVING TRIGONOMETRIC EQUATIONS THAT ARE LINEAR IN FORM

Recall that a linear equation in one variable is an equation that involves constants and **only one** variable that is raised to the first power. For example, the equation $\sqrt{2}x + 1 = 0$ is a linear equation in one variable. We solve this equation by isolating the variable x We do this by subtracting 1 from both sides and then dividing both sides by $\sqrt{2}$.

$$\sqrt{2}x + 1 = 0 \qquad \text{Write the original linear equation.}$$

$$\sqrt{2}x = -1 \qquad \text{Subtract 1 from both sides.}$$

$$x = -\frac{1}{\sqrt{2}} \qquad \text{Divide both sides by } \sqrt{2}.$$

When a trigonometric equation contains constants and **only one** trigonometric function that is raised to the first power, we say that the trigonometric equation is **linear in form**. Some examples of trigonometric equations that are linear in form are $\sin\theta = \frac{1}{2}$, $\sqrt{3}\tan\theta + 1 = 0$, $\sec\theta = -1$, $\sqrt{2}\cos 2\theta + 1 = 0$, and $\sin\frac{\theta}{2} = -\frac{\sqrt{3}}{2}$. Note that each one of these equations contains constants and only one trigonometric function that is raised to the first power. To solve these equations, we first treat each equation in much the same way as we treat a linear equation in one variable. That is, we use algebraic manipulations to isolate the trigonometric function. In the following two examples, we will follow a four-step procedure to solve each of these equations.

Steps for Solving Trigonometric Equations That Are Linear in Form

Step 1. Isolate the trigonometric function on one side of the equation.
Step 2. Determine the quadrants in which the terminal side of the argument of the function lies or determine the axis on which the terminal side of the argument of the function lies.

(continued)

Step 3. If the terminal side of the argument of the function lies within a quadrant, then determine the reference angle and the value(s) of the argument on the interval $[0, 2\pi)$.

If the terminal side of the argument of the function lies along an axis, then determine the angle associated with it on the interval $[0, 2\pi)$, choosing from $0, \dfrac{\pi}{2}, \pi,$ or $\dfrac{3\pi}{2}$.

Step 4. Use the period of the given function to determine the solutions.

My interactive video summary

⊘ **Example 1** Solving Trigonometric Equations That Are Linear in Form

Determine a general formula (or formulas) for all solutions to each equation. Then, determine the specific solutions (if any) on the interval $[0, 2\pi)$.

a. $\sin \theta = \dfrac{1}{2}$ **b.** $\sqrt{3}\tan\theta + 1 = 0$ **c.** $\sec\theta = -1$

Solution

a. Step 1. For the equation $\sin\theta = \dfrac{1}{2}$, the trigonometric function is already isolated. Therefore, we may skip to step 2.

Step 2. Because $\sin\theta = \dfrac{1}{2} > 0$, we know that the terminal side of θ must lie in Quadrant I or Quadrant II.

Step 3. All values of θ must have the same reference angle, which is $\dfrac{\pi}{6}$ because $\sin\dfrac{\pi}{6} = \dfrac{1}{2}$. It follows that

$$\theta = \text{Reference Angle} = \frac{\pi}{6} \text{ in Quadrant I and}$$

$$\theta = \pi - \text{Reference Angle} = \pi - \frac{\pi}{6} = \frac{5\pi}{6} \text{ in Quadrant II. See Figure 12.}$$

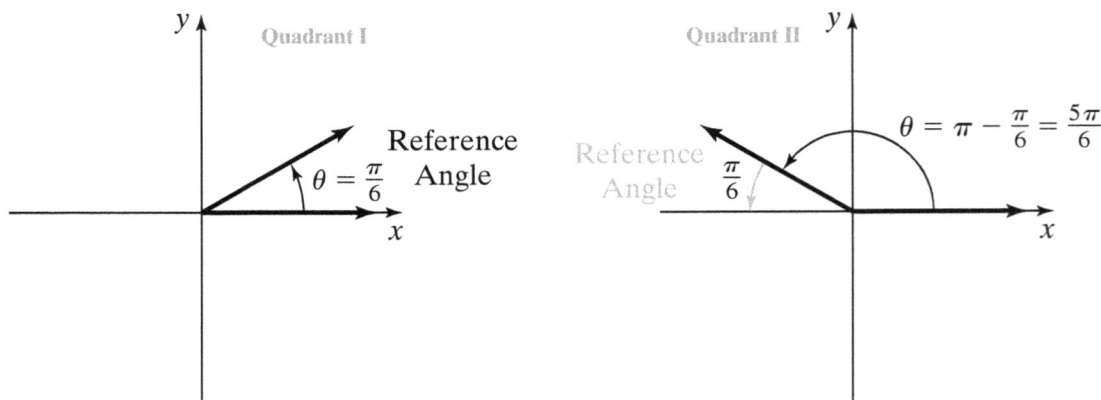

Figure 12 The reference angle is $\dfrac{\pi}{6}$. In Quadrant I, $\theta = \dfrac{\pi}{6}$. In Quadrant II, $\theta = \dfrac{5\pi}{6}$.

Step 4. The sine function has a period of 2π. Therefore, the possible general solutions are of the form

$$\theta = \frac{\pi}{6} + 2\pi k \quad \text{or} \quad \theta = \frac{5\pi}{6} - 2\pi k, \text{ where } k \text{ is any integer.}$$

These two equations produce uniquely different values and thus represent the general solutions to the equation $\sin \theta = \frac{1}{2}$.

In order to find the specific solutions between 0 and 2π, we must carefully observe the value of the angles to be certain that we have all angles in the interval and that we do not list angles outside of that interval. Using the formulas for the general solutions with $k = 1$, we get

$$\theta = \frac{\pi}{6} + 2\pi = \frac{13\pi}{6} \quad \text{or} \quad \theta = \frac{5\pi}{6} + 2\pi = \frac{17\pi}{6}$$

My interactive video summary

⊙ Both of these angles lie outside of the interval $[0, 2\pi)$. Therefore, the only solutions on the interval $[0, 2\pi)$ are $\theta = \frac{\pi}{6}$ or $\theta = \frac{5\pi}{6}$. Work through this interactive video to see the video solution to this equation.

b. Step 1. Given the equation $\sqrt{3} \tan \theta + 1 = 0$, we start by isolating the function $\tan \theta$.

$$\sqrt{3} \tan \theta + 1 = 0 \qquad \text{Write the original equation.}$$

$$\sqrt{3} \tan \theta = -1 \qquad \text{Subtract 1 from both sides.}$$

$$\tan \theta = -\frac{1}{\sqrt{3}} \qquad \text{Divide both sides by } \sqrt{3}.$$

Step 2. Because $\tan \theta = -\frac{1}{\sqrt{3}} < 0$, we know that the terminal side of θ must lie in Quadrant II or Quadrant IV.

Step 3. All values of θ must have the same reference angle of $\frac{\pi}{6}$ because $\tan \frac{\pi}{6} = \frac{1}{\sqrt{3}}$. It follows that

$$\theta = \pi - \text{Reference Angle} = \pi - \frac{\pi}{6} = \frac{5\pi}{6} \text{ or}$$

$$\theta = 2\pi - \text{Reference Angle} = 2\pi - \frac{\pi}{6} = \frac{11\pi}{6}. \text{ See Figure 13.}$$

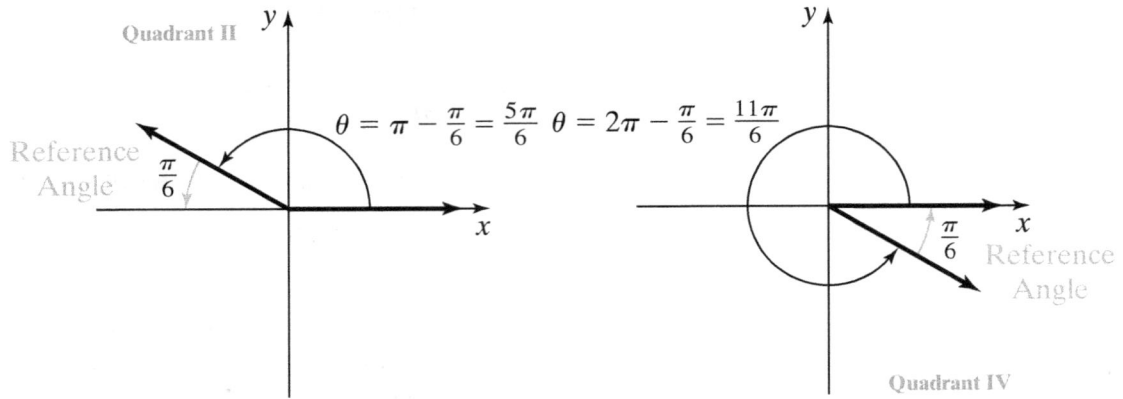

Figure 13 The reference angle is $\dfrac{\pi}{6}$. In Quadrant II, $\theta = \dfrac{5\pi}{6}$. In Quadrant IV, $\theta = \dfrac{11\pi}{6}$.

Step 4. The tangent function has a period of π. Therefore, the possible general solutions are of the form

$$\theta = \frac{5\pi}{6} + \pi k \quad \text{or} \quad \theta = \frac{11\pi}{6} + \pi k, \text{ where } k \text{ is any integer.}$$

However, these two equations do **not** produce unique values and thus using both formulas is not the most concise way to describe the general solution.

Note that the formula $\theta = \dfrac{5\pi}{6} + \pi k$ produces the exact same solutions as the formula $\theta = \dfrac{11\pi}{6} + \pi k$. Therefore, only one of these equations is needed to describe the general solution. We will choose the first equation. Thus, the general solution to the equation $\sqrt{3}\tan\theta + 1 = 0$ is $\theta = \dfrac{5\pi}{6} + \pi k$, where k is any integer.

As we observed in step 3, the only angles that satisfy the equation $\sqrt{3}\tan\theta + 1 = 0$ on the interval $[0, 2\pi)$ are $\theta = \dfrac{5\pi}{6}$ and $\theta = \dfrac{11\pi}{6}$.

My interactive video summary

⊙ Work through this interactive video to see the video solution to this equation.

c. Step 1. For the equation $\sec\theta = -1$, the trigonometric function is already isolated. Therefore, we may skip to step 2.

Step 2. To determine the quadrant in which the terminal side of θ lies or the axis on which the terminal side of θ lies, we can use the reciprocal identity $\sec\theta = \dfrac{1}{\cos\theta}$ and then use our knowledge of the more familiar cosine function.

$\sec\theta = -1$	Write the original equation.
$\dfrac{1}{\cos\theta} = -1$	Use the reciprocal identity $\sec\theta = \dfrac{1}{\cos\theta}$.
$1 = -\cos\theta$	Multiply both sides by $\cos\theta$.
$-1 = \cos\theta$	Multiply both sides by -1.

The cosine function is equal to −1 for angles that lie along the negative x-axis. Therefore, the secant function is also equal to −1 for angles that lie along the negative x-axis. Thus, the terminal side of angle θ must lie along the negative x-axis. See Figure 14.

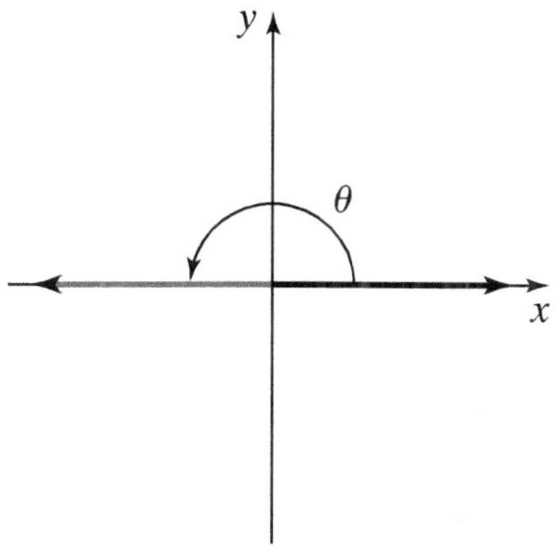

Figure 14 For the equation $\sec \theta = -1$, the terminal side of θ lies along the negative x-axis.

Step 3. The angle on the interval $[0, 2\pi)$ that lies along the negative x-axis is $\theta = \pi$.

Step 4. The secant (and cosine) function has a period of 2π. Therefore, the general solutions are of the form $\theta = \pi + 2\pi k$. The only solution to the equation $\sec \theta = -1$ on the interval $[0, 2\pi)$ is $\theta = \pi$.

My interactive video summary

Work through this interactive video to see the video solution to this equation.

You Try It Work through this You Try It problem.

Work Exercises 1–4 in this textbook or in the MyMathLab Study Plan.

In Example 1, the argument of the trigonometric function in each equation was θ. In Example 2, we look at equations in which the argument is something other than just θ.

My interactive video summary

Example 2 Solving Trigonometric Equations That Are Linear in Form

Determine a general formula (or formulas) for all solutions to each equation. Then, determine the specific solutions (if any) on the interval $[0, 2\pi)$.

a. $\sqrt{2} \cos 2\theta + 1 = 0$ **b.** $\sin \dfrac{\theta}{2} = -\dfrac{\sqrt{3}}{2}$ **c.** $\tan\left(\theta + \dfrac{\pi}{6}\right) + 1 = 0$

Solution

a. Step 1. To isolate the trigonometric function, we first subtract 1 from both sides then divide both sides by $\sqrt{2}$.

$$\sqrt{2} \cos 2\theta + 1 = 0 \qquad \text{Write the original equation.}$$
$$\sqrt{2} \cos 2\theta = -1 \qquad \text{Subtract 1 from both sides.}$$
$$\cos 2\theta = -\frac{1}{\sqrt{2}} \qquad \text{Divide both sides by } \sqrt{2}.$$

Step 2. Because $\cos 2\theta = -\dfrac{1}{\sqrt{2}} < 0$, we know that the terminal side of the argument, 2θ, must lie in Quadrant II or Quadrant III.

Step 3. All values of 2θ must have the reference angle of $\dfrac{\pi}{4}$ because

$$\cos\dfrac{\pi}{4} = \dfrac{1}{\sqrt{2}}. \text{ It follows that}$$

$$2\theta = \pi - \text{Reference Angle} = \pi - \dfrac{\pi}{4} = \dfrac{3\pi}{4} \text{ or}$$

$$2\theta = \text{Reference Angle} + \pi = \dfrac{\pi}{4} + \pi = \dfrac{5\pi}{4}. \text{ See Figure 15.}$$

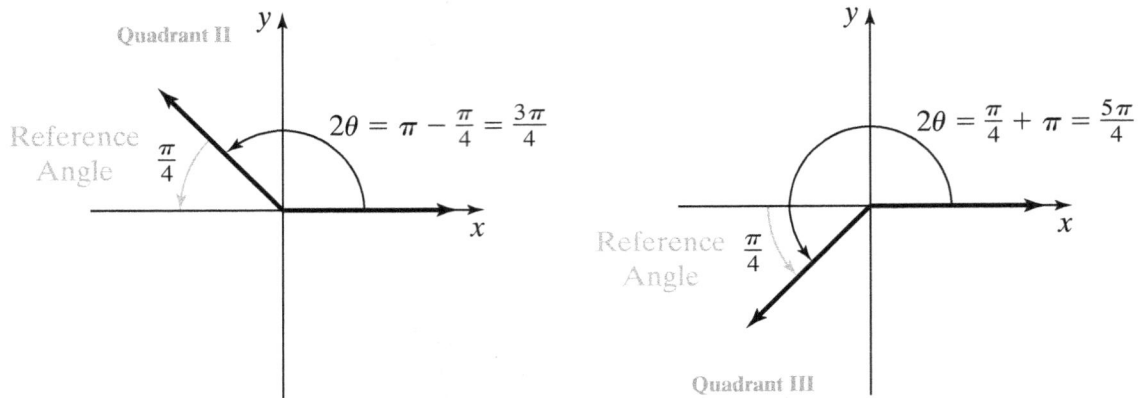

Figure 15 The reference angle is $\dfrac{\pi}{4}$. In Quadrant II, $2\theta = \dfrac{3\pi}{4}$. In Quadrant III, $2\theta = \dfrac{5\pi}{4}$.

Step 4. The cosine function has a period of 2π. Therefore, the solutions for 2θ are of the form

$$2\theta = \dfrac{3\pi}{4} + 2\pi k \quad \text{or} \quad 2\theta = \dfrac{5\pi}{4} + 2\pi k, \text{ where } k \text{ is any integer.}$$

To determine the general solutions, solve for θ by dividing both sides of each of these equations by 2.

$2\theta = \dfrac{3\pi}{4} + 2\pi k \quad \text{or} \quad 2\theta = \dfrac{5\pi}{4} + 2\pi k$ Write the two equations representing 2θ.

$\theta = \dfrac{3\pi}{8} + \pi k \quad \text{or} \quad \theta = \dfrac{5\pi}{8} + \pi k$ Divide both sides of each equation by 2.

These two equations produce uniquely different values. Thus, the formulas describing the general solutions are $\theta = \dfrac{3\pi}{8} + \pi k$ or $\theta = \dfrac{5\pi}{8} + \pi k$, where k is any integer.

In order to find the specific solutions on the interval $[0, 2\pi)$, we must carefully observe the value of the angles to be certain that we have all angles in the interval and that we do not list angles outside of that interval.

Using $k = 1$ for each formula above, we get $\theta = \dfrac{3\pi}{8} + \pi = \dfrac{11\pi}{8}$ or

$\theta = \dfrac{5\pi}{8} + \pi = \dfrac{13\pi}{8}$. Notice that both of these new solutions are within the interval $[0, 2\pi)$.

Using $k = 2$, we get $\theta = \dfrac{3\pi}{8} + 2\pi = \dfrac{19\pi}{8}$ or $\theta = \dfrac{5\pi}{8} + 2\pi = \dfrac{21\pi}{3}$ and we observe that these angles are outside of the interval $[0, 2\pi)$.

Thus, there are four specific solutions to the equation $\sqrt{2}\cos 2\theta - 1 = 0$ on the interval $[0, 2\pi)$ These solutions are $\theta = \dfrac{3\pi}{8}, \theta = \dfrac{5\pi}{8}, \theta = \dfrac{11\pi}{8}$ and $\theta = \dfrac{13\pi}{8}$.

My interactive video summary

⊙ Work through this interactive video to see the video solution to this equation.

b. Step 1. For the equation $\sin\dfrac{\theta}{2} = -\dfrac{\sqrt{3}}{2}$, the trigonometric function is already isolated. Therefore, we can skip to step 2.

Step 2. Because $\sin\dfrac{\theta}{2} = -\dfrac{\sqrt{3}}{2} < 0$, we know that the terminal side of $\dfrac{\theta}{2}$ must lie in Quadrant III or Quadrant IV.

Step 3. All values of $\dfrac{\theta}{2}$ must have the same reference angle of $\dfrac{\pi}{3}$ because $\sin\dfrac{\pi}{3} = \dfrac{\sqrt{3}}{2}$. It follows that

$$\frac{\theta}{2} = \text{Reference Angle} + \pi = \frac{\pi}{3} + \pi = \frac{4\pi}{3} \text{ or}$$

$$\frac{\theta}{2} = 2\pi - \text{Reference Angle} = 2\pi - \frac{\pi}{3} = \frac{5\pi}{3}. \text{ See Figure 16.}$$

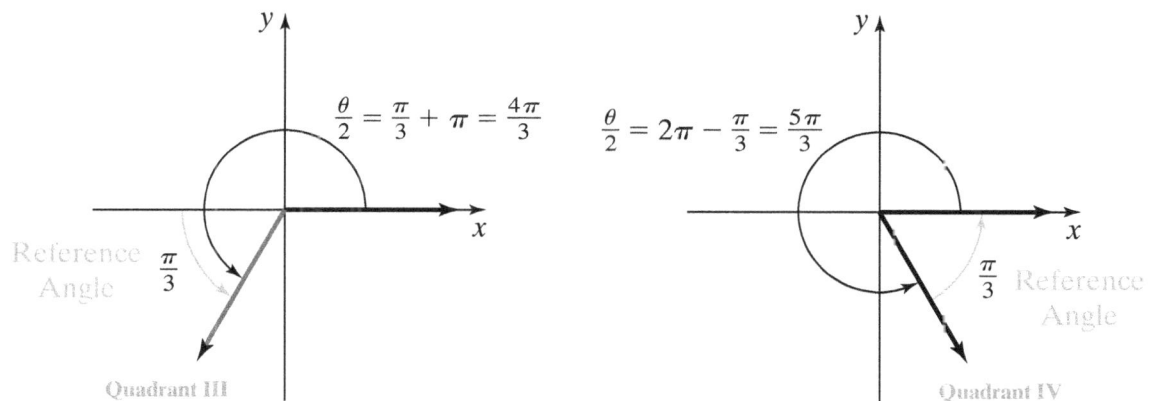

$$\frac{\theta}{2} = \frac{\pi}{3} + \pi = \frac{4\pi}{3} \qquad \frac{\theta}{2} = 2\pi - \frac{\pi}{3} = \frac{5\pi}{3}$$

Reference Angle $\dfrac{\pi}{3}$

Quadrant III

$\dfrac{\pi}{3}$ Reference Angle

Quadrant IV

Figure 16 The reference angle is $\dfrac{\pi}{3}$. In Quadrant III, $\dfrac{\theta}{2} = \dfrac{4\pi}{3}$. In Quadrant V, $\dfrac{\theta}{2} = \dfrac{5\pi}{3}$.

Step 4. The sine function has a period of 2π. Therefore, the solutions for $\dfrac{\theta}{2}$ are of the form $\dfrac{\theta}{2} = \dfrac{4\pi}{3} + 2\pi k$ or $\dfrac{\theta}{2} = \dfrac{5\pi}{3} + 2\pi k$, where k is any integer.

To determine the general solutions, solve each of these equations for θ by multiplying both sides of each equation by 2.

$$\frac{\theta}{2} = \frac{4\pi}{3} + 2\pi k \quad \text{or} \quad \frac{\theta}{2} = \frac{5\pi}{3} + 2\pi k \qquad \text{Write the two equations representing } \frac{\theta}{2}.$$

$$\theta = \frac{8\pi}{3} + 4\pi k \quad \text{or} \quad \theta = \frac{10\pi}{3} + 4\pi k \qquad \text{Multiply both sides of each equation by 2.}$$

These two equations produce uniquely different values. Thus, the general solutions are $\theta = \dfrac{8\pi}{3} + 4\pi k$ or $\theta = \dfrac{10\pi}{3} + 4\pi k$, where k is any integer.

My interactive video summary

▶ Note that for any integer value of k, the angles $\theta = \dfrac{8\pi}{3} + 4\pi k$ and $\theta = \dfrac{10\pi}{3} + 4\pi k$ lie outside of the interval $[0, 2\pi)$. Therefore, the equation $\sin\dfrac{\theta}{2} = -\dfrac{\sqrt{3}}{2}$ has **no solution** on the interval $[0, 2\pi)$. Work through this interactive video to see the video solution to this equation.

⚠ **It is possible for a trigonometric equation to have general solutions but not have a solution on the interval $[0, 2\pi)$ as in the previous example. Therefore, it is extremely important to check carefully to see if a solution exists on an indicated interval. Using a graphing utility, we can see why there is no solution to the equation $\sin\dfrac{\theta}{2} = -\dfrac{\sqrt{3}}{2}$ on the interval $[0, 2\pi)$. See Figure 17.**

c. Step 1. For the equation $\tan\left(\theta + \dfrac{\pi}{6}\right) + 1 = 0$, we isolate the trigonometric

Using Technology

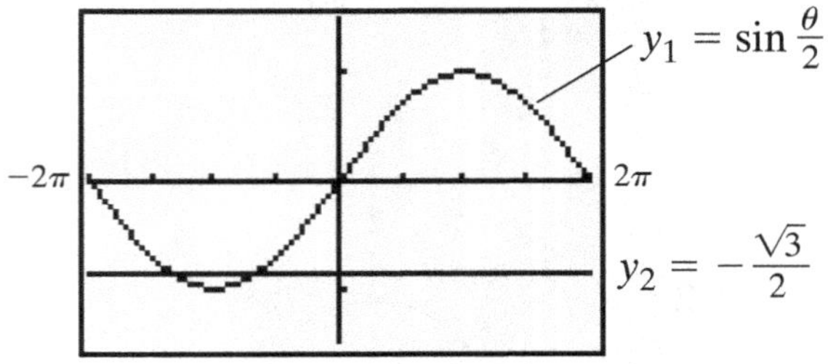

$y_1 = \sin\dfrac{\theta}{2}$

$y_2 = -\dfrac{\sqrt{3}}{2}$

Figure 17 The graphs of $y_1 = \sin\dfrac{\theta}{2}$ and $y_2 = -\dfrac{\sqrt{3}}{2}$ are shown here using a graphing utility. Notice that the two graphs do do not intersect on the interval $[0, 2\pi)$. Thus, the equation $\sin\dfrac{\theta}{2} = -\dfrac{\sqrt{3}}{2}$ has no solution on the restricted interval $[0, 2\pi)$.

function by subtracting 1 from both sides to obtain the equation $\tan\left(\theta + \dfrac{\pi}{6}\right) = -1$.

Step 2. Because $\tan\left(\theta + \dfrac{\pi}{6}\right) = -1 < 0$, we know that the terminal side of the argument, $\left(\theta + \dfrac{\pi}{6}\right)$, must lie in Quadrant II or Quadrant IV.

Step 3. All values of the angle $\left(\theta + \dfrac{\pi}{6}\right)$ must have a reference angle of $\dfrac{\pi}{4}$ because $\tan\dfrac{\pi}{4} = 1$. It follows that

$$\left(\theta + \frac{\pi}{6}\right) = \pi - \text{Reference Angle} = \pi - \frac{\pi}{4} = \frac{3\pi}{4} \text{ or}$$

$$\left(\theta + \frac{\pi}{6}\right) = 2\pi - \text{Reference Angle} = 2\pi - \frac{\pi}{4} = \frac{7\pi}{4}. \text{ See Figure 18.}$$

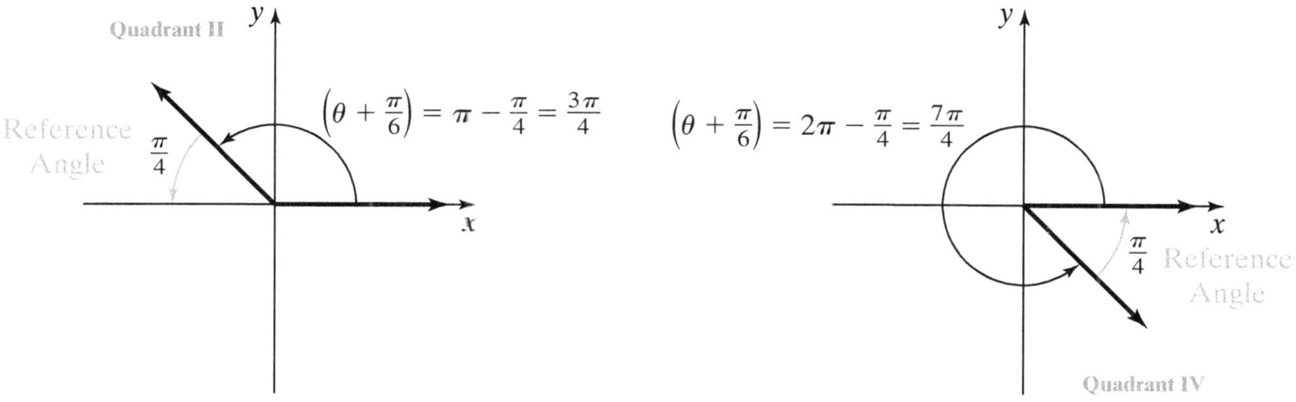

Figure 18 The reference angle is $\dfrac{\pi}{4}$. In Quadrant II, $\left(\theta + \dfrac{\pi}{6}\right) = \dfrac{3\pi}{4}$. In Quadrant IV, $\left(\theta + \dfrac{\pi}{6}\right) = \dfrac{7\pi}{4}$.

Step 4. The tangent function has a period of π. Therefore, the solutions for $\left(\theta + \dfrac{\pi}{6}\right)$ are of the form

$$\left(\theta + \frac{\pi}{6}\right) = \frac{3\pi}{4} + \pi k \quad \text{or} \quad \left(\theta + \frac{\pi}{6}\right) = \frac{7\pi}{4} + \pi k, \text{ where } k \text{ is any integer.}$$

Note that both equations produce exactly the same values and thus using both equations is not the most concise way to describe the general solution. Therefore, only one equation is needed. Thus, we will only use $\left(\theta + \dfrac{\pi}{6}\right) = \dfrac{3\pi}{4} + \pi k$. This situation also exists in **Example 1b** and often occurs when solving a trigonometric equation involving a tangent function because the period of the tangent function is π.

To determine the general solution, remove the parentheses and solve for θ.

$\theta + \dfrac{\pi}{6} = \dfrac{3\pi}{4} + \pi k$ Write the equation representing $\theta + \dfrac{\pi}{6}$.

$\theta = \dfrac{3\pi}{4} - \dfrac{\pi}{6} + \pi k$ Subtract $\dfrac{\pi}{6}$ from both sides.

$\theta = \dfrac{9\pi}{12} - \dfrac{2\pi}{12} + \pi k$ Write each fraction using a common denominator.

$\theta = \dfrac{7\pi}{12} + \pi k$ Simplify.

Thus, the general solution to the equation $\tan\left(\theta + \dfrac{\pi}{6}\right) + 1 = 0$ is $\theta = \dfrac{7\pi}{12} - \pi k$. Using the formula for the general solutions with $k = 0$ and $k = 1$, we get

$$\theta = \frac{7\pi}{12} + \pi(0) = \frac{7\pi}{12} \text{ and}$$

$$\theta = \frac{7\pi}{12} + \pi(1) = \frac{19\pi}{12}$$

Both of these angles lie in the interval $[0, 2\pi)$. Any other values of k will produce angles outside of the interval $[0, 2\pi)$. Therefore, the only solutions on the interval $[0, 2\pi)$ are $\theta = \dfrac{7\pi}{12}$ and $\theta = \dfrac{19\pi}{12}$.

*My interactive
video summary*

⊘ Work through this interactive video to see the video solution to this equation. ●

You Try It Work through this You Try It problem.

Work Exercises 5–12 in this textbook or in the MyMathLab Study Plan.

OBJECTIVE 2   SOLVING TRIGONOMETRIC EQUATIONS THAT ARE QUADRATIC IN FORM

Recall that a quadratic equation has the form $ax^2 + bx + c$, $a \neq 0$. These equations are relatively straightforward to solve because we know several methods for solving these types of equations. If $f(\theta)$ is a trigonometric function, then the equation $a(f(\theta))^2 + b(f(\theta)) + c$, $a \neq 0$ is said to be quadratic in form because we can transform it into a quadratic equation using the substitution $u = f(\theta)$. Example 3 illustrates two trigonometric equations that are quadratic in form.

Example 3 Solving a Trigonometric Equation That Is Quadratic in Form

Determine a general formula (or formulas) for all solutions to each equation. Then, determine the specific solutions (if any) on the interval $[0, 2\pi)$.

a. $\sin^2\theta - 4\sin\theta + 3 = 0$ **b.** $4\cos^2\theta - 3 = 0$

Solution

a. $\sin^2\theta - 4\sin\theta + 3 = 0$ Write the original equation.

   Let $u = \sin\theta$, then $u^2 = \sin^2\theta$.   Determine the proper substitution.

   $u^2 - 4u + 3 = 0$ Substitute u for $\sin\theta$ and u^2 for $\sin^2\theta$ into the original equation.

   $(u - 3)(u - 1) = 0$ Factor.

   $u - 3 = 0$ or $u - 1 = 0$ Use the zero product property.

   $u = 3$ or $u = 1$ Solve for u.

   $\sin\theta = 3$ or $\sin\theta = 1$ Substitute $\sin\theta$ for u.

*My interactive
video summary*

⊘ We now have two trigonometric functions that are linear in form similar to the equations from Example 1. Note that the range of $y = \sin\theta$ is $[-1, 1]$; therefore, it is impossible for the sine of an angle to equal 3. Thus, the equation $\sin\theta = 3$ has no solution. We can solve the equation $\sin\theta = 1$ using the steps for solving trigonometric equations that are linear in form discussed earlier. Try solving the equation $\sin\theta = 1$ on your own. When you're finished, check your answer or watch this interactive video to see each step of the solution process.

b. To solve the equation $4\cos^2\theta - 3 = 0$, we first make the substitution $u = \cos\theta$.

$$4\cos^2\theta - 3 = 0 \qquad \text{Write the original equation.}$$

$$\text{Let } u = \cos\theta, \text{ then } u^2 = \cos^2\theta. \qquad \text{Determine the proper substitution.}$$

$$4u^2 - 3 = 0 \qquad \text{Substitute } u^2 \text{ for } \cos^2\theta \text{ in the original equation.}$$

$$4u^2 = 3 \qquad \text{Add 3 to both sides.}$$

$$u^2 = \frac{3}{4} \qquad \text{Divide both sides by 4.}$$

$$u = \pm\sqrt{\frac{3}{4}} = \pm\frac{\sqrt{3}}{2} \qquad \text{Use the square root property.}$$

$$\cos\theta = \pm\frac{\sqrt{3}}{2} \qquad \text{Substitute } \cos\theta \text{ for } u.$$

We now see that $\cos\theta = \dfrac{\sqrt{3}}{2}$ or $\cos\theta = -\dfrac{\sqrt{3}}{2}$. To solve each equation, we follow the steps for solving trigonometric equations that are linear in form.

My interactive video summary

⊙ Follow these steps or work through this interactive video to verify the following general solutions:

General solutions to the equation $\cos\theta = \dfrac{\sqrt{3}}{2}$: $\qquad \theta = \dfrac{\pi}{6} + 2\pi k$ or $\theta = \dfrac{11\pi}{6} + 2\pi k$

General solutions to the equation $\cos\theta = -\dfrac{\sqrt{3}}{2}$: $\qquad \theta = \dfrac{5\pi}{6} + 2\pi k$ or $\theta = \dfrac{7\pi}{6} + 2\pi k$

All four of these equations could be used to describe the general solution to the original equation $4\cos^2\theta - 3 = 0$. However, this is not the most concise way to describe the general solution. Note that the two equations $\theta = \dfrac{\pi}{6} + \pi k$ and $\theta = \dfrac{5\pi}{6} + \pi k$ describe the exact same values as the four equations describe. Therefore, the most concise way to describe the general solutions to the equation $4\cos^2\theta - 3 = 0$ is $\theta = \dfrac{\pi}{6} + \pi k$ or $\theta = \dfrac{5\pi}{6} + \pi k$.

Use the general formulas $\theta = \dfrac{\pi}{6} + \pi k$ and $\theta = \dfrac{5\pi}{6} + \pi k$ using different integer values of k until we find all specific solutions to the equation $4\cos^2\theta - 3 = 0$ on the interval $[0, 2\pi)$. We start with $k = -1$ as sometimes a specific solution can be found using negative values of k. If a specific solution is found using $k = -1$, then try $k = -2$. Continue this process until the angles found are outside the interval $[0, 2\pi)$.

$k = -1$: $\quad \theta = \dfrac{\pi}{6} + \pi(-1) = \boxed{-\dfrac{5\pi}{6}}$ and $\theta = \dfrac{5\pi}{6} + \pi(-1) = \boxed{-\dfrac{\pi}{6}}$ (Neither angle lies within the interval $[0, 2\pi)$.)

There were no specific solutions found using the negative integer $k = -1$, so we now try $k = 0, k = 1$, etc.

$k = 0$: $\quad \theta = \dfrac{\pi}{6} + \pi(0) = \boxed{\dfrac{\pi}{6}}$ and $\theta = \dfrac{5\pi}{6} + \pi(0) = \boxed{\dfrac{5\pi}{6}}$ (Both angles lie within the interval $[0, 2\pi)$.)

$k = 1$: $\quad \theta = \dfrac{\pi}{6} + \pi(1) = \boxed{\dfrac{7\pi}{6}}$ and $\theta = \dfrac{5\pi}{6} + \pi(1) = \boxed{\dfrac{11\pi}{6}}$ (Both angles lie within the interval $[0, 2\pi)$.)

$k = 2$: $\quad \theta = \dfrac{\pi}{6} + \pi(2) = \boxed{\dfrac{13\pi}{6}}$ and $\theta = \dfrac{5\pi}{6} + \pi(2) = \boxed{\dfrac{17\pi}{6}}$ (Neither angle lies within the interval $[0, 2\pi)$.)

We see that when $k = 2$, the angles produced do not lie within the interval $[0, 2\pi)$. Thus, there are no more specific solutions on the interval $[0, 2\pi)$.

Therefore, $\theta = \dfrac{\pi}{6}, \theta = \dfrac{5\pi}{6}, \theta = \dfrac{7\pi}{6}$, and $\theta = \dfrac{11\pi}{6}$ are the four specific solutions to the equation $4\cos^2\theta - 3 = 0$ on the interval $[0, 2\pi)$.

My interactive video summary

⊙ Work through this interactive video to see the video solution to this equation. ●

You Try It Work through this You Try It problem.

Work Exercises 13–17 in this textbook or in the MyMathLab Study Plan.

OBJECTIVE 3 SOLVING TRIGONOMETRIC EQUATIONS USING IDENTITIES

We will often encounter trigonometric equations that cannot be solved directly. However, it might be possible to use a trigonometric identity to transform the equation into a more manageable equation. Then, we can use solving techniques already discussed. Before we look at such equations, take a moment to refresh your memory by clicking on the identity and formula links below.

Quotient Identities	Half-Angle Formulas
Reciprocal Identities	Double Angle Formulas
Pythagorean Identities	Product-to-Sum Formulas
Cofunction Identities	Sum-to-Product Formulas

My interactive video summary

⊙ **Example 4** Solving Trigonometric Equation Using Identities

Determine a general formula (or formulas) for all solutions to each equation. Then, determine the specific solutions (if any) on the interval $[0, 2\pi)$.

a. $2\sin^2\theta = 3\cos\theta + 3$ **b.** $\sin\theta\cos\theta = -\dfrac{1}{2}$

c. $\cos 2\theta + 4\sin^2\theta = 2$ **d.** $\sin 5\theta + \sin 3\theta = 0$

Solution

a. It appears that the equation $2\sin^2\theta = 3\cos\theta + 3$ might be quadratic in form. But, because there are two different trigonometric functions, there is no immediate substitution that will allow us to transform the equation into a quadratic equation. However, we can first use the Pythagorean identity of the form $\sin^2\theta = 1 - \cos^2\theta$ to produce an equation that contains all functions in terms of cosine.

$2\sin^2\theta = 3\cos\theta + 3$	Write the original equation.
$2(1 - \cos^2\theta) = 3\cos\theta + 3$	Use the substitution $\sin^2\theta = 1 - \cos^2\theta$.
$2 - 2\cos^2\theta = 3\cos\theta + 3$	Distribute.
$0 = 2\cos^2\theta + 3\cos\theta + 1$	Add $2\cos^2\theta$ to both sides and subtract 2 from both sides.

$$2 \cos^2 \theta + 3 \cos \theta + 1 = 0$$ Rewrite the equation so that 0 is on the right-hand side.

Let $u = \cos \theta$, then $u^2 = \cos^2 \theta$ Determine the proper substitution.

$$2u^2 + 3u + 1 = 0$$ Substitute u for $\cos \theta$ and u^2 for $\cos^2 \theta$.

$$(2u - 1)(u + 1) = 0$$ Factor.

$$2u + 1 = 0 \text{ or } u + 1 = 0$$ Use the zero product property.

$$u = -\frac{1}{2} \text{ or } u = -1$$ Solve for u.

$$\cos \theta = -\frac{1}{2} \text{ or } \cos \theta = -1$$ Substitute $\cos \theta$ for u.

We now see that $\cos \theta = -\frac{1}{2}$ or $\cos \theta = -1$. To solve each equation, follow the steps for solving trigonometric equations that are linear in form.

The general solutions to the equation $\cos \theta = -\frac{1}{2}$ are $\theta = \frac{2\pi}{3} + 2\pi k$ or $\theta = \frac{4\pi}{3} + 2\pi k$. The general solution to the equation $\cos \theta = -1$ is $\theta = \pi + 2\pi k$.

Therefore, the general solutions to the equation $2 \sin^2 \theta = 3 \cos \theta + 3$ are $\theta = \frac{2\pi}{3} + 2\pi k$ or $\theta = \frac{4\pi}{3} + 2\pi k$ or $\theta = \pi + 2\pi k$.

There are three specific solutions to the equation $2 \sin^2 \theta = 3 \cos \theta + 3$ on the interval $[0, 2\pi)$. These solutions are $\theta = \frac{2\pi}{3}, \theta = \pi$, and $\theta = \frac{4\pi}{3}$.

My interactive video summary

⊘ Work through this interactive video to see every step of the solution to this equation.

b. Recall that the double-angle formula for sine is given by $\sin 2\theta = 2 \sin \theta \cos \theta$. If we multiply both sides of the equation $\sin \theta \cos \theta = -\frac{1}{2}$ by 2, we will be able to use this double-angle formula to transform the equation into a trigonometric equation that is linear in form.

$$\sin \theta \cos \theta = -\frac{1}{2}$$ Write the original equation.

$$2 \sin \theta \cos \theta = -1$$ Multiply both sides by 2.

$$\sin 2\theta = -1$$ Substitute $\sin 2\theta$ for $2 \sin \theta \cos \theta$.

We now have transformed the original equation into a trigonometric equation that is linear in form. Follow the steps for solving trigonometric equations that are linear in form to solve this equation.

Step 1. For the equation $\sin 2\theta = -1$, the trigonometric function is already isolated. Therefore, we can skip to step 2.

Step 2. Because $\sin 2\theta = -1$, we know that the terminal side of the argument, 2θ, must lie along the negative y-axis.

Step 3. The angle on the interval $[0, 2\pi)$ that lies along the negative y-axis is $2\theta = \frac{3\pi}{2}$. See Figure 19.

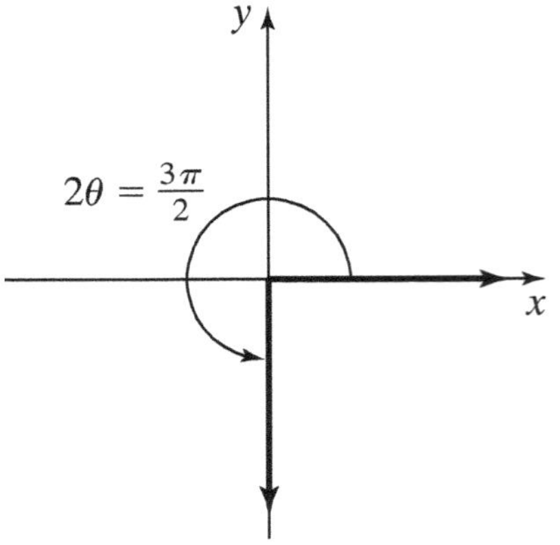

$2\theta = \dfrac{3\pi}{2}$

Figure 19 The sine function is equal to -1 for angles that lie on the negative y-axis. The angle on the interval $[0, 2\pi)$ that lies along the negative y-axis is $2\theta = \dfrac{3\pi}{2}$.

Step 4. The sine function has a period of 2π. Therefore, the solutions for 2θ are of the form $2\theta = \dfrac{3\pi}{2} + 2\pi k$, where k is any integer. We can divide both sides this equations by 2 to solve for θ.

$$2\theta = \dfrac{3\pi}{2} + 2\pi k \qquad \text{Write the equation representing } 2\theta.$$

$$\theta = \dfrac{3\pi}{4} + \pi k \qquad \text{Divide both sides by 2.}$$

Therefore, the general solution to the equation $\sin\theta \cos\theta = -\dfrac{1}{2}$ is

$$\theta = \dfrac{3\pi}{4} + \pi k, \text{ where } k \text{ is any integer.}$$

We can use the general solution $\theta = \dfrac{3\pi}{4} + \pi k$ using different values of k until we find all specific solutions on the interval $[0, 2\pi)$. We start with $k = -1$.

$k = -1$: $\theta = \dfrac{3\pi}{4} + \pi(-1) = \dfrac{3\pi}{4} - \pi = \dfrac{3\pi}{4} - \dfrac{4\pi}{4} = \left(-\dfrac{\pi}{4}\right)$ (This angle does **not** lie within the interval $[0, 2\pi)$.)

$k = 0$: $\theta = \dfrac{3\pi}{4} + \pi(0) = \left(\dfrac{3\pi}{4}\right)$ (This angle lies within the interval $[0, 2\pi)$.)

$k = 1$: $\theta = \dfrac{3\pi}{4} + \pi(1) = \dfrac{3\pi}{4} + \pi = \dfrac{3\pi}{4} + \dfrac{4\pi}{4} = \left(\dfrac{7\pi}{4}\right)$ (This angle lies within the interval $[0, 2\pi)$.)

$k = 2$: $\theta = \dfrac{3\pi}{4} + \pi(2) = \dfrac{3\pi}{4} + 2\pi = \dfrac{3\pi}{4} + \dfrac{8\pi}{4} = \left(\dfrac{11\pi}{4}\right)$ (This angle does **not** lie within the interval $[0, 2\pi)$.)

The specific solutions to the equation $\sin\theta \cos\theta = -\dfrac{1}{2}$ on the interval $[0, 2\pi)$ are $\theta = \dfrac{3\pi}{4}$ and $\theta = \dfrac{7\pi}{4}$.

My interactive video summary

▶ Work through this interactive video to every step of the solution to the equation $\sin\theta \cos\theta = -\dfrac{1}{2}$.

c. For the equation $\cos 2\theta + 4\sin^2\theta = 2$, use a **double-angle formula** for cosine to rewrite the function $\cos 2\theta$ in terms of only sine. This will transform the equation into a trigonometric equation that is quadratic in form. Try solving this equation on your own to see if you obtain the following solutions.

General solution: $\qquad\qquad\qquad\qquad\qquad\qquad \theta = \dfrac{\pi}{4} + \dfrac{\pi}{2}k$

Specific solutions on the interval $[0, 2\pi)$: $\qquad \theta = \dfrac{\pi}{4}, \theta = \dfrac{3\pi}{4}, \theta = \dfrac{5\pi}{4}$ and $\theta = \dfrac{7\pi}{4}$

My interactive video summary

⊚ Work through this interactive video to see every step of the solution to the equation $\cos 2\theta + 4 \sin^2 \theta = 2$.

d. For the equation $\sin 5\theta + \sin 3\theta = 0$, first use a sum-to-product formula. (See Section 3.4, Example 3a.) This will transform the equation into the product of two trigonometric functions that will allow us to use the zero product property.

$$\sin 5\theta + \sin 3\theta = 0 \qquad \text{Write the original equation.}$$

Use the sum-to-product formula

$$2 \sin\left(\frac{5\theta + 3\theta}{2}\right) \cos\left(\frac{5\theta - 3\theta}{2}\right) = 0 \qquad \sin \alpha + \sin \beta = 2 \sin\left(\frac{\alpha + \beta}{2}\right) \cos\left(\frac{\alpha - \beta}{2}\right).$$

$$2 \sin 4\theta \cos \theta = 0 \qquad \text{Simplify the arguments.}$$

$$\sin 4\theta \cos \theta = 0 \qquad \text{Divide both sides by 2.}$$

$$\sin 4\theta = 0 \text{ or } \cos \theta = 0 \qquad \text{Use the zero product property.}$$

My interactive video summary

⊚ Solve these two equations on your own. Check your work by watching this interactive video.

You Try It Work through this You Try It problem.

Work Exercises 18–27 in this textbook or in the MyMathLab Study Plan.

OBJECTIVE 4 SOLVING OTHER TYPES OF TRIGONOMETRIC EQUATIONS

In addition to the techniques for solving trigonometric equations discussed so far in this section, there are still other techniques that can be used. These include but are not limited to the use of factoring techniques and creative algebraic manipulations.

My interactive video summary

⊚ **Example 5** Solving Other Types of Trigonometric Equations

Determine a general formula (or formulas) for all solutions to each equation. Then, determine the specific solutions (if any) on the interval $[0, 2\pi)$.

a. $\sin 2\theta + 2 \cos \theta \sin 2\theta = 0$ $\qquad$ **b.** $\cos^2 \theta = \sin \theta \cos \theta$ $\qquad$ **c.** $\sin \theta + \cos \theta = 1$

Solution

a. Observe that the two terms of the equation $\sin 2\theta + 2 \cos \theta \sin 2\theta = 0$ on the left-hand side contain the common factor $\sin 2\theta$. We start by factoring out this common factor

$$\sin 2\theta + 2 \cos \theta \sin 2\theta = 0 \qquad \text{Write the original equation.}$$

$$\sin 2\theta(1 + 2\cos \theta) = 0 \qquad \text{Factor out the common factor of } \sin 2\theta.$$

$$\sin 2\theta = 0 \text{ or } 1 + 2 \cos \theta = 0 \qquad \text{Use the zero product property.}$$

We now solve the equations $\sin 2\theta = 0$ and $1 + 2\cos\theta = 0$ independently. Try solving these two equations on your own. When you have solved them, work through this interactive video to see if you are correct.

b. To solve the equation $\cos^2\theta = \sin\theta\cos\theta$, it is tempting to divide both sides of the equation by $\cos\theta$ and then cancel a factor of $\cos\theta$ on each side of the equation. However, if $\cos\theta = 0$, that would mean we divided by zero (which is undefined). This would result in the "loss" of any solutions that come from the equation $\cos\theta = 0$. Instead, follow the procedure below.

$$\cos^2\theta = \sin\theta\cos\theta \qquad \text{Write the original equation.}$$

$$\cos^2\theta - \sin\theta\cos\theta = 0 \qquad \text{Subtact } \sin\theta\cos\theta \text{ from both sides.}$$

$$\cos\theta\,(\cos\theta - \sin\theta) = 0 \qquad \text{Factor out the common factor of } \cos\theta.$$

$$\cos\theta = 0 \text{ or } \cos\theta - \sin\theta = 0 \qquad \text{Use the zero product property.}$$

We now solve the equations $\cos\theta = 0$ and $\cos\theta - \sin\theta = 0$ independently.

To solve the equation $\cos\theta = 0$, follow the steps for solving trigonometric equations that are linear in form. Read these steps to verify that the general solutions to the equation $\cos\theta = 0$ are of the form $\theta = \dfrac{\pi}{2} + \pi k$, where k is any integer.

To solve the equation $\cos\theta - \sin\theta = 0$, start by adding $\sin\theta$ to both sides of the equation.

$$\cos\theta - \sin\theta = 0 \qquad \text{Write the equation.}$$

$$\cos\theta = \sin\theta \qquad \text{Add } \sin\theta \text{ to both sides.}$$

$$\sin\theta = \cos\theta \qquad \text{Rearrange the order of the equation.}$$

Note that we are now able to divide both sides of the equation $\sin\theta = \cos\theta$ by $\cos\theta$ without dividing by zero and without losing any solutions. To see why this is true, read this explanation.

$$\sin\theta = \cos\theta \qquad \text{Write the equation.}$$

$$\frac{\sin\theta}{\cos\theta} = 1 \qquad \text{Divide both sides by } \cos\theta.$$

$$\tan\theta = 1 \qquad \text{Substitute } \frac{\sin\theta}{\cos\theta} = \tan\theta.$$

My interactive video summary

▶ Work through this interactive video to verify that the general solution to the equation $\tan\theta = 1$ is $\theta = \dfrac{\pi}{4} + \pi k$, where k is any integer.

Thus, the general solutions to the equation $\cos^2\theta = \sin\theta\cos\theta$ are the two formulas $\theta = \dfrac{\pi}{2} + \pi k$ or $\theta = \dfrac{\pi}{4} + \pi k$, where k is any integer.

The solutions to the equation $\cos^2\theta = \sin\theta\cos\theta$ on the interval $[0, 2\pi)$ are $\theta = \dfrac{\pi}{4}, \theta = \dfrac{\pi}{2}, \theta = \dfrac{5\pi}{4},$ and $\theta = \dfrac{3\pi}{2}.$

(!) **In the previous example it was not possible to divide both sides of the original equation $\cos^2 \theta = \sin \theta \cos \theta$ by $\cos \theta$ because for this equation, the expression $\cos \theta$ could equal 0 (and division by zero is never allowed.) However, we saw that it was possible to divide both sides of the equation $\sin \theta = \cos \theta$ by $\cos \theta$ because we verified in this equation that $\cos \theta$ could not equal 0. Therefore, be extremely careful when attempting to multiply or divide both sides of a trigonometric equation by a function. In order to perform this operation, you must first verify that the function cannot equal 0 for the given equation.**

c. At first glance it appears that the equation $\sin \theta + \cos \theta = 1$ is a Pythagorean identity. However, this is not the case because neither trigonometric function is squared. If we square both sides of this equation, it may be possible to use a Pythagorean identity.

(!) **Great care must be taken when introducing the squaring operation because doing so may produce** extraneous solutions.

$\sin \theta + \cos \theta = 1$	Write the original equation.
$(\sin \theta + \cos \theta)^2 = (1)^2$	Square both sides.
$\sin^2 \theta + 2 \sin \theta \cos \theta + \cos^2 \theta = 1$	Perform the squaring operations.
$\sin^2 \theta + \cos^2 \theta + 2 \sin \theta \cos \theta = 1$	Rearrange the terms.
$1 + 2 \sin \theta \cos \theta = 1$	Substitute $\sin^2 \theta + \cos^2 \theta = 1$.
$2 \sin \theta \cos \theta = 0$	Subtract 1 from both sides.
$\sin 2\theta = 0$	Substitute $\sin 2\theta$ for $2 \sin \theta \cos \theta$. (This is a double-angle formula.)

To solve the equation $\sin 2\theta = 0$, use the steps for solving trigonometric equations that are linear in form. Using these steps, we find that the general solution to the equation $\sin 2\theta = 0$ is $\theta = \dfrac{k\pi}{2}$, where k is *any* integer. It is possible that $\theta = \dfrac{k\pi}{2}$ is not a solution to the original equation for all integer values of k because we introduced the squaring operation during the solution process. Thus, we must check all possibilities of k for $\theta = \dfrac{k\pi}{2}$ to determine the general solution to the original equation $\sin \theta + \cos \theta = 1$.

k	$\sin\left(\dfrac{k\pi}{2}\right) + \cos\left(\dfrac{k\pi}{2}\right) = 1$	
$k = 0$:	$\sin(0) + \cos(0) = 0 + 1 = 1$	$\checkmark$
$k = 1$:	$\sin\left(\dfrac{\pi}{2}\right) + \cos\left(\dfrac{\pi}{2}\right) = 1 + 0 = 1$	$\checkmark$
$k = 2$:	$\sin(\pi) + \cos(\pi) = 0 - 1 = -1$	$\times$
$k = 3$:	$\sin\left(\dfrac{3\pi}{2}\right) + \cos\left(\dfrac{3\pi}{2}\right) = -1 + 0 = -1$	$\times$

We do not have to check any more values of k because the period of sine and cosine is 2π. In other words, $k = 4$ will produce the same result as $k = 0$. Similarly, $k = 5$ will produce the same result as $k = 1$, and so on.

We can see that $\theta = 0$ and $\theta = \dfrac{\pi}{2}$ are solutions to the equation $\sin \theta + \cos \theta = 1$.

The period of sine and cosine is 2π. Therefore, integer multiples of 2π added to each of these values must also be solutions. Thus, the general solutions to the equation $\sin \theta + \cos \theta = 1$ are of the form

$$\theta = 2\pi k \quad \text{or} \quad \theta = \frac{\pi}{2} + 2\pi k.$$

The only two solutions to the equation $\sin \theta + \cos \theta = 1$ on the interval $[0, 2\pi)$ are $\theta = 0$ and $\theta = \frac{\pi}{2}$, as illustrated by the points of intersection of the graphs in Figure 20.

My interactive video summary

⊙ Watch this interactive video to see each step of this solution and to check your work.

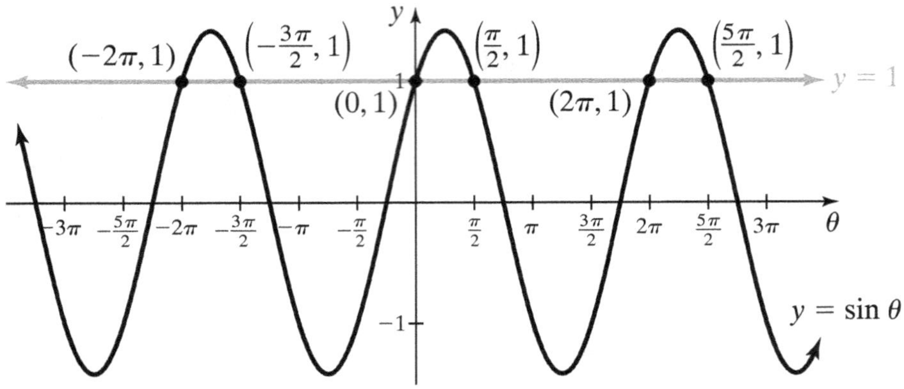

Figure 20 The solutions to the equation $\sin \theta + \cos \theta = 1$ are the first coordinates of the points of intersection of the graphs $y = \sin \theta + \cos \theta$ and $y = 1$.

The two graphs intersect at all values of $\theta = 2\pi k$ and $\theta = \frac{\pi}{2} + 2\pi k$

where k is an integer.

You Try It Work through this You Try It problem.

Work Exercises 28–32 in this textbook or in the MyMathLab Study Plan.

OBJECTIVE 5 SOLVING TRIGONOMETRIC EQUATIONS USING A CALCULATOR

In the previous examples, we were able to find an exact solution to each trigonometric equation. This is not always the case. When an exact solution cannot be obtained, it may be necessary to approximate the solution using a calculator. Before trying the following example, you should review inverse trigonometric functions that were discussed in Section 2.4.

Example 6 Solving Trigonometric Equations Using a Calculator

Approximate all solutions to the trigonometric equation $\cos \theta = -0.4$ on the interval $[0, 2\pi)$. Round your answer to four decimal places.

Solution The equation $\cos\theta = -0.4$ is a trigonometric equation that is linear in form. Therefore, we can follow the steps for solving trigonometric equations that are linear in form to solve this equation.

Step 1. For the equation $\cos\theta = -0.4$, the trigonometric function is already isolated. Therefore, we may skip to step 2.

Step 2. Because $\cos\theta = -0.4 < 0$, we know that the terminal side of θ must lie in Quadrant II or Quadrant III.

Step 3. All values of θ must have the same reference angle. The reference angle is the positive acute angle whose cosine is equal to 0.4. Thus, the reference angle is $\cos^{-1}(0.4)$. It follows that

$$\theta = \pi - \text{Reference Angle} = \pi - \cos^{-1}(0.4) \approx 1.9823 \text{ in Quadrant II and}$$

$$\theta = \pi + \text{Reference Angle} = \pi + \cos^{-1}(0.4) \approx 4.3009 \text{ in Quadrant III. See Figure 21.}$$

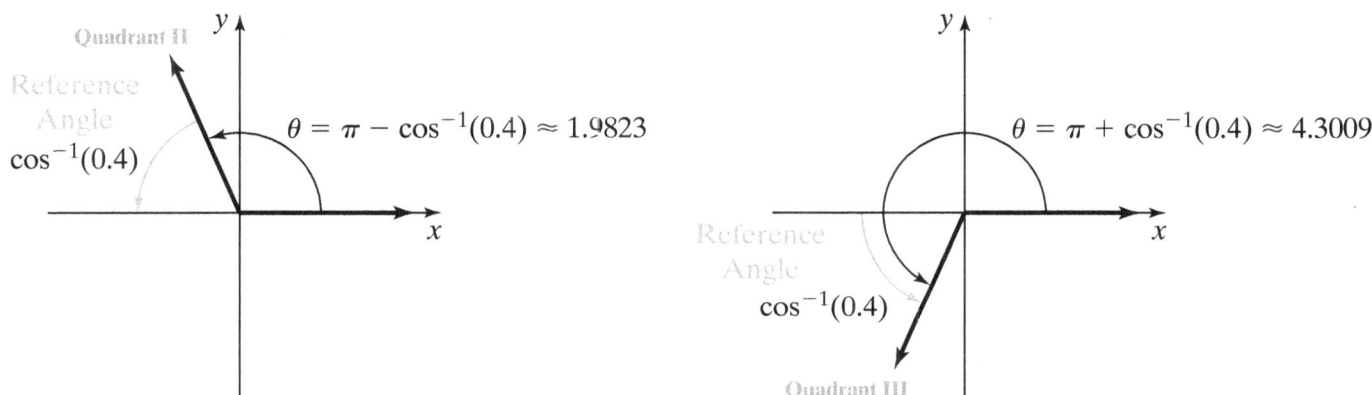

Figure 21 The reference angle is $\cos^{-1}(0.4)$. In Quadrant II, $\theta = \pi - \cos^{-1}(0.4) \approx 1.9823$. In Quadrant III, $\theta = \pi + \cos^{-1}(0.4) \approx 4.3009$.

Step 4. The period of the cosine function is 2π. Adding or subtracting any integer multiple of 2π to $\theta \approx 1.9823$ or $\theta \approx 4.3009$ will yield an angle that is outside of the interval $[0, 2\pi)$. Therefore, the only two solutions to the equation $\cos\theta = -0.4$ on the interval $[0, 2\pi)$ are $\theta \approx 1.9823$ and $\theta \approx 4.3009$.

Using the Intersect feature of a graphing utility, we can verify that our answers are correct. See Figure 22.

Using Technology

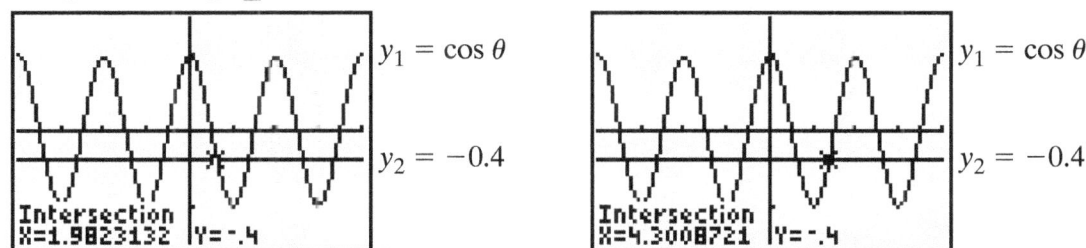

Figure 22 Using the Intersect feature of a graphing utility with $y_1 = \cos\theta$ and $y_2 = -0.4$ we see that the graphs intersect at $\theta \approx 1.9823$ and $\theta \approx 4.3009$.

⚠ **You cannot simply use an inverse trigonometric function key on your calculator to solve this trigonometric equation. Most calculators are programmed to restrict the interval of the solution of an inverse trigonometric function to the same intervals we used in Section 2.4 when we first defined** $y = \sin^{-1} x, y = \cos^{-1} x,$ **and** $y = \tan^{-1} x.$ **Therefore, using the inverse function key will never give you more than one solution, and sometimes even the one solution it does give you will not be a correct solution to the given equation.**

You Try It   Work through this You Try It problem.

Work Exercises 33–38 in this textbook or in the MyMathLab Study Plan.

3.5 Exercises

Skill Check Exercises

SCE-1. Simplify the expression $\dfrac{\pi}{4} + \dfrac{2\pi}{3}k$ for $-3, -2, -1, 0, 1,$ and 2

In Exercises SCE-2 through SCE-4, solve each equation for u.

SCE-2. $2u^2 + 3u + 1 = 0$ **SCE-3.** $u + u^2 = 0$ **SCE-4.** $2(1 - u^2) - 3u - 3 = 0$

In Exercises 1–32, determine a general formula (or formulas) for the solution to each equation. Then, determine the specific solutions (if any) on the interval $[0, 2\pi)$.

SbS **1.** $\cos \theta = \dfrac{1}{2}$ SbS **2.** $\tan \theta + \sqrt{3} = 0$ SbS **3.** $\cot \theta = 0$

SbS **4.** $5 \csc \theta - 3 = 2$ SbS **5.** $\sin 2\theta = -1$ SbS **6.** $\sqrt{2} \sin 2\theta + 1 = 0$

SbS **7.** $2 \cos 3\theta + \sqrt{2} = 0$ SbS **8.** $\tan \left(\dfrac{\theta}{2} \right) = -1$ SbS **9.** $\sin \dfrac{\theta}{2} = -\dfrac{1}{\sqrt{2}}$

SbS **10.** $\cos \dfrac{\theta}{2} = \dfrac{1}{2}$ SbS **11.** $\cos \left(\theta - \dfrac{\pi}{3} \right) - 1 = 0$ SbS **12.** $5 \sec \left(\dfrac{3\theta}{2} \right) + 10 = 0$

SbS **13.** $\tan^2 \theta = \dfrac{1}{3}$ SbS **14.** $2 \sin^2 \theta - \sin \theta - 1 = 0$ SbS **15.** $2 \cos^2 \theta - 9 \cos \theta - 5 = 0$

SbS **16.** $4 \cos^4 \theta - 7 \cos^2 \theta + 3 = 0$ SbS **17.** $2 \sin^2 \left(\dfrac{\theta}{3} - \dfrac{\pi}{4} \right) - 1 = 0$ SbS **18.** $\cos^2 \theta - \sin^2 \theta = 1 - \sin \theta$

SbS **19.** $3 \cos \theta + 3 = 2 \sin^2 \theta$ SbS **20.** $\sec^2 \theta - \tan \theta = 1$ SbS **21.** $\cos 2\theta + 3 = 5 \cos \theta$

SbS **22.** $\cos 2\theta + 10 \sin^2 \theta = 7$ SbS **23.** $\sin 2\theta \sin \theta = \cos \theta$ SbS **24.** $\tan \theta = 2 \sin \theta$

SbS **25.** $4 \cos \theta = 3 \sec \theta$ SbS **26.** $\sin (4\theta) - \sin (2\theta) = 0$ SbS **27.** $\cos (2\theta) - \cos (6\theta) = 0$

SbS **28.** $2 \cos^2 \theta + \cos \theta = 0$ SbS **29.** $3 \sin \theta = -3 \cos \theta$ SbS **30.** $\sin \theta + \cos \theta = -\sqrt{2}$

SbS **31.** $\sin \theta \cos \theta = -\dfrac{1}{2}$ SbS **32.** $\sin^2 \theta + \sqrt{3} \sin \theta \cos \theta = 0$

In Exercises 33–38, approximate the solutions to each equation on the interval $[0, 2\pi)$. Round your answer to four decimal places.

33. $\sin \theta = 0.72$

34. $\cos \theta = 0.63$

35. $\tan \theta = 0.5$

36. $\cos \theta = -0.8$

37. $\sin \theta = -0.2$

38. $\tan \theta = -4$

Brief Exercises

In Exercises 39–70, determine the specific solutions (if any) to each equation on the interval $[0, 2\pi)$.

39. $\cos \theta = \dfrac{1}{2}$

40. $\tan \theta + \sqrt{3} = 0$

41. $\cot \theta = 0$

42. $5 \csc \theta - 3 = 2$

43. $\sin 2\theta = -1$

44. $\sqrt{2} \sin 2\theta + 1 = 0$

45. $2 \cos 3\theta + \sqrt{2} = 0$

46. $\tan \left(\dfrac{\theta}{2} \right) = -1$

47. $\sin \dfrac{\theta}{2} = -\dfrac{1}{\sqrt{2}}$

48. $\cos \dfrac{\theta}{2} = \dfrac{1}{2}$

49. $\cos \left(\theta - \dfrac{\pi}{3} \right) - 1 = 0$

50. $5 \sec \left(\dfrac{3\theta}{2} \right) + 10 = 0$

51. $\tan^2 \theta = \dfrac{1}{3}$

52. $2 \sin^2 \theta - \sin \theta - 1 = 0$

53. $2 \cos^2 \theta - 9 \cos \theta - 5 = 0$

54. $4 \cos^4 \theta - 7 \cos^2 \theta + 3 = 0$

55. $2 \sin^2 \left(\dfrac{\theta}{3} - \dfrac{\pi}{4} \right) - 1 = 0$

56. $\cos^2 \theta - \sin^2 \theta = 1 - \sin \theta$

57. $3 \cos \theta + 3 = 2 \sin^2 \theta$

58. $\sec^2 \theta - \tan \theta = 1$

59. $\cos 2\theta + 3 = 5 \cos \theta$

60. $\cos 2\theta + 10 \sin^2 \theta = 7$

61. $\sin 2\theta \sin \theta = \cos \theta$

62. $\tan \theta = 2 \sin \theta$

63. $4 \cos \theta = 3 \sec \theta$

64. $\sin (4\theta) - \sin (2\theta) = 0$

65. $\cos (2\theta) - \cos (6\theta) = 0$

66. $2 \cos^2 \theta + \cos \theta = 0$

67. $3 \sin \theta = -3 \cos \theta$

68. $\sin \theta + \cos \theta = -\sqrt{2}$

69. $\sin \theta \cos \theta = -\dfrac{1}{2}$

70. $\sin^2 \theta + \sqrt{3} \sin \theta \cos \theta = 0$

CHAPTER FOUR

Applications of Trigonometry

CHAPTER FOUR CONTENTS

4.1 Right Triangle Applications

THINGS TO KNOW

Before working through this section, be sure that you are familiar with the following concepts:

VIDEO   ANIMATION   INTERACTIVE

You Try It
1. Using the Pythagorean Theorem (Section 1.3)

You Try It
2. Evaluating Trigonometric Functions Using a Calculator (Section 1.4)

OBJECTIVES

1 Solving Right Triangles

2 Applications Using Right Triangles

OBJECTIVE 1   SOLVING RIGHT TRIANGLES

Right triangles abound in our everyday world, and so it is appropriate at this point to look at just a few situations where the right triangle definitions of trigonometric functions can simplify our lives. We start with a concept called "solving right triangles." The goal of solving a right triangle is to determine the measure of all angles and the length of all sides, given certain information. In Section 4.2 and Section 4.3, we will solve triangles that do not have a right angle.

Example 1 Solving a Right Triangle

Solve the given right triangle. Round the length of side a and the measure of the two acute angles to two decimal places.

Solution Observe that we are given the following information.

Angles	Sides
$A =$ _____	$a =$ _____
$B =$ _____	$b = 11.4$ cm
$C = 90°$	$c = 12.1$ cm

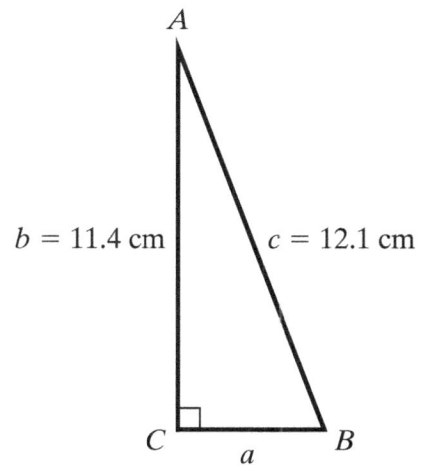

We can determine the length of side a using the Pythagorean Theorem.

$a^2 + b^2 = c^2$	Write the Pythagorean Theorem.
$a^2 + (11.4)^2 = (12.1)^2$	Substitute $b = 11.4$ and $c = 12.1$.
$a^2 + 129.96 = 146.41$	Square.
$a^2 = 16.45$	Subtract.
$a = \pm \sqrt{16.45} \approx \pm 4.06$	Solve for a.
$a \approx 4.06$	Use the positive square root since a represents the length of a side of a triangle.

We insert the new information.

Angles	Sides
$A =$ _____	$a \approx 4.06$ cm
$B =$ _____	$b = 11.4$ cm
$C = 90°$	$c = 12.1$ cm

Now that we know the lengths of all sides, we can determine the measure of the two acute angles.

It does not matter which angle to determine first. We will choose to first determine the measure of angle B. In Figure 1 we see that the length of the side opposite of angle B is 11.4 cm. The length of the hypotenuse is 12.1 cm. Therefore, we can use the trigonometric ratio, $\sin B = \dfrac{11.4}{12.1}$.

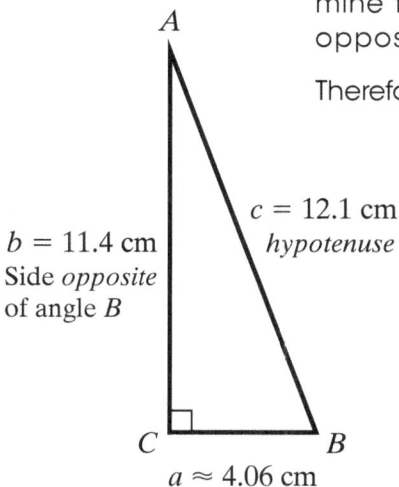

Figure 1 The length of the side opposite of B is 11.4 cm and the length of the hypotenuse is 12.1 cm. Thus, $\sin B = \dfrac{11.4}{12.1}$.

$\sin B = \dfrac{11.4}{12.1}$ The sine of angle B is equal to the ratio of the length of the side opposite angle B to the length of the hypotenuse.

$B = \sin^{-1}\left(\dfrac{11.4}{12.1}\right)$ Use the inverse sine function to solve for B.

$\approx 70.42°$ Use a calculator (in degree mode) to evaluate $\sin^{-1}\left(\dfrac{11.4}{12.1}\right)$.

Again, we insert the new information.

Angles	Sides
$A =$ _____	$a \approx$ 4.06 cm
$B \approx$ 70.42°	$b =$ 11.4 cm
$C = 90°$	$c =$ 12.1 cm

Since angle A and angle B are complementary angles, we can easily determine the measure of angle A.

$A \approx 90° - 70.42° \approx 19.58°$ Use the fact that angle A and B are complementary angles.

We have now solved the right triangle.

Angles	Sides
$A \approx$ 19.58°	$a \approx$ 4.06 cm
$B \approx$ 70.42°	$b =$ 11.4 cm
$C = 90°$	$c =$ 12.1 cm

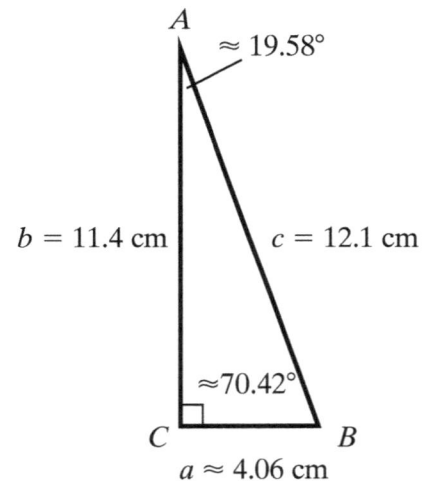

My video summary   ⊚ **Example 2** Solving a Right Triangle

Suppose that the length of the hypotenuse of a right triangle is 11 inches. If one of the acute angles is 37.34°, find the length of the two legs and the measure of the other acute angle. Round all values to two decimal places.

Solution First, draw and label a right triangle using the given information. Let $A = 37.34°$ and let $c = 11$ in.

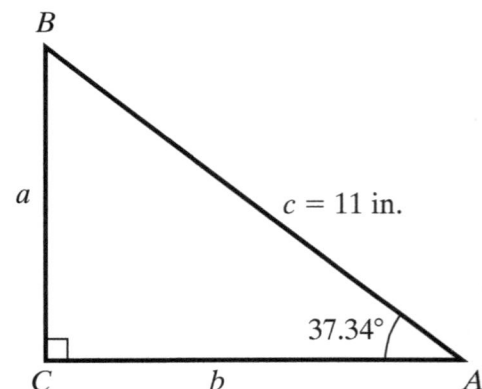

Since angle A and angle B are complementary angles, we can easily determine the measure of angle B.

$$B = 90° - 37.34° = 52.66°$$ 　　Use the fact that angle A and B are complementary angles.

Insert the information.

Angles	Sides
$A = 37.34°$	$a = $ _____
$B = \underline{52.66°}$	$b = $ _____
$C = 90°$	$c = 11$ in

Side a is opposite of angle A so we can use the sine function because we are given the length of the hypotenuse.

$$\sin 37.34° = \frac{a}{11}$$ 　　By definition, $\sin \theta = \frac{opp}{hyp}$.

$$11\sin 37.34° = a$$ 　　Multiply both sides by 11.

$$6.67 \approx a$$ 　　Use a calculator (in degree mode) to evaluate $11\sin 37.34°$.

Once again, insert the information.

Angles	Sides
$A = 37.34°$	$a \approx \underline{6.67}$ in
$B = \underline{52.66°}$	$b = $ _____
$C = 90°$	$c = 11$ in

We can determine the length of the side b in a variety of ways. Below are three different ways that can be used to solve for b.

$$(6.67)^2 + b^2 = (11)^2$$ 　　Use the Pythagorean Theorem with $a \approx 6.67$ and $c = 11$.

$$\cos 37.34° = \frac{b}{11}$$ 　　Use the cosine function, $\cos \theta = \frac{adj}{hyp}$.

$$\sin 52.66° = \frac{b}{11}$$ 　　Use the sine function, $\sin \theta = \frac{opp}{hyp}$.

You should try solving for b n each of the three equations above to verify that $b \approx 8.75$ inches. We have now solved the right triangle.

Angles	Sides
$A = 37.34°$	$a \approx \underline{6.67}$ in
$B = \underline{52.66°}$	$b \approx \underline{8.75}$ in
$C = 90°$	$c = 11$ in

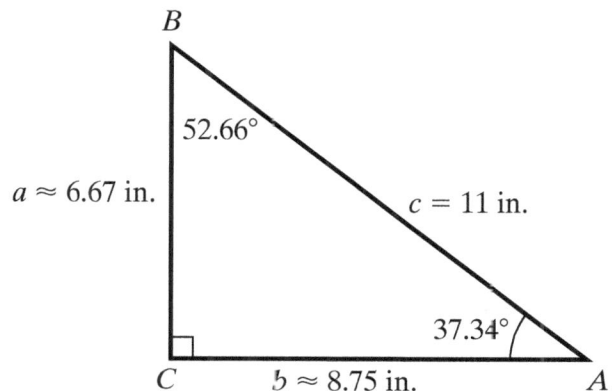

In order to solve a right triangle, we must be given at least two pieces of information. In Example 1 we were given the length of one side and the measure of an acute angle. In Example 2, we were given the length of two sides but no angles. The only two pieces of information that will not help to solve a right triangle are the measures of two angles. In this case we can easily determine the measure of the third angle but it is impossible to determine the length of the three sides. This is because there are infinitely many similar right triangles with the same three angle measures.

You Try It Work through this You Try It problem.

Work Exercises 1–7 in this textbook or in the MyMathLab **Study Plan**.

OBJECTIVE 2 APPLICATIONS USING RIGHT TRIANGLES

Applications involving right triangles often involve angles created by an observer looking up or looking down at an object. The angle between the horizontal and the line of sight of an object *above* the horizontal is called the **angle of eleva-tion**. The angle between the horizontal and the line of sight of an object below the horizontal is called the **angle of depression**. See Figure 2.

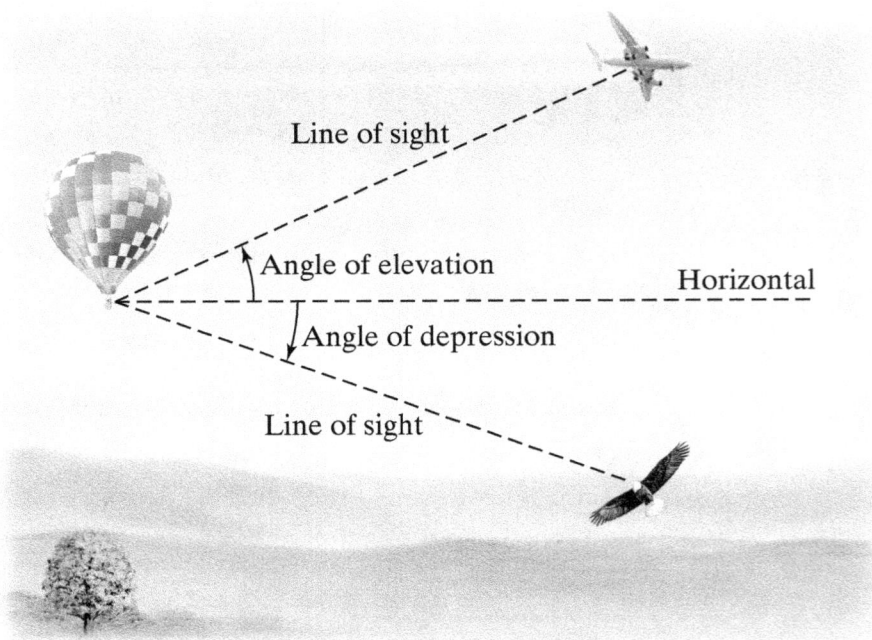

Figure 2

My video summary

⊙ **Example 3** Determining the Height of a Flagpole

The angle of elevation to the top of a flagpole measured by a digital protractor is 20° from a point on the ground 90 feet away from its base. Find the height of the flagpole. Round to two decimal places.

Solution Let h represent the height of the flagpole, as in Figure 3. Then choose the correct trigonometric function that relates the given information. Determine the value of h on your own, and then watch this video to see if you are correct.

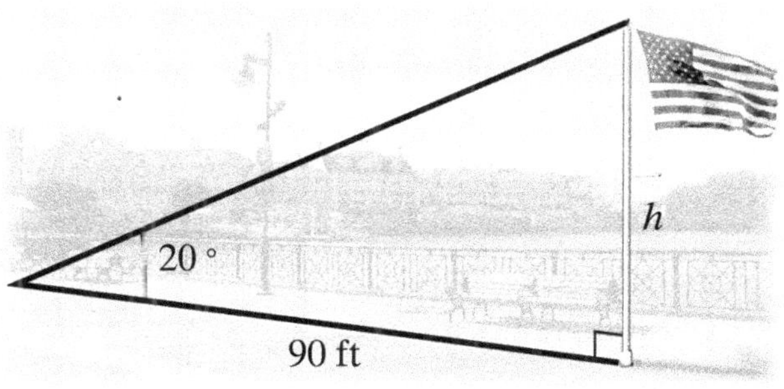

Figure 3

My video summary

⊚ **Example 4** Determining the Altitude of an Airplane

At the same instant, two observers 5 miles apart are looking at the same airplane. (See Figure 4.) The angle of elevation (from the ground) of the observer closest to the plane is 71°. The angle of elevation (from the ground) of the person furthest from the plane is 39°. Find the altitude of the plane (to the nearest foot) at the instant the two people observe the plane.

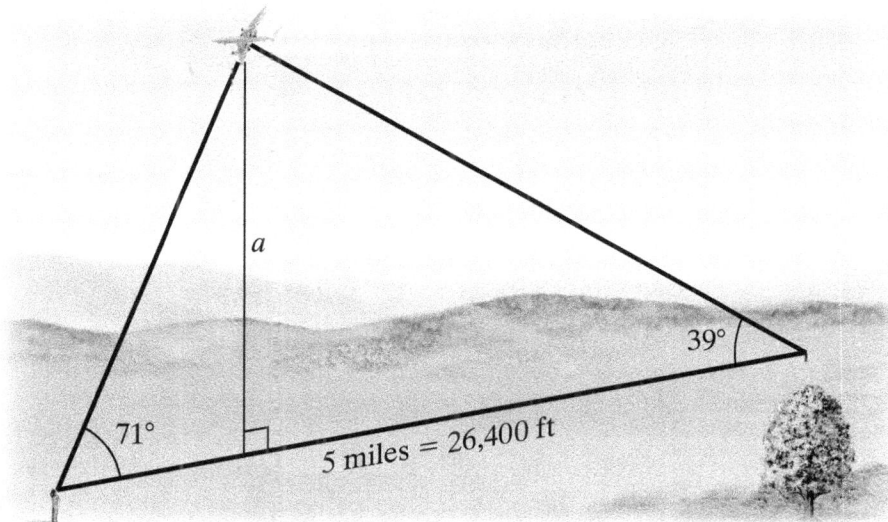

Figure 4

Solution Let a represent the altitude of the plane in feet. The 5 miles must be converted to feet to be consistent with units. There are 5280 feet in one mile. Therefore, the observers are $5 \text{ mi} \cdot 5280 \dfrac{\text{ft}}{\text{mi}} = 26{,}400$ ft apart. Let x represent the distance from the observer nearest to the plane to the base of the altitude. Then

$26,400 - x$ is the distance from the observer furthest from the plane to the base of the altitude. We can now draw and label the two right triangles shown below.

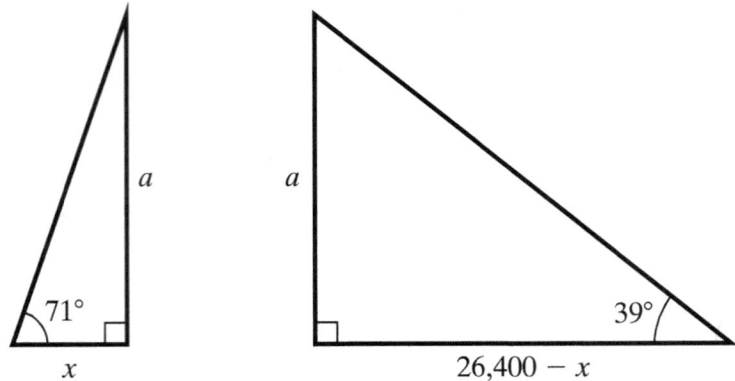

From the two right triangles, we can use the tangent function to write the following two equations:

$$\tan 71° = \frac{a}{x} \quad \tan 39° = \frac{a}{26,400 - x}$$

Multiply both sides of each equation by their respective denominators to solve each equation for a.

$$a = x \tan 71° \quad a = (26,400 - x) \tan 39°$$

The expressions $x \tan 71°$ and $(26,400 - x) \tan 39°$ both represent the altitude of the plane. Thus they are equal to each other.

$$x \tan 71° = (26,400 - x) \tan 39°$$

My video summary

⊘ Solving this equation for x, we get $x = \dfrac{26,400 \tan 39°}{\tan 71° + \tan 39°}$. Read this overview to see how to solve for x or watch this video to see the entire solution. We can solve for a by substituting $x = \dfrac{26,400 \tan 39°}{\tan 71° + \tan 39°}$ into the equation $a = x \tan 71°$.

$a = x \tan 71°$ Write an equation representing the altitude of the plane.

$a = \left(\dfrac{26,400 \tan 39°}{\tan 71° + \tan 39°} \right) \tan 71°$ Substitute $\dfrac{26,400 \tan 39°}{\tan 71° + \tan 39°}$ for x.

$a \approx 16,717 \text{ ft}$ Evaluate by using a calculator (in degree mode).

The altitude of the airplane is approximately 16,717 feet.

My video summary

⊘ **Example 5** Determining the Height of the Eiffel Tower

A tourist visiting Paris determines that the angle of elevation from point A to the top of the Eiffel tower is 12.19°. She then walks 1 km on a straight line toward the tower to point B and determines that the angle of elevation to the top of the tower is now 32.94°. Determine the height of the Eiffel Tower. Round to the nearest meter.

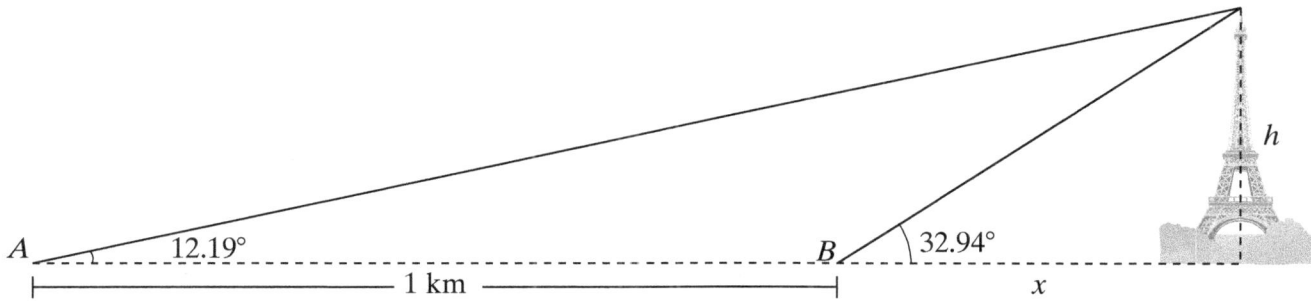

Solution We can use the two triangles in the figure to write the following equations.

$$\tan 32.94° = \frac{h}{x} \quad \text{and} \quad \tan 12.19° = \frac{h}{x+1}$$

Solving for x using the equation, $\tan 32.94° = \frac{h}{x}$, we get $x = \frac{h}{\tan 32.94°}$. Substituting $x = \frac{h}{\tan 32.94°}$ into the second equation we get $\tan 12.19° = \frac{h}{\frac{h}{\tan 32.94°} + 1}$.

Now, solve for h.

$\tan 12.19° = \dfrac{h}{\dfrac{h}{\tan 32.94°} + 1}$ Write the equation found in the previous step.

$\tan 12.19° = \dfrac{h}{\dfrac{h + \tan 32.94°}{\tan 32.94°}}$ Simplify the denominator of the complex fraction using a common denominator.

$\tan 12.19° = \dfrac{h \tan 32.94°}{h + \tan 32.94°}$ Simplify.

$\tan 12.19°(h + \tan 32.94°) = h \tan 32.94°$ Multiply both sides by $h + \tan 32.94°$.

$h \tan 12.19° + (\tan 12.19°)(\tan 32.94°) = h \tan 32.94°$ Use the distributive property.

$h \tan 12.19° - h \tan 32.94° = -(\tan 12.19°)(\tan 32.94°)$ Collect the terms with the variable h on the left-hand side.

$h(\tan 12.19° - \tan 32.94°) = -(\tan 12.19°)(\tan 32.94°)$ Factor out an h.

$h = \dfrac{-(\tan 12.19°)(\tan 32.94°)}{\tan 12.19° - \tan 32.94°} \text{ km}$ Solve for h to get an expression that describes the height in kilometers.

To determine the height in meters, we multiply this expression by 1000. Therefore,

$$h = \frac{-(\tan 12.19°)(\tan 32.94°)}{\tan 12.19° - \tan 32.94°}(1000) \approx 324 \text{ m}$$

You Try It Work through this You Try It problem.

Work Exercises 8–15 in this textbook or in the MyMathLab Study Plan.

4.1 Exercises

Skill Check Exercises

Exercises SCE-1 through SCE-3, use a calculator in degree mode to approximate the expression. Round to two decimal places.

SCE-1. $\sin^{-1}\left(\dfrac{35.1}{75.8}\right)$ **SCE-2.** $\cos^{-1}\left(\dfrac{8.1}{31.9}\right)$ **SCE-3.** $\tan^{-1}\left(\dfrac{14.8}{16.2}\right)$

In Exercises 1–5, solve each right triangle. Round the length of the sides to two decimal places and round the measure of the angles to two decimal places.

1.

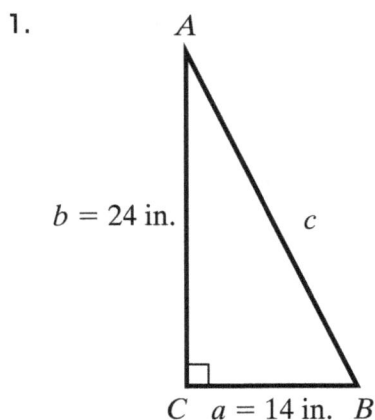

$b = 24$ in.

c

$C \quad a = 14$ in. $\quad B$

2.

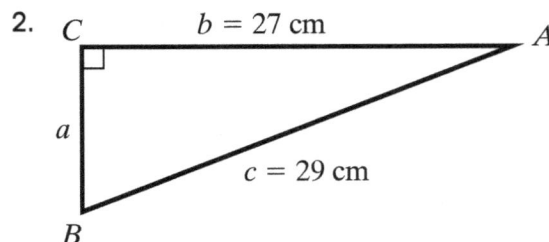

$C \qquad b = 27$ cm $\qquad A$

a

$c = 29$ cm

B

3.

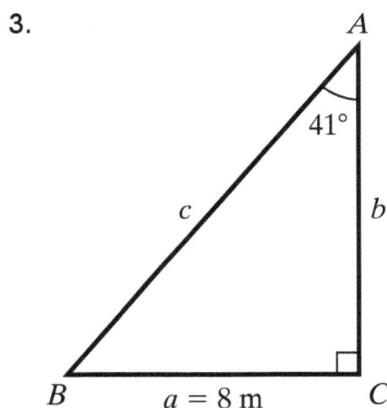

A

$41°$

$c \qquad b$

$B \qquad a = 8$ m $\qquad C$

4.

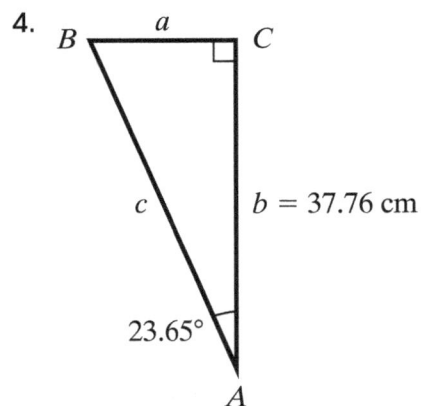

$B \qquad a \qquad C$

$c \qquad b = 37.76$ cm

$23.65°$

A

5.

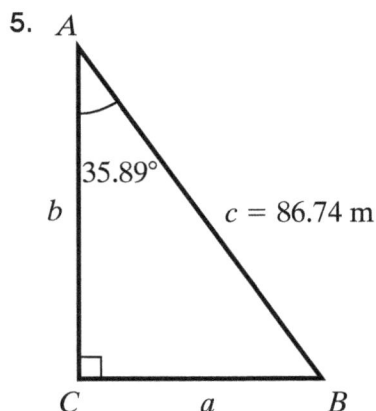

A

$35.89°$

$b \qquad c = 86.74$ m

$C \qquad a \qquad B$

6. Find the length of each leg of a right triangle given that one angle is 27° and the length of the hypotenuse is 14 inches. Round the length of the sides to two decimal places.

7. Find the length of the two missing sides of a right triangle given that one angle is 83.1° and the length of the side adjacent to the angle is 12.5 centimeters. Round the length of the sides to two decimal places.

8. The angle of elevation to the top of a flagpole measured by a digital protractor is 54.7° from a point on the ground 72 feet away from its base. Find the height of the flagpole. Round to two decimal places.

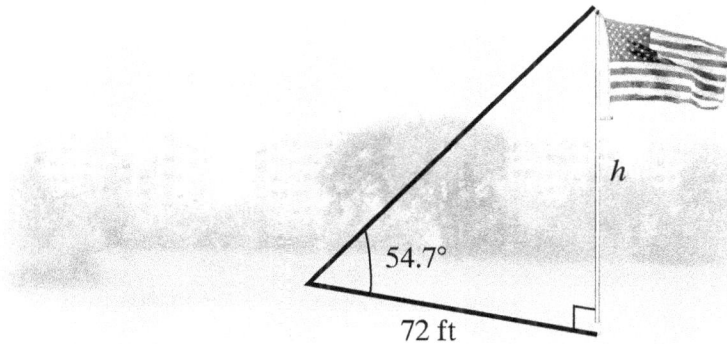

h

54.7°

72 ft

9. A hawk 330 feet above the ground spots a mouse on the ground at an angle of depression of 55.4°. Find the line of sight distance between the hawk and the mouse. Round to two decimal places.

10. Two people are standing on opposite sides of a small river. One person is located at point Q, a distance of 35 meters from a bridge. The other person is standing on the southeast corner of the bridge at point P. The angle between the bridge and the line of sight from P to Q is 72.4°. Use this information to determine the length of the bridge and the distance between the two people. Round your answers to two decimal places as needed.

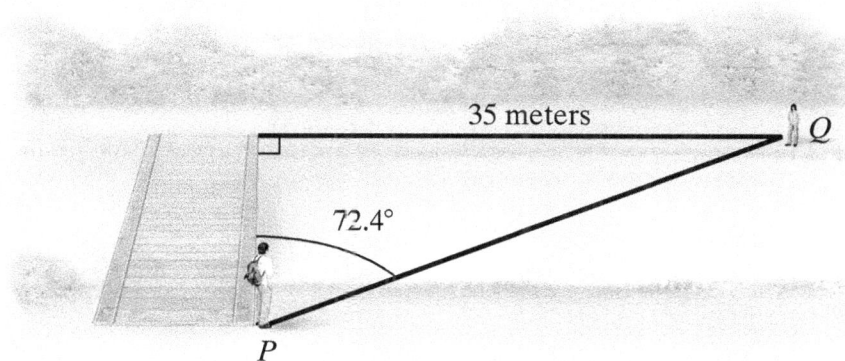

35 meters

Q

72.4°

P

11. A basketball player spots the front of the rim of a basketball hoop at an angle of elevation of 25.7°. If the player's eyes are located 6 feet above the ground, find the horizontal distance between the basketball player and the front of the rim. (Assume that the rim of the basket is 10 feet from the floor.) Round to two decimal places.

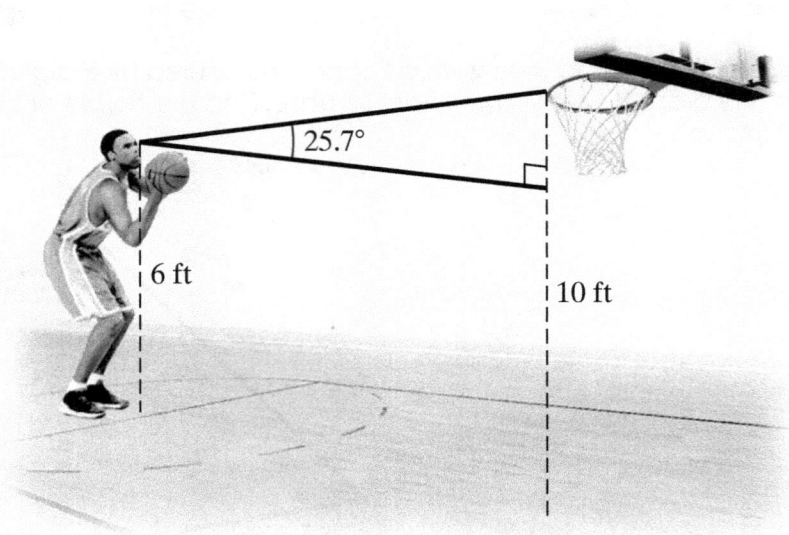

12. During the first quarter of its moon, the planet Quagnon, its moon, and its sun create a right triangle as shown. The distance between Quagnon and its moon is approximately 146,000 km. Determine the distance between the moon of Quagnon and its sun. Round to the nearest kilometer.

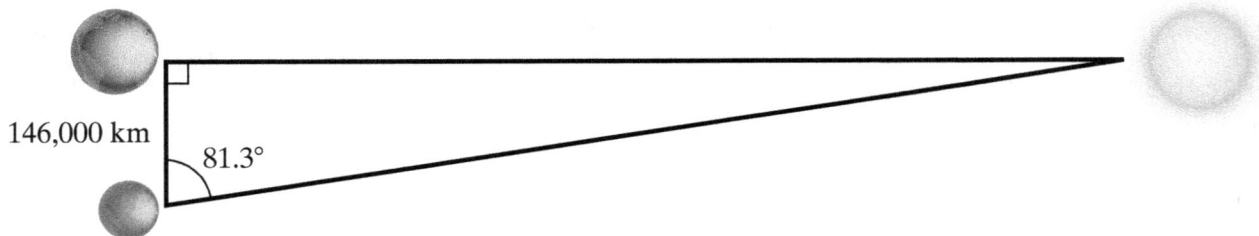

13. A mine shaft with a circular entrance has been carved into the side of a mountain. From a distance of 100 feet from the base of the mountain, the angle of elevation to the bottom of the circular opening is 27.7°. The angle of elevation to the top of the opening is 33°. Determine the diameter of the circular entrance. Round to two decimal places.

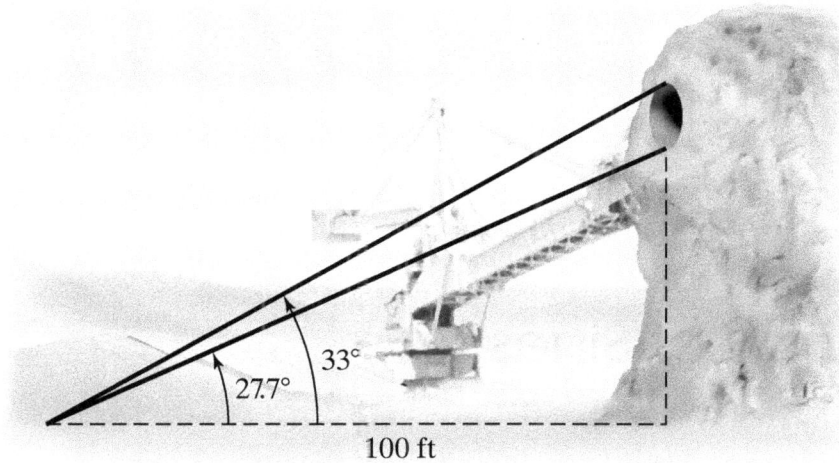

14. At the same instant, two observers 2 miles apart are looking at the same hot air balloon. See the figure. The angle of elevation (from the ground) of the observer closest to the balloon is 77°. The angle of elevation (from the ground) of the person furthest from the balloon is 43°. Find the altitude of the balloon (to the nearest foot) at the instant the two people observe the balloon.

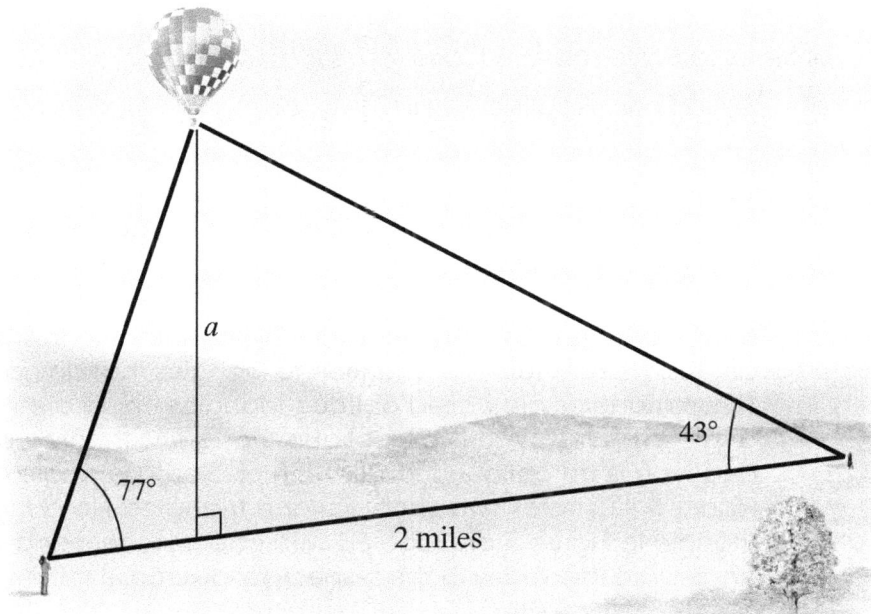

15. A surveyor wanted to determine the height of a mountain. To do this, she determined that the angle of elevation from point A to the top of the mountain is 31°. She then drove 1500 meters toward the mountain to point B where she determined that the angle of elevation to the top of the mountain was 44°. How tall is the mountain? Round to the nearest meter.

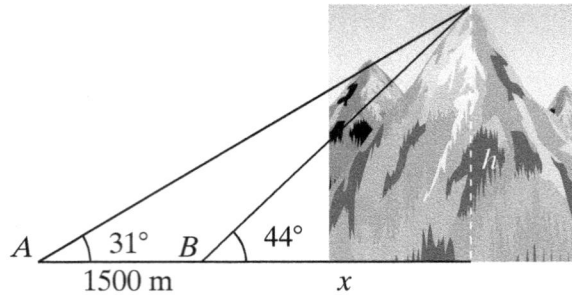

A 31° B 44°
1500 m x

4.2 The Law of Sines

THINGS TO KNOW

Before working through this section, be sure that you are familiar with the following concepts:

VIDEO   ANIMATION   INTERACTIVE

You Try It 1. Understanding the Inverse Sine Function
(Section 2.4)

OBJECTIVES

1 Determining If the Law of Sines Can Be Used to Solve an Oblique Triangle

2 Using the Law of Sines to Solve the SAA Case or ASA Case

3 Using the Law of Sines to Solve the SSA Case

4 Using the Law of Sines to Solve Applied Problems Involving Oblique Triangles

- -

OBJECTIVE 1   DETERMINING IF THE LAW OF SINES CAN BE USED TO SOLVE AN OBLIQUE TRIANGLE

Most triangles that we have worked with thus far in this text have been right triangles. We now turn our attention to triangles that do not include a right angle. These triangles are called **oblique triangles**. There are two types of oblique triangles. The first type is an oblique triangle with three acute angles. The second type is an oblique triangle with one obtuse angle and two acute angles. Figure 5 illustrates two such oblique triangles. Note that the angles in each triangle in Figure 5 are labeled with capital letters and the sides opposite the angles are labeled with the respective lowercase letter.

The goal of this section is to determine all angles and all sides of oblique triangles, given certain information. This process is called **solving oblique triangles**. In this section, we will solve (or attempt to solve) oblique triangles using the **Law of Sines**.

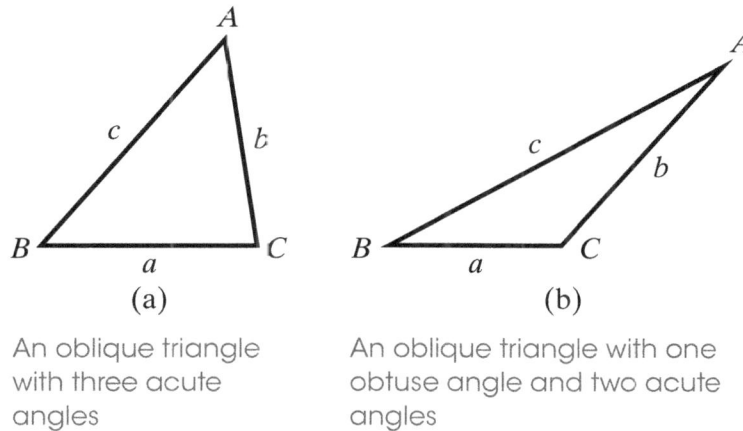

(a)

An oblique triangle with three acute angles

(b)

An oblique triangle with one obtuse angle and two acute angles

Figure 5

The Law of Sines

If A, B, and C are the measures of the angles of any triangle and if a, b, and c are the lengths of the sides opposite the corresponding angles, then

$$\frac{a}{\sin A} = \frac{b}{\sin B} = \frac{c}{\sin C} \quad \text{or} \quad \frac{\sin A}{a} = \frac{\sin B}{b} = \frac{\sin C}{c}.$$

The Law of Sines states that the ratio of the length of a side of a triangle to the sine of the angle opposite the side is equal to the ratio of the length of any other side to the sine of the angle opposite that side. We can develop a proof of the Law of Sines for the case of an oblique triangle having three acute angles by first drawing an altitude of length h from angle C to point D, as shown in Figure 6.

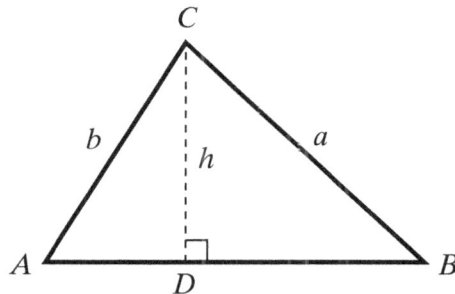

Figure 6 Oblique triangle ABC with three acute angles and an altitude of length h

Note in Figure 6 that triangle ACD and BCD are right triangles. Using the right triangle definition of the sine function, we get

$$\sin A = \frac{h}{b} \quad \text{and} \quad \sin B = \frac{h}{a}.$$

Solving both of these equations for h, we get

$$h = b \sin A \quad \text{and} \quad h = a \sin B.$$

We can now set the expressions $a \sin A$ and $b \sin B$ equal to each other because they both represent the length of the altitude of the triangle in Figure 6.

$$b \sin A = a \sin B$$ Set the expressions representing h equal to each other.

$$\frac{b \sin A}{\sin A \sin B} = \frac{a \sin B}{\sin A \sin B}$$ Divide both sides by $\sin A \sin B$.

$$\frac{b}{\sin B} = \frac{a}{\sin A}$$ Cancel common factors.

Using the same triangle from Figure 6 but drawing an altitude from angle B (see Figure 7), we can get a similar result.

$$\sin A = \frac{h}{c} \quad \text{and} \quad \sin C = \frac{h}{a}$$ Use Figure 7 and the right triangle definition of the sine function.

$$h = c \sin A \quad \text{and} \quad h = a \sin C$$ Solve both equations for h.

$$c \sin A = a \sin C$$ Set the expressions representing h equal to each other.

$$\frac{c \sin A}{\sin A \sin C} = \frac{a \sin C}{\sin A \sin C}$$ Divide both sides by $\sin A \sin C$.

$$\frac{c}{\sin C} = \frac{a}{\sin A}$$ Cancel common factors.

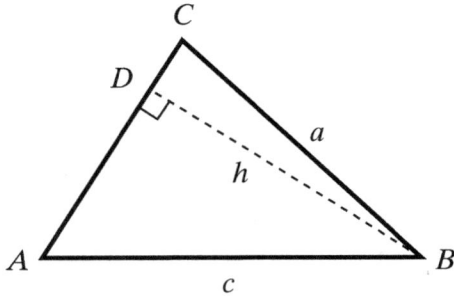

Figure 7

Thus we now have the two results,

$$\frac{b}{\sin B} = \frac{a}{\sin A} \text{ and } \frac{c}{\sin C} = \frac{a}{\sin A}.$$

Therefore, $\frac{a}{\sin A} = \frac{b}{\sin B} = \frac{c}{\sin C}$ and we have proved the Law of Sines for the case of an oblique triangle with three acute angles. The right triangle case and the oblique triangle with an obtuse angle case can be proved in a similar manner.

Before we use the Law of Sines to solve oblique triangles, it is necessary to consider when the Law of Sines can be used. To solve any oblique triangle, at least three pieces of information must be known. If S represents a known side of a triangle and if A represents a known angle of a triangle, then there are six possible situations where only three pieces of information can be known. The six possible cases are shown in Table 1.

Table 1

Triangle	Description of Case	Abbreviation of Case
	Side–Angle–Angle Two angles and a side opposite one of the angles are known.	SAA
	Angle–Side–Angle Two angles and the side between the angles are known.	ASA
	Side–Side–Angle Two sides and an angle opposite one of the sides are known.	SSA
	Side–Angle–Side Two sides and an angle between the sides are known.	SAS
	Side–Side–Side All three sides are known.	SSS
	Angle–Angle–Angle All three angles are known.	AAA

Because the Law of Sines uses proportions that involve both angles and sides, the following pieces of information are needed to solve an oblique triangle using the Law of Sines:

1. The measure of an angle must be known.

2. The length of the side opposite the known angle must be known.

3. At least one more side or one more angle must be known.

The first three cases listed in Table 1 involve situations where this information is known. Therefore, the Law of Sines can be used to solve the SAA, ASA, and SSA cases.

My video summary   ⊙ **Example 1** Determining If the Law of Sines Can Be Used to Solve an Oblique Triangle

Decide whether or not the Law of Sines can be used to solve each triangle. Do not attempt to solve the triangle.

a.

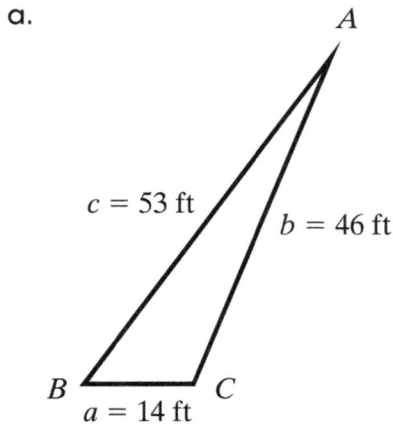

$c = 53$ ft
$b = 46$ ft
$a = 14$ ft

b.

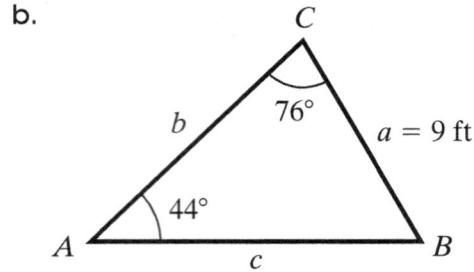

b
$76°$
$a = 9$ ft
$44°$
c

c.

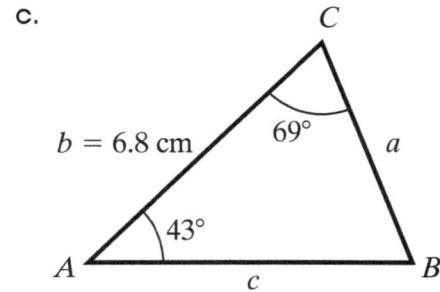

$b = 6.8$ cm
$69°$
a
$43°$
c

Solution

a. First, organize the given information.

Angles	Sides
$A = $ ____	$a = 14$ ft
$B = $ ____	$b = 46$ ft
$C = $ ____	$c = 53$ ft

The Law of Sines *cannot* be used because the first criterion is not satisfied. We are not given the measure of any angle. This is an example of the SSS case, which will be discussed in Section 4.2.

b. First, organize the given information.

Angles	Sides
$A = 44°$	$a = 9$ ft
$B = $ ____	$b = $ ____
$C = 76°$	$c = $ ____

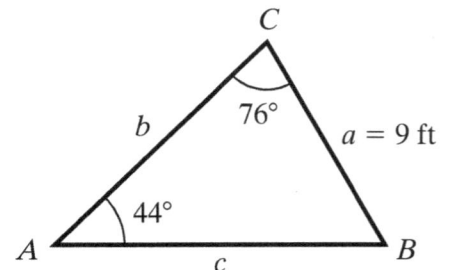

C
b
$76°$
$a = 9$ ft
$44°$
A
c
B

The Law of Sines *can* be used to solve the triangle because we are given an angle $A = 44°$ and the length of the side opposite of angle A ($a = 9$ ft). We are also given the measure of another angle $C = 76°$. This is an example of the SAA case.

c. First, organize the given information.

Angles	Sides
$A = 43°$	$a = $ ____
$B = $ ____	$b = 6.8$ cm
$C = 69°$	$c = $ ____

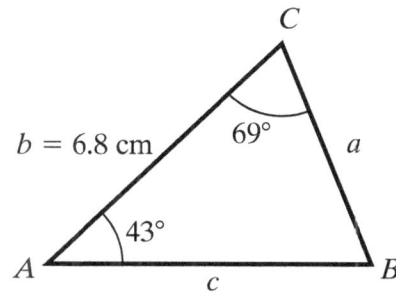

At first glance it appears that we cannot use the Law of Sines to solve this triangle because we are given the length of side b but we are not given the measure of angle B. However, because the sum of the measures of the angles of any triangle is 180°, we can find the measure of angle B.

$$B + 69° + 43° = 180°$$ The sum of the measures of the angles of any triangle is 180°.

$$B + 112° = 180°$$ Add.

$$B = 68°$$ Solve for B.

We now know the measure of an angle $B = 68°$, the length of the side opposite the angle, $b = 6.8$ cm and another angle $A = 43°$ or $C = 69°$. Therefore, the Law of Sines can be used to solve the triangle. This is an example of the ASA case.

You Try It Work through this You Try It problem.

Work Exercises 1–6 in this textbook or in the My Math Lab **Study Plan.**

OBJECTIVE 2 USING THE LAW OF SINES TO SOLVE THE SAA CASE OR ASA CASE

When the measure of any two angles of an oblique triangle are known and the length of any side is known, always start by determining the measure of the unknown angle. Then use appropriate Law of Sines proportions to solve for the lengths of the remaining unknown sides. Whenever possible, we will avoid using rounded information to solve for the remaining parts of the triangle. When this cannot be avoided, we will agree to use information rounded to one decimal place unless some other guideline is stated.

Example 2 Solving a SAA Triangle Using the Law of Sines

Solve the given oblique triangle. Round the lengths of the sides to one decimal place.

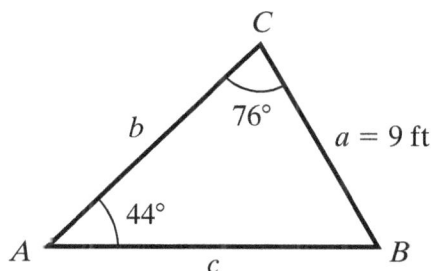

Solution Observe that we are given the following information.

Angles	Sides
$A = 44°$	$a = $ 9 ft
$B = $ ____	$b = $ ____
$C = 76°$	$c = $ ____

We start by determining the measure of angle B using the fact that the sum of the measures of the three angles of a triangle is $180°$.

$$B + 44° + 76° = 180°$$ The sum of the measures of the angles of any triangle is $180°$.

$$B + 120° = 180°$$ Add.

$$B = 60°$$ Solve for B.

We insert the new information.

Angles	Sides
$A = 44°$	$a = 9$ ft
$B = $ 60°	$b = $ ____
$C = 76°$	$c = $ ____

We will use the known ratio $\dfrac{a}{\sin A}$ to determine the lengths of the other two sides. First find b.

$$\frac{b}{\sin B} = \frac{a}{\sin A}$$ Use the Law of Sines to write a proportion.

$$\frac{b}{\sin 60°} = \frac{9}{\sin 44°}$$ Substitute the known information.

$$b = \frac{9 \sin 60°}{\sin 44°}$$ Multiply both sides by $\sin 60°$ to solve for b.

$$b \approx 11.2 \text{ ft}$$ Use a calculator in degree mode and round to one decimal place.

Again, we insert the new information.

Angles	Sides
$A = 44°$	$a = 9$ ft
$B = $ 60°	$b \approx $ 11.2 ft
$C = 76°$	$c = $ ____

Lastly, we find the length of side C.

$$\frac{c}{\sin C} = \frac{a}{\sin A} \qquad \text{Use the Law of Sines to write a proportion.}$$

$$\frac{c}{\sin 76°} = \frac{9}{\sin 44°} \qquad \text{Substitute the known information.}$$

$$c = \frac{9 \sin 76°}{\sin 44°} \qquad \text{Multiply both sides by } \sin 76° \text{ to solve for } c.$$

$$c \approx 12.6 \text{ ft} \qquad \text{Use a calculator in degree mode and round to one decimal place.}$$

We have now solved the triangle.

Angles	Sides
$A = 44°$	$a = 9$ ft
$B = \underline{60°}$	$b \approx \underline{11.2}$ ft
$C = 76°$	$c \approx \underline{12.6}$ ft

My video summary ⊘ **Example 3** Solving an ASA Triangle Using the Law of Sines

Solve oblique triangle ABC if $B = 38°$, $C = 72°$, and $a = 7.5$ cm. Round the lengths of the sides to one decimal place.

Solution Start by sketching a triangle and label the angles and sides based on the given information. See Figure 8.

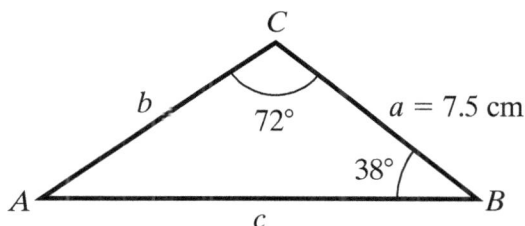

Figure 8

We can determine the measure of angle A by subtracting the measures of angle B and angle C from $180°$. Therefore, $A = 180° - 38° - 72° = 70°$.

We now organize all of the information.

Angles	Sides
$A = 70°$	$a = 7.5$ cm
$B = 38°$	$b = \underline{}$
$C = 72°$	$c = \underline{}$

Try using the Law of Sines to determine the lengths of sides b and c on your own. Work through this video to see the solution.

You Try It Work through this You Try It problem.

Work Exercises 7–12 in this textbook or in the MyMathLab Study Plan.

OBJECTIVE 3   USING THE LAW OF SINES TO SOLVE THE SSA CASE

When given the SAA case or the ASA case, as in Examples 2 and 3, we used the Law of Sines of the form $\dfrac{a}{\sin A} = \dfrac{b}{\sin B} = \dfrac{c}{\sin C}$ to determine the missing sides of a triangle. In the SAA case or the ASA case, a unique triangle is always formed.

We now want to look at the SSA case. This is the case when we are given two sides and the angle opposite one of the sides. In the SSA case, we will use the Law of Sines of the form $\dfrac{\sin A}{a} = \dfrac{\sin B}{b} = \dfrac{\sin C}{c}$ to first determine the measure of an angle. For example, suppose that we are given the measure of angle A, the length of side a, and the length of side b. In order to solve for angle B, we first use the Law of Sines and the proportion $\dfrac{\sin B}{b} = \dfrac{\sin A}{a}$ to solve for $\sin B$ as follows:

$$\frac{\sin B}{b} = \frac{\sin A}{a} \qquad \text{Use the Law of Sines to write a proportion.}$$

$$\sin B = \frac{b \sin A}{a} \qquad \text{Multiply both sides by } b \text{ to solve for } \sin B.$$

Because the values of b, a, and $\sin A$ are always positive, it follows that the value of $\sin B$ is always positive. Therefore, there are three possibilities for the value of $\sin B$.

1. $\sin B > 1$

2. $\sin B = 1$

3. $0 < \sin B < 1$

For the first possibility, recall that the sine of an angle cannot exceed a value of 1. Thus, if $\sin B > 1$, then there exists no angle B for which $\sin B > 1$ and the result is no triangle.

For the second possibility, where $\sin B = 1$, then $B = 90°$ and the result is a right triangle.

For the third possibility, where $0 < \sin B < 1$, then there are two potential solutions for B because the sine function is positive for angles having a terminal side in Quadrant I or in Quadrant II. Thus, the result is either one triangle or two triangles.

As you can see, unlike the SAA case or ASA case, where one unique triangle is always formed, the SSA case can result in zero, one, or two triangles. For this reason, the SSA case is often referred to as the **ambiguous case** of the Law of Sines.

My animation summary 　Watch this animation to see all of the possible triangles that can result when given the SSA case.

The number of triangles depends on the value of $\sin B$ and whether or not there is exactly one solution for B, two solutions for B, or no solution for B. All possible scenarios of the SSA case are outlined in Table 2 assuming that we are given the measure of angle A and the lengths of sides a and b.

Table 2

Possible Triangles	Number of Triangles	Description	The Value of sin B
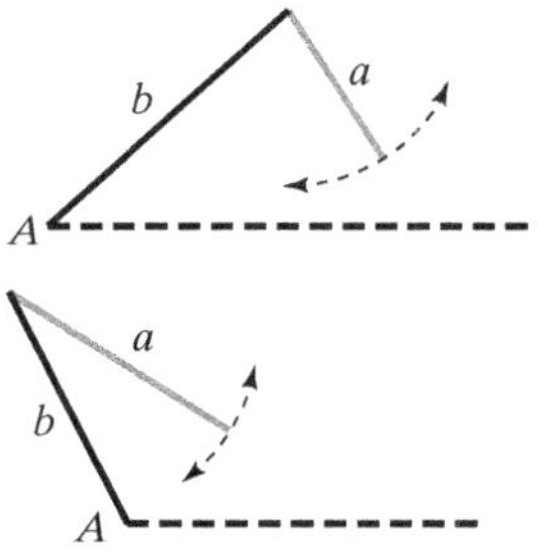	No Triangle	No angle B exists and side a is too short to reach the opposite side.	$\sin B > 1$
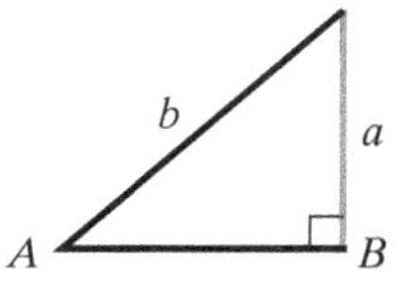	One Triangle	The measure of angle B is 90° and the result is one right triangle.	$\sin B = 1$
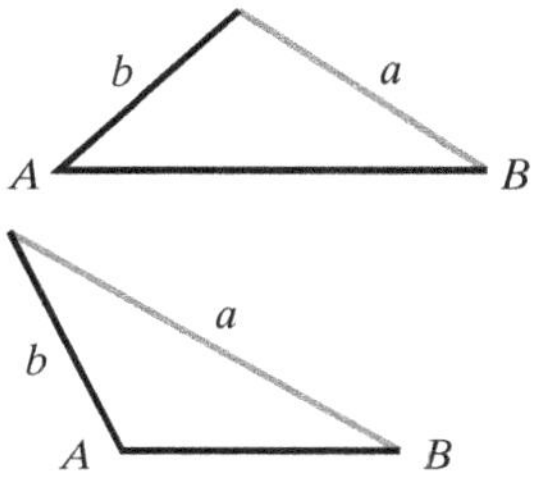	One Triangle	If there is one solution for B, then the traingle is oblique with three acute angles or the triangle is oblique with one obtuse angle.	$0 < \sin B < 1$
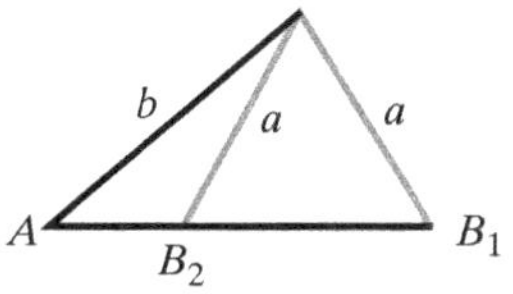	Two Triangles	If there are two solutions for B (B_1 and B_2), then the result is two triangles.	$0 < \sin B < 1$

My animation summary

▢ Watch this animation ᴛo see all of the possible triangles that can result when given the SSA case.

Example 4 Solving a SSA Triangle Using the Law of Sines (No Triangle)

Two sides and an angle are given below. Determine whether the information results in no triangle, one rght triangle, or one or two oblique triangles. Solve each resulting triangle. Rcund the measures of all angles and the lengths of all sides to one decimal place.

$$a = 10 \text{ ft}, \quad b = 28 \text{ ft}, \quad A = 29°$$

Solution First, organize the given information.

Angles	Sides
$A = 29°$	$a = 10 \text{ ft}$
$B = $ ____	$b = 28 \text{ ft}$
$C = $ ____	$c = $ ____

We use the Law of Sines to first determine the value of $\sin B$.

$$\frac{\sin B}{b} = \frac{\sin A}{a} \qquad \text{Use the Law of Sines to write a proportion.}$$

$$\frac{\sin B}{28} = \frac{\sin 29°}{10} \qquad \text{Substitute the known information.}$$

$$\sin B = \frac{28 \sin 29°}{10} \qquad \text{Multiply both sides by 28 to isolate } \sin B.$$

The value of $\sin B = \dfrac{28 \sin 29°}{10}$ is approximately 1.3575. Because the value of $\sin B$ cannot exceed 1, we know that there is no angle B. Thus, there is no triangle that satisfies the given conditions. Figure 9 illustrates that the length of side a is too short to reach the opposite side.

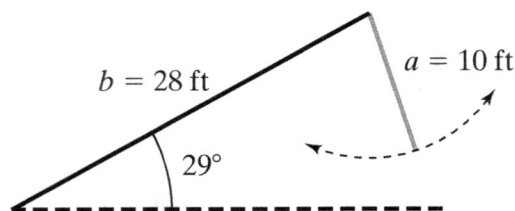

$b = 28$ ft $a = 10$ ft

$29°$

Figure 9

My video summary

◉ **Example 5** Solving a SSA Triangle Using the Law of Sines (Exactly One Triangle)

Two sides and an angle are given below. Determine whether the information results in no triangle, one right triangle, or one or two oblique triangles. Solve each resulting triangle. Round the measures of all angles and the lengths of all sides to one decimal place.

$$a = 13 \text{ cm}, \quad b = 7.8 \text{ cm}, \quad A = 67°$$

Solution First, organize the given information.

Angles	Sides
$A = 67°$	$a = 13$ cm
$B = $ ____	$b = 7.8$ cm
$C = $ ____	$c = $ ____

We use the Law of Sines to first determine the value of $\sin B$.

$$\frac{\sin B}{b} = \frac{\sin A}{a} \qquad \text{Use the Law of Sines to write a proportion.}$$

$$\frac{\sin B}{7.8} = \frac{\sin 67°}{13} \qquad \text{Substitute the known information.}$$

$$\sin B = \frac{7.8 \sin 67°}{13} \qquad \text{Multiply both sides by 7.8 to isolate } \sin B.$$

The value of $\sin B = \dfrac{7.8 \sin 67°}{13}$ is approximately 0.5523. Therefore, $0 < \sin B < 1$, and the given information results in one or two oblique triangles.

There are two possible choices for B between $0°$ and $180°$. One choice is an angle in standard position having a terminal side that lies in Quadrant I. This angle is $B_1 = \sin^{-1}\left(\dfrac{7.8 \sin 67°}{13}\right) \approx 33.5°$. The other choice is an angle in standard position having a terminal side lying in Quadrant II. This angle is $B_2 = 180° - \sin^{-1}\left(\dfrac{7.8 \sin 67°}{13}\right) \approx 146.5°$. If $B_2 \approx 146.5°$, then the sum of the measures of the angles in the triangle would be $A + B_2 + C \approx 67° + 146.5° + C \approx 213.5° + C$. This cannot happen because the sum of the measures of the three angles cannot exceed $180°$. Therefore, we exclude B_2 as a possible choice. Thus, $B_1 = B \approx 33.5°$.

We can determine the measure of angle C by subtracting the measures of angle A and angle B from $180°$. Therefore, $C \approx 180° - 67° - 33.5° \approx 79.5°$.

We now summarize the information gathered thus far.

Angles	Sides
$A = 67°$	$a = 13$ cm
$B \approx 33.5°$	$b = 7.8$ cm
$C \approx 79.5°$	$c = $ _____

To find c we use the Law of Sines. Note that we will use the approximate measure of angle C rounded to one decimal place.

$\dfrac{c}{\sin C} = \dfrac{a}{\sin A}$ Use the Law of Sines to write a proportion.

$\dfrac{c}{\sin 79.5°} \approx \dfrac{13}{\sin 67°}$ Substitute the known information.

$c \approx \dfrac{13 \sin 79.5°}{\sin 67°}$ Multiply both sides by $\sin 79.5°$ to solve for c.

$c \approx 13.9$ cm   Use a calculator in degree mode and round to one decimal place.

My video summary   ⊙ We have now solved the oblique triangle, which can be seen in Figure 10. Watch this video to see the solution to this example.

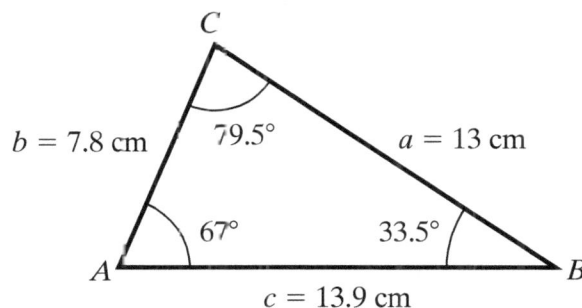

Figure 10

My video summary   ⊙ **Example 6** Solving a SSA Triangle Using the Law of Sines (Two Triangles)

Two sides and an angle are given below. Determine whether the information results in no triangle, one right triangle, or one or two oblique triangles. Solve each resulting triangle. Round the measures of all angles and the lengths of all sides to one decimal place.

$$b = 11.3 \text{ in.}, \quad c = 15.5 \text{ in.}, \quad B = 34.7°$$

Solution First organize the given information.

Angles	Sides
$A = $ ____	$a = $ ____
$B = 34.7°$	$b = 11.3$ in.
$C = $ ____	$c = 15.5$ in.

We use the Law of Sines to first determine the value of $\sin C$.

$$\frac{\sin C}{c} = \frac{\sin B}{b}$$ Use the Law of Sines to write a proportion.

$$\frac{\sin C}{15.5} = \frac{\sin 34.7°}{11.3}$$ Substitute the known information.

$$\sin C = \frac{15.5 \sin 34.7°}{11.3}$$ Multiply both sides by 15.5 to isolate $\sin C$.

The value of $\sin C = \dfrac{15.5 \sin 34°}{11.3}$ is approximately 0.7809. Thus, there are two possible

choices for C between $0°$ and $180°$. One choice is $C_1 = \sin^{-1}\left(\dfrac{15.5 \sin 34.7°}{11.3}\right) \approx 51.3°$.

The other choice is $C_2 = 180° - \sin^{-1}\left(\dfrac{15.5 \sin 34.7°}{11.3}\right) \approx 128.7°$.

If $C_1 \approx 51.3°$, then the sum of the measures of the angles in the triangle would be

$$A_1 + B + C_1 \approx A_1 + 34.7° + 51.3° \approx A_1 + 86.0°.$$

If $C_2 \approx 128.7°$, then the sum of the measures of the angles in the triangle would be

$$A_2 + B + C_2 \approx A_2 + 34.7° + 128.7° \approx A_2 + 163.4°.$$

Both C_1 and C_2 are valid choices because the sum of the measures of the three angles does not yet exceed $180°$.

We now summarize the information thus far for both triangles.

Triangle 1

Angles	Sides
$A_1 = $ ___	$a_1 = $ ___
$B = 34.7°$	$b = 11.3$ in.
$C_1 \approx 51.3°$	$c = 15.5$ in.

Triangle 2

Angles	Sides
$A_2 = $ ___	$a_2 = $ ___
$B = 34.7°$	$b = 11.3$ in.
$C_2 \approx 128.7°$	$c = 15.5$ in.

My video summary ⊘ See if you can determine the values of A_1, A_2, a_1, and a_2 on your own. Watch this video to verify that $A_1 \approx 94°$, $A_2 \approx 16.6°$, $a_1 \approx 19.8$ in., and $a_2 \approx 5.7$ in. The two triangles are displayed in Figure 11.

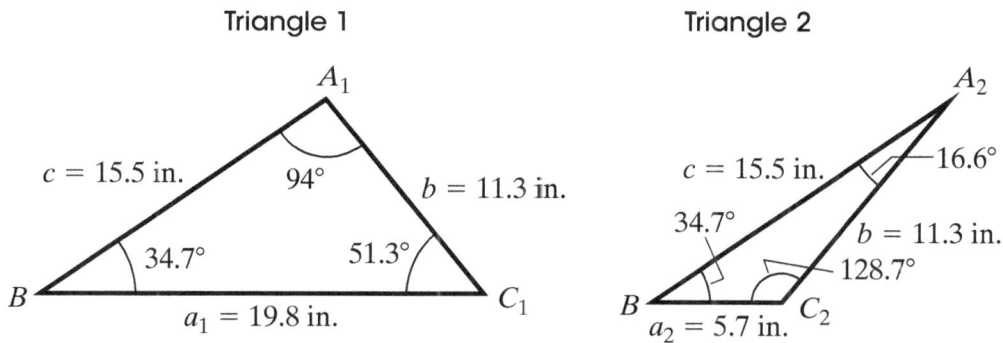

Figure 11

You Try It   Work through this You Try It problem.

Work Exercises 13–19 in this textbook or in the MyMathLab Study Plan.

We now summarize the three cases for which the Law of Sines can be used.

Using the Law of Sines

If S represents a given side of a triangle and if A represents a given angle of a triangle, then the Law of Sines can be used to solve the three oblique triangle cases shown below.

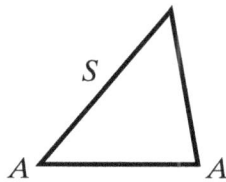
The SAA Case

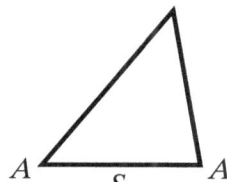
The ASA Case

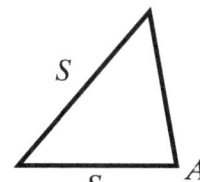
The SSA Case
(The Ambiguous Case)

Note   In Section 4.2, we will see that the Law of Cosines can be used to solve the SAS case and the SSS case. Neither the Law of Sines nor the Law of Cosines can be used to solve the AAA case because the three given angles do not define a unique triangle. Do you know why? View this explanation to see why.

OBJECTIVE 4   USING THE LAW OF SINES TO SOLVE APPLIED PROBLEMS INVOLVING OBLIQUE TRIANGLES

The Law of Sines can be a useful tool to help solve many applications that arise involving triangles that are not right triangles. Many fields, such as surveying, engineering, and navigation, require the use of the Law of Sines.

Example 7 Determining the Width of a River

To determine the width of a river, forestry workers place markers on opposite sides of the river at points A and B. A third marker is placed at point C, 200 feet away from point A, forming triangle ABC. If the angle in triangle ABC at point C is 51° and if the angle in triangle ABC at point A is 110°, then determine the width of the river rounded to the nearest tenth of a foot.

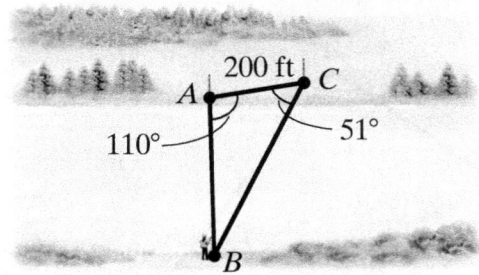

Solution First organize the given information.

Angles	Sides
$A = 110°$	$a = $ _____
$B = $ _____	$b = 200$ ft
$C = 51°$	$c = $ _____

Observe that we want to find the length of side c, which represents the width of the river.

Start by determining the measure of angle B using the fact that the sum of the measures of the three angles of a triangle is 180°.

$$B + 51° + 110° = 180°$$ The sum of the measures of the angles of any triangle is 180°.

$$B + 161° = 180°$$ Add.

$$B = 19°$$ Solve for B.

Now insert the new information.

Angles	Sides
$A = 110°$	$a = $ _____
$B = \underline{\ \ 19°\ \ }$	$b = 200$ ft
$C = 51°$	$c = $ _____

We can now use the Law of Sines of the form $\dfrac{c}{\sin C} = \dfrac{b}{\sin B}$ to solve for c.

$$\dfrac{c}{\sin C} = \dfrac{b}{\sin B}$$ Use the Law of Sines to write a proportion.

$$\dfrac{c}{\sin 51°} = \dfrac{200}{\sin 19°}$$ Substitute the known information.

$$c = \dfrac{200 \sin 51°}{\sin 19°}$$ Multiply both sides by $\sin 51°$ to solve for c.

$$c \approx 477.4 \text{ ft}$$ Use a calculator in degree mode.

Therefore, the width of the river is approximately 477.4 ft.

Many applications involving navigation use the concept of **bearing**. There are several different ways to denote a bearing. In the examples that follow, a bearing will be described as the direction that one object is from another object in relation to north, south, east, and west. Two directions and a degree measurement will be given to describe a bearing. For example, a bearing of N 45° E (read as "45 degrees east of north") can be sketched by drawing the initial side of an angle along the positive y-axis, which represents due north. The terminal side of the angle is then rotated away from the initial side in an "easterly direction" 45° toward the positive x-axis. See Figure 12a. Three other bearings all having an angle of 45° are sketched in Figures 12b–d.

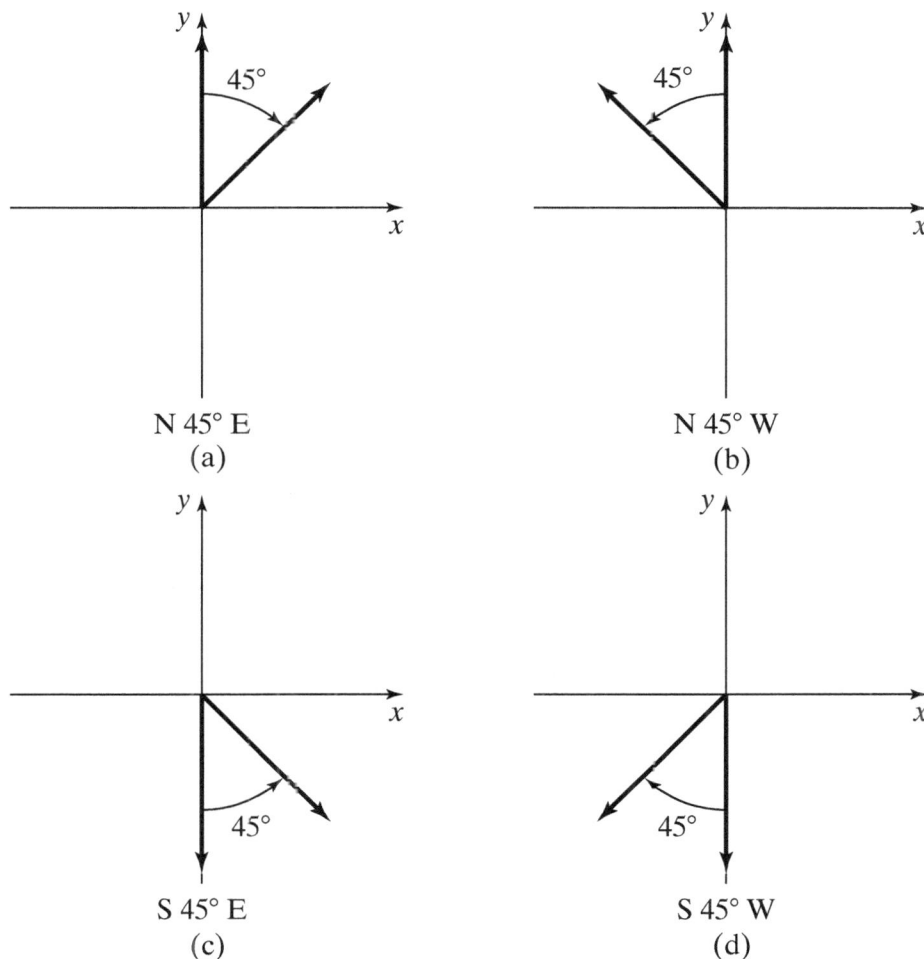

Figure 12

My video summary ⊘ **Example 8** Determining the Distance a Ship Is from Port

A ship set sail from port at a bearing of N 53° E and sailed 63 km to point B. The ship then turned and sailed an additional 69 km to point C. Determine the distance from port to point C if the ship's final bearing is N 74° E. Round to the nearest tenth of a kilometer.

Solution First, draw a diagram with the port located at the origin, as shown below in Figure 13.

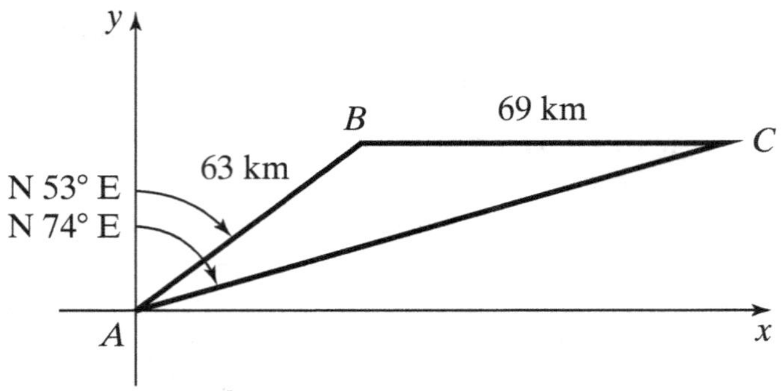

Figure 13

To determine the measure of angle A, we subtract the original bearing from the final bearing to get $A = 74° - 53° = 21°$. This gives the triangle seen in Figure 14.

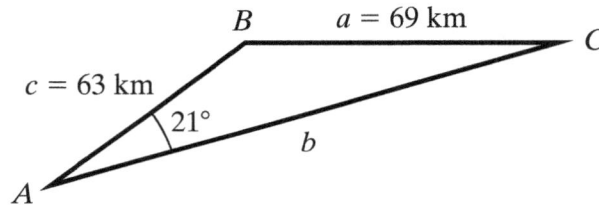

Figure 14

We now organize the information that we have thus far.

Angles	Sides
$A = 21°$	$a = 69$ km
$B = $ ____	$b = $ ____
$C = $ ____	$c = 63$ km

The goal is to find the length of side b. To do this, we must first find the measures of angles C and B. We can use the Law of Sines and the proportion $\dfrac{\sin C}{c} = \dfrac{\sin A}{a}$ to first determine the value of $\sin C$.

$$\frac{\sin C}{c} = \frac{\sin A}{a} \qquad \text{Use the Law of Sines to write a proportion.}$$

$$\frac{\sin C}{63} = \frac{\sin 21°}{69} \qquad \text{Substitute the known information.}$$

$$\sin C = \frac{63 \sin 21°}{69} \qquad \text{Multiply both sides by 63 to isolate } \sin C.$$

The value of $\sin C = \dfrac{63 \sin 21°}{69}$ is approximately 0.3272. Thus, there are two possible choices for C between 0° and 180°. One choice is an angle in standard position having a terminal side that lies in Quadrant I. This angle is $C_1 = \sin^{-1}\left(\dfrac{63 \sin 21°}{69}\right) \approx 19.1°$. The other choice is an angle in standard position having a terminal side lying in Quadrant II. This angle is $C_2 = 180° - \sin^{-1}\left(\dfrac{63 \sin 21°}{69}\right) \approx 160.9°$.

If $C_2 \approx 160.9°$, then the sum of the measures of the angles in the triangle would be $A + B + C_2 \approx 21° + B + 160.9° \approx 181.9° + B$. This cannot happen because the sum of the measures of the three angles cannot exceed 180°. Therefore, we exclude C_2 as a possible choice. Thus, $C_1 = C \approx 19.1°$. Because the sum of the three angles of a triangle is 180°, it follows that $B \approx 139.9°$.

We now insert the new information.

Angles	Sides
$A = 21°$	$a = 69$ km
$B \approx 139.9°$	$b = $ ____
$C \approx 19.1°$	$c = 63$ km

My video summary ⟩ We can now use the Law of Sines and the proportion $\dfrac{b}{\sin B} = \dfrac{a}{\sin A}$ to determine the value of b, which represents the distance from the port to the ship. Try to determine the value of b on your own. Check your answer, or watch this video to see every step of the solution to this example. ●

You Try It Work through this You Try It problem.

Work Exercises 20–25 in this textbook or in the MyMathLab Study Plan.

4.2 Exercises

Skill Check Exercises

For SCE-1 and SCE-2, use a calculator to evaluate each expression. Round to one decimal place.

SCE-1. $\dfrac{7 \sin 63°}{\sin 43°}$

SCE-2. $\dfrac{23.3 \sin 84.2°}{\sin 45.9°}$

In Exercises 1–6, decide whether or not the Law of Sines can be used to solve each triangle. Do not attempt to solve the triangle.

1.

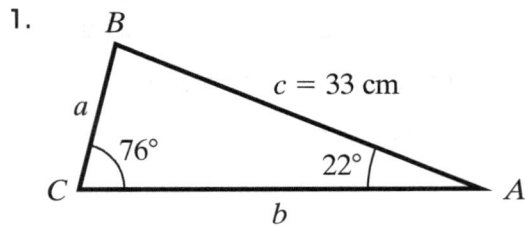

2.

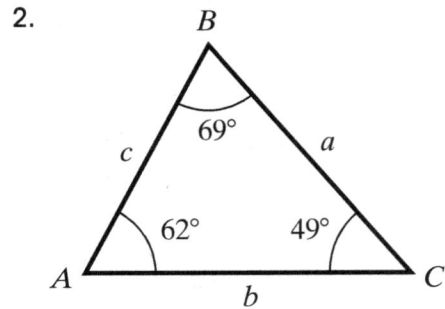

3.

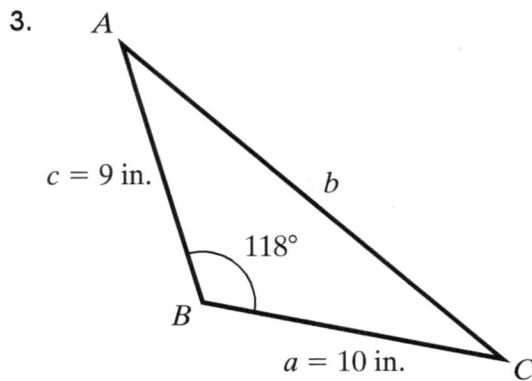

4.

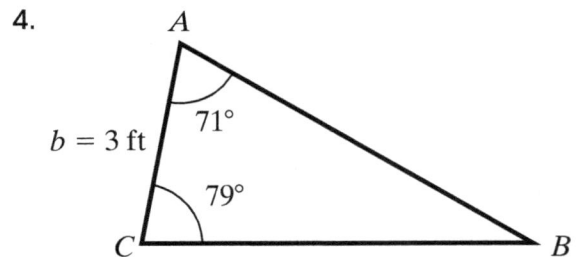

5.

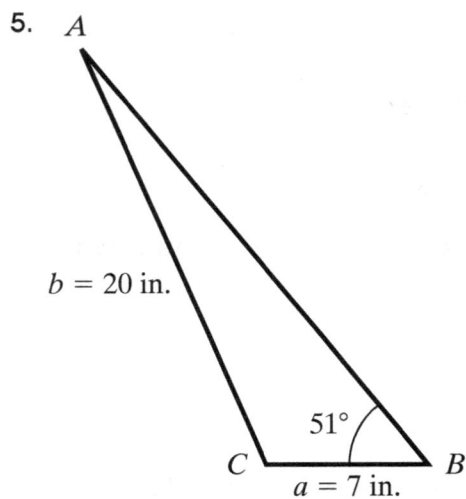

6.

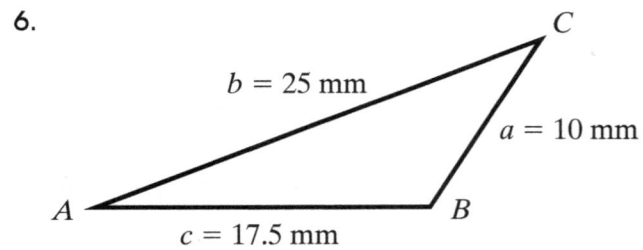

In Exercises 7–12, solve each oblique triangle. Round the measures of all angles and the lengths of all sides to one decimal place.

7.

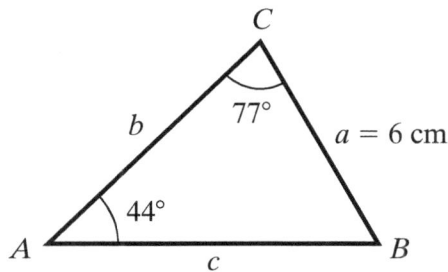

8.

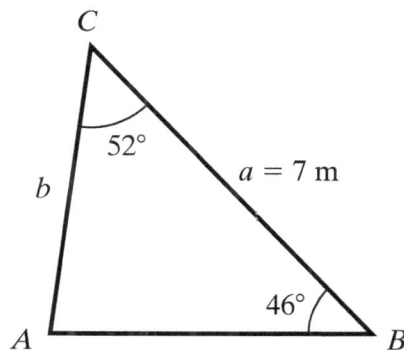

9.

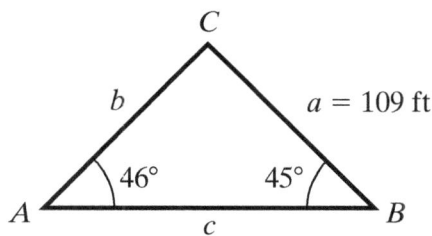

10. $A = 48°$, $B = 53°$, and $b = 6$

11. $A = 106°$, $C = 21°$, and $a = 4$

12. $B = 58°$, $C = 47°$, and $c = 9$

In Exercises 13–19, two sides and an angle are given. First determine whether the information results in no triangle, one triangle, or two triangles. Solve each resulting triangle. Round the measures of all angles and the lengths of all sides to one decimal place.

13. $a = 8.1$, $b = 7.3$, and $A = 40°$

14. $a = 21.5$, $b = 29.3$, and $A = 70.1°$

15. $a = 3.7$, $b = 7.4$, and $A = 30°$

16. $a = 14$, $b = 16.3$, and $A = 55°$

17. $a = 14.8$, $b = 15.9$, and $A = 67.7°$

18. $a = 20$, $b = 24$, anc $A = 121°$

19. $a = 13.3$, $b = 13.6$, and $A = 57.1°$

20. To determine the width of a river, forestry workers place markers or opposite sides of the river at points A and B. A third marker is placed at point C, 40 meters away from point A, forming triangle ABC. If the angle in triangle ABC at point C is 42° and if the angle in triangle ABC at point A is 112°, then determine the width of the river to the nearest tenth of a meter.

21. During World War II, the United States military led a massive assault on the beaches of Normandy in France. Omaha Beach is 5 miles long. At 5 A.M., a U.S. Navy ship was first spotted from either end of Omaha Beach. The angle made with the ship from one end of the beach was 41°. The angle made with the ship from the other end of the beach was 66°. Determine the distance from the ship to either end of the beach at the moment the ship was first spotted. Round to the nearest tenth of a mile.

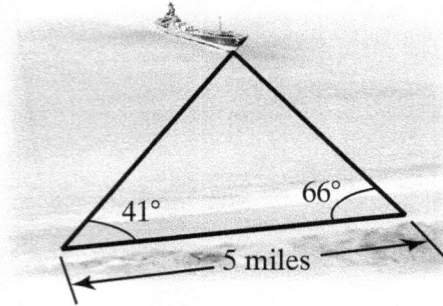

22. A ship was sighted from a lighthouse on a foggy night. When the ship was first sighted, the bearing of the ship was N 16° E. The ship then sailed for 8.6 miles due south. The new bearing was N 27° E. Find the distance between the lighthouse and the ship at each location. Round to the nearest tenth of a mile.

23. A ship set sail from port at a bearing of N 20° W and sailed 35 miles to point B. The ship then turned and sailed an additional 40 miles to point C. Determine the distance from port to the ship if the bearing from the port to point C is N 51° W. Round to the nearest tenth of a mile.

24. Forest fire lookout station Alpha is located 20 miles due west of lookout station Beta. The bearing of a fire from station Alpha is S 11° W. The bearing of the fire from station Beta is S 29° W. Determine the distance from the fire to both lookout stations. Round to the nearest tenth of a mile.

25. To determine the height of a giant Sequoia tree, researchers measure the angle of elevation to the top of the tree from two locations that are 500 feet apart. The angle of elevation from the location closest to the tree is 22°. The angle of elevation from the location farthest from the tree is 16°. What is the height of the tree to the nearest tenth of a degree?

Brief Exercises

In Exercises 26–32, two sides and an angle are given. Determine whether the information results in no triangle, one triangle, or two triangles.

26. $a = 8.1$, $b = 7.3$, and $A = 40°$

27. $a = 21.5$, $b = 29.3$, and $A = 70.1°$

28. $a = 3.7$, $b = 7.4$, and $A = 30°$

29. $a = 14$, $b = 16.3$, and $A = 55°$

30. $a = 14.8$, $b = 15.9$, and $A = 67.7°$

31. $a = 20$, $b = 24$, and $A = 121°$

32. $a = 13.3$, $b = 13.6$, and $A = 57.1°$

4.3 The Law of Cosines

THINGS TO KNOW

Before working through this section, be sure that you are familiar with the following concepts:

VIDEO   ANIMATION   INTERACTIVE

You Try It
1. Understanding the Inverse Sine Function (Section 2.4)

You Try It
2. Understanding the Inverse Cosine Function (Section 2.4)

OBJECTIVES

1 Determining If the Law of Sines or the Law of Cosines Should Be Used to Begin to Solve an Oblique Triangle

2 Using the Law of Cosines to Solve the SAS Case

3 Using the Law of Cosines to Solve the SSS Case

4 Using the Law of Cosines to Solve Applied Problems Involving Oblique Triangles

---

OBJECTIVE 1   DETERMINING IF THE LAW OF SINES OR THE LAW OF COSINES SHOULD BE USED TO BEGIN TO SOLVE AN OBLIQUE TRIANGLE

Recall that it takes at least three pieces of known information to solve an oblique triangle. In Section 4.2, we used the Law of Sines to solve oblique triangles. We saw that the Law of Sines can be used when we are given the measure of an angle, the length of the side opposite that angle, and at least one other side or one other angle. Therefore, the Law of Sines can be used for the following cases.

> **Using the Law of Sines**
>
> If S represents a given side of a triangle and if A represents a given angle of a triangle, then the Law of Sines can be used to solve the three triangle cases shown below.
>
>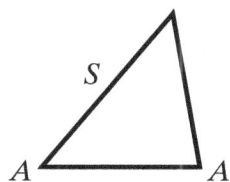
> The SAA Case
>
>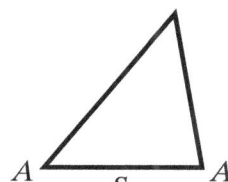
> The ASA Case
>
>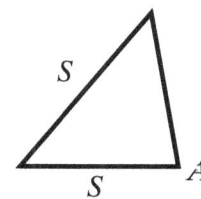
> The SSA Case
> (The Ambiguous Case)

The other cases in which three pieces of information can be known are the ASA, SSS, and AAA cases. We will ignore the AAA case because the AAA case never defines a unique triangle. Do you know why? View this explanation to see why. The Law of Sines cannot be used to begin to solve the SAS or SSS cases because in either case, no angle opposite a known side is given. However, we can use the **Law of Cosines** to obtain this needed information.

The Law of Cosines

If $A, B,$ and C are the measures of the angles of any triangle and if $a, b,$ and c are the lengths of the sides opposite the corresponding angles, then

$$a^2 = b^2 + c^2 - 2bc \cos A,$$

$$b^2 = a^2 + c^2 - 2ac \cos B, \text{ and}$$

$$c^2 = a^2 + b^2 - 2ab \cos C$$

Note that each of the three Law of Cosines equations relates all three sides of a triangle to an angle. In words, the Law of Cosines says that the square of any side of a triangle is equal to the sum of the squares of the remaining two sides, minus twice the product of the two remaining sides and the cosine of the angle between them. Note that the angle used in each of the three equations is the angle opposite the side of the triangle that is isolated on the left-hand side of the equation.

To derive the first equation, $a^2 = b^2 + c^2 - 2bc \cos A$, start by drawing a triangle so that the vertex of angle A is at the origin of a Cartesian plane. See Figure 15a.

Next, draw an **altitude** of length y from the vertex of angle C to the opposite side at point $(x, 0)$, thus creating the two right triangles shown in Figure 15b. Note that the coordinates of the vertex of angle C are (x, y). Using the **right triangle definitions** of cosine and sine, we see that $\cos A = \dfrac{x}{b}$ and $\sin A = \dfrac{y}{b}$. Therefore, we can determine that the x-coordinate of the vertex of angle C is $x = b \cos A$. Likewise, the y-coordinate of the vertex of angle C is $y = b \sin A$. See Figure 15c.

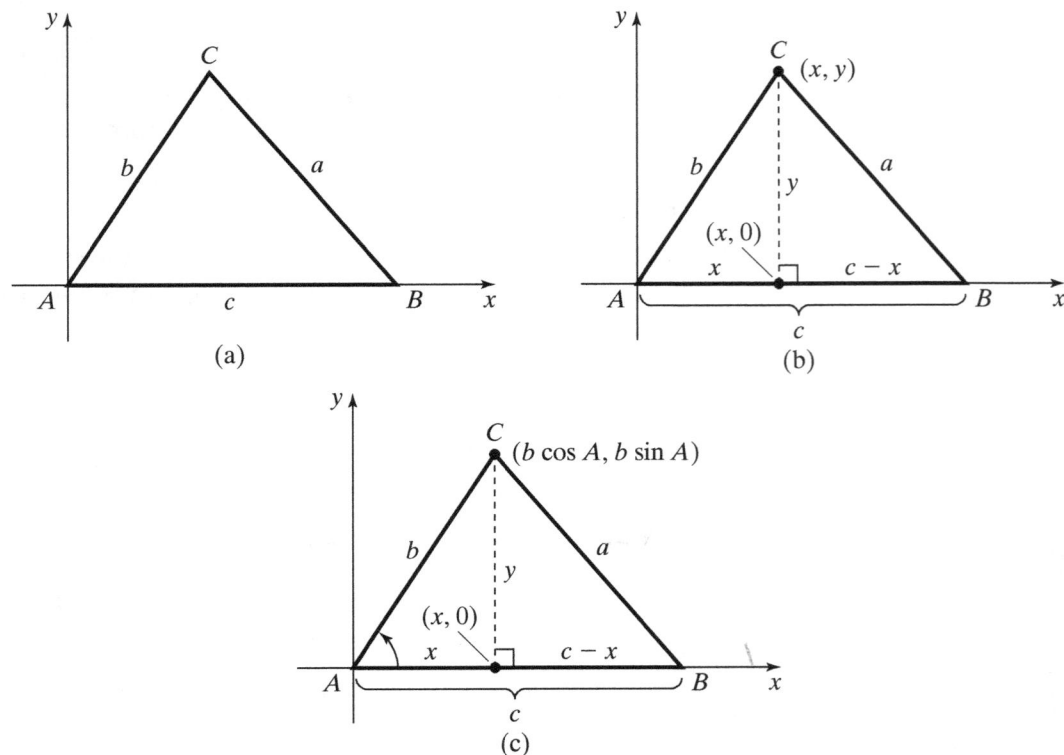

Figure 15

Now, isolate the triangle with side lengths of $c - x, y$, and a (see Figure 16) and use the Pythagorean Theorem to solve for a^2.

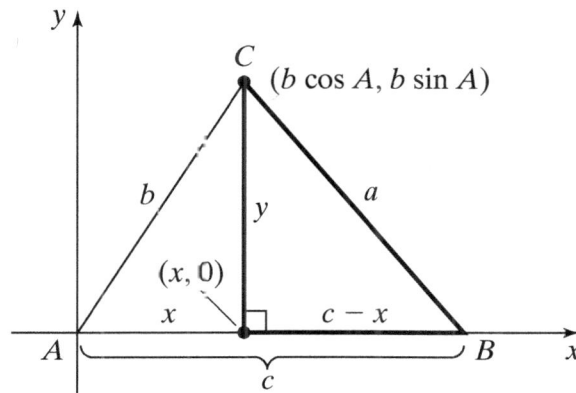

Figure 16

$$(c - x)^2 + y^2 = a^2$$ Write an equation representing the Pythagorean Theorem.

$$c^2 - 2xc + x^2 + y^2 = a^2$$ Square: $(c - x)^2 = c^2 - 2xc + x^2$.

$$c^2 - 2(b \cos A)c + (b \cos A)^2 + (b \sin A)^2 = a^2$$ Substitute $x = b \cos A$ and $y = b \sin A$.

$$c^2 - 2bc \cos A + b^2 \cos^2 A + b^2 \sin^2 A = a^2$$ Perform the operations and simplify.

$$c^2 - 2bc \cos A + b^2(\cos^2 A + \sin^2 A) = a^2$$ Factor out b^2 in the last two terms on the left-hand side.

$$c^2 - 2bc \cos A + b^2(1) = a^2$$ Use the Pythagorean identity: $\cos^2 A + \sin^2 A = 1$.

$$a^2 = b^2 + c^2 - 2bc \cos A$$ Rearrange the terms.

We have now verified the first of the three Law of Cosines equations. We can derive the other two equations in a similar manner. Note that we used an oblique triangle with three acute angles to derive the equation. This is not necessary. We could have derived the formula in a similar manner using an oblique triangle with an obtuse angle.

Solving the equation $a^2 = b^2 + c^2 - 2bc \cos A$ for $\cos A$, we get $\cos A = \dfrac{b^2 + c^2 - a^2}{2bc}$.

View this solution to verify. This leads to the following alternate form of the Law of Cosines.

The Alternate Form of the Law of Cosines

If A, B, and C are the measures of the angles of any triangle and if a, b, and c are the lengths of the sides opposite the corresponding angles, then

$$\cos A = \frac{b^2 + c^2 - a^2}{2bc},$$

$$\cos B = \frac{a^2 + c^2 - b^2}{2ac}, \text{ and}$$

$$\cos C = \frac{a^2 + b^2 - c^2}{2ab}$$

The three equations seen above are useful when determining the measure of a missing angle provided that the length of each of the three sides is known. Note that the angle on the left-hand side of the equation is opposite the side of the triangle that is squared and subtracted on the right-hand side of the equation.

My video summary

⊘ **Example 1** Determining If the Law of Sines or the Law of Cosines Should Be Used to Begin to Solve an Oblique Triangle

Decide whether the Law of Sines or the Law of Cosines should be used to begin to solve the given triangle. Do not solve the triangle.

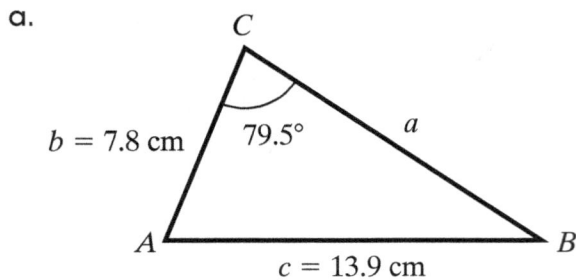

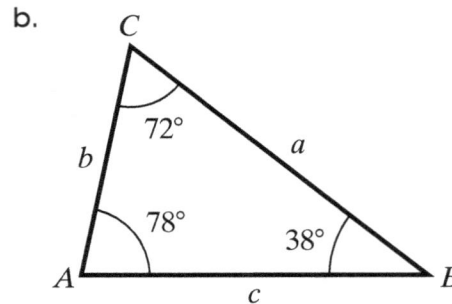

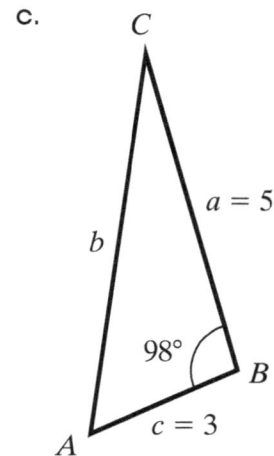

a.

$b = 7.8$ cm, $79.5°$, a, $c = 13.9$ cm

b.

$72°$, $78°$, $38°$, a, b, c

c.

$a = 5$, b, $98°$, $c = 3$

Solution

a. First, organize the given information.

Angles	Sides
$A =$ ___	$a =$ ___
$B =$ ___	$b = 7.8$ cm
$C = 79.5°$	$c = 13.9$ cm

The lengths of two sides and the measure of an angle opposite one of the sides is given. This is the SSA case. Therefore, the Law of Sines should be used to begin to solve this triangle.

b. For this triangle, we see that we are given the measure of all three angles but are not given the length of any of the sides. This is the AAA case. Neither the Law of Sines nor the Law of Cosines can be used to solve this triangle because the AAA case does not result in a unique triangle.

c. First, organize the given information.

Angles	Sides
$A =$ ___	$a = 5$
$B = 98°$	$b =$ ___
$C =$ ___	$c = 3$

The lengths of two sides and the measure of the angle between the two sides are given. This is the SAS case. Therefore, the Law of Cosines should be used to begin to solve this triangle.

You Try It   Work through this You Try It problem.

Work Exercises 1–6 in this textbook or in the MyMathLab Study Plan.

OBJECTIVE 2   USING THE LAW OF COSINES TO SOLVE THE SAS CASE

The SAS case is the situation where we are given the lengths of two sides of a triangle and the measure of the included angle. To solve such a triangle, first use the Law of Cosines to determine the length of the missing side (the side opposite the given angle.) Once we know the lengths of the three sides, we can use the Law of Sines or the Law of Cosines to determine the measure of another angle. Note that the longer the side length, the larger the angle and vice versa. This information can often be useful when solving triangles. We now summarize a technique for solving an SAS oblique triangle.

Solving an SAS Oblique Triangle

Step 1. Use the Law of Cosines to determine the length of the missing side.

Step 2. Determine the measure of the smaller of the remaining two angles using the Law of Sines or using the alternate form of the Law of Cosines.
This angle is the angle opposite of the shortest side and will always be acute.

Step 3. Use the fact that the sum of the measures of the three angles of a triangle is $180°$ to determine the measure of the remaining angle.

Whenever possible, we will avoid using rounded information to solve for the remaining parts of the triangle. When this cannot be avoided, we will agree to use information rounded to one decimal place unless some other guideline is stated.

My interactive video summary

⊙ **Example 2** Solving an SAS Triangle

Solve the given oblique triangle. Round the measures of all angles and the lengths of all sides to one decimal place.

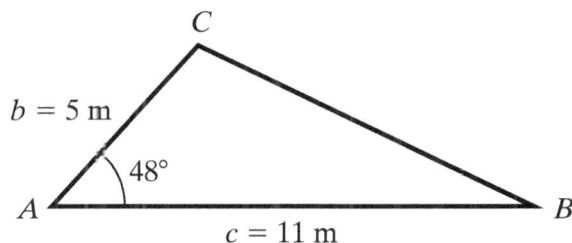

Solution Observe that we are given the following information.

Angles	Sides
$A = 48°$	$a = $ ____
$B = $ ____	$b = 5$ m
$C = $ ____	$c = 11$ m

Step 1. Use the Law of Cosines to determine the length of side a because we are given the measure of angle A.

$$a^2 = b^2 + c^2 - 2bc \cos A$$

Use the Law of Cosines equation that relates a, b, c, and A.

$$a^2 = (5)^2 + (11)^2 - 2(5)(11) \cos 48°$$

Substitute $b = 5, c = 11$, and $A = 48°$.

$$a = \sqrt{(5)^2 + (11)^2 - 2(5)(11) \cos 48°}$$

Take the square root of both sides.

$$a \approx 8.5$$

Use a calculator and round to one decimal place.

Insert the new information.

Angles	Sides
$A = 48°$	$a \approx 8.5$ m
$B = $ ____	$b = 5$ m
$C = $ ____	$c = 11$ m

Step 2. The smaller of the two missing angles is angle B because it is opposite the shortest side. We can now use either the Law of Sines or the alternate form of the Law of Cosines to determine the measure of angle B. Here, we use the alternate form of the Law of Cosines.

$$\cos B = \frac{a^2 + c^2 - b^2}{2ac}$$

Use the alternate form of the Law of Cosines equation that relates a, b, c, and $\cos B$.

$$\cos B \approx \frac{(8.5)^2 + (11)^2 - (5)^2}{2(8.5)(11)}$$

Substitute $a \approx 8.5, b = 5$, and $c = 11$.

$$\cos B \approx \frac{168.25}{187}$$

Perform the operations on the right-hand side and simplify.

Therefore, $B \approx \cos^{-1}\left(\dfrac{168.25}{187}\right) \approx 25.9°$.

Step 3. We can determine the measure of angle C by subtracting the measures of angle A and angle B from $180°$. Therefore, $C \approx 180° - 48° - 25.9° \approx 106.1°$.

My interactive video summary

⊗ We have now solved the oblique triangle, which can be seen in Figure 17. Watch this interactive video to see the solution to this example.

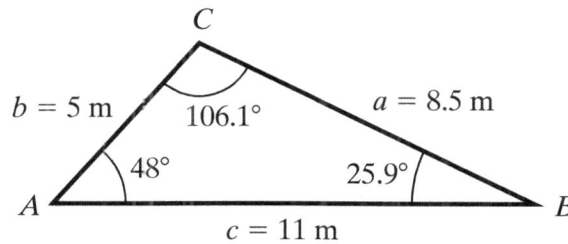

Figure 17

You Try It   Work through this You Try It problem.

Work Exercises 7–11 in this textbook or in the MyMathLab Study Plan.

OBJECTIVE 3   USING THE LAW OF COSINES TO SOLVE THE SSS CASE

The SSS case is the situation where we are given the lengths of all three sides of a triangle. Thus, we need to determine the measures of all three angles. To do this, we use the appropriate alternate form of the Law of Cosines to find the measure of the largest angle. Finding the largest angle first guarantees that the remaining two angles will always be acute. Next, use the Law of Sines or the alternate form of the Law of Cosines to find the measure of either of the two remaining acute angles. Finally, use the fact that the sum of the measures of the three angles of a triangle is $180°$ to find the measure of the third angle. We now summarize a technique for solving a SSS oblique triangle.

Solving a SSS Oblique Triangle

Step 1. Use the alternate form of the Law of Cosines to determine the measure of the largest angle. This is the angle opposite the longest side.

Step 2. Determine the measure of one of the remaining two angles using the Law of Sines or the alternate form of the Law of Cosines.
(This angle will always be acute.)

Step 3. Use the fact that the sum of the measures of the three angles of a triangle is $180°$ to determine the measure of the remaining angle.

My interactive video summary

⊗ **Example 3** Solving a SSS Triangle

Solve oblique triangle ABC f $a = 5$ ft, $b = 8$ ft, and $c = 12$ ft.

Solution

Step 1. Use the alternative form of the Law of Cosines to determine the measure of the angle C. This angle has the largest measure because it is opposite the longest side.

$$\cos C = \frac{a^2 + b^2 - c^2}{2ab}$$

Use the alternate form of the Law of Cosines equation that relates $a, b, c,$ and $\cos C$.

$$\cos C = \frac{(5)^2 + (8)^2 - (12)^2}{2(5)(8)}$$ Substitute $a = 5, b = 8$, and $c = 12$.

$$\cos C = -\frac{11}{16}$$ Perform the operations on the right-hand side and simplify.

Therefore, $C = \cos^{-1}\left(-\frac{11}{16}\right) \approx 133.4°$.

Observe that we now have the following information.

Angles	Sides
$A = $ ____	$a = 5$ ft
$B = $ ____	$b = 8$ ft
$C \approx 133.4°$	$c = 12$ ft

Step 2. We can use either the Law of Sines or the alternate form of the Law of Cosines to determine the measure of angle B. Here, we use the Law of Sines.

$$\frac{\sin B}{b} = \frac{\sin C}{c}$$ Use the Law of Sines to write a proportion.

$$\frac{\sin B}{8} \approx \frac{\sin 133.4°}{12}$$ Substitute the known information.

$$\sin B \approx \frac{8 \sin 133.4°}{12}$$ Multiply both sides by 8 to isolate $\sin B$.

Therefore, $B \approx \sin^{-1}\left(\frac{8 \sin 133.4°}{12}\right) \approx 29.0°$.

Note that there is another solution to the equation $\sin B \approx \dfrac{8 \sin 133.4°}{12}$, where $0° < B < 180°$. This solution is $B \approx 180° - \sin^{-1}\left(\dfrac{8 \sin 133.4°}{12}\right) \approx 151.0°$. However, because we determined that the largest angle was $C \approx 133.4°$, we can ignore this solution.

Step 3. To find the measure of angle A, subtract the measures of angles C and B from 180°. Thus, $A \approx 180° - 133.4° - 29.0° \approx 17.6°$ and the triangle is solved. See Figure 18.

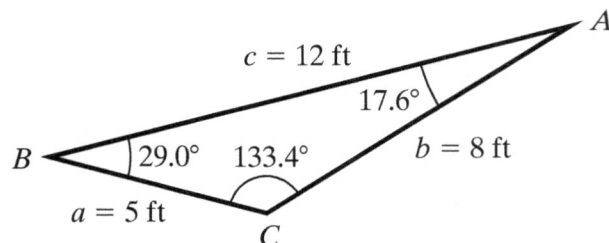

Figure 18

You Try It Work through this You Try It problem.

Work Exercises 12–16 in this textbook or in the MyMathLab Study Plan.

OBJECTIVE 4   USING THE LAW OF COSINES TO SOLVE APPLIED PROBLEMS INVOLVING OBLIQUE TRIANGLES

As with the Law of Sines, the Law of Cosines can be a useful tool to help solve many applications that arise involving triangles that are not right triangles. You may want to review the concept of bearing that was introduced in Section 4.2.

My video summary   ⊙ **Example 4** Determining the Distance between Two Airplanes

Two planes take off from different runways at the same time. One plane flies at an average speed of 350 mph with a bearing of N 21° E. The other plane flies at an average speed of 420 mph with a bearing of S 84° W. How far are the planes from each other 2 hours after takeoff? Round to the nearest tenth of a mile.

Solution   After 2 hours, the plane flying with a bearing of N 21° E is 700 miles from the airport and the plane flying with a bearing of S 84° W is 840 miles from the airport. Draw a diagram with the airport located at the origin, as shown in Figure 19.

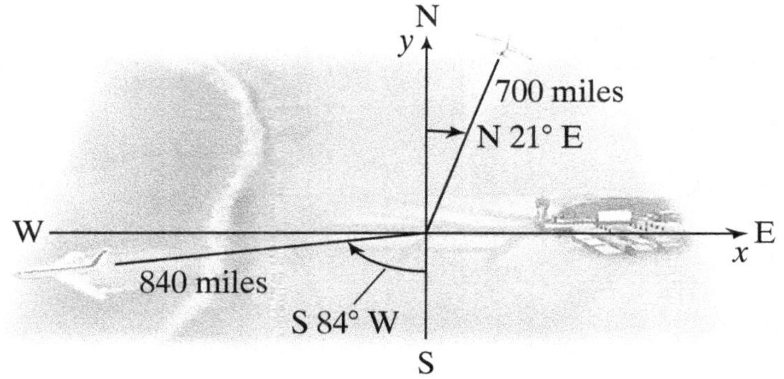

Figure 19

Let d represent the distance between the two planes. We have been given the length of two sides, $a = 700$ and $b = 840$, and the information needed to find the included angle. The measure of the included angle D is $D = 21° + 90° + 6° = 117°$. See Figure 20.

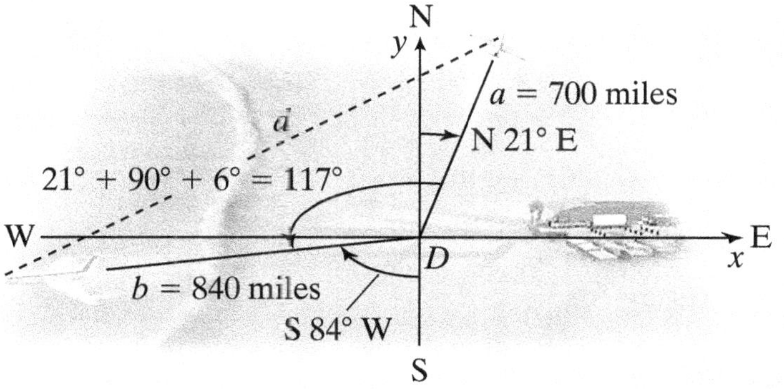

Figure 20

Now that we know the two sides and the angle between them, we use the Law of Cosines with $a = 700$, $b = 840$, and $D = 117°$ to solve for d.

$d^2 = a^2 + b^2 - 2ab \cos D$ Use the alternate form of the Law of Cosines equation that relates a, b, d, and $\cos D$.

$d^2 = (700)^2 + (840)^2 - 2(700)(840) \cos 117°$ Substitute $a = 700$, $b = 840$, and $D = 117°$.

$d = \sqrt{(700)^2 + (840)^2 - 2(700)(840) \cos 117°}$ Take the square root of both sides.

$d \approx 1315.1$ Round to the nearest tenth.

Therefore, the planes are approximately 1315.1 miles apart after 2 hours.

Example 5 Determining the Central Angle Given the Radius and the Length of a Chord

A **chord** of a circle is a line segment with endpoints that both lie on the circumference of a circle. Determine the measure of the central angle (in degrees) if the length of the chord intercepted by the central angle of a circle of radius 10 inches is 16.5 inches. Round the measure of the central angle to one decimal place.

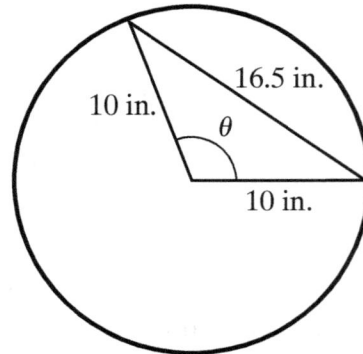

Solution To determine the measure of θ, we use the alternate form of the Law of Cosines with $a = 10$, $b = 10$, and $c = 16.5$, where c is the side opposite angle θ.

$\cos \theta = \dfrac{a^2 + b^2 - c^2}{2ab}$ Use the alternate form of the Law of Cosines equation that relates a, b, c, and $\cos \theta$.

$\cos \theta = \dfrac{(10)^2 + (10)^2 - (16.5)^2}{2(10)(10)}$ Substitute $a = 10$, $b = 10$, and $c = 16.5$.

$\cos \theta = -\dfrac{72.25}{200}$ Perform the operations on the right-hand side and simplify.

Therefore, $\theta = \cos^{-1}\left(-\dfrac{72.25}{200}\right) \approx 111.2°$.

You Try It   Work through this You Try It problem.

Work Exercises 17–25 in this textbook or in the MyMathLab Study Plan.

4.3 Exercises

Skill Check Exercises

For SCE-1 and SCE-2, use a calculator in degree mode to evaluate each expression. Round to one decimal place.

SCE-1. $\sqrt{(3)^2 + (5)^2 - 2(3)(5)\cos 36°}$

SCE-2. $\cos^{-1}\left(\dfrac{(3.1)^2 + (5)^2 - (3)^2}{2(3.1)(5)}\right)$.

In Exercises 1–6, decide whether the Law of Sines or the Law of Cosines should be used to begin to solve the given triangle. Do not solve the triangle.

1.
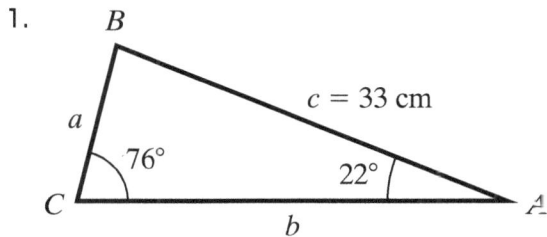

2.
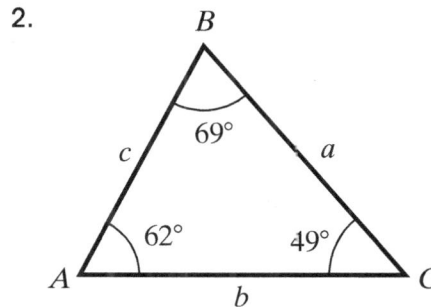

3.
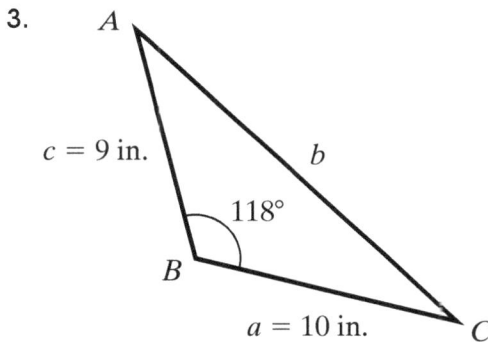

4.
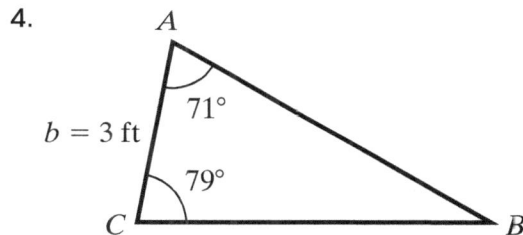

5.
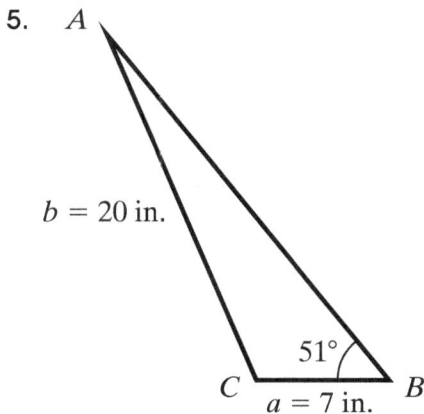

6.
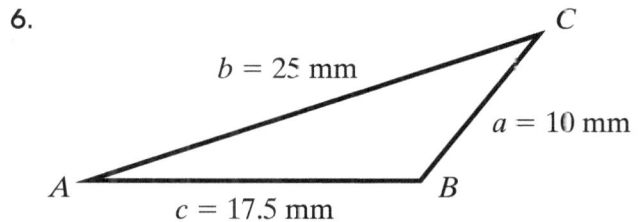

In Exercises 7–11, solve each oblique triangle. Round the measures of all angles and the lengths of all sides to one decimal place.

7.

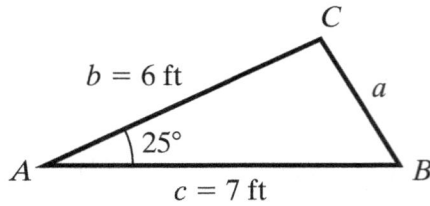

8.

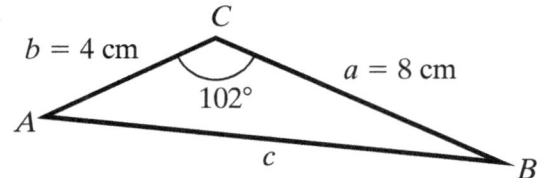

9.

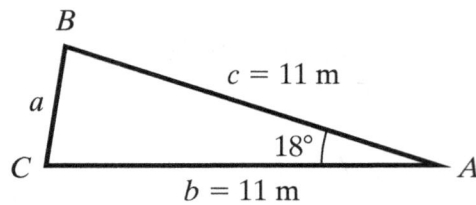

10. $a = 6, b = 8$, and $C = 20°$

11. $b = 2, c = 9$, and $A = 130°$

In Exercises 12–16, solve each triangle. Round the measures of all angles to one decimal place.

SbS **12.**

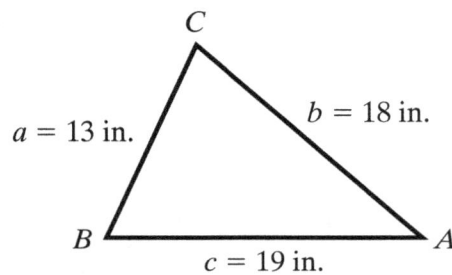

SbS **13.**

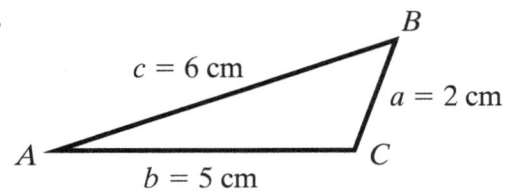

SbS **14.**

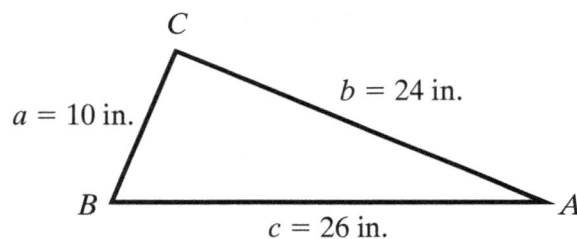

SbS **15.** $a = 11, b = 18$, and $c = 20$

SbS **16.** $a = 18, b = 14$, and $c = 13$

17. Two planes take off from the same airport at the same time using different runways. One plane flies at an average speed of 300 mph with a bearing of N 29° E. The other plane flies at an average speed of 350 mph with a bearing of S 47° W. How far are the planes from each other 3 hours after takeoff? Round to the nearest tenth of a mile.

18. Two planes take off from the same airport at the same time using different runways. One plane flies at an average speed of 500 mph with a bearing of N 32° W. The other plane flies at an average speed of 420 mph with a bearing of S 20° W. How far are the planes from each other 2 hours after takeoff? Round to the nearest tenth of a mile.

19. A cruise ship leaves port *A* on a four-day cruise and visits port *B* and port *C* as shown in the figure. On what bearing should the cruise ship navigate to return from port *C* back to port *A*? Round each angle to one decimal place.

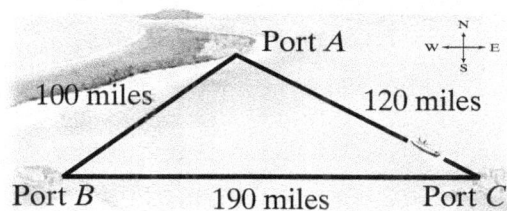

20. Determine the measure of the central angle (in degrees) if the length of the chord intercepted by the central angle of a circle of radius 35 inches is 14 inches. Round the measure of the central angle to one decimal place.

21. Determine the length of the chord intercepted by a central angle of 54° in a circle with a radius of 18 centimeters. Round to the nearest tenth of a centimeter.

22. Determine the lengths of the diagonals of a parallelogram if the adjacent sides are 13 cm and 11 cm and if the angle between them is 42°. Round the lengths of the diagonals to the nearest tenth of a centimeter.

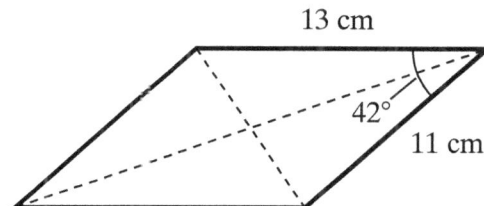

23. Three points in a plane determine a unique triangle. Solve this triangle formed by the points $A(-4, -1)$, $B(-1, 3)$, and $C(5, -1)$. Round the measures of all angles and the lengths of all sides to one decimal place.

24. Civil engineers are in the planning phases of a major tunnel project in which they plan to bore through a mountain. To determine the length of the proposed tunnel, they take measurements as shown in the figure. What is the length of the proposed tunnel to the nearest tenth of a foot?

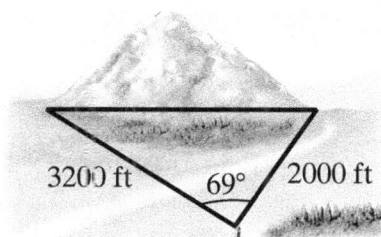

25. A common design for a mountain cabin is an A-frame cabin. A-frame cabins are fairly easy to construct, and the steeply pitched roof line is perfect for helping snow fall to the ground during heavy winterstorms. Determine the angle between the two sides of the roof of an A-frame cabin if the sides are both 22 feet long and the base of the cabin is 20 feet wide. Round to the nearest tenth of a degree.

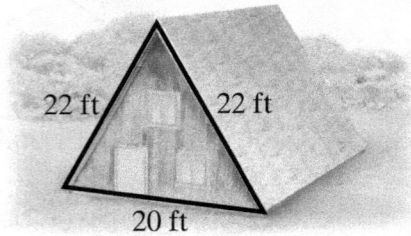

4.4 Area of Triangles

THINGS TO KNOW

Before working through this section, be sure that you are familiar with the following concepts:

	VIDEO	ANIMATION	INTERACTIVE

You Try It
1. Understanding the Inverse Sine Function (Section 2.4)

You Try It
2. Understanding the Inverse Cosine Function (Section 2.4)

You Try It
3. Using the Law of Sines to Solve the SAA Case or the ASA Case (Section 4.1)

You Try It
4. Using the Law of Cosines to Solve the SSS Case (Section 4.1)

OBJECTIVES

1 Determining the Area of Oblique Triangles

2 Using Heron's Formula to Determine the Area of an SSS Triangle

3 Solving Applied Problems Involving the Area of Triangles

....

OBJECTIVE 1   DETERMINING THE AREA OF OBLIQUE TRIANGLES

A familiar formula for the area of a triangle is $\text{Area} = \frac{1}{2}bh$, where b is the length of the base of the triangle and h is the length of the height, or **altitude**, of the triangle.

Area of a Triangle

In any triangle, the area is given by Area $= \frac{1}{2}bh$, where b is the length of the base of the triangle and h is the length of the altitude drawn to that base (or drawn to an extension of that base).

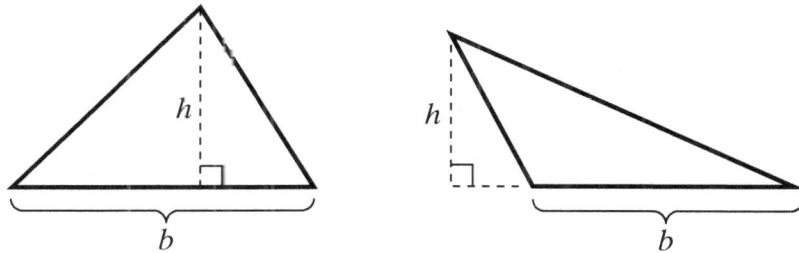

If the height of a triangle is not known, then the formula Area $= \frac{1}{2}bh$ cannot always be used readily. In this section we will develop alternate formulas to find the area of a triangle.

To develop the first set of formulas, start by drawing an oblique triangle ABC with an altitude of length h drawn from the vertex of B. See Figure 21.

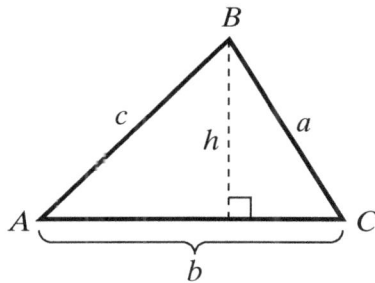

Figure 21

By the right triangle definition of the sine function, we have $\sin A = \dfrac{h}{c}$ or, equivalently,

$h = c \sin A$. Substituting $h = c \sin A$ into the area formula, we get Area $= \frac{1}{2}bh =$

$\frac{1}{2}b(c \sin A)$. Note that we could have derived this same formula had we started

with an oblique triangle with an obtuse angle. The formula Area $= \frac{1}{2}bc \sin A$ is just

one of three similar area formulas. The other two formulas can be derived the exact same way by first drawing the altitude from another vertex of the triangle.

Area of a Triangle

If A, B, and C are the measures of the angles of any triangle and if a, b, and c are the lengths of the sides opposite the corresponding angles, then the area of triangle ABC is given by

$$\text{Area} = \frac{1}{2}bc \sin A, \text{ or } \text{Area} = \frac{1}{2}ac \sin B, \text{ or } \text{Area} = \frac{1}{2}ab \sin C.$$

Note that the three area formulas above require that the lengths of two sides of a triangle are known along with the measure of the angle between them. In other words, we must have a SAS (side–angle–side) triangle before using these formulas. If not enough information is known, use the Law of Sines or the Law of Cosines to determine the needed information.

My video summary   ⊘ **Example 1** Determining the Area of an Oblique Triangle

Determine the area of each triangle. Round each answer to two decimal places.

a.

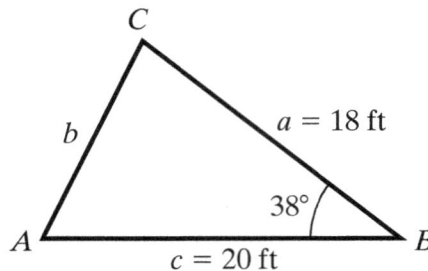

b.

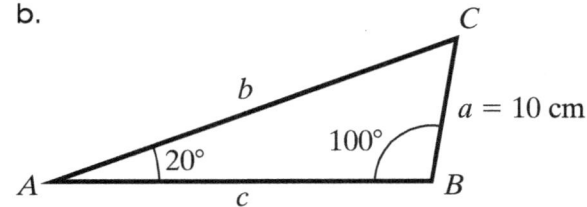

Solution

a. We are given the lengths of two sides, $a = 18$ feet and $c = 20$ feet, and the measure of the angle between them, $B = 38°$. Therefore, we use the formula $\text{Area} = \dfrac{1}{2}ac \sin B$ to determine the area.

$$\text{Area} = \frac{1}{2}ac \sin B \qquad \text{Write the appropriate area formula.}$$

$$= \frac{1}{2}(18)(20) \sin 38° \qquad \text{Substitute the known information.}$$

$$\approx 110.82 \text{ ft}^2 \qquad \text{Use a calculator in degree mode and round to two decimal places.}$$

b. Before we can use one of the area formulas, we must know the lengths of two sides and the measure of the angle between them. Because we know the length of side a and we know the measure of angle B, we need to find the length of side c, thus giving us a SAS triangle.

To determine the length of side c, first determine the measure of angle C using the fact that the sum of the measures of the three angles must be $180°$. Then, we will use the Law of Sines to find the length of side c.

$$20° + 100° + C = 180° \qquad \text{The sum of the measures of the angles of any triangle is } 180°.$$

$$120° + C = 180° \qquad \text{Add.}$$

$$C = 60° \qquad \text{Solve for } C.$$

Organize the known information.

Angles	Sides
$A = 20°$	$a = 10\text{ cm}$
$B = 100°$	$b = $ ____
$C = 60°$	$c = $ ____

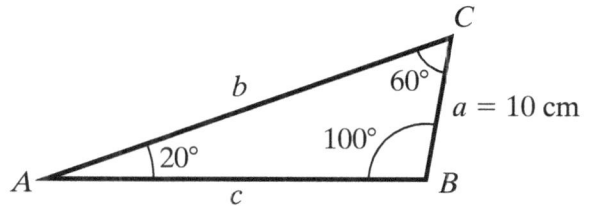

Use the Law of Sines to determine the length of side c.

$$\frac{c}{\sin C} = \frac{a}{\sin A}$$

Use the Law of Sines to write a proportion.

$$\frac{c}{\sin 60°} = \frac{10}{\sin 20°}$$

Substitute the known information.

$$c = \frac{10 \sin 60°}{\sin 20°}$$

Multiply both sides by $\sin 60°$ to solve for c.

⚠ **To avoid rounding errors, we will use *exact* values found in intermediate steps. Therefore, we will use the value $c = \dfrac{10 \sin 60°}{\sin 20°}$ when substituting into an area formula.**

Now use the formula Area $= \dfrac{1}{2} ac \sin B$ to determine the area.

$$\text{Area} = \frac{1}{2} ac \sin B$$

Write the appropriate area formula.

$$= \frac{1}{2}(10)\left(\frac{10 \sin 60°}{\sin 20°}\right) \sin 100°$$

Substitute the known information using exact values.

$$\approx 124.68 \text{ cm}^2$$

Use a calculator in degree mode and round to two decimal places.

My interactive video summary

▷ Work through this interactive video to see the entire worked-out solution. ●

You Try It Work through this You Try It problem.

Work Exercises 1–12 in this textbook or in the MyMathLab Study Plan.

OBJECTIVE 2   USING HERON'S FORMULA TO DETERMINE THE AREA
OF AN SSS TRIANGLE

Suppose that we wish to find the area of a triangle in which the lengths of the three sides are known. One way to determine this area is to use the Law of Cosines to determine the measure of any angle, then use one of the three previously stated area formulas. We illustrate this procedure in Example 2.

(eText Screens 4.4-1–4.4-19)

Example 2 Determining the Area of an SSS Oblique Triangle

Determine the area of the given triangle. Round to two decimal places.

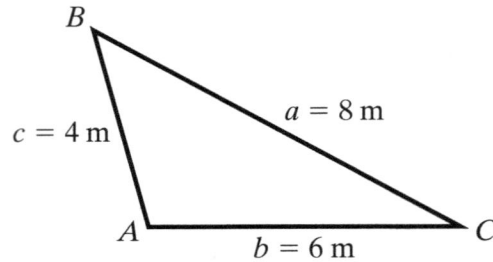

Solution To use an area formula, first determine the measure of any of the three angles. Here, we determine the measure of angle A using an alternate form of the Law of Cosines.

$$\cos A = \frac{b^2 + c^2 - a^2}{2bc}$$

Use the alternate form of the Law of Cosines equation that relates a, b, c, and $\cos A$.

$$\cos A = \frac{(6)^2 + (4)^2 - (8)^2}{2(6)(4)}$$

Substitute $a = 8$, $b = 6$, and $c = 4$.

$$\cos A = -\frac{1}{4}$$

Perform the operations on the right-hand side and simplify.

Therefore, the exact value of A is $A = \cos^{-1}\left(-\frac{1}{4}\right)$.

Once again, to avoid rounding errors, we will substitute the exact value, $A = \cos^{-1}\left(-\frac{1}{4}\right)$, when using the appropriate area formula.

Now use the formula Area $= \frac{1}{2}bc \sin A$ to determine the area.

$$\text{Area} = \frac{1}{2}bc \sin A$$

Write the appropriate area formula.

$$= \frac{1}{2}(6)(4) \sin\left(\cos^{-1}\left(-\frac{1}{4}\right)\right)$$

Substitute the known information using exact values.

$$\approx 11.62 \text{ m}^2$$

Use a calculator in degree mode and round to two decimal places.

The procedure for determining the area of an SSS triangle can be computed much more efficiently using a formula that does not require the knowledge of the measure of any angle. This formula is known as **Heron's formula** and is named after the Greek mathematician Heron of Alexandria. Heron's formula requires that we first compute the **semiperimeter**, which is exactly equal to half of the sum of the lengths of the three sides of a triangle. Heron's formula is stated below.

> **Heron's Formula**
>
> Suppose that a triangle has side lengths of a, b, and c. If the semiperimeter is $s = \dfrac{1}{2}(a + b + c)$, then the area of the triangle is
>
> $$\text{Area} = \sqrt{s(s - a)(s - b)(s - c)}.$$

My video summary ⊘ There are many ways to prove Heron's formula. One way requires the area formula $\text{Area} = \dfrac{1}{2}ab \sin C$, an a ternate form of the Law of Cosines, the semiperimeter formula, and a Pythagorean identity of the form $\sin C = \sqrt{1 - \cos^2 C}$. To see the complete proof of Heron's formula, watch this video.

In Example 3, we will use the same triangle that was used in Example 2 to illustrate the use of Heron's formula.

My video summary ⊘ **Example 3** Determining the Area of an SSS Oblique Triangle Using Heron's Formula

Use Heron's formula to determine the area of the given triangle. Round to two decimal places.

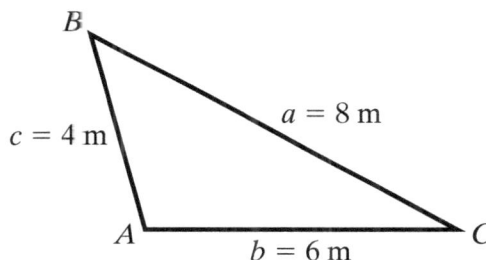

Solution First, determine the semiperimeter.

$$s = \frac{1}{2}(a + b + c) \qquad \text{Write the formula for the semiperimeter.}$$

$$= \frac{1}{2}(8 + 6 + 4) = 9 \text{ m} \qquad \text{Substitute the known information and simplify.}$$

Now use Heron's formula with $a = 8$, $b = 6$, $c = 4$, and $s = 9$.

$$\text{Area} = \sqrt{s(s - a)(s - b)(s - c)} \qquad \text{Write Heron's formula.}$$

$$= \sqrt{9(9 - 8)(9 - 6)(9 - 4)} \qquad \text{Substitute the known information.}$$

$$= \sqrt{9(1)(3)(5)} \qquad \text{Simplify inside the radical.}$$

$$\approx 11.62 \text{ m}^2 \qquad \text{Use a calculator and round to two decimal places.}$$

Notice that the area computed using Heron's formula in Example 3 is the same as the area found in Example 2!

You Try It Work through this You Try It problem.

Work Exercises 13–15 in this textbook or in the MyMathLab Study Plan.

OBJECTIVE 3   SOLVING APPLIED PROBLEMS INVOLVING THE AREA OF TRIANGLES

The formulas discussed in this section can be used to solve applications that involve finding the area of triangles.

Example 4 Determining the Area of the Cross Section of a House

A painter who is painting a house has only one side of the house left to paint. He has enough paint to cover 1200 square feet. A cross section of the unpainted side of the house is shown in the figure. What is the area of the unpainted side of the house? Does he have enough paint to finish the job?

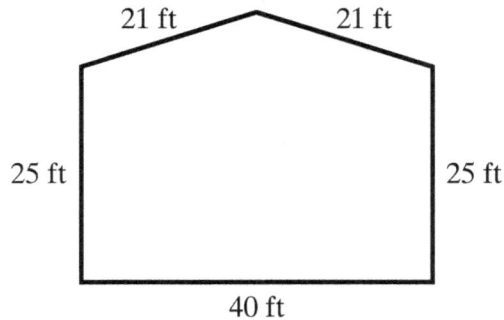

Solution First, separate the cross section of the side of the house into a rectangle and a triangle, as shown in Figure 22.

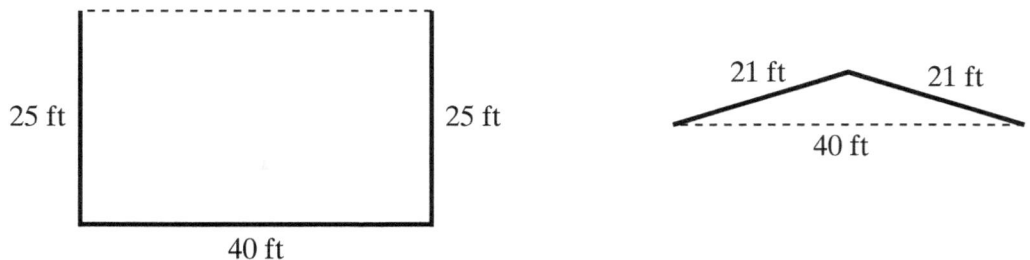

Figure 22

The total area of the cross section is the sum of the areas of the rectangular portion and the triangular portion seen in Figure 22. The area of the rectangular portion is $(25)(40) = 1000 \text{ ft}^2$.

Using Heron's formula with a semiperimeter of $s = \dfrac{1}{2}(21 + 21 + 40) = 41$ feet, we can determine the area of the triangular portion of the cross section. The area of the triangular portion of the side of the house is

$$\text{Area} = \sqrt{s(s - a)(s - b)(s - c)} = \sqrt{41(41 - 21)(41 - 21)(41 - 40)}$$
$$= \sqrt{41(20)(20)(1)} = \sqrt{16{,}400} \text{ ft}^2.$$

Thus, the total area of the unpainted side of the house is $1000 \text{ ft}^2 + \sqrt{16{,}400} \text{ ft}^2$ $\approx 1128 \text{ ft}^2$. The painter has enough paint to finish the job. ●

You Try It   Work through this You Try It problem.

Work Exercises 16–21 in this textbook or in the MyMathLab Study Plan.

4.4 Exercises

Skill Check Exercises

SCE-1. Use a calculator to evaluate the expression $\sqrt{12.5(12.5-11)(12.5-9)(12.5-5)}$. Round to two decimal places.

In Exercises 1–12, determine the area of each triangle. Round your answer to two decimal places.

1.

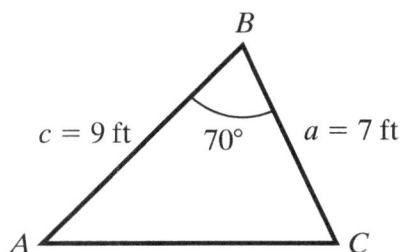

$c = 9$ ft, $70°$, $a = 7$ ft

2.

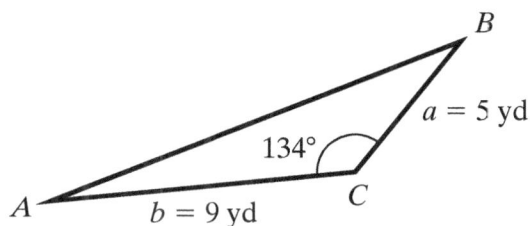

$134°$, $a = 5$ yd, $b = 9$ yd

3.

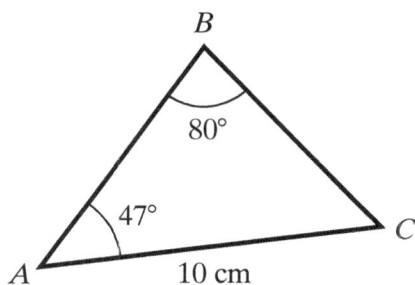

$80°$, $47°$, 10 cm

4.

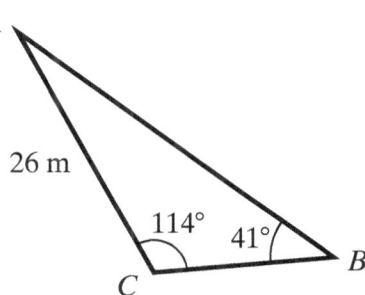

26 m, $114°$, $41°$

5.

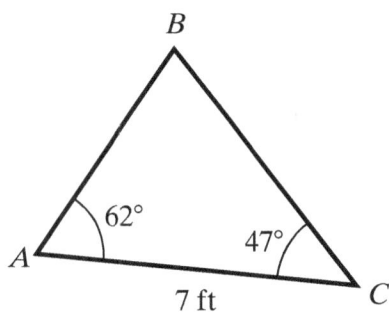

$62°$, $47°$, 7 ft

6.

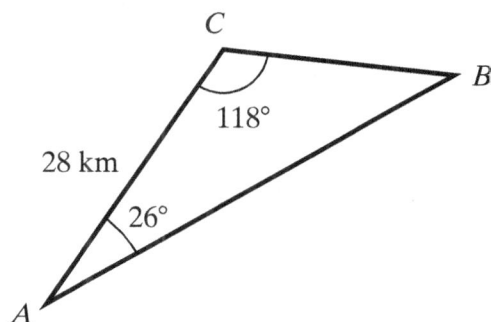

$118°$, 28 km, $26°$

7. $a = 12$ cm, $c = 15$ cm, $B = 56°$

8. $b = 11$ yd, $c = 7$ yd, $B = 102°$

9. $a = 7$ ft, $A = 72°$, $B = 51°$

10. $b = 4$ m, $B = 21°$, $C = 131°$

11. $a = 19$ km, $B = 74°$, $C = 81°$

12. $c = 20$ cm, $B = 59°$, $A = 93°$

In Exercises 13–15, use Heron's formula to determine the area of each triangle. Round your answer to two decimal places.

13.

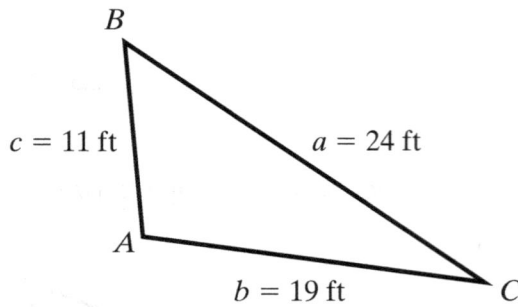

$c = 11$ ft
$a = 24$ ft
$b = 19$ ft

14.

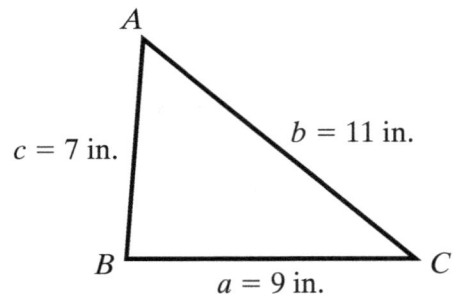

$c = 7$ in.
$b = 11$ in.
$a = 9$ in.

15.

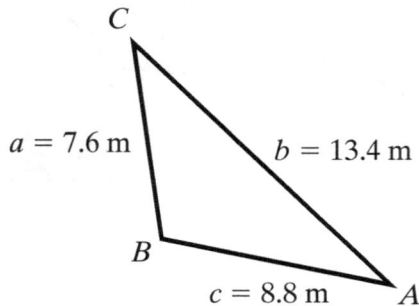

$a = 7.6$ m
$b = 13.4$ m
$c = 8.8$ m

16. Two sides of a triangular lot measure 58 feet by 72 feet, and the angle between these two sides is 65°. Find the area of the lot.

17. A triangular piece of commercial real estate is priced at $5.25 per square foot. Determine the cost of the triangular piece of land if two sides of the lot measure 180 feet by 200 feet and the angle between these two sides is 70°. Round to the nearest dollar.

18. A local merchant currently has a triangular gravel parking lot for her customers. The measurements of the lot are 125 feet by 70 feet by 65 feet. Determine the total cost to pave the lot if the price is $1.15 per square foot.

19. A painter who is painting a house has only one side of the house left to paint. He has enough paint to cover 1000 square feet. A cross section of the unpainted side of the house is shown in the figure. What is the area of the unpainted side of the house? Does the painter have enough paint to finish the job?

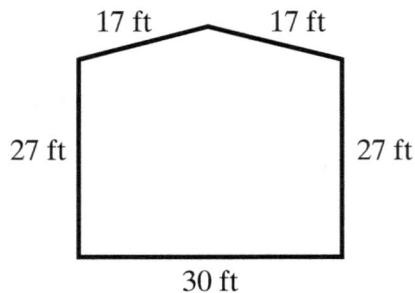

17 ft   17 ft

27 ft   27 ft

30 ft

20. Determine the area of the cross section of a house shown in the figure below.

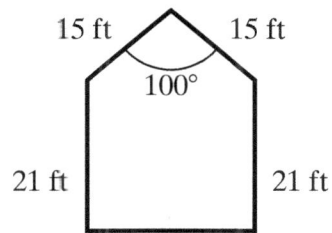

15 ft   15 ft

100°

21 ft   21 ft

21. A couple is planning on carpeting their upstairs bonus room, which has a floor in the shape of a triangle. Two of the adjacent sides of the room are 12 feet and 11 feet. The angle opposite the 11-foot side is 52°. If the cost of the carpet is $1.25 per square foot, how much, to the nearest dollar, will it cost to carpet the room?

Polar Equations, Complex Numbers, and Vectors

CHAPTER FIVE CONTENTS

5.1 Polar Coordinates and Polar Equations

THINGS TO KNOW

Before working through this section, be sure that you are familiar with the following concepts:

		VIDEO	ANIMATION	INTERACTIVE
You Try It	1. Understanding the Four Families of Special Angles (Section 1.5)			⊘
You Try It	2. Understanding the Definitions of the Trigonometric Functions of General Angles (Section 1.5)	⊘	▭	
You Try It	3. Understanding the Signs of the Trigonometric Functions (Section 1.5)	⊘		
You Try It	4. Evaluating Trigonometric Functions of Angles Belonging to the $\frac{\pi}{3}, \frac{\pi}{6},$ or $\frac{\pi}{4}$ Families (Section 1.5)			⊘
You Try It	5. Solving Trigonometric Equations That Are Linear in Form (Section 3.5)			⊘

OBJECTIVES

1 Plotting Points Using Polar Coordinates
2 Determining Different Representations of a Point (r, θ)
3 Converting a Point from Polar Coordinates to Rectangular Coordinates

OBJECTIVE 1   PLOTTING POINTS USING POLAR COORDINATES

So far in this eText we have used the rectangular coordinate system (also called the Cartesian coordinate system) when plotting points and sketching the graphs of functions. This is not the only system that can be used to plot points and sketch functions. In this section we begin our study of the **polar coordinate system.** This system is based on a fixed point called the **pole** and a ray with the vertex at the pole. This ray is called the **polar axis** and is shown in Figure 1.

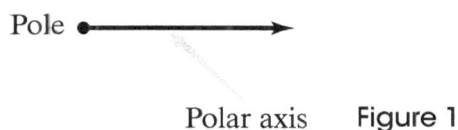

Pole •———————➤

Polar axis   **Figure 1**

A point P in the polar coordinate system is represented by the ordered pair $P(r, \theta)$. The angle θ is the angle (measured in radians or degrees) between the polar axis and the line segment from the pole to point P. This angle can be positive, negative, or zero. The value r is the **directed distance** from the pole to point P. This directed distance can be positive, negative, or zero. The following box shows the three cases of r with a positive angle θ.

Definition   Directed Distance

Given an ordered pair $P(r, \theta)$ in the polar coordinate system, the **directed distance** r can be positive, negative, or zero.

If $r > 0$, then P lies on the terminal side of angle θ.

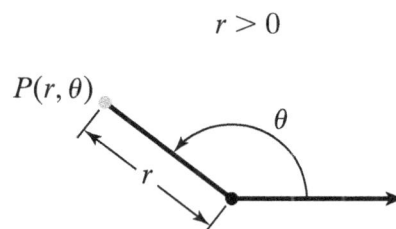

$r > 0$

If $r < 0$, then P lies on the ray opposite of the terminal side of angle θ.

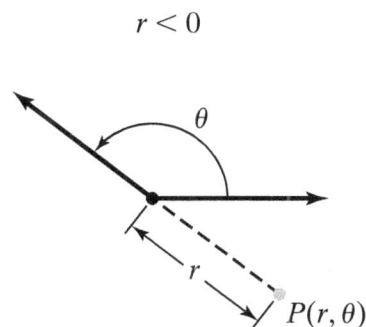

$r < 0$

If $r = 0$, then P lies on the pole regardless of the measure of angle θ.

$$r = 0$$

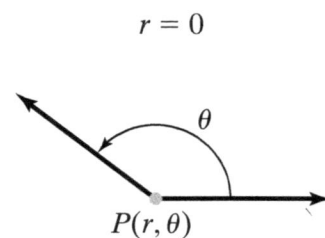

$$P(r, \theta)$$

When plotting polar coordinates and sketching polar equations, we will often use a polar grid. A polar grid consists of a series of concentric circles of different radii and pre-sketched angles in **standard position**. See Figure 2.

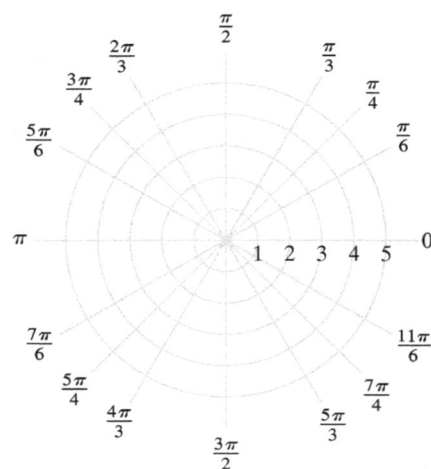

Figure 2 A polar grid

When plotting points in the polar coordinate system, we first sketch the angle in standard position. Then, we locate the point P using the appropriate directed distance for r.

📝 *My video summary*

⊙ **Example 1** Plotting Points Using Polar Coordinates

Plot the following points in a polar coordinate system.

a. $A\left(3, \dfrac{\pi}{4}\right)$ **b.** $B(-2, 120°)$ **c.** $C\left(1.5, -\dfrac{7\pi}{6}\right)$ **d.** $D\left(-3, -\dfrac{3\pi}{4}\right)$

Solution

a. To plot the point $A\left(3, \dfrac{\pi}{4}\right)$, we identify that

$r = 3$ and $\theta = \dfrac{\pi}{4}$. Thus, start by drawing the

angle $\theta = \dfrac{\pi}{4}$ on a polar grid. Because the

value of r is positive, the point A lies along the terminal side of angle θ, 3 units from the pole. See Figure 3.

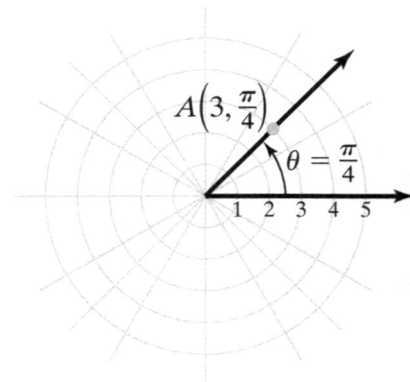

Figure 3 The point $A\left(3, \dfrac{\pi}{4}\right)$ lies along the terminal side of

$\theta = \dfrac{\pi}{4}$ a distance of 3 units from the pole.

5.1 Polar Coordinates and Polar Equations   5-3

b. To plot the point $B(-2, 120°)$, we identify that $r = -2$ and $\theta = 120°$. Thus, start by drawing the angle $\theta = 120°$ on a polar grid. Because the value of r is negative, the point B lies along the ray directly opposite from the terminal side of angle θ, 2 units from the pole. See Figure 4.

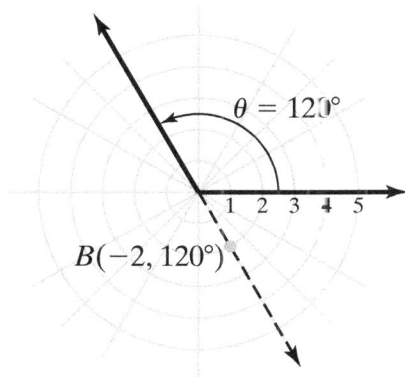

Figure 4 The point $B(-2, 120°)$ lies along the ray directly opposite the terminal side of $\theta = 120°$ a distance of 2 units from the pole.

My video summary ⊘ Try plotting the points $C\left(1.5, -\frac{7\pi}{6}\right)$ and $D\left(-3, -\frac{3\pi}{4}\right)$ on your own. Then watch this video to see how to plot these points correctly. The two points are plotted in Figure 5.

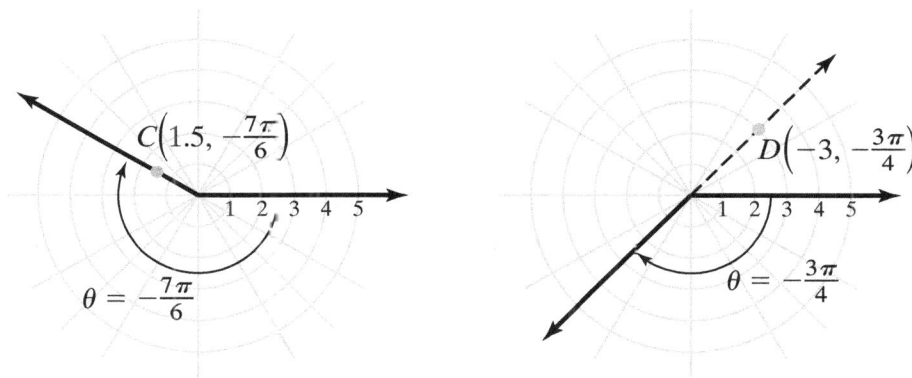

Figure 5 The points $C\left(1.5, -\frac{7\pi}{6}\right)$ and $D\left(-3, -\frac{3\pi}{4}\right)$ are shown plotted in the polar coordinate system.

You Try It Work through this You Try It problem.

Work Exercises 1–10 in this textbook or in the MyMathLab Study Plan.

OBJECTIVE 2 DETERMINING DIFFERENT REPRESENTATIONS OF THE POINT (r, θ)

My video summary ⊘ In Example 1 we plotted the points $A\left(3, \frac{\pi}{4}\right)$ and $D\left(-3, -\frac{3\pi}{4}\right)$. Watch this video to see how to plot these two points. These two points are shown again in Figure 6. Notice that points A and D are positioned at the exact same location.

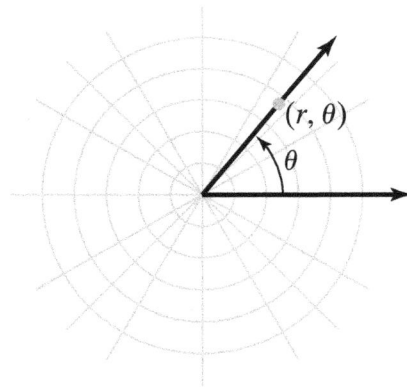

Figure 6 The points $A\left(3, \dfrac{\pi}{4}\right)$ and $D\left(-3, -\dfrac{3\pi}{4}\right)$ have the exact same location in the polar coordinate system.

Unlike points in the rectangular coordinate system, which have exactly one representation, points in the polar coordinate system can have infinitely many representations. Given a point in the polar coordinate system having coordinates (r, θ), we can determine a different representation of this point by keeping r the same and by choosing a different angle that is coterminal to the given angle. Or, we can change the sign of r and choose an angle whose terminal side is located one-half of a rotation from angle θ, as in the case of the two points shown in Figure 6. We now summarize how to obtain different representations of the same point.

Determining Different Representations of a Point (r, θ)

Use the same value of r but choose an angle coterminal to θ. The coordinates will be of the form $(r, \theta + 2\pi k)$, where k is any integer.

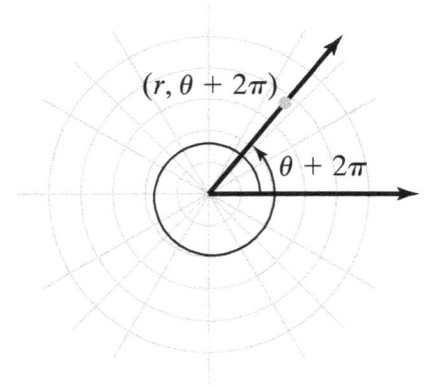

Use the opposite value of r but choose an angle coterminal to the angle located one-half of a rotation from angle θ. The coordinates will be of the form $(-r, \theta + \pi + 2\pi k)$, where k is any integer.

(continued)

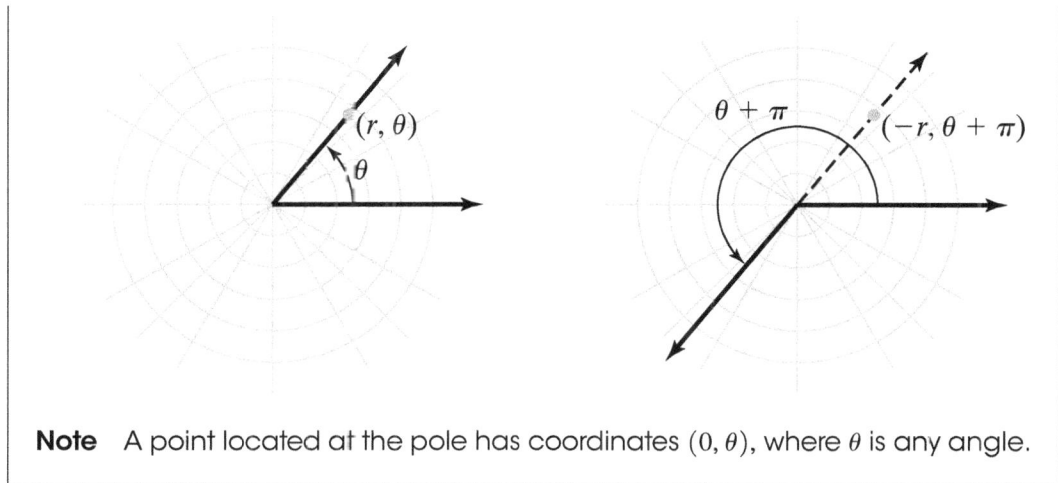

Note A point located at the pole has coordinates $(0, \theta)$, where θ is any angle.

My video summary ⊘ **Example 2** Determining Different Representations of the Same Point

The point $P\left(4, \dfrac{5\pi}{6}\right)$ is shown in Figure 7. Determine three different representations of point P that have the specified conditions.

a. $r > 0, -2\pi \le \theta < 0$ **b.** $r < 0, 0 \le \theta < 2\pi$ **c.** $r > 0, 2\pi \le \theta < 4\pi$

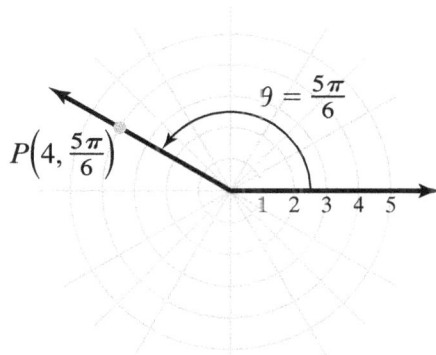

Figure 7 The point $P\left(4, \dfrac{5\pi}{6}\right)$

Solution

a. We want $r > 0$, so we use $r = 4$. The new angle must be a negative angle between -2π and 0 and must be coterminal to $\dfrac{5\pi}{6}$. Therefore, the new angle must be of the form $\dfrac{5\pi}{6} + 2\pi k$. We choose $k = -1$ to guarantee that we obtain an angle in the desired interval. The result is

$$P\left(4, \frac{5\pi}{6} + 2\pi(-1)\right) = P\left(4, \frac{5\pi}{6} - 2\pi\right) = P\left(4, \frac{5\pi}{6} - \frac{12\pi}{6}\right) = P\left(4, -\frac{7\pi}{6}\right).$$

b. We want $r < 0$, so we use $r = -4$. The new angle must be a positive angle between 0 and 2π and must be coterminal to $\dfrac{5\pi}{6} + \pi$. Therefore, the new

angle must be of the form $\dfrac{5\pi}{6} + \pi + 2\pi k$. We choose $k = 0$ to guarantee that

we obtain an angle in the desired interval. The result is

$$P\left(-4, \frac{5\pi}{6} + \pi + 2\pi(0)\right) = P\left(-4, \frac{5\pi}{6} + \pi\right) = P\left(-4, \frac{5\pi}{6} + \frac{6\pi}{6}\right) = P\left(-4, \frac{11\pi}{6}\right).$$

My video summary ⊙ Try to determine the representation of point P as described in part c on your own. Then watch this video to see if you are correct. ●

You Try It Work through this You Try It problem.

Work Exercises 11–15 in this textbook or in the MyMathLab Study Plan.

OBJECTIVE 3 CONVERTING A POINT FROM POLAR COORDINATES TO RECTANGULAR COORDINATES

We can place the pole at the origin of the rectangular coordinate system so that the polar axis coincides with the positive x-axis. See Figure 8. Therefore, we can describe point P using either rectangular coordinates or polar coordinates.

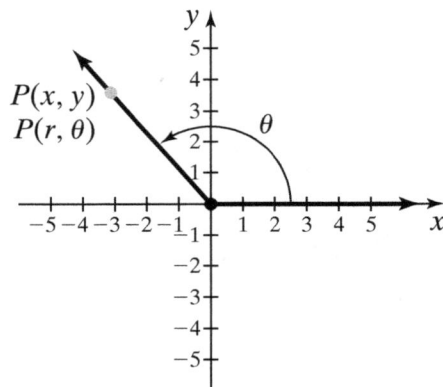

Figure 8 Rectangular and polar coordinate systems

The relationships between rectangular coordinates and polar coordinates are based on the general angle definitions of the trigonometric functions. If we consider point P seen in Figure 8, then by the general angle definitions of the trigonometric functions we know the following:

$$\cos \theta = \frac{x}{r} \quad \text{and} \quad \sin \theta = \frac{y}{r}$$

Multiplying both sides of the equations $\cos \theta = \dfrac{x}{r}$ and $\sin \theta = \dfrac{y}{r}$ by r gives $x = r \cos \theta$ and $y = r \sin \theta$. In these new forms, the value of r can be zero.

Relationships Used when Converting a Point from Polar Coordinates to Rectangular Coordinates

$x = r \cos \theta$ and $y = r \sin \theta$

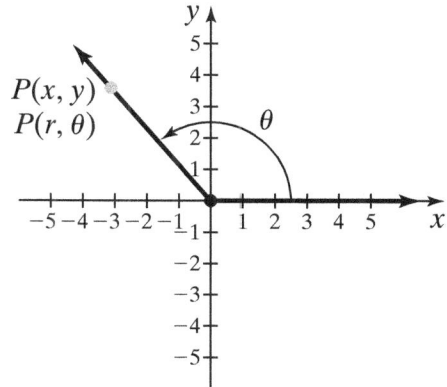

My video summary   ⊙ **Example 3** Converting a Point from Polar Coordinates to Rectangular Coordinates

Determine the rectangular coordinates for the points with the given polar coordinates.

a. $A(5, \pi)$ **b.** $B\left(-7, -\dfrac{\pi}{3}\right)$ **c.** $C\left(3\sqrt{2}, \dfrac{5\pi}{4}\right)$

Solution

a. For the point $A(5, \pi)$, we see that $r = 5$ and $\theta = \pi$. To determine the rectangular coordinates for point A, we use the equations $x = r \cos \theta$ and $y = r \sin \theta$.

$x = r \cos \theta$	Write the equation that relates x, r, and θ.	$y = r \sin \theta$	Write the equation that relates y, r, and θ.
$x = 5 \cos \pi$	Substitute $r = 5$ and $\theta = \pi$.	$y = 5 \sin \pi$	Substitute $r = 5$ and $\theta = \pi$.
$x = 5(-1)$	Evaluate $\cos \pi = -1$.	$y = 5(0)$	Evaluate $\sin \pi = 0$.
$x = -5$	Multiply.	$y = 0$	Multiply.

We see that $x = -5$ and $y = 0$. Therefore, the rectangular coordinates of $A(5, \pi)$ are $A(-5, 0)$. See Figure 9.

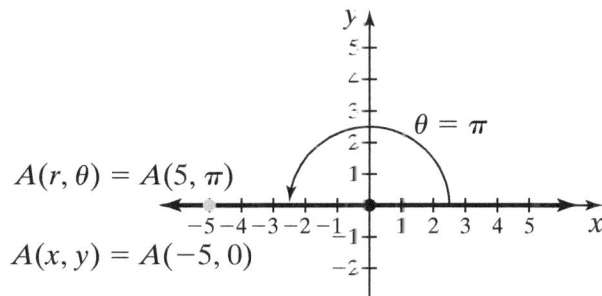

$A(r, \theta) = A(5, \pi)$

$A(x, y) = A(-5, 0)$

Figure 9 The rectangular coordinates of the point $A(5, \pi)$ are $A(-5, 0)$.

b. For the point $B\left(-7, -\dfrac{\pi}{3}\right)$, we see that $r = -7$ and $\theta = -\dfrac{\pi}{3}$. To determine the rectangular coordinates for point B, we use the equations $x = r\cos\theta$ and $y = r\sin\theta$.

$x = r\cos\theta$	Write the equation that relates x, r, and θ.	$y = r\sin\theta$	Write the equation that relates y, r, and θ.
$x = -7\cos\left(-\dfrac{\pi}{3}\right)$	Substitute $r = -7$ and $\theta = -\dfrac{\pi}{3}$.	$y = -7\sin\left(-\dfrac{\pi}{3}\right)$	Substitute $r = -7$ and $\theta = -\dfrac{\pi}{3}$.
$x = -7\left(\dfrac{1}{2}\right)$	Evaluate $\cos\left(-\dfrac{\pi}{3}\right) = \dfrac{1}{2}$.	$y = -7\left(-\dfrac{\sqrt{3}}{2}\right)$	Evaluate $\sin\left(-\dfrac{\pi}{3}\right) = -\dfrac{\sqrt{3}}{2}$.
$x = -\dfrac{7}{2}$	Multiply.	$y = \dfrac{7\sqrt{3}}{2}$	Multiply.

We see that $x = -\dfrac{7}{2}$ and $y = \dfrac{7\sqrt{3}}{2}$. Therefore, the rectangular coordinates of

$$B\left(-7, -\frac{\pi}{3}\right) \text{ are } B\left(-\frac{7}{2}, \frac{7\sqrt{3}}{2}\right).$$

My video summary

c. Try to determine the rectangular coordinates that correspond to the polar coordinates $C\left(3\sqrt{2}, \dfrac{5\pi}{4}\right)$ on your own. View the solution to see if you are correct or watch this video to see the worked-out solution.

You Try It   Work through this You Try It problem.

Work Exercises 16–25 in this textbook or in the MyMathLab Study Plan.

OBJECTIVE 4   CONVERTING A POINT FROM RECTANGULAR COORDINATES TO POLAR COORDINATES

Converting from rectangular coordinates to polar coordinates is a bit more involved than converting from polar coordinates to rectangular coordinates, because there are infinitely many representations for any given point in polar coordinates. For simplicity and consistency, we will always determine the polar coordinates with the conditions that $r \geq 0$ and $0 \leq \theta < 2\pi$.

First, consider points in the rectangular coordinate system lying along an axis and having coordinates $(a, 0)$, $(0, a)$, $(-a, 0)$, and $(0, -a)$, where $a > 0$. Each of the four cases is described in the following box.

Converting Rectangular Coordinates to Polar Coordinates for Points Lying Along an Axis

In each case, assume that $a > 0$.

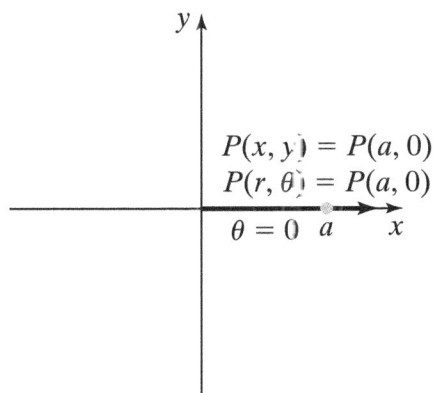

$$P(x, y) = P(a, 0)$$
$$P(r, \theta) = P(a, 0)$$
$$\theta = 0 \quad a \quad x$$

The point $P(x, y) = P(a, 0)$ lies along the positive x-axis and has polar coordinates of $P(r, \theta) = P(a, 0)$.

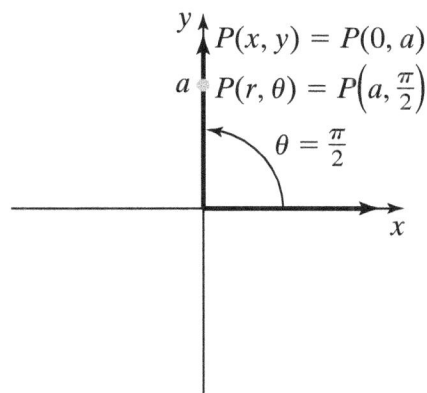

$$P(x, y) = P(0, a)$$
$$a \quad P(r, \theta) = P\left(a, \frac{\pi}{2}\right)$$
$$\theta = \frac{\pi}{2}$$

The point $P(x, y) = P(0, a)$ lies along the positive y-axis and has polar coordinates of $P(r, \theta) = P\left(a, \frac{\pi}{2}\right)$.

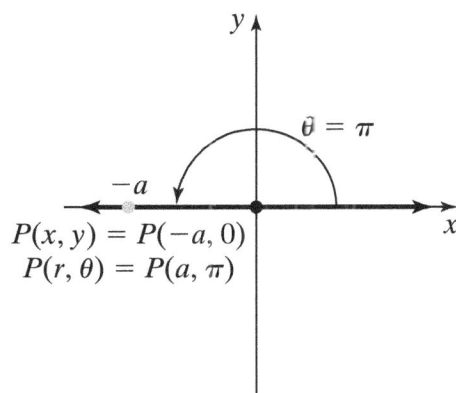

$$\theta = \pi$$
$$-a$$
$$P(x, y) = P(-a, 0)$$
$$P(r, \theta) = P(a, \pi)$$

The point $P(x, y) = P(-a, 0)$ lies along the negative x-axis and has polar coordinates of $P(r, \theta) = P(a, \pi)$.

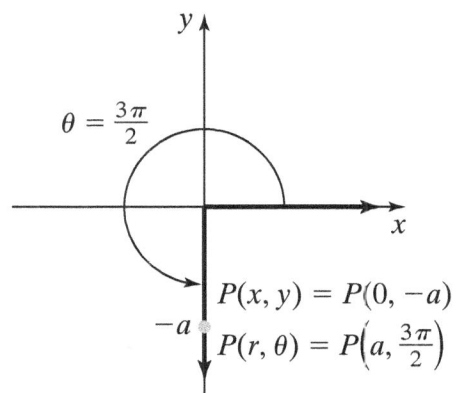

$$\theta = \frac{3\pi}{2}$$
$$P(x, y) = P(0, -a)$$
$$-a \quad P(r, \theta) = P\left(a, \frac{3\pi}{2}\right)$$

The point $P(x, y) = P(0, -a)$ lies along the negative y-axis and has polar coordinates of $P(r, \theta) = P\left(a, \frac{3\pi}{2}\right)$.

My video summary ⊙ **Example 4** Converting Rectangular Coordinates to Polar Coordinates for Points Lying Along an Axis

Determine the polar coordinates for the points with the given rectangular coordinates.

a. $A(-3.5, 0)$ **b.** $B(0, -\sqrt{7})$

Solution

a. The point $A(-3.5, 0)$ lies along the negative x-axis. Therefore, $\theta = \pi$. The distance from the origin to point A is 3.5 units, so $r = 3.5$. Thus, the polar coordinates of point A are $A(3.5, \pi)$. See Figure 10.

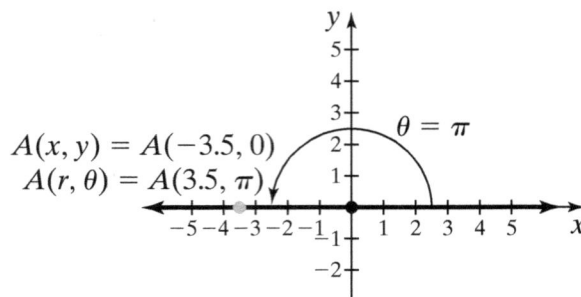

Figure 10 The polar coordinates of the point $A(-3.5, 0)$ are $A(3.5, \pi)$.

$A(x, y) = A(-3.5, 0)$
$A(r, \theta) = A(3.5, \pi)$

b. The point $B(0, -\sqrt{7})$ lies along the negative y-axis. Therefore, $\theta = \dfrac{3\pi}{2}$. The distance from the origin to point B is $\sqrt{7}$ units, so $r = \sqrt{7}$. Thus, the polar coordinates of point B are $B\left(\sqrt{7}, \dfrac{3\pi}{2}\right)$. See Figure 11.

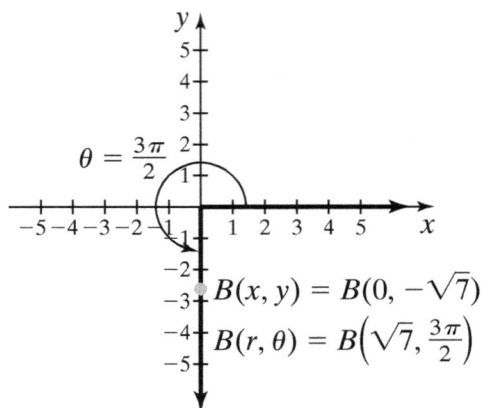

$B(x, y) = B(0, -\sqrt{7})$
$B(r, \theta) = B\left(\sqrt{7}, \dfrac{3\pi}{2}\right)$

Figure 11 The polar coordinates of the point $B(0, -\sqrt{7})$ are $B\left(\sqrt{7}, \dfrac{3\pi}{2}\right)$.

My video summary

▶ Work through this video to see the complete solution to Example 4.

You Try It Work through this You Try It problem.

Work Exercises 26–29 in this textbook or in the MyMathLab **Study Plan.**

When a point $P(x, y)$ given in rectangular coordinates does not lie along an axis, the conversion to polar coordinates is a bit more complicated. As stated earlier, we will always determine the polar coordinates $P(r, \theta)$ with the conditions that $r \geq 0$ and $0 \leq \theta < 2\pi$.

By the general angle definitions of the trigonometric functions we know the following:

$$r = \sqrt{x^2 + y^2} \quad \text{and} \quad \tan \theta = \frac{y}{x}$$

First, we will determine the value of r using the fact that $r = \sqrt{x^2 + y^2}$. Second, we will determine the value of θ, which is much more involved. In order to find θ, we first have to find the reference angle θ_R. Recall that the reference angle is an acute angle. We know that the tangent of any acute angle is a positive value. Therefore, using the fact that $\tan \theta = \dfrac{y}{x}$, we can find θ_R by solving the equation $\tan \theta_R = \left|\dfrac{y}{x}\right|$. The appropriate value of θ depends on the quadrant in which the point $P(x, y)$ lies.

Converting Rectangular Coordinates to Polar Coordinates for Points Not Lying Along an Axis

Step 1. Determine the value of r using the equation $r = \sqrt{x^2 + y^2}$.

Step 2. Plot the point and determine the quadrant in which it lies.

Step 3. Determine the value of the acute reference angle θ_R by solving the equation $\tan \theta_R = \left| \dfrac{y}{x} \right|$.

Step 4. Determine the value of θ using θ_R and the quadrant in which the point lies. Each case is outlined below.

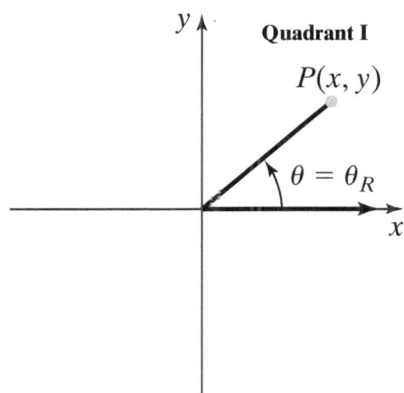

Quadrant I

If $P(x, y)$ lies in Quadrant I, then $\theta = \theta_R$.

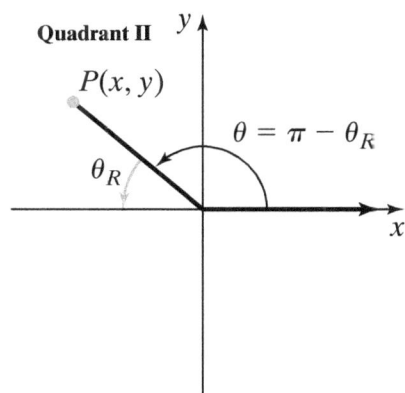

Quadrant II

If $P(x, y)$ lies in Quadrant II, then $\theta = \pi - \theta_R$.

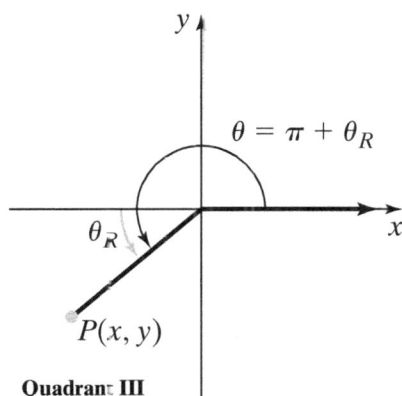

If $P(x, y)$ lies in Quadrant III, then $\theta = \pi + \theta_R$.

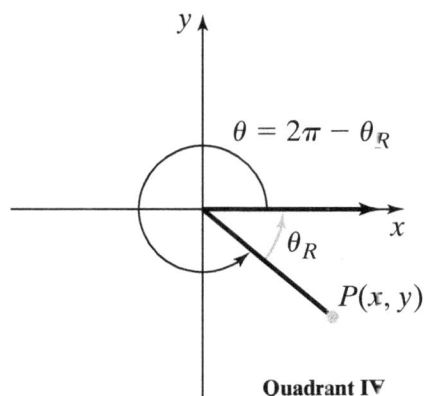

If $P(x, y)$ lies in Quadrant IV, then $\theta = 2\pi - \theta_R$.

Quadrant III

Quadrant IV

⊙ Example 5 Converting a Point from Rectangular Coordinates to Polar Coordinates

Determine the polar coordinates for the points with the given rectangular coordinates such that $r \geq 0$ and $0 \leq \theta < 2\pi$. Round the values of r and θ to two decimal places if necessary.

a. $A(-4, -4)$ **b.** $B(-2\sqrt{3}, 2)$ **c.** $C(4, -3)$

Solution

a. **Step 1.** Given the rectangular coordinates $A(-4, -4)$, find the value of r using the equation $r = \sqrt{x^2 + y^2}$.

$$r = \sqrt{x^2 + y^2} \qquad \text{Write the equation relating } r, x, \text{ and } y.$$

$$r = \sqrt{(-4)^2 + (-4)^2} \qquad \text{Substitute } x = -4 \text{ and } y = -4.$$

$$r = \sqrt{16 + 16} \qquad \text{Square each term.}$$

$$r = \sqrt{32} = 4\sqrt{2} \qquad \text{Simplify.}$$

Step 2. Plot the point $A(-4, -4)$ and recognize that the point lies in Quadrant III. See Figure 12.

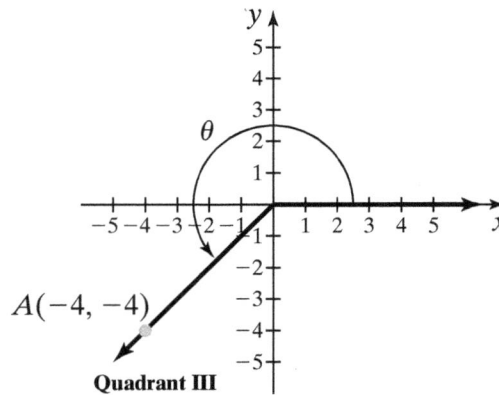

Figure 12 The point $A(-4, -4)$ lies in Quadrant III.

Step 3. To determine the value of θ, we first find the reference angle, θ_R, by solving the equation $\tan \theta_R = \left| \dfrac{y}{x} \right|$.

$$\tan \theta_R = \left| \frac{y}{x} \right| \qquad \text{Write the equation relating } \theta_R, x, \text{ and } y.$$

$$\tan \theta_R = \left| \frac{-4}{-4} \right| \qquad \text{Substitute } x = -4 \text{ and } y = -4.$$

$$\tan \theta_R = 1 \qquad \text{Simplify.}$$

The acute angle whose tangent is 1 is $\theta_R = \dfrac{\pi}{4}$.

Step 4. Because the point lies in Quadrant III, we know that $\theta = \theta_R + \pi$. Thus,

$$\theta = \frac{\pi}{4} + \pi = \frac{5\pi}{4}.$$

My interactive video summary

⊙ Therefore, the polar coordinates of point A for $r \geq 0$ and $0 \leq \theta < 2\pi$ are $A\left(4\sqrt{2}, \dfrac{5\pi}{4} \right)$. See Figure 13. Watch this interactive video to see the worked-out solution.

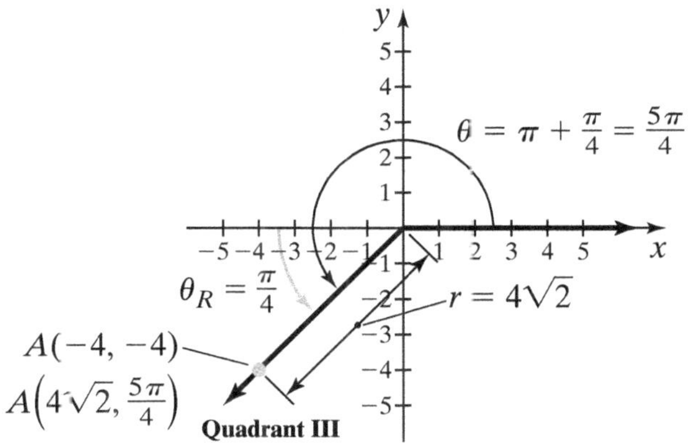

Figure 13 The polar coordinates of the point $A(-4, -4)$ for $r \geq 0$ and $0 \leq \theta < 2\pi$ are $A\left(4\sqrt{2}, \dfrac{5\pi}{4}\right)$.

b. Step 1. Given the rectangular coordinates $B(-2\sqrt{3}, 2)$, find the value of r using the equation $r = \sqrt{x^2 + y^2}$.

$$r = \sqrt{x^2 + y^2} \qquad \text{Write the equation relating } r, x, \text{ and } y.$$
$$r = \sqrt{(-2\sqrt{3})^2 + (2)^2} \qquad \text{Substitute } x = -2\sqrt{3} \text{ and } y = 2.$$
$$r = \sqrt{(4)(3) + 4} \qquad \text{Square each term.}$$
$$r = \sqrt{16} = 4 \qquad \text{Simplify.}$$

Step 2. Plot the point $B(-2\sqrt{3}, 2)$ and recognize that the point lies in Quadrant II. See Figure 14.

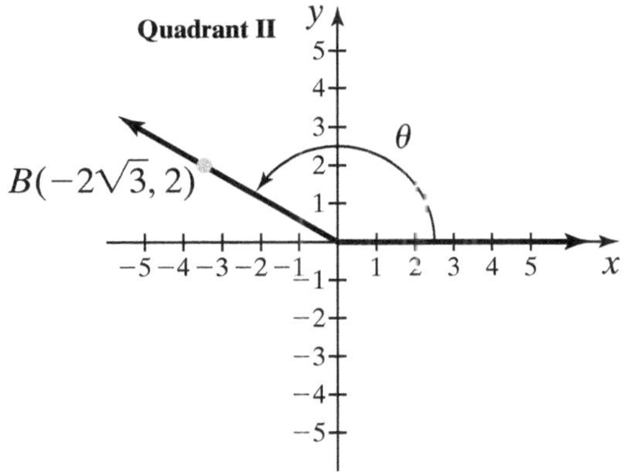

Figure 14 The point $B(-2\sqrt{3}, 2)$ lies in Quadrant II.

Step 3. Given the rectangular coordinates $B(-2\sqrt{3}, 2)$, to determine θ we first find the value of θ_R by solving the equation $\tan \theta_R = \left|\dfrac{y}{x}\right|$.

$$\tan \theta_R = \left|\frac{y}{x}\right| \qquad \text{Write the equation relating } \theta_R, x, \text{ and } y.$$
$$\tan \theta_R = \left|\frac{2}{-2\sqrt{3}}\right| \qquad \text{Substitute } y = 2 \text{ and } x = -2\sqrt{3}.$$
$$\tan \theta_R = \frac{1}{\sqrt{3}} \qquad \text{Simplify.}$$

The acute angle whose tangent is $\dfrac{1}{\sqrt{3}}$ is $\theta_R = \dfrac{\pi}{6}$.

My interactive video summary

$\blacktriangleright$ **Step 4.** Because the point lies in Quadrant II, we get $\theta = \pi - \theta_R = \pi - \dfrac{\pi}{6} = \dfrac{5\pi}{6}$. Therefore, the polar coordinates of point B for $r \geq 0$ and $0 \leq \theta < 2\pi$ are $B\left(4, \dfrac{5\pi}{6}\right)$. See Figure 15. Watch this interactive video to see the worked-out solution.

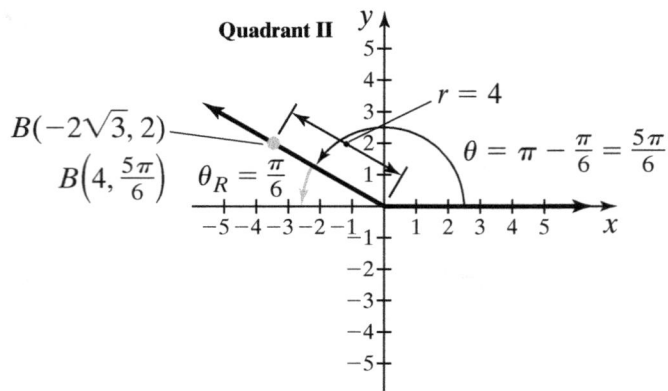

Figure 15 The polar coordinates of the point $B(-2\sqrt{3}, 2)$ for $r \geq 0$ and $0 \leq \theta < 2\pi$ are $B\left(4, \dfrac{5\pi}{6}\right)$.

c. Step 1. Given the rectangular coordinates $C(4, -3)$, find the value of r using the equation $r = \sqrt{x^2 + y^2}$.

$$r = \sqrt{x^2 + y^2} \qquad \text{Write the equation relating } r, x, \text{ and } y.$$
$$r = \sqrt{(4)^2 + (-3)^2} \qquad \text{Substitute } x = 4 \text{ and } y = -3.$$
$$r = \sqrt{16 + 9} \qquad \text{Square each term.}$$
$$r = \sqrt{25} = 5 \qquad \text{Simplify.}$$

Step 2. Plot the point $C(4, -3)$ and recognize that the point lies in Quadrant IV. See Figure 16.

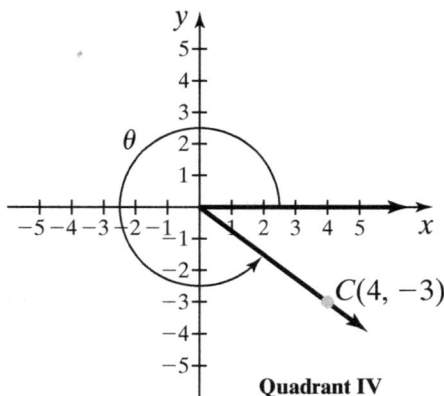

Figure 16 The point $C(4, -3)$ lies in Quadrant IV.

Step 3. Given the rectangular coordinates $C(4, -3)$, to determine θ we first find

the value of θ_R by solving the equation $\tan \theta_R = \left| \dfrac{y}{x} \right|$.

$$\tan \theta_R = \left| \frac{y}{x} \right| \qquad \text{Write the equation relating } \theta_R, x, \text{ and } y.$$

$$\tan \theta_R = \left| \frac{-3}{4} \right| \qquad \text{Substitute } x = 4 \text{ and } y = -3.$$

$$\tan \theta_R = \frac{3}{4} \qquad \text{Simplify.}$$

The angle whose tangent is $\dfrac{3}{4}$ does not belong to one of the special

angle families. However, we can use the inverse tangent function to

solve for θ_R. Thus, $\theta_R = \tan^{-1}\left(\dfrac{3}{4}\right)$.

Step 4. The point $C(4, -3)$ lies in Quadrant IV, which indicates that $\theta = 2\pi - \theta_R$.

Therefore, $\theta = 2\pi - \tan^{-1}\left(\dfrac{3}{4}\right)$.

⊙ Using a calculator, we get $\theta = 2\pi - \tan^{-1}\left(\dfrac{3}{4}\right) \approx 5.64$. Thus, the

polar coordinates of point C are approximately $C(5, 5.64)$. See Figure 17.
Watch this interactive video to see the worked-out solution.

My interactive video summary

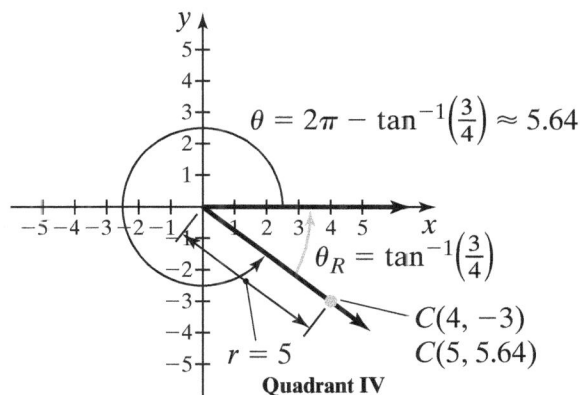

Figure 17 The polar coordinates of the point
$C(4, -3)$ for $r \geq 0$ and $0 \leq \theta < 2\pi$
are approximately $C(5, 5.64)$.

You Try It Work through this You Try It problem.

Work Exercises 30–37 in this textbook or in the MyMathLab Study Plan.

OBJECTIVE 5   CONVERTING AN EQUATION FROM RECTANGULAR FORM
TO POLAR FORM

A **polar equation** is an equation whose variables are r and θ. The independent variable, or input value, is θ and the dependent variable, or output value, is r. Below are some examples of polar equations:

$$r = 3, \quad r = 6\cos\theta, \quad \text{and} \quad r = \frac{8}{2\cos\theta - \sin\theta}$$

The equations above are said to be in polar form. To convert an equation in x and y (rectangular form) to an equation in r and θ (polar form), we must revisit some relationships previously stated. Recall that $x = r\cos\theta$, $y = r\sin\theta$, and $r = \sqrt{x^2 + y^2}$. When converting rectangular to polar coordinates, we chose, for convenience, to state our answer with $r \geq 0$. But those same points could be written using a different representation where r is negative. Therefore, we can square both sides of the equation $r = \sqrt{x^2 + y^2}$ and obtain an equation where r can take on any value. This equation is $r^2 = x^2 + y^2$. Thus, to convert an equation from rectangular form to polar form, replace x with $r\cos\theta$ and y with $r\sin\theta$. If the equation in rectangular form contains the expression $x^2 + y^2$, then replace $x^2 + y^2$ with r^2. We will always attempt to write the final polar equation in the form where either r or r^2 is isolated.

My video summary   ⊙ **Example 6** Converting an Equation from Rectangular Form to Polar Form

Convert each equation given in rectangular form into polar form.

a. $x = 7$ **b.** $2x - y = 8$ **c.** $4x^2 + 4y^2 = 3$ **d.** $x^2 + y^2 = 9y$

Solution

a. To convert the equation $x = 7$ into polar form, we replace x with $r\cos\theta$.

$\quad\quad x = 7$ Write the original equation in rectangular form.

$\quad r\cos\theta = 7$ Replace x with $r\cos\theta$.

$\quad\quad r = \dfrac{7}{\cos\theta}$ Divide both sides by $\cos\theta$.

Therefore, given the equation $x = 7$ in rectangular form, we see that the equivalent equation in polar form is $r = \dfrac{7}{\cos\theta}$. Watch this video to see the worked-out solution.

b. To convert the equation $2x - y = 8$ into polar form, we replace x with $r\cos\theta$ and replace y with $r\sin\theta$.

$\quad\quad 2x - y = 8$ Write the original equation in rectangular form.

$\quad 2r\cos\theta - r\sin\theta = 8$ Replace x with $r\cos\theta$ and replace y with $r\sin\theta$.

$\quad r(2\cos\theta - \sin\theta) = 8$ Factor.

$\quad\quad r = \dfrac{8}{2\cos\theta - \sin\theta}$ Divide both sides by $2\cos\theta - \sin\theta$.

Therefore, given the equation $2x - y = 8$ in rectangular form, we see that the equivalent equation in polar form is

$$r = \frac{8}{2\cos\theta - \sin\theta}.$$

My video summary ⊙ Watch this video to see the worked-out solution.

c. To convert the equation $4x^2 + 4y^2 = 3$ into polar form, we can factor out a 4 and then replace $x^2 + y^2$ with r^2.

$4x^2 + 4y^2 = 3$	Write the original equation in rectangular form.
$4(x^2 + y^2) = 3$	Factor.
$4r^2 = 3$	Replace $x^2 + y^2$ with r^2.
$r^2 = \dfrac{3}{4}$	Divide both sides by 4.
$r = \pm\sqrt{\dfrac{3}{4}}$	Use the square root property to take the square root of both sides.
$r = \pm\dfrac{\sqrt{3}}{2}$	Simplify.

As you will see when we graph polar equations, the graphs of $r = \dfrac{\sqrt{3}}{2}$ and $r = -\dfrac{\sqrt{3}}{2}$ are represented by the same set of points. Thus, it is unnecessary to include both equations as part of the final polar form answer. For convenience, we will only use the positive value of r. Therefore, given the equation $4x^2 + 4y^2 = 3$ in rectangular form, we see that the equivalent equation in polar form is of $r = \dfrac{\sqrt{3}}{2}$.

My video summary ⊙ Watch this video to see the worked-out solution.

d. To convert the equation $x^2 + y^2 = 9y$ into polar form, we start by replacing $x^2 + y^2$ with r^2 and y with $r\sin\theta$.

$x^2 + y^2 = 9y$	Write the original equation in rectangular form.
$r^2 = 9r\sin\theta$	Replace $x^2 + y^2$ with r^2 and replace y with $r\sin\theta$.
$r^2 - 9r\sin\theta = 0$	Subtract $9r\sin\theta$ from both sides.
$r(r - 9\sin\theta) = 0$	Factor.
$r = 0$ or $r - 9\sin\theta = 0$	Use the zero product property.
$r = 0$ or $r = 9\sin\theta$	Solve the second equation for r.

Note that for the equation $r = 9\sin\theta$, if k is an integer, then $r = 9\sin(k\pi) = 0$. Therefore, it is unnecessary to include the equation $r = 0$ as part of the final polar form answer.

Thus, the equivalent equation in polar form is $r = 9\sin\theta$.

My video summary ⊙ Watch this video to see the worked-out solution.

You Try It **Work through this You Try It problem.**

Work Exercises 38–45 in this textbook or in the My MathLab **Study Plan.**

OBJECTIVE 6 CONVERTING AN EQUATION FROM POLAR FORM TO RECTANGULAR FORM

To convert an equation in r and θ (polar form) to an equation in x and y (rectangular form), we will use the equations $x = r\cos\theta$, $y = r\sin\theta$, $r^2 = x^2 + y^2$, and $\tan\theta = \dfrac{y}{x}$ that have been previously stated. To use the equations $x = r\cos\theta$ and $y = r\sin\theta$, we may be required to first multiply both sides of the original polar equation by r. To use the equation $r^2 = x^2 + y^2$, we may need to square both sides of the original polar equation. To use the equation $\tan\theta = \dfrac{y}{x}$, we may need to rewrite the original equation in a form that involves the expression $\tan\theta$. This will allow us to substitute $\dfrac{y}{x}$ for $\tan\theta$.

My video summary ◉ **Example 7** Converting an Equation from Polar Form to Rectangular Form

Convert each equation given in polar form into rectangular form.

a. $3r\cos\theta - 4r\sin\theta = -1$ **b.** $r = 6\cos\theta$ **c.** $r = 3$ **d.** $\theta = \dfrac{\pi}{6}$

Solution

a. To convert the equation $3r\cos\theta - 4r\sin\theta = -1$ into rectangular form, we replace $r\cos\theta$ with x and replace $r\sin\theta$ with y.

$$3r\cos\theta - 4r\sin\theta = -1 \qquad \text{Write the original equation in polar form.}$$
$$3x - 4y = -1 \qquad \text{Replace } r\cos\theta \text{ with } x \text{ and } r\sin\theta \text{ with } y.$$

My video summary ◉ Therefore, given the equation $3r\cos\theta - 4r\sin\theta = -1$ in polar form, we see that the equivalent equation in rectangular form is $3x - 4y = -1$. Note that we could solve this equation for y to obtain an alternate answer of $y = \dfrac{3}{4}x + \dfrac{1}{4}$.

Watch this **video** to see the worked-out solution.

b. To convert the equation $r = 6\cos\theta$ into rectangular form, first multiply both sides of the equation by r.

$$r = 6\cos\theta \qquad \text{Write the original equation in polar form.}$$
$$r^2 = 6r\cos\theta \qquad \text{Multiply both sides by } r.$$
$$x^2 + y^2 = 6x \qquad \text{Replace } r^2 \text{ with } x^2 + y^2 \text{ and replace } r\cos\theta \text{ with } x.$$

My video summary ◉ Therefore, given the equation $r = 6\cos\theta$ in polar form, we see that the equivalent equation in rectangular form is $x^2 + y^2 = 6x$. Watch this **video** to see the worked-out solution.

c. To convert the equation $r = 3$ into rectangular form, first square both sides of the equation so that we can use the equation $r^2 = x^2 + y^2$.

$$r = 3 \qquad \text{Write the original equation in polar form.}$$
$$r^2 = 9 \qquad \text{Square both sides.}$$
$$x^2 + y^2 = 9 \qquad \text{Replace } r^2 \text{ with } x^2 + y^2.$$

My video summary ⊚ Therefore, given the equation $r = 3$ in polar form, we see that the equivalent equation in rectangular form is $x^2 + y^2 = 9$. Watch this video to see the worked-out solution.

d. To convert the equation $\theta = \dfrac{\pi}{6}$ into rectangular form, first rewrite the equation to involve the expression $\tan \theta$. This will allow us to use the equation $\tan \theta = \dfrac{y}{x}$.

$$\theta = \frac{\pi}{6}$$ Write the original equation in polar form.

$$\tan \theta = \tan \frac{\pi}{6}$$ Rewrite the original equation to involve the expression $\tan \theta$.

$$\tan \theta = \frac{1}{\sqrt{3}}$$ Evaluate $\tan \dfrac{\pi}{6} = \dfrac{1}{\sqrt{3}}$.

$$\frac{y}{x} = \frac{1}{\sqrt{3}}$$ Replace $\tan \theta$. with $\dfrac{y}{x}$.

$$y = \frac{1}{\sqrt{3}} x$$ Multiply both sides by x.

My video summary ⊚ Therefore, given the equation $\theta = \dfrac{\pi}{6}$ in polar form, we see that the equivalent equation in rectangular form is $y = \dfrac{1}{\sqrt{3}} x$. Watch this video to see the worked-out solution.

It is important to point out that in the previous example we were given polar equations and were able to convert them into recognizable rectangular equations.

Note that the graphs of the equations $3x - 4y = -1$ and $y = \dfrac{1}{\sqrt{3}} x$ are lines and the graphs of the equations $x^2 + y^2 = 6x$ and $x^2 + y^2 = 9$ are circles. In Section 5.2 we will sketch the graphs of polar equations.

You Try It   Work through this You Try It problem.

Work Exercises 46–53 in this textbook or in the MyMathLab Study Plan.

5.1 Exercises

In Exercises 1–10, plot each point in a polar coordinate system.

1. $A\left(2, \dfrac{\pi}{3}\right)$

2. $B(3, 150°)$

3. $C\left(-1, \dfrac{5\pi}{4}\right)$

4. $D(1.5, \pi)$

5. $E\left(4, -\dfrac{5\pi}{6}\right)$

6. $F(-2.5, -305°)$

7. $G\left(-2, -\dfrac{3\pi}{2}\right)$

8. $H\left(-3, -\dfrac{7\pi}{4}\right)$

9. $J\left(5, \dfrac{13\pi}{6}\right)$

10. $K\left(-4, \dfrac{11\pi}{4}\right)$

In Exercises 11–15, a point $P(r, \theta)$ is given. Plot the point and then find three different representations of point P that have the specified conditions.

a. $r > 0, -2\pi \le \theta < 0$ **b.** $r < 0, 0 \le \theta < 2\pi$ **c.** $r > 0, 2\pi \le \theta < 4\pi$

11. $P\left(5, \dfrac{\pi}{6}\right)$ **12.** $P\left(3, \dfrac{4\pi}{3}\right)$ **13.** $P\left(4, \dfrac{3\pi}{2}\right)$ **14.** $P\left(5, \dfrac{3\pi}{4}\right)$ **15.** $P\left(3, \dfrac{11\pi}{6}\right)$

In Exercises 16–25, a point $P(r, \theta)$ in polar coordinates is given. Determine the rectangular coordinates $P(x, y)$.

16. $P\left(3, \dfrac{\pi}{4}\right)$ **17.** $P\left(2, \dfrac{2\pi}{3}\right)$ **18.** $P(5, 270°)$ **19.** $P\left(1, \dfrac{7\pi}{6}\right)$ **20.** $P\left(4, \dfrac{5\pi}{3}\right)$

21. $P(-2, 30°)$ **22.** $P\left(-5, \dfrac{3\pi}{4}\right)$ **23.** $P(-7, \pi)$ **24.** $P\left(-3, \dfrac{4\pi}{3}\right)$ **25.** $P\left(-7, \dfrac{11\pi}{6}\right)$

In Exercises 26–37, a point $P(x, y)$ in rectangular coordinates is given. Determine the polar coordinates $P(r, \theta)$ such that $r \ge 0$ and $0 \le \theta < 2\pi$.

26. $P(3, 0)$ **27.** $P(0, -\sqrt{2})$ **28.** $P(-0.5, 0)$ **29.** $P(0, 1.5)$

SbS **30.** $P(3, -3)$ SbS **31.** $P(-5, 5\sqrt{3})$ SbS **32.** $P(-4\sqrt{3}, -4)$ SbS **33.** $P(\sqrt{5}, \sqrt{5})$

SbS **34.** $P(2, 2\sqrt{3})$ SbS **35.** $P(-6\sqrt{3}, 6)$ SbS **36.** $P(2, -7)$ SbS **37.** $P(\sqrt{2}, -1)$

In Exercises 38–45, convert each equation given in rectangular form into polar form.

38. $7x - 4y = 1$ **39.** $x^2 + y^2 = 25$ **40.** $x = -3$

41. $y = 5$ **42.** $x^2 + y^2 = -4y$ **43.** $x^2 + y^2 = 6x$

44. $(x + 1)^2 + y^2 = 1$ **45.** $x^2 + (y - 4)^2 = 16$

In Exercises 41–53, convert each equation given in polar form into rectangular form.

46. $2r \cos \theta + 5r \sin \theta = 3$ **47.** $r = 3 \sin \theta$ **48.** $r = 4$

49. $r = -2 \cos \theta$ **50.** $\theta = \dfrac{\pi}{3}$ **51.** $r = 6 \sin \theta - 2 \cos \theta$

52. $r = 2 \sec \theta$ **53.** $r = -3 \csc \theta$

Brief Exercises

In Exercises 54–59, a point $P(x, y)$ in rectangular coordinates is given. Determine the polar coordinates $P(r, \theta)$ such that $r \ge 0$ and $0 \le \theta < 2\pi$.

54. $P(2, 2)$ **55.** $P(-5, 5\sqrt{3})$ **56.** $P(-2\sqrt{3}, 2)$

57. $P(-\sqrt{5}, \sqrt{5})$ **58.** $P(-1, -\sqrt{3})$ **59.** $P(-6\sqrt{3}, -6)$

5.2 Graphing Polar Equations

THINGS TO KNOW

Before working through this section, be sure that you are familiar with the following concepts:

You Try It
1. Evaluating Trigonometric Functions of Angles Belonging to the $\frac{\pi}{3}, \frac{\pi}{6}$, or $\frac{\pi}{4}$ Families (Section 1.5)

You Try It
2. Solving Trigonometric Equations That Are Linear in Form (Section 3.5)

You Try It
3. Plotting Points Using Polar Coordinates (Section 5.1)

You Try It
4. Converting an Equation from Polar Form to Rectangular Form (Section 5.1)

OBJECTIVES

1 Sketching Equations of the Form $\theta = \alpha$, $r \cos \theta = a$, $r \sin \theta = a$, and $ar \cos \theta + br \sin \theta = c$

2 Sketching Equations of the Form $r = a$, $r = a \sin \theta$, and $r = a \cos \theta$

3 Sketching Equations of the Form $r = a + b \sin \theta$ and $r = a + b \cos \theta$

4 Sketching Equations of the Form $r = a \sin n\theta$ and $r = a \cos n\theta$

5 Sketching Equations of the Form $r^2 = a^2 \sin 2\theta$ and $r^2 = a^2 \cos 2\theta$

Introduction to Section 5.2

In our study of algebra, we learned to recognize the shape of the graph of some equations just by looking at the equation rather than just plotting a myriad of points. We know that the equation $x^2 + y^2 = 4$ is represented by the graph of a circle with center at the origin and radius of 2, that the equation $x = 2$ is represented by the graph of a vertical line 2 units to the right of the y-axis, and that the equation $y = -3$ is represented by the graph of a horizontal line 3 units below the x-axis. Recognizing these shapes and considering other factors specific to the shape reduces the work needed to get an accurate sketch of the graph.

As with equations in rectangular form, plotting points to sketch a polar graph does not always satisfactorily yield an accurate picture of the graph. Therefore, we will learn to recognize the shapes that represent various polar equations to reduce the work needed to accurately sketch their graphs. Other factors that may also be considered are symmetry, finding the greatest value r attains and the associated values of θ, and finding the value(s) of θ when $r = 0$. Though some polar equations may look quite complex at first, knowing this information can make their graphs easier to sketch using the polar coordinate system than graphing in the rectangular coordinate system.

OBJECTIVE 1   SKETCHING EQUATIONS OF THE FORM $\theta = \alpha$, $r \cos \theta = a$, $r \sin \theta = a$, AND $ar \cos \theta + br \sin \theta = c$

We begin our study of polar equations by looking at equations of the form $\theta = \alpha$, $r \cos \theta = a$, $r \sin \theta = a$, and $ar \cos \theta + br \sin \theta = c$. We will see that these equations all have something in common. We start by analyzing the graph of an equation of the form $\theta = \alpha$.

The graph of $\theta = \alpha$ is the set of all ordered pairs (r, α), where r is any value. When $r > 0$, the point lies along the terminal side of angle α. When $r < 0$, the point lies along the ray opposite the terminal side of angle α. The graph of $\theta = \alpha$ is a line through the pole that makes an angle of α with the polar axis. Figure 18 illustrates the graph of $\theta = \alpha$ for $0 < \alpha < \dfrac{\pi}{2}$.

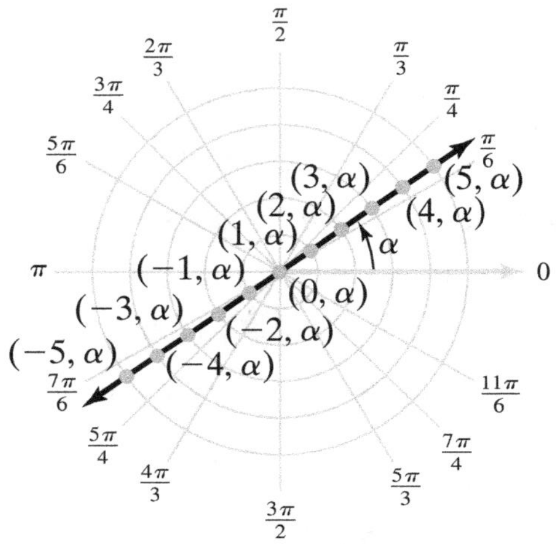

Figure 18 The graph of $\theta = a$ for $0 < \alpha < \dfrac{\pi}{2}$

Note that it is possible to convert the equation $\theta = \alpha$ in polar form into rectangular form, but there is no advantage in doing so.

My video summary ⊘ **Example 1** Sketching the Graph of a Polar Equation of the Form $\boldsymbol{\theta = \alpha}$

Sketch the graph of the polar equation $\theta = \dfrac{2\pi}{3}$.

Solution We recognize that the equation is of the form $\theta = \alpha$, where $\alpha = \dfrac{2\pi}{3}$.

Thus, the graph of $\theta = \dfrac{2\pi}{3}$ is a line passing through the pole that makes an angle of $\dfrac{2\pi}{3}$ with the polar axis.

We now consider equations of the form $r \cos \theta = a$ as in Example 2.

(eText Screens 5.2-1–5.2-50)

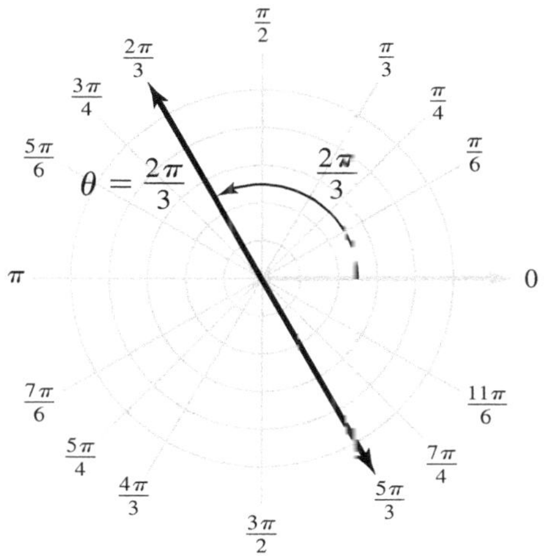

Figure 19 The graph of $\theta = \frac{2\pi}{3}$

Example 2 Sketching the Graph of an Equation of the Form $r\cos\theta = a$

Sketch the graph of the polar equation $r\cos\theta = 2$.

Solution To sketch the graph of $r\cos\theta = 2$, we begin by arbitrarily choosing values of the independent variable, θ, and then determining the corresponding values of the dependent variable, r. To do this, first solve the equation $r\cos\theta = 2$ for r by dividing both sides by $\cos\theta$ to obtain the equation $r = \frac{2}{\cos\theta}$. We now create a table of values using special angles between 0 and $\frac{\pi}{2}$ and connect the points with a smooth curve.

θ	$r = \frac{2}{\cos\theta}$	(r, θ)
0	$r = \frac{2}{\cos 0} = \frac{2}{1} = 2$	$(2, 0)$
$\frac{\pi}{6}$	$r = \frac{2}{\cos\frac{\pi}{6}} = \frac{2}{\frac{\sqrt{3}}{2}} = \frac{4}{\sqrt{3}}$	$\left(\frac{4}{\sqrt{3}}, \frac{\pi}{6}\right)$
$\frac{\pi}{4}$	$r = \frac{2}{\cos\frac{\pi}{4}} = \frac{2}{\frac{1}{\sqrt{2}}} = 2\sqrt{2}$	$\left(2\sqrt{2}, \frac{\pi}{4}\right)$
$\frac{\pi}{3}$	$r = \frac{2}{\cos\frac{\pi}{3}} = \frac{2}{\frac{1}{2}} = 4$	$\left(4, \frac{\pi}{3}\right)$
$\frac{\pi}{2}$	$r = \frac{2}{\cos\frac{\pi}{2}} = \frac{2}{0} =$ undefined	There is no point that corresponds to $\theta = \frac{\pi}{2}$.

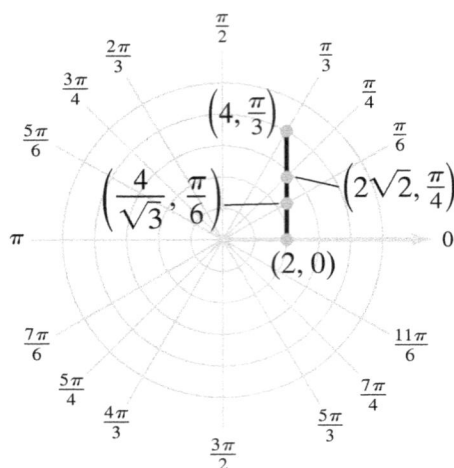

Figure 20 A partial graph of $r\cos\theta = 2$ for special angles between 0 and $\dfrac{\pi}{2}$

The partial graph of $r\cos\theta = 2$ shown in Figure 20 appears to be taking on the shape of a straight line. This should not be surprising at all. Recall from Section 5.1 that one of the relationships used when converting an equation from polar form to rectangular form is $x = r\cos\theta$. Therefore, rather than plotting points, we could have used this relationship to convert the original equation into rectangular coordinates.

$$r\cos\theta = 2 \qquad \text{Write the original equation in polar form.}$$

$$x = 2 \qquad \text{Replace } r\cos\theta \text{ with } x.$$

The equation $r\cos\theta = 2$ in polar form is equivalent to the equation $x = 2$ in rectangular form. We should recognize that the graph of the equation $x = 2$ is a vertical line 2 units to the right of the y-axis. Thus, the graph of the equation $r\cos\theta = 2$ must also be a vertical line 2 units to the right of the vertical line $\theta = \dfrac{\pi}{2}$ on a polar grid. See Figure 21.

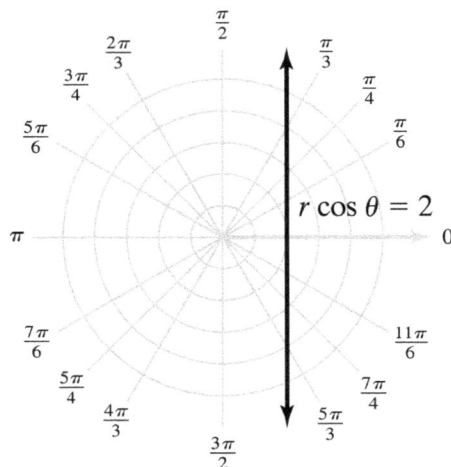

Figure 21 The graph of $r\cos\theta = 2$ is a vertical line located two units to the right of the vertical line $\theta = \dfrac{\pi}{2}$.

When a is a constant, the graph of the equation $r\cos\theta = a$ is a vertical line.

We now analyze the graph of an equation of the form $r\sin\theta = a$. We can convert the equation $r\sin\theta = a$ from polar form to rectangular form.

$$r\sin\theta = a \qquad \text{Write the original equation in polar form.}$$

$$y = a \qquad \text{Replace } r\sin\theta \text{ with } y.$$

The graph of the equation $y = a$ in rectangular coordinates is a horizontal line. Thus, when a is a constant, the graph of the equation $r \sin \theta = a$ in polar form is a horizontal line.

My video summary ◉ **Example 3** Sketching the Graph of an Equation of the Form $r \sin \theta = a$

Sketch the graph of the polar equation $r \sin \theta = -3$.

Solution There is no need to plot points to determine the graph of the equation $r \sin \theta = -3$ because we should immediately recognize that this equation is equivalent to the equation $y = -3$ in rectangular form. Thus, the graph of $r \sin \theta = -3$ is a horizontal line located 3 units below the line $\theta = 0$ on a polar grid. See Figure 22.

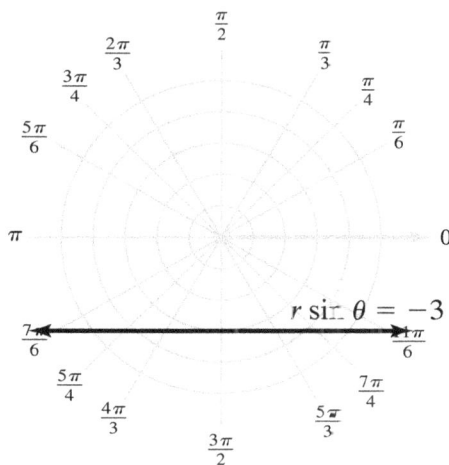

$r \sin \theta = -3$

Figure 22 The graph of $r \sin \theta = -3$ is a horizontal line located three units below the line $\theta = 0$.

We now analyze polar equations containing both $r \cos \theta$ and $r \sin \theta$. These equations are represented by graphs of lines that are neither vertical nor horizontal.

My video summary ◉ **Example 4** Sketching the Graph of a Polar Equation of the Form $ar \cos \theta + br \sin \theta = c$

Sketch the graph of the polar equation $3r \cos \theta - 2r \sin \theta = -6$.

Solution We see that this equation contains the expressions $r \cos \theta$ and $r \sin \theta$. Therefore, we can replace $r \cos \theta$ with x and replace $r \sin \theta$ with y to convert the equation $3r \cos \theta - 2r \sin \theta = -6$ into rectangular form.

$$3r \cos \theta - 2r \sin \theta = -6 \qquad \text{Write the original equation in polar form.}$$
$$3x - 2y = -6 \qquad \text{Replace } r \cos \theta \text{ with } x \text{ and } r \sin \theta \text{ with } y.$$

The graph of the equation $3x - 2y = -6$ is a line. We can solve this equation for y to write the equation in slope-intercept form as $y = \frac{3}{2}x + 3$, where $m = \frac{3}{2}$ and $b = 3$. We can therefore sketch the graph of the equation $3r \cos \theta - 2r \sin \theta = -6$ in polar form by sketching the graph of the linear equation $y = \frac{3}{2}x + 3$ in rectangular form. Watch this video to see how to sketch the graph shown in Figure 23.

When a, b, and c are constants, the graph of the equation $ar\cos\theta + br\sin\theta = c$ is a line with slope $m = -\dfrac{a}{b}$ and y-intercept $\dfrac{c}{b}$.

We now summarize some polar equations whose graphs are lines.

Figure 23 The graph of
$$3r\cos\theta - 2r\sin\theta = -6$$

Graphs of Polar Equations of the Form $\theta = \alpha$, $r\cos\theta = a$, $r\sin\theta = a$, and $ar\cos\theta + br\sin\theta = c$, where a, b, and c are Constants

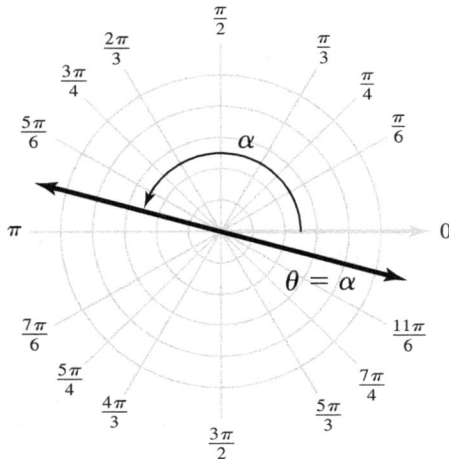

The graph of $\theta = \alpha$ is a line through the pole that makes an angle of α with the polar axis.

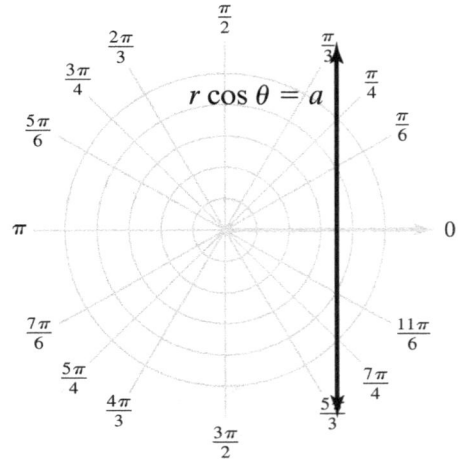

The graph of $r\cos\theta = a$ is a vertical line.

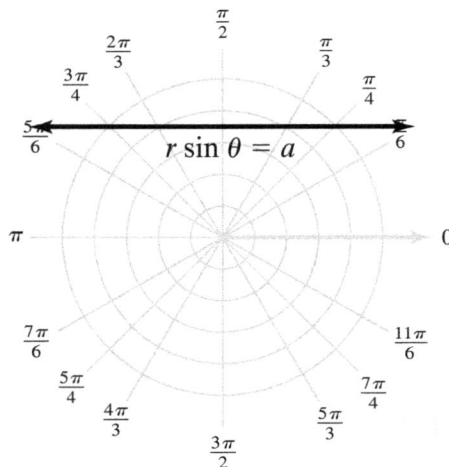

The graph of $r\sin\theta = a$ is a horizontal line.

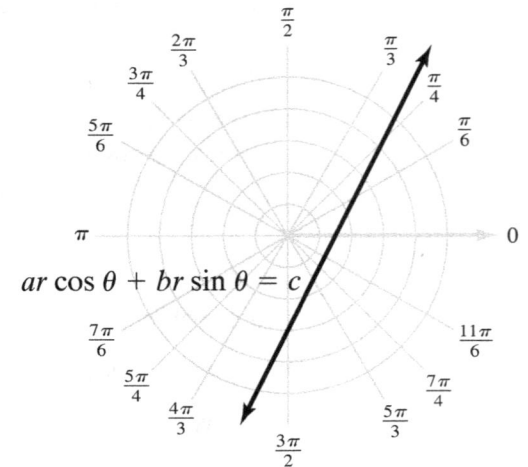

The graph of $ar\cos\theta + br\sin\theta = c$ is a line with slope $m = -\dfrac{a}{b}$ and y-intercept $\dfrac{c}{b}$.

You Try It   Work through this You Try It problem.

Work Exercises 1–8 in this textbook or in the MyMath Lab **Study Plan.**

OBJECTIVE 2   SKETCHING EQUATIONS OF THE FORM $r = a, r = a \sin \theta$, AND $r = a \cos \theta$

Consider the polar equation $r = a$, where a is a constant. Note that this equation does not contain the independent variable θ. This means that regardless of angle θ, the value of r is always equal to a. Thus, the graph of $r = a$ is a circle centered at the pole with a radius of length $|a|$. If $a = 0$, then the graph is a single point located at the pole.

We can verify that the graph of $r = a$ is a circle by converting the equation into rectangular form.

$$r = a \qquad \text{Write the original equation in polar form.}$$
$$r^2 = a^2 \qquad \text{Square both sides.}$$
$$x^2 + y^2 = a^2 \qquad \text{Replace } r^2 \text{ with } x^2 + y^2.$$

The equation $x^2 + y^2 = a^2$ in rectangular form is the equation of a circle centered at the origin with a radius of length $|a|$.

Example 5 Sketching the Graph of a Polar Equation of the Form $r = a$

Sketch the graph of the polar equation $r = -3$.

Solution

The graph is a circle centered at the pole with a radius of length $|-3| = 3$. See Figure 24.

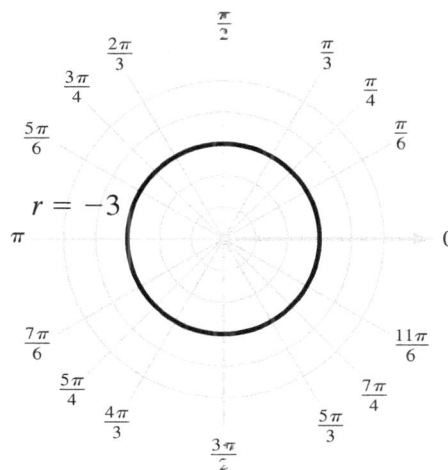

Figure 24   The graph of $r = -3$

We now consider two more equations whose graphs are circles. These equations have the form $r = a \sin \theta$ and $r = a \cos \theta$. (Remember that if $a = 0$, then the graph is a single point located at the pole.) To sketch the graphs of equations of this form, we can convert the equations into rectangular form as in Example 6.

My interactive video summary

⊘ **Example 6** Sketching the Graph of a Polar Equation of the Form $r = a \sin \theta$ and $r = a \cos \theta$

Sketch the graph of each polar equation.

a. $r = 4 \sin \theta$ **b.** $r = -2 \cos \theta$

Solution

a. First, convert the equation $r = 4 \sin \theta$ from polar form to rectangular form.

$r = 4 \sin \theta$	Write the original equation in polar form.
$r^2 = 4r \sin \theta$	Multiply both sides by r.
$x^2 + y^2 = 4y$	Replace r^2 with $x^2 + y^2$ and replace $r \sin \theta$ with y.

Now, write the equation $x^2 + y^2 = 4y$ in standard form by subtracting $4y$ from both sides and completing the square.

$x^2 + y^2 = 4y$	Write the equation in rectangular form.
$x^2 + y^2 - 4y = 0$	Subtract $4y$ from both sides.
$x^2 + y^2 - 4y + 4 = 4$	Complete the square on the y-terms by adding 4 to both sides.
$x^2 + (y - 2)^2 = 4$	Factor.

We see that the equation $r = 4 \sin \theta$ in polar form is equivalent to the equation $x^2 + (y - 2)^2 = 4$ in rectangular form. The graph of the equation $x^2 + (y - 2)^2 = 4$ is a circle centered at the point $(0, 2)$ with a radius of 2 units. Therefore, the equation $r = 4 \sin \theta$ is a circle with a center located 2 units directly above the pole with a radius of 2 units. See Figure 25.

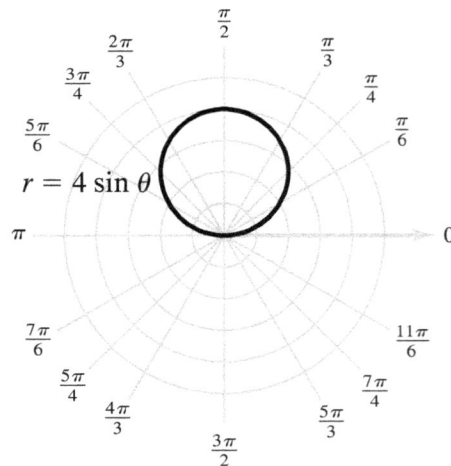

Figure 25   The graph of $r = 4 \sin \theta$ is a circle centered two units above the pole with a radius of 2 units.

My interactive video summary

⊘ **b.** Try sketching the graph of $r = -2 \cos \theta$ on your own. Watch this **interactive video** to see the complete solution or view the **graph**.

We now summarize the graphs of $r = a$, $r = a \sin \theta$, and $r = a \cos \theta$.

Graphs of Polar Equations of the Form $r = a$, $r = a \sin \theta$, and $r = a \cos \theta$, where $a \neq 0$ is a Constant

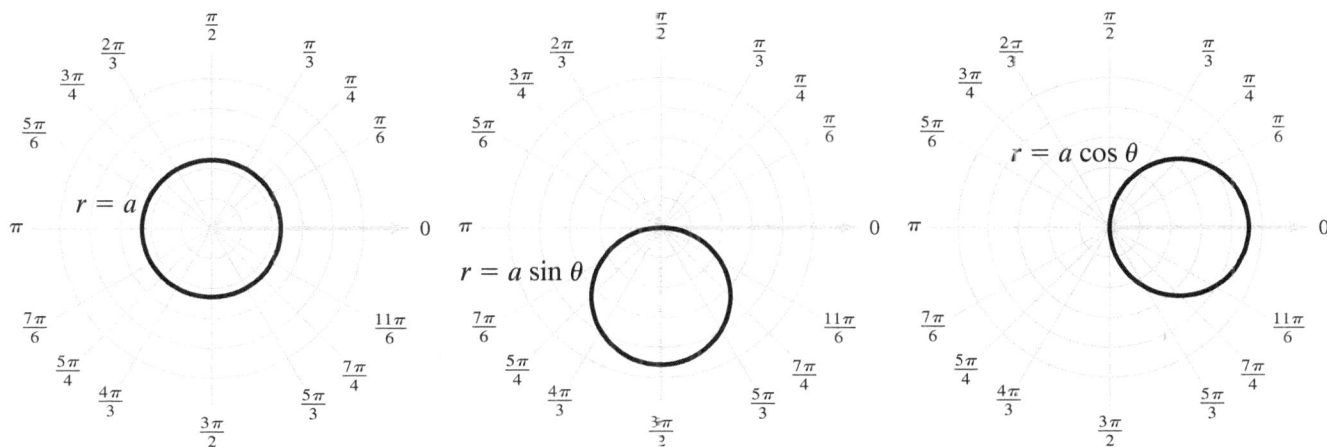

The graph of $r = a$ is a circle centered at the pole with radius of length $|a|$.

The graph of $r = a \sin \theta$ is a circle centered along the line $\theta = \dfrac{\pi}{2}$. The center is located $\dfrac{|a|}{2}$ units from the pole. The radius is $\dfrac{|a|}{2}$ units.

The graph of $r = a \cos \theta$ is a circle centered along the line $\theta = 0$. The center is located $\dfrac{|a|}{2}$ units from the pole. The radius is $\dfrac{|a|}{2}$ units.

You Try It   Work through this You Try It problem.

Work Exercises 9–14 in this textbook or in the MyMathLab Study Plan.

OBJECTIVE 3   SKETCHING EQUATIONS OF THE FORM $r = a + b \sin \theta$ AND $r = a + b \cos \theta$

So far we have sketched polar equations whose graphs had familiar shapes, namely lines and circles. We now begin to explore polar equations whose graphs may be a bit unfamiliar. The next category of graphs that we will study are called **limacons**. The word *limacon* is derived from the Latin word *limax*, which means "snail." All limacons have the form $r = a + b \sin \theta$ or $r = a + b \cos \theta$, where a and b are constants. Although we can convert these equations into rectangular coordinates, the rectangular form will not help to sketch the graph. We will see that the ratio $\left|\dfrac{a}{b}\right|$ will determine the shape of the limacon. We start with an example where the ratio $\left|\dfrac{a}{b}\right|$ is equal to 1.

Example 7 Sketching the Graph of a Polar Equation

of the Form $r = a + b \sin \theta$, where $\left| \dfrac{a}{b} \right| = 1$

Sketch the graph of the polar equation $r = 3 - 3 \sin \theta$.

Solution To sketch the graph of $r = 3 - 3 \sin \theta$, begin by choosing values of θ and then determine the corresponding values of r. We start by choosing the quadrantal angles $\theta = 0, \theta = \dfrac{\pi}{2}, \theta = \pi$, and $\theta = \dfrac{3\pi}{2}$. See Figure 26.

θ	$r = 3 - 3 \sin \theta$	(r, θ)
0	$r = 3 - 3 \sin 0 = 3 - 0 = 3$	$(3, 0)$
$\dfrac{\pi}{2}$	$r = 3 - 3 \sin \dfrac{\pi}{2} = 3 - 3 = 0$	$\left(0, \dfrac{\pi}{2}\right)$
π	$r = 3 - 3 \sin \pi = 3 - 0 = 3$	$(3, \pi)$
$\dfrac{3\pi}{2}$	$r = 3 - 3 \sin \dfrac{3\pi}{2} = 3 + 3 = 6$	$\left(6, \dfrac{3\pi}{2}\right)$

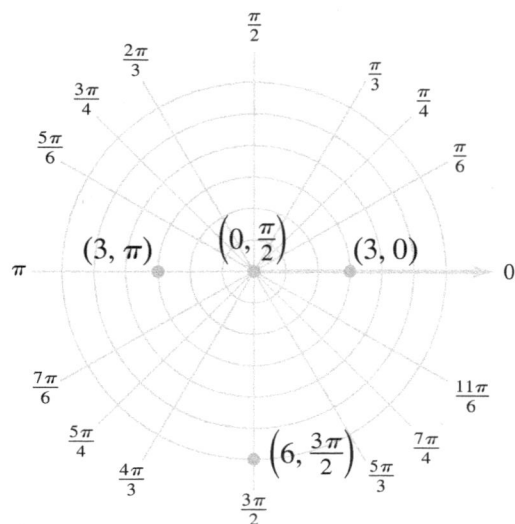

Figure 26 Points lying on the graph of $r = 3 - 3 \sin \theta$ whose angles are quadrantal angles

If we were to connect the points plotted in Figure 26, it would appear that the graph has the shape of a "T." This is not the actual shape of the graph, so we continue to plot several more points. We now choose the remaining special angles between 0 and π and connect the points with a smooth curve. See Figure 27.

θ	$r = 3 - 3 \sin \theta$	(r, θ)
$\dfrac{\pi}{6}$	$r = 3 - 3 \sin \dfrac{\pi}{6} = 3 - \dfrac{3}{2} = 1.5$	$\left(1.5, \dfrac{\pi}{6}\right)$
$\dfrac{\pi}{4}$	$r = 3 - 3 \sin \dfrac{\pi}{4} = 3 - \dfrac{3}{\sqrt{2}} \approx 0.88$	$\left(0.88, \dfrac{\pi}{4}\right)$
$\dfrac{\pi}{3}$	$r = 3 - 3 \sin \dfrac{\pi}{3} = 3 - \dfrac{3\sqrt{3}}{2} \approx 0.40$	$\left(0.40, \dfrac{\pi}{3}\right)$
$\dfrac{2\pi}{3}$	$r = 3 - 3 \sin \dfrac{2\pi}{3} = 3 - \dfrac{3\sqrt{3}}{2} \approx 0.40$	$\left(0.40, \dfrac{2\pi}{3}\right)$
$\dfrac{3\pi}{4}$	$r = 3 - 3 \sin \dfrac{3\pi}{4} = 3 - \dfrac{3}{\sqrt{2}} \approx 0.88$	$\left(0.88, \dfrac{3\pi}{4}\right)$
$\dfrac{5\pi}{6}$	$r = 3 - 3 \sin \dfrac{5\pi}{6} = 3 - \dfrac{3}{2} = 1.5$	$\left(1.5, \dfrac{5\pi}{6}\right)$

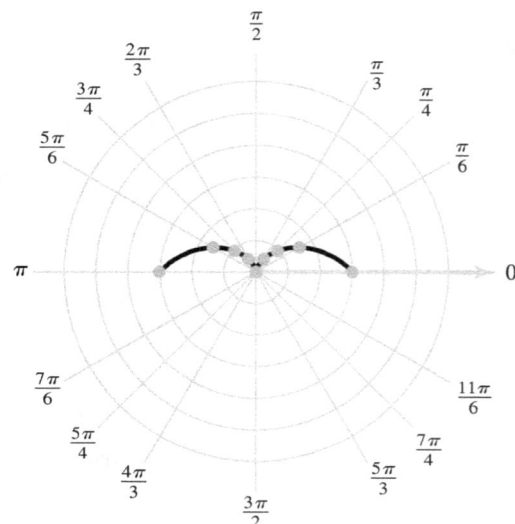

Figure 27 The graph of $r = 3 - 3 \sin \theta$ for $0 \le \theta \le \pi$

As you can see in Figure 27, the graph of $r = 3 - 3 \sin \theta$ appears to be symmetric about the vertical line $\theta = \dfrac{\pi}{2}$. Every equation of the form $r = a + b \sin \theta$ will be symmetric about this vertical line. Thus, in order to complete the graph, we need only to plot points having values of θ between π and $\dfrac{3\pi}{2}$ and then use symmetry to finish the graph. See Figure 28.

θ	$r = 3 - 3 \sin \theta$	(r, θ)
$\dfrac{7\pi}{6}$	$r = 3 - 3 \sin \dfrac{7\pi}{6} = 3 + \dfrac{3}{2} = 4.5$	$\left(4.5, \dfrac{7\pi}{6}\right)$
$\dfrac{5\pi}{4}$	$r = 3 - 3 \sin \dfrac{5\pi}{4} = 3 + \dfrac{3}{\sqrt{2}} \approx 5.12$	$\left(5.12, \dfrac{5\pi}{4}\right)$
$\dfrac{4\pi}{3}$	$r = 3 - 3 \sin \dfrac{4\pi}{3} = 3 + \dfrac{3\sqrt{3}}{2} \approx 5.60$	$\left(5.60, \dfrac{4\pi}{3}\right)$

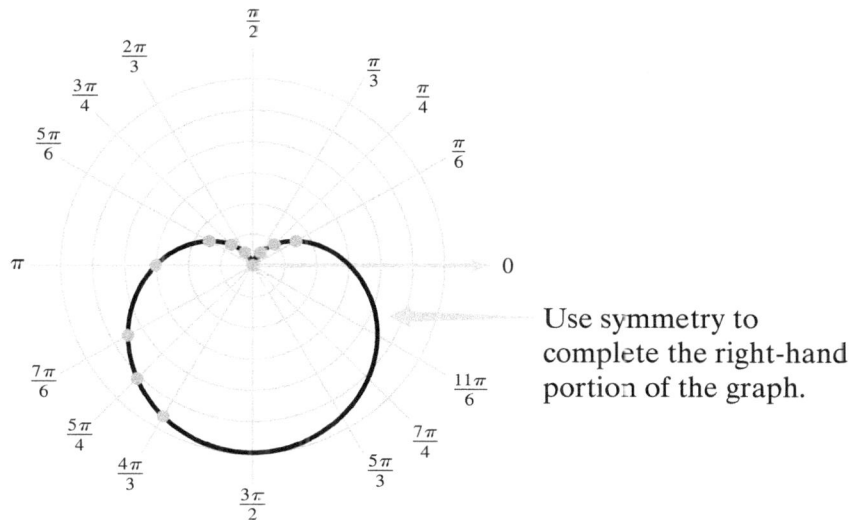

Use symmetry to complete the right-hand portion of the graph.

Figure 28 The complete graph of $r = 3 - 3 \sin \theta$

Observe that the graph of $r = 3 - 3 \sin \theta$ has a "heart shape." For this reason, graphs with this shape are called **cardioids**. All polar equations of the form $r = a + b \sin \theta$ or $r = a + b \cos \theta$, where $\left|\dfrac{a}{b}\right| = 1$, have graphs that are cardioids. Equations of this form involving sine will have graphs that are symmetric about the vertical line $\theta = \dfrac{\pi}{2}$. Equations of this form involving cosine will have graphs that are symmetric about the horizontal line $\theta = 0$. In fact, the graphs of all equations of the form $r = a + b \sin \theta$ or $r = a + b \cos \theta$ follow these symmetric properties. We now investigate equations of the form $r = a + b \sin \theta$ or $r = a + b \cos \theta$, where $\left|\dfrac{a}{b}\right| < 1$.

Example 8 Sketching the Graph of a Polar Equation of the Form $r = a + b \cos \theta$, where $\left| \dfrac{a}{b} \right| < 1$

Sketch the graph of the polar equation $r = -1 + 2 \cos \theta$.

Solution The equation is of the form $r = a + b \cos \theta$, so the graph will be symmetric about the horizontal line $\theta = 0$. Plotting points for values of $0 \leq \theta \leq \pi$, we get exactly "one-half" of the graph. See Figure 29.

θ	0	$\dfrac{\pi}{6}$	$\dfrac{\pi}{4}$	$\dfrac{\pi}{3}$	$\dfrac{\pi}{2}$	$\dfrac{2\pi}{3}$	$\dfrac{3\pi}{4}$	$\dfrac{5\pi}{6}$	π
r	1	0.73	0.41	0	-1	-2	-2.41	-2.73	-3
(r, θ)	$(1, 0)$	$\left(0.73, \dfrac{\pi}{6}\right)$	$\left(0.41, \dfrac{\pi}{4}\right)$	$\left(0, \dfrac{\pi}{3}\right)$	$\left(-1, \dfrac{\pi}{2}\right)$	$\left(-2, \dfrac{2\pi}{3}\right)$	$\left(-2.41, \dfrac{3\pi}{4}\right)$	$\left(-2.73, \dfrac{5\pi}{6}\right)$	$(-3, \pi)$

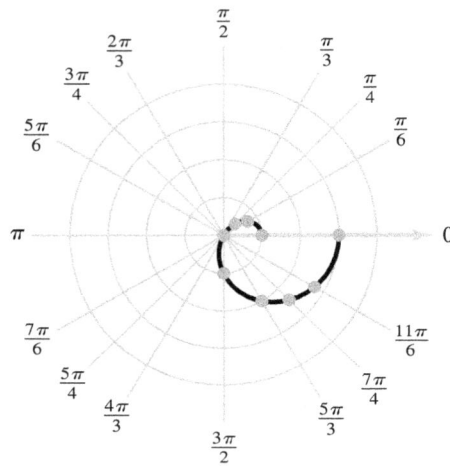

Figure 29 The graph of $r = -1 + 2 \cos \theta$ for $0 \leq \theta \leq \pi$

Because the graph is symmetric about the horizontal line $\theta = 0$, the graph can be completed by drawing the reflection of Figure 29 about the line $\theta = 0$. See Figure 30.

Figure 30 The complete graph of $r = -1 + 2 \cos \theta$

Notice that the graph has an "inner loop" that intersects the pole. The graph actually intersects the pole twice; once at the point $\left(0, \dfrac{\pi}{3}\right)$ and again at the point $\left(0, \dfrac{5\pi}{3}\right)$.

Graphs of the form $r = a + b \sin \theta$ or $r = a + b \cos \theta$ for $\left|\dfrac{a}{b}\right| < 1$ are limacons that have an inner loop that intersects the pole twice.

The next example illustrates the graph of $r = a + b \sin \theta$ for the case where $1 < \left|\dfrac{a}{b}\right| < 2$.

Example 9 Sketching the Graph of a Polar Equation of the Form $r = a + b \sin \theta$, where $1 < \left|\dfrac{a}{b}\right| < 2$

Sketch the graph of the polar equation $r = 3 + 2 \sin \theta$.

Solution The equation is of the form $r = a + b \sin \theta$, so the graph will be symmetric about the vertical line $\theta = \dfrac{\pi}{2}$. We start by plotting points for values of $0 \le \theta \le \dfrac{\pi}{2}$. We can then reflect this portion of the graph about the vertical line $\theta = \dfrac{\pi}{2}$ to obtain the graph of $r = 3 + 2 \sin \theta$ for $0 \le \theta \le \pi$. See Figure 31.

$\boldsymbol{\theta}$	0	$\dfrac{\pi}{6}$	$\dfrac{\pi}{4}$	$\dfrac{\pi}{3}$	$\dfrac{\pi}{2}$
$\boldsymbol{r}$	3	4	4.41	4.73	5
$\boldsymbol{(r, \theta)}$	$(3, 0)$	$\left(4, \dfrac{\pi}{6}\right)$	$\left(4.41, \dfrac{\pi}{4}\right)$	$\left(4.73, \dfrac{\pi}{3}\right)$	$\left(5, \dfrac{\pi}{2}\right)$

Use symmetry to determine the upper left-hand portion of the graph.

Figure 31 The graph of $r = 3 + 2 \sin \theta$ for $0 \le \theta \le \pi$

We can now plot points for values of $\pi < \theta \le \dfrac{3\pi}{2}$ and then once again use symmetry to complete the graph. See Figure 32.

θ	$\dfrac{7\pi}{6}$	$\dfrac{5\pi}{4}$	$\dfrac{4\pi}{3}$	$\dfrac{3\pi}{2}$
r	2	1.59	1.27	1
(r, θ)	$\left(2, \dfrac{7\pi}{6}\right)$	$\left(1.59, \dfrac{5\pi}{4}\right)$	$\left(1.27, \dfrac{4\pi}{3}\right)$	$\left(1, \dfrac{3\pi}{2}\right)$

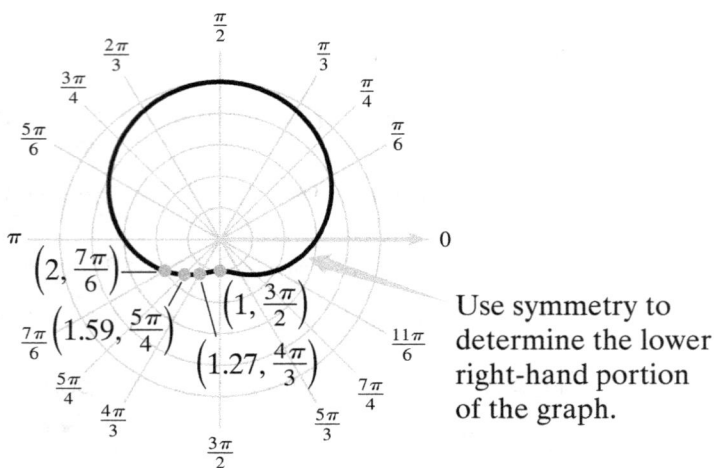

Use symmetry to determine the lower right-hand portion of the graph.

Figure 32 The complete graph of $r = 3 + 2\sin\theta$

The graph of $r = 3 + 2\sin\theta$ is known as a limacon with a dimple. All equations of the form $r = a + b\sin\theta$ or $r = a + b\cos\theta$ for $1 < \left|\dfrac{a}{b}\right| < 2$ have graphs that are limacons with dimples.

The final type of limacon has no inner loop and no dimple. These equations are of the form $r = a + b\sin\theta$ or $r = a + b\cos\theta$ for $\left|\dfrac{a}{b}\right| \ge 2$. Example 10 shows such an equation.

We now summarize each of the four types of limacons and create a step-by-step procedure for sketching their graphs.

Graphs of Polar Equations of the Form $r = a + b \sin \theta$ and $r = a + b \cos \theta$, where $a \neq 0$ and $b \neq 0$ Are Constants

The graph is a cardioid if $\left| \dfrac{a}{b} \right| = 1$.

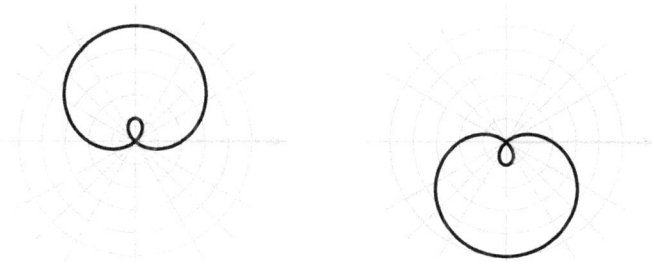

The graph is a limacon with an inner loop if $\left| \dfrac{a}{b} \right| < 1$.

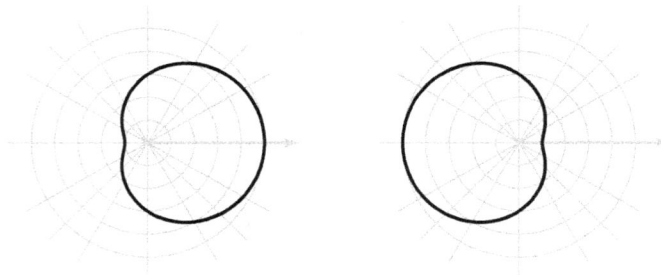

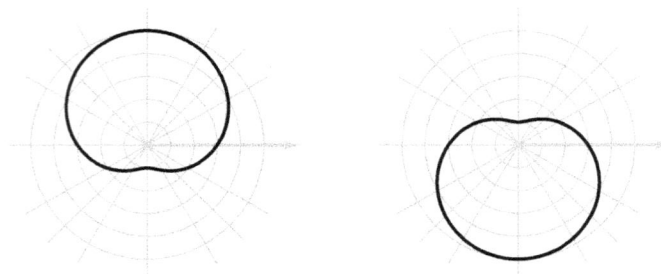

The graph is a limacon with a dimple if $1 < \left|\dfrac{a}{b}\right| < 2.$

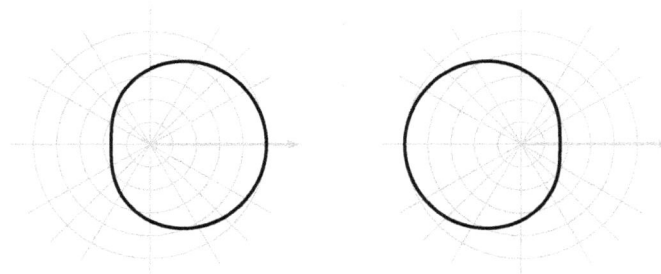

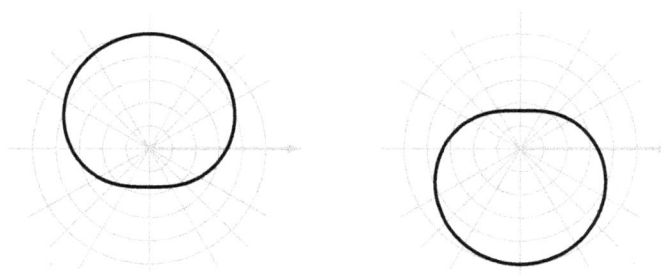

The graph is a limacon with no inner loop and no dimple if $\left|\dfrac{a}{b}\right| \geq 2.$

Steps for Sketching Polar Equations (Limacons) of the Form
$r = a + b \sin \theta$ **and** $r = a + b \cos \theta$

Step 1. Identify the general shape using the ratio $\left| \dfrac{a}{b} \right|$.

If $\left| \dfrac{a}{b} \right| = 1$, then the graph is a cardioid.

If $\left| \dfrac{a}{b} \right| < 1$, then the graph is a limacon with an inner loop that intersects the pole.

If $1 < \left| \dfrac{a}{b} \right| < 2$, then the graph is a limacon with a dimple.

If $\left| \dfrac{a}{b} \right| \geq 2$, then the graph is a limacon with no inner loop and no dimple.

Step 2. Determine the symmetry.

If the equation is of the form $r = a + b \sin \theta$, then the graph must be symmetric about the line $\theta = \dfrac{\pi}{2}$.

If the equation is of the form $r = a + b \cos \theta$, then the graph must be symmetric about the line $\theta = 0$.

Step 3. Plot the points corresponding to the quadrantal angles $\theta = 0, \theta = \dfrac{\pi}{2}, \theta = \pi$, and $\theta = \dfrac{3\pi}{2}$.

Step 4. If necessary, plot a few more points until symmetry can be used to complete the graph.

My interactive video summary

⊘ **Example 10** Sketching the Graph of a Polar Equation of the Form $r = a + b \sin \theta$ and $r = a + b \cos \theta$

Sketch the graph of each polar equation.

a. $r = 4 - 3 \cos \theta$ **b.** $r = 2 + \sin \theta$ **c.** $r = -2 + 2 \cos \theta$ **d.** $r = 3 - 4 \sin \theta$

Solution Follow the four-step process to sketch each graph. Carefully work through this interactive video to see how to use this four-step process, or, once you have sketched a graph, click on the corresponding link below to see if your graph is correct.

View the graph of $r = 4 - 3 \cos \theta$.
View the graph of $r = 2 + \sin \theta$.
View the graph of $r = -2 + 2 \cos \theta$.
View the graph of $r = 3 - 4 \sin \theta$.

You Try It Work through this You Try It problem.

Work Exercises 15–26 in this textbook or in the MyMathLab Study Plan.

OBJECTIVE 4   SKETCHING EQUATIONS OF THE FORM $r = a \sin n\theta$ AND $r = a \cos n\theta$

We now consider polar equations of the form $r = a \sin n\theta$ and $r = a \cos n\theta$, where $a \neq 0$ is a constant and $n \neq 1$ is a positive integer. The graphs of these equations are called roses because the shapes of the graphs resemble the petals of a flower. The number of petals depends on the value of n. Again, converting these types of equations into rectangular form would be of no advantage. We start by sketching a graph of the form $r = a \sin n\theta$, where n is even.

Example 11 Sketching the Graph of a Polar Equation of the Form $r = a \sin n\theta$

Sketch the graph of $r = 3 \sin 2\theta$.

Solution The equation $r = 3 \sin 2\theta$ is of the form $r = a \sin n\theta$, where $a = 3$ and $n = 2$. We start by plotting several points to complete one petal of a rose. We will use values of θ such that the angle 2θ will yield a special angle, as seen in the table below. Note that decimal values of r are approximations.

θ	0	$\dfrac{\pi}{12}$	$\dfrac{\pi}{6}$	$\dfrac{\pi}{4}$	$\dfrac{\pi}{3}$	$\dfrac{5\pi}{12}$	$\dfrac{\pi}{2}$
2θ	0	$\dfrac{\pi}{6}$	$\dfrac{\pi}{3}$	$\dfrac{\pi}{2}$	$\dfrac{2\pi}{3}$	$\dfrac{5\pi}{6}$	π
r	0	$\dfrac{3}{2}$	2.6	3	2.6	$\dfrac{3}{2}$	0
(r, θ)	$(0, 0)$	$\left(\dfrac{3}{2}, \dfrac{\pi}{12}\right)$	$\left(2.6, \dfrac{\pi}{6}\right)$	$\left(3, \dfrac{\pi}{4}\right)$	$\left(2.6, \dfrac{\pi}{3}\right)$	$\left(\dfrac{3}{2}, \dfrac{5\pi}{12}\right)$	$\left(0, \dfrac{\pi}{2}\right)$

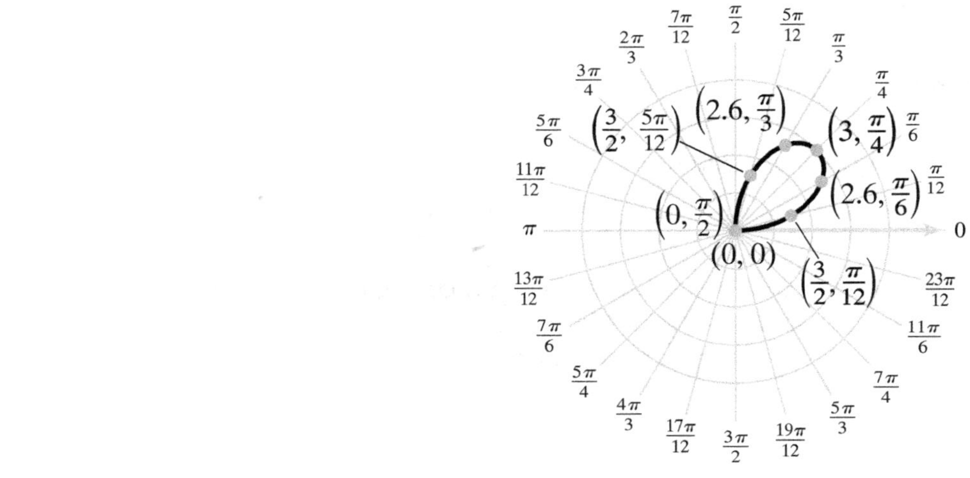

Figure 33 One petal of the graph of $r = 3 \sin 2\theta$

It is not necessary to continue to plot more points to complete the graph. Each petal of the rose will have the exact same shape as the petal sketched in Figure 33. Note that the length of the petal is 3 units. If we can determine the endpoints of the remaining petals, then we can easily complete the graph. These endpoints occur for values of θ for which $r = 3$ or $r = -3$. Thus, we must solve the two trigonometric equations $3 \sin 2\theta = 3$ and $3 \sin 2\theta = -3$. These equations are equivalent to the equations $\sin 2\theta = 1$ and $\sin 2\theta = -1$. Both of these equations can be solved using the steps for solving trigonometric equations that are linear in form that were discussed in Section 3.5.

The solutions to the equation **$\sin 2\theta = 1$** on the interval $[0, 2\pi)$ are $\theta = \dfrac{\pi}{4}$ and $\theta = \dfrac{5\pi}{4}$. The solutions to the equation **$\sin 2\theta = -1$** on the interval $[0, 2\pi)$ are $\theta = \dfrac{3\pi}{4}$ and $\theta = \dfrac{7\pi}{4}$. Therefore, the endpoints of the four petals occur at the points and $\left(3, \dfrac{\pi}{4}\right)$, $\left(3, \dfrac{5\pi}{4}\right)$, $\left(-3, \dfrac{3\pi}{4}\right)$, and $\left(-3, \dfrac{7\pi}{4}\right)$. See Figure 34. Now that we know the endpoints of the four petals, we can complete the graph of $r = 3 \sin 2\theta$. See Figure 35.

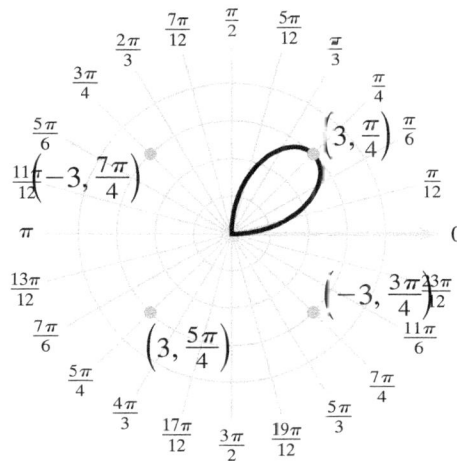

Figure 34 The endpoints of the four petals of the graph of $r = 3 \sin 2\theta$ occur at the points $\left(3, \dfrac{\pi}{4}\right)$, $\left(3, \dfrac{5\pi}{4}\right)$, $\left(-3, \dfrac{3\pi}{4}\right)$, and $\left(-3, \dfrac{7\pi}{4}\right)$.

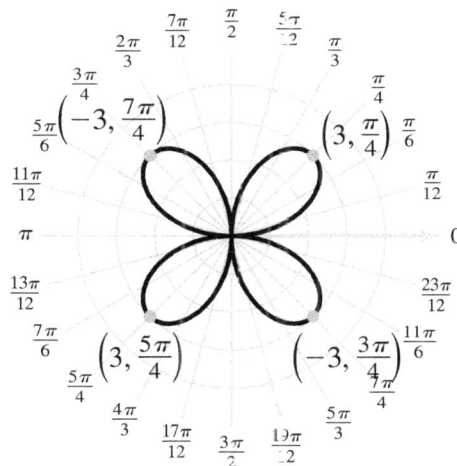

Figure 35 The complete graph of $r = 3 \sin 2\theta$

Note that the graph of the equation $r = 3 \sin 2\theta$ has 4 petals. The length of each petal is 3 units. In general, equations of the form $r = a \sin n\theta$ and $r = a \cos n\theta$ have petals of length $|a|$ units. After further examination we will be able to generalize the following:

If n is even, then there will be $2n$ petals.

If n is odd, there will be n petals.

We now summarize each type of rose and create a step-by-step procedure for sketching their graphs.

Graphs of Polar Equations of the Form $r = a \sin n\theta$ and $r = a \cos n\theta$, where $a \neq 0$ Is a Constant and $n \neq 1$ Is a Positive Integer

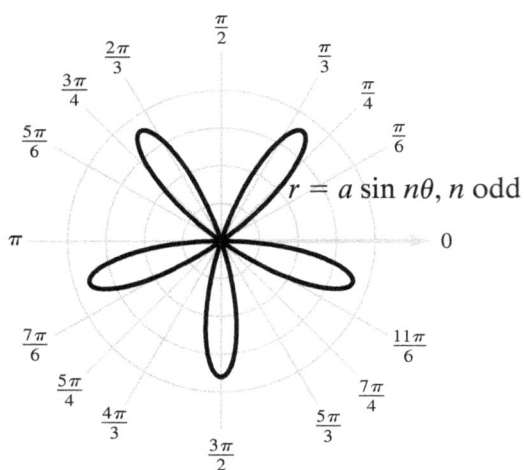

$r = a \sin n\theta$, n odd

The graph is a rose with n petals. The endpoint of one petal lies along the vertical line $\theta = \dfrac{\pi}{2}$.

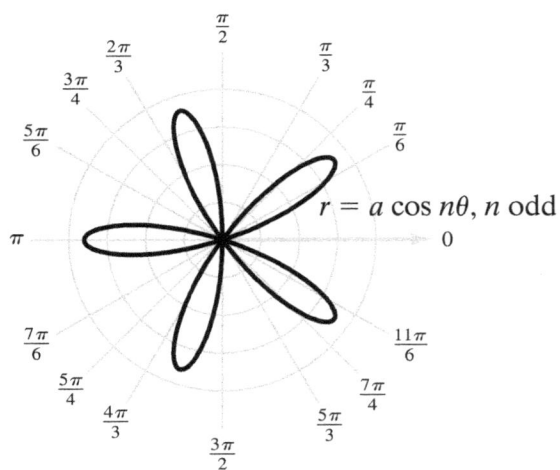

$r = a \cos n\theta$, n odd

The graph is a rose with n petals. The endpoint of one petal lies along the line $\theta = 0$.

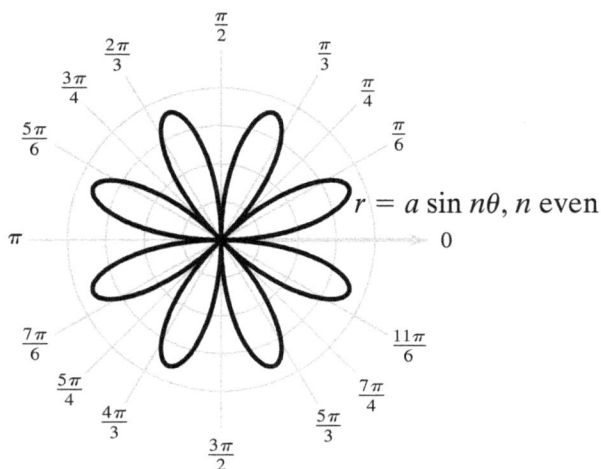

$r = a \sin n\theta$, n even

The graph is a rose with $2n$ petals. None of the petals has an endpoint lying on either the line $\theta = 0$ or the line $\theta = \dfrac{\pi}{2}$.

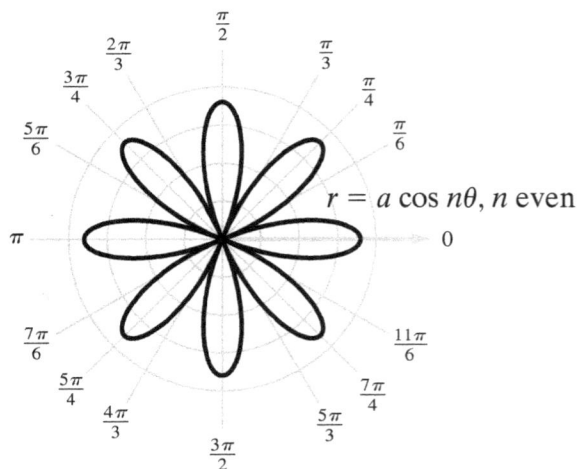

$r = a \cos n\theta$, n even

The graph is a rose with $2n$ petals. Two petals have endpoints lying along the vertical line $\theta = \dfrac{\pi}{2}$. Two petals have endpoints lying along the horizontal line $\theta = 0$.

Steps for Sketching Polar Equations (Roses) of the Form $r = a \sin n\theta$ and $r = a \cos n\theta$, where $a \neq 0$ and $n \neq 1$ Is a Positive Integer

Step 1. Identify the number of petals.

If n is even, then there are $2n$ petals.

If n is odd, then there are n petals.

Step 2. Determine the length of each petal.

The length of each petal is $|a|$ units.

Step 3. Determine all angles where an endpoint of a petal lies.

If the equation is of the form $r = a \sin n\theta$, then the endpoints occur for angles on the interval $[0, 2\pi)$* that satisfy the equations $\sin n\theta = 1$ and $\sin n\theta = -1$.

If the equation is of the form $r = a \cos n\theta$, then the endpoints occur for angles on the interval $[0, 2\pi)$* that satisfy the equations $\cos n\theta = 1$ and $\cos n\theta = -1$.

*Note that when n is odd, it is only necessary to consider angles on the interval $[0, \pi)$. A complete graph is obtained on this interval because the graph will completely traverse itself on the interval $[\pi, 2\pi)$.

Step 4. Substitute each angle determined in Step 3 back into the original equation to obtain the appropriate values of r for each angle. The ordered pairs obtained represent the endpoints of the rose petals. Plot these points on the graph.

Step 5. Determine angles where the graph passes through the pole. These angles serve as a guide when sketching the width of a petal.

If the equation is of the form $r = a \sin n\theta$, then the graph passes through the pole when $\sin n\theta = 0$.

If the equation is of the form $r = a \cos n\theta$, then the graph passes through the pole when $\cos n\theta = 0$.

Step 6. Draw each petal to complete the graph.

My interactive video summary

⊘ **Example 12** Sketching the Graph of a Polar Equation of the Form $r = a \sin n\theta$ and $r = a \cos n\theta$

Sketch the graph of each polar equation.

a. $r = -4 \cos 3\theta$ **b.** $r = -2 \sin 5\theta$ **c.** $r = 5 \cos 4\theta$

Solution

a. To sketch the graph of $r = -4 \cos 3\theta$, we follow the five-step process for sketching roses.

Step 1. For the equation $r = -4 \cos 3\theta$, $n = 3$, which is odd. So there are 3 petals.

Step 2. The length of each petal is $|a| = |-4| = 4$ units.

Step 3. Solve the equations $\cos 3\theta = 1$ and $\cos 3\theta = -1$ to determine all angles where an endpoint of a petal lies. View these steps to verify that the solutions to the equation $\cos 3\theta = 1$ on the interval $[0, 2\pi)$ are

$\theta = 0, \theta = \dfrac{2\pi}{3}$ and $\theta = \dfrac{4\pi}{3}$. View these **steps** to verify that the solutions

to the equation $\cos 3\theta = -1$ on the interval $[0, 2\pi)$ are $\theta = \dfrac{\pi}{3}, \theta = \pi,$

and $\theta = \dfrac{5\pi}{3}.$

Step 4. Although there are only three petals, we found six values of θ in step 3. This is because when n is odd, we obtain a complete graph on the interval $[0, \pi)$. The graph will completely traverse itself on the interval $[\pi, 2\pi)$.

For this reason, we choose the three values of θ from step 3 that lie on the interval $[0, \pi)$. These values are $\theta = 0, \theta = \dfrac{\pi}{3}$, and $\theta = \dfrac{2\pi}{3}.$

Thus, the coordinates of the three endpoints of the petals are $(-4, 0), \left(4, \dfrac{\pi}{3}\right),$ and $\left(-4, \dfrac{2\pi}{3}\right).$ See Figure 36.

Step 5. The graph passes through the pole when $\cos 3\theta = 0$. View these **steps** to verify that these angles on the interval $[0, \pi)$ are $\theta = \dfrac{\pi}{6}, \theta = \dfrac{\pi}{2},$ and $\theta = \dfrac{5\pi}{6}.$ These three angles can be used as a guide when determining the width of each petal.

Step 6. We now draw the three petals of the rose, making certain that the graph never intersects the lines $\theta = \dfrac{\pi}{6}, \theta = \dfrac{\pi}{2},$ and $\theta = \dfrac{5\pi}{6}$ except at the pole. See the complete graph in Figure 37.

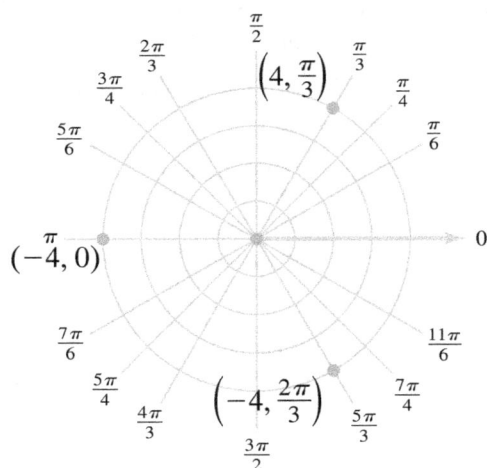

Figure 36 The endpoints of the 3 petals of the graph of $r = -4\cos 3\theta$

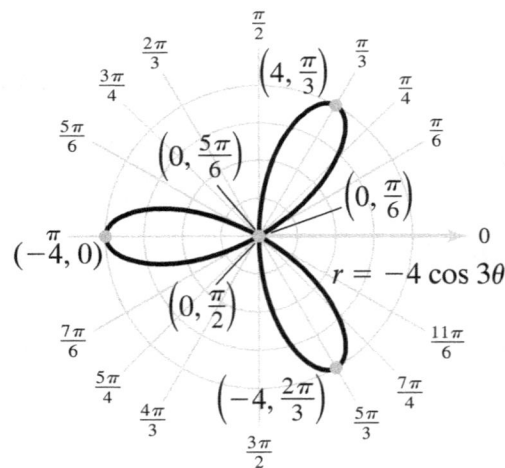

Figure 37 The complete graph of $r = -4\cos 3\theta$

*My interactive
video summary*

⊘ For parts b and c, carefully work through this interactive video to see how to sketch each graph or click on the corresponding link below to see if your graph is correct.

View the graph of $r = -2 \sin 5\theta$.

View the graph of $r = 5 \cos 4\theta$.

You Try It   Work through this You Try It problem.

Work Exercises 27–34 in this textbook or in the MyMathLab Study Plan.

OBJECTIVE 5   SKETCHING EQUATIONS OF THE FORM $r^2 = a^2 \sin 2\theta$ AND $r^2 = a^2 \cos 2\theta$

The final type of polar graph that we will consider is called a **lemniscate**. Lemniscates resemble a Figure 8, or a propeller, and have equations of the form $r^2 = a^2 \sin 2\theta$ and $r^2 = a^2 \cos 2\theta$. The endpoints of the two loops of lemniscates of the form $r^2 = a^2 \sin 2\theta$ occur when $\theta = \dfrac{\pi}{4}$ and $\theta = \dfrac{5\pi}{4}$, whereas the endpoints of the two loops of lemniscates of the form $r^2 = a^2 \cos 2\theta$ occur when $\theta = 0$ and $\theta = \pi$. The length of each loop of a lemniscate is $|a|$ units. The two types of lemniscates are shown on the following page.

Graphs of Polar Equations of the Form $r^2 = a^2 \sin 2\theta$, and $r^2 = a^2 \cos 2\theta$, where $a \neq 0$ Is a Constant

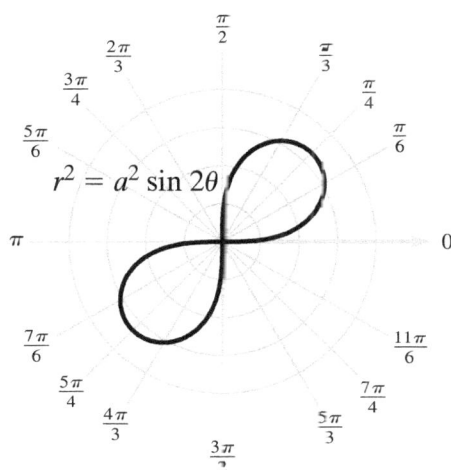

$r^2 = a^2 \sin 2\theta$

$r^2 = a^2 \cos 2\theta$

The graph is a lemniscate symmetric about the pole and the line $\theta = \dfrac{\pi}{4}$. The endpoints of the two loops occur when $\theta = \dfrac{\pi}{4}$ and $\theta = \dfrac{5\pi}{4}$.

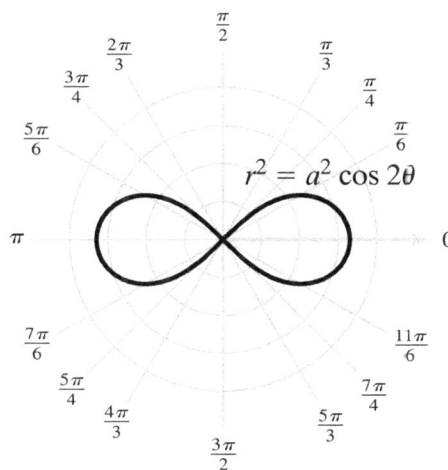

The graph is a lemniscate symmetric about the pole, the horizontal line $\theta = 0$, and the vertical line $\theta = \dfrac{\pi}{2}$. The endpoints of the two loops occur when $\theta = 0$ and $\theta = \pi$.

My video summary   ⊙ **Example 13** Sketching the Graph of a Polar Equation of the Form $r^2 = a^2 \sin 2\theta$ and $r^2 = a^2 \cos 2\theta$

Sketch the graph of each polar equation.

a. $r^2 = 9 \cos 2\theta$ **b.** $r^2 = 16 \sin 2\theta$

Solution

a. The equation is of the form $r^2 = a^2 \cos 2\theta$, where $a^2 = 9$. So the graph is a lemniscate. The length of each loop is 3 units. The graph is symmetric about the pole, the horizontal line $\theta = 0$, and the vertical line $\theta = \dfrac{\pi}{2}$. See Figure 38.

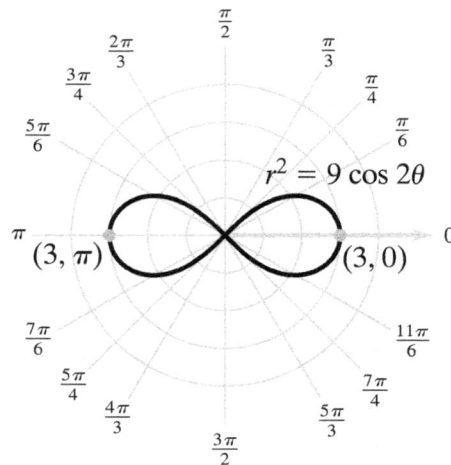

Figure 38 The complete graph of $r^2 = 9 \cos 2\theta$

b. Try sketching the graph of the equation $r^2 = 16 \sin 2\theta$ on your own. When you have finished sketching the graph, view the completed **graph** to see if you are correct or watch this **video**.

You Try It   Work through this You Try It problem.

Work Exercises 35 and 36 in this textbook or in the My MathLab **Study Plan**.

5.2 Exercises

Skill Check Exercises

SCE-1. Simplify the expression $\dfrac{\pi}{4} + \dfrac{2\pi}{3}k$ for $k = -3, -2, -1, 0, 1,$ and 2.

In Exercises 1–36, sketch the graph of the polar equation.

SbS **1.** $r \cos \theta = 5$ SbS **2.** $r \cos \theta = -2$ SbS **3.** $r \sin \theta = 3$ SbS **4.** $r \sin \theta = -4$

SbS **5.** $2r \cos \theta + 5r \sin \theta = 10$ SbS **6.** $3r \cos \theta - 4r \sin \theta = -5$

7. $\theta = -\dfrac{\pi}{4}$ **8.** $\theta = \dfrac{5\pi}{6}$ **9.** $r = 3$ **10.** $r = -2$

11. $r = 2\cos\theta$ **12.** $r = -3\sin\theta$ **13.** $r = -\cos\theta$ **14.** $r = 5\sin\theta$

15. $r = 1 + \sin\theta$ **16.** $r = 3 - 4\cos\theta$ **17.** $r = 4 + 3\sin\theta$ **18.** $r = 3 - 3\cos\theta$

19. $r = 4 + 2\sin\theta$ **20.** $r = 1 + 2\sin\theta$ **21.** $r = -4 - 4\sin\theta$ **22.** $r = 5 - 3\cos\theta$

23. $r = -2 + 5\cos\theta$ **24.** $r = -2 + 2\cos\theta$ **25.** $r = -2 - 3\sin\theta$ **26.** $r = 3 - \cos\theta$

27. $r = -4\cos 3\theta$ **28.** $r = 2\sin 2\theta$ **29.** $r = 5\cos 4\theta$ **30.** $r = -3\sin 5\theta$

31. $r = 5\sin 3\theta$ **32.** $r = 3\cos 5\theta$ **33.** $r = -4\sin 4\theta$ **34.** $r = -2\cos 2\theta$

35. $r^2 = 16\sin 2\theta$ **36.** $r^2 = 9\cos 2\theta$

Brief Exercises

In Exercises 37–72, identify the type of graph of each polar equation.

37. $r\cos\theta = 5$ **38.** $r\cos\theta = -2$ **39.** $r\sin\theta = 3$ **40.** $r\sin\theta = -4$

41. $2r\cos\theta + 5r\sin\theta = 10$ **42.** $3r\cos\theta - 4r\sin\theta = -5$

43. $\theta = -\dfrac{\pi}{4}$ **44.** $\theta = \dfrac{5\pi}{6}$ **45.** $r = 3$ **46.** $r = -2$

47. $r = 2\cos\theta$ **48.** $r = -3\sin\theta$ **49.** $r = -\cos\theta$ **50.** $r = 5\sin\theta$

51. $r = 1 + \sin\theta$ **52.** $r = 3 - 4\cos\theta$ **53.** $r = 4 + 3\sin\theta$ **54.** $r = 3 - 3\cos\theta$

55. $r = 4 + 2\sin\theta$ **56.** $r = 1 + 2\sin\theta$ **57.** $r = -4 - 4\sin\theta$ **58.** $r = 5 - 3\cos\theta$

59. $r = -2 + 5\cos\theta$ **60.** $r = -2 + 2\cos\theta$ **61.** $r = -2 - 3\sin\theta$ **62.** $r = 3 - \cos\theta$

63. $r = -4\cos 3\theta$ **64.** $r = 2\sin 2\theta$ **65.** $r = 5\cos 4\theta$ **66.** $r = -3\sin 5\theta$

67. $r = 5\sin 3\theta$ **68.** $r = 3\cos 5\theta$ **69.** $r = -4\sin 4\theta$ **70.** $r = -2\cos 2\theta$

71. $r^2 = 16\sin 2\theta$ **72.** $r^2 = 9\cos 2\theta$

In Exercises 73–108, sketch the graph of the polar equation.

73. $r\cos\theta = 5$ **74.** $r\cos\theta = -2$ **75.** $r\sin\theta = 3$ **76.** $r\sin\theta = -4$

77. $2r\cos\theta + 5r\sin\theta = 10$ **78.** $3r\cos\theta - 4r\sin\theta = -5$

79. $\theta = -\dfrac{\pi}{4}$ **80.** $\theta = \dfrac{5\pi}{6}$ **81.** $r = 3$ **82.** $r = -2$

83. $r = 2\cos\theta$ **84.** $r = -3\sin\theta$ **85.** $r = -\cos\theta$ **86.** $r = 5\sin\theta$

87. $r = 1 + \sin\theta$ **88.** $r = 3 - 4\cos\theta$ **89.** $r = 4 + 3\sin\theta$ **90.** $r = 3 - 3\cos\theta$

91. $r = 4 + 2 \sin \theta$ **92.** $r = 1 + 2 \sin \theta$ **93.** $r = -4 - 4 \sin \theta$ **94.** $r = 5 - 3 \cos \theta$

95. $r = -2 + 5 \cos \theta$ **96.** $r = -2 + 2 \cos \theta$ **97.** $r = -2 - 3 \sin \theta$ **98.** $r = 3 - \cos \theta$

99. $r = -4 \cos 3\theta$ **100.** $r = 2 \sin 2\theta$ **101.** $r = 5 \cos 4\theta$ **102.** $r = -3 \sin 5\theta$

103. $r = 5 \sin 3\theta$ **104.** $r = 3 \cos 5\theta$ **105.** $r = -4 \sin 4\theta$ **106.** $r = -2 \cos 2\theta$

107. $r^2 = 16 \sin 2\theta$ **108.** $r^2 = 9 \cos 2\theta$

5.3 Complex Numbers in Polar Form; De Moivre's Theorem

THINGS TO KNOW

Before working through this section, be sure that you are familiar with the following concepts:

                                                                 VIDEO   ANIMATION   INTERACTIVE

You Try It   **1.** Understanding the Four Families of Special Angles (Section 1.5)   *(Interactive)*

You Try It   **2.** Understanding the Definitions of the Trigonometric Functions of General Angles (Section 1.5)   *(Video) (Animation)*

You Try It   **3.** Evaluating Trigonometric Functions of Angles Belonging to the $\frac{\pi}{3}, \frac{\pi}{6}$, or $\frac{\pi}{4}$ Families (Section 1.5)   *(Interactive)*

You Try It   **4.** Solving Trigonometric Equations That Are Linear in Form (Section 3.5)   *(Interactive)*

You Try It   **5.** Plotting Points Using Polar Coordinates (Section 5.1)   *(Video)*

You Try It   **6.** Converting a Point from Rectangular Coordinates to Polar Coordinates (Section 5.1)   *(Interactive)*

OBJECTIVES

1 Understanding the Rectangular Form of a Complex Number

2 Understanding the Polar Form of a Complex Number

3 Converting a Complex Number from Polar Form to Rectangular Form

4 Converting a Complex Number from Rectangular Form to Standard Polar Form

5 Determining the Product or Quotient of Complex Numbers in Polar Form

6 Using De Moivre's Theorem to Raise a Complex Number to a Power

7 Using De Moivre's Theorem to Find the Roots of a Complex Number

OBJECTIVE 1   UNDERSTANDING THE RECTANGULAR FORM OF
A COMPLEX NUMBER

Complex numbers are numbers of the form $a + bi$, where a and b are real numbers and i is the imaginary unit. We often use the variable z to denote a complex number. Complex numbers of the form $z = a + bi$ are said to be in **rectangular form**.

Definition   The Rectangular Form of a Complex Number

A complex number is said to be in **rectangular form** if it is written as $z = a + bi$, where a and b are real numbers and i is the imaginary unit.

Complex numbers have no order. That is, there is no way to compare two complex numbers and determine if one is greater than the other. However, we can characterize complex numbers by representing them geometrically on a graph.

Every complex number can be represented by a point in the **complex plane**. The complex plane looks similar to a rectangular coordinate system. Just like in the rectangular coordinate system, the point of intersection of the two axes is called the **origin** of the complex plane. However, in the complex plane the horizontal axis is called the **real** axis and the vertical axis is called the **imaginary axis**. We plot the complex number $z = a + bi$ in the same way as we would plot the point (a, b) in the rectangular coordinate system. See Figure 39.

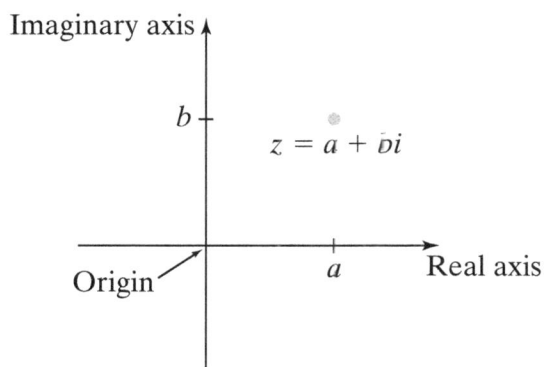

Figure 39 A point in the complex plane

Recall that the absolute value of a real number a is defined as its distance from 0 on a number line and is denoted as $|a|$. Similarly, **the absolute value of a complex number** $z = a - bi$ is the distance from the origin to the point z in the complex plane and is denoted by $|z|$. Therefore, we can use the distance formula to determine the absolute value of a complex number by finding the distance between the origin and the point (a, b) in a rectangular coordinate system.

Definition   The Absolute Value of a Complex Number

The **absolute value of a complex number** $z = a + bi$ is denoted by $|z|$ and is the distance from the origin to z in the complex plane and is given by $|z| = \sqrt{a^2 + b^2}$.

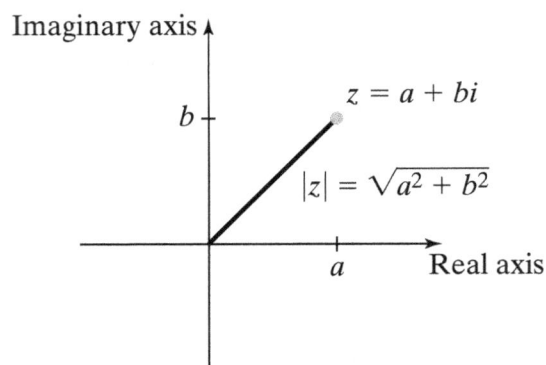

My video summary   ⊗ **Example 1 Plotting Complex Numbers in the Complex Plane**

Plot each complex number in the complex plane and determine its absolute value.

a. $z_1 = 3 - 4i$ **b.** $z_2 = -2 + 5i$ **c.** $z_3 = 3$ **d.** $z_4 = -2i$

Solution Each complex number is plotted in Figure 40.

The absolute values of each number are as follows:

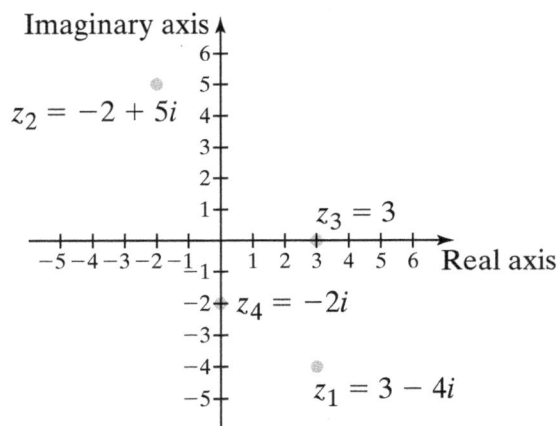

Figure 40

a. For $z_1 = 3 - 4i$, $a = 3$ and $b = -4$. Thus, $|z_1| = \sqrt{3^2 + (-4)^2} = \sqrt{9 + 16} = \sqrt{25} = 5$.

b. For $z_2 = -2 + 5i$, $a = -2$ and $b = 5$. Thus, $|z_2| = \sqrt{(-2)^2 + (5)^2} = \sqrt{4 + 25} = \sqrt{29}$.

c. For $z_3 = 3$, $a = 3$ and $b = 0$. Thus, $|z_3| = \sqrt{3^2 + 0^2} = \sqrt{9} = 3$.

d. For $z_4 = -2i$, $a = 0$ and $b = -2$. Thus, $|z_4| = \sqrt{0^2 + (-2)^2} = \sqrt{4} = 2$.

You Try It Work through this You Try It problem.

Work Exercises 1–4 in this textbook or in the MyMathLab **Study Plan**.

OBJECTIVE 2 UNDERSTANDING THE PCLAR FORM OF A COMPLEX NUMBER

Suppose that we represent a complex number $z = a + bi$ using the polar coordinates (r, θ). Notice in Figure 41 that the absolute value of z, $|z| = \sqrt{a^2 + b^2}$, is the same as the value of r. Therefore, $|z| = r = \sqrt{a^2 + b^2}$.

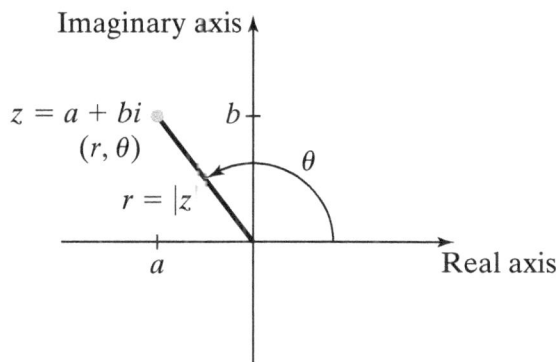

Figure 41

Considering the point sketched in Figure 41 and using the general angle definitions of the trigonometric functions, we know the following:

$$\cos \theta = \frac{a}{r}, \ \sin \theta = \frac{b}{r}, \text{ and } \tan \theta = \frac{b}{a}$$

Multiplying both sides of the equations $\cos \theta = \dfrac{a}{r}$ and $\sin \theta = \dfrac{b}{r}$ by r gives $a = r \cos \theta$ and $b = r \sin \theta$. We can use these two equations to write the number $z = a + bi$ in an alternate form.

$z = a + bi$	Write the complex number in rectangular form.
$z = r \cos \theta + (r \sin \theta)i$	Substitute $r \cos \theta$ for a and $r \cos \theta$ for b.
$z = r(\cos \theta + i \sin \theta)$	Factor out a common factor of r.

We say that the complex number $z = r(\cos \theta + i \sin \theta)$ is written in **polar form**. The value of r is called the **modulus** of z (or the magnitude of z) and the angle θ is called the **argument** of z.

Definition   The Polar Form of a Complex Number

A complex number $z = a + bi$ is said to be in **polar form** if it is written as

$z = r(\cos \theta + i \sin \theta)$, where $a = r \cos \theta$, $b = r \sin \theta$, $\tan \theta = \dfrac{b}{a}$, and $r = \sqrt{a^2 + b^2}$.

Because of the periodic nature of sine and cosine, there are infinitely many representations of every complex number in polar form. A complex number $z = r(\cos \theta + i \sin \theta)$ is equivalent to a complex number of the form $z = r(\cos(\theta + 2\pi k) + i \sin(\theta + 2\pi k))$, where k is any integer. However, we will always write the polar form of a complex number with $0 \le \theta < 2\pi$. Complex numbers in polar form, where $0 \le \theta < 2\pi$ (or $0 \le \theta < 360°$), are said to be in **standard polar form**.

Definition   The Standard Polar Form of a Complex Number

A complex number $z = r(\cos\theta + i\sin\theta)$ is said to be in **standard polar form** when $0 \leq \theta < 2\pi$ (or $0 \leq \theta < 360°$).

My video summary   ⊙ **Example 2** Using the Standard Polar Form of a Complex Number

Rewrite the complex number in standard polar form, plot the number in the complex plane, and determine the quadrant in which the point lies or the axis on which the point lies.

a. $z = 3\left(\cos\dfrac{5\pi}{8} + i\sin\dfrac{5\pi}{8}\right)$

b. $z = 2\left(\cos\dfrac{23\pi}{4} + i\sin\dfrac{23\pi}{4}\right)$

c. $z = 4(\cos(-3\pi) + i\sin(-3\pi))$

Solution

a. The complex number $z = 3\left(\cos\dfrac{5\pi}{8} + i\sin\dfrac{5\pi}{8}\right)$ is already written in standard polar form because $\theta = \dfrac{5\pi}{8}$ and $0 \leq \dfrac{5\pi}{8} < 2\pi$. To plot z, we plot a point that lies on the terminal side of $\theta = \dfrac{5\pi}{8}$ and is a distance of $r = 3$ units from the origin. We observe that this complex number lies in Quadrant II. See Figure 42.

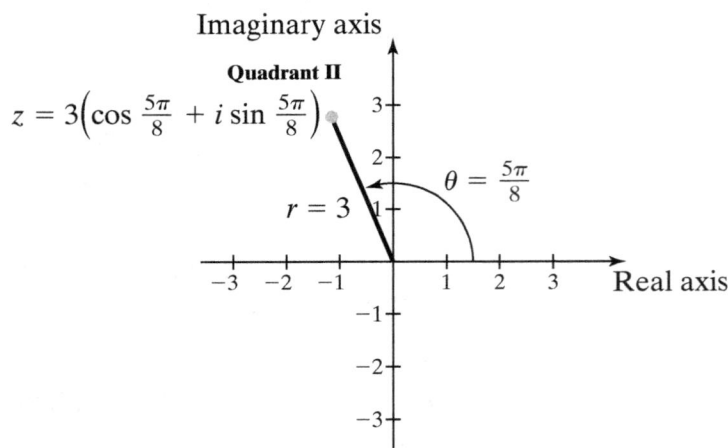

Figure 42 The complex number $z = 3\left(\cos\dfrac{5\pi}{8} + i\sin\dfrac{5\pi}{8}\right)$ is located in Quadrant II in the complex plane.

b. To write the complex number $z = 2\left(\cos\dfrac{23\pi}{4} + i\sin\dfrac{23\pi}{4}\right)$ in standard polar form, we must determine the angle on the interval $[0, 2\pi)$ that is coterminal with $\dfrac{23\pi}{4}$. This angle is $\theta = \dfrac{23\pi}{4} - 4\pi = \dfrac{23\pi}{4} - \dfrac{16\pi}{4} = \dfrac{7\pi}{4}$. Therefore, the standard polar form is $z = 2\left(\cos\dfrac{7\pi}{4} + i\sin\dfrac{7\pi}{4}\right)$. To plot z, we plot a point that lies on the

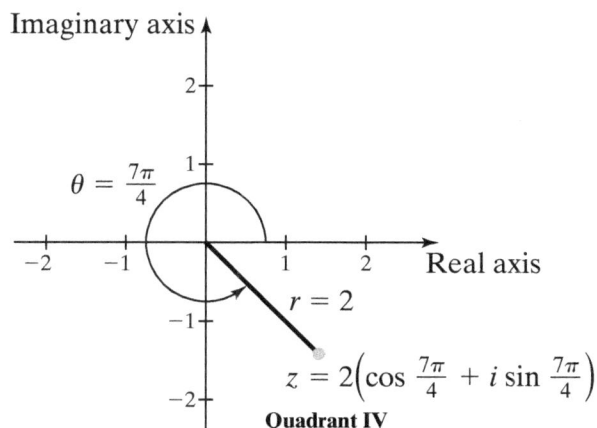

$$z = 2\left(\cos \frac{7\pi}{4} + i \sin \frac{7\pi}{4}\right)$$

Quadrant IV

Figure 43 The complex number

$z = 2\left(\cos \dfrac{23\pi}{4} + i \sin \dfrac{23\pi}{4}\right)$ written in

standard polar form is

$z = 2\left(\cos \dfrac{7\pi}{4} + i \sin \dfrac{7\pi}{4}\right)$ and is located

in Quadrant IV in the complex plane.

terminal side of $\theta = \dfrac{7\pi}{4}$ and is a distance of $r = 2$ units from the origin. We observe that this complex number lies in Quadrant IV. Figure 43.

c. To write the complex number $z = 4(\cos(-3\pi) + i \sin(-3\pi))$ in standard polar form, we must determine the angle on the interval $[0, 2\pi)$ that is coterminal with -3π. This angle is $\theta = -3\pi + 4\pi = \pi$. Therefore, the standard polar form is $z = 4(\cos \pi + i \sin \pi)$. To plot z, we plot a point that lies on the terminal side of $\theta = \pi$ and is a distance of $r = 4$ units from the origin. We observe that this complex number lies along the real axis. See Figure 44.

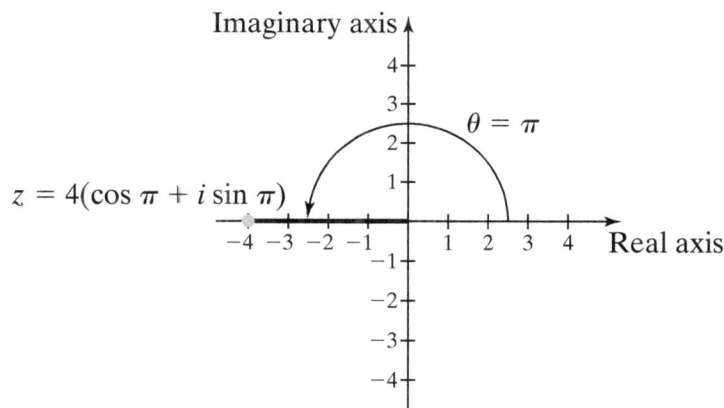

Figure 44 The complex number $z = 4(\cos(-3\pi) +$
$i \sin(-3\pi))$ written in standard polar form is
$z = 4(\cos \pi + i \sin \pi)$ and is located along
the real axis in the complex plane.

You Try It Work through this You Try It problem.

Work Exercises 5–10 in this textbook or in the MyMathLab Study Plan.

OBJECTIVE 3 CONVERTING A COMPLEX NUMBER FROM POLAR FORM TO RECTANGULAR FORM

If we are given a complex number in polar form, it is very easy to convert the complex number into rectangular form. We simply evaluate $\cos \theta$ and $\sin \theta$ within the expression $z = r(\cos \theta + i \sin \theta)$.

My video summary   ⊘ **Example 3** Converting a Complex Number from Polar Form to Rectangular Form

Write each complex number in rectangular form using exact values if possible. Otherwise, round to two decimal places.

a. $z = 3\left(\cos\dfrac{7\pi}{4} + i\sin\dfrac{7\pi}{4}\right)$ 　　　　**b.** $z = 4(\cos 80° + i\sin 80°)$

Solution

a. To convert $z = 3\left(\cos\dfrac{7\pi}{4} + i\sin\dfrac{7\pi}{4}\right)$ to rectangular form, we must evaluate

$\cos\dfrac{7\pi}{4}$ and $\sin\dfrac{7\pi}{4}$.

$z = 3\left(\cos\dfrac{7\pi}{4} + i\sin\dfrac{7\pi}{4}\right)$ 　　　Write the original complex number in polar form.

$z = 3\left(\dfrac{1}{\sqrt{2}} + i\left(-\dfrac{1}{\sqrt{2}}\right)\right)$ 　　　Evaluate $\cos\dfrac{7\pi}{4} = \dfrac{1}{\sqrt{2}}$ and $\sin\dfrac{7\pi}{4} = -\dfrac{1}{\sqrt{2}}$.

$z = \dfrac{3}{\sqrt{2}} - \dfrac{3}{\sqrt{2}}i$ 　　　Multiply.

Therefore, the rectangular form of the complex number $z = 3\left(\cos\dfrac{7\pi}{4} + i\sin\dfrac{7\pi}{4}\right)$

is $z = \dfrac{3}{\sqrt{2}} - \dfrac{3}{\sqrt{2}}i$. See Figure 45.

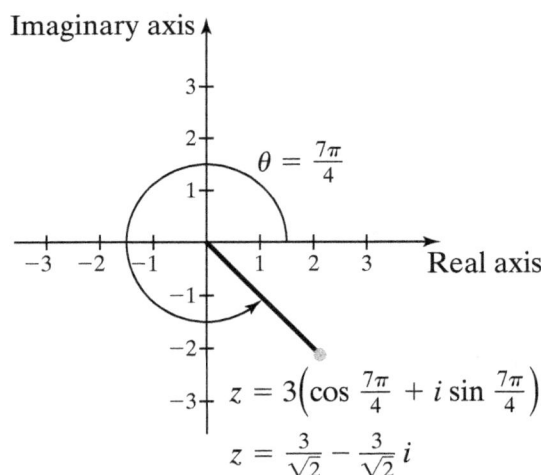

Figure 45 The rectangular form of $z = 3\left(\cos\dfrac{7\pi}{4} + i\sin\dfrac{7\pi}{4}\right)$ is $z = \dfrac{3}{\sqrt{2}} - \dfrac{3}{\sqrt{2}}i.$

b. To convert $z = 4(\cos 80° + i\sin 80°)$ to rectangular form, we observe that $\theta = 80°$ does not belong to one of the special angle families. Thus, we will use a calculator set in degree mode to approximate values.

$z = 4(\cos 80° + i\sin 80°)$ 　　　Write the original complex number in polar form.

$z = 4\cos 80° + 4i\sin 80°$ 　　　Multiply.

$z = 0.69 + 3.94i$ 　　　Approximate $4\cos 80° \approx 0.69$ and $4\sin 80° \approx 3.94$.

Therefore, the rectangular form of the complex number $z = 4(\cos 80° + i\sin 80°)$ is approximately $z = 0.69 + 3.94i$. See Figure 46.

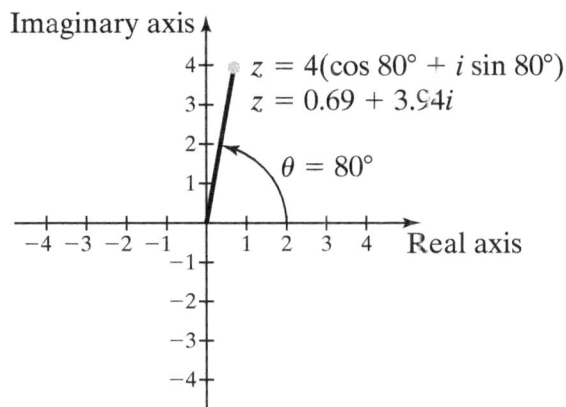

Figure 46 The rectangular form of $z = 4(\cos 80° + i \sin 80°)$ is approximately $z = 0.69 + 3.94i$.

You Try It Work through this You Try It problem.

Work Exercises 11–16 in this textbook or in the MyMathLab Study Plan.

OBJECTIVE 4 CONVERTING A COMPLEX NUMBER FROM RECTANGULAR FORM TO STANDARD POLAR FORM

Converting a complex number from rectangular form to standard polar form is similar to the process of converting a point in rectangular coordinates to polar coordinates. This skill was introduced in Section 5.1.

First consider a complex number of the form $z = a + bi$ where $a = 0$ or $b = 0$. Complex numbers of this form are located along the real axis or the imaginary axis in the complex plane. Each case is outlined below.

Converting Complex Numbers From Rectangular Form to Standard Polar Form for Complex Numbers Lying Along the Real Axis or Imaginary Axis

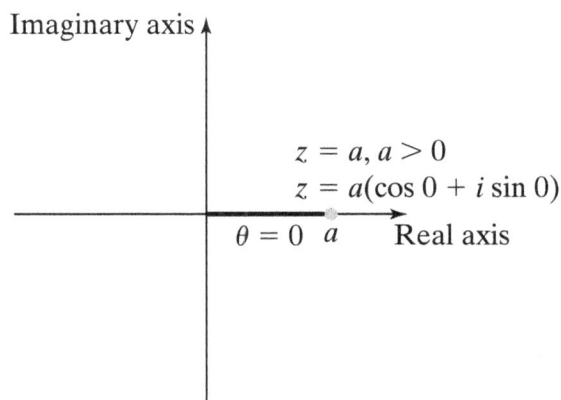

The complex number $z = a, a > 0$ lies along the positive real axis so $\theta = 0$. The standard polar form is $z = a(\cos 0 + i \sin 0)$.

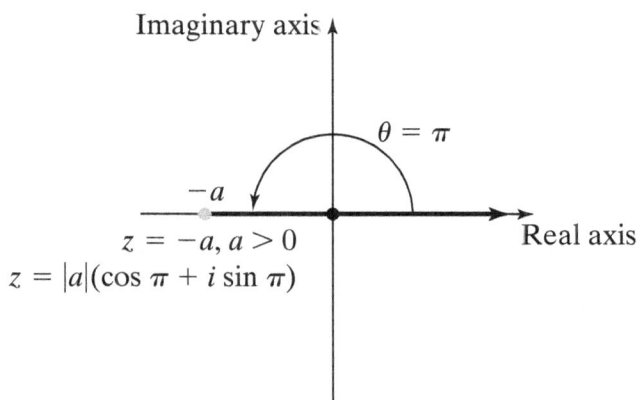

The complex number $z = -a, a > 0$ lies along the negative real axis so $\theta = \pi$. The standard polar form is $z = |a|(\cos \pi + i \sin \pi)$.

Imaginary axis

$z = bi, b > 0$

$z = b\left(\cos\dfrac{\pi}{2} + i\sin\dfrac{\pi}{2}\right)$

$\theta = \dfrac{\pi}{2}$

Real axis

Imaginary axis

$\theta = \dfrac{3\pi}{2}$

Real axis

$z = -bi, b > 0$

$z = |b|\left(\cos\dfrac{3\pi}{2} + i\sin\dfrac{3\pi}{2}\right)$

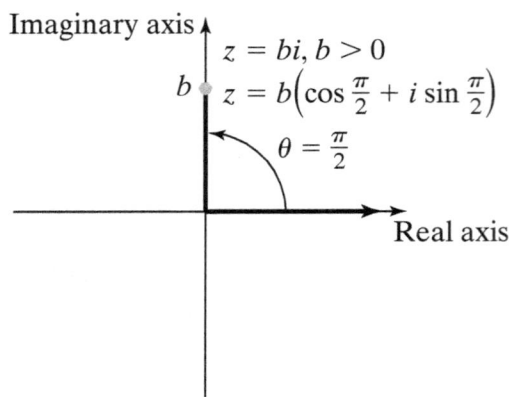

The complex number $z = bi, b > 0$ lies along the positive imaginary axis so $\theta = \dfrac{\pi}{2}$. The standard polar form is $z = b\left(\cos\dfrac{\pi}{2} + i\sin\dfrac{\pi}{2}\right)$.

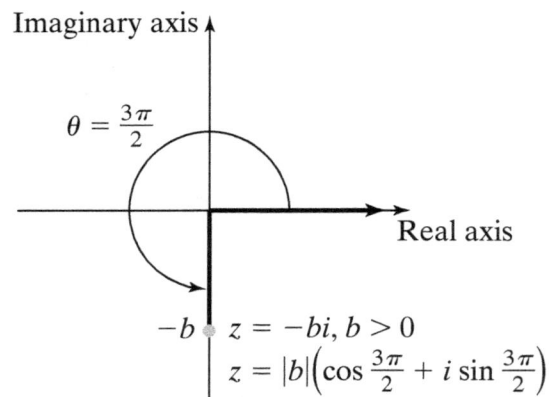

The complex number $z = -bi, b > 0$ lies along the negative imaginary axis so $\theta = \dfrac{3\pi}{2}$. The standard polar form is $z = |b|\left(\cos\dfrac{3\pi}{2} + i\sin\dfrac{3\pi}{2}\right)$.

My video summary ⊙ **Example 4** Converting a Complex Number from Rectangular Form to Standard Polar Form

Determine the standard polar form of each complex number. Write the argument using radians.

a. $z = 5$ **b.** $z = -3i$ **c.** $z = -\sqrt{7}$ **d.** $z = \dfrac{7}{2}i$

Solution Work through the video to verify the following:

a. The standard polar form of $z = 5$ is $z = 5(\cos 0 + i\sin 0)$.

b. The standard polar form of $z = -3i$ is $z = 3\left(\cos\dfrac{3\pi}{2} + i\sin\dfrac{3\pi}{2}\right)$.

c. The standard polar form of $z = -\sqrt{7}$ is $z = \sqrt{7}\,(\cos\pi + i\sin\pi)$.

d. The standard polar form of $z = \dfrac{7}{2}i$ is $z = \dfrac{7}{2}\left(\cos\dfrac{\pi}{2} + i\sin\dfrac{\pi}{2}\right)$.

In order to convert a complex number $z = a + bi$ into the form $z = r(\cos\theta + i\sin\theta)$, for $a \neq 0$ and $b \neq 0$, we must determine the values of r and θ. Determining the value of the modulus r is straightforward because we know that $r = \sqrt{a^2 + b^2}$. Determining the value of θ is much more involved. In order to find θ, we first have to find the reference angle θ_R. Recall that the reference angle is an acute angle. We know that the tangent of any acute angle is a positive value. Therefore, using the fact that $\tan\theta = \dfrac{b}{a}$, we can find θ_R by solving the equation $\tan\theta_R = \left|\dfrac{b}{a}\right|$. The appropriate value of θ depends on the quadrant in which the complex number $z = a + bi$ lies in the complex plane.

Converting a Complex Number from Rectangular Form to Standard Polar Form for $a \neq 0$ and $b \neq 0$

Step 1. Determine the value of r using the equation $r = \sqrt{a^2 + b^2}$.

Step 2. Plot the point and determine the quadrant in which it lies.

Step 3. Determine the value of the acute reference angle θ_R by solving the equation $\tan \theta_R = \left| \dfrac{b}{a} \right|$.

Step 4. Determine the value of θ using θ_R and the quadrant in which the point lies. Each case is outlined below.

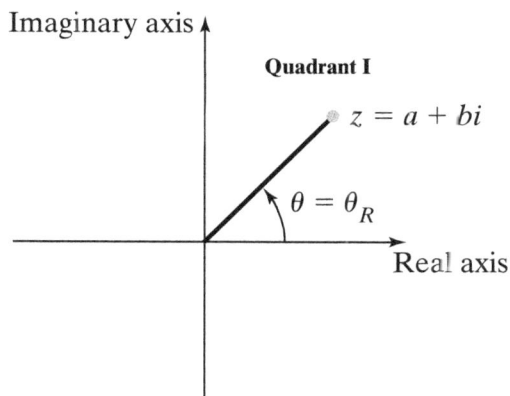

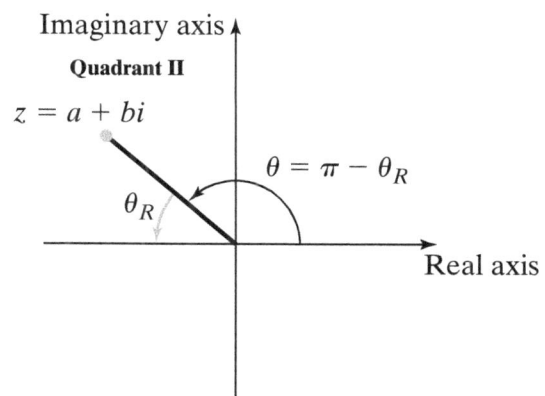

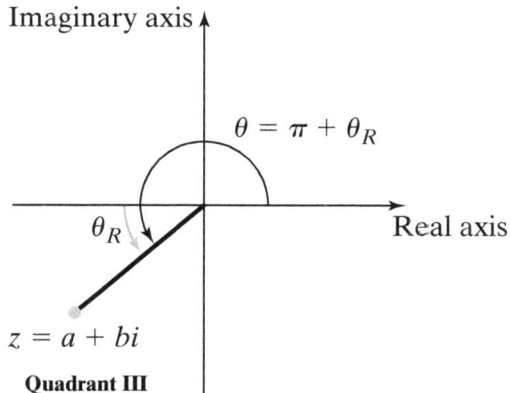

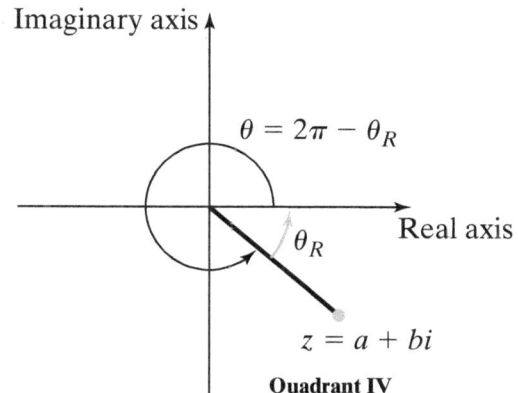

Example 5 Converting a Complex Number from Rectangular Form to Standard Polar Form

My interactive video summary

Determine the standard polar form of each complex number. Write the argument in radians using exact values if possible. Otherwise, round the argument to two decimal places.

a. $z = -2\sqrt{3} + 2i$ **b.** $z = 4 - 3i$

Solution

a. Step 1. To find the value of r, use the equation $r = \sqrt{a^2 + b^2}$.

$$r = \sqrt{a^2 + b^2} \qquad \text{Write the equation relating } r, a, \text{ and } b.$$

$$r = \sqrt{(-2\sqrt{3})^2 + (2)^2} \qquad \text{Substitute } a = -2\sqrt{3} \text{ and } b = 2.$$

$$r = \sqrt{12 + 4} \qquad \text{Square each term.}$$

$$r = \sqrt{16} = 4 \qquad \text{Simplify.}$$

Step 2. Plot the complex number $z = -2\sqrt{3} + 2i$ and recognize that it lies in Quadrant II, a distance of 4 units from the origin. See Figure 47.

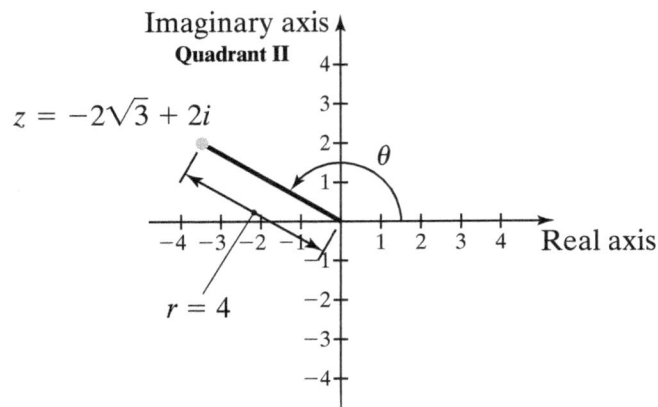

Figure 47 The complex number $z = -2\sqrt{3} + 2i$ lies in Quadrant II, a distance of four units from the origin.

Step 3. Given the rectangular form $z = -2\sqrt{3} + 2i$, to determine θ we first find the value of θ_R by solving the equation $\tan \theta_R = \left| \dfrac{b}{a} \right|$.

$$\tan \theta_R = \left| \frac{b}{a} \right| \qquad \text{Write the equation relating } \theta_R, a, \text{ and } b.$$

$$\tan \theta_R = \left| \frac{2}{-2\sqrt{3}} \right| \qquad \text{Substitute } a = -2\sqrt{3} \text{ and } b = 2.$$

$$\tan \theta_R = \frac{1}{\sqrt{3}} \qquad \text{Simplify.}$$

The acute angle whose tangent is $\dfrac{1}{\sqrt{3}}$ is $\theta_R = \dfrac{\pi}{6}$.

Step 4. Because the point lies in Quadrant II, we know that $\theta = \pi - \theta_R$.

Thus, $\theta = \pi - \dfrac{\pi}{6} = \dfrac{5\pi}{6}$.

My interactive video summary

▶ Therefore, the standard polar form of the complex number is $z = -2\sqrt{3} + 2i$ is $z = 4\left(\cos \dfrac{5\pi}{6} + i \sin \dfrac{5\pi}{6} \right)$. See Figure 48. Watch this interactive video to see the worked-out solution.

$z = -2\sqrt{3} + 2i$

$z = 4\left(\cos\dfrac{5\pi}{6} - i\sin\dfrac{5\pi}{6}\right)$

Imaginary axis

Quadrant II

$\theta = \pi - \dfrac{\pi}{6} = \dfrac{5\pi}{6}$

$\theta_R = \dfrac{\pi}{6}$

Real axis

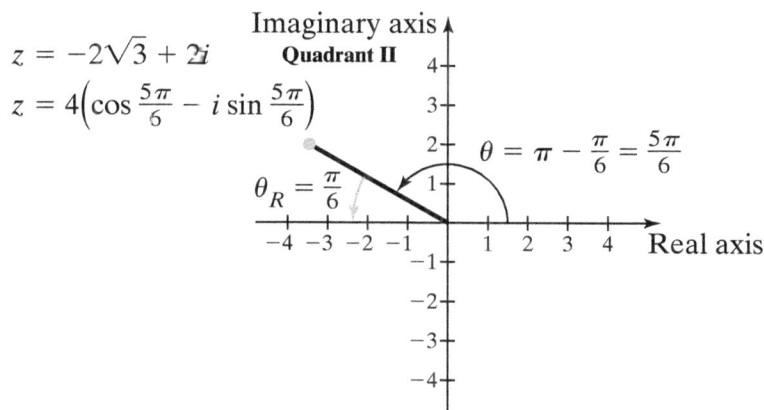

Figure 48 The polar form of the complex number

$$z = -2\sqrt{3} + 2i \text{ is } z = 4\left(\cos\frac{5\pi}{6} + i\sin\frac{5\pi}{6}\right).$$

b. Step 1. Given the rectangular form $z = 4 - 3i$, find the value of r using the equation $r = \sqrt{a^2 + b^2}$.

$r = \sqrt{a^2 + b^2}$ Write the equation relating r, a, and b.

$r = \sqrt{(4)^2 + (-3)^2}$ Substitute $a = 4$ and $b = -3$.

$r = \sqrt{16 + 9}$ Square each term.

$r = \sqrt{25} = 5$ Simplify.

Step 2. Plot the complex number $z = 4 - 3i$ and recognize that it lies in Quadrant IV, a distance of 5 units from the origin. See Figure 49.

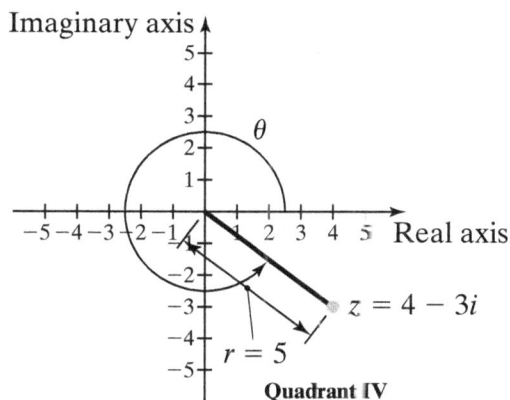

Imaginary axis

θ

Real axis

$z = 4 - 3i$

$r = 5$

Quadrant IV

Figure 49 The complex number $z = 4 - 3i$ lies in Quadrant IV, a distance of five units from the origin.

Step 3. Given the rectangular form $z = 4 - 3i$, to determine θ we first find the value of θ_R by solving the equation $\tan\theta_R = \left|\dfrac{b}{a}\right|$.

$\tan\theta_R = \left|\dfrac{b}{a}\right|$ Write the equation relating θ_R, a, and b.

$\tan\theta_R = \left|\dfrac{-3}{4}\right|$ Substitute $a = 4$ and $b = -3$.

$\tan\theta_R = \dfrac{3}{4}$ Simplify.

The angle whose tangent is $\frac{3}{4}$ is not an angle that is a member of one of the special families of angles. However, using the inverse tangent function, we get $\theta_R = \tan^{-1}\left(\frac{3}{4}\right)$.

Step 4. The complex number $z = 4 - 3i$ lies in Quadrant IV, which indicates that $\theta = 2\pi - \theta_R$. Therefore, $\theta = 2\pi - \tan^{-1}\left(\frac{3}{4}\right)$. Using a calculator set in radian mode, we get $\theta = 2\pi - \tan^{-1}\left(\frac{3}{4}\right) \approx 5.64$.

My interactive video summary

⊘ Therefore, using $r = 5$ and an approximate value of $\theta \approx 5.64$, we see that the standard polar form of the complex number $z = 4 - 3i$ is approximately $z = 4(\cos 5.64 + i \sin 5.64)$. See Figure 50. Watch this interactive video to see the worked-out solution.

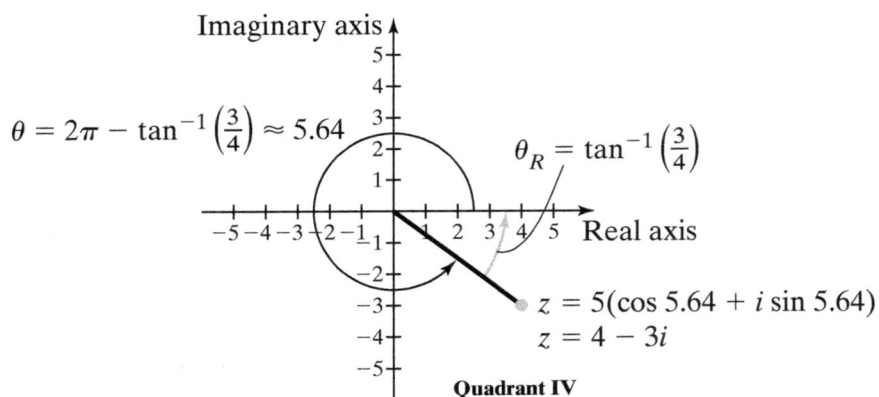

Figure 50 The standard polar form of the complex number $z = 4 - 3i$ is approximately $z = 5(\cos 5.64 + i \sin 5.64)$.

You Try It Work through this You Try It problem.

Work Exercises 17–23 in this textbook or in the MyMathLab Study Plan.

OBJECTIVE 5 DETERMINING THE PRODUCT OR QUOTIENT OF COMPLEX NUMBERS IN POLAR FORM

My video summary

⊘ Given two complex numbers of the form $z_1 = r_1(\cos \theta_1 + i \sin \theta_1)$ and $z_2 = r_2(\cos \theta_2 + i \sin \theta_2)$, we wish to determine a formula for the product $z_1 z_2$ and the quotient $\frac{z_1}{z_2}$. To derive these formulas, we must recall the sum and difference formulas for sine and cosine that were first introduced in Section 3.2. Watch this video to see how to derive the product and quotient of complex numbers whose formulas are given below.

The Product and Quotient of Two Complex Numbers Written in Polar Form

If $z_1 = r_1(\cos\theta_1 + i\sin\theta_1)$ and $z_2 = r_2(\cos\theta_2 + i\sin\theta_2)$ are two complex numbers, then

$$z_1 z_2 = r_1 r_2(\cos(\theta_1 + \theta_2) + i\sin(\theta_1 + \theta_2))$$

and

$$\frac{z_1}{z_2} = \frac{r_1}{r_2}[\cos(\theta_1 - \theta_2) + i\sin(\theta_1 - \theta_2)]$$

Simply stated, we can say that to find the product of complex numbers we "multiply the r's and add the angles." Similarly, to divide complex numbers we "divide the r's and subtract the angles."

My video summary ⊘ **Example 6** Finding the Product and Quotient of Two Complex Numbers

Let $z_1 = 4\left(\cos\dfrac{2\pi}{3} + i\sin\dfrac{2\pi}{3}\right)$ and $z_2 = 5\left(\cos\dfrac{11\pi}{6} + i\sin\dfrac{11\pi}{6}\right)$. Find $z_1 z_2$ and $\dfrac{z_1}{z_2}$ and write the answers in standard polar form.

Solution First, we identify that $r_1 = 4$, $r_2 = 5$, $\theta_1 = \dfrac{2\pi}{3}$, and $\theta_2 = \dfrac{11\pi}{6}$.
The product of r_1 and r_2 is $r_1 r_2 = 4 \cdot 5 = 20$ and the sum of θ_1 and θ_2 is

$$\theta_1 + \theta_2 = \frac{2\pi}{3} + \frac{11\pi}{6} = \frac{15\pi}{6} = \frac{5\pi}{2}.$$

To write the answer in standard polar form, we find the angle on the interval $[0, 2\pi)$ that is coterminal with $\dfrac{5\pi}{2}$ is $\dfrac{\pi}{2}$. Thus, the product written in standard polar form is

$$z_1 z_2 = 20\left(\cos\frac{\pi}{2} + i\sin\frac{\pi}{2}\right).$$

The quotient of r_1 and r_2 is $\dfrac{r_1}{r_2} = \dfrac{4}{5}$ and the difference of θ_1 and θ_2 is

$$\theta_1 - \theta_2 = \frac{2\pi}{3} - \frac{11\pi}{6} = -\frac{7\pi}{6}.$$

The angle on the interval $[0, 2\pi)$ that is coterminal with $-\dfrac{7\pi}{6}$ is $\dfrac{5\pi}{6}$.

Therefore, the quotient written in standard polar form is $\dfrac{z_1}{z_2} = \dfrac{4}{5}\left(\cos\dfrac{5\pi}{6} + i\sin\dfrac{5\pi}{6}\right)$.

Note that because the arguments of the product and quotient are both angles that belong to a special family of angles, it is easy to determine the corresponding rectangular form. We could have rewritten the product and quotient as

$$z_1 z_2 = 20\left(\cos\frac{\pi}{2} + i\sin\frac{\pi}{2}\right) = 20(0 + i(1)) = 20i$$

and

$$\frac{z_1}{z_2} = \frac{4}{5}\left(\cos\frac{5\pi}{6} + i\sin\frac{5\pi}{6}\right) = \frac{4}{5}\left(-\frac{\sqrt{3}}{2} + i\left(\frac{1}{2}\right)\right) = -\frac{2\sqrt{3}}{5} + \frac{2}{5}i.$$

You Try It   Work through this You Try It problem.

Work Exercises 24–27 in this textbook or in the MyMathLab Study Plan.

OBJECTIVE 6   USING DE MOIVRE'S THEOREM TO RAISE A COMPLEX NUMBER TO A POWER

If $z = r(\cos\theta + i\sin\theta)$ is a complex number, then we can use the product of two complex numbers formula to find z^2.

$z^2 = z \cdot z = r(\cos\theta + i\sin\theta) \cdot r(\cos\theta + i\sin\theta)$ Write z^2 as the product of $z = r(\cos\theta + i\sin\theta)$ and $z = r(\cos\theta + i\sin\theta)$.

$\quad = r \cdot r(\cos(\theta + \theta) + i\sin(\theta + \theta))$ Use the formula for the product of complex numbers written in polar form.

$z^2 = r^2(\cos 2\theta + i\sin 2\theta)$ Write $r \cdot r$ as r^2 and $\theta + \theta$ as 2θ.

We can use this result to find z^3.

$z^3 = z^2 \cdot z = r^2(\cos 2\theta + i\sin 2\theta) \cdot r(\cos\theta + i\sin\theta)$ Write z^3 as the product of $z^2 = r^2(\cos 2\theta + i\sin 2\theta)$ and $z = r(\cos\theta + i\sin\theta)$.

$\quad = r^2 \cdot r(\cos(2\theta + \theta) + i\sin(2\theta + \theta))$ Use the formula for the product of complex numbers written in polar form.

$z^3 = r^3(\cos 3\theta + i\sin 3\theta)$ Write $r^2 \cdot r$ as r^3 and $2\theta + \theta$ as 3θ.

We can use this same procedure to get

$$z^4 = r^4(\cos 4\theta + i\sin 4\theta) \text{ and } z^5 = r^5(\cos 5\theta + i\sin 5\theta).$$

In general, if $z = r(\cos\theta + i\sin\theta)$ is a complex number and if n is a positive integer, then $z^n = r^n(\cos n\theta + i\sin n\theta)$. This formula for the nth power of a complex number is known as **De Moivre's Theorem**, named after the French mathematician Abraham de Moivre.

De Moivre's Theorem for Finding Powers of a Complex Number

If $z = r(\cos\theta + i\sin\theta)$ is a complex number written in polar form and if n is a positive integer, then z^n is given by the formula

$$z^n = [r(\cos\theta + i\sin\theta)]^n = r^n(\cos n\theta + i\sin n\theta).$$

---

*My interactive
video summary*

⊘ **Example 7** Using De Moivre's Theorem to Raise a Complex Number to a Power

a. Find $\left[5\left(\cos\dfrac{3\pi}{4} + i\sin\dfrac{3\pi}{4}\right)\right]^3$ and write your answer in standard polar form.

b. Find $(\sqrt{3} - i)^4$ and write your answer in rectangular form.

Solution

a. Apply De Moivre's Theorem with $r = 5, \theta = \dfrac{3\pi}{4}$, and $n = 3$.

$$5\left[\left(\cos\frac{3\pi}{4} + i\sin\frac{3\pi}{4}\right)\right]^3$$ Write the original expression.

$$= 5^3\left(\cos\left(3\cdot\frac{3\pi}{4}\right) + i\sin\left(3\cdot\frac{3\pi}{4}\right)\right)$$ Use De Moivre's Theorem with $r = 5, \theta = \dfrac{3\pi}{4}$, and $n = 3$.

$$= 125\left(\cos\frac{9\pi}{4} + i\sin\frac{9\pi}{4}\right)$$ Multiply.

$$= 125\left(\cos\frac{\pi}{4} + i\sin\frac{\pi}{4}\right)$$ Substitute the angle $\dfrac{\pi}{4}$ for its coterminal angle $\dfrac{9\pi}{4}$ to write in standard polar form.

b. To evaluate the expression $(\sqrt{3} - i)^4$, first write the complex number $z = \sqrt{3} - i$ in standard polar form. View these steps to verify that the standard polar form of

$$z = \sqrt{3} - i \text{ is } z = 2\left(\cos\frac{11\pi}{6} + i\sin\frac{11\pi}{6}\right).$$

Next, apply De Moivre's Theorem with $r = 2, \theta = \dfrac{11\pi}{6}$, and $n = 4$.

$$\left[2\left(\cos\frac{11\pi}{6} + i\sin\frac{11\pi}{6}\right)\right]^4$$ Write the original expression.

$$= 2^4\left(\cos\left(4\cdot\frac{11\pi}{6}\right) + i\sin\left(4\cdot\frac{11\pi}{6}\right)\right)$$ Use De Moivre's Theorem with $r = 2, \theta = \dfrac{11\pi}{6}$, and $n = 4$.

$$= 16\left(\cos\frac{44\pi}{6} + i\sin\frac{44\pi}{6}\right)$$ Multiply.

$$= 16\left(\cos\frac{22\pi}{3} + i\sin\frac{22\pi}{3}\right)$$ Simplify: $\dfrac{44\pi}{6} = \dfrac{22\pi}{3}$.

$$= 16\left(\cos\frac{4\pi}{3} + i\sin\frac{4\pi}{3}\right)$$ Substitute the angle $\dfrac{4\pi}{3}$ for its coterminal angle $\dfrac{22\pi}{3}$ to write in standard polar form.

$$= 16\left(-\frac{1}{2} + i\left(-\frac{\sqrt{3}}{2}\right)\right)$$ Evaluate $\cos\dfrac{4\pi}{3}$ and $\sin\dfrac{4\pi}{3}$.

$$= -8 - 8\sqrt{3}i$$ Multiply.

You Try It Work through this You Try It problem.

Work Exercises 28–37 in this textbook or in the MyMathLab Study Plan.

OBJECTIVE 7 USING DE MOIVRE'S THEOREM TO FIND THE ROOTS OF A COMPLEX NUMBER

We now want to investigate the nth roots of a complex number. There are exactly n distinct nth roots of any nonzero complex number. For example, there are 2 distinct 2nd roots (square roots) of a complex number, there are 3 distinct 3rd roots (cube roots) of a complex number, there are 4 distinct 4th roots of a complex number, and so on.

To derive a formula for the nth roots of a complex number, start with De Moivre's Theorem for positive integer powers of n and substitute $\frac{1}{n}$ for n.

$$z^n = [r(\cos\theta + i\sin\theta)]^n = r^n(\cos n\theta + i\sin n\theta)$$

Write De Moivre's Theorem for integer powers of a complex number.

$$z^{\frac{1}{n}} = [r(\cos\theta + i\sin\theta)]^{\frac{1}{n}} = r^{\frac{1}{n}}\left(\cos\frac{1}{n}\theta + i\sin\frac{1}{n}\theta\right)$$

Substitute $\frac{1}{n}$ for n.

Writing $z^{\frac{1}{n}}$ as $\sqrt[n]{z}$ and $r^{\frac{1}{n}}$ as $\sqrt[n]{r}$, we get the formula $\sqrt[n]{z} = \sqrt[n]{r}\left(\cos\frac{1}{n}\theta + i\sin\frac{1}{n}\theta\right)$.

However, this formula only determines *one* of the n distinct nth roots. To find all n roots, we start overusing the fact that $z = r(\cos\theta + i\sin\theta)$ is equivalent to

$$z = r(\cos(\theta + 2\pi k) + i\sin(\theta + 2pk)), \text{ where } k \text{ is any integer.}$$

$$z^n = [r(\cos(\theta + 2\pi k) + i\sin(\theta + 2\pi k))]^n = r^n(\cos(n(\theta + 2\pi k)) + i\sin(n(\theta + 2\pi k)))$$

Write De Moivre's Theorem for integer powers of a complex number using $\theta + 2\pi k$ as the argument.

$$z^{\frac{1}{n}} = [r(\cos(\theta + 2\pi k) + i\sin(\theta + 2\pi k))]^{\frac{1}{n}} = r^{\frac{1}{n}}\left[\cos\left(\frac{1}{n}(\theta + 2\pi k)\right) + i\sin\left(\frac{1}{n}(\theta + 2\pi k)\right)\right]$$

Substitute $\frac{1}{n}$ for n.

We can write this formula as $\sqrt[n]{z} = \sqrt[n]{r}\left[\cos\left(\frac{\theta + 2\pi k}{n}\right) + i\sin\left(\frac{\theta + 2\pi k}{n}\right)\right]$.

This formula determines the n distinct nth roots of a complex number for $k = 0, 1, 2, \cdots, (n-1)$.

De Moivre's Theorem for Finding the nth Roots of a Complex Number

If $z = r(\cos\theta + i\sin\theta)$ is a complex number written in polar form and if n is a positive integer, then z has exactly n complex roots determined by the formula

$$\sqrt[n]{z} = \sqrt[n]{r}\left[\cos\left(\frac{\theta + 2\pi k}{n}\right) + i\sin\left(\frac{\theta + 2\pi k}{n}\right)\right]$$

or

$$\sqrt[n]{z} = \sqrt[n]{r}\left[\cos\left(\frac{\theta + (360°)k}{n}\right) + i\sin\left(\frac{\theta + (360°)k}{n}\right)\right]$$

for $k = 0, 1, 2, \cdots, (n-1)$. We will label the n roots as $z_0, z_1, z_2, \cdots, z_{n-1}$.

Example 8 Finding Complex Roots of a Real Number

Find the complex cube roots of 8. Write your answers in rectangular form.

Solution First write 8 in polar form as $8 = 8(\cos 0 + i \sin 0)$.

Next, apply De Moivre's Theorem for finding nth roots with $r = 8$, $\theta = 0$, and $n = 3$.

$$\sqrt[3]{8} = \sqrt[3]{8(\cos 0 + i \sin 0)} = \sqrt[3]{8}\left[\cos\left(\frac{0 + 2\pi k}{3}\right) + i \sin\left(\frac{0 + 2\pi k}{3}\right)\right] = \sqrt[3]{8}\left[\cos\left(\frac{2\pi k}{3}\right) + i \sin\left(\frac{2\pi k}{3}\right)\right] \text{ for } k = 0, 1, 2.$$

If $k = 0$: $z_0 = \sqrt[3]{8}\left[\cos\left(\frac{2\pi \cdot 0}{3}\right) + i \sin\left(\frac{2\pi \cdot 0}{3}\right)\right] = 2[\cos 0 + i \sin 0] = 2(1 + i \cdot 0) = 2$

If $k = 1$: $z_1 = \sqrt[3]{8}\left[\cos\left(\frac{2\pi \cdot 1}{3}\right) + i \sin\left(\frac{2\pi \cdot 1}{3}\right)\right] = 2\left[\cos\frac{2\pi}{3} + i \sin\frac{2\pi}{3}\right] = 2\left(-\frac{1}{2} + i \cdot \frac{\sqrt{3}}{2}\right) = -1 + \sqrt{3}i$

If $k = 2$: $z_2 = \sqrt[3]{8}\left[\cos\left(\frac{2\pi \cdot 2}{3}\right) + i \sin\left(\frac{2\pi \cdot 2}{3}\right)\right] = 2\left[\cos\frac{4\pi}{3} + i \sin\frac{4\pi}{3}\right] = 2\left(-\frac{1}{2} + i \cdot \left(-\frac{\sqrt{3}}{2}\right)\right) = -1 - \sqrt{3}i$

If we plot the three distinct cube roots of 8, we notice that the three roots are equally spaced around a circle centered at the origin of radius 2. See Figure 51. Each successive root is located exactly $\frac{2\pi}{3}$ radians away from the previous root.

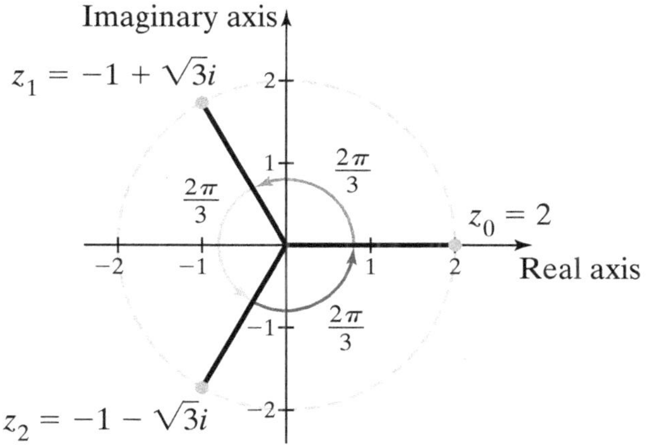

$z_1 = -1 + \sqrt{3}i$

$z_2 = -1 - \sqrt{3}i$

$z_0 = 2$

Figure 51 The 3 cube roots of 8 lie on the graph of a circle of radius 2 in the complex plane. Each successive root is located $\frac{2\pi}{3}$ radians from the previous root.

In Example 8 we saw that the three complex cube roots of 8 were all located on a circle of radius 2, where each root was exactly $\frac{2\pi}{3}$ radians from the previous root. See Figure 51. All complex nth roots will have this same graphical representation. If we can find z_0, then we can easily determine the remaining $n - 1$ roots by adding $\frac{2\pi}{n}$ $\left(\text{or } \frac{360°}{n}\right)$ to each successive argument. We demonstrate this in Example 9.

My video summary ⊙ **Example 9 Finding Complex Roots**

a. Find the complex fourth roots of $z = 81\left(\cos\frac{3\pi}{5} + i \sin\frac{3\pi}{5}\right)$. Write your answers in standard polar form.

b. Find the complex square roots of $z = -2\sqrt{3} + 2i$. Write your answer in rectangular form with each part rounded to two decimal places.

Solution

a. We start by finding z_0 using De Moivre's Theorem for finding nth roots with $r = 81, \theta = \dfrac{3\pi}{5}$, and $n = 4$.

$$z_0 = \sqrt[4]{81}\left[\cos\left(\frac{\frac{3\pi}{5} + 2\pi \cdot 0}{4}\right) + i\sin\left(\frac{\frac{3\pi}{5} + 2\pi \cdot 0}{4}\right)\right] = 3\left[\cos\frac{3\pi}{20} + i\sin\frac{3\pi}{20}\right]$$

To find z_1, z_2, and z_3, we must add $\dfrac{2\pi}{4} = \dfrac{\pi}{2}$ to the argument of the previous root.

Therefore, we get

$$z_1 = 3\left[\cos\left(\frac{3\pi}{20} + \frac{\pi}{2}\right) + i\sin\left(\frac{3\pi}{20} + \frac{\pi}{2}\right)\right] = 3\left[\cos\frac{13\pi}{20} + i\sin\frac{13\pi}{20}\right]$$

$$z_2 = 3\left[\cos\left(\frac{13\pi}{20} + \frac{\pi}{2}\right) + i\sin\left(\frac{13\pi}{20} + \frac{\pi}{2}\right)\right] = 3\left[\cos\frac{23\pi}{20} + i\sin\frac{23\pi}{20}\right]$$

$$z_3 = 3\left[\cos\left(\frac{23\pi}{20} + \frac{\pi}{2}\right) + i\sin\left(\frac{23\pi}{20} + \frac{\pi}{2}\right)\right] = 3\left[\cos\frac{33\pi}{20} + i\sin\frac{33\pi}{20}\right]$$

My interactive video summary

▶ The four roots are shown in Figure 52. Notice that each successive root is located exactly $\dfrac{\pi}{2}$ radians away from the previous root in the complex plane. Watch this interactive video to see the solution to this example.

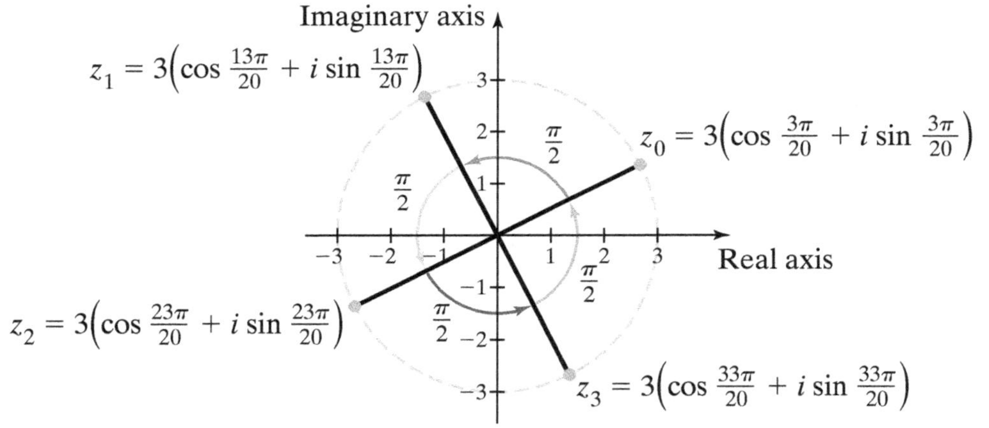

Figure 52 The 4 fourth roots of $z = 81\left(\cos\dfrac{3\pi}{5} + i\sin\dfrac{3\pi}{5}\right)$ lie on the graph of a circle of radius 3. Each successive root is located $\dfrac{2\pi}{4} = \dfrac{\pi}{2}$ radians from the previous root in the complex plane.

b. To find the two square roots of $z = -2\sqrt{3} + 2i$, first write the complex number in standard polar form. See Example 5 to verify that the standard polar form of $z = -2\sqrt{3} + 2i$ is $z = 4\left(\cos\dfrac{5\pi}{6} + i\sin\dfrac{5\pi}{6}\right)$. We find z_0 using De Moivre's Theorem for finding nth roots with $r = 4, \theta = \dfrac{5\pi}{6}$, and $n = 2$.

$$z_0 = \sqrt{4}\left[\cos\left(\frac{\frac{5\pi}{6} + 2\pi \cdot 0}{2}\right) + i\sin\left(\frac{\frac{5\pi}{6} + 2\pi \cdot 0}{2}\right)\right] = 2\left(\cos\frac{5\pi}{12} + i\sin\frac{5\pi}{12}\right)$$

To find z_1, we add $\frac{2\pi}{2} = \pi$ to the argument of z_0.

$$z_1 = 2\left[\cos\left(\frac{5\pi}{12} + \pi\right) + i\sin\left(\frac{5\pi}{12} + \pi\right)\right] = 2\left(\cos\frac{17\pi}{12} + i\sin\frac{17\pi}{12}\right)$$

Using a calculator set in radian mode, we can evaluate $\cos\frac{5\pi}{12}$, $\sin\frac{5\pi}{12}$, $\cos\frac{17\pi}{12}$, and $\sin\frac{17\pi}{12}$ to approximate the rectangular form of z_0 and z_1. Thus, the approximate rectangular forms for the complex square roots of $z = -2\sqrt{3} + 2i$ are

$$z_0 = 2\left(\cos\frac{5\pi}{12} + i\sin\frac{5\pi}{12}\right) \approx 0.52 + 1.93i$$

and

$$z_1 = 2\left(\cos\frac{17\pi}{12} + i\sin\frac{17\pi}{12}\right) \approx -0.52 - 1.93i$$

Note Because the argument of each root has a denominator of 12, it is actually possible to use the sum formulas for sine and cosine discussed in Section 3.2 to determine the exact values of $\cos\frac{5\pi}{12}$, $\sin\frac{5\pi}{12}$, $\cos\frac{17\pi}{12}$, and $\sin\frac{17\pi}{12}$. Therefore, we can determine the exact rectangular form of z_0 and z_1. View these steps to see how to show that the exact rectangular form of z_0 and z_1 is

$$z_0 = 2\left(\cos\frac{5\pi}{12} + i\sin\frac{5\pi}{12}\right) = \frac{\sqrt{3} - 1}{\sqrt{2}} + \frac{\sqrt{3} + 1}{\sqrt{2}}i$$

and

$$z_1 = 2\left(\cos\frac{17\pi}{12} + i\sin\frac{17\pi}{12}\right) = -\left(\frac{\sqrt{3} - 1}{\sqrt{2}}\right) - \left(\frac{\sqrt{3} + 1}{\sqrt{2}}\right)i$$

You Try It Work through this You Try It problem.

Work Exercises 38–46 in this textbook or in the MyMathLab Study Plan.

5.3 Exercises

In Exercises 1–4, plot each complex number in the complex plane and determine its absolute value.

1. $z = -1 + 2i$ **2.** $z = -5 - 4i$ **3.** $z = 5$ **4.** $z = -3i$

In Exercises 5–10, rewrite the complex number in standard polar form, plot the number in the complex plane, and determine the quadrant in which the point lies or the axis on which the point lies.

5. $z = 3\left(\cos\frac{11\pi}{7} + i\sin\frac{11\pi}{7}\right)$ **6.** $z = 2(\cos 280° + i\sin 280°)$

SbS **7.** $z = 5(\cos \pi + i \sin \pi)$

SbS **8.** $z = 3\left(\cos \dfrac{25\pi}{3} + i \sin \dfrac{25\pi}{3}\right)$

SbS **9.** $z = 2(\cos 390° + i \sin 390°)$

SbS **10.** $z = 5\left(\cos \dfrac{5\pi}{2} + i \sin \dfrac{5\pi}{2}\right)$

In Exercises 11–16, write each complex number in rectangular form using exact values if possible. Otherwise, round to two decimal places.

11. $z = 5\left(\cos \dfrac{7\pi}{4} + i \sin \dfrac{7\pi}{4}\right)$

12. $z = 4(\cos 240° + i \sin 240°)$

13. $z = 8\left(\cos \dfrac{3\pi}{2} + i \sin \dfrac{3\pi}{2}\right)$

14. $z = 10\left(\cos \dfrac{5\pi}{3} + i \sin \dfrac{5\pi}{3}\right)$

15. $z = 10\left(\cos \dfrac{\pi}{10} + i \sin \dfrac{\pi}{10}\right)$

16. $z = 18(\cos 208° + i \sin 208°)$

In Exercises 17–23, write each complex number in standard polar form. Write the argument in radians using exact values if possible. Otherwise, round the argument to two decimal places.

SbS **17.** $z = -4 + 4i$ SbS **18.** $z = 5\sqrt{3} - 5i$ SbS **19.** $z = -2 - 2\sqrt{3}i$ SbS **20.** $z = -5$

SbS **21.** $z = 7i$ SbS **22.** $z = -3 + 2i$ SbS **23.** $z = 1 - 4i$

24. Find $z_1 z_2$ and $\dfrac{z_1}{z_2}$ if $z_1 = 5\left(\cos \dfrac{\pi}{6} + i \sin \dfrac{\pi}{6}\right)$ and $z_2 = 4\left(\cos \dfrac{\pi}{7} + i \sin \dfrac{\pi}{7}\right)$. Write your answers in standard polar form with the argument in radians.

25. Find $z_1 z_2$ and $\dfrac{z_1}{z_2}$ if $z_1 = 7(\cos 80° + i \sin 80°)$ and $z_2 = 6(\cos 30° + i \sin 30°)$. Write your answers in standard polar form with the argument in degrees.

26. Find $z_1 z_2$ and $\dfrac{z_1}{z_2}$ if $z_1 = 2\left(\cos \dfrac{3\pi}{4} + i \sin \dfrac{3\pi}{4}\right)$ and $z_2 = 3\left(\cos \dfrac{11\pi}{6} + i \sin \dfrac{11\pi}{6}\right)$. Write your answers in standard polar form with the argument in radians.

27. Find $z_1 z_2$ and $\dfrac{z_1}{z_2}$ if $z_1 = 11(\cos 150° + i \sin 150°)$ and $z_2 = 9(\cos 310° + i \sin 310°)$. Write your answers in standard polar form with the argument in degrees.

In Exercises 28–30, determine the given complex number raised to a power and write the answer in standard polar form.

28. $\left[2\left(\cos \dfrac{\pi}{7} + i \sin \dfrac{\pi}{7}\right)\right]^5$

29. $[3(\cos 71° + i \sin 71°)]^4$

30. $\left[7\left(\cos \dfrac{16\pi}{9} + i \sin \dfrac{16\pi}{9}\right)\right]^2$

In Exercises 31–37, determine the given complex number raised to a power and write the answer in rectangular form.

31. $\left[3\left(\cos \dfrac{\pi}{9} + i \sin \dfrac{\pi}{9}\right)\right]^3$

32. $[2(\cos 80° + i \sin 80°)]^3$

33. $\left[4\left(\cos \dfrac{\pi}{5} + i \sin \dfrac{\pi}{5}\right)\right]^5$

34. $\left[\sqrt{3}(\cos 20° + i \sin 20°)\right]^6$ **35.** $\left[\sqrt{5}\left(\cos \dfrac{7\pi}{6} + i \sin \dfrac{7\pi}{6}\right)\right]^4$ **36.** $(-1 - i)^7$ **37.** $(3 - 3\sqrt{3}i)^6$

38. Find all complex square roots of $z = 36\left(\cos \dfrac{\pi}{6} + i \sin \dfrac{\pi}{6}\right)$. Write each root in standard polar form with the argument in radians.

39. Find all complex cube roots of $z = 27(\cos 250° + i \sin 250°)$. Write each root in standard polar form with the argument in degrees.

40. Find all complex fourth roots of $z = -16$. Write each root in standard polar form with the argument in radians.

41. Find all complex fifth roots of $z = -4i$. Write each root in standard polar form with the argument in radians.

42. Find all complex cube roots of $z = 6 + 6\sqrt{3}i$. Write each root in standard polar form with the argument in radians.

43. Find all complex fourth roots of $z = 256$. Write each root in rectangular form.

44. Find all complex cube roots of $z = -125$. Write each root in rectangular form.

45. Find all complex fourth roots of $z = 64\left(\cos \dfrac{5\pi}{3} + i \sin \dfrac{5\pi}{3}\right)$. Write each root in rectangular form, rounding to two decimal places.

46. Find all complex cube roots of $z = 1 - 4i$. Write each root in rectangular form, rounding to two decimal places.

5.4 Vectors

THINGS TO KNOW

Before working through this section, be sure that you are familiar with the following concepts:

VIDEO ANIMATION INTERACTIVE

You Try It 1. Converting a Point from Polar Coordinates to Rectangular Coordinates (Section 5.1)

You Try It 2. Converting a Point from Rectangular Coordinates to Polar Coordinates (Section 5.1)

You Try It 3. Converting a Complex Number from Rectangular Form to Standard Polar Form (Section 5.3)

OBJECTIVES

1 Understanding the Geometric Representation of a Vector

2 Understanding Operations on Vectors Represented Geometrically

3 Understanding Vectors in Terms of Components

4 Understanding Vectors in Terms of **i** and **j**

5 Finding a Unit Vector

6 Determining the Direction Angle of a Vector

7 Representing a Vector in Terms of **i** and **j** Given Its Magnitude and Direction Angle

8 Using Vectors to Solve Applications Involving Velocity

9 Using Vectors to Solve Applications Involving Force

Introduction to Section 5.4

Many physical quantities can be described by a single real number. For example, the length of a fence, the circumference of a circle, the temperature of a room, or the area of a farmer's field can all be described by a single real number. The size of these numbers is called the **magnitude** of the quantity. Quantities that can be described using only magnitude are called **scalar quantities**. Some quantities, such as velocity and force, require that we know both the magnitude of the quantity and the direction of the quantity. For example, the two cars in Figure 53 left from the same initial point and are both traveling at a speed of 60 mph. However, one car is moving in a Northwest direction and the other is moving in a Northeast direction. To describe the position of a car, it is not enough to know the speed of the car (the magnitude); it is also necessary to know the direction in which the car is moving.

Figure 53

The speed of both cars in Figure 53 is the same (magnitude of 60 mph), but they are traveling in different directions. Quantities that are described using both magnitude and direction are called **vector quantities** or **vectors**. In this section we will study vectors and we will learn how to represent vectors in three different ways. We start by introducing a geometric representation of vectors.

OBJECTIVE 1 UNDERSTANDING THE GEOMETRIC REPRESENTATION OF A VECTOR

We can represent a vector geometrically using an arrow of finite length drawn in a particular direction. Vectors can be thought of as a directed line segment that begins at an **initial point** (sometimes called the tail of the vector) and ends at a **terminal point** (sometimes called the head or tip of the vector).

Definition   Geometric Representation of a Vector

A **vector v** can be represented geometrically as a directed line segment⁺ in a plane having an initial point P and a terminal point Q.

The vector can be denoted as the boldface letter $\mathbf{v}$, $\overrightarrow{PQ}$, or $\vec{v}$.

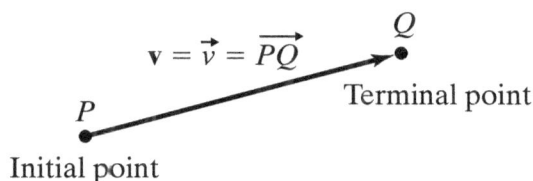

$$\mathbf{v} = \vec{v} = \overrightarrow{PQ}$$

Terminal point Q

P Initial point

Throughout Sections 5.4 and 5.5, we will denote vectors using a boldface letter. Because we cannot write boldface letters using pencil and paper, vectors in all videos will be denoted by an arrow above a single letter.

The magnitude of a vector represented geometrically is the length of the arrow. A vector whose magnitude is 0 is called the **zero vector** and is denoted by $\vec{0}$ or $\mathbf{0}$. If we know the coordinates of the initial point and the terminal point of a vector in a rectangular coordinate system, then we can use the distance formula to determine the magnitude, as stated in the following definition.

Definition   The Magnitude of a Vector Represented Geometrically

The **magnitude** of a vector $\mathbf{v}$ is the distance between the initial point and the terminal point and is denoted by $\|\mathbf{v}\|$. If the initial point has coordinates $P(x_1, y_1)$ and the terminal point has coordinates $Q(x_2, y_2)$, then

$$\|\mathbf{v}\| = \sqrt{(x_2 - x_1)^2 + (y_2 - y_1)^2}.$$

My video summary

⊙ **Example 1** Determining the Magnitude of a Vector Represented Geometrically

Determine the magnitude of $\mathbf{v}$.

Solution Vector **v** has an initial point $(x_1, y_1) = (-3, -1)$ and a terminal point $(x_2, y_2) = (2, 3)$. Therefore, the magnitude of **v** is

$$\|\mathbf{v}\| = \sqrt{(x_2 - x_1)^2 + (y_2 - y_1)^2} = \sqrt{(2 - (-3))^2 + (3 - (-1))^2}$$
$$= \sqrt{5^2 + 4^2} = \sqrt{41}.$$

You Try It Work through this You Try It problem.

Work Exercises 1–3 in this textbook or in the MyMathLab Study Plan.

OBJECTIVE 2 UNDERSTANDING OPERATIONS ON VECTORS REPRESENTED GEOMETRICALLY

SCALAR MULTIPLICATION

My video summary ◉ Recall that a quantity described using only magnitude is called a **scalar quantity**. When working in the context of vectors, real numbers are called **scalars**. Consider the two vectors in Figure 54a. If we multiply vector **v** by a scalar k, we obtain the product $k\mathbf{v}$. Multiplying a vector by a scalar is called **scalar multiplication**. Suppose that k is a positive scalar such that $k \neq 1$. Then the vector $k\mathbf{v}$ has the same direction as **v** but has a magnitude that is k times the magnitude of **v**. See Figure 54b. Similarly, if vector **u** is multiplied by $k < 0$, then $k\mathbf{u}$ is a vector in the exact opposite direction as **u** with a magnitude of $|k|\|\mathbf{u}\|$. See Figure 54c. Watch this video to see a description of scalar multiplication.

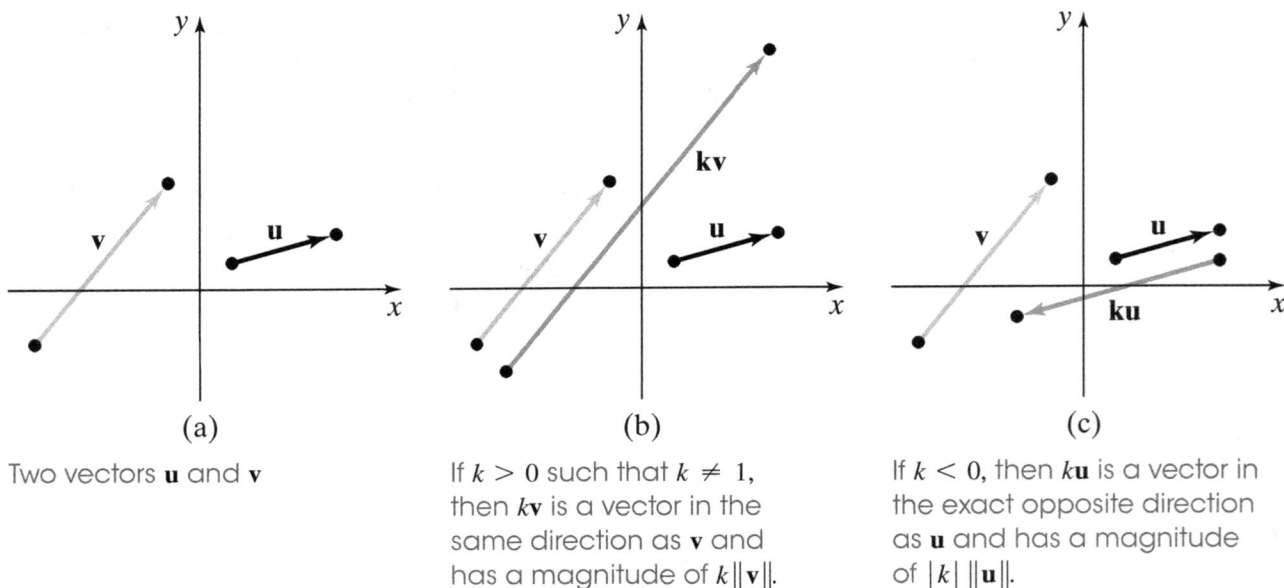

(a)	(b)	(c)
Two vectors **u** and **v**	If $k > 0$ such that $k \neq 1$, then $k\mathbf{v}$ is a vector in the same direction as **v** and has a magnitude of $k\|\mathbf{v}\|$.	If $k < 0$, then $k\mathbf{u}$ is a vector in the exact opposite direction as **u** and has a magnitude of $|k|\|\mathbf{u}\|$.

Figure 54

Note that if $k = 0$, then $k\mathbf{v}$ is the zero vector, **0**.

Example 2 Multiplying a Vector by a Scalar

Given the vector **v**, draw the vectors $2\mathbf{v}$ and $-\dfrac{1}{2}\mathbf{v}$.

Solution The vector $2\mathbf{v}$ is twice as long as vector $\mathbf{v}$ and is in the same direction as $\mathbf{v}$. The vector $-\dfrac{1}{2}\mathbf{v}$ is half as long as vector $\mathbf{v}$ and is in the exact opposite direction as $\mathbf{v}$.

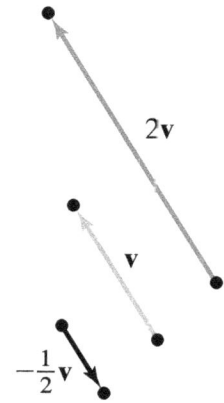

Note that the three vectors in Example 2 are **parallel** to each other. Any nonzero vector $\mathbf{u}$ that is a scalar multiple of a vector $\mathbf{v}$ is said to be parallel to $\mathbf{v}$.

Definition   Parallel Vectors

Two vectors $\mathbf{u}$ and $\mathbf{v}$ are **parallel vectors** if there is a nonzero scalar k such that $\mathbf{u} = k\mathbf{v}$.

My video summary   ▶ **VECTOR ADDITION**

Suppose that we wish to add the two nonzero vectors $\mathbf{u}$ and $\mathbf{v}$ shown in Figure 55a. The sum of $\mathbf{u}$ and $\mathbf{v}$ is denoted by $\mathbf{u} + \mathbf{v}$ and is called the **resultant vector**. To add the two vectors geometrically, we start by drawing an exact copy of vector $\mathbf{v}$ so that the initial point of $\mathbf{v}$ coincides with the terminal point of vector $\mathbf{u}$. See Figure 55b. The resultant vector $\mathbf{u} + \mathbf{v}$ is the vector that shares the initial point with $\mathbf{u}$ extending to and sharing the terminal point with $\mathbf{v}$. You can see in Figure 55c why vector addition is sometimes referred to as the **triangle law**. Watch this video to see a description of vector addition.

(a)

Two vectors $\mathbf{u}$ and $\mathbf{v}$

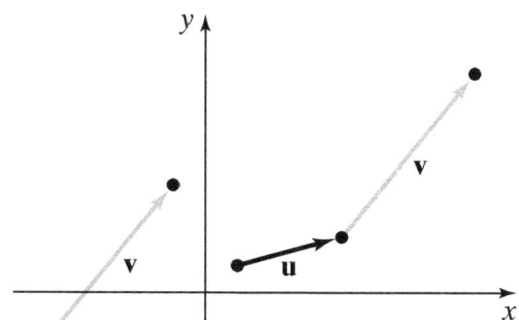

(b)

To find $\mathbf{u} + \mathbf{v}$, start by placing the initial point of $\mathbf{v}$ on the terminal point of $\mathbf{u}$.

Figure 55 Continues

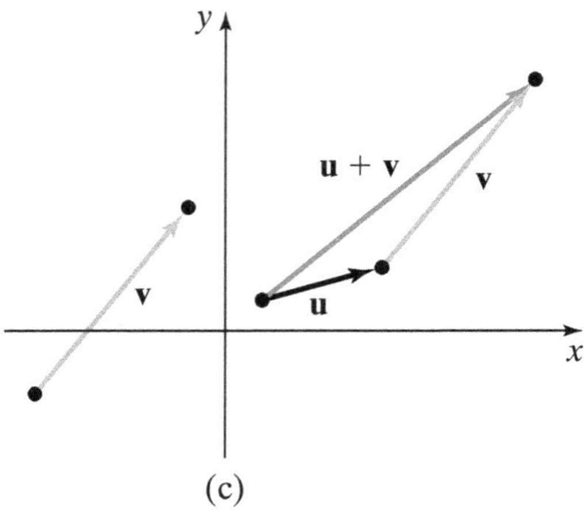

(c)

The resultant vector **u** + **v** extends from the initial point of **u** to the terminal point of **v**. The three vectors form a triangle.

Figure 55

My video summary

⊙ THE PARALLELOGRAM LAW FOR VECTOR ADDITION

Another way to illustrate vector addition is to think of the resultant vector **u** + **v** as the diagonal of a parallelogram. We do this by starting with the vectors **u** and **v**, as shown in Figure 56a. We can complete a parallelogram by drawing copies of vectors **u** and **v** on opposite sides of each other. As you can see in Figure 56b, we can represent the resultant vector **u** + **v** as the diagonal of a parallelogram. This is known as the **parallelogram law for vector addition**. The parallelogram law is also a nice way to illustrate that **vector addition is commutative**, which means that **u** + **v** = **v** + **u**. The commutative property of vector addition is just one of many properties of vectors. We will state several more properties later in this section.

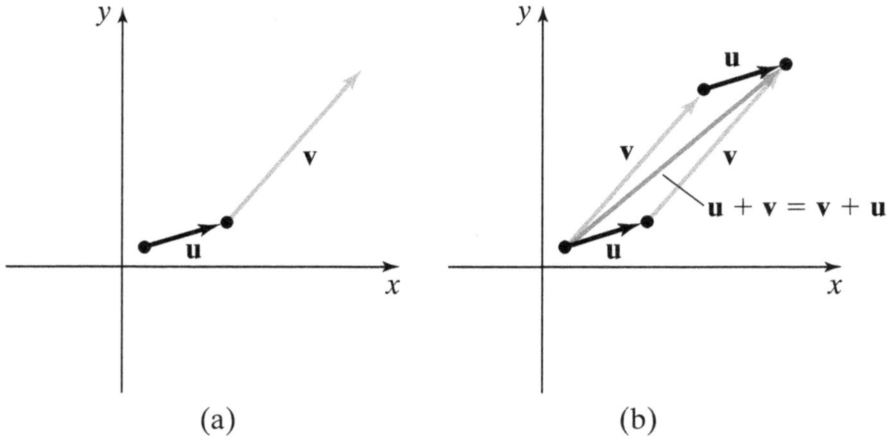

(a)

Start by placing the initial point of **v** on the terminal point of **u**.

(b)

Construct a parallelogram. The diagonal is the vector **u** + **v** = **v** + **u**.

Figure 56

My video summary

⊙ VECTOR SUBTRACTION

The difference of two vectors **u** − **v** is defined as **u** − **v** = **u** + (−**v**), where −**v** is obtained by multiplying **v** by the scalar −1. Therefore, to find the difference of two vectors **u** − **v**, first draw the vector −**v** in such a way that the initial point of −**v** coincides with the terminal point of **u**. Then use vector addition to add −**v** to **u**. See Figure 57b. The resultant vector **u** − **v** = **u** + (−**v**) is the vector that shares the initial point, with **u** extending to and sharing the terminal point with **v**. See Figure 57c. Watch this video to see a description of vector subtraction.

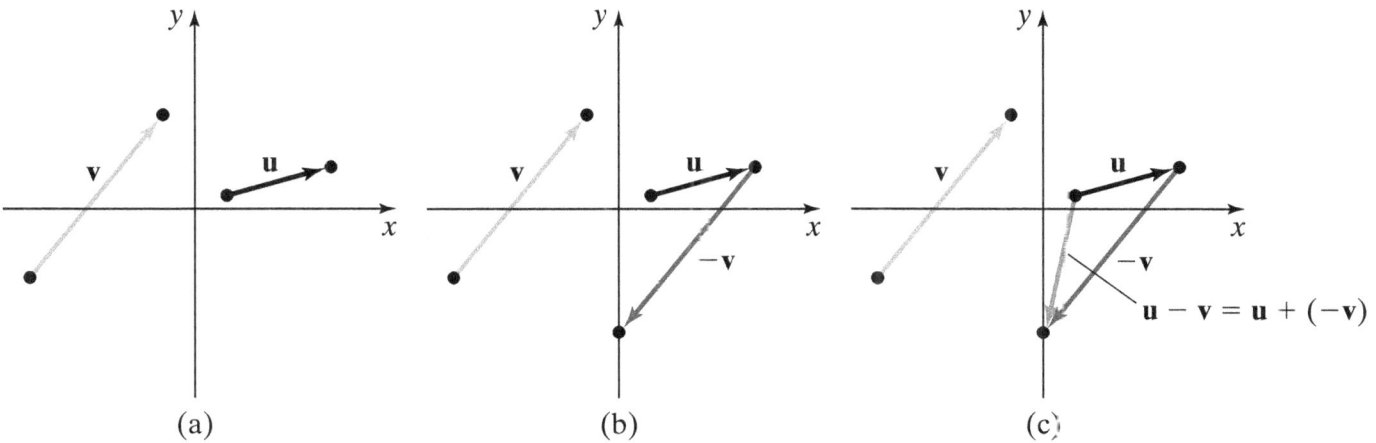

(a) | (b) | (c)

Two vectors **u** and **v** | To find **u** − **v**, start by creating −**v** and place −**v** on the terminal point of **u**. | The resultant vector **u** − **v** extends from the initial point of **u** to the terminal point of −**v**.

Figure 57

My video summary ⊘ **Example 3** Performing Operations on Vectors Represented Geometrically

Given the vectors **u**, **v**, and **w**, draw each of the following vectors.

a. **w** + **v**   b. **v** − **u**   c. **v** + 2**w** − **u**

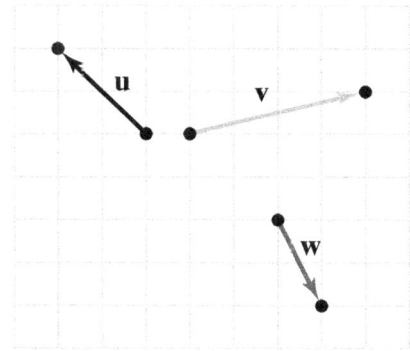

Solution Each vector is illustrated in green below. Watch this video to see a complete description of how to sketch each vector.

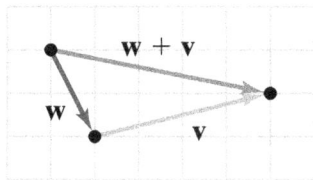

a. **w** + **v**

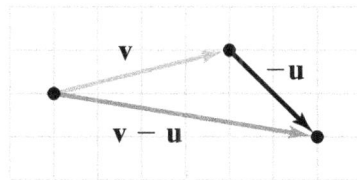

b. **v** − **u**

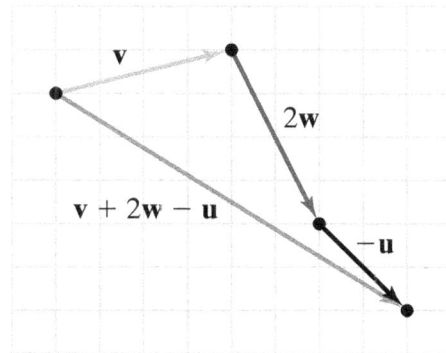

c. **v** + 2**w** − **u**

You Try It Work through this You Try It problem.

Work Exercises 4–10 in this textbook or in the MyMathLab Study Plan.

OBJECTIVE 3   UNDERSTANDING VECTORS IN TERMS OF COMPONENTS

Every vector has magnitude and direction. Two vectors in a plane are **equal vectors** if they have the same magnitude and the same direction, even if they do not share the same initial point and terminal point.

Definition   Equal Vectors

Two vectors in a plane are **equal vectors** if they have the same magnitude and the same direction.

Consider the three vectors in Figure 58, which are sketched in a rectangular coordinate system. Note that all three vectors are equal (they have the same magnitude and same direction) and they can all be drawn by starting at the initial point and then moving 3 units right followed by moving 4 units up.

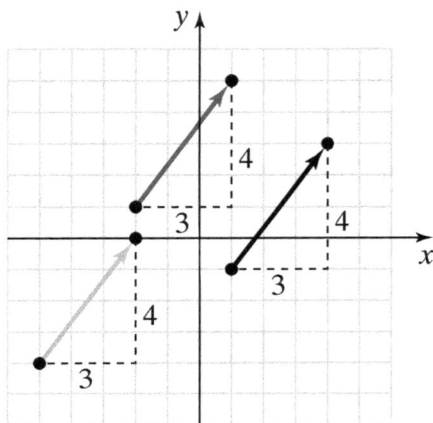

Figure 58 Each vector can be obtained by moving horizontally to the right of the initial point 3 units followed by moving vertically up 4 units.

The three equal vectors in Figure 58 can each be thought of as the "3 units right and 4 units up vector." This suggests another way to represent vectors. For each of the vectors in Figure 58, we can use the number 3 to describe the horizontal displacement from the initial point and the number 4 to describe the vertical displacement. These displacement values are called the **components** of the vector. Vectors represented using components a and b can be written in the form $\mathbf{v} = \langle a, b \rangle$, where a is the horizontal component and b is the vertical component. Each vector in Figure 58 can be represented as $\mathbf{v} = \langle 3, 4 \rangle$.

Definition   Representing a Vector in Terms
of Components a and b

If the initial point of a vector $\mathbf{v}$ is $P(x_1, y_1)$ and if the terminal point is $Q(x_2, y_2)$, then $\mathbf{v}$ is a **vector represented by components a and b**, where $\mathbf{v} = \langle x_2 - x_1, y_2 - y_1 \rangle = \langle a, b \rangle$.

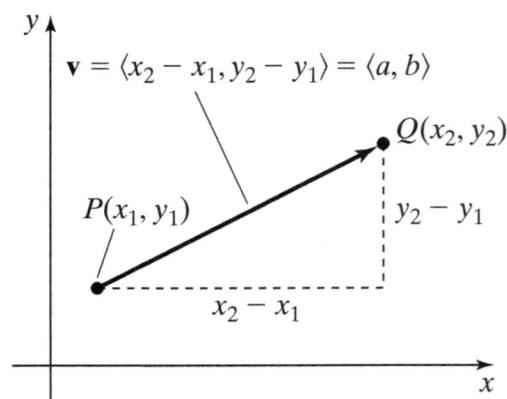

Example 4 Determining the Components and Magnitude of a Vector

Determine the component representation and the magnitude of a vector **v** having an initial point $P(5, 3)$ and a terminal point $Q(-6, 5)$.

Solution The component representation is $\mathbf{v} = \langle x_2 - x_1, y_2 - y_1 \rangle = \langle -6 - 5, 5 - 3 \rangle = \langle -11, 2 \rangle$ and the magnitude of **v** is $\|\mathbf{v}\| = \|\langle -11, 2 \rangle\| = \sqrt{(-11)^2 + 2^2} = \sqrt{121 + 4} = \sqrt{125} = 5\sqrt{5}$. See Figure 59.

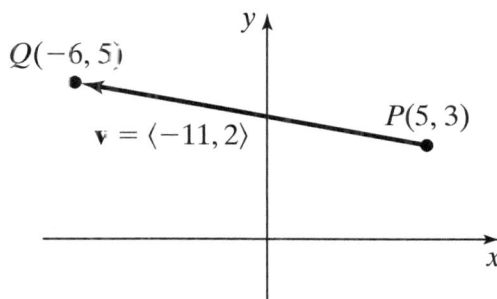

Figure 59 The vector $\mathbf{v} = \langle -11, 2 \rangle$

You Try It Work through this You Try It problem.

Work Exercises 11–13 in this textbook or in the My Math**Lab Study Plan.**

We have previously stated that two vectors are equal if they have the same magnitude and direction. Notice that the vector $\mathbf{v} = \langle -11, 2 \rangle$ in Example 4 is equal to the vector **u** that has an initial point at the origin and a terminal point of $(-11, 2)$. See Figure 60.

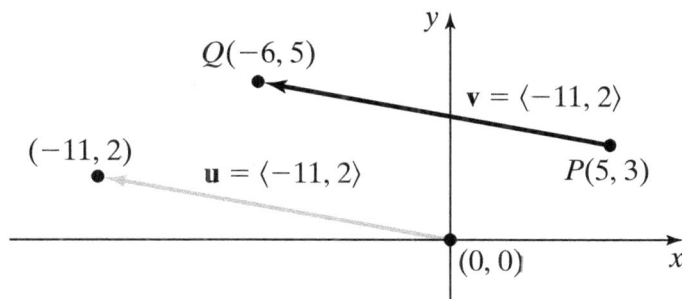

Figure 60 The vector $\mathbf{v} = \langle -11, 2 \rangle$ with initial point $P(5, 3)$ and terminal point $Q(-6, 5)$ is equal to the vector **u** with initial point at the origin and terminal point at $(-11, 2)$.

For any given vector, we can always find an equivalent vector whose initial point is at the origin. Vectors having an initial point at the origin are said to be in **standard position**.

Definition   A Vector in Standard Position

A vector $\mathbf{v} = \langle a, b \rangle$ is in **standard position** if the initial point coincides with the origin. The magnitude is $\|\mathbf{v}\| = \|\langle a, b \rangle\| = \sqrt{a^2 + b^2}$.

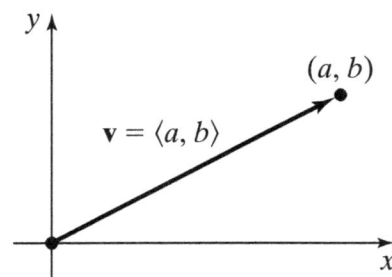

For simplicity, unless otherwise stated, we will consider all vectors discussed in the remainder of this section to be in standard position. This is possible because any vector can be represented in the form $\mathbf{v} = \langle a, b \rangle$, where the initial point is the origin and the terminal point is (a, b).

OBJECTIVE 4   UNDERSTANDING VECTORS REPRESENTED IN TERMS OF i AND j

So far we have seen a geometric representation of vectors and vectors represented using components. There is one other very useful way to represent vectors. Before we discuss this third representation, we must introduce the concept of unit vectors.

A **unit vector** is a vector that has a magnitude of 1 unit.

Definition   Unit Vector

A **unit vector** is a vector that has a magnitude of 1 unit.

There are two unit vectors that are particularly important. The first is the unit vector whose direction is along the positive x-axis. This vector is named $\mathbf{i}$, where $\mathbf{i} = \langle 1, 0 \rangle$. The other is the unit vector whose direction is along the positive y-axis. This vector is named $\mathbf{j}$, where $\mathbf{j} = \langle 0, 1 \rangle$.

Definition   Unit Vectors i and j

The unit vector whose direction is along the positive x-axis is $\mathbf{i} = \langle 1, 0 \rangle$.

The unit vector whose direction is along the positive y-axis is $\mathbf{j} = \langle 0, 1 \rangle$.

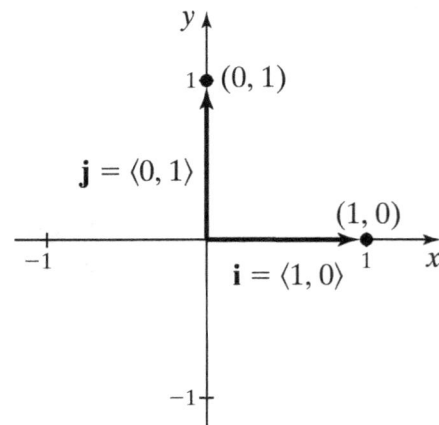

To sketch the vector $\mathbf{v} = a\mathbf{i} + b\mathbf{j}$, start by placing the initial point of $b\mathbf{j}$ on the terminal point of $a\mathbf{i}$. See Figure 61a. The resultant vector $\mathbf{v} = a\mathbf{i} + b\mathbf{j}$ extends from the origin to the terminal point of $b\mathbf{j}$. See Figure 61b.

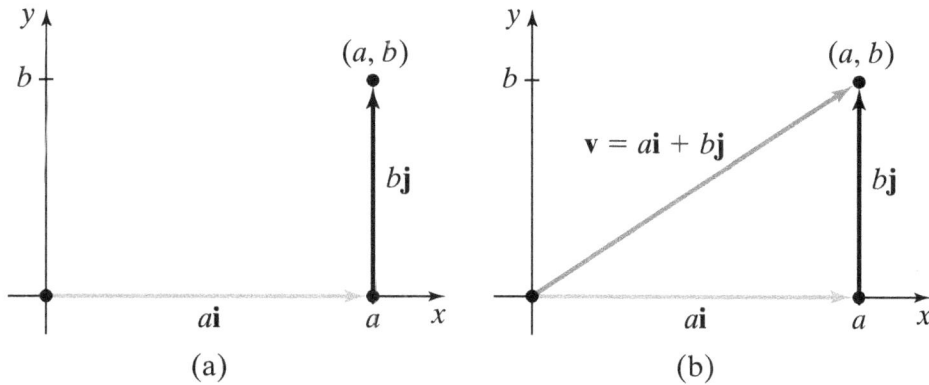

(a)

Place the initial point of $b\mathbf{j}$ on the terminal point of $a\mathbf{i}$.

(b)

The resultant vector $\mathbf{v} = a\mathbf{i} + b\mathbf{j}$ extends from the origin to the terminal point of $b\mathbf{j}$.

Figure 61

As you can see in Figure 61b, the vector $\mathbf{v} = a\mathbf{i} + b\mathbf{j}$ is equivalent to the vector $\mathbf{v} = \langle a, b \rangle$. Therefore, any vector can be written in terms of unit vectors $\mathbf{i}$ and $\mathbf{j}$, which gives us a third way of representing vectors.

Definition   A Vector Represented in Terms of $\mathbf{i}$ and $\mathbf{j}$

Any vector $\mathbf{v} = \langle a, b \rangle$ can be **represented in terms of the unit vectors $\mathbf{i}$ and $\mathbf{j}$**, where $\mathbf{v} = \langle a, b \rangle = a\mathbf{i} + b\mathbf{j}$.

The magnitude is $\|\mathbf{v}\| = \sqrt{a^2 + b^2}$.

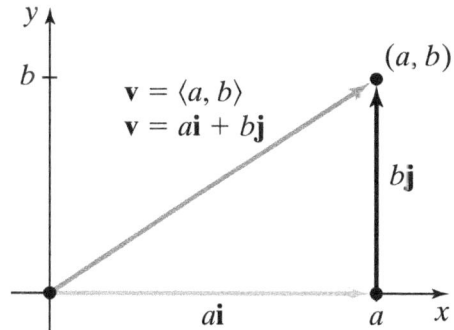

Example 5 Writing a Vector in Terms of $\mathbf{i}$ and $\mathbf{j}$

Write the vector $\mathbf{v} = \langle -5, 2 \rangle$ in terms of the unit vectors $\mathbf{i}$ and $\mathbf{j}$.

Solution We write $\mathbf{v}$ in terms of the unit vectors $\mathbf{i}$ and $\mathbf{j}$ as $\mathbf{v} = \langle -5, 2 \rangle = -5\mathbf{i} + 2\mathbf{j}$. See Figure 62.

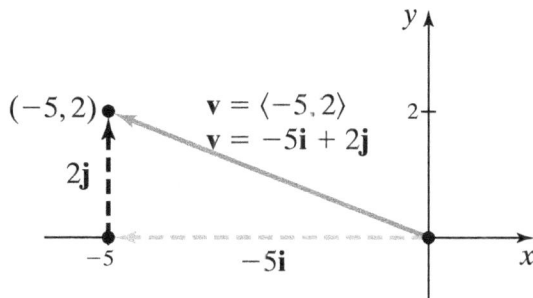

Figure 62 The vector $\mathbf{v} = \langle -5, 2 \rangle = -5\mathbf{i} + 2\mathbf{j}$

You Try It   Work through this You Try It problem.

Work Exercises 14–16 in this textbook or in the My MathLab **Study Plan.**

When vectors are represented in terms of **i** and **j**, then the operations of addition, subtraction, and scalar multiplication are very straightforward.

> ### Operations with Vectors in Terms of i and j
>
> If $\mathbf{u} = a\mathbf{i} + b\mathbf{j}$, $\mathbf{v} = c\mathbf{i} + d\mathbf{j}$, and k is a scalar, then
>
> $$\mathbf{u} + \mathbf{v} = (a + c)\mathbf{i} + (b + d)\mathbf{j}$$
>
> $$\mathbf{u} - \mathbf{v} = (a - c)\mathbf{i} + (b - d)\mathbf{j}$$
>
> $$k\mathbf{u} = (ka)\mathbf{i} + (kb)\mathbf{j}$$

My video summary

⊘ **Example 6** Performing Operations on Vectors in Terms of **i** and **j**

Let $\mathbf{u} = -3\mathbf{i} + 7\mathbf{j}$ and $\mathbf{v} = 5\mathbf{i} - \mathbf{j}$. Find each vector in terms of **i** and **j** and determine the magnitude of each vector.

a. $-\dfrac{1}{2}\mathbf{u}$ b. $\mathbf{u} + \mathbf{v}$ c. $\mathbf{u} - \mathbf{v}$ d. $3\mathbf{u} - 5\mathbf{v}$

My video summary

⊘ **Solution**   Work through this video to verify the following:

a. $-\dfrac{1}{2}\mathbf{u} = \left(\left(-\dfrac{1}{2}\right)(-3)\right)\mathbf{i} + \left(\left(-\dfrac{1}{2}\right)(7)\right)\mathbf{j} = \dfrac{3}{2}\mathbf{i} - \dfrac{7}{2}\mathbf{j};$

$\left\|-\dfrac{1}{2}\mathbf{u}\right\| = \sqrt{\left(\dfrac{3}{2}\right)^2 + \left(-\dfrac{7}{2}\right)^2} = \sqrt{\dfrac{9}{4} + \dfrac{49}{4}} = \sqrt{\dfrac{58}{4}} = \dfrac{\sqrt{58}}{2}$

b. $\mathbf{u} - \mathbf{v} = (-3 - 5)\mathbf{i} + (7 - (-1))\mathbf{j} = -8\mathbf{i} + 8\mathbf{j};$

$\|\mathbf{u} - \mathbf{v}\| = \sqrt{(-8)^2 + 8^2} = \sqrt{64 + 64} = \sqrt{128} = 8\sqrt{2}$

My video summary   ⊘ Work through this video to see the solutions to parts b and d.

You Try It   Work through this You Try It problem.

Work Exercises 17–22 in this textbook or in the My MathLab **Study Plan.**

We have previously used the parallelogram law to introduce the fact that vector addition is commutative. We now restate the commutative property and list several other properties of vectors.

> ### Properties of Vectors
>
> If $\mathbf{u}$, $\mathbf{v}$, and $\mathbf{w}$ are vectors and if A and B are scalars, then the following properties are true.
>
> 1. $\mathbf{u} + \mathbf{v} = \mathbf{v} + \mathbf{u}$ Commutative Property for Vector Addition
>
> 2. $(\mathbf{u} + \mathbf{v}) + \mathbf{w} = \mathbf{u} + (\mathbf{v} + \mathbf{w})$ Associative Property for Vector Addition

(continued)

3. $\mathbf{u} + \mathbf{0} = \mathbf{u}$ Additive Identity Property

4. $\mathbf{u} + (-\mathbf{u}) = \mathbf{0}$ Additive Inverse Property

5. $(AB)\mathbf{u} = A(B\mathbf{u})$ Associative Property for Scalar Multiplication

6. $A(\mathbf{u} + \mathbf{v}) = A\mathbf{u} + A\mathbf{v}$ Distributive Property

7. $(A + B)\mathbf{u} = A\mathbf{u} + B\mathbf{u}$ Distributive Property

8. $1\mathbf{u} = \mathbf{u}$ Multiplicative Identity Property

9. $0\mathbf{u} = \mathbf{0}$ Zero Multiplication Property

10. $\|A\mathbf{u}\| = |A|\,\|\mathbf{u}\|$ Magnitude of Scalar Multiplication Property

OBJECTIVE 5   FINDING A UNIT VECTOR

Recall that a unit vector has a magnitude of 1 unit. Given a vector $\mathbf{v} = a\mathbf{i} + b\mathbf{j}$, it is often useful to find the unit vector $\mathbf{u}$ that has the same direction as $\mathbf{v}$. We can find this unit vector $\mathbf{u}$ by dividing $\mathbf{v}$ by its magnitude $\|\mathbf{v}\|$.

Definition   The Unit Vector in the Same Direction of a Given Vector

Given a nonzero vector $\mathbf{v} = a\mathbf{i} + b\mathbf{j}$, the **unit vector** $\mathbf{u}$ in the same direction as $\mathbf{v}$ is $\mathbf{u} = \dfrac{\mathbf{v}}{\|\mathbf{v}\|}$.

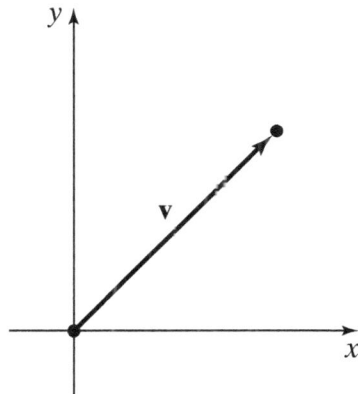

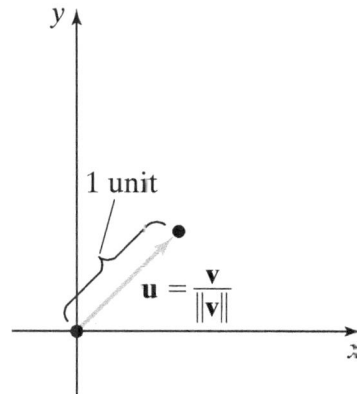

My video summary   ▷ **Example 7** Finding a Unit Vector

Find the unit vector that has the same direction as $\mathbf{v} = 6\mathbf{i} - 8\mathbf{j}$.

Solution First, determine the magnitude of $\mathbf{v}$.

$$\|\mathbf{v}\| = \|6\mathbf{i} - 8\mathbf{j}\| = \sqrt{6^2 + (-8)^2} = \sqrt{36 - 64} = \sqrt{100} = 10$$

The unit vector in the same direction as $\mathbf{v}$ is $\mathbf{u} = \dfrac{\mathbf{v}}{\|\mathbf{v}\|} = \dfrac{6\mathbf{i} - 8\mathbf{j}}{10} = \dfrac{6}{10}\mathbf{i} - \dfrac{8}{10}\mathbf{j} = \dfrac{3}{5}\mathbf{i} - \dfrac{4}{5}\mathbf{j}$.

We can verify that the length of $\mathbf{u}$ is 1 unit by determining the magnitude of $\mathbf{u}$.

$$\|\mathbf{u}\| = \left\|\tfrac{3}{5}\mathbf{i} - \tfrac{4}{5}\mathbf{j}\right\| = \sqrt{\left(\tfrac{3}{5}\right)^2 + \left(\tfrac{-4}{5}\right)^2} = \sqrt{\tfrac{9}{25} + \tfrac{16}{25}} = \sqrt{\tfrac{25}{25}} = \sqrt{1} = 1$$

You Try It   Work through this You Try It problem.

Work Exercises 23–26 in this textbook or in the MyMathLab Study Plan.

OBJECTIVE 6   DETERMINING THE DIRECTION ANGLE OF A VECTOR

Recall that every vector has magnitude and direction. We can describe the direction using the positive angle formed between the positive x-axis and a vector in standard position. This positive angle is called the **direction angle**.

Definition   The Direction Angle of a Vector

Given a vector $\mathbf{v} = \langle a, b \rangle = a\mathbf{i} + b\mathbf{j}$ in standard position, the **direction angle of v** is the positive angle θ between the positive x-axis and vector that satisfies the equation $\tan \theta = \dfrac{b}{a}$, where $a \neq 0$.

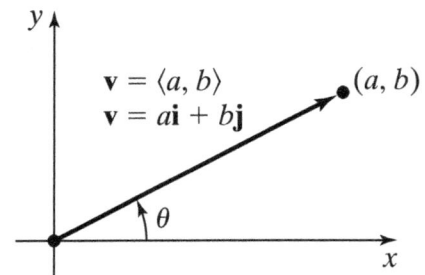

My video summary

⊙ Example 8 Determining the Direction Angle of a Vector

Determine the direction angle of the vector $\mathbf{v} = -3\mathbf{i} + 2\mathbf{j}$.

Solution To find the direction angle θ, start by determining the reference angle θ_R. Using the fact that $\tan \theta = \dfrac{b}{a}$, we can find θ_R by solving the equation $\tan \theta_R = \left| \dfrac{b}{a} \right|$. This technique was first introduced in Section 5.1.

$$\tan \theta_R = \left| \frac{b}{a} \right| \qquad \text{Write the equation relating } \theta_R, a, \text{ and } b.$$

$$\tan \theta_R = \left| \frac{2}{-3} \right| \qquad \text{Substitute } a = -3 \text{ and } b = 2.$$

$$\tan \theta_R = \frac{2}{3} \qquad \text{Simplify.}$$

The angle whose tangent is $\dfrac{2}{3}$ does not belong to one of the special angle families. However, we can use the inverse tangent function to solve for θ_R. Thus,

$$\theta_R = \tan^{-1}\left(\frac{2}{3} \right).$$

The terminal side of θ lies in Quadrant II, which indicates that $\theta = 180° - \theta_R$. Using a calculator set in degree mode, we get $\theta = 180° - \tan^{-1}\left(\dfrac{2}{3} \right) \approx 146.3°$. See Figure 63.

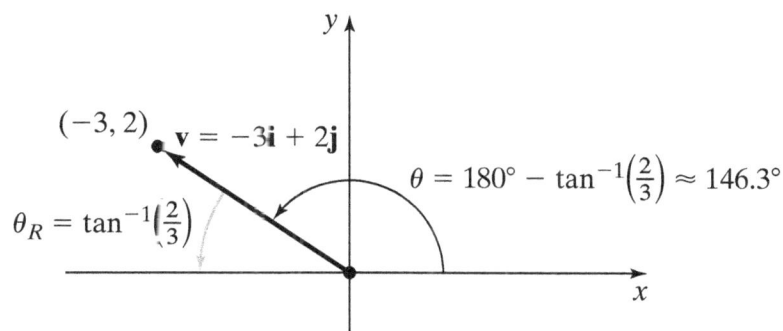

Figure 63 The vector $v = -3i + 2j$ has a direction angle of $\theta \approx 146.3°$.

You Try It   Work through this You Try It problem.

Work Exercises 27–31 in this textbook or in the MyMathLab Study Plan.

OBJECTIVE 7   REPRESENTING A VECTOR IN TERMS OF **i** AND **j** GIVEN ITS MAGNITUDE AND DIRECTION ANGLE

If we know the magnitude and the direction angle of a vector, we can find the horizontal component a and the vertical component b. We do this in much the same way as we did when we converted polar coordinates to rectangular coordinates.

Consider the nonzero vector **v** in Figure 64 with magnitude $\|v\|$ and direction angle θ. The goal is to determine the values of a and b.

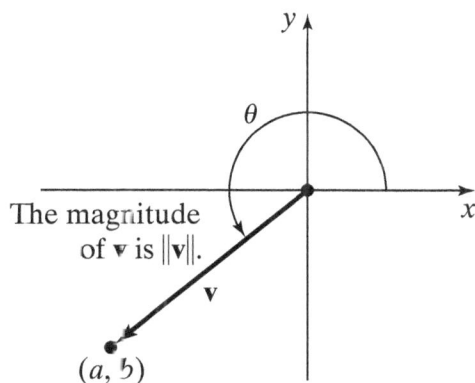

Figure 64

By the general angle definition of sine and cosine, we know that $\cos \theta = \dfrac{a}{\|v\|}$ and $\sin \theta = \dfrac{b}{\|v\|}$. Multiplying both sides of these two equations by $\|v\|$, we get $a = \|v\| \cos \theta$ and $b = \|v\| \sin \theta$. Therefore, we can write **v** as follows:

$v = ai + bj$ Write the vector in terms of unit vectors **i** and **j**.

$v = (\|v\| \cos \theta)i + (\|v\| \sin \theta)j$ Substitute $b = \|v\| \cos \theta$ and $b = \|v\| \sin \theta$.

Representing a Vector in Terms of **i** and **j** Given its Magnitude and Direction Angle

If θ is the direction angle measured from the positive x-axis to a vector **v**, then the vector can be expressed as $\mathbf{v} = (\|\mathbf{v}\| \cos \theta)\mathbf{i} + (\|\mathbf{v}\| \sin \theta)\mathbf{j}$.

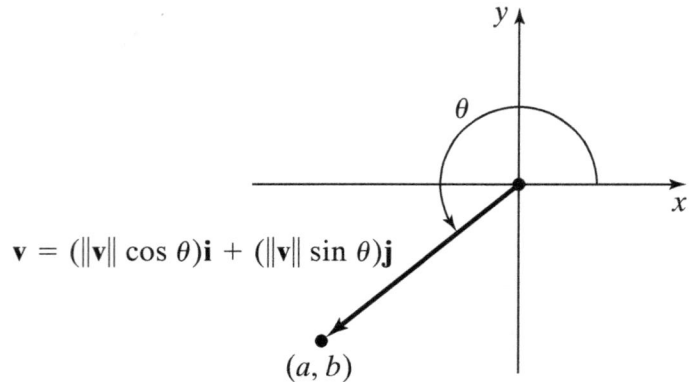

$$\mathbf{v} = (\|\mathbf{v}\| \cos \theta)\mathbf{i} + (\|\mathbf{v}\| \sin \theta)\mathbf{j}$$

(a, b)

My video summary

⊙ Example 9 Representing a Vector in Terms of **i** and **j** Given Its Magnitude and Direction Angle

The vector **v** has a magnitude of 20 units and direction angle of $\theta = 50°$. Represent this vector in the form $\mathbf{v} = a\mathbf{i} + b\mathbf{j}$. Round a and b to two decimal places.

Solution First sketch the vector.

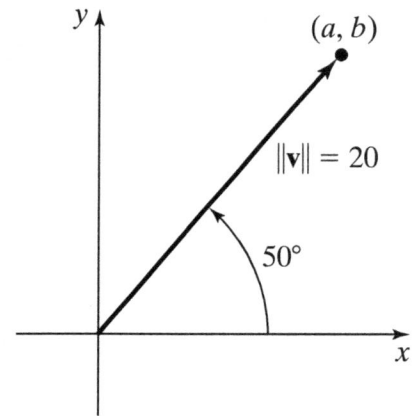

(a, b)

$\|\mathbf{v}\| = 20$

$50°$

The horizontal component of the vector is $a = \|\mathbf{v}\| \cos \theta = 20 \cos 50° \approx 12.86$.
The vertical component of the vector is $b = \|\mathbf{v}\| \sin \theta = 20 \sin 50° \approx 15.32$.
Thus, the vector is approximately $\mathbf{v} = 12.86\mathbf{i} + 15.32\mathbf{j}$.

You Try It Work through this **You Try It** problem.

Work Exercises 32–36 in this textbook or in the MyMathLab **Study Plan.**

OBJECTIVE 8 USING VECTORS TO SOLVE APPLICATIONS INVOLVING VELOCITY

Velocity has both magnitude (speed) and direction. Thus, given the speed and direction of an object in motion, we can use a vector to represent the velocity of the object.

My video summary   ⊙ **Example 10** Using a Vector to Represent the Velocity of an Airplane

An airplane takes off from a runway at a speed of 190 mph at an angle of 11°. Express the velocity of the plane at takeoff as a vector in terms of **i** and **j**. Round a and b to two decimal places.

Solution Let **v** be the vector that represents the velocity of the plane at takeoff. The magnitude is $\|\mathbf{v}\| = 190$ and the direction angle is $\theta = 11°$. To determine **v**, we use the formula $\mathbf{v} = (\|\mathbf{v}\| \cos \theta)\mathbf{i} + (\|\mathbf{v}\| \sin \theta)\mathbf{j}$.

$$\mathbf{v} = (\|\mathbf{v}\| \cos \theta)\mathbf{i} - (\|\mathbf{v}\| \sin \theta)\mathbf{j} = (190 \cos 11°)\mathbf{i} + (190 \sin 11°)\mathbf{j}$$

Using a calculator in degree mode, we get $190 \cos 11° \approx 186.51$ and $190 \sin 11° \approx 36.25$. Therefore, the vector representing the velocity of the plane is approximately $\mathbf{v} = 186.51\mathbf{i} + 36.25\mathbf{j}$.

You Try It   Work through this You Try It problem.

Work Exercises 37 and 38 in this textbook or in the MyMathLab Study Plan.

Many applications give the direction of a vector in terms of its bearing. The concept of a bearing was first introduced in Section 4.1. Figure 65 illustrates a review of how to sketch the bearing of an object. Each object has a bearing of 45° in one of four directions.

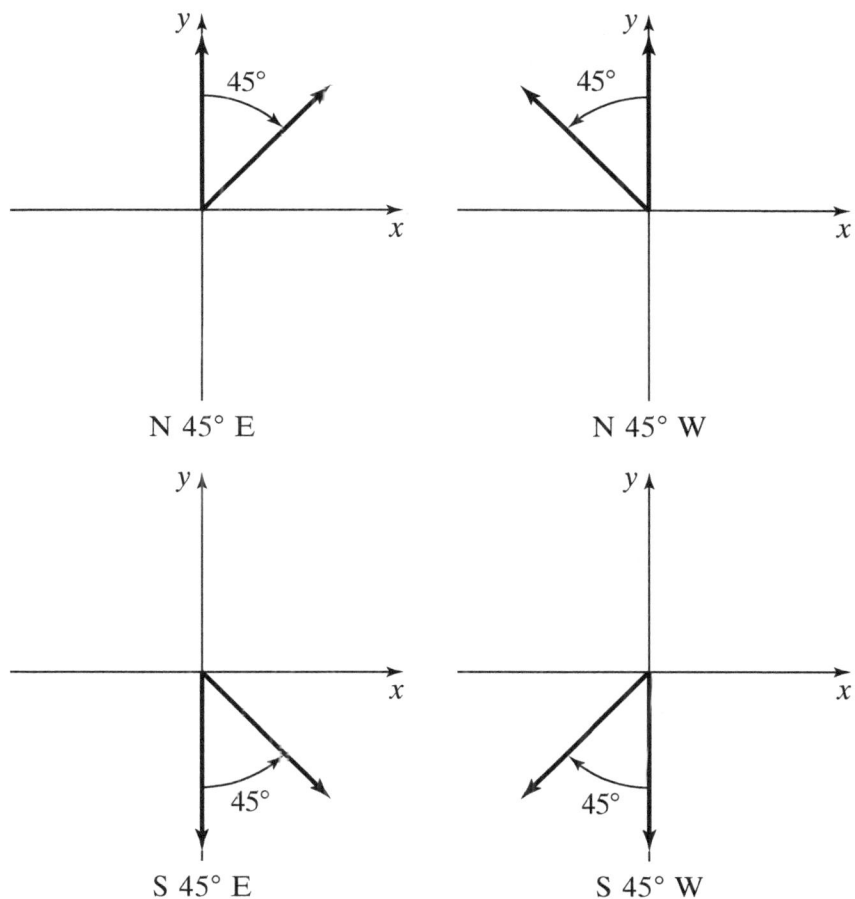

N 45° E   N 45° W

S 45° E   S 45° W   Figure 65

Recall that a bearing of N 45° E is read as "45° east of north."

For the rest of the application examples in this section, the vectors drawn in the corresponding figures will not include the initial point or terminal point. Remember that all vectors have finite length and that the arrows do not mean that the vector extends indefinitely. The arrowheads in the figures simply represent the direction of each vector.

My video summary ⊙ **Example 11** Using a Vector to Represent Wind Velocity

The wind is blowing at a speed of 35 mph in a direction of N 30° E. Express the velocity of the wind as a vector in terms of **i** and **j**.

Solution Let **w** be the vector having a magnitude of $\|\mathbf{w}\| = 35$ and a direction angle of $\theta = 60°$. See Figure 66.

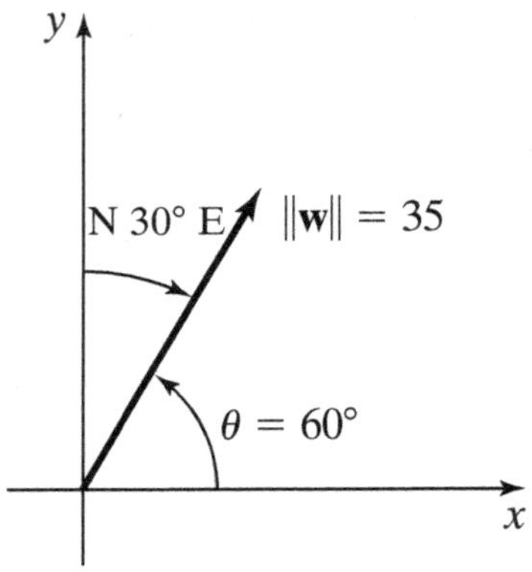

Figure 66

To represent the velocity of the wind in terms of **i** and **j**, we use the formula $\mathbf{w} = (\|\mathbf{w}\| \cos\theta)\mathbf{i} + (\|\mathbf{w}\| \sin\theta)\mathbf{j}$.

$$\mathbf{w} = (\|\mathbf{w}\| \cos\theta)\mathbf{i} + (\|\mathbf{w}\| \sin\theta)\mathbf{j} = (35 \cos 60°)\mathbf{i} + (35 \sin 60°)\mathbf{j} = \left(35 \cdot \frac{1}{2}\right)\mathbf{i} + \left(35 \cdot \frac{\sqrt{3}}{2}\right)\mathbf{j} = \frac{35}{2}\mathbf{i} + \frac{35\sqrt{3}}{2}\mathbf{j}$$ ●

You Try It Work through this You Try It problem.

Work Exercises **39** and **40** in this textbook or in the My Math Lab **Study Plan.**

My video summary ⊙ **Example 12** Using Vectors to Determine the Ground Speed and Direction of an Airplane

A 747 jet was heading due east at 520 mph in still air and encountered a 60 mph headwind blowing in the direction N 40° W. Determine the resulting ground speed of the plane and its new bearing. Round the ground speed to the nearest hundredth.

Solution Let **v** represent the velocity of the plane in still air and let **w** represent the velocity of the wind; then $\|\mathbf{v}\| = 520$ and $\|\mathbf{w}\| = 60$. The resultant vector $\mathbf{v} + \mathbf{w}$ is the diagonal of the parallelogram shown in Figure 67. The ground speed of the airplane is the magnitude of the resultant vector $\|\mathbf{v} + \mathbf{w}\|$.

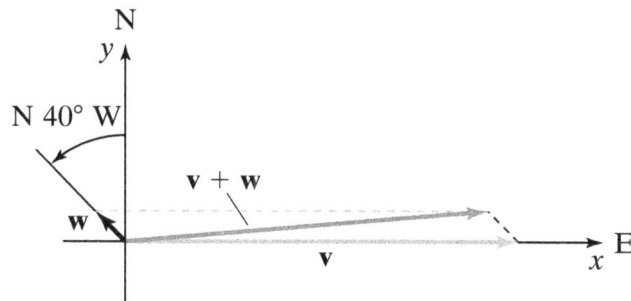

Figure 67

We first find **v** using the formula $\mathbf{v} = (\|\mathbf{v}\| \cos\theta)\mathbf{i} + (\|\mathbf{v}\| \sin\theta)\mathbf{j}$, where $\theta = 0°$ (the plane was heading due east) and $\|\mathbf{v}\| = 520$. Thus, $\mathbf{v} = (520\cos 0°)\mathbf{i} + (520\sin 0°)\mathbf{j} = 520(1)\mathbf{i} + 520(0)\mathbf{j} = 520\mathbf{i}$.

To find **w**, we use the formula $\mathbf{w} = (\|\mathbf{w}\| \cos\theta)\mathbf{i} + (\|\mathbf{w}\| \sin\theta)\mathbf{j}$, where $\theta = 40° + 90° = 130°$ and $\|\mathbf{w}\| = 60$. Therefore, $\mathbf{w} = (60\cos 130°)\mathbf{i} + (60\sin 130°)\mathbf{j}$.

The resultant vector is $\mathbf{v} + \mathbf{w} = 520\mathbf{i} + (60\cos 130°)\mathbf{i} + (60\sin 130°)\mathbf{j} = (520 + 60\cos 130°)\mathbf{i} + (60\sin 130°)\mathbf{j}$.

Using the formula for the magnitude of a vector, we see that the ground speed of the plane is $\|\mathbf{v} + \mathbf{w}\| = \sqrt{(520 + 60\cos 130°)^2 + (60\sin 130°)^2} \approx 483.62$ mph.

The direction angle is $\theta = \tan^{-1}\dfrac{60\sin 130°}{520 + 60\cos 130°} \approx 5.45°$. The bearing of the plane is measured clockwise from the y-axis to $\mathbf{v} + \mathbf{w}$. Thus we must determine the complement of the direction angle, which is $90° - 5.45° = 84.55°$. Therefore, the final bearing of the plane is N 84.55° E. See Figure 68.

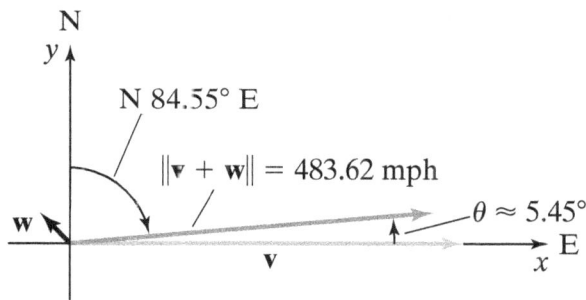

Figure 68

You Try It   Work through this You Try It problem.

Work Exercises 41 and 42 in this textbook or in the MyMathLab Study Plan.

OBJECTIVE 9   USING VECTORS TO SOLVE APPLICATIONS INVOLVING FORCE

Vectors can be used to solve many applied problems involving the concept of **force**. This is because a force has *both* magnitude and direction. You can think of force as the push or pull on an object that causes the object to move. Examples include the pushing of a chair across a room or the downward gravitational pull on a ball. If more than one force is acting on an object, then the resultant force experienced by the object is the vector sum of these forces.

Isaac Newton's second law of motion states that for every action, there is an equal and opposite reaction. In the context of force, we can think of this law as saying that "forces always come in pairs." For example, when you place a 5-pound bag of groceries on your kitchen table, there is a 5-pound force of gravity acting on the bag of groceries and pulling it downward. We can represent this force using the vector $F_1 = -5j$. However, the bag does not move because the kitchen table stops it. This is because there is a force of equal magnitude (5 pounds) exerted by the table in the exact opposite direction.

We can represent the force of the table acting on the bag by the vector $F_2 = 5j$. The resultant force is $F_1 + F_2 = -5j + 5j = 0$, where 0 is the zero vector. Thus, the magnitude of the total force acting on the grocery bag is 0, indicating that the bag will not move. See Figure 69.

$F_2 = 5j$

$F_1 = -5j$

Figure 69 A 5-pound bag of groceries on a kitchen table does not move because the downward force of gravity is equal to the upward force of the table acting on the bag.

When the net result of the total force acting on an object is the zero vector, we say that the object is in **static equilibrium**. In applied problems involving force, it is often necessary to determine the forces acting on an object that will result in static equilibrium.

Definition   Static Equilibrium

If $F_1, F_2, \cdots, F_n$ are forces acting on an object, then the object is in **static equilibrium** if the resultant force $F = F_1 + F_2 + \cdots + F_n$ is equal to 0 (the zero vector).

My video summary

⊙ **Example 13** Determining the Force Necessary for an Object to Be in Static Equilibrium

The forces $F_1 = 6i - 8j$ and $F_2 = 3i + 2j$ are acting on an object. What additional force is required for the object to be in static equilibrium?

Solution The resultant force acting on the object is $F = F_1 + F_2 = (6 + 3)i + (-8 + 2)j = 9i - 6j$. The additional force F_3 required for the object to be in equilibrium is $F_3 = -F = -(9i - 6j) = -9i + 6j$. ●

You Try It Work through this You Try It problem.

Work Exercises 43–45 in this textbook or in the MyMathLab Study Plan.

If the sum of the forces acting on an object is not 0, then the resultant forces cause the object to move.

My video summary

⊚ Example 14 Finding the Magnitude and Bearing of a Ship Being Towed

Two tugboats are towing a large ship out of port and into the open sea. One tugboat exerts a force of $\|\mathbf{F_1}\| = 2000$ pounds in a direction N 35° W. The other tugboat pulls with a force of $\|\mathbf{F_2}\| = 1400$ pounds in a direction S 55° W. Find the magnitude of the resultant force and the bearing of the ship.

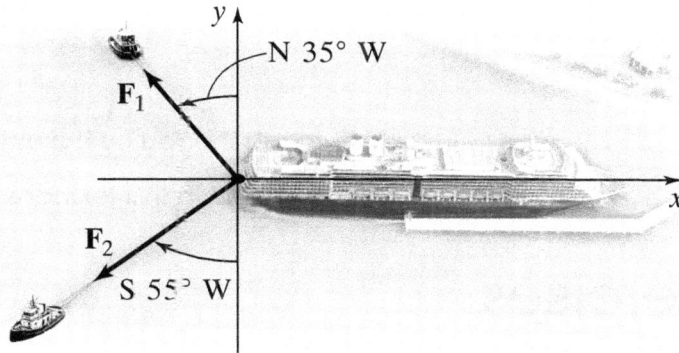

Solution We first find $\mathbf{F_1}$ using the formula $\mathbf{F_1} = (\|\mathbf{F_1}\| \cos\theta)\mathbf{i} + (\|\mathbf{F_1}\| \sin\theta)\mathbf{j}$, where $\|\mathbf{F_1}\| = 2000$ and $\theta = 90° + 35° = 125°$. Thus, $\mathbf{F_1} = (2000\cos 125°)\mathbf{i} + (2000\sin 125°)\mathbf{j}$.

Similarly, we find $\mathbf{F_2}$ using the formula $\mathbf{F_2} = (\|\mathbf{F_2}\| \cos\theta)\mathbf{i} + (\|\mathbf{F_2}\| \sin\theta)\mathbf{j}$, where $\|\mathbf{F_2}\| = 1400$ and $\theta = 270° - 55° = 215°$. Thus, $\mathbf{F_2} = (1400\cos 215°)\mathbf{i} + (1400\sin 215°)\mathbf{j}$.

The resultant force is $\mathbf{F} = \mathbf{F_1} + \mathbf{F_2} = (2000\cos 125° + 1400\cos 215°)\mathbf{i} + (2000\sin 125° + 1400\sin 215°)\mathbf{j} \approx -2293.97\mathbf{i} + 835.30\mathbf{j}$.

The magnitude of the resultant force is approximately

$$\|\mathbf{F}\| = \sqrt{(-2293.97)^2 + (835.30)^2} \approx 2441.32.$$

To determine the ship's bearing, we must first determine the direction angle θ for $\mathbf{F} = -2293.97\mathbf{i} + 835.30\mathbf{j}$.

To find the direction angle θ, start by determining the reference angle θ_R. Using the fact that $\tan\theta = \dfrac{b}{a}$, we can find θ_R by solving the equation $\tan\theta_R = \left|\dfrac{b}{a}\right|$.

$$\tan\theta_R = \left|\frac{b}{a}\right| \qquad \text{Write the equation relating } \theta_R, a, \text{ and } b.$$

$$\tan\theta_R = \left|\frac{835.30}{-2293.97}\right| \qquad \text{Substitute } a = -2293.97 \text{ and } b = 835.30.$$

$$\tan\theta_R = \frac{835.30}{2293.97} \qquad \text{Simplify.}$$

We can use the inverse tangent function to solve for θ_R. Thus,

$$\theta_R = \tan^{-1}\left(\frac{835.30}{2293.97}\right) \approx 20.00°.$$

The bearing of the ship is measured from the positive y-axis to $\mathbf{F}$. Thus we must determine the complement of θ_R, which is $90° - 20.00° \approx 70.00°$. Therefore, the ship's bearing is N 70.00° W. See Figure 70.

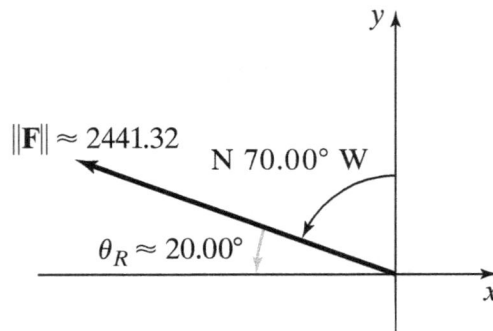

Figure 70

You Try It Work through this You Try It problem.

Work Exercises 46 and 47 in this textbook or in the MyMathLab Study Plan.

5.4 Exercises

In Exercises 1–3, determine the magnitude of **v**.

1.

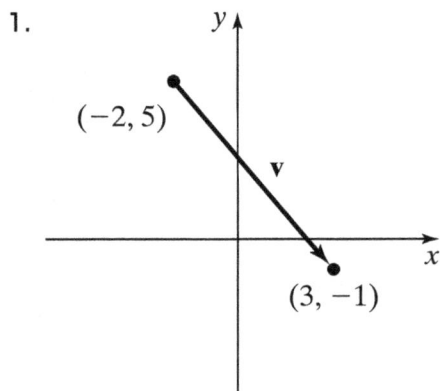

$(-2, 5)$

v

$(3, -1)$

2.

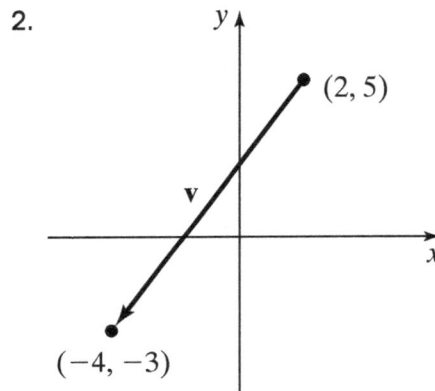

$(2, 5)$

v

$(-4, -3)$

3.

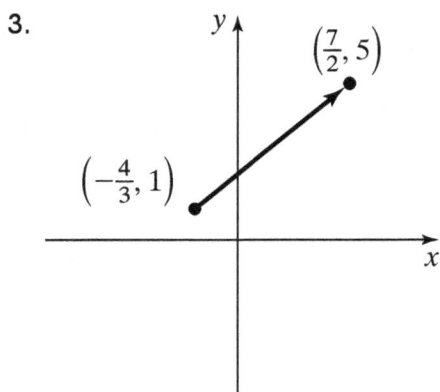

$\left(\frac{7}{2}, 5\right)$

$\left(-\frac{4}{3}, 1\right)$

In Exercises 4–10, use the given vectors **u**, **v**, and **w** to draw each vector.

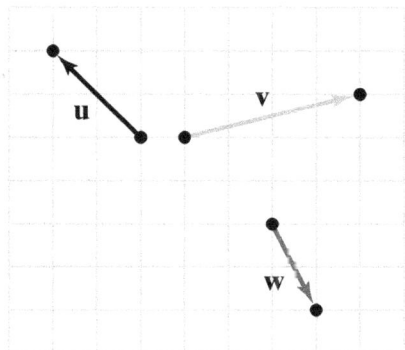

4. $2\mathbf{w}$

5. $\mathbf{u} + \mathbf{v}$

6. $-2\mathbf{v}$

7. $\mathbf{w} - \mathbf{v}$

8. $\dfrac{1}{2}\mathbf{u}$

9. $\mathbf{v} - 2\mathbf{u}$

10. $2\mathbf{w} + 2\mathbf{u} - \mathbf{v}$

In Exercises 11–13, determine the component representation and magnitude of the vector having the initial point P and the terminal point Q.

11. $P(2, 3)$ and $Q(6, 6)$

12. $P(-2, 1)$ and $Q(1, 4)$

13. $P(1, -4)$ and $Q(-5, 7)$

In Exercises 14–16, write each vector in terms of the unit vectors **i** and **j**.

14. $\mathbf{v} = \langle 3, 5 \rangle$

15. $\mathbf{v} = \langle -4, 1 \rangle$

16. $\mathbf{v} = \langle -6, -7 \rangle$

In Exercises 17–22, if $\mathbf{u} = -3\mathbf{i} + 7\mathbf{j}$ and $\mathbf{v} = 5\mathbf{i} - \mathbf{j}$, then find the indicated vectors in terms of **i** and **j** and find the magnitude of each vector.

17. $3\mathbf{u}$ **18.** $\mathbf{u} + \mathbf{v}$ **19.** $\mathbf{u} - \mathbf{v}$ **20.** $4\mathbf{u} + 3\mathbf{v}$ **21.** $-\dfrac{1}{3}\mathbf{u}$ **22.** $\dfrac{1}{2}\mathbf{u} - \dfrac{1}{4}\mathbf{v}$

In Exercises 23–26, find the unit vector in the same direction as the given vector.

23. $\mathbf{v} = -2\mathbf{j}$

24. $\mathbf{v} = 4\mathbf{i} - 3\mathbf{j}$

25. $\mathbf{v} = 5\mathbf{i} - 2\mathbf{j}$

26. $\mathbf{v} = \mathbf{i} - 4\mathbf{j}$

In Exercises 27–31, determine the direction angle for each vector.

27. $\mathbf{v} = 3\mathbf{i} - 3\mathbf{j}$

28. $\mathbf{v} = -5\mathbf{i} + 5\sqrt{3}\mathbf{j}$

29. $\mathbf{v} = 2\sqrt{3}\mathbf{i} + 2\mathbf{j}$

30. $\mathbf{v} = 2\mathbf{i} + 5\mathbf{j}$

31. $\mathbf{v} = 2\mathbf{i} - 7\mathbf{j}$

In Exercises 32–36, represent the vector **v** in the form $v = a\mathbf{i} + b\mathbf{j}$ whose magnitude and direction angle are given.

32. $\|\mathbf{v}\| = 15; \theta = 150°$

33. $\|\mathbf{v}\| = 22; \theta = 315°$

34. $\|\mathbf{v}\| = 7; \theta = 300°$

35. $\|\mathbf{v}\| = 10; \theta = 100°$ (Round a and b to two decimal places.)

36. $\|\mathbf{v}\| = \dfrac{2}{3}; \theta = 212°$ (Round a and b to two decimal places.)

37. An outfielder releases a baseball with a speed of 100 feet per second at an angle of 30° with the horizontal. Express the velocity of the ball as a vector in terms of $\mathbf{i}$ and $\mathbf{j}$.

38. An airplane approaches a runway at 180 miles per hour at an angle of 6° with the runway. Express the velocity of the plane as a vector in terms of $\mathbf{i}$ and $\mathbf{j}$. Round a and b to two decimal places.

39. The wind is blowing at a speed of 27 mph in a direction of N 50° E. Express the velocity of the wind as a vector in terms of $\mathbf{i}$ and $\mathbf{j}$.

40. The wind is blowing at a speed of 40 mph in a direction of S 20° W. Express the velocity of the wind as a vector in terms of $\mathbf{i}$ and $\mathbf{j}$.

41. An airplane was heading due east at 280 mph in still air and encountered a 30 mph headwind blowing in the direction N 40° W. Determine the resulting ground speed of the plane and its new bearing. Round the ground speed to the nearest hundredth.

42. An airplane was heading due west at 280 mph in still air and encountered a 50 mph tailwind blowing in the direction N 40° W. Determine the ground speed of the plane and its new bearing. Round the ground speed to the nearest hundredth.

43. The forces $\mathbf{F_1} = -5\mathbf{i} + 4\mathbf{j}$ and $\mathbf{F_2} = 6\mathbf{i} - 11\mathbf{j}$ are acting on an object. What additional force is required for the object to be in static equilibrium?

44. The forces $\mathbf{F_1} = 8\mathbf{i} - 11\mathbf{j}$, $\mathbf{F_2} = -12\mathbf{i} - 13\mathbf{j}$, and $\mathbf{F_3} = -7\mathbf{i} + 8\mathbf{j}$ are acting on an object. What additional force is required for the object to be in static equilibrium?

45. At the county Fourth of July fair, a local strong man challenges contestants two at a time to a tug-of-war contest. Contestant A can tug with a force of 200 pounds. Contestant B can tug with a force of 350 pounds. The angle between the ropes of the two contestants is 40°. With how much force must the local strong man tug so that the rope does not move?

46. Two tugboats are towing a Naval ship out of port and into the open sea. One tugboat exerts a force of $\|\mathbf{F_1}\| = 1200$ pounds in a direction N 45° W. The other tugboat pulls with a force of $\|\mathbf{F_2}\| = 800$ pounds in a direction S 60° W. Find the magnitude of the resultant force and bearing of the Naval ship.

47. Two tugboats are towing a grain barge out of dangerous shallow water. One tugboat exerts a force of $\|\mathbf{F_1}\| = 500$ pounds in a direction N 25° E. The other tugboat pulls with a force of $\|\mathbf{F_2}\| = 900$ pounds in a direction S 20° E. Find the magnitude of the resultant force and bearing of the grain barge.

5.5 The Dot Product

THINGS TO KNOW

Before working through this section, be sure that you are
familiar with the following concepts:

VIDEO   ANIMATION   INTERACTIVE

You Try It
1. Understanding Vectors in Terms of **i** and **j** (Section 5.4)

You Try It
2. Finding a Unit Vector (Section 5.4)

You Try It
3. Determining the Direction Angle of a Vector (Section 5.4)

You Try It
4. Representing a Vector in Terms of **i** and **j** Given Its Magnitude and Direction (Section 5.4)

OBJECTIVES

1 Understanding the Dot Product and Its Properties

2 Using the Dot Product to Determine the Angle between Two Vectors

3 Using the Dot Product to Determine If Two Vectors Are Orthogonal or Parallel

4 Decomposing a Vector into Two Orthogonal Vectors

5 Solving Applications Involving Forces on an Inclined Plane

6 Solving Applications Involving Work

In Section 5.4 we introduced three different ways to represent vectors. A vector **v** can be represented geometrically by an arrow of finite length in a plane, it can be represented using the component notation $\mathbf{v} = \langle a, b \rangle$, or it can be represented using the unit vectors **i** and **j** as $\mathbf{v} = a\mathbf{i} + b\mathbf{j}$. We will use this third representation of vectors throughout this section.

OBJECTIVE 1   UNDERSTANDING THE DOT PRODUCT AND ITS PROPERTIES

In Section 5.4 we learned how to add and subtract two vectors. We also learned how to multiply a vector by a scalar. We now introduce an operation called the **dot product** of two vectors. The dot product of two vectors results in a scalar. As you will see at the end of this section, the dot product is useful in many applications.

Definition   The Dot Product of Two Vectors

If $\mathbf{u} = a\mathbf{i} + b\mathbf{j}$ and $\mathbf{v} = c\mathbf{i} + d\mathbf{j}$, then the **dot product** is $\mathbf{u} \cdot \mathbf{v} = ac + bd.$

Note that the dot product is often referred to as the **inner product** or the **scalar product** of two vectors.

Example 1 Determining the Dot Product of Two Vectors

If $\mathbf{u} = -3\mathbf{i} + 5\mathbf{j}$ and $\mathbf{v} = 7\mathbf{i} - 4\mathbf{j}$, then find $\mathbf{u} \cdot \mathbf{v}$ and $\mathbf{v} \cdot \mathbf{u}$.

Solution $\mathbf{u} \cdot \mathbf{v} = (-3)(7) + (5)(-4) = -21 - 20 = -41$

$\mathbf{v} \cdot \mathbf{u} = (7)(-3) + (-4)(5) = -21 - 20 = -41$

It is no coincidence that the values of $\mathbf{u} \cdot \mathbf{v}$ and $\mathbf{v} \cdot \mathbf{u}$ in Example 1 are equal. In fact, for any two vectors $\mathbf{u}$ and $\mathbf{v}$, $\mathbf{u} \cdot \mathbf{v} = \mathbf{v} \cdot \mathbf{u}$. This is just one of five properties of the dot product.

Dot Product Properties

If $\mathbf{u}$, $\mathbf{v}$, and $\mathbf{w}$ are vectors and if k is a scalar, then the following properties are true.

1. $\mathbf{u} \cdot \mathbf{v} = \mathbf{v} \cdot \mathbf{u}$ 2. $\mathbf{u} \cdot (\mathbf{v} + \mathbf{w}) = \mathbf{u} \cdot \mathbf{v} + \mathbf{u} \cdot \mathbf{w}$ 3. $k(\mathbf{u} \cdot \mathbf{v}) = (k\mathbf{u}) \cdot \mathbf{v} = \mathbf{u} \cdot (k\mathbf{v})$

4. $\mathbf{0} \cdot \mathbf{v} = 0$ 5. $\mathbf{v} \cdot \mathbf{v} = \|\mathbf{v}\|^2$

My video summary   ⊙ **Example 2** Finding Dot Products

If $\mathbf{u} = 4\mathbf{i} + 6\mathbf{j}$, $\mathbf{v} = -2\mathbf{i} + 8\mathbf{j}$, and $\mathbf{w} = -3\mathbf{i} - \mathbf{j}$, then find each of the following.

a. $\mathbf{u} \cdot \mathbf{v}$ b. $\mathbf{u} \cdot (\mathbf{v} + \mathbf{w})$ c. $\mathbf{u} \cdot (-5\mathbf{v})$ d. $\|\mathbf{w}\|^2$

Solution

a. $\mathbf{u} \cdot \mathbf{v} = (4)(-2) + (6)(8) = -8 + 48 = 40$

b. $\mathbf{u} \cdot (\mathbf{v} + \mathbf{w})$ Write the original expression.

$\quad = \mathbf{u} \cdot \mathbf{v} + \mathbf{u} \cdot \mathbf{w}$ Use Property 2.

$\quad = 40 + (4)(-3) + (6)(-1)$ Use the result from part a and determine $\mathbf{u} \cdot \mathbf{w}$.

$\quad = 22$ Simplify.

c. $\mathbf{u} \cdot (-5\mathbf{v})$ Write the original expression.

$\quad = -5(\mathbf{u} \cdot \mathbf{v})$ Use Property 3.

$\quad = -5(40)$ Use the result from part a.

$\quad = -200$ Simplify.

d. $\|\mathbf{w}\|^2$ Write the original expression.

$\quad = \mathbf{w} \cdot \mathbf{w}$ Use Property 5.

$\quad = (-3)(-3) + (-1)(-1)$ Substitute $a = -3$, $c = -3$, $b = -1$, and $d = -1$ into the dot product formula.

$\quad = 10$ Simplify

You Try It Work through this You Try It problem.

Work Exercises 1–8 in this textbook or in the MyMathLab Study Plan.

OBJECTIVE 2 USING THE DOT PRODUCT TO DETERMINE THE ANGLE BETWEEN TWO VECTORS

One of the important applications of the dot product is that it is useful in determining the angle between two non-zero vectors. Suppose that $\mathbf{u}$ and $\mathbf{v}$ are non-zero vectors that share the same initial point. Let θ be the angle between the two vectors such that $0 \leq \theta \leq 180°$. We can determine the angle between the two vectors using the following formula.

The Angle between Two Vectors

If $\mathbf{u}$ and $\mathbf{v}$ are non-zero vectors and if θ is the angle between $\mathbf{u}$ and $\mathbf{v}$, then $\cos \theta = \dfrac{\mathbf{u} \cdot \mathbf{v}}{\|\mathbf{u}\|\|\mathbf{v}\|}$.

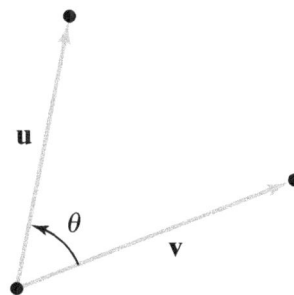

To prove the formula $\cos \theta = \dfrac{\mathbf{u} \cdot \mathbf{v}}{\|\mathbf{u}\|\|\mathbf{v}\|}$, start by coinciding the initial point of vector $-\mathbf{v}$ with the terminal point of $\mathbf{u}$. The angle between $\mathbf{u}$ and $-\mathbf{v}$ is θ because $\mathbf{v}$ and $-\mathbf{v}$ are parallel and alternate interior angles are congruent. See Figure 71a. The vector that extends from the initial point of $\mathbf{u}$ to the terminal point of $-\mathbf{v}$ is $\mathbf{u} - \mathbf{v}$. See Figure 71b. The triangle formed in Figure 71b has sides of lengths $\|\mathbf{u}\|$, $\|-\mathbf{v}\| = \|\mathbf{v}\|$, and $\|\mathbf{u} - \mathbf{v}\|$. See Figure 71c.

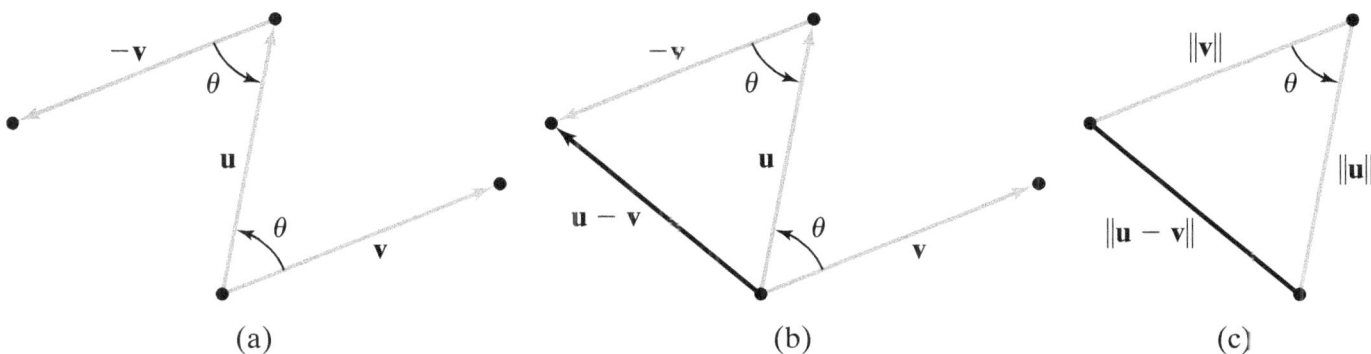

(a)   (b)   (c)

Figure 71

My video summary ◎ We use the triangle in Figure 71c and the Law of Cosines to get $\|\mathbf{u} - \mathbf{v}\|^2 = \|\mathbf{u}\|^2 + \|\mathbf{v}\|^2 - 2\|\mathbf{u}\|\|\mathbf{v}\|\cos \theta$. Carefully watch this video to see how to use dot product properties to rewrite this equation as $\cos \theta = \dfrac{\mathbf{u} \cdot \mathbf{v}}{\|\mathbf{u}\|\|\mathbf{v}\|}$.

My video summary ◎ **Example 3** Determining the Angle between Two Vectors

Determine the angle between each pair of vectors. Give the angle in degrees rounded to the nearest hundredth of a degree.

a. $\mathbf{u} = \mathbf{i} + 4\mathbf{j}$, $\mathbf{v} = -2\mathbf{i} + 5\mathbf{j}$ **b.** $\mathbf{u} = 3\mathbf{i} - 2\mathbf{j}$, $\mathbf{v} = 4\mathbf{i} + 6\mathbf{j}$

Solution

a. To determine the angle between $\mathbf{u} = \mathbf{i} + 4\mathbf{j}$ and $\mathbf{v} = -2\mathbf{i} + 5\mathbf{j}$, we use the formula $\cos\theta = \dfrac{\mathbf{u} \cdot \mathbf{v}}{\|\mathbf{u}\|\,\|\mathbf{v}\|}$.

$$\cos\theta = \frac{(1)(-2) + (4)(5)}{\sqrt{17}\sqrt{29}} = \frac{18}{\sqrt{493}}$$

To find θ, we use a calculator set in degree mode and use the inverse cosine function. Recall that the range of the inverse cosine function is $0 \leq \theta \leq 180°$.

$$\theta = \cos^{-1}\left(\frac{18}{\sqrt{493}}\right) \approx 35.84°$$

b. To determine the angle between $\mathbf{u} = 3\mathbf{i} - 2\mathbf{j}$ and $\mathbf{v} = 4\mathbf{i} + 6\mathbf{j}$, we use the formula $\cos\theta = \dfrac{\mathbf{u} \cdot \mathbf{v}}{\|\mathbf{u}\|\,\|\mathbf{v}\|}$.

$$\cos\theta = \frac{(3)(4) + (-2)(6)}{\sqrt{13}\sqrt{52}} = \frac{0}{\sqrt{676}} = 0$$

The only angle on the interval $0 \leq \theta \leq 180°$ such that $\cos\theta = 0$ is $\theta = 90°$.

Note that if $\mathbf{u}$ and $\mathbf{v}$ are non-zero vectors, then we can multiply both sides of the equation $\cos\theta = \dfrac{\mathbf{u} \cdot \mathbf{v}}{\|\mathbf{u}\|\,\|\mathbf{v}\|}$ by $\|\mathbf{u}\|\,\|\mathbf{v}\|$ to obtain the equivalent equation $\|\mathbf{u}\|\,\|\mathbf{v}\| \cos\theta = \mathbf{u} \cdot \mathbf{v}$. This is an alternate form of the dot product.

Definition   The Alternate Form of the Dot Product of Two Vectors

If $\mathbf{u}$ and $\mathbf{v}$ are non-zero vectors, then the **alternate form of the dot product** is

$$\mathbf{u} \cdot \mathbf{v} = \|\mathbf{u}\|\,\|\mathbf{v}\| \cos\theta.$$

You Try It   Work through this You Try It problem.

Work Exercises 9–13 in this textbook or in the MyMathLab Study Plan.

OBJECTIVE 3   USING THE DOT PRODUCT TO DETERMINE IF TWO VECTORS ARE ORTHOGONAL OR PARALLEL

ORTHOGONAL VECTORS

In Example 3b, we see that the angle between the vectors $\mathbf{u} = 3\mathbf{i} - 2\mathbf{j}$ and $\mathbf{v} = 4\mathbf{i} + 6\mathbf{j}$ is $\theta = 90°$. When the angle between two non-zero vectors is $\theta = 90°$, then the vectors are said to be **orthogonal** vectors.

Definition   Orthogonal Vectors

Two non-zero vectors $\mathbf{u}$ and $\mathbf{v}$ are **orthogonal** if the angle between them is $\theta = 90°$.

Because we know from the definition of orthogonal vectors that the angle between **u** and **v** is $\theta = 90°$, then by the alternate definition of the dot product we get $\mathbf{u} \cdot \mathbf{v} = \|\mathbf{u}\|\,\|\mathbf{v}\|\cos 90° = \|\mathbf{u}\|\,\|\mathbf{v}\|(0) = 0$. This gives us the following test to determine if two non-zero vectors are orthogonal.

Test for Orthogonal Vectors

Two non-zero vectors **u** and **v** are orthogonal if and only if $\mathbf{u} \cdot \mathbf{v} = 0$.

Note: Every vector is orthogonal to the zero vector.

Example 4 Finding a Component of an Orthogonal Vector

Determine the value of b so that the vectors $\mathbf{u} = \mathbf{i} + b\mathbf{j}$ and $\mathbf{v} = 3\mathbf{i} + 10\mathbf{j}$ are orthogonal.

Solution We know that when two vectors are orthogonal, then the dot product must be zero. Thus, we are looking for the value of b such that $\mathbf{u} \cdot \mathbf{v} = 0$. The dot product of **u** and **v** is $\mathbf{u} \cdot \mathbf{v} = (1)(3) + (b)(10) = 3 + 10b$.

We now set the expression $3 + 10b$ equal to zero to obtain $b = -\dfrac{3}{10}$.

We can verify that the vectors $\mathbf{u} = \mathbf{i} - \dfrac{3}{10}\mathbf{j}$ and $\mathbf{v} = 3\mathbf{i} + 10\mathbf{j}$ are orthogonal by showing that $\mathbf{u} \cdot \mathbf{v} = 0$.

$$\mathbf{u} \cdot \mathbf{v} = (1)(3) + \left(-\frac{3}{10}\right)(10) = 3 - 3 = 0$$

You Try It Work through this You Try It problem.

Work Exercises 14 and 15 in this textbook or in the MyMathLab Study Plan.

PARALLEL VECTORS

Recall that two non-zero vectors **u** and **v** are parallel vectors if there is a non-zero scalar k such that $\mathbf{v} = k\mathbf{u}$. To establish a test to determine if two vectors are parallel, we can examine the angle between them. We start by finding the dot product.

$\mathbf{u} \cdot \mathbf{v}$ Start with the dot product of **u** and **v**.

$\mathbf{u} \cdot (k\mathbf{u})$ Vector **v** must be a scalar multiple of **u**.

$= k(\mathbf{u} \cdot \mathbf{u})$ Use Dot Product Property 3.

$= k\|\mathbf{u}\|^2$ Use Dot Product Property 5.

Therefore, the cosine of the angle between parallel vectors **u** and $\mathbf{v} = k\mathbf{u}$ is

$$\cos\theta = \frac{\mathbf{u} \cdot k\mathbf{u}}{\|\mathbf{u}\|\,\|k\mathbf{u}\|} = \frac{k\|\mathbf{u}\|^2}{|k|\,\|\mathbf{u}\|\,\|\mathbf{u}\|} = \frac{k\|\mathbf{u}\|^2}{|k|\,\|\mathbf{u}\|^2} = \frac{k}{|k|}.$$

If $k > 0$, then $\cos\theta = \dfrac{k}{|k|} = \dfrac{k}{k} = 1$ and hence $\theta = 0°$.

If $k < 0$, then $\cos\theta = \dfrac{k}{|k|} = \dfrac{k}{-k} = -1$ and hence $\theta = 180°$.

This gives us the following test to determine if two non-zero vectors are parallel.

Test for Parallel Vectors

Two non-zero vectors **u** and **v** are parallel if $\dfrac{\mathbf{u} \cdot \mathbf{v}}{\|\mathbf{u}\|\,\|\mathbf{v}\|} = 1$ or $\dfrac{\mathbf{u} \cdot \mathbf{v}}{\|\mathbf{u}\|\,\|\mathbf{v}\|} = -1$.

Note that the angle between parallel vectors is $\theta = 0°$ if $\dfrac{\mathbf{u} \cdot \mathbf{v}}{\|\mathbf{u}\|\,\|\mathbf{v}\|} = 1$.

The angle between parallel vectors is $\theta = 180°$ if $\dfrac{\mathbf{u} \cdot \mathbf{v}}{\|\mathbf{u}\|\,\|\mathbf{v}\|} = -1$.

My video summary

◉ **Example 5** Determining If Two Vectors Are Orthogonal, Parallel, or Neither

Determine if $\mathbf{u} = -\dfrac{1}{2}\mathbf{i} - \mathbf{j}$ and $\mathbf{v} = 2\mathbf{i} + 4\mathbf{j}$ are orthogonal, parallel, or neither.

Solution We start by determining the dot product $\mathbf{u} \cdot \mathbf{v}$. The dot product of **u** and **v** is

$$\mathbf{u} \cdot \mathbf{v} = \left(-\frac{1}{2}\right)(2) + (-1)(4) = -1 - 4 = -5$$

The vectors are not orthogonal because the value of $\mathbf{u} \cdot \mathbf{v}$. is non-zero.

Next, determine the value of $\dfrac{\mathbf{u} \cdot \mathbf{v}}{\|\mathbf{u}\|\,\|\mathbf{v}\|}$.

$$\frac{\mathbf{u} \cdot \mathbf{v}}{\|\mathbf{u}\|\,\|\mathbf{v}\|} = \frac{-5}{\sqrt{\left(\frac{-1}{2}\right)^2 + (-1)^2}\sqrt{2^2 + 4^2}} = \frac{-5}{\sqrt{\frac{5}{4}}\sqrt{20}} = \frac{-5}{\sqrt{\frac{100}{4}}} = \frac{-5}{\sqrt{25}} = \frac{-5}{5} = -1$$

Therefore, the vectors are parallel. Furthermore, the angle between them is $\theta = 180°$ because $\cos\theta = \dfrac{\mathbf{u} \cdot \mathbf{v}}{\|\mathbf{u}\|\,\|\mathbf{v}\|} = -1$.

You Try It Work through this You Try It problem.

Work Exercises 16–19 in this textbook or in the MyMathLab Study Plan.

OBJECTIVE 4 DECOMPOSING A VECTOR INTO TWO ORTHOGONAL VECTORS

An important use of dot products is the idea of **projections**. Consider the non-zero vectors **v** and **w** that share the same initial point P. See Figure 72a. Draw a line segment from the terminal point of **v** that is perpendicular to **w**. Label this point Q, thus creating a right triangle. See Figure 72b. The green vector $\overrightarrow{PQ}$, which is the side of the right triangle adjacent to θ in Figure 72c, is called the **vector projection of v onto w**. We will denote the vector projection of **v** onto **w** as $\text{proj}_{\mathbf{w}}\mathbf{v}$.

Our goal is to represent the vector $\text{proj}_{\mathbf{w}}\mathbf{v}$ in terms of **v** and **w**. We start by determining the magnitude of $\text{proj}_{\mathbf{w}}\mathbf{v}$. Looking at the right triangle in Figure 72c, we see that the length of the hypotenuse of the right triangle is the magnitude of **v** and the

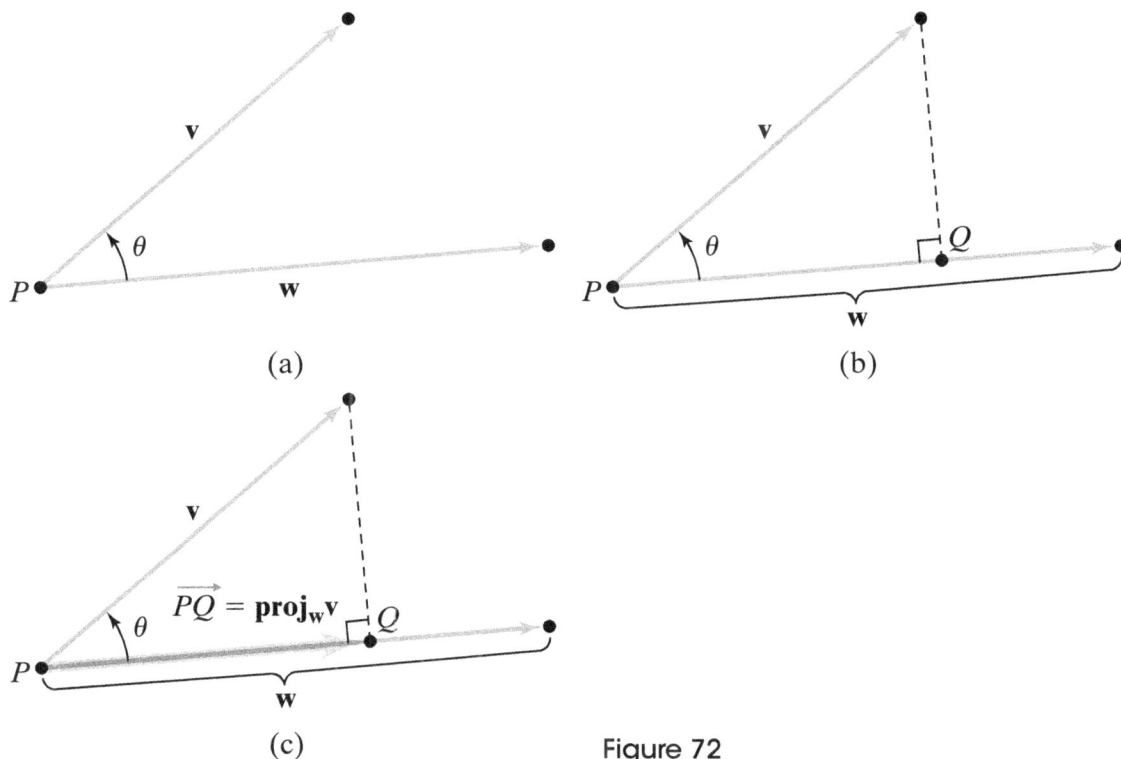

(a)

(b)

(c)

Figure 72

length of the side adjacent to θ is the magnitude of $\text{proj}_w\mathbf{v}$. Using the right triangle definition of the cosine function, we get

$$\cos \theta = \frac{adj}{hyp} = \frac{\|\text{proj}_w\mathbf{v}\|}{\|\mathbf{v}\|}.$$

We can multiply both sides of this equation by $\|\mathbf{v}\|$ because $\mathbf{v}$ is a non-zero vector. The result is $\|\text{proj}_w\mathbf{v}\| = \|\mathbf{v}\| \cos \theta$, which is called the **scalar component of $\mathbf{v}$ in the direction of $\mathbf{w}$**.

Definition The Scalar Component of $\mathbf{v}$ in the Direction of $\mathbf{w}$

If θ is the angle between $\mathbf{v}$ and $\mathbf{w}$, then the **scalar component of $\mathbf{v}$ in the direction of $\mathbf{w}$** is $\|\text{proj}_w\mathbf{v}\| = \|\mathbf{v}\| \cos \theta$.

Now that we know the magnitude of $\text{proj}_w\mathbf{v}$, we can develop a formula that represents $\text{proj}_w\mathbf{v}$ in terms of $\mathbf{v}$ and $\mathbf{w}$. Recall that the unit vector in the direction of $\mathbf{w}$ is defined as $\dfrac{\mathbf{w}}{\|\mathbf{w}\|}$. The vector $\text{proj}_w\mathbf{v}$ is equal to its magnitude (given by $\|\mathbf{v}\| \cos \theta$) times the unit vector in the direction of $\mathbf{w}$ (given by $\dfrac{\mathbf{w}}{\|\mathbf{w}\|}$). We can thus represent the vector $\text{proj}_w\mathbf{v}$ as

$$\text{proj}_w\mathbf{v} = \|\mathbf{v}\| \cos \theta \frac{\mathbf{w}}{\|\mathbf{w}\|}.$$

We will multiply the right-hand side of $\text{proj}_{\mathbf{w}}\mathbf{v} = \|\mathbf{v}\| \cos\theta \dfrac{\mathbf{w}}{\|\mathbf{w}\|}$ by $1 = \dfrac{\|\mathbf{w}\|}{\|\mathbf{w}\|}$ and then use the alternate definition of the dot product to represent the vector in terms of $\mathbf{v}$ and $\mathbf{w}$.

$$\text{proj}_{\mathbf{w}}\mathbf{v} = \|\mathbf{v}\| \cos\theta \frac{\mathbf{w}}{\|\mathbf{w}\|}$$

Start with the vector representation of $\text{proj}_{\mathbf{w}}\mathbf{v} = \|\mathbf{v}\| \cos\theta \dfrac{w}{\|\mathbf{w}\|}$.

$$= \|\mathbf{v}\| \cos\theta \frac{\mathbf{w}}{\|\mathbf{w}\|} \frac{\|\mathbf{w}\|}{\|\mathbf{w}\|}$$

Multiply the right-hand side by $1 = \dfrac{\|\mathbf{w}\|}{\|\mathbf{w}\|}$.

$$= \|\mathbf{v}\|\|\mathbf{w}\| \cos\theta \frac{\mathbf{w}}{\|\mathbf{w}\|^2}$$

Rearrange the order of the factors and write $\|\mathbf{w}\|\|\mathbf{w}\|$ as $\|\mathbf{w}\|^2$.

$$= \frac{\mathbf{v} \cdot \mathbf{w}}{\|\mathbf{w}\|^2} \mathbf{w}$$

Substitute $\mathbf{v} \cdot \mathbf{w}$ for $\|\mathbf{v}\|\|\mathbf{w}\| \cos\theta$ using the alternate definition of the dot product.

We now define $\text{proj}_{\mathbf{w}}\mathbf{v}$, the **vector projection of v onto w**, in terms of $\mathbf{v}$ and $\mathbf{w}$.

Definition The Vector Projection of **v** onto **w**

Given two non-zero vectors $\mathbf{v}$ and $\mathbf{w}$, the **vector projection of v onto w** is given by $\text{proj}_{\mathbf{w}}\mathbf{v} = \dfrac{\mathbf{v} \cdot \mathbf{w}}{\|\mathbf{w}\|^2} \mathbf{w}$.

Example 6 Determining the Vector Projection of **v** onto **w**

If $\mathbf{v} = \mathbf{i} + 2\mathbf{j}$ and $\mathbf{w} = 4\mathbf{i} + \mathbf{j}$, determine the vector $\text{proj}_{\mathbf{w}}\mathbf{v}$.

Solution To determine the vector $\text{proj}_{\mathbf{w}}\mathbf{v}$, we use the formula $\text{proj}_{\mathbf{w}}\mathbf{v} = \dfrac{\mathbf{v} \cdot \mathbf{w}}{\|\mathbf{w}\|^2}\mathbf{w}$.

$$\text{proj}_{\mathbf{w}}\mathbf{v} = \frac{\mathbf{v} \cdot \mathbf{w}}{\|\mathbf{w}\|^2} \mathbf{w} = \frac{(1)(4) + (2)(1)}{(4)^2 + (1)^2}(4\mathbf{i} + \mathbf{j}) = \frac{6}{17}(4\mathbf{i} + \mathbf{j}) = \frac{24}{17}\mathbf{i} + \frac{6}{17}\mathbf{j}$$

The vectors $\mathbf{v}$, $\mathbf{w}$, and $\text{proj}_{\mathbf{w}}\mathbf{v}$ are shown in Figure 73.

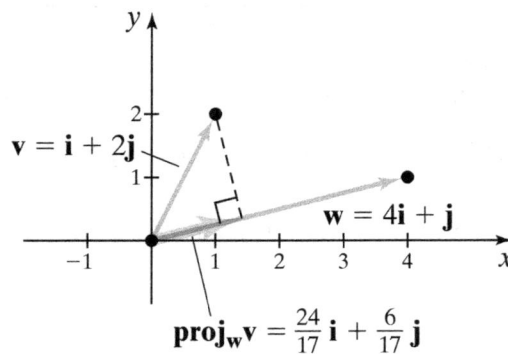

$$\text{proj}_{\mathbf{w}}\mathbf{v} = \frac{24}{17}\mathbf{i} + \frac{6}{17}\mathbf{j}$$

Figure 73

Every non-zero vector can be written as the sum of two orthogonal vectors. To illustrate this, consider the vectors $\mathbf{v}$, $\mathbf{w}$, $\mathbf{v}_1 = \text{proj}_\mathbf{w}\mathbf{v}$, and $\mathbf{v}_2$, where $\mathbf{v}_1$ is orthogonal to $\mathbf{v}_2$. See Figure 74.

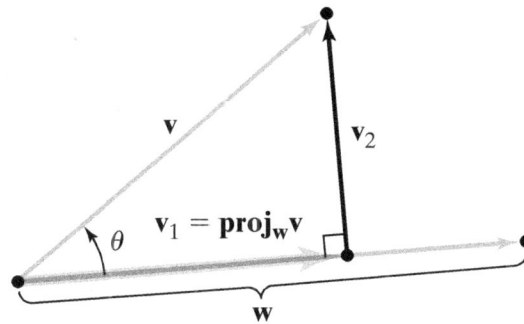

Figure 74

Using the addition of vectors represented geometrically, we see that $\mathbf{v} = \mathbf{v}_1 + \mathbf{v}_2$, where $\mathbf{v}_1 = \text{proj}_\mathbf{w}\mathbf{v}$ is parallel to $\mathbf{w}$ and $\mathbf{v}_2$ is orthogonal to $\mathbf{w}$. Note that $\mathbf{v}_2 = \mathbf{v} - \mathbf{v}_1$. The vector sum $\mathbf{v} = \mathbf{v}_1 + \mathbf{v}_2$ is called the **vector decomposition of $\mathbf{v}$ into orthogonal components**.

Definition   The Vector Decomposition of $\mathbf{v}$ into Orthogonal Components

Let $\mathbf{v}$ and $\mathbf{w}$ be non-zero vectors. Then $\mathbf{v}$ can be written as the sum of orthogonal vectors $\mathbf{v}_1$ and $\mathbf{v}_2$, where $\mathbf{v}_1$ is parallel to $\mathbf{w}$ and $\mathbf{v}_2$ is orthogonal to $\mathbf{w}$ such that

$$\mathbf{v}_1 = \text{proj}_\mathbf{w}\mathbf{v} = \frac{\mathbf{v}\cdot\mathbf{w}}{\|\mathbf{w}\|^2}\,\mathbf{w} \text{ and } \mathbf{v}_2 = \mathbf{v} - \mathbf{v}_1.$$

The process of expressing $\mathbf{v}$ as the sum of $\mathbf{v}_1$ and $\mathbf{v}_2$ is called the **vector decomposition of $\mathbf{v}$ into orthogonal components $\mathbf{v}_1$ and $\mathbf{v}_2$.**

My video summary   ⊙ **Example 7** Decomposing a Vector into Orthogonal Components

Let $\mathbf{v} = \mathbf{i} + 2\mathbf{j}$ and $\mathbf{w} = 4\mathbf{i} - \mathbf{j}$. Determine the vector decomposition of $\mathbf{v}$ into orthogonal components $\mathbf{v}_1$ and $\mathbf{v}_2$, where $\mathbf{v}_1$ is parallel to $\mathbf{w}$ and $\mathbf{v}_2$ is orthogonal to $\mathbf{w}$.

Solution   In Example 6 we determined that $\text{proj}_\mathbf{w}\mathbf{v} = \frac{\mathbf{v}\cdot\mathbf{w}}{\|\mathbf{w}\|^2}\,\mathbf{w} = \frac{24}{17}\mathbf{i} + \frac{6}{17}\mathbf{j}$.

Therefore, $\mathbf{v}_1 = \frac{24}{17}\mathbf{i} + \frac{6}{17}\mathbf{j}$ and $\mathbf{v}_2 = \mathbf{v} - \mathbf{v}_1 = (\mathbf{i} + 2\mathbf{j}) - \left(\frac{24}{17}\mathbf{i} + \frac{6}{17}\mathbf{j}\right) =$

$\left(\frac{17}{17}\mathbf{i} + \frac{34}{17}\mathbf{j}\right) - \left(\frac{24}{17}\mathbf{i} + \frac{6}{17}\mathbf{j}\right) = -\frac{7}{17}\mathbf{i} + \frac{28}{17}\mathbf{j}$. See Figure 75.

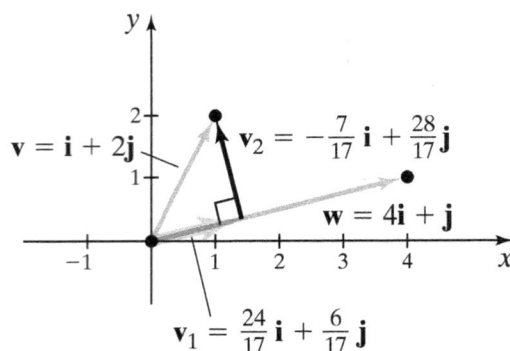

Figure 75

Note that $v_1 \cdot v_2 = \left(\dfrac{24}{17}\right)\left(-\dfrac{7}{17}\right) + \left(\dfrac{6}{17}\right)\left(\dfrac{28}{17}\right) = -\dfrac{168}{289} + \dfrac{168}{289} = 0$, which verifies that v_1 and v_2 are orthogonal.

You Try It Work through this You Try It problem.

Work Exercises 20–22 in this textbook or in the MyMathLab Study Plan.

OBJECTIVE 5 SOLVING APPLICATIONS INVOLVING FORCES ON AN INCLINED PLANE

Projections and the decomposition of vectors into orthogonal components can be used to solve applications when the force of gravity is acting on an object that is on an inclined plane, such as a hill or a ramp. The steeper the incline, the more force is required to pull or push the object up the incline.

Suppose that an object is placed on a ramp that has an incline angle of α. Let **w** represent a vector parallel to the incline. Let **G** represent the force due to gravity pulling the object directly downward. There are two other forces involved, $\mathbf{F}_1$ and $\mathbf{F}_2$, where $\mathbf{F}_1$ represents the force exerted by the object pushing down the incline and $\mathbf{F}_2$ represents the force of the object pushing against the ramp at a right angle. See Figure 76.

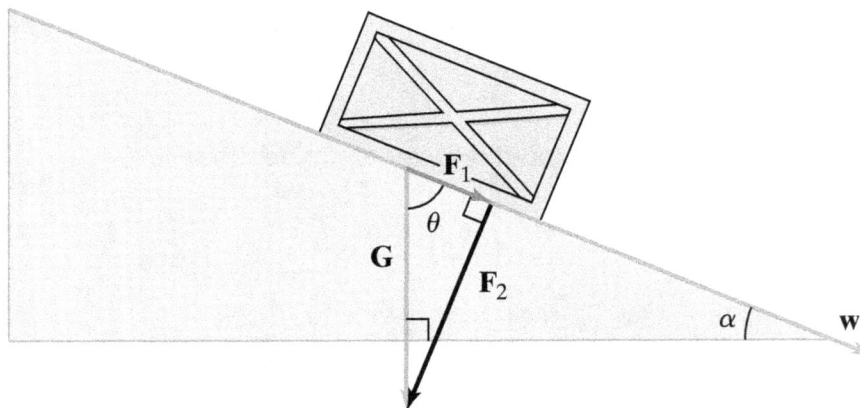

Figure 76

Note that a portion of the vector **G** forms a right angle with the base of the incline. The right triangle formed contains the complementary angles α and θ. The force of gravity **G** can be written as the vector decomposition of the orthogonal components $\mathbf{F}_1$ and $\mathbf{F}_2$. Thus, $\mathbf{G} = \mathbf{F}_1 + \mathbf{F}_2$, where $\mathbf{F}_1 = \text{proj}_{\mathbf{w}}\mathbf{G}$. Therefore, the force required to prevent the object from sliding down the ramp must be equal in magnitude to F_1. The magnitude of F_1 is precisely the scalar component of G in the direction of w, which is

$$\|F_1\| = \|\text{proj}_w G\| = \|G\| \cos \theta.$$

My video summary   ⊙ **Example 8** Determining the Magnitude of Forces of an Object on an Incline

A 200-pound object is placed on a ramp that is inclined at 22°. What is the magnitude of the force needed to hold the box in a stationary position to prevent the box from sliding down the ramp? What is the magnitude of the force pushing against the ramp?

Solution We can draw a diagram with vectors w, G, F_1, and F_2, where w is a vector positioned parallel to the incline, $G = -200j$ is the force due to gravity, $F_1 = \text{proj}_w G$ is the force pushing down the incline, and F_2 is the force of the object pushing against the ramp at a right angle. The angle between G and w is 68°. See Figure 77.

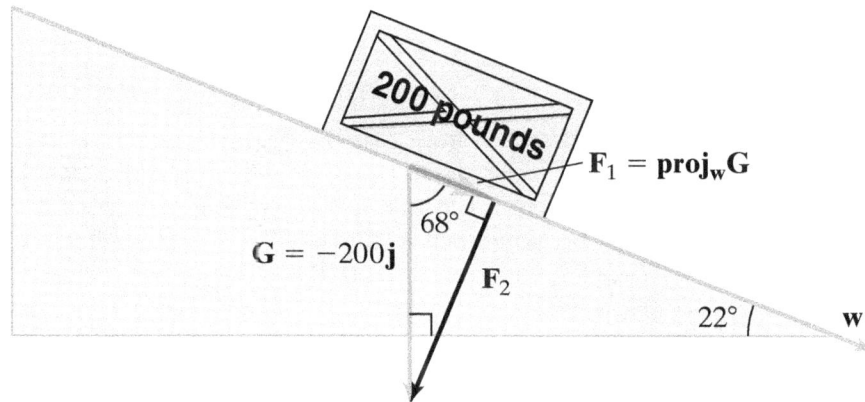

Figure 77

The force needed to hold the object in a stationary position is the magnitude of F_1, which is the scalar component of G in the direction of w. So, the magnitude of the needed force is

$$\|F_1\| = \|\text{proj}_w G\| = \|G\| \cos \theta = \sqrt{0^2 + (-200)^2} \cos 68° = 200 \cos 68° \approx 74.92 \text{ pounds.}$$

From the Pythagorean Theorem, we get the equation $\|G\|^2 = \|F_1\|^2 + \|F_2\|^2$, which will allow us to solve for $\|F_2\|$.

$\|G\|^2 = \|F_1\|^2 + \|F_2\|^2$	Write the Pythagorean Theorem.
$\|F_2\| = \sqrt{\|G\|^2 - \|F_1\|^2}$	Solve for $\|F_2\|$.
$\|F_2\| = \sqrt{(200)^2 - (74.92)^2}$	Substit $\|G\| = 200$ and $\|F_1\| = 74.92$.
≈ 185.44	Simplify.

Therefore, the magnitude of the force pushing against the ramp is approximately 185.44 pounds.

You Try It Work through this You Try It problem.

Work Exercises 23–26 in this textbook or in the MyMathLab Study Plan.

OBJECTIVE 6   SOLVING APPLICATIONS INVOLVING WORK

In physics, the concept of **work** refers to the transfer of energy from one object to another that causes the object to move a certain distance. So, work is done by a force whenever the force moves an object. The simplest formula for work is $W = Fd$, where F is the magnitude of the force and d is the distance traveled. However, this formula is true only when the force is applied parallel to the line of motion. See Figure 78.

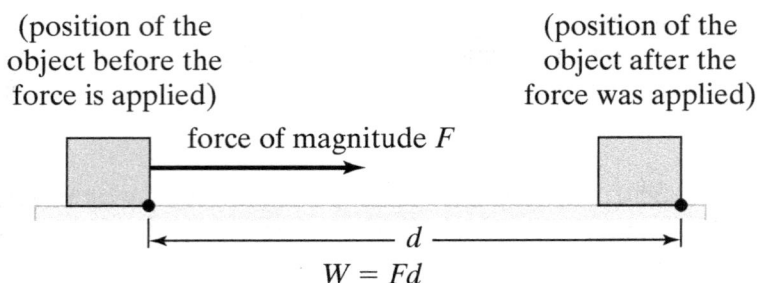

(position of the
object before the
force is applied)

(position of the
object after the
force was applied)

force of magnitude F

d

$W = Fd$

Figure 78

For example, the amount of work required to lift a 10-pound weight vertically 3 feet is

$$W = Fd = (10 \text{ lbs})(3 \text{ feet}) = 30 \text{ foot-pounds}.$$

Note that the units used for work in this scenario is foot-pounds.

Suppose that the constant force applied to an object is not parallel to the line of motion but instead is applied to the object at an angle θ between the object and the horizontal. In this situation, let $\mathbf{F}$ represent the force of an object with a direction angle θ relative to the line of motion. If the object moves from a point P to a point Q, then let $\mathbf{D} = \overrightarrow{PQ}$. The force of $\mathbf{F}$ in the direction of $\mathbf{D}$ is the vector $\text{proj}_\mathbf{D}\mathbf{F}$. See Figure 79.

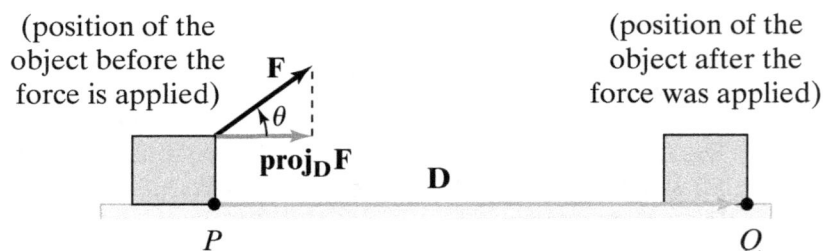

(position of the
object before the
force is applied)

$\mathbf{F}$

θ

$\text{proj}_\mathbf{D}\mathbf{F}$

$\mathbf{D}$

(position of the
object after the
force was applied)

P Q **Figure 79**

The work done by $\mathbf{F}$ is equal to the magnitude of the force in the direction of $\mathbf{D}$ times the distance from P to Q. This gives the following.

$W = $ (Magnitude of the force in the direction of the line of motion)(Distance from P to Q)

$= $ (Scalar Projection of $\mathbf{F}$ onto $\mathbf{D}$)(Magnitude of $\mathbf{D}$)

$= (\|\text{proj}_\mathbf{D}\mathbf{F}\|)(\|\mathbf{D}\|)$ Substitute the appropriate expressions.

$= (\|\mathbf{F}\| \cos\theta)(\|\mathbf{D}\|)$ Use the definition of $\text{proj}_\mathbf{D}\mathbf{F}$.

$= \|\mathbf{F}\|\|\mathbf{D}\| \cos\theta$ Rearrange the order of the factors.

$= \mathbf{F} \cdot \mathbf{D}$ Substitute using the alternate definition of the dot product.

Thus, the work is equal to the dot product of **F** and **D**. If the angle of direction is given, then the most direct way to compute work is to use the alternate form of the dot product $\mathbf{F} \cdot \mathbf{D} = \|\mathbf{F}\|\,\|\mathbf{D}\| \cos \theta$.

Example 9 Calculating Work

A horse is pulling a plow with a force of 400 pounds. The angle between the harness and the ground is 20°. How much work is done to pull the plow 50 feet?

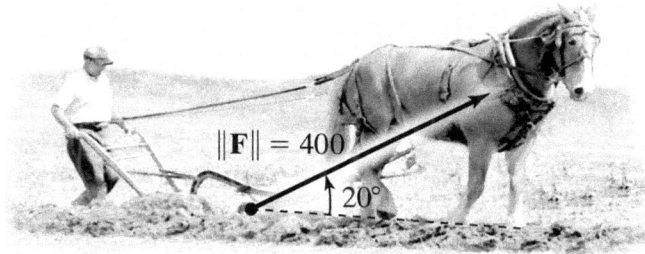

$\|\mathbf{F}\| = 400$
$20°$

Solution We can calculate the work done as follows.

$$W = \mathbf{F} \cdot \mathbf{D}$$

$$= \|\mathbf{F}\|\,\|\mathbf{D}\| \cos \theta$$

$$= (400)(50) \cos 20° \approx 18{,}793.85 \text{ foot-pounds.}$$

You Try It Work through this You Try It problem.

Work Exercises 27–30 in this textbook or in the MyMathLab Study Plan.

5.5 Exercises

In Exercises 1–8, find the value of each scalar given that $\mathbf{u} = 5\mathbf{i} - 2\mathbf{j}$, $\mathbf{v} = 6\mathbf{i} + \mathbf{j}$, and $\mathbf{w} = -\mathbf{i} - 7\mathbf{j}$.

1. $\mathbf{u} \cdot \mathbf{v}$ **2.** $\mathbf{v} \cdot \mathbf{w}$ **3.** $(\mathbf{u} \cdot \mathbf{v})\|\mathbf{w}\|$ **4.** $\mathbf{u} \cdot (\mathbf{v} + \mathbf{w})$

5. $(\mathbf{u} + \mathbf{v}) \cdot \mathbf{w}$ **6.** $(-7\mathbf{u}) \cdot \mathbf{w}$ **7.** $\|\mathbf{w}\|^2$ **8.** $\|\mathbf{u}\|\,\|\mathbf{v}\|$

In Exercises 9–13, determine the angle between **u** and **v**. Give the angle in degrees rounded to the nearest hundredth of a degree.

9. $\mathbf{u} = 5\mathbf{i} + 2\mathbf{j}$, $\mathbf{v} = \mathbf{i} + 4\mathbf{j}$ **10.** $\mathbf{u} = 2\mathbf{i} + 3\mathbf{j}$, $\mathbf{v} = -\mathbf{i} + 2\mathbf{j}$ **11.** $\mathbf{u} = 7\mathbf{i} + 3\mathbf{j}$, $\mathbf{v} = -5\mathbf{i} + 2\mathbf{j}$

12. $\mathbf{u} = -4\mathbf{i} + 4\mathbf{j}$, $\mathbf{v} = 3\mathbf{i} - 6\mathbf{j}$ **13.** $\mathbf{u} = \dfrac{1}{2}\mathbf{i} - \dfrac{1}{3}\mathbf{j}$, $\mathbf{v} = -3\mathbf{i} + 2\mathbf{j}$

14. Determine the value of b so that the vectors $\mathbf{u} = 4\mathbf{i} + b\mathbf{j}$ and $\mathbf{v} = -\mathbf{i} + 2\mathbf{j}$ are orthogonal.

15. Determine the value of a so that the vectors $\mathbf{u} = -5\mathbf{i} - 7\mathbf{j}$ and $\mathbf{v} = a\mathbf{i} + 3\mathbf{j}$ are orthogonal.

In Exercises 16–19, determine if the given vectors are orthogonal, parallel, or neither.

SbS 16. $u = 2i - 3j, v = 6i + 4j$ SbS 17. $u = -i + 2j, v = 2i + 4j$

SbS 18. $u = -i + \sqrt{3}j, v = \sqrt{3}i - 3j$ SbS 19. $u = -\sqrt{2}i + j, v = \sqrt{2}i + 3j$

In Exercises 20–22, non-zero vectors v and w are given. Determine the vector decomposition of v into orthogonal components v_1 and v_2, where v_1 is parallel to w and v_2 is orthogonal to w.

20. $v = -5i - 9j; w = -i + j$

21. $v = 3i + 4j; w = -2i + 7j$

22. $v = \dfrac{1}{2}i - 8j; w = 5i + \dfrac{1}{3}j$

23. How much force is required to keep a 1000-pound block from sliding down an incline of 36°?

24. A mover in a moving truck is using a rope to pull a 400-pound box up a ramp that has an incline of 25°. What is the force needed to hold the box in a stationary position to prevent the box from sliding down the ramp? What is the magnitude of the force pushing against the ramp?

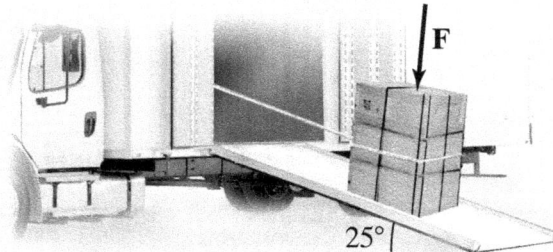

25. A car is being winched out of a ditch at an angle of 40°. Determine the weight of the car if the winch is pulling with a maximum capacity of 1200 pounds.

26. An object weighing 800 pounds is being hauled up a ramp. The object is held still by a force of 350 pounds. Determine the ramp's incline angle.

27. How much work is required to lift a 50-pound weight vertically 4 feet?

28. A horse is pulling a plow with a force of 500 pounds. The angle between the harness and the ground is 25°. How much work is done to pull the plow 60 feet?

29. A car is being winched out of a ditch at an angle of 40°. The winch exerts a constant force of 1000 pounds. What is the total work done by the winch in order to pull the car a horizontal distance of 35 feet?

30. Ten thousand foot-pounds of work was done by a force of 200 pounds to move the object 75 feet. If the force was exerted on the object at an angle θ between the object and the horizontal, then determine the angle θ. Write the angle in degrees rounded to the nearest tenth of a degree.

Glossary

Acronym An acronym is a word (real or made up) formed by abbreviating the name of an organization or a multiword statement.

Acute Angles An acute angle is an angle with measure between 0 and 90 degrees.

Algebraic Expression An algebraic expression consists of one or more terms that may include variables, constants, and operating symbols such as $+$ and $-$.

Alternate Form of the Dot Product of Two Vectors
If $\mathbf{u}$ and $\mathbf{v}$ are non-zero vectors, then the alternate form of the dot product is $\mathbf{u} \cdot \mathbf{v} = \|\mathbf{u}\| \|\mathbf{v}\| \cos \theta$.

Alternate Form of the Law of Cosines If A, B, and C are the measures of the angles of any triangle and if a, b, and c are the lengths of the sides opposite the corresponding angles, then $\cos A = \dfrac{b^2 + c^2 - a^2}{2bc}$, $\cos B = \dfrac{a^2 + c^2 - b^2}{2ac}$, and $\cos C = \dfrac{a^2 + b^2 - c^2}{2ab}$.

Altitude An altitude of a triangle is a line segment from a vertex of a triangle perpendicular to the (possibly extended) opposite side. The length of the altitude is the distance between the vertex and the base.

Argument The argument of a function is the variable, term, or expression on which a function operates.

Central Angle An angle whose vertex is at the center of a circle is called a central angle.

Closed Interval A closed interval is an interval of real numbers that include two endpoints.

Complementary Angles Two angles are called complementary angles if the sum of their measures is 90°
(or $\dfrac{\pi}{2}$ radians).

Concentric Circles Concentric circles are circles that share a common center. When concentric circles intersect, their points of intersection form a hyperbola.

Coterminal Angles Angles in standard position having the same terminal side are called coterminal angles.

Dependent Variable A dependent variable is the variable whose value is determined by the value assumed by an independent variable.

Equation of a Circle in Standard Form Centered at the Origin A circle with radius r centered at the origin has equation $x^2 + y^2 = r^2$.

Even A function $y = f(x)$ is even if for every x in the domain, $f(-x) = f(x)$. Even functions are symmetric about the y-axis.

Even Functions A function f is even if $f(-x) = f(x)$ for all x in the domain of f.

Extraneous Solution An extraneous solution is a solution obtained through algebraic manipulations that is *not* a solution to the original equation.

Four Quadrants The axes of a rectangular coordinate system divide the rectangular coordinate system into four quadrants.

Horizontal Asymptote A horizontal line $y = H$ is a horizontal asymptote of a function f if the values of $f(x)$ approach some fixed number H as the values of x approach ∞ or $-\infty$.

Horizontal Line Test Given the graph of a function f, if every horizontal line intersects the graph of f at most once, then f is a one-to-one function.

Independent Variable An independent variable is the variable whose value determines the value of the other variables.

Inverse Cosine Function The inverse cosine function, denoted as $y = \cos^{-1} x$ is the inverse of $y = \cos x$, $0 \le x \le \pi$.

Inverse Function Let f be a one-to-one function with domain A and range B. Then f^{-1} is the inverse function of f with domain B and range A.

Inverse Sine Function The inverse sine function, denoted as $y = \sin^{-1} x$ is the inverse of $y = \sin x$, $-\dfrac{\pi}{2} \le x \le \dfrac{\pi}{2}$.

Inverse Tangent Function The inverse tangent function, denoted as $y = \tan^{-1} x$ is the inverse of $y = \tan x$, $-\dfrac{\pi}{2} \le x \le \dfrac{\pi}{2}$.

Law of Cosines If A, B, and C are the measures of the angles of any triangle and if a, b, and c are the lengths of the sides opposite the corresponding angles, then $a^2 = b^2 + c^2 - 2bc \cos A$, $b^2 = a^2 + c^2 - 2ac \cos B$, and $c^2 = a^2 + b^2 - 2ab \cos C$.

Law of Sines If A, B, and C are the measures of the angles of any triangle and if a, b, and c are the lengths of the sides opposite the corresponding angles, then $\dfrac{a}{\sin A} = \dfrac{b}{\sin B} = \dfrac{c}{\sin C}$ or $\dfrac{\sin A}{a} = \dfrac{\sin B}{b} = \dfrac{\sin C}{c}$.

Magnitude of a Vector The magnitude of a vector of the form $\mathbf{v} = a\mathbf{i} + b\mathbf{j}$ is $\|\mathbf{v}\| = \sqrt{a^2 + b^2}$

Obtuse Angles An obtuse angle is an angle with measure between 90 and 180 degrees.

Odd Functions A function f is odd if $f(-x) = -f(x)$ for all x in the domain of f.

Origin The origin is the point in the Cartesian plane where the x-axis and y-axis intersect. The origin has coordinates $(0, 0)$.

Periodic Function A function is said to be periodic if there is a positive number P such that $f(x + P) = f(x)$ for all x in the domain of f.

Proportion A proportion is an equation involving two ratios in which the first ratio is equal to the second ratio.

Pythagorean Theorem Pythagorean Theorem states that given any right triangle, the sum of the squares of the lengths of the legs is equal to the square of the length of the hypotenuse.

Quadrantal Angle   A quadrantal angle is an angle in standard position whose terminal side lies on an axis.

Rationalizing a Denominator   The process of removing radicals so that the denominator contains only a rational number is called rationalizing the denominator.

Ray   A ray consists of an endpoint called the vertex and a half-line extending indefinitely in one direction.

Reciprocal   A reciprocal is a number or algebraic expression related to another number or algebraic expression in such a way that when multiplied together their product is 1.

Relative Maximum   When a function $y = f(x)$ changes from increasing to decreasing at a point $(c, f(c))$, then f is said to have a relative maximum at $x = c$.

Relative Minimum   When a function $y = f(x)$ changes from decreasing to increasing at a point $(c, f(c))$, then f is said to have a relative minimum at $x = c$.

Right Angle   A right angle is an angle that has a measure of exactly $\dfrac{\pi}{2}$ radians (or 90°)

Square Root Property   The solution to the quadratic equation $x^2 - c = 0$ or equivalently $x^2 = c$ is $x^2 = \pm\sqrt{c}$.

Standard Position   An angle is in standard position if the vertex is at the origin of a rectangular coordinate system and the initial side of the angle is along the positive x-axis.

Symmetry About the Origin   The unit circle is symmetric about the origin. Thus, for any point (a, b) that lies on the graph, the point $(-a, -b)$ must also lie on the graph.

Symmetry About the x-Axis   The unit circle is symmetric about the x-axis. Thus, for any point (a, b) that lies on the graph, the point $(a, -b)$ must also lie on the graph.

Symmetry About the y-Axis   The unit circle is symmetric about the y-axis. Thus, for any point (a, b) that lies on the graph, the point $(-a, b)$ must also lie on the graph.

Unit Vector in the Same Direction of a Given Vector   Given a nonzero vector $\mathbf{v} = a\mathbf{i} - b\mathbf{j}$, the unit vector $\mathbf{u}$ in the same direction as $\mathbf{v}$ is $\mathbf{u} = \dfrac{\mathbf{v}}{\|\mathbf{v}\|}$

Vertical Asymptote   A vertical line $x = a$ is a vertical asymptote of a function f if the values of $f(x)$ approach ∞ or $-\infty$ as the values of x approach a.

Zero   A zero of a function is a value of x for which $f(x) = 0$.

Zero of a Function   A zero of a function $f(x)$ is a value of x for which $f(x) = 0$.

Answers

CHAPTER 1

1.1 Exercises

SCE-1. $\dfrac{7\pi}{5}$ **SCE-2.** 150   **SCE-3.** 4   **SCE-4.** 10   **SCE-5.** 16π **SCE-6.** -30π

1. ; Quadrant I   **2.** ; y-axis   **3.** ; Quadrant III   **4.** ; Quadrant II

5. ; Quadrant II   **6.** $128°$ **7.** $98°$ **8.** $309°, -411°$ **9.** $-990°, -270°, 90°, 450°$ **10.** ; x-axis

11. ; Quadrant III   **12.** ; y-axis   **13.** ; Quadrant IV   **14.** ; Quadrant I

15. ; Quadrant III   **16.** $\dfrac{\pi}{12}$ **17.** $\dfrac{5\pi}{6}$ **18.** -3π **19.** $\dfrac{8\pi}{15}$ **20.** $-\dfrac{119\pi}{60}$ **21.** $15°$ **22.** $108°$ **23.** $-450°$

24. $205.71°$ **25.** $-348°$ **26.** $171.89°$ **27.** $\dfrac{11\pi}{8}$ **28.** $\dfrac{5\pi}{6}$ **29.** $\dfrac{\pi}{3}$ **30.** $\dfrac{139\pi}{70}$ **31.** $\dfrac{13\pi}{12}, -\dfrac{35\pi}{12}$ **32.** $-\dfrac{11\pi}{2}$, $-\dfrac{3\pi}{2}, \dfrac{\pi}{2}, \dfrac{5\pi}{2}$

1.2 Exercises

1. $6\,\text{cm}^2$ **2.** $\approx 282.74\,\text{in.}^2$ **3.** $\approx 33.51\,\text{m}^2$ **4.** $\approx 17.87\,\text{in.}^2$ **5.** $\dfrac{60}{49}\,\text{rad}$ **6.** $\approx 70.03\,\text{ft}$ **7.** $\approx 9.42\,\text{in.}$ **8.** $\approx 226.19\,\text{cm}$

9. $\approx 81.16\,\text{m}$ **10.** $\approx 212.93\,\text{km}$ **11.** $\dfrac{8}{5}\,\text{rad}$ **12.** $1516\,\text{miles}$ **13.** $\approx 11.67°$ **14.** $122°$ **15.** $12\pi\,\text{rad/min}$

16. $100\pi\,\text{rad/min}$ **17.** $\approx 142.8\,\text{mph}$ **18.** $\approx 7.85\,\text{m/min}$ **19.** $\approx 64.71\,\text{mph}$ **20.** $\approx 3.22\,\text{cm}$ **21.** $\approx 85.68\,\text{mph}$

1.3 Exercises

1. obtuse, scalene   **2.** right, scalene   **3.** acute, scalene   **4.** $\sqrt{193}$ **5.** 30   **6.** $\sqrt{137}$ **7.** 3   **8.** $\sqrt{7}$ **9.** $\approx 84.85\,\text{ft}$

10. $\approx 8.77\,\text{m}$ **11.** 8 in. by 15 in.   **12.** $\approx 5.74\,\text{ft}$ **13.** $k = 2, RS = 7, QS = 10$ **14.** $k = \dfrac{5}{2}, HK = 10, QT = \dfrac{16}{5}$

15. $k = \dfrac{2}{\sqrt{3}}, LV = \dfrac{2\sqrt{5}}{\sqrt{3}}, LZ = 2\sqrt{3}$ **16.** $k = \dfrac{5}{\sqrt{2}}, DJ = 5\sqrt{2}, PT = \dfrac{12}{5}$ **17.** $AB = 25, XZ = 48, YZ = 14$

18. $RT = \dfrac{35}{2}, ST = \dfrac{37}{2}, KM = 35$ **19.** $DV = \dfrac{9}{\sqrt{5}}, LV = \dfrac{6}{\sqrt{5}}, BF = 4$ **20.** $HN = 3\sqrt{15}, BG = \dfrac{77}{3\sqrt{15}}, GX = \dfrac{112}{3\sqrt{15}}$

21. $10, 10\sqrt{2}$ **22.** $\sqrt{2}, 2\sqrt{2}$ **23.** $\dfrac{9}{2}, \dfrac{9\sqrt{3}}{2}$ **24.** $4\sqrt{2}, 4\sqrt{2}$ **25.** $2, 2\sqrt{3}$ **26.** $\dfrac{\sqrt{10}}{\sqrt{3}}, \dfrac{2\sqrt{10}}{\sqrt{3}}$ **27.** $\sqrt{2}, 2$

28. $\dfrac{\sqrt{5}}{\sqrt{3}}, \dfrac{2\sqrt{5}}{\sqrt{3}}$ **29.** $\dfrac{\sqrt{7}}{\sqrt{2}}, \dfrac{\sqrt{7}}{\sqrt{2}}$ **30.** $2\sqrt{15}, 4\sqrt{5}$ **31.** $202.5\,\text{ft}$ **32.** $\approx 34.64\,\text{ft}, 20\,\text{ft}$ **33.** $\approx 2.67\,\text{ft}$ **34.** $\approx 16{,}970.56\,\text{ft}$

35. length of bridge: 20.21 ft, distance between the two people: 40.41 ft   **36.** 272 ft

1.4 Exercises

SCE-1. $-8, 8$ **SCE-2.** $-\sqrt{22}, \sqrt{22}$ **SCE-3.** $-7, 7$ **SCE-4.** $-\sqrt{31}, \sqrt{31}$ **SCE-5.** $-10, 10$

SCE-6. $-\sqrt{11}, \sqrt{11}$ **SCE-7.** $\dfrac{\pi}{3}$ **SCE-8.** $\dfrac{\pi}{12}$

1. $\sin\theta = \dfrac{3}{5}, \cos\theta = \dfrac{4}{5}, \tan\theta = \dfrac{3}{4}, \csc\theta = \dfrac{5}{3}, \sec\theta = \dfrac{5}{4}, \cot\theta = \dfrac{4}{3}$ **2.** $\sin\theta = \dfrac{12}{13}, \cos\theta = \dfrac{5}{13}, \tan\theta = \dfrac{12}{5}, \csc\theta = \dfrac{13}{12},$

$\sec\theta = \dfrac{13}{5}, \cot\theta = \dfrac{5}{12}$ **3.** $\sin\theta = \dfrac{4}{\sqrt{137}}, \cos\theta = \dfrac{11}{\sqrt{137}}, \tan\theta = \dfrac{4}{11}, \csc\theta = \dfrac{\sqrt{137}}{4}, \sec\theta = \dfrac{\sqrt{137}}{11}, \cot\theta = \dfrac{11}{4}$

4. $\sin\theta = \dfrac{3}{\sqrt{13}}, \cos\theta = \dfrac{2}{\sqrt{13}}, \tan\theta = \dfrac{3}{2}, \csc\theta = \dfrac{\sqrt{13}}{3}, \sec\theta = \dfrac{\sqrt{13}}{2}, \cot\theta = \dfrac{2}{3}$ **5.** $\sin\theta = \dfrac{1}{2}, \cos\theta = \dfrac{\sqrt{3}}{2}, \tan\theta = \dfrac{1}{\sqrt{3}},$

$\csc\theta = 2, \sec\theta = \dfrac{2}{\sqrt{3}}, \cot\theta = \sqrt{3}$ **6.** $\cos\theta = \dfrac{12}{13}, \tan\theta = \dfrac{5}{12}, \csc\theta = \dfrac{13}{5}, \sec\theta = \dfrac{13}{12}, \cot\theta = \dfrac{12}{5}$ **7.** $\sin\theta = \dfrac{11}{61},$

$\tan\theta = \dfrac{11}{60}, \csc\theta = \dfrac{61}{11}, \sec\theta = \dfrac{61}{60}, \cot\theta = \dfrac{60}{11}$ **8.** $\sin\theta = \dfrac{1}{\sqrt{5}}, \cos\theta = \dfrac{2}{\sqrt{5}}, \csc\theta = \sqrt{5}, \sec\theta = \dfrac{\sqrt{5}}{2}, \cot\theta = 2$

9. $\sin\theta = \dfrac{1}{\sqrt{6}}, \cos\theta = \dfrac{\sqrt{5}}{\sqrt{6}}, \tan\theta = \dfrac{1}{\sqrt{5}}, \sec\theta = \dfrac{\sqrt{6}}{\sqrt{5}}, \cot\theta = \sqrt{5}$ **10.** $\sin\theta = \dfrac{2\sqrt{11}}{7}, \cos\theta = \dfrac{\sqrt{5}}{7}, \tan\theta = \dfrac{2\sqrt{11}}{\sqrt{5}},$

$\csc\theta = \dfrac{7}{2\sqrt{11}}, \cot\theta = \dfrac{\sqrt{5}}{2\sqrt{11}}$ **11.** $\sin\theta = \dfrac{1}{\sqrt{10}}, \cos\theta = \dfrac{3}{\sqrt{10}}, \tan\theta = \dfrac{1}{3}, \csc\theta = \sqrt{10}, \sec\theta = \dfrac{\sqrt{10}}{3}$ **12.** $\dfrac{1}{2}$ **13.** $\sqrt{2}$

14. $\dfrac{2}{\sqrt{3}}$ **15.** $\sqrt{2} - \sqrt{3}$ **16.** $2\sqrt{3} - 2$ **17.** $\dfrac{10}{3}$ **18.** $\dfrac{3}{4}$ **19.** $\dfrac{\pi}{6}$ **20.** $\dfrac{\pi}{6}$ **21.** $\dfrac{\pi}{4}$ **22.** $\tan\theta = \dfrac{7}{24}, \csc\theta = \dfrac{25}{7},$

$\sec\theta = \dfrac{25}{24}, \cot\theta = \dfrac{24}{7}$ **23.** $\tan\theta = \dfrac{3}{\sqrt{2}}, \csc\theta = \dfrac{\sqrt{11}}{3}, \sec\theta = \dfrac{\sqrt{11}}{\sqrt{2}}, \cot\theta = \dfrac{\sqrt{2}}{3}$ **24.** 0 **25.** 1 **26.** 0 **27.** $\csc\theta$

28. $\sec\theta$ **29.** $\cos\left(-\dfrac{3\pi}{14}\right)$ **30.** $\tan 51°$ **31.** -1 **32.** $\tan\theta$ **33.** $\cot\left(105° - \theta\right)$ **34.** 0.7660 **35.** 1.0125

36. 1.7604 **37.** 0.8746

1.5 Exercises

SCE-1. Quadrant IV   **SCE-2.** x-axis

1. $0°, 90°, 180°, 270°$ **2.** $\dfrac{\pi}{6}, \dfrac{5\pi}{6}, \dfrac{7\pi}{6}, \dfrac{11\pi}{6}$ **3.** quadrantal family, $, \theta_c = \dfrac{\pi}{2}$ **4.** $\dfrac{\pi}{3}$ family,   ,

$\theta_c = \dfrac{2\pi}{3}$ **5.** $\dfrac{\pi}{3}$ family, $, \theta_c = \dfrac{2\pi}{3}$ **6.** quadrantal family, $, \theta_c = \dfrac{3\pi}{2}$

7. $\dfrac{\pi}{4}$ family, $, \theta_c = \dfrac{\pi}{4}$ **8.** quadrantal family, $, \theta_c = 0$ **9.** $\dfrac{\pi}{6}$ family, $, \theta_c = \dfrac{7\pi}{6}$

10. $\dfrac{\pi}{3}$ family, $\theta_c = \dfrac{5\pi}{3}$ **11.** $\dfrac{\pi}{4}$ family, $\theta_c = \dfrac{3\pi}{4}$ **12.** $\dfrac{\pi}{3}$ family, $\theta_c = \dfrac{\pi}{3}$

13. $\sin\theta = \dfrac{3}{5}, \cos\theta = \dfrac{4}{5}, \tan\theta = \dfrac{3}{4}, \csc\theta = \dfrac{5}{3}, \sec\theta = \dfrac{5}{4}, \cot\theta = \dfrac{4}{3}$ **14.** $\sin\theta = \dfrac{12}{13}, \cos\theta = -\dfrac{5}{13}, \tan\theta = -\dfrac{12}{5},$
$\csc\theta = \dfrac{13}{12}, \sec\theta = -\dfrac{13}{5}, \cot\theta = -\dfrac{5}{12}$ **15.** $\sin\theta = -\dfrac{5}{\sqrt{29}}, \cos\theta = -\dfrac{2}{\sqrt{29}}, \tan\theta = \dfrac{5}{2}, \csc\theta = -\dfrac{\sqrt{29}}{5}, \sec\theta = -\dfrac{\sqrt{29}}{2},$
$\cot\theta = \dfrac{2}{5}$ **16.** $\sin\theta = -\dfrac{4}{\sqrt{17}}, \cos\theta = \dfrac{1}{\sqrt{17}}, \tan\theta = -4, \csc\theta = -\dfrac{\sqrt{17}}{4}, \sec\theta = \sqrt{17}, \cot\theta = -\dfrac{1}{4}$ **17.** $\sin\theta = \dfrac{\sqrt{3}}{2},$
$\cos\theta = -\dfrac{1}{2}, \tan\theta = -\sqrt{3}, \csc\theta = \dfrac{2}{\sqrt{3}}, \sec\theta = -2, \cot\theta = -\dfrac{1}{\sqrt{3}}$ **18.** $\sin\theta = -\dfrac{1}{\sqrt{2}}, \cos\theta = -\dfrac{1}{\sqrt{2}}, \tan\theta = 1,$
$\csc\theta = -\sqrt{2}, \sec\theta = -\sqrt{2}, \cot\theta = 1$ **19.** 0 **20.** 0 **21.** undefined   **22.** undefined   **23.** -1 **24.** undefined
25. 1 **26.** 0 **27.** undefined   **28.** undefined   **29.** 1 **30.** 0 **31.** Quadrants I and II   **32.** Quadrants I and IV

33. Quadrants II and IV   **34.** Quadrants III and IV   **35.** Quadrant IV   **36.** Quadrant I   **37.** Quadrant I   **38.** $\dfrac{25}{24}$

39. $-\dfrac{1}{\sqrt{17}}$ **40.** $-\dfrac{5}{\sqrt{21}}$ **41.** $\dfrac{\pi}{3}$ **42.** $\dfrac{\pi}{4}$ **43.** $\dfrac{\pi}{3}$ **44.** $\dfrac{\pi}{4}$ **45.** $\dfrac{\pi}{3}$ **46.** $\dfrac{2\pi}{9}$ **47.** $\dfrac{3\pi}{7}$ **48.** $30°$ **49.** $45°$

50. $19°$ **51.** $\dfrac{1}{2}$ **52.** $-\dfrac{1}{\sqrt{2}}$ **53.** $\dfrac{1}{\sqrt{3}}$ **54.** -2 **55.** $\sqrt{3}$ **56.** $-\dfrac{\sqrt{3}}{2}$ **57.** $\sqrt{2}$ **58.** -1 **59.** $\dfrac{1}{2}$ **60.** $-\sqrt{3}$

61. $\dfrac{2}{\sqrt{3}}$ **62.** $-\dfrac{2}{\sqrt{3}}$ **63.** $\dfrac{1}{\sqrt{2}}$ **64.** $-\dfrac{1}{2}$ **65.** $\dfrac{1}{\sqrt{3}}$ **66.** -2 **67.** 0 **68.** 0 **69.** undefined   **70.** undefined

71. -1 **72.** undefined   **73.** 1 **74.** 0 **75.** undefined   **76.** undefined   **77.** 1 **78.** 0 **79.** $\dfrac{1}{2}$ **80.** $-\dfrac{1}{\sqrt{2}}$

81. $\dfrac{1}{\sqrt{3}}$ **82.** -2 **83.** $\sqrt{3}$ **84.** $-\dfrac{\sqrt{3}}{2}$ **85.** $\sqrt{2}$ **86.** -1 **87.** $\dfrac{1}{2}$ **88.** $-\sqrt{3}$ **89.** $\dfrac{2}{\sqrt{3}}$ **90.** $-\dfrac{2}{\sqrt{3}}$

91. $\dfrac{1}{\sqrt{2}}$ **92.** $-\dfrac{1}{2}$ **93.** $\dfrac{1}{\sqrt{3}}$ **94.** -2

1.6 Exercises

SCE-1. $-\dfrac{3}{5}, \dfrac{3}{5}$ **SCE-2.** $-\dfrac{1}{2}, \dfrac{1}{2}$ **SCE-3.** $-\dfrac{2\sqrt{10}}{7}, \dfrac{2\sqrt{10}}{7}$ **SCE-4.** $-\dfrac{1}{\sqrt{2}}, \dfrac{1}{\sqrt{2}}$

1. $-\dfrac{\sqrt{5}}{3}$ **2.** $\dfrac{2\sqrt{6}}{5}$ **3.** $-\dfrac{4\sqrt{3}}{7}$ **4.** $-\dfrac{\sqrt{77}}{9}$ **5.** $-\dfrac{2\sqrt{11}}{3\sqrt{5}}$ **6.** $\left(\dfrac{1}{7}\right)^2 + \left(\dfrac{4\sqrt{3}}{7}\right)^2 = \dfrac{1}{49} + \dfrac{48}{49} = 1; \left(-\dfrac{1}{7}, \dfrac{4\sqrt{3}}{7}\right),$
$\left(\dfrac{1}{7}, -\dfrac{4\sqrt{3}}{7}\right), \left(-\dfrac{1}{7}, -\dfrac{4\sqrt{3}}{7}\right)$ **7.** $\left(-\dfrac{1}{2}\right)^2 + \left(-\dfrac{\sqrt{3}}{2}\right)^2 = \dfrac{1}{4} + \dfrac{3}{4} = 1; \left(\dfrac{1}{2}, \dfrac{\sqrt{3}}{2}\right), \left(-\dfrac{1}{2}, \dfrac{\sqrt{3}}{2}\right), \left(\dfrac{1}{2}, -\dfrac{\sqrt{3}}{2}\right)$

8. $\left(-\dfrac{4}{\sqrt{17}}\right)^2 + \left(-\dfrac{1}{\sqrt{17}}\right)^2 = \dfrac{16}{17} + \dfrac{1}{17} = 1; \left(\dfrac{4}{\sqrt{17}}, \dfrac{1}{\sqrt{17}}\right), \left(-\dfrac{4}{\sqrt{17}}, \dfrac{1}{\sqrt{17}}\right), \left(\dfrac{4}{\sqrt{17}}, -\dfrac{1}{\sqrt{17}}\right)$ **9.** $\sin t = -\dfrac{4\sqrt{3}}{7},$
$\cos t = \dfrac{1}{7}, \tan t = -4\sqrt{3}, \csc t = -\dfrac{7}{4\sqrt{3}}, \sec t = 7, \cot t = -\dfrac{1}{4\sqrt{3}}$ **10.** $\sin t = -\dfrac{1}{\sqrt{17}}, \cos t = -\dfrac{4}{\sqrt{17}}, \tan t = \dfrac{1}{4},$
$\csc t = -\sqrt{17}, \sec t = -\dfrac{\sqrt{17}}{4}, \cot t = 4$ **11.** $\sin t = -\dfrac{\sqrt{3}}{2}, \cos t = -\dfrac{1}{2}, \tan t = \sqrt{3}, \csc t = -\dfrac{2}{\sqrt{3}}, \sec t = -2, \cot t = \dfrac{1}{\sqrt{3}}$

12. 1 **13.** 0 **14.** undefined   **15.** -1 **16.** undefined   **17.** 0 **18.** 0 **19.** -1 **20.** $\dfrac{1}{2}$ **21.** $-\dfrac{1}{\sqrt{2}}$ **22.** $\dfrac{1}{\sqrt{3}}$

23. $-\sqrt{2}$ **24.** $\sqrt{3}$ **25.** $-\dfrac{\sqrt{3}}{2}$ **26.** $\sqrt{2}$ **27.** -1 **28.** $\dfrac{1}{2}$ **29.** $-\sqrt{3}$ **30.** $\dfrac{2}{\sqrt{3}}$ **31.** $-\dfrac{2}{\sqrt{3}}$ **32.** $\dfrac{\sqrt{3}}{2}$

33. $-\dfrac{1}{2}$ **34.** $\dfrac{1}{\sqrt{3}}$ **35.** $-\dfrac{2}{\sqrt{3}}$ **36.** 1 **37.** 0 **38.** undefined   **39.** -1 **40.** undefined   **41.** 0 **42.** 0 **43.** -1

44. $\dfrac{1}{2}$ **45.** $-\dfrac{1}{\sqrt{2}}$ **46.** $\dfrac{1}{\sqrt{3}}$ **47.** $-\sqrt{2}$ **48.** $\sqrt{3}$ **49.** $-\dfrac{\sqrt{3}}{2}$ **50.** $\sqrt{2}$ **51.** -1 **52.** $\dfrac{1}{2}$ **53.** $-\sqrt{3}$

54. $\dfrac{2}{\sqrt{3}}$ **55.** $-\dfrac{2}{\sqrt{3}}$ **56.** $\dfrac{\sqrt{3}}{2}$ **57.** $-\dfrac{1}{2}$ **58.** $\dfrac{1}{\sqrt{3}}$ **59.** $-\dfrac{2}{\sqrt{3}}$

CHAPTER 2

2.1 Exercises

SCE-1. $\dfrac{\pi}{10}$ **SCE-2.** 6

1. ; the following properties apply: The function is an odd function. The function is decreasing on the interval $\left(\pi, \dfrac{3\pi}{2}\right)$.

The domain is $(-\infty, \infty)$. The range is $[-1, 1]$. The y-intercept is 0. The zeros are of the form n, where n is any integer. The function obtains a relative maximum at $x = \dfrac{\pi}{2} + 2n\pi$, where n is an integer. The function obtains a relative minimum at $x = \dfrac{3\pi}{2} + 2n\pi$, where n is an integer. **2.** $-\dfrac{11\pi}{6}, -\dfrac{7\pi}{6}, \dfrac{\pi}{6}, \dfrac{5\pi}{6}, \dfrac{13\pi}{6}$ **3.** $-\dfrac{7\pi}{4}, -\dfrac{5\pi}{4}, \dfrac{\pi}{4}$ **4.** $-\dfrac{8\pi}{3}, -\dfrac{7\pi}{3}, -\dfrac{2\pi}{3}, -\dfrac{\pi}{6}, \dfrac{4\pi}{3}$

5. iii **6.** i **7.** ii **8.** i **9.** ; the following properties apply: The function is an even function. The function is

increasing on the interval $\left(-\pi, -\dfrac{\pi}{2}\right)$. The domain is $(-\infty, \infty)$. The range is $[-1, 1]$. The y-intercept is 1. The zeros are of

the form $\dfrac{(2n + 1)\pi}{2}$, where n is any integer. The function obtains a relative maximum at $x = 2\pi n$, where n is an integer.

10. $-\dfrac{5\pi}{2}, -\dfrac{3\pi}{2}, -\dfrac{\pi}{2}, \dfrac{\pi}{2}$ **11.** $-\dfrac{9\pi}{4}, -\dfrac{7\pi}{4}, -\dfrac{\pi}{4}, \dfrac{\pi}{4}$ **12.** $-\dfrac{19\pi}{6}, -\dfrac{17\pi}{6}, -\dfrac{7\pi}{6}, -\dfrac{5\pi}{6}, \dfrac{5\pi}{6}, \dfrac{7\pi}{6}$ **13.** iii **14.** i **15.** ii

16. i **17.** amplitude: 3; range: $[-3, 3]$; **18.** amplitude: 4; range: $[-4, 4]$; **19.** amplitude: $\dfrac{1}{3}$;

range: $\left[-\dfrac{1}{3}, \dfrac{1}{3}\right]$; **20.** amplitude: $\dfrac{1}{4}$; range: $\left[-\dfrac{1}{4}, \dfrac{1}{4}\right]$; **21.** amplitude: $\dfrac{3}{2}$; range: $\left[-\dfrac{3}{2}, \dfrac{3}{2}\right]$;

22. amplitude: 2; range: $[-2, 2]$; **23.** amplitude: 5; range: $[-5, 5]$;

24. amplitude: $\dfrac{3}{4}$; range: $\left[-\dfrac{3}{4}, \dfrac{3}{4}\right]$; **25.** amplitude: $\dfrac{1}{2}$; range: $\left[-\dfrac{1}{2}, \dfrac{1}{2}\right]$; **26.** amplitude: $\dfrac{5}{4}$;

range: $\left[-\dfrac{5}{4}, \dfrac{5}{4}\right]$;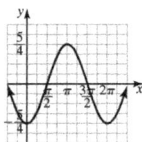

27. $\dfrac{2\pi}{3}$;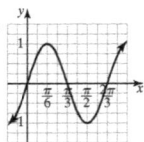

28. 6π;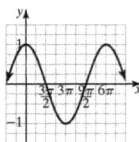

29. $\dfrac{4\pi}{3}$;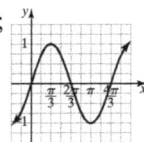

30. $\dfrac{2}{3}$;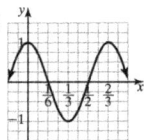

31. 6;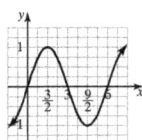

32. $\dfrac{\pi}{2}$;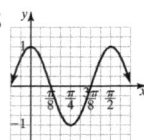

33. 8π;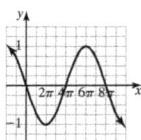

34. $\dfrac{4\pi}{5}$;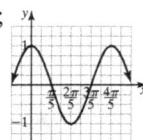

35. $\dfrac{2}{3}$;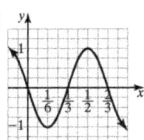

36. 4;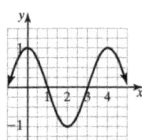

37. amplitude: 2; range: $[-2, 2]$; period: $\dfrac{2\pi}{3}$;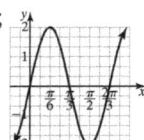

38. amplitude: $\dfrac{5}{3}$; range: $\left[-\dfrac{5}{3}, \dfrac{5}{3}\right]$; period: $\dfrac{\pi}{2}$;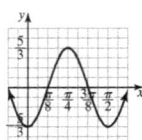

39. amplitude: $\dfrac{3}{4}$; range: $\left[-\dfrac{3}{4}, \dfrac{3}{4}\right]$; period: π;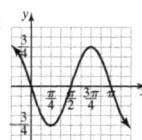

40. amplitude: 4; range: $[-4, 4]$; period: $\dfrac{2\pi}{3}$;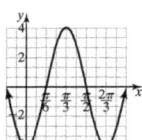

41. amplitude: 5; range: $[-5, 5]$; period: 4π;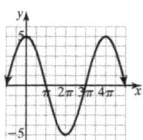

42. amplitude: $\dfrac{1}{2}$; range: $\left[-\dfrac{1}{2}, \dfrac{1}{2}\right]$; period: 8π;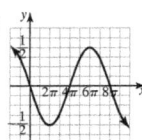

43. amplitude: 1; range: $[-1, 1]$; period: 10π;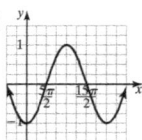

44. amplitude: 6; range: $[-6, 6]$; period: $\dfrac{4\pi}{3}$;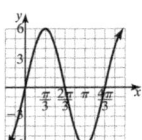

45. amplitude: 2; range: $[-2, 2]$; period: $\dfrac{3\pi}{2}$;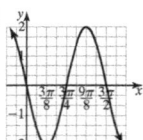

46. amplitude: 2; range: $[-2, 2]$; period: $\dfrac{8\pi}{5}$;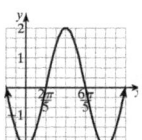

47. amplitude: 1; range: $[-1, 1]$; period: $\dfrac{4\pi}{5}$;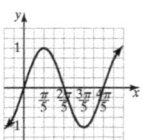

48. amplitude: 3; range: $[-3, 3]$; period: 4;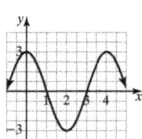

49. amplitude: 4; range: $[-4, 4]$; period: 6;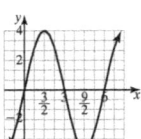

50. amplitude: $\dfrac{5}{4}$; range: $\left[-\dfrac{5}{4}, \dfrac{5}{4}\right]$; period: 2;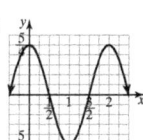

51. amplitude: $\dfrac{1}{2}$; range: $\left[-\dfrac{1}{2}, \dfrac{1}{2}\right]$; period: 2;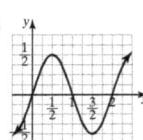

52. amplitude: 6; range: $[-6, 6]$; period: 3;

53. $y = 3 \cos x$　**54.** $y = -4 \sin x$　**55.** $y = \dfrac{1}{2} \sin x$

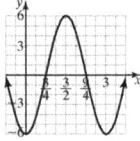

56. $y = -\dfrac{1}{3} \cos x$　**57.** $y = \cos 2x$　**58.** $y = \sin\left(\dfrac{1}{5}x\right)$　**59.** $y = 3 \sin 4x$　**60.** $y = -5 \cos 3x$　**61.** $y = 4 \cos\left(\dfrac{1}{5}x\right)$

62. $y = -2 \sin\left(\dfrac{1}{4}x\right)$　**63.** 3　**64.** 1　**65.** 2　**66.** $[-3, 3]$　**67.** $[-1, 1]$　**68.** $[-2, 2]$　**69.** 2π　**70.** $\dfrac{2\pi}{3}$

71. $\dfrac{8\pi}{5}$　**72.** $(0, 0), \left(\dfrac{\pi}{2}, -3\right), (\pi, 0), \left(\dfrac{3\pi}{2}, 3\right), (2\pi, 0)$　**73.** $(0, 1), \left(\dfrac{\pi}{6}, 0\right), \left(\dfrac{\pi}{3}, -1\right), \left(\dfrac{\pi}{2}, 0\right), \left(\dfrac{2\pi}{3}, 1\right)$

74. $(0, -2), \left(\dfrac{2\pi}{5}, 0\right), \left(\dfrac{4\pi}{5}, 2\right), \left(\dfrac{6\pi}{5}, 0\right), \left(\dfrac{8\pi}{5}, -2\right)$　**75.**

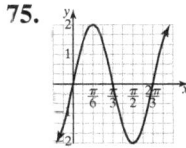

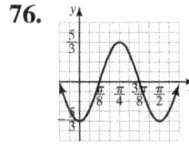

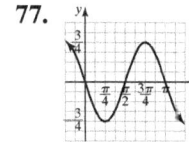

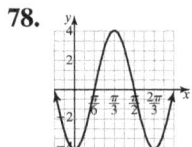

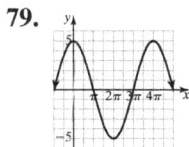

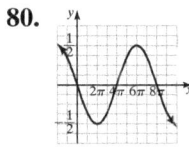

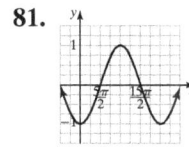

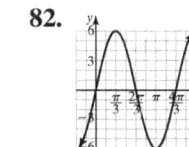

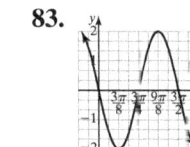

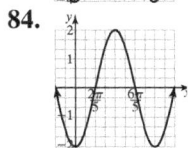

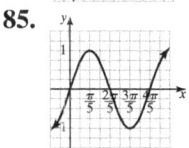

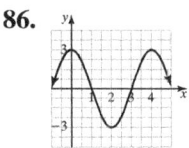

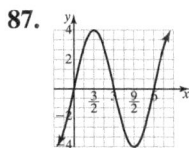

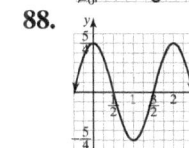

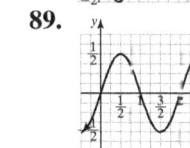

　76.

　77.

78.

　79.

　80.

　81.

　82.

　83.

84.

　85.

　86.

　87.

　88.

　89.

90.

　91. $y = 3 \cos x$　**92.** $y = -4 \sin x$　**93.** $y = \dfrac{1}{2} \sin x$　**94.** $y = -\dfrac{1}{3} \cos x$　**95.** $y = \cos 2x$

96. $y = \sin\left(\dfrac{1}{5}x\right)$　**97.** $y = 3 \sin 4x$　**98.** $y = -5 \cos 3x$　**99.** $y = 4 \cos\left(\dfrac{1}{5}x\right)$　**100.** $y = -2 \sin\left(\dfrac{1}{4}x\right)$

2.2 Exercises

SCE-1. $\left[\dfrac{\pi}{2}, \dfrac{3\pi}{2}\right]$　**SCE-2.** $\left[\dfrac{\pi}{18}, \dfrac{13\pi}{18}\right]$　**SCE-3.** $\left[\dfrac{3}{\pi}, 2 + \dfrac{3}{\pi}\right]$　**SCE-4.** $\dfrac{\pi}{4}$　**SCE-5.** $\dfrac{\pi}{8}$　**SCE-6.** $4\left(x - \dfrac{\pi}{4}\right)$

SCE-7. $5\left(x - \dfrac{\pi}{15}\right)$　**SCE-8.** $-3\left(x - \dfrac{\pi}{3}\right)$　**SCE-9.** $-4\left(x - \dfrac{\pi}{12}\right)$　**SCE-10.** $-2\left(x - \dfrac{\pi}{2}\right)$　**SCE-11.** $\pi\left(x - \dfrac{3}{\pi}\right)$

1. amplitude: 1; range: $[-1, 1]$; period: 2π; phase shift: π;

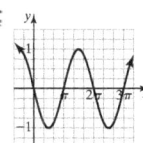

2. amplitude: 1; range: $[-1, 1]$; period: 2π; phase

shift: $-\dfrac{\pi}{2}$;

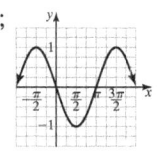

　3. amplitude: 1; range: $[-1, 1]$; period: 2π; phase shift: $\dfrac{\pi}{6}$;

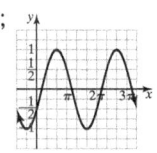

　4. amplitude: 1; range:

$[-1, 1]$; period: 2π; phase shift: $-\dfrac{2\pi}{3}$;   **5.** amplitude: 1; range: $[-1, 1]$; period: 2π; phase shift: 2;

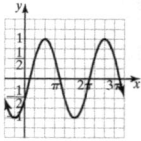

6. amplitude: 1; range: $[-1, 1]$; period: $\dfrac{2\pi}{3}$; phase shift: $-\dfrac{\pi}{3}$;   **7.** amplitude: 1; range: $[-1, 1]$; period: π;

phase shift: $\dfrac{\pi}{2}$;   **8.** amplitude: 3; range: $[-3, 3]$; period: $\dfrac{\pi}{2}$; phase shift: $\dfrac{\pi}{4}$;   **9.** amplitude: 2;

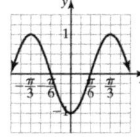

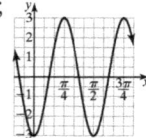

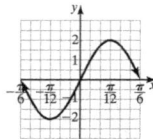

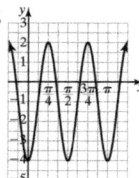

range: $[-2, 2]$; period: $\dfrac{\pi}{3}$; phase shift: $-\dfrac{\pi}{6}$;   **10.** amplitude: 4; range: $[-4, 4]$; period: π; phase

shift: $-\dfrac{\pi}{2}$;   **11.** amplitude: 3; range: $[-3, 3]$; period: π; phase shift: $-\dfrac{\pi}{4}$;

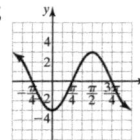

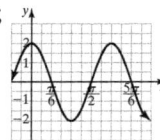

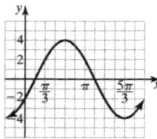

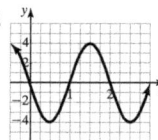

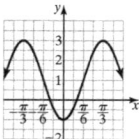

12. amplitude: 2; range: $[-2, 2]$; period: $\dfrac{2\pi}{3}$; phase shift: $\dfrac{\pi}{6}$;   **13.** amplitude: 4; range: $[-4, 4]$; period: 2π;

phase shift: $-\dfrac{\pi}{3}$;   **14.** amplitude: 1; range: $[-1, 1]$; period: π; phase shift: $\dfrac{\pi}{2}$;

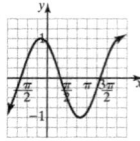

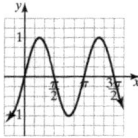

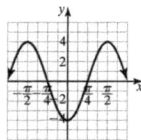

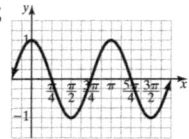

15. amplitude: 4; range: $[-4, 4]$; period: 2; phase shift: $\dfrac{3}{\pi}$;   **16.** amplitude: 2; range: $[-1, 3]$; period: $\dfrac{2\pi}{3}$;

phase shift: $-\dfrac{\pi}{3}$;   **17.** amplitude: 1; range: $[-3, -1]$; period: π; phase shift: $\dfrac{\pi}{2}$;

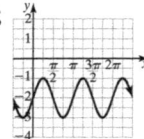

18. amplitude: 3; range: $[-4, 2]$; period: $\dfrac{\pi}{2}$; phase shift: $\dfrac{\pi}{4}$;   **19.** amplitude: 2; range: $[-1, 3]$; period: $\dfrac{\pi}{3}$; phase shift: $-\dfrac{\pi}{6}$;   **20.** amplitude: 4; range: $[-1, 7]$; period: π; phase shift: $-\dfrac{\pi}{2}$;   **21.** amplitude: 3; range:

$[-7, -1]$; period: π; phase shift: $\dfrac{\pi}{8}$;   **22.** amplitude: 2; range: $[3, 7]$; period: $\dfrac{2\pi}{3}$; phase shift: $\dfrac{\pi}{6}$;

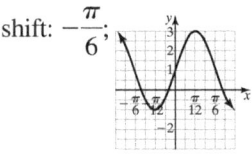

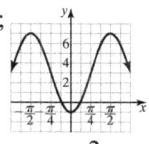

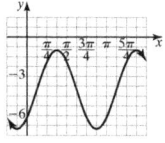

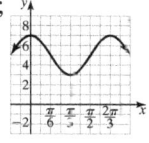

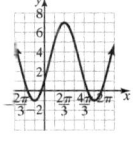

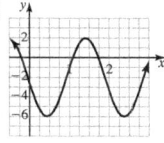

23. amplitude: 4; range: $[-1, 7]$; period: 2π; phase shift: $-\dfrac{\pi}{3}$;   **24.** amplitude: 1; range: $[2, 4]$ period: π; phase

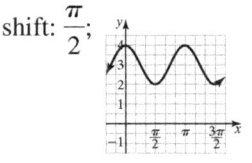

shift: $\dfrac{\pi}{2}$;   **25.** amplitude: 4; range: $[-6, 2]$; period: 2; phase shift: $\dfrac{3}{\pi}$;   **26.** $y = \sin\left(x - \dfrac{\pi}{3}\right)$

27. $y = \cos\left(x + \dfrac{2\pi}{3}\right)$ **28.** $y = -3\cos(x + \pi)$ **29.** $y = -4\sin\left(x + \dfrac{\pi}{6}\right)$ **30.** $y = 2\sin\left(x - \dfrac{5\pi}{6}\right)$

31. $y = -5\cos(3x - \pi)$ **32.** $y = -3\cos\left(x - \dfrac{\pi}{4}\right) - 1$ **33.** $y = -2\sin\left(3x - \dfrac{\pi}{2}\right) + 4$ **34.** 1   **35.** 3   **36.** 2

37. $[-1, 1]$ **38.** $[-3, 3]$ **39.** $[-4, 0]$ **40.** 2π **41.** π **42.** 2π **43.** $\dfrac{\pi}{6}$ **44.** $-\dfrac{\pi}{2}$ **45.** $\dfrac{\pi}{3}$

46. $\left(\dfrac{\pi}{6}, 0\right), \left(\dfrac{2\pi}{3}, 1\right), \left(\dfrac{7\pi}{6}, 0\right), \left(\dfrac{5\pi}{3}, -1\right), \left(\dfrac{13\pi}{6}, 0\right)$ **47.** $\left(-\dfrac{\pi}{2}, -3\right), \left(-\dfrac{\pi}{4}, 0\right), (0, 3), \left(\dfrac{\pi}{4}, 0\right), \left(\dfrac{\pi}{2}, -3\right)$

48. $\left(\dfrac{\pi}{3}, -2\right), \left(\dfrac{5\pi}{6}, -4\right), \left(\dfrac{4\pi}{3}, -2\right), \left(\dfrac{11\pi}{6}, 0\right), \left(\dfrac{7\pi}{3}, -2\right)$ **49.**   **50.**   **51.**

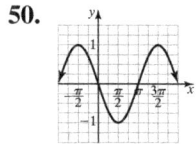

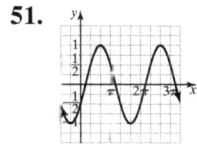

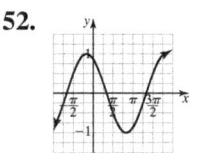

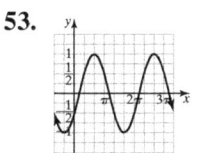

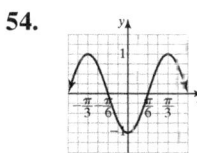

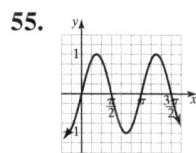

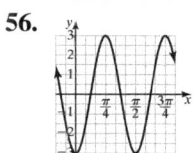

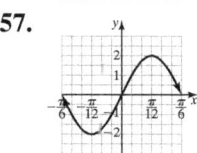

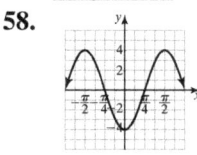

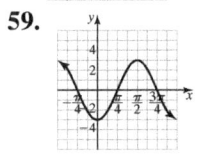

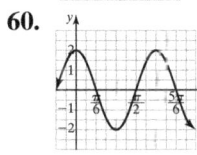

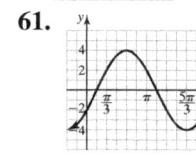

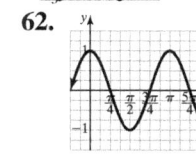

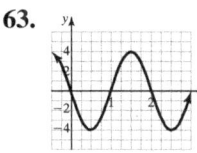

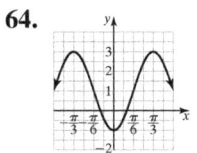

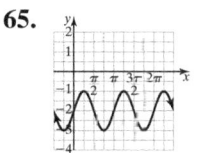

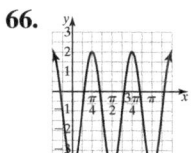

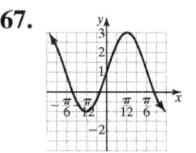

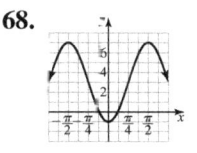

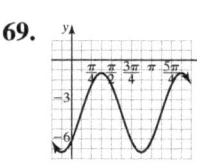

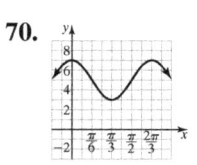

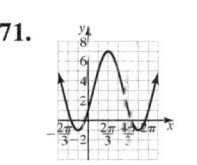

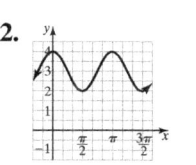

52.   **53.**   **54.**   **55.**   **56.**   **57.**

58.   **59.**   **60.**   **61.**   **62.**

63.   **64.**   **65.**   **66.**   **67.**   **68.**

69.   **70.**   **71.**   **72.**   **73.** $y = \sin\left(x - \dfrac{\pi}{3}\right)$

74. $y = \cos\left(x + \dfrac{2\pi}{3}\right)$ **75.** $y = -3\cos(x + \pi)$ **76.** $y = -4\sin\left(x + \dfrac{\pi}{6}\right)$ **77.** $y = 2\sin\left(x - \dfrac{5\pi}{6}\right)$

78. $y = -5\cos(3x - \pi)$ **79.** $y = -3\cos\left(x - \dfrac{\pi}{4}\right) - 1$ **80.** $y = -2\sin\left(3x - \dfrac{\pi}{2}\right) + 4$

2.3 Exercises

SCE-1. $\left[\dfrac{\pi}{4}, \dfrac{3\pi}{4}\right]$ **SCE-2.** $\left[-\dfrac{\pi}{3}, 0\right]$ **SCE-3.** $\left[-\dfrac{3\pi}{2}, \dfrac{3\pi}{2}\right]$ **SCE-4.** $[0, 5\pi]$

1. ; the function is an odd function. The domain is all real numbers except odd integer multiples of $\dfrac{\pi}{2}$. The

range is $(-\infty, \infty)$. Every halfway point has a y-coordinate of -1 or 1. The interval for the graph of the principal cycle

is $\left(-\dfrac{\pi}{2}, \dfrac{\pi}{2}\right)$. The y-intercept is 0. The function has a period of $P = \pi$. The zeros are of the form $n\pi$, where n is an integer.

2. $\left(-\dfrac{5\pi}{3}, \sqrt{3}\right), \left(-\dfrac{2\pi}{3}, \sqrt{3}\right), \left(\dfrac{\pi}{3}, \sqrt{3}\right), \left(\dfrac{4\pi}{3}, \sqrt{3}\right), \left(\dfrac{7\pi}{3}, \sqrt{3}\right)$ **3.** $\left(-\dfrac{13\pi}{6}, -\dfrac{1}{\sqrt{3}}\right), \left(-\dfrac{7\pi}{6}, -\dfrac{1}{\sqrt{3}}\right), \left(-\dfrac{\pi}{6}, -\dfrac{1}{\sqrt{3}}\right)$

4. $\left(-\dfrac{7\pi}{4}, 1\right), \left(-\dfrac{5\pi}{4}, -1\right), \left(-\dfrac{3\pi}{4}, 1\right), \left(-\dfrac{\pi}{4}, -1\right), \left(\dfrac{\pi}{4}, 1\right), \left(\dfrac{3\pi}{4}, -1\right)$ **5.** iii   **6.** principal cycle: $\left(-\dfrac{3\pi}{4}, \dfrac{\pi}{4}\right)$; period: π;

vertical asymptotes: $x = -\dfrac{3\pi}{4}, x = \dfrac{\pi}{4}$; center point: $\left(-\dfrac{\pi}{4}, 0\right)$; halfway points: $\left(-\dfrac{\pi}{2}, -1\right), (0, 1)$;

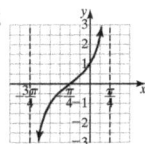

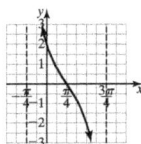

7. principal cycle: $\left(-\dfrac{\pi}{3}, \dfrac{2\pi}{3}\right)$; period: π; vertical asymptotes: $x = -\dfrac{\pi}{3}, x = \dfrac{2\pi}{3}$; center point: $\left(\dfrac{\pi}{6}, 0\right)$; halfway points:

$\left(-\dfrac{\pi}{12}, -3\right), \left(\dfrac{5\pi}{12}, 3\right)$;   **8.** principal cycle: $\left(-\dfrac{\pi}{4}, \dfrac{3\pi}{4}\right)$; period: π; vertical asymptotes: $x = -\dfrac{\pi}{4}, x = \dfrac{3\pi}{4}$;

center point: $\left(\dfrac{\pi}{4}, 0\right)$; halfway points: $(0, 2), \left(\dfrac{\pi}{2}, -2\right)$;   **9.** principal cycle: $\left(-\dfrac{\pi}{2}, -\dfrac{\pi}{6}\right)$; period: $\dfrac{\pi}{3}$;

vertical asymptotes: $x = -\dfrac{\pi}{2}, x = -\dfrac{\pi}{6}$; center point: $\left(-\dfrac{\pi}{3}, 0\right)$; halfway points: $\left(-\dfrac{5\pi}{12}, -1\right), \left(-\dfrac{\pi}{4}, 1\right)$;

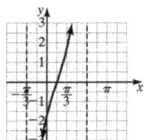

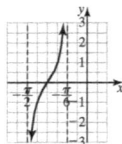

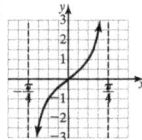

10. principal cycle: $\left(-\dfrac{\pi}{4}, \dfrac{\pi}{4}\right)$; period: $\dfrac{\pi}{2}$; vertical asymptotes: $x = -\dfrac{\pi}{4}, x = \dfrac{\pi}{4}$; center point: $(0, 0)$; halfway points:

$\left(-\dfrac{\pi}{8}, -1\right), \left(\dfrac{\pi}{8}, 1\right)$;   **11.** principal cycle: $(-\pi, \pi)$; period: 2π; vertical asymptotes: $x = -\pi, x = \pi$;

center point: $(0, 0)$; halfway points: $\left(-\dfrac{\pi}{2}, -1\right)$, $\left(\dfrac{\pi}{2}, 1\right)$;

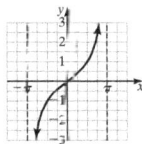

12. principal cycle: $\left(-\dfrac{\pi}{6}, \dfrac{\pi}{6}\right)$; period: $\dfrac{\pi}{3}$;

vertical asymptotes: $x = -\dfrac{\pi}{6}$, $x = \dfrac{\pi}{6}$; center point: $(0, 0)$; halfway points: $\left(-\dfrac{\pi}{12}, 2\right)$, $\left(\dfrac{\pi}{12}, -2\right)$;

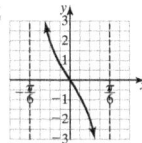

13. principal cycle: $(-3\pi, -\pi)$; period: 2π; vertical asymptotes: $x = -3\pi$, $x = -\pi$; center point: $(-2\pi, 0)$; halfway points: $\left(-\dfrac{5\pi}{2}, -1\right)$, $\left(-\dfrac{3\pi}{2}, 1\right)$;

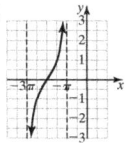

14. principal cycle: $\left(\dfrac{\pi}{4}, \dfrac{3\pi}{4}\right)$; period: $\dfrac{\pi}{2}$; vertical asymptotes: $x = \dfrac{\pi}{4}$, $x = \dfrac{3\pi}{4}$;

center point: $\left(\dfrac{\pi}{2}, 0\right)$; halfway points: $\left(\dfrac{3\pi}{8}, 1\right)$, $\left(\dfrac{5\pi}{8}, -1\right)$;

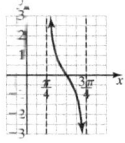

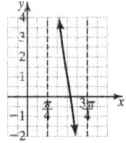

15. principal cycle: $\left(-\dfrac{3\pi}{4}, -\dfrac{\pi}{4}\right)$; period: $\dfrac{\pi}{2}$;

vertical asymptotes: $x = -\dfrac{3\pi}{4}$, $x = -\dfrac{\pi}{4}$; center point: $\left(-\dfrac{\pi}{2}, -1\right)$; halfway points: $\left(-\dfrac{5\pi}{8}, -4\right)$, $\left(-\dfrac{3\pi}{8}, 2\right)$;

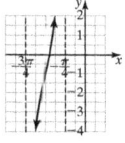

16. principal cycle: $\left(\dfrac{\pi}{4}, \dfrac{3\pi}{4}\right)$; period: $\dfrac{\pi}{2}$; vertical asymptotes: $x = \dfrac{\pi}{4}$, $x = \dfrac{3\pi}{4}$; center point: $\left(\dfrac{\pi}{2}, 1\right)$; halfway points:

$\left(\dfrac{3\pi}{8}, 5\right)$, $\left(\dfrac{5\pi}{8}, -3\right)$;

17. principal cycle: $\left(-\dfrac{\pi}{6}, \dfrac{\pi}{6}\right)$; period: $\dfrac{\pi}{3}$; vertical asymptotes: $x = -\dfrac{\pi}{6}$, $x = \dfrac{\pi}{6}$;

center point: $(0, 4)$; halfway points: $\left(-\dfrac{\pi}{12}, \dfrac{9}{2}\right)$, $\left(\dfrac{\pi}{12}, \dfrac{7}{2}\right)$;

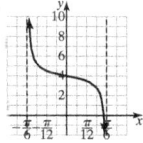

18. principal cycle: $(0, \pi)$; period: π;

vertical asymptotes: $x = 0$, $x = \pi$; center point: $\left(\dfrac{\pi}{2}, 1\right)$; halfway points: $\left(\dfrac{\pi}{4}, 2\right)$, $\left(\dfrac{3\pi}{4}, 0\right)$;

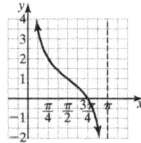

19. principal cycle: $\left(\dfrac{\pi}{4}, \dfrac{3\pi}{4}\right)$; period: $\dfrac{\pi}{2}$; vertical asymptotes: $x = \dfrac{\pi}{4}$, $x = \dfrac{3\pi}{4}$; center point: $\left(\dfrac{\pi}{2}, -1\right)$; halfway points:

$\left(\dfrac{3\pi}{8}, -4\right)$, $\left(\dfrac{5\pi}{8}, 2\right)$;

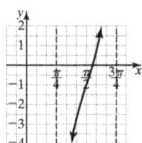

20. ; the function is an odd function. The range is $(-\infty, \infty)$. The zeros are

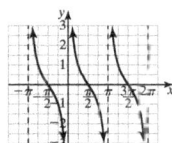

of the form $\dfrac{n\pi}{2}$, where n is an odd integer. Every halfway point has a y-coordinate of -1 or 1. The domain is all real numbers except integer multiples of π. The function has a period of $P = \pi$. The interval for the graph of the principal cycle is $(0, \pi)$.

21. $\left(-\dfrac{11\pi}{6}, \sqrt{3}\right), \left(-\dfrac{5\pi}{6}, \sqrt{3}\right), \left(\dfrac{\pi}{6}, \sqrt{3}\right), \left(\dfrac{7\pi}{6}, \sqrt{3}\right), \left(\dfrac{13\pi}{6}, \sqrt{3}\right)$ **22.** $\left(-\dfrac{7\pi}{3}, -\dfrac{1}{\sqrt{3}}\right), \left(-\dfrac{4\pi}{3}, -\dfrac{1}{\sqrt{3}}\right),$
$\left(-\dfrac{\pi}{3}, -\dfrac{1}{\sqrt{3}}\right)$ **23.** $\left(-\dfrac{7\pi}{4}, 1\right), \left(-\dfrac{5\pi}{4}, -1\right), \left(-\dfrac{3\pi}{4}, 1\right), \left(-\dfrac{\pi}{4}, -1\right), \left(\dfrac{\pi}{4}, 1\right), \left(\dfrac{3\pi}{4}, -1\right)$ **24.** i **25.** principal cycle:
$\left(-\dfrac{\pi}{4}, \dfrac{3\pi}{4}\right)$; period: π; vertical asymptotes: $x = -\dfrac{\pi}{4}, x = \dfrac{3\pi}{4}$; center point: $\left(\dfrac{\pi}{4}, 0\right)$; halfway points: $(0, 1), \left(\dfrac{\pi}{2}, -1\right)$;

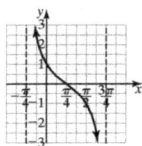

26. principal cycle: $\left(\dfrac{\pi}{2}, \dfrac{3\pi}{2}\right)$; period: π; vertical asymptotes: $x = \dfrac{\pi}{2}, x = \dfrac{3\pi}{2}$; center point: $(\pi, 0)$;

halfway points: $\left(\dfrac{3\pi}{4}, 3\right), \left(\dfrac{5\pi}{4}, -3\right)$; **27.** principal cycle: $\left(\dfrac{\pi}{3}, \dfrac{4\pi}{3}\right)$; period: π; vertical asymptotes:

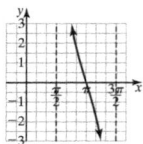

$x = \dfrac{\pi}{3}, x = \dfrac{4\pi}{3}$; center point: $\left(\dfrac{5\pi}{6}, 0\right)$; halfway points: $\left(\dfrac{7\pi}{12}, -2\right), \left(\dfrac{13\pi}{12}, 2\right)$; 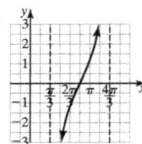 **28.** principal cycle:

$\left(-\dfrac{\pi}{3}, 0\right)$; period: $\dfrac{\pi}{3}$; vertical asymptotes: $x = -\dfrac{\pi}{3}, x = 0$; center point: $\left(-\dfrac{\pi}{6}, 0\right)$; halfway points: $\left(-\dfrac{\pi}{4}, 1\right), \left(-\dfrac{\pi}{12}, -1\right)$;

29. principal cycle: $\left(0, \dfrac{\pi}{2}\right)$; period: $\dfrac{\pi}{2}$; vertical asymptotes: $x = 0, x = \dfrac{\pi}{2}$; center point: $\left(\dfrac{\pi}{4}, 0\right)$;

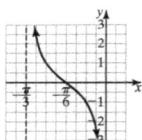

halfway points: $\left(\dfrac{\pi}{8}, 1\right), \left(\dfrac{3\pi}{8}, -1\right)$; 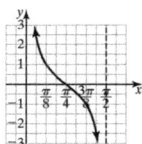 **30.** principal cycle: $(0, 3\pi)$; period: 3π; vertical asymptotes:

$x = 0, x = 3\pi$; center point: $\left(\dfrac{3\pi}{2}, 0\right)$; halfway points: $\left(\dfrac{3\pi}{4}, 1\right), \left(\dfrac{9\pi}{4}, -1\right)$; 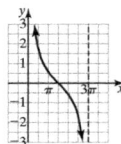 **31.** principal cycle: $\left(0, \dfrac{\pi}{3}\right)$;

period: $\dfrac{\pi}{3}$; vertical asymptotes: $x = 0, x = \dfrac{\pi}{3}$; center point: $\left(\dfrac{\pi}{6}, 0\right)$; halfway points: $\left(\dfrac{\pi}{12}, -2\right), \left(\dfrac{\pi}{4}, 2\right)$;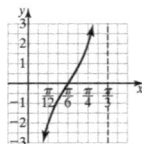

32. principal cycle: $\left(\dfrac{\pi}{2}, \pi\right)$; period: $\dfrac{\pi}{2}$; vertical asymptotes: $x = \dfrac{\pi}{2}, x = \pi$; center point: $\left(\dfrac{3\pi}{4}, 0\right)$; halfway points: $\left(\dfrac{5\pi}{8}, 3\right),$

$\left(\dfrac{7\pi}{8}, -3\right)$;

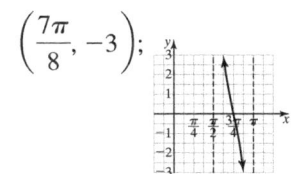

33. principal cycle: $\left(\dfrac{\pi}{2}, \pi\right)$; period: $\dfrac{\pi}{2}$; vertical asymptotes: $x = \dfrac{\pi}{2}$, $x = \pi$; center point: $\left(\dfrac{3\pi}{4}, 0\right)$;

halfway points: $\left(\dfrac{5\pi}{8}, -1\right), \left(\dfrac{7\pi}{8}, 1\right)$;

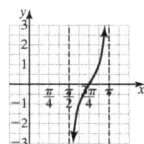

34. principal cycle: $(-2\pi, 0)$; period: 2π; vertical asymptotes:

$x = -2\pi$, $x = 0$; center point: $(-\pi, 0)$; halfway points: $\left(-\dfrac{3\pi}{2}, 1\right), \left(-\dfrac{\pi}{2}, -1\right)$;

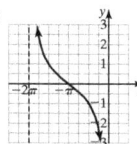

35. principal cycle: $\left(-\dfrac{\pi}{2}, 0\right)$;

period: $\dfrac{\pi}{2}$; vertical asymptotes: $x = -\dfrac{\pi}{2}$, $x = 0$; center point: $\left(-\dfrac{\pi}{4}, -1\right)$; halfway points: $\left(-\dfrac{3\pi}{8}, 2\right), \left(-\dfrac{\pi}{8}, -4\right)$;

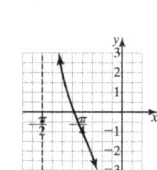

36. principal cycle: $\left(0, \dfrac{\pi}{3}\right)$; period: $\dfrac{\pi}{3}$; vertical asymptotes: $x = 0$, $x = \dfrac{\pi}{3}$; center point: $\left(\dfrac{\pi}{6}, 4\right)$;

halfway points: $\left(\dfrac{\pi}{12}, \dfrac{7}{2}\right), \left(\dfrac{\pi}{4}, \dfrac{9}{2}\right)$;

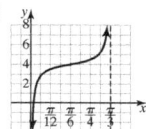

37. principal cycle: $\left(\dfrac{\pi}{2}, \dfrac{3\pi}{2}\right)$; period: π; vertical asymptotes:

$x = \dfrac{\pi}{2}$, $x = \dfrac{3\pi}{2}$; center point: $(\pi, 1)$; halfway points: $\left(\dfrac{3\pi}{4}, 0\right), \left(\dfrac{5\pi}{4}, 2\right)$;

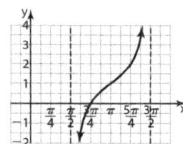

38. principal cycle: $\left(\dfrac{\pi}{2}, \pi\right)$;

period: $\dfrac{\pi}{2}$; vertical asymptotes: $x = \dfrac{\pi}{2}$, $x = \pi$; center point: $\left(\dfrac{3\pi}{4}, -1\right)$; halfway points: $\left(\dfrac{5\pi}{8}, 2\right), \left(\dfrac{7\pi}{8}, -4\right)$;

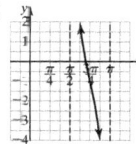

39. $y = 2 \tan (x - \pi)$ or $y = -2 \cot \left(x - \dfrac{\pi}{2}\right)$ **40.** $y = 3 \tan \left(x - \dfrac{\pi}{6}\right)$ or $y = -3 \cot \left(x + \dfrac{\pi}{3}\right)$

41. $y = -4 \tan \left(x + \dfrac{\pi}{2}\right) + 1$ or $y = 4 \cot (x + \pi) + 1$ **42.** $y = -5 \tan \left(x - \dfrac{3\pi}{4}\right) + 2$ or $y = 5 \cot \left(x - \dfrac{\pi}{4}\right) + 2$

43. $y = 2 \tan \left(3x + \dfrac{\pi}{2}\right)$ or $y = -2 \cot (3x + \pi)$ **44.** $y = 4 \tan (2x + \pi) - 3$ or $y = -4 \cot \left(2x + \dfrac{3\pi}{2}\right) - 3$

45.

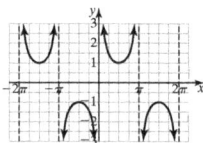

; the function is an odd function. The domain is all real numbers except integer multiples of π. The

function has a period of $P = 2\pi$. The vertical asymptotes are of the form $x = n\pi$, where n is an integer. The function obtains

a relative maximum at $x = -\dfrac{\pi}{2} + 2\pi n$, where n is an integer. The range is $(-\infty, -1] \cup [1, \infty)$.

46. ; the function is an even function. The function has a period of $P = 2\pi$. The domain is all real numbers

except odd integer multiples of $\dfrac{\pi}{2}$. The range is $(-\infty, -1] \cup [1, \infty)$. The function obtains a relative maximum at $x = n\pi$,

where n is an odd integer. The vertical asymptotes are of the form $x = \dfrac{n\pi}{2}$, where n is an odd integer.   **47.** two cycles on the

interval $\left[-\dfrac{3\pi}{2}, \dfrac{5\pi}{2}\right]$; vertical asymptotes: $x = -\dfrac{3\pi}{2}, x = -\dfrac{\pi}{2}, x = \dfrac{\pi}{2}, x = \dfrac{3\pi}{2}, x = \dfrac{5\pi}{2}$; relative maximums: $(0, -2), (2\pi, -2)$;

relative minimums: $(-\pi, 2), (\pi, 2)$;   **48.** two cycles on the interval $\left[-\dfrac{3\pi}{2}, \dfrac{5\pi}{2}\right]$; vertical asymptotes:

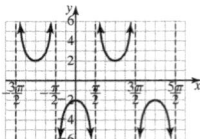

$x = -\dfrac{3\pi}{2}, x = -\dfrac{\pi}{2}, x = \dfrac{\pi}{2}, x = \dfrac{3\pi}{2}, x = \dfrac{5\pi}{2}$; relative maximums: $(-\pi, -2), (\pi, -2)$; relative minimums: $(0, 2), (2\pi, 2)$;

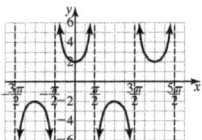

     49. two cycles on the interval $\left[-\dfrac{7\pi}{4}, \dfrac{9\pi}{4}\right]$; vertical asymptotes: $x = -\dfrac{7\pi}{4}, x = -\dfrac{3\pi}{4}, x = \dfrac{\pi}{4}, x = \dfrac{5\pi}{4},$

$x = \dfrac{9\pi}{4}$; relative maximums: $\left(-\dfrac{\pi}{4}, -1\right), \left(\dfrac{7\pi}{4}, -1\right)$; relative minimums: $\left(-\dfrac{5\pi}{4}, 1\right), \left(\dfrac{3\pi}{4}, 1\right)$;

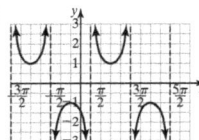

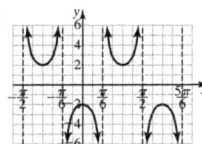

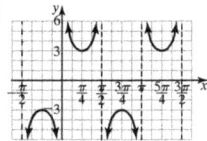

50. two cycles on the interval $\left[-\dfrac{7\pi}{4}, \dfrac{9\pi}{4}\right]$; vertical asymptotes: $x = -\dfrac{7\pi}{4}, x = -\dfrac{3\pi}{4}, x = \dfrac{\pi}{4}, x = \dfrac{5\pi}{4}, x = \dfrac{9\pi}{4}$;

relative maximums: $\left(-\dfrac{5\pi}{4}, -1\right), \left(\dfrac{3\pi}{4}, -1\right)$; relative minimums: $\left(-\dfrac{\pi}{4}, 1\right), \left(\dfrac{7\pi}{4}, 1\right)$;

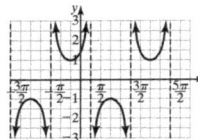

51. two cycles on the interval $\left[-\dfrac{\pi}{2}, \dfrac{5\pi}{6}\right]$; vertical asymptotes: $x = -\dfrac{\pi}{2}, x = -\dfrac{\pi}{6}, x = \dfrac{\pi}{6}, x = \dfrac{\pi}{2}, x = \dfrac{5\pi}{6}$; relative maximums:

$(0, -2), \left(\dfrac{2\pi}{3}, -2\right)$; relative minimums: $\left(-\dfrac{\pi}{3}, 2\right), \left(\dfrac{\pi}{3}, 2\right)$;   **52.** two cycles on the interval $\left[-\dfrac{\pi}{2}, \dfrac{3\pi}{2}\right]$;

vertical asymptotes: $x = -\dfrac{\pi}{2}, x = 0, x = \dfrac{\pi}{2}, x = \pi, x = \dfrac{3\pi}{2}$; relative maximums: $\left(-\dfrac{\pi}{4}, -3\right), \left(\dfrac{3\pi}{4}, -3\right)$; relative minimums:

$\left(\dfrac{\pi}{4}, 3\right), \left(\dfrac{5\pi}{4}, 3\right)$;   **53.** two cycles on the interval $\left[-\dfrac{3\pi}{4}, \dfrac{5\pi}{4}\right]$; vertical asymptotes: $x = -\dfrac{3\pi}{4}, x = -\dfrac{\pi}{4},$

$x = \dfrac{\pi}{4}, x = \dfrac{3\pi}{4}, x = \dfrac{5\pi}{4}$; relative maximums: $(0, -1), (\pi, -1)$; relative minimums: $\left(-\dfrac{\pi}{2}, 7\right), \left(\dfrac{\pi}{2}, 7\right)$;

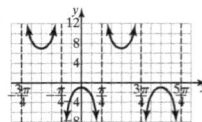

54. two cycles on the interval $\left[-\dfrac{7\pi}{4}, \dfrac{9\pi}{4}\right]$; vertical asymptotes: $x = -\dfrac{7\pi}{4}$, $x = -\dfrac{3\pi}{4}$, $x = \dfrac{\pi}{4}$, $x = \dfrac{5\pi}{4}$, $x = \dfrac{9\pi}{4}$;

relative maximums: $\left(-\dfrac{5\pi}{4}, -7\right)$, $\left(\dfrac{3\pi}{4}, -7\right)$; relative minimums $\left(-\dfrac{\pi}{4}, -1\right)$, $\left(\dfrac{7\pi}{4}, -1\right)$;

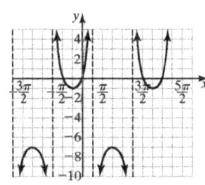

55. two cycles on the interval $\left[-\dfrac{3\pi}{2}, \dfrac{5\pi}{2}\right]$; vertical asymptotes: $x = -\dfrac{3\pi}{2}$, $x = -\dfrac{\pi}{2}$, $x = \dfrac{\pi}{2}$, $x = \dfrac{3\pi}{2}$, $x = \dfrac{5\pi}{2}$; relative maximums:

$(-\pi, 3)$, $(\pi, 3)$; relative minimums: $(0, 7)$, $(2\pi, 7)$;   **56.** two cycles on the interval $\left[-\dfrac{\pi}{2}, \dfrac{3\pi}{2}\right]$;

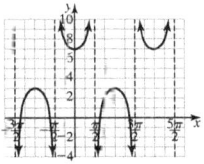

vertical asymptotes: $x = -\dfrac{\pi}{2}$, $x = 0$, $x = \dfrac{\pi}{2}$, $x = \pi$, $x = \dfrac{3\pi}{2}$; relative maximums: $\left(-\dfrac{\pi}{4}, 2\right)$, $\left(\dfrac{3\pi}{4}, 2\right)$; relative minimums:

$\left(\dfrac{\pi}{4}, 4\right)$, $\left(\dfrac{5\pi}{4}, 4\right)$;   **57.** $\left(-\dfrac{\pi}{3}, \dfrac{2\pi}{3}\right)$ **58.** $\left(0, \dfrac{\pi}{2}\right)$ **59.** $\left(-\dfrac{\pi}{2}, -\dfrac{\pi}{6}\right)$ **60.** $\left(\dfrac{\pi}{2}, \pi\right)$

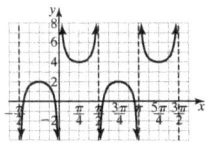

61. $\left(\dfrac{\pi}{4}, \dfrac{3\pi}{4}\right)$ **62.** π **63.** $\dfrac{\pi}{2}$ **64.** $\dfrac{\pi}{3}$ **65.** $\dfrac{\pi}{2}$ **66.** $\dfrac{\pi}{2}$ **67.** $x = -\dfrac{\pi}{3}$, $x = \dfrac{2\pi}{3}$ **68.** $x = 0$, $x = \dfrac{\pi}{2}$

69. $x = -\dfrac{\pi}{2}$, $x = -\dfrac{\pi}{6}$ **70.** $x = \dfrac{\pi}{2}$, $x = \pi$ **71.** $x = \dfrac{\pi}{4}$, $x = \dfrac{3\pi}{4}$ **72.** $\left(\dfrac{\pi}{6}, 0\right)$ **73.** $\left(\dfrac{\pi}{4}, 0\right)$ **74.** $\left(-\dfrac{\pi}{3}, 0\right)$

75. $\left(\dfrac{3\pi}{4}, 0\right)$ **76.** $\left(\dfrac{\pi}{2}, 1\right)$ **77.** $\left(-\dfrac{\pi}{12}, -3\right)$, $\left(\dfrac{5\pi}{12}, 3\right)$ **78.** $\left(\dfrac{\pi}{8}, 1\right)$, $\left(\dfrac{3\pi}{8}, -1\right)$ **79.** $\left(-\dfrac{5\pi}{12}, -1\right)$, $\left(-\dfrac{\pi}{4}, 1\right)$

80. $\left(\dfrac{5\pi}{8}, -1\right)$, $\left(\dfrac{7\pi}{8}, 1\right)$ **81.** $\left(\dfrac{3\pi}{8}, -2\right)$, $\left(\dfrac{5\pi}{8}, 4\right)$ **82.** $y = 2\tan(x - \pi)$ or $y = -2\cot\left(x - \dfrac{\pi}{2}\right)$

83. $y = 3\tan\left(x - \dfrac{\pi}{6}\right)$ or $y = -3\cot\left(x + \dfrac{\pi}{3}\right)$ **84.** $y = -4\tan\left(x + \dfrac{\pi}{2}\right) + 1$ or $y = 4\cot(x + \pi) + 1$

85. $y = -5\tan\left(x - \dfrac{3\pi}{4}\right) + 2$ or $y = 5\cot\left(x - \dfrac{\pi}{4}\right) + 2$ **86.** $y = 2\tan\left(3x + \dfrac{\pi}{2}\right)$ or $y = -2\cot(3x + \pi)$

87. $y = 4\tan(2x + \pi) - 3$ or $y = -4\cot\left(2x + \dfrac{3\pi}{2}\right) - 3$ **88.** $x = -\dfrac{\pi}{2}$, $x = \dfrac{\pi}{2}$, $x = \dfrac{3\pi}{2}$, $x = \dfrac{5\pi}{2}$

89. $x = -\dfrac{\pi}{2}$, $x = 0$, $x = \dfrac{\pi}{2}$, $x = \pi$, $x = \dfrac{3\pi}{2}$ **90.** $(-\pi, 2)$, $(\pi, 2)$, $(3\pi, 2)$ **91.** $\left(-\dfrac{\pi}{4}, 5\right)$, $\left(\dfrac{3\pi}{4}, 5\right)$

92. $(0, -2)$, $(2\pi, -2)$ **93.** $\left(\dfrac{\pi}{4}, -3\right)$, $\left(\dfrac{5\pi}{4}, -3\right)$ **94.**   **95.**   **96.**

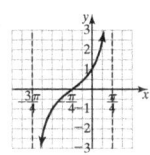 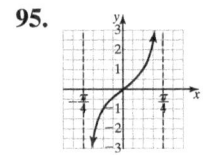 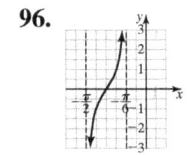

97.   **98.**   **99.**   **100.**   **101.**

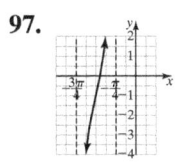 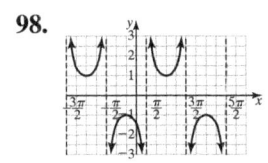 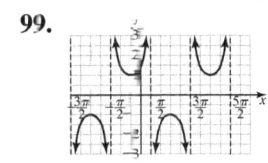 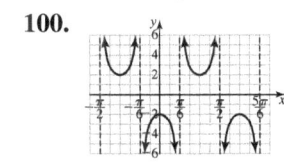 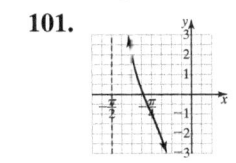

102. 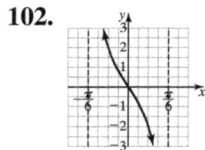   **103.** 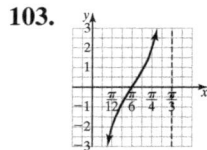   **104.** 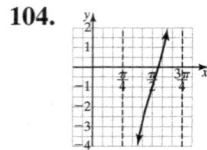   **105.** 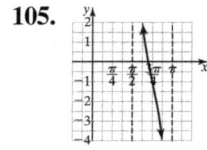   **106.**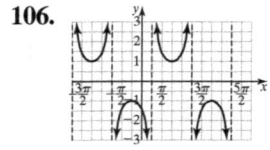

107. 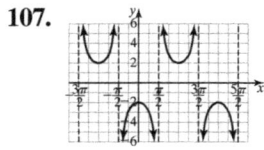   **108.** 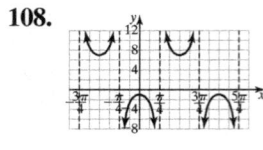   **109.**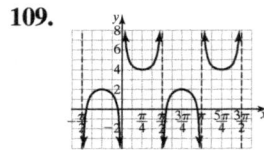

2.4 Exercises

1. $\dfrac{\pi}{3}$ **2.** $-\dfrac{\pi}{4}$ **3.** $\dfrac{\pi}{2}$ **4.** .3363 rad   **5.** -1.0552 rad   **6.** does not exist   **7.** $\dfrac{\pi}{6}$ **8.** $\dfrac{3\pi}{4}$ **9.** 0   **10.** 1.2132 rad

11. 2.7389 rad   **12.** does not exist   **13.** $\dfrac{\pi}{6}$ **14.** $-\dfrac{\pi}{3}$ **15.** $\dfrac{\pi}{4}$ **16.** .6499 rad   **17.** 1.3877 rad   **18.** $\left[-\dfrac{\pi}{2}, \dfrac{\pi}{2}\right]$

19. $[0, \pi]$ **20.** $\left(-\dfrac{\pi}{2}, \dfrac{\pi}{2}\right)$ **21.** Quadrant I   **22.** Quadrant IV   **23.** y-axis   **24.** Quadrant I   **25.** Quadrant II

26. x-axis   **27.** Quadrant I   **28.** Quadrant IV   **29.** Quadrant I   **30.** $\dfrac{\pi}{3}$ **31.** $-\dfrac{\pi}{4}$ **32.** $\dfrac{\pi}{2}$ **33.** $\dfrac{\pi}{6}$ **34.** $\dfrac{3\pi}{4}$

35. 0   **36.** $\dfrac{\pi}{6}$ **37.** $-\dfrac{\pi}{3}$ **38.** $\dfrac{\pi}{4}$

2.5 Exercises

1. $\dfrac{1}{\sqrt{2}}$ **2.** $\dfrac{\sqrt{3}}{2}$ **3.** $\dfrac{1}{\sqrt{3}}$ **4.** $-\dfrac{\sqrt{3}}{2}$ **5.** $-\dfrac{1}{2}$ **6.** $-\sqrt{3}$ **7.** $\dfrac{7}{9}$ **8.** does not exist   **9.** 35.4   **10.** $\dfrac{\pi}{7}$ **11.** $\dfrac{\pi}{5}$

12. $\dfrac{\pi}{3}$ **13.** $-\dfrac{\pi}{4}$ **14.** $\dfrac{3\pi}{4}$ **15.** $\dfrac{\pi}{6}$ **16.** $\dfrac{\pi}{3}$ **17.** $\dfrac{\pi}{4}$ **18.** $\dfrac{\pi}{2}$ **19.** 0   **20.** $-\dfrac{\pi}{3}$ **21.** $-\dfrac{\pi}{4}$ **22.** $\dfrac{5\pi}{6}$ **23.** $\dfrac{2\pi}{5}$

24. $\dfrac{\pi}{7}$ **25.** $\dfrac{\pi}{9}$ **26.** $-\dfrac{\pi}{12}$ **27.** $\dfrac{7\pi}{10}$ **28.** $\dfrac{\sqrt{3}}{2}$ **29.** does not exist   **30.** $\dfrac{1}{2}$ **31.** $\dfrac{2}{\sqrt{3}}$ **32.** 2   **33.** $\dfrac{1}{\sqrt{2}}$ **34.** $\dfrac{\sqrt{3}}{2}$

35. -1 **36.** $\dfrac{7}{4}$ **37.** $\dfrac{\sqrt{74}}{7}$ **38.** $\dfrac{\sqrt{3}}{\sqrt{13}}$ **39.** $\dfrac{\sqrt{11}}{\sqrt{7}}$ **40.** $\dfrac{\pi}{6}$ **41.** does not exist   **42.** $-\dfrac{\pi}{4}$ **43.** does not exist

44. $\dfrac{3\pi}{4}$ **45.** $\dfrac{\pi}{4}$ **46.** $\dfrac{9\pi}{22}$ **47.** $\dfrac{7\pi}{18}$ **48.** $\dfrac{5\pi}{6}$ **49.** $\dfrac{\pi}{6}$ **50.** $-\dfrac{\pi}{3}$ **51.** 0.0666 rad   **52.** does not exist   **53.** 0.3285 rad

54. $\sqrt{1 - u^2}$ **55.** $\dfrac{\sqrt{1 - 4u^2}}{2u}$ **56.** $\dfrac{\sqrt{u^2 - 9}}{u}$ **57.** $\dfrac{\sqrt{11}}{u}$ **58.** $\dfrac{\sqrt{u^2 + 121}}{11}$ **59.** $\dfrac{1}{\sqrt{2}}$ **60.** $\dfrac{\sqrt{3}}{2}$ **61.** $\dfrac{1}{\sqrt{3}}$

62. $-\dfrac{\sqrt{3}}{2}$ **63.** $-\dfrac{1}{2}$ **64.** $-\sqrt{3}$ **65.** $\dfrac{7}{9}$ **66.** does not exist   **67.** 35.4   **68.** $\dfrac{\pi}{7}$ **69.** $\dfrac{\pi}{5}$ **70.** $\dfrac{\pi}{3}$ **71.** $-\dfrac{\pi}{4}$

72. $\dfrac{3\pi}{4}$ **73.** $\dfrac{\pi}{6}$ **74.** $\dfrac{\pi}{3}$ **75.** $\dfrac{\pi}{4}$ **76.** $\dfrac{\pi}{2}$ **77.** 0   **78.** $-\dfrac{\pi}{3}$ **79.** $-\dfrac{\pi}{4}$ **80.** $\dfrac{5\pi}{6}$ **81.** $\dfrac{2\pi}{5}$ **82.** $\dfrac{\pi}{7}$ **83.** $\dfrac{\pi}{9}$

84. $-\dfrac{\pi}{12}$ **85.** $\dfrac{7\pi}{10}$ **86.** $\dfrac{\sqrt{3}}{2}$ **87.** does not exist   **88.** $\dfrac{1}{2}$ **89.** $\dfrac{2}{\sqrt{3}}$ **90.** 2   **91.** $\dfrac{1}{\sqrt{2}}$ **92.** $\dfrac{\sqrt{3}}{2}$ **93.** -1 **94.** $\dfrac{7}{4}$

95. $\dfrac{\sqrt{74}}{7}$ **96.** $\dfrac{\sqrt{3}}{\sqrt{13}}$ **97.** $\dfrac{\sqrt{11}}{\sqrt{7}}$ **98.** $\dfrac{\pi}{6}$ **99.** does not exist   **100.** $-\dfrac{\pi}{4}$ **101.** does not exist   **102.** $\dfrac{3\pi}{4}$ **103.** $\dfrac{\pi}{4}$

104. $\dfrac{9\pi}{22}$ **105.** $\dfrac{7\pi}{18}$

CHAPTER 3

3.1 Exercises

1. $\cos\theta$ **2.** $\cot\theta$ **3.** $\sin\theta$ **4.** $\cos\theta$ **5.** $\sin^2\theta$ **6.** $\tan^2\theta;\ 1$ **7.** $-\tan^2\theta$

8.
$$1 + \cot^2(-\theta) \overset{?}{=} \csc^2\theta$$
$$1 + (\cot(-\theta))^2 \overset{?}{=} \csc^2\theta$$
$$1 + (-\cot\theta)^2 \overset{?}{=} \csc^2\theta$$
$$1 + \cot^2\theta \overset{?}{=} \csc^2\theta$$
$$\csc^2\theta = \csc^2\theta$$

9.
$$\frac{\cot^2\beta + 1}{\csc\beta} \overset{?}{=} \csc\beta$$
$$\frac{\csc^2\beta}{\csc\beta} \overset{?}{=} \csc\beta$$
$$\csc\beta = \csc\beta$$

10.
$$\tan^2 4x + \csc^2 4x - \cot^2 4x \overset{?}{=} \sec^2 4x$$
$$\tan^2 4x + 1 + \cot^2 4x - \cot^2 4x \overset{?}{=} \sec^2 4x$$
$$\tan^2 4x + 1 \overset{?}{=} \sec^2 4x$$
$$\sec^2 4x = \sec^2 4x$$

11.
$$(3\cos\theta - 4\sin\theta)^2 + (4\cos\theta + 3\sin\theta)^2 \overset{?}{=} 25$$
$$9\cos^2\theta - 24\cos\theta\sin\theta + 16\sin^2\theta + 16\cos^2\theta + 24\cos\theta\sin\theta + 9\sin^2\theta \overset{?}{=} 25$$
$$9\cos^2\theta + 16\sin^2\theta + 16\cos^2\theta + 9\sin^2\theta \overset{?}{=} 25$$
$$9(\cos^2\theta + \sin^2\theta) + 16(\sin^2\theta + \cos^2\theta) \overset{?}{=} 25$$
$$9(1) + 16(1) \overset{?}{=} 25$$
$$25 = 25$$

12.
$$7\sin^2\theta + 4\cos^2\theta \overset{?}{=} 4 + 3\sin^2\theta$$
$$3\sin^2\theta + 4\sin^2\theta + 4\cos^2\theta \overset{?}{=} 4 + 3\sin^2\theta$$
$$3\sin^2\theta + 4(\sin^2\theta + \cos^2\theta) \overset{?}{=} 4 + 3\sin^2\theta$$
$$3\sin^2\theta + 4(1) \overset{?}{=} 4 + 3\sin^2\theta$$
$$4 + 3\sin^2\theta = 4 + 3\sin^2\theta$$

13.
$$\cos\theta\csc\theta \overset{?}{=} \cot\theta$$
$$\cos\theta\cdot\frac{1}{\sin\theta} \overset{?}{=} \cot\theta$$
$$\frac{\cos\theta}{\sin\theta} \overset{?}{=} \cot\theta$$
$$\cot\theta = \cot\theta$$

14.
$$\tan(-x)\cos x \overset{?}{=} -\sin x$$
$$-\tan x\cos x \overset{?}{=} -\sin x$$
$$-\frac{\sin x}{\cos x}\cdot\cos x \overset{?}{=} -\sin x$$
$$-\sin x = -\sin x$$

15.
$$\cos\theta\tan\theta\csc\theta \overset{?}{=} 1$$
$$\cos\theta\cdot\frac{\sin\theta}{\cos\theta}\cdot\frac{1}{\sin\theta} \overset{?}{=} 1$$
$$\frac{\cos\theta}{\cos\theta}\cdot\frac{\sin\theta}{\sin\theta} \overset{?}{=} 1$$
$$1 = 1$$

16.
$$\sec\theta - \cos\theta \overset{?}{=} \sin\theta\tan\theta$$
$$\frac{1}{\cos\theta} - \cos\theta \overset{?}{=} \sin\theta\tan\theta$$
$$\frac{1}{\cos\theta} - \cos\theta\cdot\frac{\cos\theta}{\cos\theta} \overset{?}{=} \sin\theta\tan\theta$$
$$\frac{1}{\cos\theta} - \frac{\cos^2\theta}{\cos\theta} \overset{?}{=} \sin\theta\tan\theta$$
$$\frac{1 - \cos^2\theta}{\cos\theta} \overset{?}{=} \sin\theta\tan\theta$$
$$\frac{\sin^2\theta}{\cos\theta} \overset{?}{=} \sin\theta\tan\theta$$
$$\frac{\sin\theta\cdot\sin\theta}{\cos\theta} \overset{?}{=} \sin\theta\tan\theta$$
$$\sin\theta\cdot\frac{\sin\theta}{\cos\theta} \overset{?}{=} \sin\theta\tan\theta$$
$$\sin\theta\tan\theta = \sin\theta\tan\theta$$

17.
$$\csc t\sin t - \sin^2 t \overset{?}{=} \cos^2 t$$
$$\frac{1}{\sin t}\cdot\sin t - \sin^2 t \overset{?}{=} \cos^2 t$$
$$1 - \sin^2 t \overset{?}{=} \cos^2 t$$
$$\cos^2 t = \cos^2 t$$

18.
$$\frac{\csc\theta}{\sin\theta} - \cos(-\theta)\sec(-\theta) \overset{?}{=} \cot^2\theta$$
$$\frac{1}{\sin\theta}\cdot\csc\theta - \cos\theta\sec\theta \overset{?}{=} \cot^2\theta$$
$$\csc\theta\cdot\csc\theta - \cos\theta\cdot\frac{1}{\cos\theta} \overset{?}{=} \cot^2\theta$$
$$\csc^2\theta - 1 \overset{?}{=} \cot^2\theta$$
$$\cot^2\theta = \cot^2\theta$$

19.
$$\sec x + \tan^2 x\sec x \overset{?}{=} \sec^3 x$$
$$\sec x(1 + \tan^2 x) \overset{?}{=} \sec^3 x$$
$$\sec x(\sec^2 x) \overset{?}{=} \sec^3 x$$
$$\sec^3 x = \sec^3 x$$

20.
$$\frac{\sin^2\beta - \cos^2\beta}{\sin\beta - \cos\beta} \overset{?}{=} \sin\beta + \cos\beta$$
$$\frac{(\sin\beta - \cos\beta)(\sin\beta + \cos\beta)}{\sin\beta - \cos\beta} \overset{?}{=} \sin\beta + \cos\beta$$
$$\sin\beta + \cos\beta = \sin\beta + \cos\beta$$

21.
$$\frac{\cot^3\theta + 1}{\cot\theta + 1} \overset{?}{=} \csc^2\theta - \cot\theta$$
$$\frac{(\cot\theta + 1)(\cot^2\theta - \cot\theta + 1)}{\cot\theta + 1} \overset{?}{=} \csc^2\theta - \cot\theta$$
$$\cot^2\theta - \cot\theta + 1 \overset{?}{=} \csc^2\theta - \cot\theta$$
$$\cot^2\theta + 1 - \cot\theta \overset{?}{=} \csc^2\theta - \cot\theta$$
$$\csc^2\theta - \cot\theta = \csc^2\theta - \cot\theta$$

22.
$$\frac{6\csc^2\theta - 7\csc\theta - 3}{1 + 3\csc\theta} \overset{?}{=} 2\csc\theta - 3$$
$$\frac{(2\csc\theta - 3)(3\csc\theta + 1)}{3\csc\theta + 1} \overset{?}{=} 2\csc\theta - 3$$
$$2\csc\theta - 3 = 2\csc\theta - 3$$

23.
$$\sin^4\theta - \cos^4\theta \overset{?}{=} 2\sin^2\theta - 1$$
$$(\sin^2\theta - \cos^2\theta)(\sin^2\theta + \cos^2\theta) \overset{?}{=} 2\sin^2\theta - 1$$
$$(\sin^2\theta - (1 - \sin^2\theta))(1) \overset{?}{=} 2\sin^2\theta - 1$$
$$2\sin^2\theta - 1 = 2\sin^2\theta - 1$$

24.
$$\frac{1 + 2\sec\theta}{\sec\theta} \overset{?}{=} 2 + \cos\theta$$
$$\frac{1}{\sec\theta} + \frac{2\sec\theta}{\sec\theta} \overset{?}{=} 2 + \cos\theta$$
$$\cos\theta + 2 \overset{?}{=} 2 + \cos\theta$$
$$2 + \cos\theta = 2 + \cos\theta$$

25.
$$\frac{\cot\theta + 1}{\csc\theta} \overset{?}{=} \sin\theta + \cos\theta$$
$$\frac{\cot\theta}{\csc\theta} + \frac{1}{\csc\theta} \overset{?}{=} \sin\theta + \cos\theta$$
$$\cot\theta \cdot \frac{1}{\csc\theta} + \sin\theta \overset{?}{=} \sin\theta + \cos\theta$$
$$\frac{\cos\theta}{\sin\theta} \cdot \sin\theta + \sin\theta \overset{?}{=} \sin\theta + \cos\theta$$
$$\cos\theta + \sin\theta \overset{?}{=} \sin\theta + \cos\theta$$
$$\sin\theta + \cos\theta = \sin\theta + \cos\theta$$

26.
$$\frac{\tan\alpha + \cot\alpha}{\tan\alpha} - \frac{\cot\alpha + \tan\alpha}{\cot\alpha} \overset{?}{=} \cot^2\alpha - \tan^2\alpha$$
$$\frac{\tan\alpha}{\tan\alpha} + \frac{\cot\alpha}{\tan\alpha} - \frac{\cot\alpha}{\cot\alpha} - \frac{\tan\alpha}{\cot\alpha} \overset{?}{=} \cot^2\alpha - \tan^2\alpha$$
$$1 + \cot\alpha \cdot \frac{1}{\tan\alpha} - 1 - \tan\alpha \cdot \frac{1}{\cot\alpha} \overset{?}{=} \cot^2\alpha - \tan^2\alpha$$
$$\cot\alpha \cdot \cot\alpha - \tan\alpha \cdot \tan\alpha \overset{?}{=} \cot^2\alpha - \tan^2\alpha$$
$$\cot^2\alpha - \tan^2\alpha = \cot^2\alpha - \tan^2\alpha$$

27.
$$\frac{\sin x + \tan x + 1}{\cos x} \overset{?}{=} \sec x + \tan x + \sin x \sec^2 x$$
$$\frac{\sin x}{\cos x} + \frac{\tan x}{\cos x} + \frac{1}{\cos x} \overset{?}{=} \sec x + \tan x + \sin x \sec^2 x$$
$$\tan x + \tan x \cdot \frac{1}{\cos x} + \sec x \overset{?}{=} \sec x + \tan x + \sin x \sec^2 x$$
$$\tan x + \frac{\sin x}{\cos x} \cdot \frac{1}{\cos x} + \sec x \overset{?}{=} \sec x + \tan x + \sin x \sec^2 x$$
$$\tan x + \sin x \cdot \frac{1}{\cos^2 x} + \sec x \overset{?}{=} \sec x + \tan x + \sin x \sec^2 x$$
$$\tan x + \sin x \sec^2 x + \sec x \overset{?}{=} \sec x + \tan x + \sin x \sec^2 x$$
$$\sec x + \tan x + \sin x \sec^2 x = \sec x + \tan x + \sin x \sec^2 x$$

28.
$$\frac{1 - \cos\theta}{\sin\theta} + \frac{\sin\theta}{1 - \cos\theta} \overset{?}{=} 2\csc\theta$$
$$\frac{1 - \cos\theta}{\sin\theta} \cdot \frac{1 - \cos\theta}{1 - \cos\theta} + \frac{\sin\theta}{1 - \cos\theta} \cdot \frac{\sin\theta}{\sin\theta} \overset{?}{=} 2\csc\theta$$
$$\frac{1 - 2\cos\theta + \cos^2\theta}{\sin\theta(1 - \cos\theta)} + \frac{\sin^2\theta}{\sin\theta(1 - \cos\theta)} \overset{?}{=} 2\csc\theta$$
$$\frac{1 - 2\cos\theta + \cos^2\theta + \sin^2\theta}{\sin\theta(1 - \cos\theta)} \overset{?}{=} 2\csc\theta$$
$$\frac{1 - 2\cos\theta + 1}{\sin\theta(1 - \cos\theta)} \overset{?}{=} 2\csc\theta$$
$$\frac{2 - 2\cos\theta}{\sin\theta(1 - \cos\theta)} \overset{?}{=} 2\csc\theta$$
$$\frac{2(1 - \cos\theta)}{\sin\theta(1 - \cos\theta)} \overset{?}{=} 2\csc\theta$$
$$\frac{2}{\sin\theta} \overset{?}{=} 2\csc\theta$$
$$2\csc\theta = 2\csc\theta$$

29.
$$\frac{\sin t}{1 + \cos t} + \cot t \overset{?}{=} \csc t$$
$$\frac{\sin t}{1 + \cos t} + \frac{\cos t}{\sin t} \overset{?}{=} \csc t$$
$$\frac{\sin t}{1 + \cos t} \cdot \frac{\sin t}{\sin t} + \frac{\cos t}{\sin t} \cdot \frac{1 + \cos t}{1 + \cos t} \overset{?}{=} \csc t$$
$$\frac{\sin^2 t}{(1 + \cos t)\sin t} + \frac{\cos t + \cos^2 t}{\sin t(1 + \cos t)} \overset{?}{=} \csc t$$
$$\frac{\sin^2 t + \cos t + \cos^2 t}{(1 + \cos t)\sin t} \overset{?}{=} \csc t$$
$$\frac{1 + \cos t}{(1 + \cos t)\sin t} \overset{?}{=} \csc t$$
$$\frac{1}{\sin t} \overset{?}{=} \csc t$$
$$\csc t = \csc t$$

30.

$$\dfrac{\sec\beta}{\sin\beta} - \dfrac{\sin\beta}{\sec\beta} \overset{?}{=} \dfrac{\tan^2\beta + \cos^2\beta}{\tan\beta}$$

$$\dfrac{\dfrac{\sec\beta}{\sin\beta} \cdot \dfrac{\sec\beta}{\sec\beta} - \dfrac{\sin\beta}{\sec\beta} \cdot \dfrac{\sin\beta}{\sin\beta}} \overset{?}{=} \dfrac{\tan^2\beta + \cos^2\beta}{\tan\beta}$$

$$\dfrac{\dfrac{\sec^2\beta}{\sin\beta\sec\beta} - \dfrac{\sin^2\beta}{\sec\beta\sin\beta}} \overset{?}{=} \dfrac{\tan^2\beta + \cos^2\beta}{\tan\beta}$$

$$\dfrac{\sec^2\beta - \sin^2\beta}{\sin\beta\sec\beta} \overset{?}{=} \dfrac{\tan^2\beta + \cos^2\beta}{\tan\beta}$$

$$\dfrac{1 + \tan^2\beta - (1 - \cos^2\beta)}{\sin\beta \cdot \dfrac{1}{\cos\beta}} \overset{?}{=} \dfrac{\tan^2\beta + \cos^2\beta}{\tan\beta}$$

$$\dfrac{1 + \tan^2\beta - 1 + \cos^2\beta}{\dfrac{\sin\beta}{\cos\beta}} \overset{?}{=} \dfrac{\tan^2\beta + \cos^2\beta}{\tan\beta}$$

$$\dfrac{\tan^2\beta + \cos^2\beta}{\tan\beta} = \dfrac{\tan^2\beta + \cos^2\beta}{\tan\beta}$$

31.

$$\dfrac{\sin\theta}{1 - \cos\theta} \overset{?}{=} \dfrac{1 + \cos\theta}{\sin\theta}$$

$$\dfrac{\sin\theta}{1 - \cos\theta} \cdot \dfrac{1 + \cos\theta}{1 + \cos\theta} \overset{?}{=} \dfrac{1 + \cos\theta}{\sin\theta}$$

$$\dfrac{\sin\theta\,(1 + \cos\theta)}{1 - \cos^2\theta} \overset{?}{=} \dfrac{1 + \cos\theta}{\sin\theta}$$

$$\dfrac{\sin\theta\,(1 + \cos\theta)}{\sin^2\theta} \overset{?}{=} \dfrac{1 + \cos\theta}{\sin\theta}$$

$$\dfrac{1 + \cos\theta}{\sin\theta} = \dfrac{1 + \cos\theta}{\sin\theta}$$

32.

$$\dfrac{\sec t - 1}{\tan t} \overset{?}{=} \dfrac{\tan t}{\sec t + 1}$$

$$\dfrac{\sec t - 1}{\tan t} \cdot \dfrac{\sec t + 1}{\sec t + 1} \overset{?}{=} \dfrac{\tan t}{\sec t + 1}$$

$$\dfrac{\sec^2 t - 1}{\tan t\,(\sec t + 1)} \overset{?}{=} \dfrac{\tan t}{\sec t + 1}$$

$$\dfrac{\tan^2 t}{\tan t\,(\sec t + 1)} \overset{?}{=} \dfrac{\tan t}{\sec t + 1}$$

$$\dfrac{\tan t}{\sec t + 1} = \dfrac{\tan t}{\sec t + 1}$$

33.
$$\cot^2 3x + \sec^2 3x - \tan^2 3x \overset{?}{=} \csc^2 3x$$
$$\cot^2 3x + 1 \overset{?}{=} \csc^2 3x$$
$$\csc^2 3x = \csc^2 3x$$

34.
$$1 + \cot^2(-\theta) \overset{?}{=} \csc^2\theta$$
$$1 + (\cot(-\theta))^2 \overset{?}{=} \csc^2\theta$$
$$1 + (-\cot\theta)^2 \overset{?}{=} \csc^2\theta$$
$$1 + \cot^2\theta \overset{?}{=} \csc^2\theta$$
$$\csc^2\theta = \csc^2\theta$$

35.
$$\tan(-x)\cos x \overset{?}{=} -\sin x$$
$$\tan(-x)\cos x \overset{?}{=} -\sin x$$
$$-\tan x\cos x \overset{?}{=} -\sin x$$
$$-\dfrac{\sin x}{\cos x} \cdot \cos x \overset{?}{=} -\sin x$$
$$-\sin x = -\sin x$$

36.

$$\dfrac{\csc\theta + 1}{\cot\theta} \overset{?}{=} \sec\theta + \tan\theta$$

$$\dfrac{\csc\theta}{\cot\theta} + \dfrac{1}{\cot\theta} \overset{?}{=} \sec\theta + \tan\theta$$

$$\csc\theta \cdot \dfrac{1}{\cot\theta} + \tan\theta \overset{?}{=} \sec\theta + \tan\theta$$

$$\csc\theta \cdot \tan\theta + \tan\theta \overset{?}{=} \sec\theta + \tan\theta$$

$$\dfrac{1}{\sin\theta} \cdot \dfrac{\sin\theta}{\cos\theta} + \tan\theta \overset{?}{=} \sec\theta + \tan\theta$$

$$\dfrac{1}{\cos\theta} + \tan\theta \overset{?}{=} \sec\theta + \tan\theta$$

$$\sec\theta + \tan\theta \overset{?}{=} \sec\theta + \tan\theta$$

37.

$$\dfrac{1 - \sec\theta}{\tan\theta} - \dfrac{\tan\theta}{1 - \sec\theta} \overset{?}{=} 2\cot\theta$$

$$\dfrac{1 - \sec\theta}{\tan\theta}\cdot\dfrac{1 - \sec\theta}{1 - \sec\theta} - \dfrac{\tan\theta}{1 - \sec\theta}\cdot\dfrac{\tan\theta}{\tan\theta} \overset{?}{=} 2\cot\theta$$

$$\dfrac{1 - 2\sec\theta + \sec^2\theta}{\tan\theta\,(1 - \sec\theta)} - \dfrac{\tan^2\theta}{(1 - \sec\theta)\tan\theta} \overset{?}{=} 2\cot\theta$$

$$\dfrac{1 - 2\sec\theta + \sec^2\theta - \tan^2\theta}{\tan\theta\,(1 - \sec\theta)} \overset{?}{=} 2\cot\theta$$

$$\dfrac{1 - 2\sec\theta + 1}{\tan\theta\,(1 - \sec\theta)} \overset{?}{=} 2\cot\theta$$

$$\dfrac{2 - 2\sec\theta}{\tan\theta\,(1 - \sec\theta)} \overset{?}{=} 2\cot\theta$$

$$\dfrac{2(1 - \sec\theta)}{\tan\theta\,(1 - \sec\theta)} \overset{?}{=} 2\cot\theta$$

$$\dfrac{2}{\tan\theta} \overset{?}{=} 2\cot\theta$$

$$2\cot\theta = 2\cot\theta$$

38.

$$\dfrac{\csc\theta}{\sec\theta} + \dfrac{4\cos\theta}{\sin\theta} \overset{?}{=} 5\cot\theta$$

$$\csc\theta\cos\theta + 4\cot\theta \overset{?}{=} 5\cot\theta$$

$$\dfrac{1}{\sin\theta}\cdot\cos\theta + 4\cot\theta \overset{?}{=} 5\cot\theta$$

$$\cot\theta + 4\cot\theta \overset{?}{=} 5\cot\theta$$

$$5\cot\theta = 5\cot\theta$$

39.

$$\dfrac{\cos^2 t + 3\cos t - 10}{\cos t + 5} \overset{?}{=} \dfrac{1 - 2\sec t}{\sec t}$$

$$\dfrac{(\cos t + 5)(\cos t - 2)}{\cos t + 5} \overset{?}{=} \dfrac{1 - 2\sec t}{\sec t}$$

$$\dfrac{\cos t - 2}{1} \overset{?}{=} \dfrac{1 - 2\sec t}{\sec t}$$

$$\dfrac{\cos t - 2}{1}\cdot\dfrac{\sec t}{\sec t} \overset{?}{=} \dfrac{1 - 2\sec t}{\sec t}$$

$$\dfrac{\cos t\cdot\sec t - 2\sec t}{\sec t} \overset{?}{=} \dfrac{1 - 2\sec t}{\sec t}$$

$$\dfrac{1 - 2\sec t}{\sec t} = \dfrac{1 - 2\sec t}{\sec t}$$

40.

$$1 + \dfrac{1 - \cot^2 x}{1 + \cot^2 x} \overset{?}{=} 2\sin^2 x$$

$$\dfrac{1 + \cot^2 x}{1 + \cot^2 x} + \dfrac{1 - \cot^2 x}{1 + \cot^2 x} \overset{?}{=} 2\sin^2 x$$

$$\dfrac{1 + \cot^2 x + 1 - \cot^2 x}{1 + \cot^2 x} \overset{?}{=} 2\sin^2 x$$

$$\dfrac{2}{1 + \cot^2 x} \overset{?}{=} 2\sin^2 x$$

$$\dfrac{2}{\csc^2 x} \overset{?}{=} 2\sin^2 x$$

$$2\sin^2 x = 2\sin^2 x$$

41.

$$\sec^4\theta - \tan^4\theta \overset{?}{=} 2\sec^2\theta - 1$$

$$(\sec^2\theta + \tan^2\theta)(\sec^2\theta - \tan^2\theta) \overset{?}{=} 2\sec^2\theta - 1$$

$$(\sec^2\theta + \sec^2\theta - 1)(1) \overset{?}{=} 2\sec^2\theta - 1$$

$$2\sec^2\theta - 1 = 2\sec^2\theta - 1$$

42.

$$\dfrac{\tan(-\theta)}{\cot(-\theta)} - \sin\theta\csc(-\theta) \overset{?}{=} \sec^2\theta$$

$$\dfrac{-\tan\theta}{-\cot\theta} - \sin\theta\,(-\csc\theta) \overset{?}{=} \sec^2\theta$$

$$\dfrac{\tan\theta}{\cot\theta} + \sin\theta\,(\csc\theta) \overset{?}{=} \sec^2\theta$$

$$\tan\theta\cdot\dfrac{1}{\cot\theta} + \sin\theta\cdot\dfrac{1}{\sin\theta} \overset{?}{=} \sec^2\theta$$

$$\tan\theta\cdot\tan\theta + 1 \overset{?}{=} \sec^2\theta$$

$$\tan^2\theta + 1 \overset{?}{=} \sec^2\theta$$

$$\sec^2\theta \overset{?}{=} \sec^2\theta$$

43.

$$\dfrac{\cos^3\theta + \sin^3\theta}{\cos\theta + \sin\theta} \overset{?}{=} 1 - \cos\theta\sin\theta$$

$$\dfrac{(\cos\theta + \sin\theta)(\cos^2\theta - \cos\theta\sin\theta + \sin^2\theta)}{\cos\theta + \sin\theta} \overset{?}{=} 1 - \cos\theta\sin\theta$$

$$\cos^2\theta - \cos\theta\sin\theta + \sin^2\theta \overset{?}{=} 1 - \cos\theta\sin\theta$$

$$\cos^2\theta + \sin^2\theta - \cos\theta\sin\theta \overset{?}{=} 1 - \cos\theta\sin\theta$$

$$1 - \cos\theta\sin\theta = 1 - \cos\theta\sin\theta$$

3.2 Exercises

SCE-1. $\dfrac{1 + \sqrt{3}}{2\sqrt{2}}$ **SCE-2.** $-\dfrac{10}{\sqrt{2501}}$ **SCE-3.** $2 + \sqrt{3}$

1. $\dfrac{1 - \sqrt{3}}{2\sqrt{2}}$ **2.** $\dfrac{\sqrt{3} - 1}{2\sqrt{2}}$ **3.** $\dfrac{\sqrt{3} + 1}{2\sqrt{2}}$ **4.** $\dfrac{1 + \sqrt{3}}{2\sqrt{2}}$ **5.** $\dfrac{\sqrt{3} - 1}{2\sqrt{2}}$ **6.** $\dfrac{\sqrt{3} + 1}{2\sqrt{2}}$ **7.** $\dfrac{\sqrt{3} + 1}{2\sqrt{2}}$ **8.** $\dfrac{1 - \sqrt{3}}{2\sqrt{2}}$

9. $\dfrac{1}{2}$ **10.** $\dfrac{\sqrt{3}}{2}$ **11.** $\dfrac{1 - \sqrt{3}}{2\sqrt{2}}$ **12.** $\dfrac{\sqrt{3} - 1}{2\sqrt{2}}$ **13.** $\sqrt{2}(\sqrt{3} - 1)$ **14.** $\dfrac{45 + 8\sqrt{2}}{17\sqrt{11}}$ **15.** $\dfrac{-45 + 8\sqrt{7}}{68}$

16. $\dfrac{1 + \sqrt{3}}{2\sqrt{2}}$ **17.** $\dfrac{1 + \sqrt{3}}{2\sqrt{2}}$ **18.** $\dfrac{\sqrt{3} - 1}{2\sqrt{2}}$ **19.** $\dfrac{\sqrt{3} - 1}{2\sqrt{2}}$ **20.** $\dfrac{1 + \sqrt{3}}{2\sqrt{2}}$ **21.** $\dfrac{1 - \sqrt{3}}{2\sqrt{2}}$ **22.** $\dfrac{\sqrt{3} - 1}{2\sqrt{2}}$

23. $\dfrac{-\sqrt{3} - 1}{2\sqrt{2}}$ **24.** $\dfrac{1}{\sqrt{2}}$ **25.** $\dfrac{1}{2}$ **26.** $\dfrac{\sqrt{3} - 1}{2\sqrt{2}}$ **27.** $\dfrac{\sqrt{3} + 1}{2\sqrt{2}}$ **28.** $-\sqrt{2}(1 + \sqrt{3})$ **29.** $\dfrac{24 - 15\sqrt{2}}{17\sqrt{11}}$

30. $\dfrac{55}{\sqrt{5141}}$ **31.** $2 + \sqrt{3}$ **32.** $-2 - \sqrt{3}$ **33.** $2 - \sqrt{3}$ **34.** $\sqrt{3} - 2$ **35.** $2 + \sqrt{3}$ **36.** $2 - \sqrt{3}$

37. $-2 + \sqrt{3}$ **38.** $-2 - \sqrt{3}$ **39.** $\sqrt{3}$ **40.** $-\sqrt{3}$ **41.** $2 - \sqrt{3}$ **42.** $2 + \sqrt{3}$ **43.** $-2 + \sqrt{3}$

44. $\dfrac{192 - 25\sqrt{15}}{119}$ **45.** $\dfrac{390 + 61\sqrt{42}}{102}$ **46.**

$$\sin\left(\frac{\pi}{2} + \theta\right) \overset{?}{=} \cos\theta$$
$$\sin\left(\frac{\pi}{2}\right)\cos\theta + \sin\theta\cos\left(\frac{\pi}{2}\right) \overset{?}{=} \cos\theta$$
$$(1)\cos\theta + \sin\theta\,(0) \overset{?}{=} \cos\theta$$
$$\cos\theta = \cos\theta$$

47.

$$\cos(\alpha + \beta) + \cos(\alpha - \beta) \overset{?}{=} 2\cos\alpha\cos\beta$$
$$\cos\alpha\cos\beta - \sin\alpha\sin\beta + \cos\alpha\cos\beta + \sin\alpha\sin\beta \overset{?}{=} 2\cos\alpha\cos\beta$$
$$2\cos\alpha\cos\beta = 2\cos\alpha\cos\beta$$

48.

$$\frac{\cos(\alpha + \beta)}{\sin\alpha\cos\beta} \overset{?}{=} \cot\alpha - \tan\beta$$
$$\frac{\cos\alpha\cos\beta - \sin\alpha\sin\beta}{\sin\alpha\cos\beta} \overset{?}{=} \cot\alpha - \tan\beta$$
$$\frac{\cos\alpha\cos\beta}{\sin\alpha\cos\beta} - \frac{\sin\alpha\sin\beta}{\sin\alpha\cos\beta} \overset{?}{=} \cot\alpha - \tan\beta$$
$$\frac{\cos\alpha}{\sin\alpha} - \frac{\sin\beta}{\cos\beta} \overset{?}{=} \cot\alpha - \tan\beta$$
$$\cot\alpha - \tan\beta = \cot\alpha - \tan\beta$$

49.

$$\frac{\cos(\alpha - \beta)}{\cos(\alpha + \beta)} \overset{?}{=} \frac{1 + \tan\alpha\tan\beta}{1 - \tan\alpha\tan\beta}$$
$$\frac{\cos\alpha\cos\beta + \sin\alpha\sin\beta}{\cos\alpha\cos\beta - \sin\alpha\sin\beta} \overset{?}{=} \frac{1 + \tan\alpha\tan\beta}{1 - \tan\alpha\tan\beta}$$
$$\frac{\cos\alpha\cos\beta + \sin\alpha\sin\beta}{\cos\alpha\cos\beta - \sin\alpha\sin\beta} \cdot \frac{\dfrac{1}{\cos\alpha\cos\beta}}{\dfrac{1}{\cos\alpha\cos\beta}} \overset{?}{=} \frac{1 + \tan\alpha\tan\beta}{1 - \tan\alpha\tan\beta}$$
$$\frac{\dfrac{\cos\alpha\cos\beta + \sin\alpha\sin\beta}{\cos\alpha\cos\beta}}{\dfrac{\cos\alpha\cos\beta - \sin\alpha\sin\beta}{\cos\alpha\cos\beta}} \overset{?}{=} \frac{1 + \tan\alpha\tan\beta}{1 - \tan\alpha\tan\beta}$$
$$\frac{\dfrac{\cos\alpha\cos\beta}{\cos\alpha\cos\beta} + \dfrac{\sin\alpha\sin\beta}{\cos\alpha\cos\beta}}{\dfrac{\cos\alpha\cos\beta}{\cos\alpha\cos\beta} - \dfrac{\sin\alpha\sin\beta}{\cos\alpha\cos\beta}} \overset{?}{=} \frac{1 + \tan\alpha\tan\beta}{1 - \tan\alpha\tan\beta}$$
$$\frac{1 + \tan\alpha\tan\beta}{1 - \tan\alpha\tan\beta} = \frac{1 + \tan\alpha\tan\beta}{1 - \tan\alpha\tan\beta}$$

50.
$$\tan(\alpha - \beta) \stackrel{?}{=} \frac{\cot\beta - \cot\alpha}{\cot\alpha\cot\beta + 1}$$
$$\frac{\tan\alpha - \tan\beta}{1 + \tan\alpha\tan\beta} \stackrel{?}{=} \frac{\cot\beta - \cot\alpha}{\cot\alpha\cot\beta + 1}$$
$$\frac{\tan\alpha - \tan\beta}{1 + \tan\alpha\tan\beta} \cdot \frac{\cot\alpha\cot\beta}{\cot\alpha\cot\beta} \stackrel{?}{=} \frac{\cot\beta - \cot\alpha}{\cot\alpha\cot\beta + 1}$$
$$\frac{\tan\alpha\cot\alpha\cot\beta - \tan\beta\cot\alpha\cot\beta}{\cot\alpha\cot\beta + \tan\alpha\tan\beta\cot\alpha\cot\beta} \stackrel{?}{=} \frac{\cot\beta - \cot\alpha}{\cot\alpha\cot\beta + 1}$$
$$\frac{(\tan\alpha\cot\alpha)\cot\beta - (\tan\beta\cot\beta)\cot\alpha}{\cot\alpha\cot\beta + (\tan\alpha\cot\alpha)(\tan\beta\cot\beta)} \stackrel{?}{=} \frac{\cot\beta - \cot\alpha}{\cot\alpha\cot\beta + 1}$$
$$\frac{(1)\cot\beta - (1)\cot\alpha}{\cot\alpha\cot\beta + (1)(1)} \stackrel{?}{=} \frac{\cot\beta - \cot\alpha}{\cot\alpha\cot\beta + 1}$$
$$\frac{\cot\beta - \cot\alpha}{\cot\alpha\cot\beta + 1} = \frac{\cot\beta - \cot\alpha}{\cot\alpha\cot\beta + 1}$$

51.
$$\csc(\alpha - \beta) \stackrel{?}{=} \frac{\csc\alpha\csc\beta}{\cot\beta - \cot\alpha}$$
$$\frac{1}{\sin(\alpha - \beta)} \stackrel{?}{=} \frac{\csc\alpha\csc\beta}{\cot\beta - \cot\alpha}$$
$$\frac{1}{\sin\alpha\cos\beta - \sin\beta\cos\alpha} \stackrel{?}{=} \frac{\csc\alpha\csc\beta}{\cot\beta - \cot\alpha}$$
$$\frac{1}{\sin\alpha\cos\beta - \sin\beta\cos\alpha} \cdot \frac{\dfrac{1}{\sin\alpha\sin\beta}}{\dfrac{1}{\sin\alpha\sin\beta}} \stackrel{?}{=} \frac{\csc\alpha\csc\beta}{\cot\beta - \cot\alpha}$$
$$\frac{\dfrac{1}{\sin\alpha\sin\beta}}{\dfrac{\sin\alpha\cos\beta}{\sin\alpha\sin\beta} - \dfrac{\sin\beta\cos\alpha}{\sin\alpha\sin\beta}} \stackrel{?}{=} \frac{\csc\alpha\csc\beta}{\cot\beta - \cot\alpha}$$
$$\frac{\dfrac{1}{\sin\alpha} \cdot \dfrac{1}{\sin\beta}}{\dfrac{\cos\beta}{\sin\beta} - \dfrac{\cos\alpha}{\sin\alpha}} \stackrel{?}{=} \frac{\csc\alpha\csc\beta}{\cot\beta - \cot\alpha}$$
$$\frac{\csc\alpha\csc\beta}{\cot\beta - \cot\alpha} = \frac{\csc\alpha\csc\beta}{\cot\beta - \cot\alpha}$$

52. $\dfrac{1 + \sqrt{3}}{2\sqrt{2}}$ **53.** -1 **54.** $\dfrac{-3 - 10\sqrt{14}}{42}$ **55.** $\sqrt{3} - 2$

3.3 Exercises

SCE-1. $\dfrac{\sqrt{2 - \sqrt{2}}}{2}$ **SCE-2.** $\dfrac{\sqrt{2 + \sqrt{2}}}{2}$ **SCE-3.** $-\dfrac{\sqrt{3}}{\sqrt{11}}$ **SCE-4.** $\dfrac{\sqrt{9}}{\sqrt{14}}$ **SCE-5.** $\sqrt{3 + 2\sqrt{2}}$

1. $\sin\dfrac{\pi}{4} = \dfrac{1}{\sqrt{2}}$ **2.** $\cos 150° = -\dfrac{\sqrt{3}}{2}$ **3.** $\tan\dfrac{7\pi}{4} = -1$ **4.** $\cos\left(-\dfrac{5\pi}{6}\right) = -\dfrac{\sqrt{3}}{2}$ **5.** $\sin 405° = \dfrac{1}{\sqrt{2}}$

6. $\cos\left(-\dfrac{5\pi}{4}\right) = -\dfrac{1}{\sqrt{2}}$ **7.** $\tan(-135°) = 1$ **8.** $\cos 210° = -\dfrac{\sqrt{3}}{2}$ **9.** $\cos(-315°) = \dfrac{1}{\sqrt{2}}$

10. $\cos\left(-\dfrac{9\pi}{4}\right) = \dfrac{1}{\sqrt{2}}$ **11.** $\sin 2\theta = -\dfrac{4\sqrt{165}}{169}$; $\cos 2\theta = \dfrac{161}{169}$; $\tan 2\theta = -\dfrac{4\sqrt{165}}{161}$ **12.** $\sin 2\theta = -\dfrac{840}{841}$; $\cos 2\theta = \dfrac{41}{841}$;

$\tan 2\theta = -\dfrac{840}{41}$ **13.** $\sin 2\theta = \dfrac{240}{289}$; $\cos 2\theta = \dfrac{161}{289}$; $\tan 2\theta = \dfrac{240}{161}$ **14.** $\sin 2\theta = \dfrac{20}{101}$; $\cos 2\theta = \dfrac{99}{101}$; $\tan 2\theta = \dfrac{20}{99}$

15. $\sin\theta = \dfrac{5}{\sqrt{26}}$; $\cos\theta = \dfrac{1}{\sqrt{26}}$; $\tan\theta = 5$ **16.** $\sin\theta = \dfrac{1}{\sqrt{5}}$; $\cos\theta = -\dfrac{2}{\sqrt{5}}$; $\tan\theta = -\dfrac{1}{2}$ **17.** $\sin\theta = \dfrac{\sqrt{53 + 7\sqrt{53}}}{\sqrt{106}}$;

$\cos\theta = -\dfrac{\sqrt{53 - 7\sqrt{53}}}{\sqrt{106}}$; $\tan\theta = -\dfrac{\sqrt{53} + 7}{2}$ **18.** $\sin\theta = \dfrac{\sqrt{7 - 2\sqrt{6}}}{\sqrt{14}}$; $\cos\theta = -\dfrac{\sqrt{7 + 2\sqrt{6}}}{\sqrt{14}}$;

$\tan\theta = \dfrac{2\sqrt{6} - 7}{5}$ **19.** $f(x) = -\dfrac{3}{2} + \dfrac{3}{2}\cos 2x$ **20.** $f(x) = \dfrac{7}{2} + \dfrac{7}{2}\cos 2x$ **21.** $f(x) = \dfrac{3}{4} - \cos 2x + \dfrac{1}{4}\cos 4x$

22. $f(x) = -\dfrac{9}{8} - \dfrac{3}{2}\cos 2x - \dfrac{3}{8}\cos 4x$ **23.** $f(x) = -\dfrac{1}{4} + \dfrac{1}{4}\cos 4x$ **24.** $-\sqrt{\dfrac{1 + \dfrac{1}{\sqrt{2}}}{2}}$ **25.** $\sqrt{\dfrac{1 + \dfrac{1}{\sqrt{2}}}{2}}$

26. $\dfrac{-\dfrac{1}{\sqrt{2}}}{1 + \dfrac{1}{\sqrt{2}}}$ **27.** $\sqrt{\dfrac{1 + \dfrac{1}{\sqrt{2}}}{2}}$ **28.** $\sqrt{\dfrac{1 + \dfrac{\sqrt{3}}{2}}{2}}$ **29.** $\dfrac{\dfrac{1}{2}}{1 - \dfrac{\sqrt{3}}{2}}$ **30.** $-\dfrac{1}{\sqrt{\dfrac{1 - \dfrac{1}{\sqrt{2}}}{2}}}$ **31.** $-\dfrac{1}{\sqrt{\dfrac{1 + \dfrac{\sqrt{3}}{2}}{2}}}$

32. $-\dfrac{\sqrt{2+\sqrt{2}}}{2}$ **33.** $\dfrac{\sqrt{2+\sqrt{2}}}{2}$ **34.** $1-\sqrt{2}$ **35.** $\dfrac{\sqrt{2+\sqrt{2}}}{2}$ **36.** $\dfrac{\sqrt{2+\sqrt{3}}}{2}$ **37.** $2+\sqrt{3}$

38. $-\sqrt{4+2\sqrt{2}}$ **39.** $-2\sqrt{2-\sqrt{3}}$ **40.** $\sin\left(\dfrac{\alpha}{2}\right)=\dfrac{3}{\sqrt{13}}; \cos\left(\dfrac{\alpha}{2}\right)=\dfrac{2}{\sqrt{13}}; \tan\left(\dfrac{\alpha}{2}\right)=\dfrac{3}{2}$ **41.** $\sin\left(\dfrac{\alpha}{2}\right)=\dfrac{5}{\sqrt{29}};$

$\cos\left(\dfrac{\alpha}{2}\right)=\dfrac{2}{\sqrt{29}}; \tan\left(\dfrac{\alpha}{2}\right)=\dfrac{5}{2}$ **42.** $\sin\left(\dfrac{\alpha}{2}\right)=\dfrac{3}{\sqrt{13}}; \cos\left(\dfrac{\alpha}{2}\right)=-\dfrac{2}{\sqrt{13}}; \tan\left(\dfrac{\alpha}{2}\right)=-\dfrac{3}{2}$ **43.** $\sin\left(\dfrac{\alpha}{2}\right)=\dfrac{3}{\sqrt{10}};$

$\cos\left(\dfrac{\alpha}{2}\right)=-\dfrac{1}{\sqrt{10}}; \tan\left(\dfrac{\alpha}{2}\right)=-3$ **44.** $\sin\left(\dfrac{\alpha}{2}\right)=\dfrac{\sqrt{10+\sqrt{10}}}{2\sqrt{5}}; \cos\left(\dfrac{\alpha}{2}\right)=-\dfrac{\sqrt{10-\sqrt{10}}}{2\sqrt{5}};$

$\tan\left(\dfrac{\alpha}{2}\right)=-\dfrac{\sqrt{10}+1}{3}$ **45.** $\sin\left(\dfrac{\alpha}{2}\right)=\dfrac{\sqrt{3}}{\sqrt{11}}; \cos\left(\dfrac{\alpha}{2}\right)=-\dfrac{4}{\sqrt{22}}; \tan\left(\dfrac{\alpha}{2}\right)=-\dfrac{\sqrt{6}}{4}$

46.
$$\dfrac{\tan^2\theta-1}{1+\tan^2\theta}\overset{?}{=}-\cos 2\theta$$
$$\dfrac{\tan^2\theta-1}{\sec^2\theta}\overset{?}{=}-\cos 2\theta$$
$$(\tan^2\theta-1)\cdot\dfrac{1}{\sec^2\theta}\overset{?}{=}-\cos 2\theta$$
$$(\tan^2\theta-1)\cdot\cos^2\theta\overset{?}{=}-\cos 2\theta$$
$$\tan^2\theta\cos^2\theta-\cos^2\theta\overset{?}{=}-\cos 2\theta$$
$$\sin^2\theta-\cos^2\theta\overset{?}{=}-\cos 2\theta$$
$$-(\cos^2\theta-\sin^2\theta)\overset{?}{=}-\cos 2\theta$$
$$-\cos 2\theta=-\cos 2\theta$$

47.
$$\cot 2\theta\overset{?}{=}\dfrac{1}{2}\sec\theta\csc\theta-\tan\theta$$
$$\dfrac{\cos 2\theta}{\sin 2\theta}\overset{?}{=}\dfrac{1}{2}\sec\theta\csc\theta-\tan\theta$$
$$\dfrac{1-2\sin^2\theta}{2\sin\theta\cos\theta}\overset{?}{=}\dfrac{1}{2}\sec\theta\csc\theta-\tan\theta$$
$$\dfrac{1}{2\sin\theta\cos\theta}-\dfrac{2\sin^2\theta}{2\sin\theta\cos\theta}\overset{?}{=}\dfrac{1}{2}\sec\theta\csc\theta-\tan\theta$$
$$\dfrac{1}{2}\cdot\dfrac{1}{\sin\theta}\cdot\dfrac{1}{\cos\theta}-\dfrac{\sin\theta}{\cos\theta}\overset{?}{=}\dfrac{1}{2}\sec\theta\csc\theta-\tan\theta$$
$$\dfrac{1}{2}\sec\theta\csc\theta-\tan\theta=\dfrac{1}{2}\sec\theta\csc\theta-\tan\theta$$

48.
$$\sec 2\theta\overset{?}{=}\dfrac{\csc^2\theta}{\csc^2\theta-2}$$
$$\dfrac{1}{\cos 2\theta}\overset{?}{=}\dfrac{\csc^2\theta}{\csc^2\theta-2}$$
$$\dfrac{1}{1-2\sin^2\theta}\overset{?}{=}\dfrac{\csc^2\theta}{\csc^2\theta-2}$$
$$\dfrac{1}{1-2\sin^2\theta}\cdot\dfrac{\csc^2\theta}{\csc^2\theta}\overset{?}{=}\dfrac{\csc^2\theta}{\csc^2\theta-2}$$
$$\dfrac{\csc^2\theta}{\csc^2\theta-2\sin^2\theta\csc^2\theta}\overset{?}{=}\dfrac{\csc^2\theta}{\csc^2\theta-2}$$
$$\dfrac{\csc^2\theta}{\csc^2\theta-2}=\dfrac{\csc^2\theta}{\csc^2\theta-2}$$

49.
$$\tan\theta\overset{?}{=}\dfrac{1-\cos 2\theta}{\sin 2\theta}$$
$$\tan\theta\overset{?}{=}\dfrac{1-(1-2\sin^2\theta)}{2\sin\theta\cos\theta}$$
$$\tan\theta\overset{?}{=}\dfrac{2\sin^2\theta}{2\sin\theta\cos\theta}$$
$$\tan\theta\overset{?}{=}\dfrac{\sin\theta}{\cos\theta}$$
$$\tan\theta=\tan\theta$$

50.
$$\sin^2\dfrac{\theta}{2}\overset{?}{=}\dfrac{\tan\theta-\sin\theta}{2\tan\theta}$$
$$\dfrac{1-\cos\theta}{2}\overset{?}{=}\dfrac{\tan\theta-\sin\theta}{2\tan\theta}$$
$$\dfrac{1-\cos\theta}{2}\cdot\dfrac{\tan\theta}{\tan\theta}\overset{?}{=}\dfrac{\tan\theta-\sin\theta}{2\tan\theta}$$
$$\dfrac{\tan\theta-\cos\theta\tan\theta}{2\tan\theta}\overset{?}{=}\dfrac{\tan\theta-\sin\theta}{2\tan\theta}$$
$$\dfrac{\tan\theta-\sin\theta}{2\tan\theta}=\dfrac{\tan\theta-\sin\theta}{2\tan\theta}$$

51.
$$\tan\dfrac{\theta}{2}\overset{?}{=}\dfrac{1}{\csc\theta+\cot\theta}$$
$$\dfrac{\sin\theta}{1+\cos\theta}\overset{?}{=}\dfrac{1}{\csc\theta+\cot\theta}$$
$$\dfrac{\sin\theta}{1+\cos\theta}\cdot\dfrac{\frac{1}{\sin\theta}}{\frac{1}{\sin\theta}}\overset{?}{=}\dfrac{1}{\csc\theta+\cot\theta}$$
$$\dfrac{\frac{\sin\theta}{\sin\theta}}{\frac{1}{\sin\theta}+\frac{\cos\theta}{\sin\theta}}\overset{?}{=}\dfrac{1}{\csc\theta+\cot\theta}$$
$$\dfrac{1}{\csc\theta+\cot\theta}=\dfrac{1}{\csc\theta+\cot\theta}$$

52.
$$\cot\dfrac{\theta}{2}-\cot\theta\overset{?}{=}\csc\theta$$
$$\dfrac{1}{\tan\frac{\theta}{2}}-\cot\theta\overset{?}{=}\csc\theta$$
$$\dfrac{1}{\tan\frac{\theta}{2}}-\cot\theta\overset{?}{=}\csc\theta$$
$$\dfrac{1+\cos\theta}{\sin\theta}-\cot\theta\overset{?}{=}\csc\theta$$
$$\dfrac{1}{\sin\theta}+\dfrac{\cos\theta}{\sin\theta}-\cot\theta\overset{?}{=}\csc\theta$$
$$\csc\theta+\cot\theta-\cot\theta\overset{?}{=}\csc\theta$$
$$\csc\theta=\csc\theta$$

53. $\dfrac{\sqrt{3}}{2}$ **54.** 0 **55.** $\dfrac{20}{29}$ **56.** $\dfrac{31}{81}$

57. $\sqrt{\dfrac{1+\dfrac{1}{2}}{2}}$ **58.** $\sqrt{\dfrac{1+\dfrac{1}{\sqrt{2}}}{2}}$ **59.** $\sqrt{\dfrac{1+\dfrac{3}{7}}{2}}$ **60.** $\sqrt{\dfrac{1+\dfrac{5}{11}}{2}}$ **61.** $-\sqrt{\dfrac{1-\dfrac{1}{\sqrt{2}}}{2}}$ **62.** $\sqrt{\dfrac{1+\dfrac{\sqrt{3}}{2}}{2}}$

63. $\sqrt{\dfrac{1-\dfrac{7}{\sqrt{74}}}{2}}$ **64.** $\sqrt{\dfrac{1+\dfrac{1}{\sqrt{26}}}{2}}$ **65.** $\dfrac{\sqrt{3}}{2}$ **66.** $\dfrac{\sqrt{2+\sqrt{2}}}{2}$ **67.** $\dfrac{\sqrt{5}}{\sqrt{7}}$ **68.** $\dfrac{2\sqrt{2}}{\sqrt{11}}$ **69.** $-\dfrac{\sqrt{2-\sqrt{2}}}{2}$

70. $\dfrac{\sqrt{2+\sqrt{3}}}{2}$ **71.** $\dfrac{\sqrt{74-7\sqrt{74}}}{2\sqrt{37}}$ **72.** $\dfrac{\sqrt{26+\sqrt{26}}}{2\sqrt{13}}$ **73.** $-\sqrt{\dfrac{1+\dfrac{1}{\sqrt{2}}}{2}}$ **74.** $\sqrt{\dfrac{1+\dfrac{1}{\sqrt{2}}}{2}}$ **75.** $\dfrac{-\dfrac{1}{\sqrt{2}}}{1+\dfrac{1}{\sqrt{2}}}$

76. $\sqrt{\dfrac{1+\dfrac{1}{\sqrt{2}}}{2}}$ **77.** $\sqrt{\dfrac{1+\dfrac{\sqrt{3}}{2}}{2}}$ **78.** $\dfrac{\dfrac{1}{2}}{1-\dfrac{\sqrt{3}}{2}}$ **79.** $-\dfrac{1}{\sqrt{\dfrac{1-\dfrac{1}{\sqrt{2}}}{2}}}$ **80.** $-\dfrac{1}{\sqrt{\dfrac{1+\dfrac{\sqrt{3}}{2}}{2}}}$ **81.** $-\dfrac{\sqrt{2+\sqrt{2}}}{2}$

82. $\dfrac{\sqrt{2+\sqrt{2}}}{2}$ **83.** $1-\sqrt{2}$ **84.** $\dfrac{\sqrt{2+\sqrt{2}}}{2}$ **85.** $\dfrac{\sqrt{2+\sqrt{3}}}{2}$ **86.** $2+\sqrt{3}$ **87.** $-\sqrt{4+2\sqrt{2}}$

88. $-2\sqrt{2-\sqrt{3}}$ **89.** $\sqrt{\dfrac{1+\dfrac{1}{2}}{2}}$ **90.** $\sqrt{\dfrac{1+\dfrac{1}{\sqrt{2}}}{2}}$ **91.** $\sqrt{\dfrac{1+\dfrac{3}{7}}{2}}$ **92.** $\sqrt{\dfrac{1+\dfrac{5}{11}}{2}}$ **93.** $-\sqrt{\dfrac{1-\dfrac{1}{\sqrt{2}}}{2}}$

94. $\sqrt{\dfrac{1+\dfrac{\sqrt{3}}{2}}{2}}$ **95.** $\sqrt{\dfrac{1-\dfrac{7}{\sqrt{74}}}{2}}$ **96.** $\sqrt{\dfrac{1+\dfrac{1}{\sqrt{26}}}{2}}$ **97.** $\dfrac{\sqrt{3}}{2}$ **98.** $\dfrac{\sqrt{2+\sqrt{2}}}{2}$ **99.** $\dfrac{\sqrt{5}}{\sqrt{7}}$ **100.** $\dfrac{2\sqrt{2}}{\sqrt{11}}$

101. $-\dfrac{\sqrt{2-\sqrt{2}}}{2}$ **102.** $\dfrac{\sqrt{2+\sqrt{3}}}{2}$ **103.** $\dfrac{\sqrt{74-7\sqrt{74}}}{2\sqrt{37}}$ **104.** $\dfrac{\sqrt{26+\sqrt{26}}}{2\sqrt{13}}$

3.4 Exercises

1. $\dfrac{1}{2}\cos(4\theta)-\dfrac{1}{2}\cos(12\theta)$ **2.** $\dfrac{1}{2}\sin(10\theta)+\dfrac{1}{2}\sin(4\theta)$ **3.** $\dfrac{1}{2}\cos(4\theta)+\dfrac{1}{2}\cos(6\theta)$ **4.** $\dfrac{1}{2}\sin(3\theta)+\dfrac{1}{2}\sin(2\theta)$

5. $\dfrac{1+\sqrt{2}}{4}$ **6.** $\dfrac{1}{4}$ **7.** $\dfrac{1-\sqrt{2}}{2\sqrt{2}}$ **8.** $\dfrac{\sqrt{2}-1}{2\sqrt{2}}$ **9.** $2\sin\theta\cos(8\theta)$ **10.** $2\cos(4\theta)\cos\theta$ **11.** $2\sin(10\theta)\cos\theta$

12. $2\sin(4\theta)\sin\left(\dfrac{3\theta}{2}\right)$ **13.** $\dfrac{1}{\sqrt{2}}$ **14.** 0 **15.** $-\dfrac{1}{\sqrt{2}}$ **16.** $-\sqrt{2}$ **17.**

$$\dfrac{\sin\theta-\sin 3\theta}{\cos\theta+\cos 3\theta}\overset{?}{=}-\tan\theta$$

$$\dfrac{2\sin\left(\dfrac{\theta-3\theta}{2}\right)\cos\left(\dfrac{\theta+3\theta}{2}\right)}{2\cos\left(\dfrac{\theta+3\theta}{2}\right)\cos\left(\dfrac{\theta-3\theta}{2}\right)}\overset{?}{=}-\tan\theta$$

$$\dfrac{2\sin(-\theta)\cos 2\theta}{2\cos 2\theta\cos(-\theta)}\overset{?}{=}-\tan\theta$$

$$\dfrac{\sin(-\theta)}{\cos(-\theta)}\overset{?}{=}-\tan\theta$$

$$\dfrac{-\sin\theta}{\cos\theta}\overset{?}{=}-\tan\theta$$

$$-\tan\theta=-\tan\theta$$

18.

$$\frac{\cos\theta + \cos 3\theta}{\sin\theta + \sin 3\theta} \overset{?}{=} \cot 2\theta$$

$$\frac{2\cos\left(\dfrac{\theta + 3\theta}{2}\right)\cos\left(\dfrac{\theta - 3\theta}{2}\right)}{2\sin\left(\dfrac{\theta + 3\theta}{2}\right)\cos\left(\dfrac{\theta - 3\theta}{2}\right)} \overset{?}{=} \cot 2\theta$$

$$\frac{2\cos 2\theta\cos(-\theta)}{2\sin 2\theta\cos(-\theta)} \overset{?}{=} \cot 2\theta$$

$$\frac{\cos 2\theta}{\sin 2\theta} \overset{?}{=} \cot 2\theta$$

$$\cot 2\theta = \cot 2\theta$$

19.

$$\frac{\sin x + \sin y}{\cos x + \cos y} \overset{?}{=} \tan\frac{x + y}{2}$$

$$\frac{2\sin\dfrac{x + y}{2}\cos\dfrac{x - y}{2}}{2\cos\dfrac{x + y}{2}\cos\dfrac{x - y}{2}} \overset{?}{=} \tan\frac{x + y}{2}$$

$$\frac{\sin\dfrac{x + y}{2}}{\cos\dfrac{x + y}{2}} \overset{?}{=} \tan\frac{x + y}{2}$$

$$\tan\frac{x + y}{2} = \tan\frac{x + y}{2}$$

20.

$$\frac{\cos x + \cos y}{\cos x - \cos y} \overset{?}{=} -\cot\frac{x + y}{2}\cot\frac{x - y}{2}$$

$$\frac{2\cos\dfrac{x + y}{2}\cos\dfrac{x - y}{2}}{-2\sin\dfrac{x + y}{2}\sin\dfrac{x - y}{2}} \overset{?}{=} -\cot\frac{x + y}{2}\cot\frac{x - y}{2}$$

$$-\frac{\cos\dfrac{x + y}{2}}{\sin\dfrac{x + y}{2}} \cdot \frac{\cos\dfrac{x - y}{2}}{\sin\dfrac{x - y}{2}} \overset{?}{=} -\cot\frac{x + y}{2}\cot\frac{x - y}{2}$$

$$-\cot\frac{x + y}{2}\cot\frac{x - y}{2} = -\cot\frac{x + y}{2}\cot\frac{x - y}{2}$$

21.

$$\frac{\sin(6\theta) + \sin(8\theta)}{\sin(6\theta) - \sin(8\theta)} \overset{?}{=} -\frac{\tan(7\theta)}{\tan\theta}$$

$$\frac{2\sin\left(\dfrac{6\theta + 8\theta}{2}\right)\cos\left(\dfrac{6\theta - 8\theta}{2}\right)}{2\sin\left(\dfrac{6\theta - 8\theta}{2}\right)\cos\left(\dfrac{6\theta + 8\theta}{2}\right)} \overset{?}{=} -\frac{\tan(7\theta)}{\tan\theta}$$

$$\frac{2\sin(7\theta)\cos(-\theta)}{2\sin(-\theta)\cos(7\theta)} \overset{?}{=} -\frac{\tan(7\theta)}{\tan\theta}$$

$$\frac{\sin(7\theta)\cos(-\theta)}{\sin(-\theta)\cos(7\theta)} \overset{?}{=} -\frac{\tan(7\theta)}{\tan\theta}$$

$$\frac{\sin(7\theta)\cos\theta}{-\sin\theta\cos(7\theta)} \overset{?}{=} -\frac{\tan(7\theta)}{\tan\theta}$$

$$-\frac{\sin(7\theta)}{\cos(7\theta)}\cdot\frac{\cos\theta}{\sin\theta} \overset{?}{=} -\frac{\tan(7\theta)}{\tan\theta}$$

$$-\tan(7\theta)\cdot\cot\theta \overset{?}{=} -\frac{\tan(7\theta)}{\tan\theta}$$

$$-\tan(7\theta)\cdot\frac{1}{\tan\theta} \overset{?}{=} -\frac{\tan(7\theta)}{\tan\theta}$$

$$-\frac{\tan(7\theta)}{\tan\theta} = -\frac{\tan(7\theta)}{\tan\theta}$$

3.5 Exercises

SCE-1. $-\dfrac{7\pi}{4}, -\dfrac{13\pi}{12}, -\dfrac{5\pi}{12}, \dfrac{\pi}{4}, \dfrac{11\pi}{12}, \dfrac{19\pi}{12}$ **SCE-2.** $-1, -\dfrac{1}{2}$ **SCE-3.** $-1, 0$ **SCE-4.** $-1, -\dfrac{1}{2}$

1. $\theta = \dfrac{\pi}{3} + 2\pi k$ or $\theta = \dfrac{5\pi}{3} + 2\pi k$, where k is any integer; $\theta = \dfrac{\pi}{3}$, $\theta = \dfrac{5\pi}{3}$ **2.** $\theta = \dfrac{2\pi}{3} + \pi k$, where k is any integer;

$\theta = \dfrac{2\pi}{3}, \theta = \dfrac{5\pi}{3}$ **3.** $\theta = \dfrac{\pi}{2} + \pi k$, where k is any integer; $\theta = \dfrac{\pi}{2}, \theta = \dfrac{3\pi}{2}$ **4.** $\theta = \dfrac{\pi}{2} + 2\pi k$, where k is any integer;

$\theta = \dfrac{\pi}{2}$ **5.** $\theta = \dfrac{3\pi}{4} + \pi k$, where k is any integer; $\theta = \dfrac{3\pi}{4}, \theta = \dfrac{7\pi}{4}$ **6.** $\theta = \dfrac{5\pi}{8} + \pi k$ or $\theta = \dfrac{7\pi}{8} + \pi k$,

where k is any integer; $\theta = \dfrac{5\pi}{8}, \theta = \dfrac{7\pi}{8}, \theta = \dfrac{13\pi}{8}, \theta = \dfrac{15\pi}{8}$ **7.** $\theta = \dfrac{\pi}{4} + \dfrac{2\pi k}{3}$ or $\theta = \dfrac{5\pi}{12} + \dfrac{2\pi k}{3}$, where k is any integer;

$\theta = \dfrac{\pi}{4}, \theta = \dfrac{5\pi}{12}, \theta = \dfrac{11\pi}{12}, \theta = \dfrac{13\pi}{12}, \theta = \dfrac{19\pi}{12}, \theta = \dfrac{21\pi}{12}$ **8.** $\theta = \dfrac{3\pi}{2} + 2\pi k$, where k is any integer; $\theta = \dfrac{3\pi}{2}$

9. $\theta = \dfrac{5\pi}{2} + 4\pi k$ or $\theta = \dfrac{7\pi}{2} + 4\pi k$, where k is any integer; no solution   **10.** $\theta = \dfrac{2\pi}{3} + 4\pi k$ or $\theta = \dfrac{10\pi}{3} + 4\pi k$,

where k is any integer; $\theta = \dfrac{2\pi}{3}$ **11.** $\theta = \dfrac{\pi}{3} + 2\pi k$, where k is any integer; $\theta = \dfrac{\pi}{3}$ **12.** $\theta = \dfrac{4\pi}{9} + \dfrac{4\pi k}{3}$ or $\theta = \dfrac{8\pi}{9} + \dfrac{4\pi k}{3}$, where k is any integer; $\theta = \dfrac{4\pi}{9}$, $\theta = \dfrac{8\pi}{9}$, $\theta = \dfrac{16\pi}{9}$ **13.** $\theta = \dfrac{\pi}{6} + \pi k$ or $\theta = \dfrac{5\pi}{6} + \pi k$, where k is any integer; $\theta = \dfrac{\pi}{6}$, $\theta = \dfrac{5\pi}{6}$, $\theta = \dfrac{7\pi}{6}$, $\theta = \dfrac{11\pi}{6}$ **14.** $\theta = \dfrac{\pi}{2} + 2\pi k$, $\theta = \dfrac{7\pi}{6} + 2\pi k$, or $\theta = \dfrac{11\pi}{6} + 2\pi k$, where k is any integer; $\theta = \dfrac{\pi}{2}$, $\theta = \dfrac{7\pi}{6}$, $\theta = \dfrac{11\pi}{6}$ **15.** $\theta = \dfrac{2\pi}{3} + 2\pi k$ or $\theta = \dfrac{4\pi}{3} + 2\pi k$, where k is any integer; $\theta = \dfrac{2\pi}{3}$, $\theta = \dfrac{4\pi}{3}$ **16.** $\theta = \pi k$, $\theta = \dfrac{\pi}{6} + \pi k$, or $\theta = \dfrac{5\pi}{6} + \pi k$, where k is any integer; $\theta = 0$, $\theta = \dfrac{\pi}{6}$, $\theta = \dfrac{5\pi}{6}$, $\theta = \pi$, $\theta = \dfrac{7\pi}{6}$, $\theta = \dfrac{11\pi}{6}$ **17.** $\theta = \dfrac{3\pi k}{2}$, where k is any integer; $\theta = 0$, $\theta = \dfrac{3\pi}{2}$ **18.** $\theta = \pi k$, $\theta = \dfrac{\pi}{6} + 2\pi k$, or $\theta = \dfrac{5\pi}{6} + 2\pi k$, where k is any integer $\theta = 0$, $\theta = \dfrac{\pi}{6}$, $\theta = \dfrac{5\pi}{6}$, $\theta = \pi$

19. $\theta = \dfrac{2\pi}{3} + 2\pi k$, $\theta = \pi + 2\pi k$, or $\theta = \dfrac{4\pi}{3} + 2\pi k$, where k is any integer; $\theta = \dfrac{2\pi}{3}$, $\theta = \pi$, $\theta = \dfrac{4\pi}{3}$

20. $\theta = \pi k$ or $\theta = \dfrac{\pi}{4} + \pi k$, where k is any integer; $\theta = 0$, $\theta = \dfrac{\pi}{4}$, $\theta = \pi$, $\theta = \dfrac{5\pi}{4}$ **21.** $\theta = \dfrac{\pi}{3} + 2\pi k$ or $\theta = \dfrac{5\pi}{3} + 2\pi k$, where k is any integer; $\theta = \dfrac{\pi}{3}$, $\theta = \dfrac{5\pi}{3}$ **22.** $\theta = \dfrac{\pi}{3} + \pi k$ or $\theta = \dfrac{2\pi}{3} + \pi k$, where k is any integer; $\theta = \dfrac{\pi}{3}$, $\theta = \dfrac{2\pi}{3}$, $\theta = \dfrac{4\pi}{3}$, $\theta = \dfrac{5\pi}{3}$ **23.** $\theta = \dfrac{\pi}{2} + \pi k$ or $\theta = \dfrac{\pi}{4} + \dfrac{\pi k}{2}$, where k is any integer; $\theta = \dfrac{\pi}{4}$, $\theta = \dfrac{\pi}{2}$, $\theta = \dfrac{3\pi}{4}$, $\theta = \dfrac{5\pi}{4}$, $\theta = \dfrac{3\pi}{2}$, $\theta = \dfrac{7\pi}{4}$

24. $\theta = \pi k$, $\theta = \dfrac{\pi}{3} + 2\pi k$, or $\theta = \dfrac{5\pi}{3} + 2\pi k$, where k is any integer; $\theta = 0$, $\theta = \dfrac{\pi}{3}$, $\theta = \pi$, $\theta = \dfrac{5\pi}{3}$ **25.** $\theta = \dfrac{\pi}{6} + \pi k$ or $\theta = \dfrac{5\pi}{6} + \pi k$, where k is any integer; $\theta = \dfrac{\pi}{6}$, $\theta = \dfrac{5\pi}{6}$, $\theta = \dfrac{7\pi}{6}$, $\theta = \dfrac{11\pi}{6}$ **26.** $\theta = \pi k$ or $\theta = \dfrac{\pi}{6} + \dfrac{\pi k}{3}$, where k is any integer; $\theta = 0$, $\theta = \dfrac{\pi}{6}$, $\theta = \dfrac{\pi}{2}$, $\theta = \dfrac{5\pi}{6}$, $\theta = \pi$, $\theta = \dfrac{7\pi}{6}$, $\theta = \dfrac{3\pi}{2}$, $\theta = \dfrac{11\pi}{6}$ **27.** $\theta = \dfrac{\pi k}{4}$, where k is any integer; $\theta = 0$, $\theta = \dfrac{\pi}{4}$, $\theta = \dfrac{\pi}{2}$, $\theta = \dfrac{3\pi}{4}$, $\theta = \pi$, $\theta = \dfrac{5\pi}{4}$, $\theta = \dfrac{3\pi}{2}$, $\theta = \dfrac{7\pi}{4}$ **28.** $\theta = \dfrac{\pi}{2} + \pi k$, $\theta = \dfrac{2\pi}{3} + 2\pi k$, or $\theta = \dfrac{4\pi}{3} + 2\pi k$, where k is any integer; $\theta = \dfrac{\pi}{2}$, $\theta = \dfrac{2\pi}{3}$, $\theta = \dfrac{4\pi}{3}$, $\theta = \dfrac{3\pi}{2}$ **29.** $\theta = \dfrac{3\pi}{4} + \pi k$, where k is any integer; $\theta = \dfrac{3\pi}{4}$, $\theta = \dfrac{7\pi}{4}$ **30.** $\theta = \dfrac{5\pi}{4} + 2\pi k$, where k is any integer; $\theta = \dfrac{5\pi}{4}$ **31.** $\theta = \dfrac{3\pi}{4} + \pi k$, where k is any integer; $\theta = \dfrac{3\pi}{4}$, $\theta = \dfrac{7\pi}{4}$ **32.** $\theta = \pi k$ or $\theta = \dfrac{2\pi}{3} + k\pi$, where k is any integer; $\theta = 0$, $\theta = \dfrac{2\pi}{3}$, $\theta = \pi$, $\theta = \dfrac{5\pi}{3}$ **33.** $\theta \approx 0.8038$, $\theta \approx 2.3378$ **34.** $\theta \approx 0.8892$, $\theta \approx 5.3939$

35. $\theta \approx 0.4636$, $\theta \approx 3.6052$ **36.** $\theta \approx 2.4981$, $\theta \approx 3.7851$ **37.** $\theta \approx 3.3430$, $\theta \approx 6.0818$ **38.** $\theta \approx 1.8158$, $\theta \approx 4.9574$

39. $\theta = \dfrac{\pi}{3}$, $\theta = \dfrac{5\pi}{3}$ **40.** $\theta = \dfrac{2\pi}{3}$, $\theta = \dfrac{5\pi}{3}$ **41.** $\theta = \dfrac{\pi}{2}$, $\theta = \dfrac{3\pi}{2}$ **42.** $\theta = \dfrac{\pi}{2}$ **43.** $\theta = \dfrac{3\pi}{4}$, $\theta = \dfrac{7\pi}{4}$ **44.** $\theta = \dfrac{5\pi}{8}$, $\theta = \dfrac{7\pi}{8}$, $\theta = \dfrac{13\pi}{8}$, $\theta = \dfrac{15\pi}{8}$ **45.** $\theta = \dfrac{\pi}{4}$, $\theta = \dfrac{5\pi}{12}$, $\theta = \dfrac{11\pi}{12}$, $\theta = \dfrac{13\pi}{12}$, $\theta = \dfrac{19\pi}{12}$, $\theta = \dfrac{21\pi}{12}$ **46.** $\theta = \dfrac{3\pi}{2}$ **47.** no solution

48. $\theta = \dfrac{2\pi}{3}$ **49.** $\theta = \dfrac{\pi}{3}$ **50.** $\theta = \dfrac{4\pi}{9}$, $\theta = \dfrac{8\pi}{9}$, $\theta = \dfrac{16\pi}{9}$ **51.** $\theta = \dfrac{\pi}{6}$, $\theta = \dfrac{5\pi}{6}$, $\theta = \dfrac{7\pi}{6}$, $\theta = \dfrac{11\pi}{6}$ **52.** $\theta = \dfrac{\pi}{2}$, $\theta = \dfrac{7\pi}{6}$, $\theta = \dfrac{11\pi}{6}$ **53.** $\theta = \dfrac{2\pi}{3}$, $\theta = \dfrac{4\pi}{3}$ **54.** $\theta = 0$, $\theta = \dfrac{\pi}{6}$, $\theta = \dfrac{5\pi}{6}$, $\theta = \pi$, $\theta = \dfrac{7\pi}{6}$, $\theta = \dfrac{11\pi}{6}$ **55.** $\theta = 0$, $\theta = \dfrac{3\pi}{2}$

56. $\theta = 0$, $\theta = \dfrac{\pi}{6}$, $\theta = \dfrac{5\pi}{6}$, $\theta = \pi$ **57.** $\theta = \dfrac{2\pi}{3}$, $\theta = \pi$, $\theta = \dfrac{4\pi}{3}$ **58.** $\theta = 0$, $\theta = \dfrac{\pi}{4}$, $\theta = \pi$, $\theta = \dfrac{5\pi}{4}$ **59.** $\theta = \dfrac{\pi}{3}$, $\theta = \dfrac{5\pi}{3}$ **60.** $\theta = \dfrac{\pi}{3}$, $\theta = \dfrac{2\pi}{3}$, $\theta = \dfrac{4\pi}{3}$, $\theta = \dfrac{5\pi}{3}$ **61.** $\theta = \dfrac{\pi}{4}$, $\theta = \dfrac{\pi}{2}$, $\theta = \dfrac{3\pi}{4}$, $\theta = \dfrac{5\pi}{4}$, $\theta = \dfrac{3\pi}{2}$, $\theta = \dfrac{7\pi}{4}$ **62.** $\theta = 0$, $\theta = \dfrac{\pi}{3}$, $\theta = \pi$, $\theta = \dfrac{5\pi}{3}$ **63.** $\theta = \dfrac{\pi}{6}$, $\theta = \dfrac{5\pi}{6}$, $\theta = \dfrac{7\pi}{6}$, $\theta = \dfrac{11\pi}{6}$ **64.** $\theta = 0$, $\theta = \dfrac{\pi}{6}$, $\theta = \dfrac{\pi}{2}$, $\theta = \dfrac{5\pi}{6}$, $\theta = \pi$, $\theta = \dfrac{7\pi}{6}$, $\theta = \dfrac{3\pi}{2}$, $\theta = \dfrac{11\pi}{6}$ **65.** $\theta = 0$, $\theta = \dfrac{\pi}{4}$, $\theta = \dfrac{\pi}{2}$, $\theta = \dfrac{3\pi}{4}$, $\theta = \pi$, $\theta = \dfrac{5\pi}{4}$, $\theta = \dfrac{3\pi}{2}$, $\theta = \dfrac{7\pi}{4}$ **66.** $\theta = \dfrac{\pi}{2}$, $\theta = \dfrac{2\pi}{3}$, $\theta = \dfrac{4\pi}{3}$, $\theta = \dfrac{3\pi}{2}$

67. $\theta = \dfrac{3\pi}{4}$, $\theta = \dfrac{7\pi}{4}$ **68.** $\theta = \dfrac{5\pi}{4}$ **69.** $\theta = \dfrac{3\pi}{4}$, $\theta = \dfrac{7\pi}{4}$ **70.** $\theta = 0$, $\theta = \dfrac{2\pi}{3}$, $\theta = \pi$, $\theta = \dfrac{5\pi}{3}$

CHAPTER 4

4.1 Exercises

SCE-1. 27.58°   **SCE-2.** 75.29°   **SCE-3.** 42.41°

1. $c = 27.78$ in.; $A = 30.26°$; $B = 59.74°$ **2.** $a = 10.58$ cm; $A = 21.4°$; $B = 68.6°$ **3.** $b = 9.2$ m; $c = 12.19$ m; $B = 49°$
4. $a = 16.54$ cm; $c = 41.22$ cm; $B = 66.35°$ **5.** $a = 50.85$ in.; $b = 70.27$ in.; $B = 54.11°$ **6.** 6.36 in.; 12.47 in.   **7.** 103.29 cm;
104.05 cm   **8.** 101.69 ft   **9.** 400.91 ft   **10.** length of bridge: 11.1 m; distance between the two people: 36.72 m   **11.** 8.31 ft
12. 965,220 km   **13.** 12.44 ft   **14.** 8103 ft   **15.** 2386 m

4.2 Exercises

SCE-1. 9.1   **SCE-2.** 32.3

1. yes   **2.** no   **3.** yes   **4.** yes   **5.** yes   **6.** no   **7.** $B = 59°, b \approx 7.4, c \approx 8.4$ **8.** $A = 82°, b \approx 5.1, c \approx 5.6$
9. $C = 89°, b \approx 107.1, c \approx 151.5$ **10.** $C = 79°, a \approx 5.6, c \approx 7.4$ **11.** $B = 53°, b \approx 3.3, c \approx 1.5$ **12.** $A = 75°$,
$a \approx 11.9, b \approx 10.4$ **13.** one triangle; $B \approx 35.4°, C \approx 104.6°, c \approx 12.2$ **14.** no triangle   **15.** one triangle; $B = 90°$,
$C = 60°, c \approx 6.4$ **16.** two triangles; $B_1 \approx 72.5°, C_1 \approx 52.5°, c_1 \approx 13.6; B_2 \approx 107.5°, C_2 \approx 17.5°, c_2 \approx 5.1$ **17.** two
triangles; $B_1 \approx 83.7°, C_1 \approx 28.6°, c_1 \approx 7.7; B_2 \approx 96.3, C_2 \approx 16.0°, c_2 \approx 4.4$ **18.** no triangle   **19.** two triangles;
$B_1 \approx 59.2°, C_1 \approx 63.7°, c_1 \approx 14.2; B_2 \approx 120.8, C_2 \approx 2.1°, c_2 \approx .6$ **20.** 61.1 m   **21.** 4.8 mi and 3.4 mi   **22.** 20.5 mi
and 12.4 mi   **23.** 65.7 mi   **24.** distance from Alpha: 56.6 mi, distance from Beta: 63.5 mi   **25.** 493.9 ft   **26.** one triangle
27. no triangle   **28.** one triangle   **29.** two triangles   **30.** two triangles   **31.** no triangle   **32.** two triangles

4.3 Exercises

SCE-1. 3.11   **SCE-2.** 34.3°

1. Law of Sines   **2.** The triangle cannot be solved.   **3.** Law of Cosines   **4.** Law of Sines   **5.** Law of Sines   **6.** Law of
Cosines   **7.** $a \approx 3.0$ ft, $B \approx 58.4°, C \approx 96.6°$ **8.** $c \approx 9.7$ cm, $A \approx 54.1°, B \approx 23.9°$ **9.** $a \approx 3.4$ m, $B = 81°, C = 81°$
10. $c \approx 3.1, A \approx 41.0°, C \approx 119.0°$ **11.** $a \approx 10.4, B \approx 8.5°, C \approx 41.5°$ **12.** $A \approx 41.0°, B \approx 65.4°, C \approx 73.6°$
13. $A \approx 18.2°, B \approx 51.3°, C \approx 110.5°$ **14.** $A \approx 22.6°, B \approx 67.4°, C = 90°$ **15.** $A \approx 33.1°, B \approx 63.4°, C \approx 83.5°$
16. $A \approx 83.5°, B \approx 50.6°, C \approx 45.9°$ **17.** 1926.1 mi   **18.** 1655.3 mi   **19.** N 62.6° W   **20.** 23.1°   **21.** 16.3 cm
22. 8.8 cm and 22.4 cm   **23.** $a \approx 7.2, b = 9, c = 5, A \approx 53.1°, B \approx 93.2°, C \approx 33.7°$ **24.** 3106.9 ft   **25.** 54.1°

4.4 Exercises

SCE-1. 22.19

1. 29.60 ft^2 **2.** 16.19 yd^2 **3.** 29.65 cm^2 **4.** 198.91 m^2 **5.** 16.73 ft^2 **6.** 258.13 km^2 **7.** 74.61 cm^2 **8.** 24.49 yd^2
9. 16.71 ft^2 **10.** 7.91 m^2 **11.** 405.50 km^2 **12.** 364.66 cm^2 **13.** 101.82 cm^2 **14.** 31.42 in.^{2} **15.** 31.55 m^2
16. 1892 ft^2 **17.** \$88,801   **18.** \$1831   **19.** 930 ft^2; yes   **20.** 593.4 ft^2 **21.** \$77

CHAPTER 5

5.1 Exercises

1.   **2.**   **3.**   **4.**   **5.**   **6.**   **7.**

8.   **9.**   **10.**   **11.**   **a.** $P\left(5, -\dfrac{11\pi}{6}\right)$ **b.** $P\left(-5, \dfrac{7\pi}{6}\right)$ **c.** $P\left(5, \dfrac{13\pi}{6}\right)$

12.   **a.** $P\left(3, -\dfrac{2\pi}{3}\right)$ **b.** $P\left(-3, \dfrac{\pi}{3}\right)$ **c.** $P\left(3, \dfrac{10\pi}{3}\right)$ **13.**   **a.** $P\left(4, -\dfrac{\pi}{2}\right)$ **b.** $P\left(-4, \dfrac{\pi}{2}\right)$

c. $P\left(4, \dfrac{7\pi}{2}\right)$ **14.** •   **a.** $P\left(5, -\dfrac{5\pi}{4}\right)$ **b.** $P\left(-5, \dfrac{7\pi}{4}\right)$ **c.** $P\left(5, \dfrac{11\pi}{4}\right)$ **15.**   **a.** $P\left(3, -\dfrac{\pi}{6}\right)$

b. $P\left(-3, \dfrac{5\pi}{6}\right)$ **c.** $P\left(3, \dfrac{23\pi}{6}\right)$ **16.** $P\left(\dfrac{3}{\sqrt{2}}, \dfrac{3}{\sqrt{2}}\right)$ **17.** $P(-1, \sqrt{3})$ **18.** $P(0, -5)$ **19.** $P\left(-\dfrac{\sqrt{3}}{2}, \dfrac{1}{2}\right)$

20. $P(2, -2\sqrt{3})$ **21.** $P(-\sqrt{3}, -1)$ **22.** $P\left(\dfrac{5}{\sqrt{2}}, -\dfrac{5}{\sqrt{2}}\right)$ **23.** $P(7, 0)$ **24.** $P\left(\dfrac{3}{2}, \dfrac{3\sqrt{3}}{2}\right)$ **25.** $P\left(-\dfrac{7\sqrt{3}}{2}, \dfrac{7}{2}\right)$

26. $P(3, 0)$ **27.** $P\left(\sqrt{2}, \dfrac{3\pi}{2}\right)$ **28.** $P(0.5, \pi)$ **29.** $P\left(1.5, \dfrac{\pi}{2}\right)$ **30.** $P\left(3\sqrt{2}, \dfrac{7\pi}{4}\right)$ **31.** $P\left(10, \dfrac{2\pi}{3}\right)$

32. $P\left(8, \dfrac{7\pi}{6}\right)$ **33.** $P\left(\sqrt{10}, \dfrac{\pi}{4}\right)$ **34.** $P\left(4, \dfrac{\pi}{3}\right)$ **35.** $P\left(12, \dfrac{5\pi}{6}\right)$ **36.** $P\left(\sqrt{53}, 2\pi - \tan^{-1}\dfrac{7}{2}\right)$

37. $P\left(\sqrt{3}, 2\pi - \tan^{-1}\dfrac{1}{\sqrt{2}}\right)$ **38.** $r = \dfrac{1}{7\cos\theta - 4\sin\theta}$ **39.** $r = 5$ **40.** $r = -3\sec\theta$ **41.** $r = 5\csc\theta$

42. $r = -4\sin\theta$ **43.** $r = 6\cos\theta$ **44.** $r = -2\cos\theta$ **45.** $r = 8\sin\theta$ **46.** $2x + 5y = 3$ **47.** $x^2 + y^2 = 3y$

48. $x^2 + y^2 = 16$ **49.** $x^2 + y^2 = -2x$ **50.** $y = \sqrt{3}x$ **51.** $x^2 + y^2 = 6y - 2x$ **52.** $x = 2$ **53.** $y = -3$

54. $P\left(2\sqrt{2}, \dfrac{\pi}{4}\right)$ **55.** $P\left(10, \dfrac{2\pi}{3}\right)$ **56.** $P\left(4, \dfrac{5\pi}{6}\right)$ **57.** $P\left(\sqrt{10}, \dfrac{3\pi}{4}\right)$ **58.** $P\left(2, \dfrac{4\pi}{3}\right)$ **59.** $P\left(12, \dfrac{7\pi}{6}\right)$

5.2 Exercises

SCE-1. $-\dfrac{7\pi}{4}, -\dfrac{13\pi}{12}, -\dfrac{5\pi}{12}, \dfrac{\pi}{4}, \dfrac{11\pi}{12}, \dfrac{19\pi}{12}$ **1.**   **2.**   **3.**   **4.**   **5.**

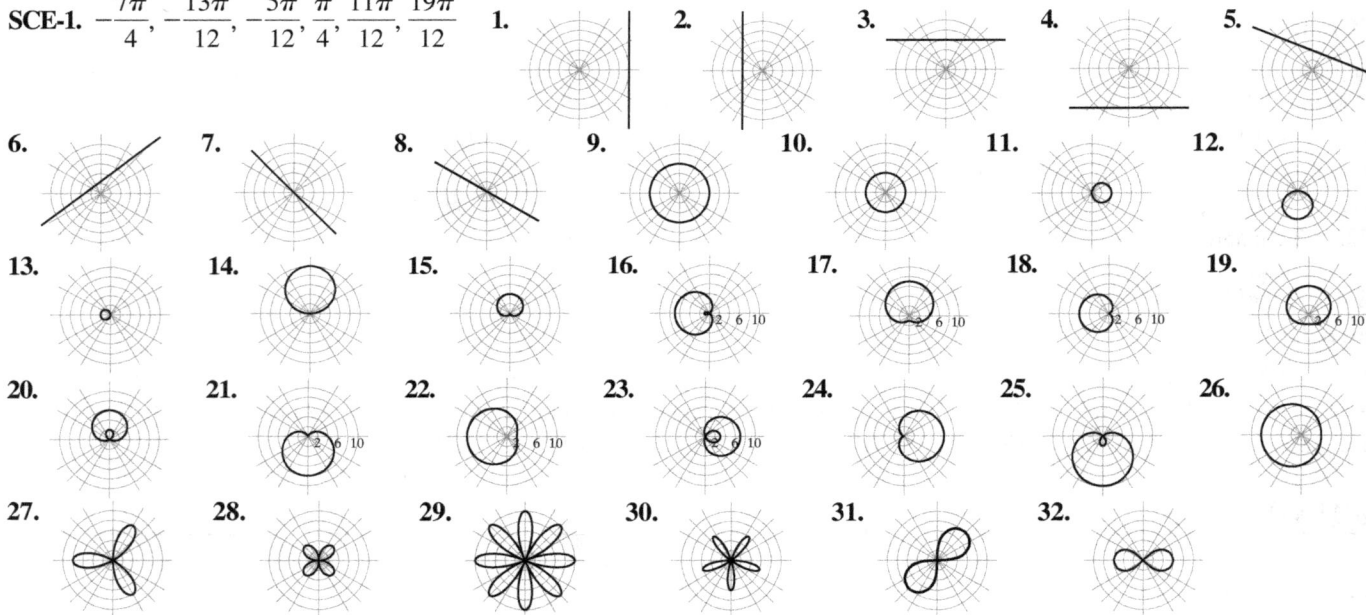

6.   **7.**   **8.**   **9.**   **10.**   **11.**   **12.**

13.   **14.**   **15.**   **16.**   **17.**   **18.**   **19.**

20.   **21.**   **22.**   **23.**   **24.**   **25.**   **26.**

27.   **28.**   **29.**   **30.**   **31.**   **32.**

33. vertical line   **34.** vertical line   **35.** horizontal line   **36.** horizontal line   **37.** line   **38.** line
39. line through the pole   **40.** line through the pole   **41.** circle centered at the pole   **42.** circle centered at the pole
43. circle   **44.** circle   **45.** circle   **46.** circle   **47.** cardioid   **48.** limacon with an inner loop   **49.** limacon with a dimple
50. cardioid   **51.** limacon with no inner loop and no dimple   **52.** limacon with an inner loop   **53.** cardioid
54. limacon with a dimple   **55.** limacon with an inner loop   **56.** cardioid   **57.** limacon with an inner loop   **58.** limacon
with no inner loop and no dimple   **59.** three-petal rose   **60.** four-petal rose   **61.** eight-petal rose   **62.** five-petal rose
63. lemniscate   **64.** lemniscate   **65.**   **66.**   **67.**   **68.**

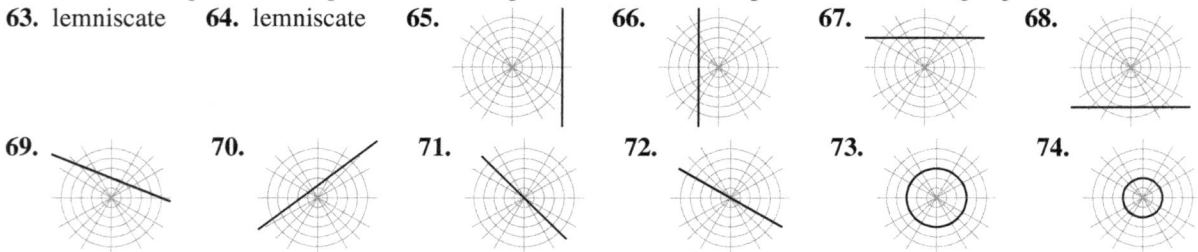

69.   **70.**   **71.**   **72.**   **73.**   **74.**

75. 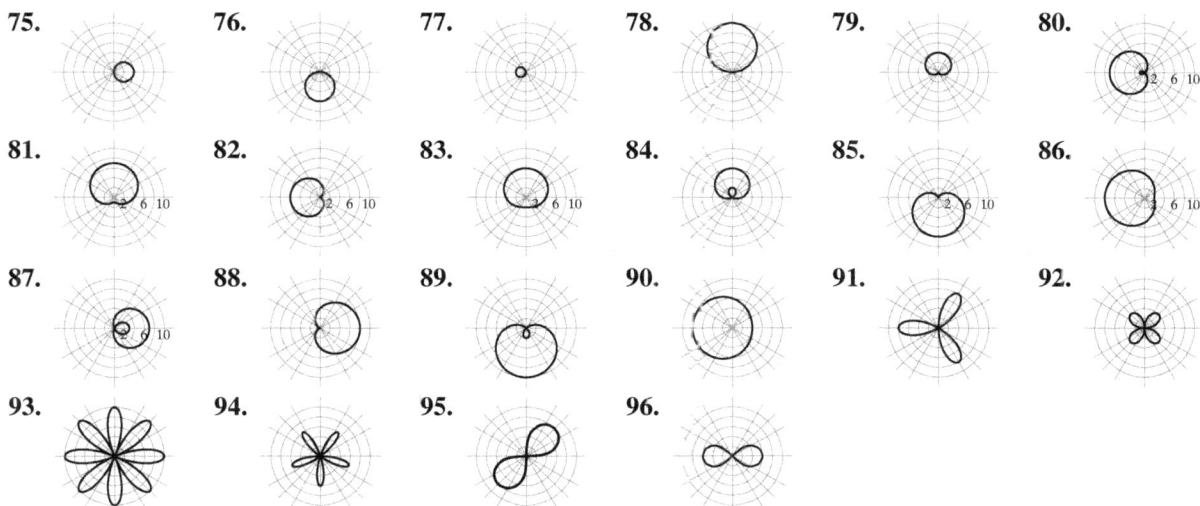 **76.** **77.** **78.** **79.** **80.**

81. **82.** **83.** **84.** **85.** **86.**

87. **88.** **89.** **90.** **91.** **92.**

93. **94.** **95.** **96.**

5.3 Exercises

1. 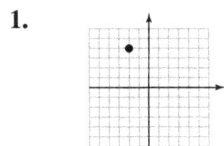 **2.** 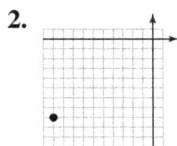 **3.** 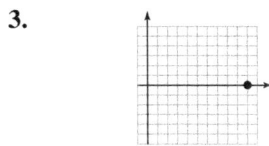 **4.**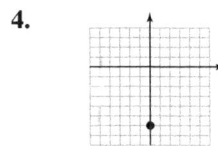

5. $z = 3\left(\cos\dfrac{11\pi}{7} + i\sin\dfrac{11\pi}{7}\right)$, ; Quadrant IV   **6.** $z = 2(\cos 280° + i\sin 280°)$, ;

Quadrant IV   **7.** $z = 5(\cos\pi + i\sin\pi)$, ; real axis   **8.** $z = 3\left(\cos\dfrac{\pi}{3} + i\sin\dfrac{\pi}{3}\right)$, ;

Quadrant I   **9.** $z = 2(\cos 30° + i\sin 30°)$, 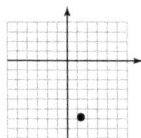 ; Quadrant I   **10.** $z = 5\left(\cos\dfrac{\pi}{2} + i\sin\dfrac{\pi}{2}\right)$,

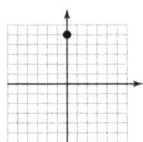 ; imaginary axis   **11.** $z = \dfrac{5}{\sqrt{2}} - \dfrac{5}{\sqrt{2}}i$ **12.** $z = -2 - 2\sqrt{3}i$ **13.** $z = -8i$ **14.** $z = 5 - 5\sqrt{3}i$

15. $z = 9.51 + 3.09i$ **16.** $z = -15.89 - 8.45i$ **17.** $z = 4\sqrt{2}\left(\cos\dfrac{3\pi}{4} + i\sin\dfrac{3\pi}{4}\right)$ **18.** $z = 10\left(\cos\dfrac{11\pi}{6} + i\sin\dfrac{11\pi}{6}\right)$

19. $z = 4\left(\cos\dfrac{4\pi}{3} + i\sin\dfrac{4\pi}{3}\right)$ **20.** $z = 5(\cos\pi + i\sin\pi)$ **21.** $z = 7\left(\cos\dfrac{\pi}{2} + i\sin\dfrac{\pi}{2}\right)$

22. $z = \sqrt{13}(\cos 2.55 + i\sin 2.55)$ **23.** $z = \sqrt{17}(\cos 4.96 + i\sin 4.96)$ **24.** $z_1 z_2 = 20\left(\cos\dfrac{13\pi}{42} + i\sin\dfrac{13\pi}{42}\right)$,

$\dfrac{z_1}{z_2} = \dfrac{5}{4}\left(\cos\dfrac{\pi}{42} + i\sin\dfrac{\pi}{42}\right)$ **25.** $z_1 z_2 = 42(\cos 110° + i\sin 110°)$, $\dfrac{z_1}{z_2} = \dfrac{7}{6}(\cos 50° + i\sin 50°)$

26. $z_1 z_2 = 6\left(\cos\dfrac{7\pi}{12} + i\sin\dfrac{7\pi}{12}\right), \dfrac{z_1}{z_2} = \dfrac{2}{3}\left(\cos\dfrac{11\pi}{12} + i\sin\dfrac{11\pi}{12}\right)$ **27.** $z_1 z_2 = 99(\cos 100° + i\sin 100°)$,

$\dfrac{z_1}{z_2} = \dfrac{11}{9}(\cos 200° + i\sin 200°)$ **28.** $32\left(\cos\dfrac{5\pi}{7} + i\sin\dfrac{5\pi}{7}\right)$ **29.** $81(\cos 284° + i\sin 284°)$ **30.** $49\left(\cos\dfrac{14\pi}{9} + i\sin\dfrac{14\pi}{9}\right)$

31. $\dfrac{27}{2} + \dfrac{27\sqrt{3}}{2}i$ **32.** $-4 - 4\sqrt{3}i$ **33.** -1024 **34.** $-\dfrac{27}{2} + \dfrac{27\sqrt{3}}{2}i$ **35.** $-\dfrac{25}{2} + \dfrac{25\sqrt{3}}{2}i$ **36.** $-8 + 8i$

37. $46{,}656$ **38.** $6\left(\cos\dfrac{\pi}{12} + i\sin\dfrac{\pi}{12}\right), 6\left(\cos\dfrac{13\pi}{12} + i\sin\dfrac{13\pi}{12}\right)$ **39.** $3\left(\cos\dfrac{250°}{3} + i\sin\dfrac{250°}{3}\right), 3\left(\cos\dfrac{610°}{3} + i\sin\dfrac{610°}{3}\right),$

$3\left(\cos\dfrac{970°}{3} + i\sin\dfrac{970°}{3}\right)$ **40.** $2\left(\cos\dfrac{\pi}{4} + i\sin\dfrac{\pi}{4}\right), 2\left(\cos\dfrac{3\pi}{4} + i\sin\dfrac{3\pi}{4}\right), 2\left(\cos\dfrac{5\pi}{4} + i\sin\dfrac{5\pi}{4}\right),$

$2\left(\cos\dfrac{7\pi}{4} + i\sin\dfrac{7\pi}{4}\right)$ **41.** $\sqrt[5]{4}\left(\cos\dfrac{3\pi}{10} + i\sin\dfrac{3\pi}{10}\right), \sqrt[5]{4}\left(\cos\dfrac{7\pi}{10} + i\sin\dfrac{7\pi}{10}\right), \sqrt[5]{4}\left(\cos\dfrac{11\pi}{10} + i\sin\dfrac{11\pi}{10}\right),$

$\sqrt[5]{4}\left(\cos\dfrac{3\pi}{2} + i\sin\dfrac{3\pi}{2}\right), \sqrt[5]{4}\left(\cos\dfrac{19\pi}{10} + i\sin\dfrac{19\pi}{10}\right)$ **42.** $\sqrt[3]{12}\left(\cos\dfrac{\pi}{9} + i\sin\dfrac{\pi}{9}\right), \sqrt[3]{12}\left(\cos\dfrac{7\pi}{9} + i\sin\dfrac{7\pi}{9}\right),$

$\sqrt[3]{12}\left(\cos\dfrac{13\pi}{9} + i\sin\dfrac{13\pi}{9}\right)$ **43.** $4, 4i, -4, -4i$ **44.** $\dfrac{5}{2} + \dfrac{5\sqrt{3}}{2}i, -5, \dfrac{5}{2} - \dfrac{5\sqrt{3}}{2}i$ **45.** $0.73 + 2.73i, -2.73 + 0.73i,$

$-0.73 - 2.73i, 2.73 - 0.73i$ **46.** $-0.13 + 1.60i, -1.32 - 0.91i, 1.45 - 0.69i$

5.4 Exercises

1. $\sqrt{61}$ **2.** 10 **3.** $\dfrac{\sqrt{1417}}{6}$ **4.** **5.** **6.** **7.** **8.**

9. **10.** **11.** $<4, 3>, 5$ **12.** $<4, 3>, 5$ **13.** $<-6, 11>, \sqrt{157}$

14. $3\mathbf{i} + 5\mathbf{j}$ **15.** $-4\mathbf{i} + \mathbf{j}$ **16.** $-6\mathbf{i} - 7\mathbf{j}$ **17.** $-9\mathbf{i} + 21\mathbf{j}, 3\sqrt{58}$ **18.** $2\mathbf{i} + 6\mathbf{j}, 2\sqrt{10}$ **19.** $-8\mathbf{i} + 8\mathbf{j}, 8\sqrt{2}$

20. $3\mathbf{i} + 25\mathbf{j}, \sqrt{634}$ **21.** $\mathbf{i} - \dfrac{7}{3}\mathbf{j}, \dfrac{\sqrt{58}}{3}$ **22.** $-\dfrac{11}{4}\mathbf{i} + \dfrac{15}{4}\mathbf{j}, \dfrac{\sqrt{346}}{4}$ **23.** $-\mathbf{j}$ **24.** $\dfrac{4}{5}\mathbf{i} - \dfrac{3}{5}\mathbf{j}$ **25.** $\dfrac{5}{\sqrt{29}}\mathbf{i} - \dfrac{2}{\sqrt{29}}\mathbf{j}$

26. $\dfrac{1}{\sqrt{17}}\mathbf{i} - \dfrac{4}{\sqrt{17}}\mathbf{j}$ **27.** $\theta = 315°$ **28.** $\theta = 120°$ **29.** $\theta = 30°$ **30.** $\theta \approx 68.20°$ **31.** $\theta \approx 285.95°$

32. $-\dfrac{15\sqrt{3}}{2}\mathbf{i} + \dfrac{15}{2}\mathbf{j}$ **33.** $11\sqrt{2}\mathbf{i} - 11\sqrt{2}\mathbf{j}$ **34.** $\dfrac{7}{2}\mathbf{i} - \dfrac{7\sqrt{3}}{2}\mathbf{j}$ **35.** $-1.74\mathbf{i} + 9.85\mathbf{j}$ **36.** $-0.57\mathbf{i} - 0.35\mathbf{j}$

37. $50\sqrt{3}\mathbf{i} + 50\mathbf{j}$ **38.** $179.01\mathbf{i} - 18.82\mathbf{j}$ **39.** $20.68\mathbf{i} + 17.36\mathbf{j}$ **40.** $-13.68\mathbf{i} - 37.59\mathbf{j}$ **41.** 261.73 mph, N84.96°E

42. 314.48 mph, N83°W **43.** $-\mathbf{i} + 7\mathbf{j}$ **44.** $11\mathbf{i} + 16\mathbf{j}$ **45.** 519.37 pounds **46.** 1605.28 pounds, N73.78°W

47. 650.85 pounds, S52.90°E

5.5 Exercises

1. 28 **2.** -13 **3.** $140\sqrt{2}$ **4.** 37 **5.** -4 **6.** -63 **7.** 50 **8.** $\sqrt{1073}$ **9.** $54.16°$ **10.** $60.26°$ **11.** $135°$

12. $161.57°$ **13.** $180°$ **14.** 2 **15.** $-\dfrac{21}{5}$ **16.** orthogonal **17.** neither **18.** parallel **19.** neither **20.** $\mathbf{v}_1 = 2\mathbf{i} - 2\mathbf{j}$,

$\mathbf{v}_2 = -7\mathbf{i} - 7\mathbf{j}$ **21.** $\mathbf{v}_1 = -\dfrac{44}{53}\mathbf{i} + \dfrac{154}{53}\mathbf{j}, \mathbf{v}_2 = \dfrac{203}{53}\mathbf{i} + \dfrac{58}{53}\mathbf{j}$ **22.** $\mathbf{v}_1 = -\dfrac{15}{452}\mathbf{i} - \dfrac{1}{452}\mathbf{j}, \mathbf{v}_2 = \dfrac{241}{452}\mathbf{i} - \dfrac{3615}{452}\mathbf{j}$

23. 587.79 pounds **24.** 169.05 pounds, 362.52 pounds **25.** 1866.87 pounds **26.** 25.94° **27.** 200 foot-pounds

28. 27,189.23 foot-pounds **29.** 26,811.56 foot-pounds **30.** 48.2°

Index